RADIOACTIVITY IN GEOLOGY
Principles and Applications

ELLIS HORWOOD SERIES IN GEOLOGY

Editors: D. T. DONOVAN, Professor of Geology, University College London and J. W. MURRAY, Professor of Geology, University of Exeter

This series aims to build up a library of books on geology which will include student texts and also more advanced works of interest to professional geologists and to industry. The series will include translation of important books recently published in Europe, and also books specially commissioned.

A GUIDE TO CLASSIFICATION IN GEOLOGY
J. W. MURRAY, Professor of Geology, University of Exeter
THE CENOZOIC ERA: Tertiary and Quaternary
C. POMEROL, Professor, University of Paris VI
Translated by D. W. HUMPHRIES, Department of Geology, University of Sheffield, and E. E. HUMPHRIES
Edited by Professor D. CURRY and D. T. DONOVAN, University College London
INTRODUCTION TO PALAEOBIOLOGY: GENERAL PALAEONTOLOGY
B. ZIEGLER, Professor of Geology and Palaeontology, University of Stuttgart, and Director of the State Museum for Natural Science, Stuttgart
FAULT AND FOLD TECTONICS
W. JAROSZEWSKI, Faculty of Geology, University of Warsaw
RADIOACTIVITY IN GEOLOGY: Principles and Applications
E. M. DURRANCE, Department of Geology, University of Exeter

BRITISH MICROPALAEONTOLOGICAL SOCIETY SERIES

This series, published by Ellis Horwood Limited for the British Micropalaeontological Society, aims to gather together knowledge of a particular faunal group for specialist and non-specialist geologists alike. The original series of Stratigraphic Atlas or Index volumes ultimately will cover all groups and will describe and illustrate the common elements of the microfauna through time (whether index or long-ranging species) thus enabling the reader to identify characteristic species in addition to those of restricted stratigraphic range. The series has now been enlarged to include the reports of conferences, organized by the Society, and collected essays on specialist themes.

The synthesis of knowledge presented in the series will reveal its strengths and prove its usefulness to the practising micropalaeontologist, and to those teaching and learning the subject. By identifying some of the gaps in this knowledge, the series will, it is believed, promote and stimulate further active research and investigation.

STRATIGRAPHICAL ATLAS OF FOSSIL FORAMINIFERA
Editors: D. G. JENKINS, The Open University, and J. W. MURRAY, Professor of Geology, University of Exeter
MICROFOSSILS FROM RECENT AND FOSSIL SHELF SEAS
Editors: J. W. NEALE, Professor of Micropalaeontology, University of Hull, and M. D. BRASIER, Lecturer in Geology, University of Hull
FOSSIL AND RECENT OSTRACODS
Editors: R. H. BATE, Stratigraphic Services International, Guildford, E. ROBINSON, Department of Geology, University College London, and L. SHEPPARD, Stratigraphic Services International, Guildford
A STRATIGRAPHICAL INDEX OF CALCAREOUS NANNOFOSSILS
Editor: A. R. LORD, Department of Geology, University College London
A STRATIGRAPHICAL INDEX OF CONODONTS
Editors: A. C. HIGGINS, Geological Survey of Canada, Calgary, and R. L. AUSTIN, Department of Geology, University of Southampton
CONODONTS: Investigative Techniques and Applications
Editor: R. L. AUSTIN, Department of Geology, University of Southampton
PALAEOBIOLOGY OF CONODONTS
Editor: R. J. ALDRIDGE, Department of Geology, University of Nottingham
MICROPALAEONTOLOGY OF CARBONATE ENVIRONMENTS
Editor: M. B. HART, Professor of Micropalaeontology and Head of Department of Geological Studies, Plymouth Polytechnic

ELLIS HORWOOD SERIES IN APPLIED GEOLOGY

The books listed below are motivated by the up-to-date applications of geology to a wide range of industrial and environmental factors: they are practical, for use by the professional and practising geologist or engineer, for use in the field, for study, and for reference.

A GUIDE TO PUMPING TESTS
F. C. BRASSINGTON, Principal Hydrogeologist, North West Water Authority
QUATERNARY GEOLOGY: Processes and Products
JOHN A. CATT, Rothamsted Experimental Station, Harpenden, UK
PRACTICAL PEDOLOGY: Manual of Soil Formation, Description and Mapping
S. G. McRAE and C. P. BURNHAM, Department of Environmental Studies and Countryside Planning, Wye College (University of London)

RADIOACTIVITY IN GEOLOGY

Principles and Applications

ERIC MICHAEL DURRANCE
Senior Lecturer in Geology
University of Exeter

ELLIS HORWOOD LIMITED
Publishers · Chichester

Halsted Press: a division of
JOHN WILEY & SONS
New York · Chichester · Brisbane · Toronto

First published in 1986
ELLIS HORWOOD LIMITED
Market Cross House, Cooper Street,
Chichester, West Sussex, PO19 1EB, England
The publisher's colophon is reproduced from James Gillison's drawing of the ancient Market Cross, Chichester.

Distributors:
Australia and New Zealand:
JACARANDA WILEY LIMITED
GPO Box 859, Brisbane, Queensland 4001, Australia
Canada:
JOHN WILEY & SONS CANADA LIMITED
22 Worcester Road, Rexdale, Ontario, Canada
Europe and Africa:
JOHN WILEY & SONS LIMITED
Baffins Lane, Chichester, West Sussex, England
North and South America and the rest of the world:
Halsted Press: a division of
JOHN WILEY & SONS
605 Third Avenue, New York, NY 10158, USA

British Library Cataloguing in Publication Data
Durrance, E.M.
Radioactivity in geology: principles and applications.
— (Ellis Horwood series in geology)
1. Radioactivity
I. Title
539.7'5'024553 QC795

Library of Congress CIP Data also available

ISBN 0–85312–761–1 (Ellis Horwood Limited)
ISBN 0–470–20389–7 (Halsted Press)

Phototypeset in Times by Ellis Horwood Limited
Printed in Great Britain by Butler & Tanner, Frome, Somerset

Table of contents

Table of Contents

Table of Contents

Preface

This book is an outline of the way in which radioactivity relates to the work of the geologist. Indeed, because of the physical nature of radioactivity and the controls exercised by chemical environments on the distribution of radionuclides within the Earth, the geophysicist and geochemist are also closely concerned with many aspects of natural radioactivity.

The study of natural radioactivity in the Earth is a very young science, radioactivity itself having only been discovered in 1896. Yet today, the energy released by radioactive decay is considered to be one of the most important sources of heat within the Earth, and thus to be a primary cause of tectonism and volcanism. Radioactivity, too, has provided the means to date geological processes over time scales which range up to several billion years and to trace the evolution of geological processes.

Since the advent of the 'atomic age' and the increasing use of nuclear energy for both armaments and peaceful purposes, the geologist has become involved in the study of artificially produced radionuclides as well as of natural radioactivity. In particular, proposals for disposal within the geological environment of radioactive waste produced by nuclear fission power stations have presented the geologist with a wide range of problems that must be investigated and solved. Again, the nature of the system means that the geophysicist and geochemist are closely involved in different aspects of the work and the expertise offered by the engineering geologist is of considerable importance.

Moreover, the rapid increase in the use of nuclear energy over the last 40 years has meant that the demand for U ore as a raw material has expanded enormously in this period. Although there are fluctuations in demand which reflect national economic situations and political decisions, the long-term depletion of fossil fuels and the, as yet, still awaited development of nuclear fusion reactors mean that considerable reserves of U will be needed in the future to supply fuel for electricity generation. The greatest demand will certainly be from those large industrialised economies which cannot be serviced by small-scale renewable energy sources such as solar and wind power. Here, too, the geologist, geophysicist and geochemist will be increasingly involved in the exploration for, and working of, economic deposits.

The development of nuclear fission reactors is often regarded as one area where mankind's activities on Earth may be strictly termed artificial, but that is not the case. Natural nuclear fission within the Earth is now known to have occurred about 2000 Ma ago, when a favourable combination of circumstances in the geological environment combined with the then greater abundance of the fissionable isotope ^{235}U, to produce self-sustaining nuclear chain reactions. This fact is not just of academic interest. The phenomenon can be viewed as an analogue of the problems that may arise in the subterranean disposal of radioactive waste and thus provide unparalleled opportunities for study of the way in which radionuclides formed by U fission migrate through the geological environment. Also, of course, if natural fission chain reactions have occurred within a U ore body in the past, then the ore is now depleted in fissionable ^{235}U and so less valuable as an energy source.

The danger to health of radioactivity has clearly been recognised for a good number of years now, especially in those fields of work which involve the use of artificial sources that have high activity, but the problems which can arise from natural radioactivity must not be considered minimal. In fact, of the average radiation dose received by the population of the UK, by far the greatest proportion (about 87%) is of natural origin. Indeed, some 43% of that dose is just due to the inhalation of Rn (the only naturally occurring radioactive gas) and its daughters. This arises from the presence of U and Th in the rocks and soils close to the Earth's surface. The distribution of radionuclides in different rocks, the take-up of these elements from the soil by plants which can then enter the food chain and find their way to man, and the nature and design of dwellings which may unwittingly enhance natural radiation are, therefore, of concern to the geologist interested in the effects of radioactivity. Even a beneficial activity such as the use of fertilizers relates to radioactivity in geology, in that phosphates are obtained from phosphatic rocks which often contain appreciable quantities of U.

Finally, in considering the scope of radioactivity in geology, we need to recognise that artificial radioactive sources are used in a wide variety of geological investigations. Directly, because of the great sensitivity with which they can be detected, radionuclides are often employed as tracers which can be added to a system to follow the movement of materials through various chemical or physical processes. Alternatively, radioactive sources are used in mineralogical investigations both in the laboratory and in the field. Indeed, they are widely used in geophysical borehole logging. By emitting radiations which penetrate and interact with their surroundings in a way which is determined by the physical and chemical nature of the material, non-destructive and *in situ* analyses can be carried out.

Radioactivity in geology is thus an enormously wide field. No single work can adequately cover all aspects, and it is not the intention of this book to attempt such a task. Instead, the book hopes to provide for students an introduction to some of the more important fields in which the geologist is likely to encounter the effects of radioactivity and its measurement. As such, it partly reflects the author's own interests, experience and prejudices. The material is presented in three main sections. The first covers the general physical and chemical background needed to appreciate the nature of radioactivity in the geological environment, while the second section then deals with the distribution and effects of radionuclides in the Earth. The third section is devoted to a description of the measurement and uses of radioactivity in geology, and of the environmental impact of both natural and man-made radioactivity.

In order to ease the flow of text, references have only been included within the body of the chapters where they are historical markers or point to specific studies or reviews. However, each chapter is followed by a bibliography of source material which it is hoped will allow the reader to pursue any specific aspect in greater detail. Within the bibliographies are publications which are of a general nature as well as articles presenting new data and interpretations. Also included are references which offer alternative views to those given herein. It is fair to point out, however, that the bibliographies are purposely directed towards publications in the English language, and thus the large body of French and Russian texts that has been published is not given due weight. A strong bias also exists towards references with a geological rather than a physical or chemical emphasis. Sources of information contained in the figures and tables are given in each caption where appropriate and grateful thanks are offered for the various permissions to reproduce material. To avoid too much repetition of references from one chapter to the next, most entries are only presented in terms of their main relevance. However, sources of information contained in figures and tables are given at the end of each appropriate chapter.

The use of radioactivity in geology owes much to the development of instrumentation and data-processing methods. Only a relatively brief discussion of these aspects has been possible in this book, but undoubtedly the future will see great advances in the measurement and interpretation of data both in the laboratory and in the field.

I should like to thank all those friends and colleagues who have given help and advice during the preparation of this book. In particular I would like to record my gratitude to Dr T. K. Ball, Dr S. Bouda, Mr P. J. Bourke, Professor S. H. U. Bowie, Dr A. V. Bromley, Professor G. C. Brown, Dr B. M. R. Green, Mr J. Hunter, Dr M. Ivanovich, Dr N. L. Jefferies, Mr U.McL. Michie, Mr D. Ostle, Dr R. C. Selley and Mr P. R. Simpson. Much of the production and editing of the typescript was carried out by Dr M. J. Heath to whom I owe a considerable debt of gratitude, and my thanks also go to Dr R. G. Gregory who drew the figures. Photographs were produced by Mr J. Jones. Finally, my thanks are due to the University of Exeter for providing the facilities without which this book could not have been written, and to my wife, Eunice, for her help and support.

1

Introduction

1.1 IONISING AND NON-IONISING RADIATION

Radiation is present in every environment of the Earth's surface, beneath the Earth and in the atmosphere. In general terms, radiation may be considered as either ionising or non-ionising depending on whether or not it possesses the ability to cause neutral atoms or molecules to acquire an electrical charge. Non-ionising radiations are typically part of the electromagnetic spectrum, with wavelengths greater than about 1.25×10^{-6} m, that is, frequencies less than 2.4×10^{14} Hz or energies less than 1 eV (1 eV = $1.602\,19 \times 10^{-19}$ J). Some ionising radiations are also electromagnetic, but with higher energies. These are the X-rays and γ-rays which typically have energies greater than 0.4 keV. Other ionising radiations are caused by beams of particles, including α-particles, β-particles, neutrons and cosmic ray particles. This account is concerned solely with ionising radiations. Indeed, discussion will be confined to those ionising radiations that arise naturally within the Earth, although there is no difference in either kind or effect between natural and artificial sources.

Ionising radiations are intensely harmful to human beings and people must be protected from unnecessary or excessive exposure. The risks associated with artificial sources, such as fall-out from nuclear weapons tests or radioactive waste from the nuclear power industry, are generally appreciated by the non-scientist. What is surpising to many people, however, is that, on average, ionising radiations of natural origin account for the largest proportion of the total exposure to the human population.

Although ionising radiations are dangerous, they can also be extremely useful. The use of X-ray systems in hospitals, for example, is probably the best-known application of artificial sources. Ionising radiations are also used therapeutically in the external treatment of cancer, where beams of high energy X-rays or γ-rays are used to irradiate malignant tissues and so to prevent the tumour cells from functioning. The property of certain materials to emit ionising radiations naturally is also utilised in medical tracer investigations (such as brain and body scans) and in the internal treatment of some cancers. Extensive applications also exist in industrial scanning and tracer work. Here, sources are usually artificially produced so that the material has the optimum characteristics for any particular application. As natural sources cannot be designed to specified limits, they are not used in medical or industrial materials. Their worth lies in what can be learned about the environment and materials in which they occur by measuring their activity and identifying their origin. It is this aspect of naturally radioactive materials with which this book is mainly concerned.

1.2 HISTORY OF RESEARCH

1.2.1 Discovery of radioactivity

Ionising radiations are more conveniently referred to as 'radioactivity'. This term, which is in common use today, was first introduced by

Marie Curie in 1898 to encompass all external manifestations of the processes occurring in materials which give rise to ionising radiations. However, the history of the study of radioactivity really began with the discovery of X-rays by Röntgen in 1895. Röntgen found that X-rays (or Röntgen rays as they were known for a while) were produced in a discharge tube by the action of cathode rays on the glass walls of the tube. He also discovered that, as X-rays were produced, a mild luminescence of the glass tube developed. There was reason to think that this might be related to the fluorescence shown by some natural materials which appear to glow. So, in 1896, Becquerel (Fig. 1.1) investigated some fluorescent compounds and found that one containing U did indeed produce natural ionising radiation. Like X-rays, the radiation was found to penetrate 'solid' matter, and so at first Becquerel considered the link established. However, later work by Becquerel showed that, whereas the fluorescence of U-bearing compounds was not always present depending upon whether or not they were crystallised and upon the growth environment of the crystallised form, they always emitted ionising radiation. Becquerel concluded that the radiation was not related to the fluorescence and that its emission was independent of the form of the U compound and of the environ-

Fig. 1.1 — Henri Becquerel (1852–1908).
From Romer 1964 (reproduced with permission).

ment in which it was found. He considered that the energy was somehow trapped in the U and resorted to calling it a form of phosphorescence, but one which apparently did not need to be replenished and which did not decrease with the passage of time. In fact, Becquerel anticipated the term radioactivity by referring to this phenomenon as 'radiations actives'. U at first remained unique as a repository of such energy but in 1898 Schmidt found that Th showed the same property. Once it was recognised that specific elements were radioactive, Marie Curie started to investigate the radioactivity of U minerals in detail. Surprisingly, she discovered that some minerals, notably pitchblende, torbernite, autunite and carnotite, were more radioactive than the quantity of U which they contained warranted. The activity was still too great even when the presence of Th was allowed for. Marie Curie therefore concluded that these minerals must contain some other radioactive element. From knowledge of the activities of U and Th, she also concluded that this element must be many times more active than U itself.

Pierre and Marie Curie then set about isolating this radioactive element using material from the pitchblende mines at Jachymov in the Krusne Hory mountains of Czechoslovakia (then known as Joachimsthal in the Erzgebirge). In fact, the Curies found two new radioactive elements, one they named polonium (after Poland, the country of Marie Curie's birth) and the other radium (which was discovered in collaboration with Bermont). In 1903, Becquerel and the Curies were awarded the Nobel Prize in Physics for their work on radioactivity, and in 1911 the Nobel Prize in Chemistry was awarded to Marie Curie for isolating radium (Pierre Curie having died in 1906). Marie Curie died in 1934 at the age of 67 years as a result of prolonged exposure to the radioactivity of Ra.

1.2.2 Transformation of elements

In 1899, Rutherford and Owens found that the radiation from Th varied with the passage of a current of air over Th-bearing materials. Rutherford then discovered that the air was continually ionised as though it were carrying some radioactive substance with the properties of a gas. This he called 'the emanation' and found that its activity decreased rapidly with time, falling by almost half with every minute that passed. Ra was also found to give off a radioactive gas, but in this case the activity persisted for much longer than that of the emanation. Rutherford also discovered an 'excited radioactivity' which seemed to come from a solid deposit laid down on surfaces where the emanation had been. To add further to the mystery, Crookes found in 1900 that he could apparently eliminate the radioactivity from U by removal of a highly radioactive substance which he named 'uranium X'. However, after a while the radioactivity of the U returned and no matter how often the U was purged of uranium X the radioactivity was regenerated (Figs 1.2 and 1.3). With Th too, Rutherford and Soddy found in 1902 that the radioactivity was lost with removal of a substance called 'thorium X' and that it was thorium X which produced the emanation (Fig. 1.4). Like U, the purged Th also regenerated its radioactivity with the passage of time and regained the power to produce the emanation, clearly by the new formation of thorium X. Hence as there was nothing but Th to produce thorium X, and as thorium X was chemically quite different from Th, it was clear to Rutherford and Soddy that transformation of the elements must be occurring to account for these observations. A similar hypothesis had been proposed in 1901 by both Becquerel and Perrin but was severely criticised by the Curies in 1902.

Thus the link between atomic transformation and the emission of energy was established. Detailed observation of the radiations by Rutherford had distinguished α-rays, which are strongly ionising but have little penetrating power, passing through only a few metres of air and being stopped by a sheet of paper, and β-rays, which cause little ionisation but are more penetrating, passing through several milli-

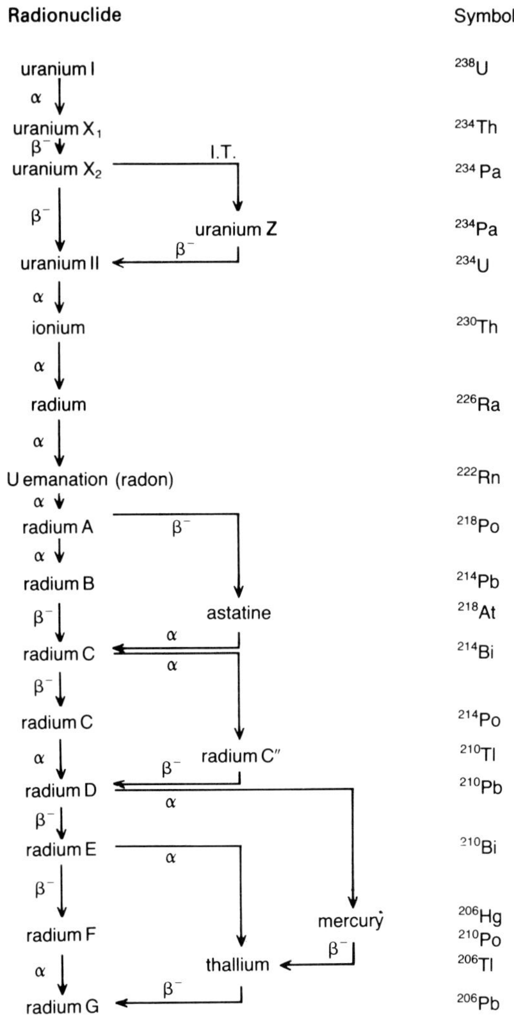

Fig. 1.2 — The classical ^{238}U decay series.

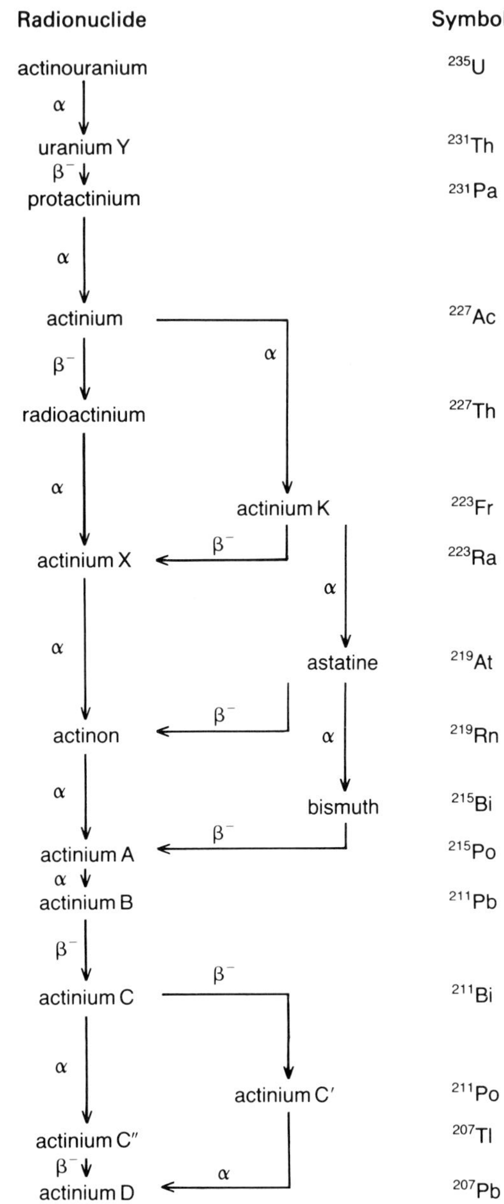

Fig. 1.3 — The classical ^{235}U decay series.

metres of Al sheet. In fact, the experiments which claimed to show that U and Th lose their radioactivity after the removal of uranium X and thorium X only allowed detection of the β-rays, the purged material still emitting α-rays. In 1900, Villard found a new type of radiation, very weakly ionising but much more penetrating than the β-rays. He called this type γ-radiation. Also in 1900, Becquerel found that β-rays are streams of high energy particles with a negative charge and a charge/mass ratio the same as that of cathode rays. These particles travel at speeds up to that of light (3×10^8m s^{-1}) and have an energy spectrum which controls their absorption profile. In 1902, Rutherford found that α-rays are also streams of

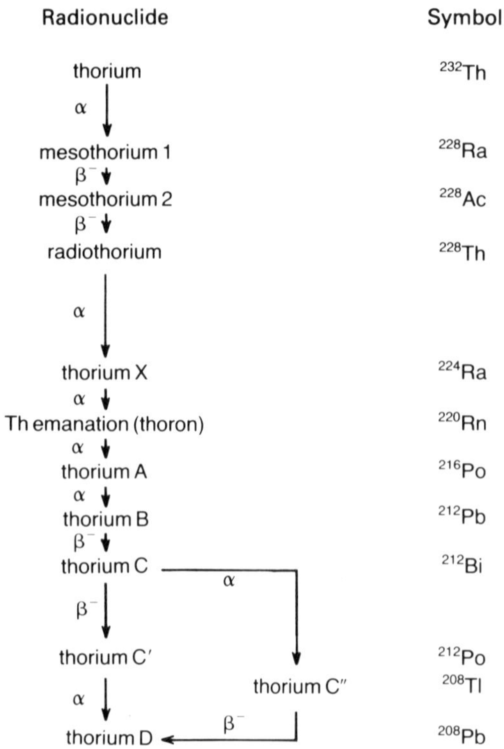

Fig. 1.4 — The classical ^{232}Th decay series.

particles but, unlike β-particles, α-particles are positively charged and are much larger and more massive than electrons. In fact, Rutherford realised that the energy of α-particles meant that they must account for almost all the energy released in the form of radiation during atomic transformations and, with Soddy, calculated that Ra releases approximately 2.5 J kg^{-1} $year^{-1}$. In contrast with the energy spectrum of β-particles, α-particles were found to have discrete velocities, travelling at about 2.5×10^7 m s^{-1}. The charge/mass ratio for the α-particle was also found to be about 1000 times that of the β-particle. So, to be sure that α-particles were the main means of energy transfer, Rutherford in 1903 studied the heat produced in a Ra sample from which all the emanation had been purged and in the emanation itself. He found that the heat produced was directly proportional to the intensity of their α-activity. Also in 1903, Ramsay and Soddy attempted to collect and identify the emanation from Ra. Instead of finding a new element, however, they discovered that part of the collected gas consisted of He. Since the Ra bromide which they used had been in solution and since He is a chemically inert gas, this was clear proof of the transformation of elements, as the He could only be formed from the Ra. The radioactive gas contained in the emanation was later identified as a separate element and named radon (Rn).

From work on the charge/mass ratio of H^+ ions, it was considered that, if the charge on the α-particle is equal to the charge of the H^+ ion, the mass of the α-particle is equal to about twice the mass of the H atom. However, if the charge on the α-particle is twice that of the H^+ ion, the mass of the α-particle will approximately equal that of a He atom. Rutherford preferred the latter interpretation as it agreed with the observations of Ramsay and Soddy, but support for this option only came in 1908 when Rutherford and Geiger made minute measurements of the charge on the α-particle and found it to be twice that of the H^+ ion. The conclusion was that α-particles, after losing their charge, become He atoms, a conclusion which was proved correct by Rutherford and Royds in 1909 when it was shown that the α-activity of the emanation caused the formation of He.

Becquerel, in 1903, had considered γ-rays to be similar to X-rays. It was not until 1912 that they were found to be diffracted by the atomic lattice of NaCl crystals and, therefore, to be electromagnetic in nature and indeed similar to light and X-rays but with a shorter wavelength than X-rays. In general, however, the theory of the transformation of elements accompanied by radioactive emission had been worked out by Rutherford for the Bakerian Lecture delivered to the Royal Society in May 1904 and published in 1905. A number of 'parent' radioactive elements were recognised (Ra, Th, U and Ac), each of which give rise to radioactive products.

These products, in turn, give rise to more radioactive products until a final stable product is formed. The possibility that Ra formed from U, Th or Ac was considered by Rutherford, and in a final footnote he mentioned that experimental work by Boltwood produced results which could support the idea that Ra formed from U. Rutherford also demonstrated that radioactivity decays with time and that in a transformation series the products are produced as fast as they decay.

Since Rutherford's Bakerian Lecture, considerable advances have been made in acquiring an understanding of the nature and details of radioactive emission and element transformation, and this is discussed in Chapter 3. However, it was during the period 1876–1904 that the foundations were laid.

Today, U and Th, with their daughter products (Figs 1.5, 1.6, and 1.7), are recognised as the two main sources of natural radiation. Usually in small quantities, these elements are

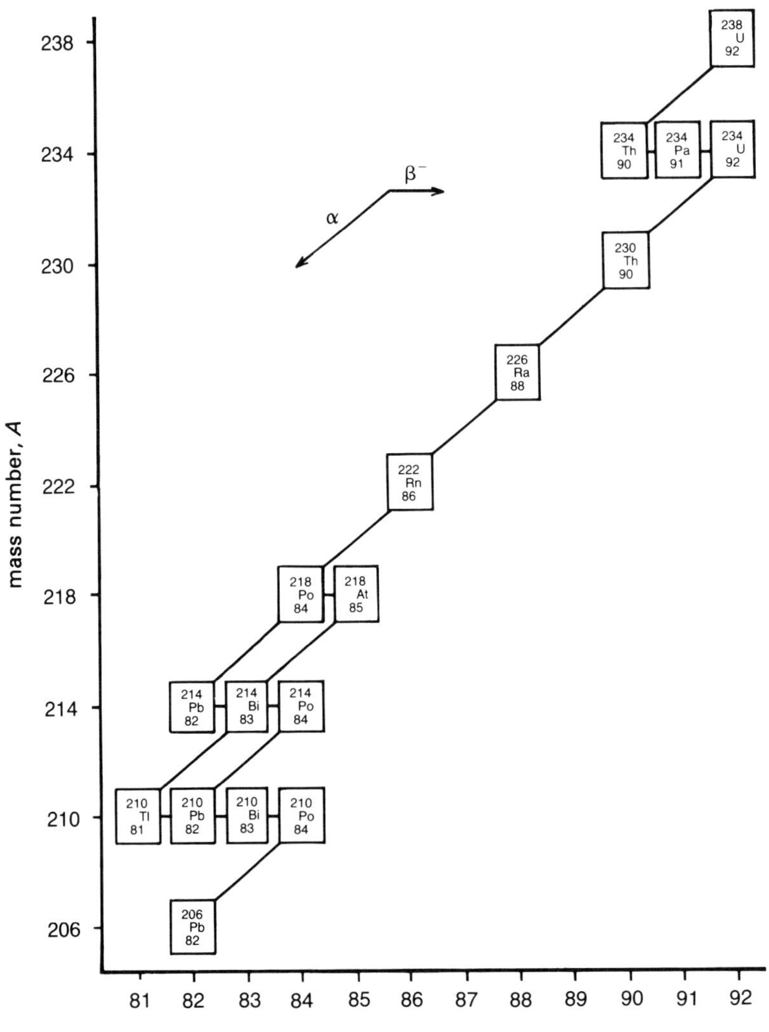

Fig. 1.5 — The ^{238}U decay series.

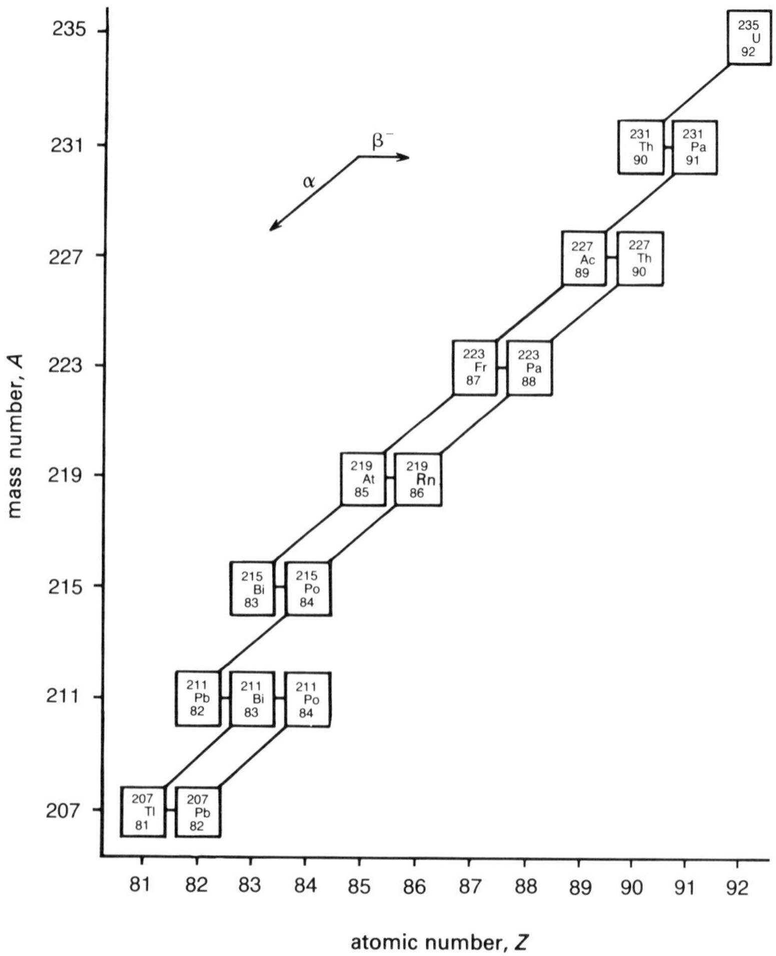

Fig. 1.6 — The ^{235}U decay series.

widely distributed in rocks and mineral deposits. In addition to these two main series, there have now been recognised a number of other naturally occurring radioactive elements. Chief amongst these in terms of activity is K, which at just over 2% is the seventh most abundant element in crustal rocks. Although K is a much weaker source of radiation than the U and Th series, because it is so abundant it is found to be an equal contributor to U and Th in the natural radiation flux.

Elements which undergo spontaneous transformation are said to be unstable or to decay. Interestingly, K is found to occur in both stable and unstable forms. This is also true of a number of other elements: for example, Rb and V. The explanation for this phenomenon was also discovered early in the history of the study of radioactivity. By 1911, some 40 radioactive substances had been detected. Many of them were identical in their chemical properties, and perhaps only a dozen or so quite separate elements in the chemical sense could be identified. Thus, in 1913, Soddy recognised that materials

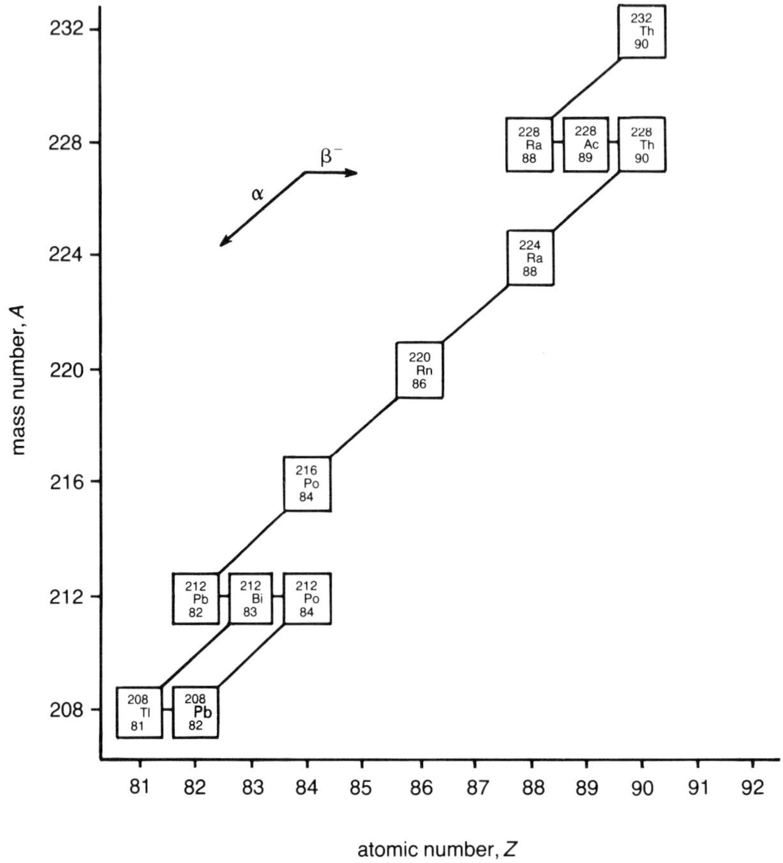

Fig. 1.7 — The ^{232}Th decay series.

exist which are single elements in their chemical behaviour but possess different physical origins. Following a suggestion by Todd, he called these variations 'isotopes'.

1.2.3 Rays and particles

The idea that an atom could be represented as a nucleus surrounded by a zone of negatively charged particles was put forward by Rutherford in 1911 as a result of his experiments on the scattering of α-particles. Until about 1930, it was believed that only three small subatomic particles existed: the electron, with a small mass and negative electrical charge (it was streams of this particle that produced cathode rays and β-rays); the proton, with a mass much larger than the electron but a positive charge equal to that of the negative charge of the electron (it is this particle which forms the nucleus of the H atom); the α-particle or He nucleus (which was supposed to be constructed of 4 protons and 2 electrons). Heavier atoms were thought to be constructed in some way from protons and electrons, which were thus the fundamental units from which the Universe was built. However, in 1931 it was found that, when certain light elements are bombarded with α-particles, a very penetrating non-ionising radiation is emitted. When this radiation encounters materials containing H, protons with very high velocities are ejected. Chadwick, in 1932, suggested that this radiation consisted of uncharged particles with the same mass as the proton and called the particles 'neutrons'.

Neutrons are now recognised as combining with protons to form the α-particle (2 neutrons and 2 protons) and to provide the explanation for the occurrence of isotopes of elements. Thus the chemical behaviour of an atom appeared to be governed by the number of protons in its nucleus and therefore defined the various elements, while the number of neutrons in the nucleus of an atom could vary and this variation defined the isotopes of the element.

Unfortunately, the term 'neutron' was originally suggested by Pauli in 1931 for a theoretical particle that would be of very small mass and no charge and that could be formed in nuclear transitions to account for an energy imbalance. The neutron, as discussed by Chadwick, has too large a mass to fit Pauli's description, and so the small particle was subsequently described as a 'neutrino' by Fermi. The existence of the neutrino as a separate particle was demonstrated in the late 1940s. β-particles are now recognised as forming from the breakdown of a neutron to give a proton, an electron and a neutrino (strictly speaking, an antineutrino).

The positive electron (β^+) or positron was discovered in 1933 by Anderson following a prediction of its existence by Dirac in 1931. Radioactive materials can emit a stream of positrons to bring about nuclear transformations. However, nuclear transformations with essentially the same results can also be due to the nuclear capture of an orbital electron, the transfer of charge number being identical and mass transfer only involving the mass of an electron. Emission of β^+-particles is accompanied by neutrino (*sensu stricto*) emission.

Following Chadwick's discovery of the neutron, Heisenberg proposed in 1932 that the elementary constituents of the nucleus are protons and neutrons. Because nuclear electrons (negatrons) are emitted as β^--particles during radioactive decay, it is possible that neutrons may be converted into protons by the emission of a β^--particle. It is also convenient to think of positrons as being generated at the time of their emission. It is now recognised that the atomic number of an element is the number of protons in the nucleus of an atom of that element and that an atom is electrically neutral, with the same number of extranuclear electrons as protons in the nucleus determining the chemical properties of the element. When atoms combine together to form larger uncharged units, the term 'molecule' is applied. Atoms and molecules may gain or lose extranuclear electrons and so acquire an electrical charge and become ions. The mass number of an atom is the number of protons plus neutrons (nucleons).

A strong nuclear force must exist which exceeds the electrostatic repulsion of individual protons and holds the nucleus together. Most species of atom can be characterised by the atomic number and the mass number and are called 'nuclides'. Nuclides of an element that have different numbers of neutrons are isotopes of that element. Some nuclides are stable but many are not. An unstable nuclide emits radiation as a result of its transformation or decay and so is radioactive. Such nuclides are called 'radionuclides'. Clearly, in radionuclides the strong force keeping the nucleus together is overcome.

More recent studies of atomic structures have resulted in the recognition of a much larger number of subatomic particles originating within the nucleus than the neutron–proton (heavy particle or 'hadron') association suggested by Heisenberg. For example, a family known as the 'strange hadrons' comprises the K^+, K^-, K^0, $\bar{K}^0$, Λ^0, Σ^+, Σ^-, Ξ^- and Ξ^0 particles. In any interaction between these, charge Q is conserved, and the total number of protons and neutrons does not change, even though the number of protons and neutrons in the nucleus may decrease because of the emission of α-particles. This conservation is considered to be related to the 'baryon number' B of the particle. Neutrons and protons have a baryon number of 1, but for leptons (electrons, positrons, neutrinos, etc.) and photons it is 0. However, not all hadrons have a baryon number of 1; pions, for example, are found in their interactions to have a value of 0. Hadrons with a baryon number of 0 are grouped together as 'mesons', while those with a value of 1 are called 'baryons'. In strong interactions of hadrons other than neutrons, protons and pions, it is also found that another property has to be conserved to account for observed reactions. This is called 'strangeness', which can have a value from −2 to +1. Strange hadrons are those of which the strangeness number *S* is other than 0 (that is, −2, −1 or +1). However, in some cases the decay of strange hadrons does not conserve strangeness (this is known as 'weak decay') although, in some weak decays, strangeness may be conserved (Table 1.1). These three quantum numbers, *B, Q* and *S*, and the control which they exercise on interactions, led Gell-Mann to develop in 1962 a classification system of subatomic particles which allowed him successfully to predict the existence of previously undiscovered particles. No new combination of *B, Q* and *S* properties of subatomic particles have been discovered since 1964. The reason for the occurrence of these

Table 1.1 — Properties of some representative hadrons.

Symbol	Name	Mass (MeV c^{-2})	*B*	*Q*	*S*	Half-life (s)	Main decay
π^0	Pi zero	135	0	0	0	0.6×10^{-16}	$\pi^0 \rightarrow \gamma + \gamma$
π^+	Pi plus	140	0	+1	0	1.8×10^{-8}	$\pi^+ \rightarrow \mu^+ + \nu$
π^-	Pi minus	140	0	−1	0	1.8×10^{-8}	$\pi^- \rightarrow \pi^- + \nu$
K^+	K plus	494	0	+1	+1	0.9×10^{-8}	$K^+ \rightarrow \mu^+ + \nu$
K^-	K minus	494	0	−1	−1	0.9×10^{-8}	$K^- \rightarrow \mu^- + \nu$
K^0	K zero	498	0	0	+1	0.6×10^{-10}	$K^0 \rightarrow \pi^+ + \pi^-$
$\bar{K}^0$	K zero bar	498	0	0	−1	0.6×10^{-10}	$\bar{K}^0 \rightarrow \pi^+ + \pi^-$
η^0	Eta zero	549	0	0	0	0.5×10^{-18}	$\eta^0 \rightarrow \gamma + \gamma$
p	Proton	938	1	+1	0	STABLE	NONE
n	Neutron	940	1	0	0	0.6×10^3	$n \rightarrow p + e^- + \nu$
Λ^0	Lambda zero	1116	1	0	−1	1.8×10^{-10}	$\Lambda^0 \rightarrow p + \pi^-$
Σ^+	Sigma plus	1189	1	+1	−1	0.6×10^{-10}	$\Sigma^+ \rightarrow p + \pi^0$
Σ^0	Sigma zero	1192	1	0	−1	$\sim 10^{-19}$	$\Sigma^0 \rightarrow \Lambda^0 + \gamma$
Σ^-	Sigma minus	1197	1	−1	−1	1.0×10^{-10}	$\Sigma^- \rightarrow n + \pi^-$
Δ^{++}	Delta double plus	1235	1	+2	0	0.4×10^{-23}	$\Delta^{++} \rightarrow p + \pi^+$
Δ^+	Delta plus	1235	1	+1	0	0.4×10^{-23}	$\Delta^+ \rightarrow p + \pi^0$
Δ^0	Delta zero	1235	1	0	0	0.4×10^{-23}	$\Delta^0 \rightarrow n + \pi^0$
Δ^-	Delta minus	1235	1	−1	0	0.4×10^{-23}	$\Delta^- \rightarrow n + \pi^-$
Ξ^0	Xi zero	1315	1	0	−2	2.1×10^{-10}	$\Xi^0 \rightarrow \Lambda^0 + \pi^0$
Ξ^-	Xi minus	1321	1	−1	−2	1.1×10^{-10}	$\Xi^- \rightarrow \Lambda^0 + \pi^-$
Σ^{*+}	Sigma star plus	1385	1	+1	−1	1.3×10^{-23}	$\Sigma^{*+} + \Lambda^0 + \pi^+$
Σ^{*0}	Sigma star zero	1385	1	0	−1	1.3×10^{-23}	$\Sigma^{*0} \rightarrow \Lambda^0 + \pi^0$
Σ^{*-}	Sigma star minus	1385	1	−1	−1	1.3×10^{-23}	$\Sigma^{*-} \rightarrow \Lambda^0 + \pi^-$
Ξ^{*0}	Xi star zero	1530	1	0	−2	0.5×10^{-22}	$\Xi^{*0} \rightarrow \Xi^- + \pi^+$
Ξ^{*-}	Xi star minus	1530	1	−1	−2	0.5×10^{-22}	$\Xi^{*-} \rightarrow \Xi^0 + \pi^+$

B, baryon number; *Q*, charge; *S*, strangeness; γ, photon; ν, neutrino; μ, muon; e, electron.
($\mu^+ \rightarrow e^+ + \nu + \nu$; $\mu^- \rightarrow e^- + \nu + \nu$). Mass M given in terms of energy E by $E = Mc^2$ where c is the speed of light in a vacuum.
(Reproduced, with permission, from Open University (1979); copyright (1979) The Open University Press.)

combinations was in 1964 considered by Gell-Mann, and independently by Zweig, to be due to their being composed of even simpler particles. Zweig called these 'aces' but Gell-Mann's suggestion of 'quark' (from *Finnegan's Wake* by James Joyce, a nonsense word) has become accepted. Despite the results of Fairbank's experiments in 1977, it is unlikely that quarks have yet been detected, and so their physical existence is still in doubt. Nevertheless, the idea that hadrons are made of quarks led to the prediction of a new property of hadrons called 'charm' C. This is a fourth quantum number and has the value 1, 0 or -1. Although protons and neutrons, for example, have $C=0$, charmed hadrons have recently been discovered. Like strangeness, charm is conserved in strong interactions but not in weak interactions. The existence of four quantum numbers for subatomic particles is in good agreement with predictions of a four-quark model. However, clear evidence supporting the existence of six quarks, known as 'up', 'down', 'strange', 'charm', 'beauty' and 'truth', is now available. Each of these is considered to have three specific properties known as 'colours'. Thus the fundamental particles are considered to be quarks and leptons. The existence of other particles ('gauge particles'), associated with the forces between quarks and leptons, is shown by the evidence of weak force exchange particles, known as 'W' and 'Z' particles, obtained in 1982 and 1983 by Rubbia. Another gauge particle is the photon, which imparts electromagnetic force. Finally, it is important to note that the discovery of the positron as, effectively, a positively charged electron was perhaps the first hint that particles are accompanied by antiparticles of opposite quantum number. Thus the proton has its opposite in the antiproton, first detected in 1955, while other pairs are K^+, K^-, π^+, π^-, etc. If a particular quantum number is 0 for a particle, it is also 0 for the antiparticle. It therefore follows that, if quarks exist, the existence of antiquarks is not impossible, and is indeed required by the quark model of hadrons.

When the weak force breaks down, each neutron in the nucleus of a radioactive atom changes into a proton, an electron and an antineutrino, but the strong force which holds the nucleus together is much greater and is carried by gauge particles, as yet unidentified, called 'gluons'. The strong force is about 10^2 greater than electromagnetic forces and about 10^7 greater than the weak force.

1.3 USES OF RADIOACTIVITY

1.3.1 Radioactivity as a geological tracer

Natural radioactivity has been utilised in many ways in geological investigations. As all rocks are radioactive to a greater or lesser degree, it was soon recognised that, by mapping the variation in intensity of the natural radioactivity in an area, a preliminary assessment of its geological structure could be obtained. This can be extremely rapid if the survey work is carried out from an aircraft. Preliminary geological maps can thus be prepared at very little cost. The radioactivity of rocks and minerals is widely used in the dating of geological materials, as the rate of radioactive decay is independent of the chemical and physical environments and changes in those environments. The rate of decay for a particular nuclide is governed by the half-life of the decay which can be defined as the time taken for a given number N of atoms of that radionuclide to decay to half that number, that is, $N/2$. To enable dating to be carried out accurately, half-lives that are of the same order as the age of the material to be dated are required. Fortunately, natural radionuclides have half-lives which range from less than 1 s to more than 10^{10} years, and so a very wide range of dating is possible, including estimates of the age of the Earth and the Solar System.

The radioactivity of certain minerals permits their easy identification and their use as tracers of geological processes, such as differentiation sequences in igneous rocks. The unravelling of complex geological histories becomes possible by studying the mobility of radionuclides within minerals and rocks. The transfer of radionuclides according to their geochemical properties has also been used to trace the evolution of the crust and mantle of the Earth, to follow the hydrological cycle and to account for some aspects of the composition of the atmosphere.

The energy liberated during radioactive decay is now considered to be one of the main sources of heat within the Earth. Indeed, the driving force for lithospheric plate movements, mountain building and vulcanicity may have an origin in the presence of radionuclides. Chemical transfer of elements and the evolution of the mantle and crust can be traced by, and owe their origin to, the presence of radionuclides. Even on a local scale, the presence of radionuclides is considered to give rise to areas of high heat flow in crustal rocks. This in turn causes convective circulation of groundwater which can lead to the formation of hydrothermal systems and associated mineralisation.

Many of these applications are of more than academic interest, but the main use of natural radioactivity is in the detection and evaluation of U-bearing and, to a lesser extent, Th-bearing ore deposits.

1.3.2 Industrial uses

U has become an important raw material since the discovery that its naturally occurring isotopes do not decay just by emission of α-particles and β-particles but also by splitting of the nucleus into two nuclides with approximately equal mass numbers: Ba and Kr, for example. This phenomenon is known as nuclear fission. Accompanying the disintegration of the nucleus is the release of a large quantity of energy. It is this process in ^{235}U which is utilised as the energy source for an atomic bomb or a nuclear power station. Fission may be made to occur artificially in materials other than U, but U is still the main raw material.

Before the discovery of radioactivity, U ores were used only as a pigmentation agent but, when Ra became of therapeutic importance in the treatment of cancer, U compounds were produced as a byproduct of Ra extraction, although these were commodities for which there was little initial demand. The chief use for U oxides remained as a colouring agent for pottery and glass, with smaller amounts used in the textile industry, photography and analytical chemistry. In the ceramics industry, U salts have been used to produce a fine glaze, yellow, orange, brown, green or black in colour. Uranium yellow, which is either the Na, K or ammonium diuranate, gives glass an opalescent yellow colour and the glazes are highly fluorescent. Ferro-uranium has been tried as an addition to special steels. It is said to impart toughness and hardness to tool steels. Some U was also used in the manufacture of luminous paints. Finally, it is interesting to note that U compounds have been incorporated in false teeth in carefully controlled amounts to emulate the fluorescence of natural teeth. It is especially important to judge the amount correctly as with too much fluorescence the teeth appear to glow in artifical light, while with too little they appear black. However, this use of radionuclides is now discouraged on radiological safety grounds as non-radioactive substances giving the same effect are now available.

Unlike U, naturally occurring Th is not an important fissionable element but it is a source of fertile material because it can be made to produce fissile decay products by artificial neutron bombardment. Th has been used for the manufacture of gas mantles and for medical purposes. Fissionable nuclides can be produced from naturally occurring Th in breeder reactors.

1.4 HEALTH HAZARDS

The health hazard usually associated with radioactive material is the effect of irradiation itself (Fig. 1.8). However, many of the naturally occurring radioactive elements are also chemically toxic in the same way that non-radioactive heavy metals, such as Pb, As and Sb, are toxic if ingested or inhaled as dust. Perhaps more insidious, however, is the fact that both U and Th are responsible for the evolution of radioactive gases. Inhalation of

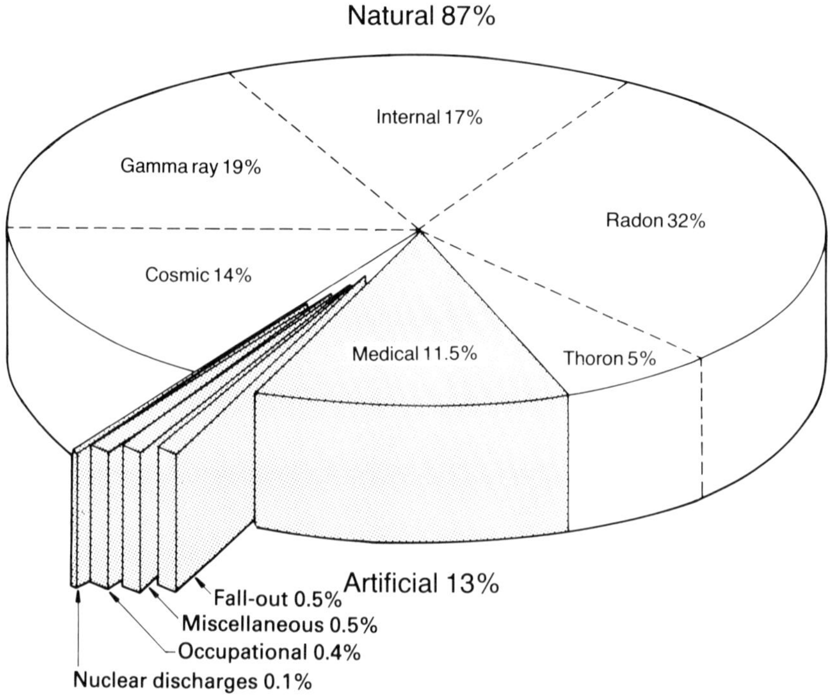

Fig. 1.8 — The average annual radiation dose to the population of the UK (National Radiological Protection Board, 1986, reproduced with permission.)

these gases is an almost unavoidable occurrence in the natural environment.

Ingestion of toxic elements should, of course, be avoided at all times, whether they are radioactive or not. Sensible precautions such as careful washing after handling such materials and before eating should always be taken. When dusty atmospheres are present, facemasks are a sensible means of preventing inhalation of dust. It is also good practice not to smoke while handling radioactive materials. If the problems of toxicity encountered with the stable nuclides are a cause for concern, ingestion and inhalation of radionuclides are a double risk, for in addition to the chemical toxicity is the effect of irradiation. Where the inhalation of toxic dust is concerned, whereas soluble dust can be hazardous in terms of chemical toxicity, it is insoluble dust which is more dangerous in terms of radiation because the particles may be deposited more or less permanently in the lungs. However, where soluble dust contains radionuclides which are absorbed and stored by the body, as is the case with Ra, then this too is hazardous. For example, Ra is chemically similar to Ca and is laid down in bone. This results in irradiation of bone marrow and an increase in the chance of contracting bone cancer. It is well documented that American Ra luminisers, who were in the habit of licking their paint brushes to obtain a fine point for painting small figures on dials, had a higher than normal incidence of bone cancer (United Nations Scientific Committee on the Effects of Atomic Radiation, 1982).

When radiation enters tissue, energy is transferred to that tissue. As α-particles and β-particles possess electrical charges, the energy transfer to the atoms or molecules in the tissue takes the form of an electrical interaction. The result may be two-fold. Firstly, the atoms in the tissue may be converted to positively charged ions by the ejection of electrons which in turn, may then ionise other atoms in the tissue, producing a cascading effect. Secondly, although the electrons may receive energy from the radiation, it may be insufficient to cause their ejection, simply leaving them in an excited state; this excitation energy is then dissipated as heat. Although β-particles are more penetrating than α-particles, the ionising and excitation effects produced by α-particles are greater than those for β-particles. However, because their penetration is so limited, external sources of α-particles are not particularly dangerous as even the dead outer layer of skin is only penetrated with difficulty. In contrast, β-particles may penetrate a centimetre or so into the living tissue; external sources can, therefore, be hazardous to the superficial part of the body, although internal organs are not at risk from external sources of these radiations. Both α-particle and β-particle sources, however, become extremely dangerous if taken inside the body where the short penetration of the α-particles can render sensitive organs susceptible to damage. The lungs and intestines are particularly vulnerable (Tables 1.2 and 1.3).

The main effect of the ionisation produced by radiation is the breakdown of water molecules into H^+ ions and OH^- ions. The OH^- ion is particularly reactive and, as the cells of tissue are largely composed of water, the resulting effect can be very significant. Of particular importance is deoxyribonucleic acid (DNA) which controls the structure and function of the cell and determines the inheritance of cell characteristics during cell reproduction. DNA molecules can be damaged by being directly ionised by radiation (producing abnormal chemical reactions) or by reacting with OH^- ions produced elsewhere by radiation. In both cases the chemical results may have a harmful biological effect, particularly if abnormal multiplication of cells occurs, causing cancer.

γ-radiation is only very weakly ionising, but it is much more penetrating that α-particles and β-particles and so all the internal organs of the body are at risk, even from external sources. The results of interaction of γ-rays with tissue are very similar to those produced by α-particles and β-particles. Electrons are liberated from the atoms and molecules of tissue and cause abnormal chemical reactivity. Most of the external radiation received by ordinary men

Table 1.2 — Main harmful effects of radiation

Early appearance	
	Very high dose and dose rate to
Death	much of body
Erythema	area of skin
Sterility	testes and ovaries
Late appearance	
Malignant diseases	Any dose or dose rate, probability depends on dose. May be manifested long after exposure
Non-malignant changes	Very high dose. May be manifested at various times
Developmental changes	Irradiation of embryo. Manifested at birth
Hereditary defects	Any dose or dose rate, probability depends on dose. Manifested in descendants

(National Radiological Protection Board, 1981, reproduced with permission.)

Table 1.3 — Risk of serious hereditary defects in the first two generations and of fatal cancers from irradiation of various tissues and organs.

Tissue or organ	Risk (Sv^{-1})
Gonads (hereditary)	1 in 250
Breast	1 in 400
Red bone marrow	1 in 500
Lung	1 in 500
Thyroid	1 in 2000
Bone surfaces	1 in 2000
Remainder of body	1 in 200
Whole body	1 in 60

(National Radiological Protection Board, 1981, reproduced with permission.)

and women is γ-radiation from natural sources, the radionuclides in the Earth irradiating the whole body more or less uniformly. Since most building materials are either natural or derived from natural materials, they too are usually radioactive to some degree. Therefore, radiation is received indoors as well as in the open air.

The amount of natural radiation received from rocks and building materials varies from one locality to another with the geology and the types of material used in house building. Because the U, Th and K contents of the granites of southwest England are all high, anyone living on one of these granites will receive more radiation than if they lived on the Chalk Downs of Hampshire. Indeed, if the house on the granite also happened to be built of granite blocks, the γ-radiation received would be even higher. The temptation may be to live in wooden houses on the Chalk Downs (Fig. 1.9), but the increase experienced by most people in a granite area is small and rarely exceeds double the average for the UK as a whole.

Clearly, however, some selection of building materials is sensible, and it is interesting to note in this context the abandonment of a possible source of relatively inexpensive $CaSO_4$ (gypsum) for plasterboards because of a high radioactivity caused by the presence of Ra. In this case the $CaSO_4$ was a byproduct (known as phosphogypsum) of the production of inorganic fertiliser by treatment of natural phosphatic rocks with H_2SO_4. Phosphatic rocks tend to have high U contents and hence high levels of Ra.

The avoidance of natural radiation is, in fact, impossible unless you live in a Pb-lined box several thousand metres underground for, even if α-particles, β-particles and γ-rays from the Earth are excluded, all of us are subject to bombardment by cosmic rays. This radiation consists of high energy charged particles, some of which originate in the Sun while others appear to originate outside the Solar System. Those of solar origin are more numerous at times of solar flares. These particles pass through even the most dense materials. Their intensity decreases with decreasing altitude and, because they are affected by the Earth's magnetic field, there is also a decrease from the polar regions towards the equator.

In terms of natural sources of danger, α-particles, β-particles and γ-rays originating from rocks and building materials, and cosmic rays, are much less hazardous than the radioactive gases produced by U and Th. Outdoors, concentrations of these gases in the atmosphere are negligible as the gas given off by the ground is rapidly dispersed in the air. However, when the gases enter a building, either from the walls or through the floor, the concentrations of gas can build up, especially if the building is poorly ventilated. Although the radioactivity of the gases is not particularly dangerous itself (even though they are α-particle emitters), the gases decay with short half-lives to form solid daughter radionuclides. When one of these gases is inhaled, some decay will take place within the lung, producing solid daughters which may not be exhaled. As these solid daughters often attach themselves to dust particles in the air, they may also be inhaled directly and may be retained within the lung. Thus the lung becomes directly exposed to radiation. It is considered by the National Radiological Protection Board that this is the

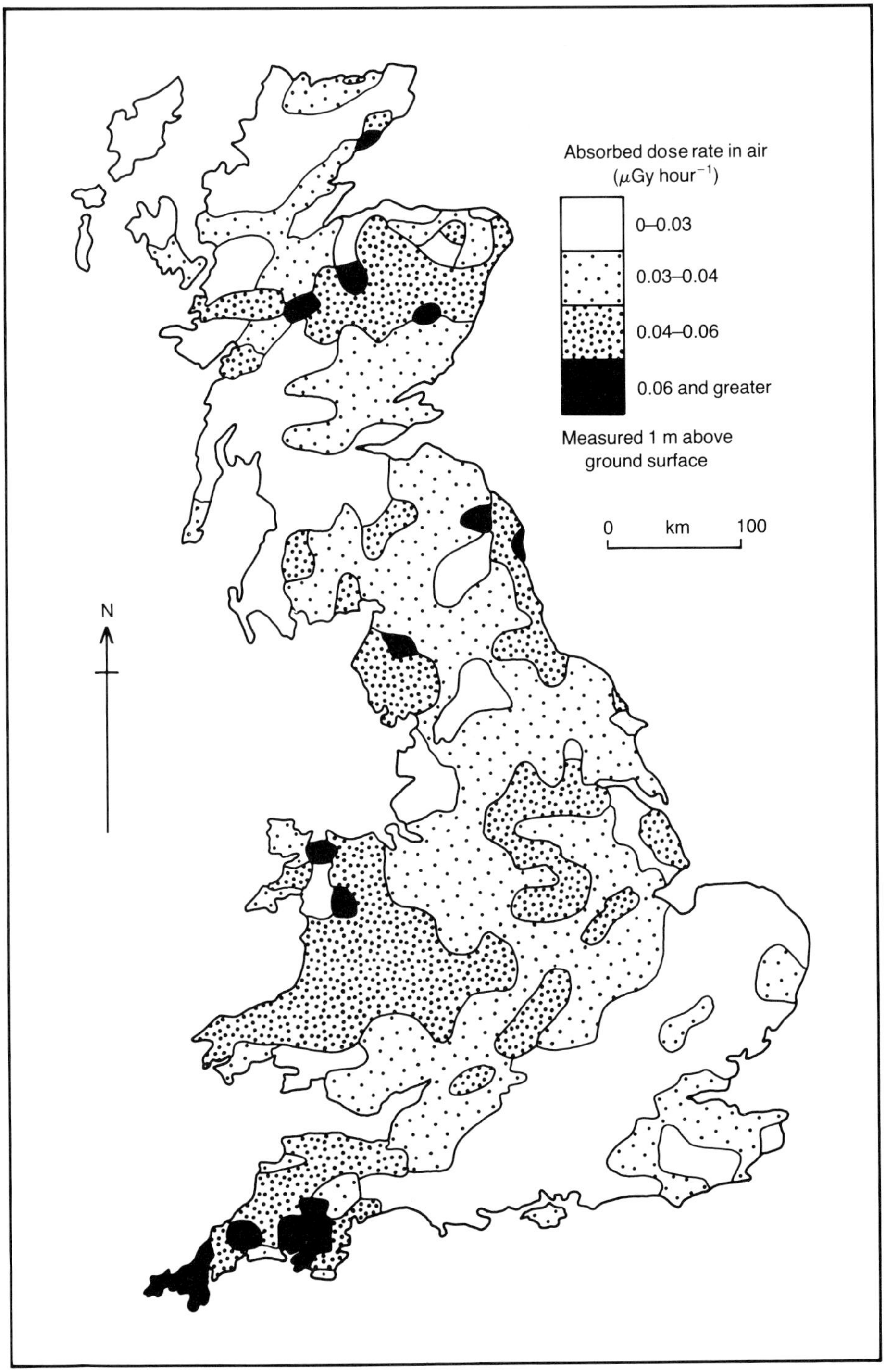

Fig. 1.9 — Terrestrial γ-ray dose rates out of doors in Great Britain. (Bradley and Green, 1984, redrawn with permission.)

main source of the natural radiation dose received in the UK. Moreover, this exposure is very variable when individuals are considered. In the UK, measurements have shown that doses to some individuals can be more than two orders of magnitude greater than the average exposure for the population as a whole. Even so, when compared with other risks to life, radiation exposure presents only a small danger (Table 1.4).

Table 1.4 — Average annual risk of death in the UK

Cause	Risk
Smoking ten cigarettes a day	1 in 250
Natural causes at 40 years of age	1 in 850
Road accidents	1 in 9500
Accidents in the home	1 in 26 000
Accidents at work	1 in 43 500
Radiation exposure of 1 mSv $year^{-1}$	1 in 80 000

(After National Radiological Protection Board, 1986.)

BIBLIOGRAPHY

Anderson, C. D. (1933) The positive electron. *Phys. Rev.* **43**, 491.

Auerbach, C. (1956) *Genetics in the atomic age.* Edinburgh, Oliver and Boyd.

Becquerel, A. H. (1896) Sur les radiations émises par phosphorescence. *C. R. Acad. Sci.* **122**, 420.

Becquerel, A. H. (1896) Sur les radiations invisibles émises par les corps phosphorescents. *C. R. Acad. Sci.* **122**, 501.

Becquerel, A. H. (1896) Sur les radiations invisibles émises par les sels d'uranium. *C. R. Acad. Sci.* **122**, 689.

Becquerel, A. H. (1896) Sur les propriétés différentes des radiations invisibles émises par les sels d'uranium, et du rayonnement de la paroi anticathodique d'un tube Crookes. *C. R. Acad. Sci.* **122**, 762.

Becquerel, A. H. (1896) Emission de radiations nouvelles par l'uranium métallique. *C. R. Acad. Sci.* **122**, 1086.

Becquerel, A. H. (1899) Note sur quelques proprietes du rayonnement de l'uranium et des corps radio-actifs. *C. R. Acad. Sci.* **128**, 771.

Bennet, D. J. (1972) *The elements of nuclear power.* London, Longmans.

Boslough, J. (1985) Worlds within the atom. *Natl. Geog.* **167**, 634.

Bowie, S. H. U. (1976) Whither uranium. *Trans. Inst. Min. Metall. B* **85**, 163.

Bradley, E. J. and Green, B. M. R., 1984. Outdoor γ-ray dose rates in Great Britain: preliminary results. *Radiat. Prot. Bull.* **59**.

Bridges, B. A. (1976) *Bacterial reaction to radiation.* Shildon, Durham, Meadowfield.

Callow, R. J. (1967) *The industrial chemistry of the lanthanons, yttrium, thorium and uranium.* Oxford, Pergamon.

Chadwick, J. (1932) The existence of the neutron. *Proc. R. Soc. Lond. A* **136**, 692.

Cheng, D. C. (1979) *Elementary particle physics: an introduction.* Reading, Massachusetts, Addison-Wesley.

Close, F. E. (1979) *An introduction to quarks and partons.* London, Academic Press.

Coggle, J. E. (1971) *Biological effect of radiation.* London, Wykeham.

Crookes, W. (1900) Radio-activity of uranium. *Proc. R. Soc. Lond. A* **66**, 409.

Curie, P., and Curie, S. (1898) Sur une substance nouvelle radio-active contenue dans la pechblende. *C. R. Acad. Sci.* **127**, 175.

Curie, P., Curie, M., and Bemont, G. (1898) Sur une nouvelle substance fortement radioactive, contenue dans la pechblende. *C. R. Acad. Sci.* **127**, 1215.

Curie, P., and Curie, M. (1902) Sur les corps radio-actifs. *C. R. Acad. Sci.* **134**, 85.

Curie, Sklodowska (1898) Rayons émis par les composés de l'uranium et du thorium. *C. R. Acad. Sci.* **126**, 1101.

Darnley, A. G. (1975) Geophysics in uranium exploration. In Williams, R. M. (ed.), Uranium to 2000, an exploration challenge. *Paper Geol. Surv. Can.* **75–26**, 13.

Dyck, W. (1975) Geochemistry applied to uranium exploration. In Williams, R. M. (ed.), Uranium to 2000, an exploration challenge. *Paper Geol. Surv. Can.* **75–26**, 33.

Fleck, A. (1957) Frederick Soddy. *Biogr. Mem. Fellows R. Soc.* **3**, 202.

Haddow, A. (1952) *Biological hazards of atomic energy.* Oxford, Clarendon.

Halnan, K. E. (1957) *Atomic energy in medicine.* London, Butterworths.

Henderson, P. (1982) *Inorganic geochemistry.* Oxford, Pergamon.

Hughes, I. S. (1972) *Elementary particles.* Harmondsworth, Penguin.

Hughes, J. S., and Roberts, G. C. (1984) The radiation exposure of the U.K. population. *NRPB* **R-173**. London, National Radiological Protection Board.

Hyde, E. K. (1971) *Fission phenomena.* New York, Dover Publications.

International Atomic Energy Agency (1977) *Isotopes in day-to-day life.* Vienna, International Atomic Energy Agency.

Kaufman, W. J. (1980) *Particles and fields.* San Francisco, California, Freeman.

Kriz, J. A. (1970) *Mesons.* Brno, Grafia.

Lakshmanan, V. I., and Lawson, G. J. (1975) Radiotracers in mineral engineering. In Nichol, A. W. (ed.), *Physicochemical methods of mineral analysis.* New York, Plenum.

Leader, E., and Predazzie, E. (1982) *An introduction to gauge theories and the 'new physics'.* Cambridge, Cambridgeshire, Cambridge University Press.

Mann, W. B., Ayres, R. L., and Garfinkel, S. B. (1980) *Radioactivity and its measurement.* Oxford, Pergamon.

Martin, A. D., and Spearman, T. D. (1970) *Elementary particle theory.* Amsterdam, North-Holland.

Ministry of Labour (1965) *Code of practice for the protection of persons exposed to ionising radiations.* London, HMSO.

Morgan, K. Z., and Turner, J. E. (1967) *Principles of radiation protection: a textbook of health physics.* New York, Wiley.

National Radiological Protection Board (1981) *Living with radiation.* London, HMSO.

National Radiological Protection Board (1986) *Living with radiation.* London, HMSO.

Oehme, F. W. (1979) *Toxicity of heavy metals in the environment.* New York, Dekker.

Open University (1979) The search for fundamental particles. *S101 Unit 31* Milton Keynes, Open University.

Organisation for Economic Cooperation and Development (1975) *Uranium: resources, production, demand.* Paris, Organisation for Economic Cooperation and Development.

Organisation for Economic Cooperation and Development (1980) *World uranium: geology and resource potential.* Vienna, International Atomic Energy Agency.

Owens, R. B. (1899) Thorium radiation. *Phil. Mag.* **48**, 360.

Plant, J., Ostle, D., and Miller, J. (1983) Natural levels of uranium in Britain — economic and environmental significance. *Rep. Inst. Geol. Sci. Lond.* **83-1**, 32.

Ramsay, W., and Soddy, F. (1903) Experiments in radioactivity and the production of helium from radium. *Proc. R. Soc. Lond. A* **72**, 204.

Regato, J. A. del (1967) *Proceedings of the Conference on radiobiology and radiotherapy, Monogr. 24.* National Cancer Institute.

Romer, A. (1964) *The discovery of radioactivity and transmutation.* New York, Dover Publications.

Röntgen, W. C. (1895) Uber eine neue Art von Strahlen. *Sitsungsber. Phys. Med. Ges. Wurzberg,* 132.

Rutherford, E. (1900) A radioactive substance emitted by thorium compounds. *Phil. Mag.* **49**, 1.

Rutherford, E. (1903) The magnetic and electric deviation of the easily absorbed rays from radium. *Phil. Mag.* **5**, 177.

Rutherford, E. (1904) The succession of changes in radioactive bodies. *Phil. Trans. R. Soc. Land. A* **204**, 169.

Rutherford, E. (1904) Slow transformation products of radium. *Phil. Mag.* **8**, 636.

Rutherford, E. (1911) The scattering of α and β particles by matter and the structure of the atom. *Phil. Mag.* **21**, 669.

Rutherford, E., and Geiger, H. (1908) An electrical method of counting the number of α particles from radio-active substances. *Proc. R. Soc. Lond. A* **81**, 141.

Rutherford, E., and Owens, R. B. (1899) Thorium and uranium radiation. *Trans. R. Soc. Can.* **5**, 9.

Rutherford, E., and Royds, T. (1909) The nature of the α-particle from radioactive substances. *Phil. Mag.* **17**, 281.

Rutherford, E., and Soddy, F. (1903) Radioactive change. *Phil. Mag.* **5**, 576.

Segré, E. G. (1977) *An introduction to nuclear and subnuclear physics.* Reading, Massachusetts, Benjamin.

Soddy, F. (1913) The radio-elements and the periodic law. *Chem. News* **107**, 97.

Spriggs, M. J., and Casteel, K. D. (1978) *Uranium supply and demand.* London, Mining Journal.

Trefil, J. S. (1980) *From atoms to quarks: an introduction to the strange world of particle physics.* London, Athlone.

United Kingdom Atomic Energy Authority (1976) *Evidence submitted to the Royal Commission on environmental pollution.* London, HMSO.

United Nations Scientific Committee on the Effects of Atomic Radiation (1982) *Ionising radiation: sources and biological effects.* New York, United Nations.

Wang, C. H., and Wills, D. I. (1965) *Radiotracer methodology in biological science.* Englewood Cliffs, New Jersey, Prentice-Hall.

Yuan, L. C. L. (1971) *Elementary particles. Science, technology and Society.* New York, Academic Press.

2

An introduction to geochemistry

2.1 ATOMIC STRUCTURES

An electrically neutral atom must have equal numbers of protons and electrons. The number of protons in the nucleus is the atomic number which defines the element, while the electronic configuration of the atom controls its chemical characteristics. The presentation of the various elements is in the form of a periodic table, in which those elements with similar chemical characteristics are placed together. The chief difficulty with the Rutherford model of the atom, in which a positively charged nucleus is surrounded by negatively charged electrons, was to explain how such an arrangement could be stable. If the electrons were at rest, they would be attracted to the nucleus and, if orbiting the nucleus, they would be continuously accelerated towards it and radiating energy (according to the Maxwell theory that an accelerated charge radiates energy) until they eventually lost all their energy and fell into the nucleus. An explanation was offered in 1913 by Bohr who proposed that there were certain orbits in which electrons could rotate while emitting no radiation at all. These were quantum-related orbits for which the angular momentum of the electron was given by integral multiples of $h/2\pi$ where h is Planck's constant. Later, de Broglie modified this in terms of the wave relationship $\lambda = h/p$, where λ is the wavelength and p the momentum, for $2\pi r = n\lambda$, where r is the radius of a particular orbit. The angular momentum of an electronic particle is pr and so, in an orbit for which $2\pi r = n\lambda$, this must equal $nh/2\pi$. In this system a quantum of energy would be emitted when an electron moved from one of these orbits to another, the size of the quantum being equal to the difference between the energies of the electrons in the initial and final states (Fig. 2.1).

The characteristic separation between electrons means that, as the number of electrons required to give electrical neutrality to an atom increases, the packing arrangement becomes more complex. In fact, the packing takes place in a series of electron shells, named K, L, M, N, O, P and Q, which can hold 2, 8, 18, 32, 18, 18 and 2 electrons respectively (as far as naturally occurring elements are concerned). This is the basis of the seven periods.

Within these shells are subshell levels, termed s, p, d and f, capable of holding 2, 6, 10 and 14 electrons respectively. Thus the L shell has subshells s and p while the N shell has subshells s, p, d and f. In the first period of the periodic table there is only one electronic subshell, and this is occupied either by 1 electron (in H) or 2 electrons (in He); the second and third periods have s and p subshells and the elements possess between 3 electrons (Li) and 10 electrons (Ne) in the second period, and between 22 electrons (Na) and 18 electrons (O) in the third period. In fact, each period can be seen to be built upon a filled core of noble gas structure: the second period builds on a He core (2 electrons), the third period on a Ne core (10 electrons), the fourth period on an Ar core (18 electrons), the fifth period on a Kr core (36 electrons), the sixth period on a Xe core (54 electrons) and the seventh period on a Rn core (86 electrons). This is shown in Tables 2.1 and 2.2.

Generally, the outer subshells are occupied after the inner subshells have been filled. There are exceptions, however, 4s being occupied

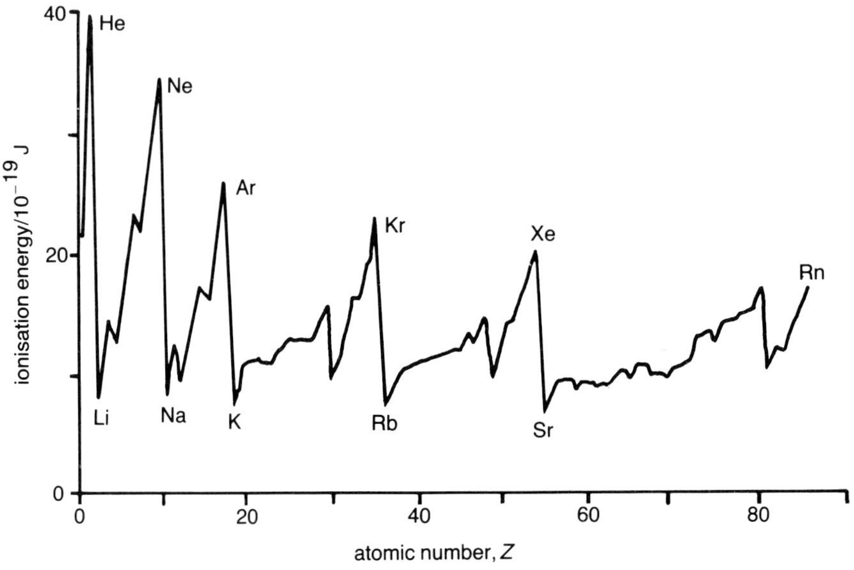

Fig. 2.1 — First ionisation energies for the elements.

before 3d, 5s before 4d, 6s before 4f, 4f before 5d, 7s before 5f, and 5f before 6d (Fig. 2.2).

The grouping of elements of similar chemical character in the periodic table is based upon the number of vacancies in the electron subshells. The noble or inert gases have full configurations with 8 electrons in their outer subshells. Where the number of electrons in the outer subshell is close to 8, the tendency is to acquire electrons in chemical reactions and to produce negatively charged ions (group VII elements). S, with 6 electrons (a group VI element), however, may either acquire 2 electrons (to form H_2S, for example, when it is a negatively charged ion) or shed 6 electrons (to form, say, SO_3 when it is a positively charged ion). When the number of electrons is much less than 8, the elements tend to lose electrons and so form positively charged ions.

The electronegativity of an element is a measure of its ability to attract electrons and so to become negatively charged. Pauling, in 1959, devised a scale for electronegativity between 0 and 4 which is dimensionless. The inert gases, which have stable outer subshells and exert very little influence on other electrons, have low values. Elements with much less than 8 electrons in their outer subshells tend to lose electrons and also to exhibit low electronegativity. In contrast, the group VII elements (with 7 electrons in their outer subshells) are strongly electronegative. Electronegativity therefore increases across the periodic table (Fig. 2.3). There is also a general decrease down through the periods with increasing electron density around the nucleus. However, this last effect is not observed in all groups as the screening effect of the closely packed electron subshells in relation to the size of the nucleus is not simple and, between groups II and III, electronegativity may increase down through the periods. Generally, however, the lowest electronegativities are found in the seventh period, group I, and the highest in the second period, group VII.

The representation used for ions is X^{n+} or X^{n-}, where X is the element involved and n is an integer bearing the sign of the electrical

Table 2.1 — Character of electron shells

K shell	$1s^2$			
L shell	$2s^2$	$2p^6$		
M shell	$3s^2$	$3p^6$	$3d^{10}$	
N shell	$4s^2$	$4p^6$	$4d^{10}$	$4f^{14}$
O shell	$5s^2$	$5p^6$	$5d^{10}$	$5f^{14}$
P shell	$6s^2$	$6p^6$	$6d^{10}$	
Q shell	$7s^2$	$7p^6$		
He shell	$1s^2$			
Ne shell	$2s^2$	$2p^6$		
Ar shell	$3s^2$	$3p^6$		
Kr shell	$3d^{10}$	$4s^2$	$4p^6$	
Xe shell	$4d^{10}$	$5s^2$	$5p^6$	
Rn shell	$4f^{14}$	$5d^{10}$	$6s^2$	$6p^6$
Eka-Rn shell	$5f^{14}$	$6d_{10}$	$7s^2$	$7p^6$

charge. The value of this integer is also known as the valency (Fig. 2.4). Positively charged ions are known as 'cations' (as they are attracted to the negatively charged plate, or cathode, of an electrolytic cell) while negatively charged ions are termed 'anions' (as they are attracted to the positive plate, or anode).

2.2 MOLECULAR STRUCTURES: BONDING AND PERIODICITY

There are three important types of bond between ions: ionic, covalent and metallic. Ionic bonds are the result of strong electrostatic attraction between oppositely charged ions. Typically, it is elements from groups I and VII which combine in this way. Similarly, elements in groups II and VI may combine ionically to form oxides, and in groups II and VII to form halides. In either case, the elements involved display a strong contrast in electronegativities. Covalent bonds are characterised by electron sharing. This is found particularly in elements from groups III, IV and V, which tend to retain their electrons more effectively than those from groups I and II. The elements forming the bond thus have similar electronegativities of intermediate values. Metallic bonds are those in which positively charged ions are arranged in a closely packed regular structure, with the electrons surrounding that structure. The electrostatic attraction between the ions and the electrons holds the structure together, but some electrons are only weakly held, making the material a good conductor of electricity. Elements with high electronegativities, including many of the transition elements, particularly those in groups V and VI, tend to form this type of bond. Finally, in the formation of complex ions the bonding together of elements of mainly covalent character produces a compound which favours ionic bonding.

Individual elements do not form just one type of bond, but each is capable of forming different types. The ability to form either ionic or covalent bonds is, in fact, shared by many elements. Here, control over which type of bond forms appears to be exerted by the difference in electronegativity between the pair of elements, large differences favouring the formation of ionic bonds, and small differences favouring covalent bonds. A difference in electronegativity of about 2 marks the boundary between mainly ionic and mainly covalent bonding. The transition elements, in particular, are liable to form either type of bond and, indeed, some may readily form metallic bonds.

Many elements, too, may form more than one ion. Thus Fe can be present as Fe^{2+} or Fe^{3+}, depending on whether 2 or 3 electrons have been lost from the outer subshell. Similarly, U can be present as U^{6+} or U^{4+}. These ionic charges can be taken to represent the oxidation state of the element. This is because, the greater the amount of O^{2-} that can combine with the ion, the greater the oxidation state. So, Fe_2O_3, containing $2Fe^{3+}$ in combination with $3O^{2-}$ is more oxidised than FeO in which Fe^{2+} is in combination with O^{2-}.

2.3 THE ORIGIN OF THE ELEMENTS AND COSMIC ABUNDANCES

In the Solar System, elements are generated in the Sun. Nuclear fusion takes place, converting H to He. The process appears to be a step function:

Table 2.2 — Electron configuration of the elements

	He	Ne		Ar			Kr			Xe				Rn			Eka-Rn		
	1s	2s	2p	3s	3p	3d	4s	4p	4d	5s	5p	4f	5d	6s	6p	5f	6d	7s	7p
1 H	1																		
2 He	2																		
3 Li	2	1																	
4 Be	2	2																	
5 B	2	2	1																
6 C	2	2	2																
7 N	2	2	3																
8 O	2	2	4																
9 F	2	2	5																
10 Ne	2	2	6																
11 Na			Ne core	1															
12 Mg				2															
13 Al				2	1														
14 Si				2	2														
15 P				2	3														
16 S				2	4														
17 Cl				2	5														
18 Ar	2	2	6	2	6														
19 K			Ar core				1												
20 Ca							2												
21 Sc						1	2												
22 Ti						2	2												
23 V						3	2												
24 Cr						5	1												
25 Mn						5	2												
26 Fe						6	2												
27 Co						7	2												
28 Ni						8	2												
29 Cu						10	1												
30 Zn						10	2												
31 Ga						10	2	1											
32 Ge						10	2	2											
33 As						10	2	3											
34 Se						10	2	4											
35 Br						10	2	5											
36 Kr	2	2	6	2	6	10	2	6											
37 Rb			Kr core							1									
38 Sr										2									
39 Y									1	2									
40 Zr									2	2									
41 Nb									4	1									
42 Mo									5	1									
43 Tc									5	2									
44 Ru									7	1									
45 Rh									8	1									
46 Pd									10										
47 Ag									10	1									
48 Cd									10	2									
49 In									10	2	1								
50 Sn									10	2	2								
51 Sb									10	2	3								
52 Te									10	2	4								
53 I									10	2	5								
54 Xe	2	2	6	2	6	10	2	6	10	2	6								
55 Cs			Xe core											1					
56 Ba														2					
57 La													1	2					
58 Ce												1	1	2					
59 Pr												2	1	2					
60 Nd												3	1	2					
61 Pm												4	1	2					
62 Sm												5	1	2					
63 Eu												6	1	2					
64 Gd												7	1	2					
65 Tb												8	1	2					
66 Dy												9	1	2					
67 Ho												10	1	2					
68 Er												11	1	2					
69 Tm												12	1	2					
70 Yb												13	1	2					
71 Lu												14	1	2					
72 Hf												14	2	2					
73 Ta												14	3	2					
74 W												14	4	2					
75 Re												14	5	2					
76 Os												14	6	2					
77 Ir												14	7	2					
78 Pt												14	9	1					
79 Au												14	10	1					
80 Hg												14	10	2					
81 Tl												14	10	2	1				
82 Pb												14	10	2	2				
83 Bi												14	10	2	3				
84 Po												14	10	2	4				
85 At												14	10	2	5				
86 Rn	2	2	6	2	6	10	2	6	10	2	6	14	10	2	6				
87 Fr			Rn core															1	
88 Ra																		2	
89 Ac																	1	2	
90 Th																	2	2	
91 Pa																	3	2	
92 U																	4	2	

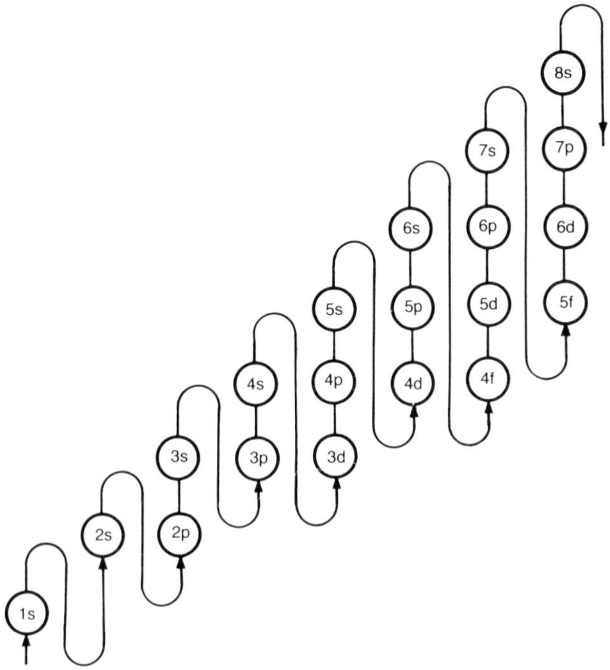

Fig. 2.2 — The order in which electron subshells are filled. (Redrawn, with permission, from The Open University (1979); copyright © 1979 The Open University Press.)

period \ group	I	II	transition elements, lanthanides and actinides										III	IV	V	VI	VII	inert gases
1	H 2.1																	He —
2	Li 1.0	Be 1.5											B 2.0	C 2.5	N 3.0	O 3.5	F 4.0	Ne —
3	Na 0.9	Mg 1.2											Al 1.5	Si 1.8	P 2.1	S 2.5	Cl 3.0	Ar —
4	K 0.8	Ca 1.0	Sc 1.3	Ti 1.5	V 1.6	Cr 1.6	Mn 1.5	Fe 1.8	Co 1.8	Ni 1.8	Cu 1.9	Zn 1.6	Ga 1.6	Ge 1.8	As 2.0	Se 2.4	Br 2.8	Kr —
5	Rb 0.8	Sr 1.0	Y 1.2	Zr 1.4	Nb 1.6	Mo 1.8	Tc 1.9	Ru 2.2	Rh 2.2	Pd 2.2	Ag 1.9	Cd 1.7	In 1.7	Sn 1.8	Sb 1.9	Te 2.1	I 2.5	Xe —
6	Cs 0.7	Ba 0.9	La-Lu 1.1–1.2	Hf 1.3	Ta 1.5	W 1.7	Re 1.9	Os 2.2	Ir 2.2	Pt 2.2	Au 2.4	Hg 1.9	Tl 1.8	Pb 1.8	Bi 1.9	Po 2.0	At 2.2	Rn —
7	Fr 0.7	Ra 0.7	Ac 1.1	Th 1.3	Pa 1.5	U 1.7	Np-Lr 1.3											

Fig. 2.3 — The periodic table of the elements and their electronegativities.

$$ {}^{1}_{1}\mathrm{H} + {}^{1}_{1}\mathrm{H} \rightarrow {}^{2}_{1}\mathrm{H} + \beta^{+} + \nu, $$

$$ {}^{2}_{1}\mathrm{H} + {}^{2}_{1}\mathrm{H} \rightarrow {}^{3}_{2}\mathrm{He} + \mathrm{n}, $$

$$ {}^{3}_{2}\mathrm{He} + {}^{3}_{2}\mathrm{He} \rightarrow {}^{4}_{2}\mathrm{He} + {}^{1}_{1}\mathrm{H} + {}^{1}_{1}\mathrm{H}, $$

and energy is released (as discussed in Chapter 3). The formation of elements of more complex structure in the Sun is not possible because temperatures are too low. In some massive stars, however, where internal temperatures are higher than in the Sun, the combination of ${}^{4}_{4}\mathrm{He}$ to form ${}^{12}_{6}\mathrm{C}$ is thought to take place:

$$ {}^{4}_{2}\mathrm{He} + {}^{4}_{2}\mathrm{He} + {}^{4}_{2}\mathrm{He} \rightarrow {}^{12}_{6}\mathrm{C}. $$

Further addition of ${}^{4}_{2}\mathrm{He}$ to the structures will give elements ${}^{16}_{8}\mathrm{O}$, ${}^{20}_{10}\mathrm{Ne}$, etc., up to ${}^{56}_{26}\mathrm{Fe}$, the number of excess neutrons increasing with increasing atomic number as protons are converted to neutrons. Other elements up to mass number 56 can be produced by nuclear reactions involving the addition, in various combinations, of protons and neutrons to elements with mass numbers that are multiples of 4. It is possible that a cycle of reactions also occurs in which, in addition to producing He, the presence of ${}^{12}_{6}\mathrm{C}$ leads to the production of N and O. This is the C–N–O carbon–nitrogen–oxygen cycle:

$$ {}^{12}_{6}\mathrm{C} + {}^{1}_{1}\mathrm{H} \rightarrow {}^{13}_{7}\mathrm{N}, $$

$$ {}^{13}_{7}\mathrm{N} \rightarrow {}^{13}_{6}\mathrm{C} + \beta^{+} + \nu, $$

$$ {}^{13}_{6}\mathrm{C} + {}^{1}_{1}\mathrm{H} \rightarrow {}^{14}_{7}\mathrm{N}, $$

$$ {}^{14}_{7}\mathrm{N} + {}^{1}_{1}\mathrm{H} \rightarrow {}^{15}_{8}\mathrm{O}, $$

$$ {}^{15}_{8}\mathrm{O} \rightarrow {}^{15}_{7}\mathrm{N} + \beta^{+} + \nu, $$

$$ {}^{15}_{7}\mathrm{N} + {}^{1}_{1}\mathrm{H} \rightarrow {}^{12}_{6}\mathrm{C} + {}^{4}_{2}\mathrm{He}. $$

The limiting mass number of 56 is governed by

group / period	I	II	transition elements, lanthanides and actinides										III	IV	V	VI	VII	inert gases
1	H 0.21/1−																	He —
2	Li 0.08/1+	Be 0.04/2+											B 0.02/3+	C 0.01/4+	N 0.01/5+ 0.17/2−	O 0.13/2−	F 0.13/1−	Ne —
3	Na 0.11/1+	Mg 0.08/2+											Al 0.06/3+	Si 0.03/4+	P 0.02/5+	S 0.02/6+ 0.17/2−	Cl 0.17/1−	Ar —
4	K 0.15/1+	Ca 0.10/2+	Sc 0.08/3+	Ti 0.08/3+ 0.07/4+	V 0.07/3+ 0.06/5+	Cr 0.07/3+ 0.04/6+	Mn 0.09/2+ 0.06/4+	Fe 0.09/2+ 0.07/3+	Co 0.08/2+	Ni 0.08/2+	Cu 0.10/1+ 0.08/2+	Zn 0.08/2+	Ga 0.07/3+	Ge 0.06/4+	As 0.06/3+ 0.06/5+	Se 0.04/6+ 0.19/2−	Br 0.18/1−	Kr —
5	Rb 0.16/1+	Sr 0.12/2+	Y 0.10/3+	Zr 0.08/4+	Nb 0.07/5+	Mo 0.07/4+ 0.07/6+	Tc –	Ru 0.07/3+ 0.06/4+	Rh 0.09/2+	Pd 0.09/2+	Ag 0.12/1+	Cd 0.10/2+	In 0.09/3+	Sn 0.13/2+ 0.08/4+	Sb 0.09/3+ 0.07/5+	Te 0.06/6+ 0.22/2−	I 0.10/5+ 0.21/1−	Xe — –
6	Cs 0.18/1+	Ba 0.14/2+	La-Lu 0.11/3+	Hf 0.08/4+	Ta 0.07/5+	W 0.07/6+	Re 0.07/4+ 0.06/7+	Os 0.07/4+	Ir 0.07/4+	Pt 0.05/2+	Au 0.14/1+	Hg 0.11/2+	Tl 0.16/1+ 0.10/3+	Pb 0.13/2+	Bi 0.11/3+	Po	At –	Rn –
7	Fr 0.19/1+	Ra 0.16/2+	Ac 0.12/3+	Th 0.11/4+	Pa 0.11/3+ 0.09/4+	U 0.11/4+ 0.08/6+	Np-Lr 0.10/3+											

Fig. 2.4 — The periodic table of the elements with radii (10^{-9} m) and ionic charges.

the energy exchange characteristics of the reactions. Up to mass number 56, all the fusion reactions are exothermic (Fig. 2.5) and thus self-sustaining. As structures become more complex, however, less energy is released, as governed by the binding fraction of the nucleus (see Section 3.2). Also, the temperatures required for the processes involved greatly increase as the complexity increases; for example, $^{1}_{1}H + ^{1}_{1}H$ reactions occur at about 2×10^8 °C, while a temperature of 5×10^8 °C is needed for the C–N–O cycle, and a temperature of about 4×10^9 °C would be needed for $^{28}_{14}Si + ^{28}_{14}Si \rightarrow ^{56}_{28}Ni \rightarrow ^{56}_{26}Fe + 2\beta^+$.

As these nuclear fusion reactions are all endothermic, a large input of energy is required to produce elements with mass numbers greater than 56. This cannot take place in stars unless they are very massive and evolve through the red giant stage to become supernovae. This critical stage occurs when the star has a Fe-rich core at a temperature of about 10^{10} °C. At this temperature, fission of the Fe takes place. This rapidly absorbs energy and the star contracts, bringing into high temperature regions the outer cooler (and lighter) parts of the star. Rapid nuclear fusion processes then occur in these lighter elements and an enormous amount of energy is generated very rapidly, causing the outer part of the star to be explosively ejected. The rapid rise in temperature caused by this massive liberation of energy probably leads to the synthesis of all the heavy nuclides. Thus the elements with mass numbers greater than 56 are believed to be formed in the 'death throes' of a red giant. Starting with $^{56}_{26}Fe$ nuclei, the addition of protons and neutrons can account for the formation of elements of any mass number. However, beyond Pb the nuclei are unstable, and the release of energy occurs spontaneously by fission and radioactive decay.

The cosmic abundances of the elements

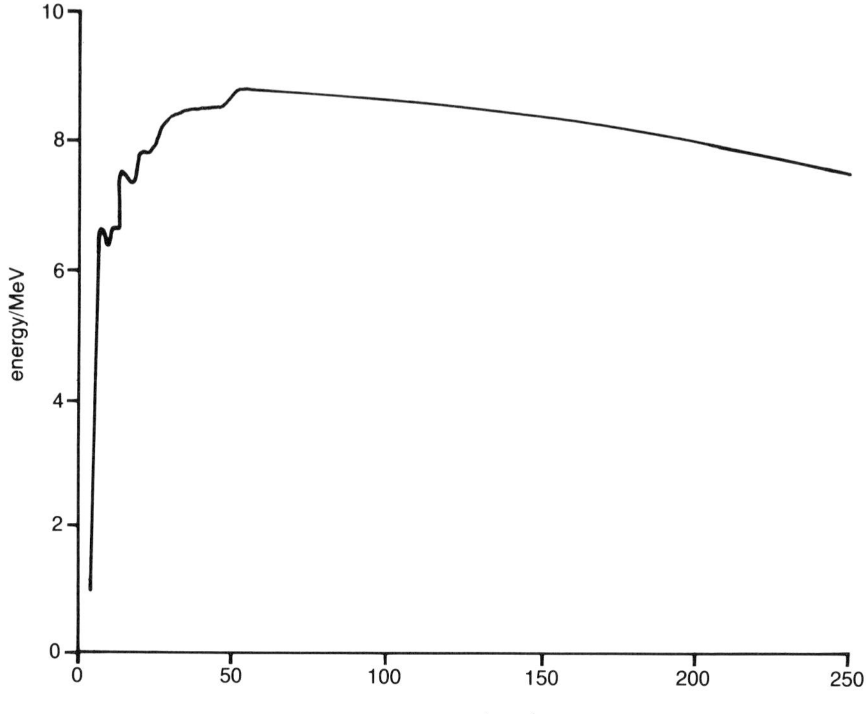

Fig. 2.5 — Nuclear unbinding energies (average energy per nucleon).

obtained by spectrographic analysis of the Sun and stars, expressed in relation to the abundance of Si, clearly show the overwhelming importance of H and He. Next in order of abundance are elements with mass numbers up to 56, of which those which are multiples of 4 are the most common. The rare heavy elements are the least abundant (Table 2.3).

The Sun is probably a relatively young star, about 5 Ga old. It formed 10–15 Ga after the 'big bang' which is believed to have produced the primordial H and He in the Universe. In this interval of time, stars would have formed by accretion of the primordial matter. Some of these stars would undoubtedly have been large, would have caused formation of elements up to $^{56}_{26}Fe$ and would then have become supernovae to form the heavier elements. Thus within interstellar space the abundance of the elements gradually built up until eventually, in the part of space that we call the Solar System, the Sun formed from a contracting cloud of material (or nebula) that already contained the array of elements that we now find. Some details of the pre-Solar System history of the elements can be obtained from the study of the radionuclides that were formed in these supernovae. Indeed, many unstable transuranic elements also appear to have been formed but, as they had short half-lives, we now only find their decay products.

Table 2.3 — Abundance of elements in the cosmos (relative to Si = 10^6)

Element	Amount	Element	Amount
H	4×10^{10}	Ru	1.49
He	3.08×10^9	Rh	2.14×10^{-1}
Li	10^2	Pd	6.75×10^{-1}
Be	2.0×10	Ag	2.6×10^{-1}
B	2.4×10	Cd	8.9×10^{-1}
C	3.5×10^6	In	1.1×10^{-1}
N	6.6×10^6	Sn	1.33
O	2.15×10^7	Sb	2.46×10^{-1}
F	1.6×10^3	Te	4.67
Ne	8.6×10^6	I	8×10^{-1}
Na	4.38×10^4	Xe	4.0
Mg	9.12×10^5	Cs	4.56×10^{-1}
Al	9.48×10^4	Ba	3.66
Si	10^6	La	2.0
P	10^4	Ce	2.26
S	3.75×10^5	Pr	4×10^{-1}
Cl	8.5×10^3	Nd	1.44
Ar	1.5×10^5	Pm	—
K	3.16×10^3	Sm	6.64×10^{-1}
Ca	4.9×10^4	Eu	1.87×10^{-1}
Sc	2.8×10	Gd	6.84×10^{-1}
Ti	2.44×10^3	Tb	9.5×10^{-2}
V	2.2×10^2	Dy	5.56×10^{-1}
Cr	7.8×10^3	Ho	1.18×10^{-1}
Mn	6.85×10^3	Er	3.16×10^{-1}
Fe	6×10^5	Tm	3.1×10^{-2}
Co	1.8×10^3	Yb	2.2×10^{-1}
Ni	2.74×10^4	Lu	5×10^{-2}
Cu	2.12×10^2	Hf	4.38×10^{-1}
Zn	4.86×10^2	Ta	6.5×10^{-2}
Ga	1.14×10	W	4.9×10^{-1}
Ge	5.05×10	Re	1.35×10^{-1}
As	4.0	Os	1.0
Se	6.76×10	Ir	8.24×10^{-1}
Br	1.34×10	Pt	1.625
Kr	5.13×10	Au	1.45×10^{-1}
Rb	6.5	Hg	2.84×10^{-1}
Sr	1.89×10	Tl	1.08×10^{-1}
Y	8.9	Pb	4.7×10^{-1}
Zr	5.45×10	Bi	1.44×10^{-1}
Nb	1.0	Th	2.7×10^{-2}
Mo	2.42	U	10^{-2}
Tc	—		

(After Rösler and Lange, 1972.)

2.4 THE INTERNAL STRUCTURE OF THE EARTH

Seismological data show that body waves (those that pass through the Earth, in contrast with surface waves which travel around the outside of the Earth) may be separated into 'primary' or P waves and 'secondary' or S waves. P waves are compressional elastic waves of which the vibrational motion is in the direction of propagation of the wave, while S waves are shear waves of which the vibrational motion is at right angles to the direction of propagation. P waves are transmitted at a speed given by $(\gamma/\rho)^{\frac{1}{2}}$, where γ is the axial modulus of the transmitting material and ρ its density. The S wave transmission speed is given by $(\mu/\rho)^{\frac{1}{2}}$, where μ is the modulus of rigidity of the material. In any material, γ is greater than μ; P waves, therefore, travel faster than S waves. As the value of μ for liquids is zero, S waves will not pass through a liquid medium.

When body waves pass through the Earth they are strongly refracted by a marked increase in velocity known as the 'mohorovicic discontinuity', or Moho, which varies in depth between about 5 and 11 km beneath oceans, and between 25 and 90 km beneath continents. This discontinuity marks the boundary between the crust and the mantle (Fig. 2.6) and probably reflects a major compositional change. The difference in the depth of the Moho between oceans and continents also reflects a difference between the composition of oceanic and that of the continental crust. By and large, the oceanic crust can be considered to be basaltic and the continental crust granitic–dioritic in composition. The underlying mantle, also, is not homogeneous, exhibiting different physical properties beneath oceans from those beneath continents, but in general it is thought to be composed essentially of Fe and Mg silicates and oxides and to have a peridotitic composition at least in its upper part. Within the mantle, a number of layers can be recognised on the basis of changes in the transmission speeds of P and S waves. The upper mantle extends to a depth of about 1000 km with a lower section about 600 km thick identified as a transition zone. The lower mantle then extends to a depth of about 2900 km. These layers probably owe their origin to phase changes in the minerals present in the mantle; the boundary between the upper mantle and the transition zone may result from the formation of olivine and stishovite from pyroxene, spinel from olivine, ilmenite from spinel and stishovite, and ilmenite and periclase from spinel. These reactions proceed in order

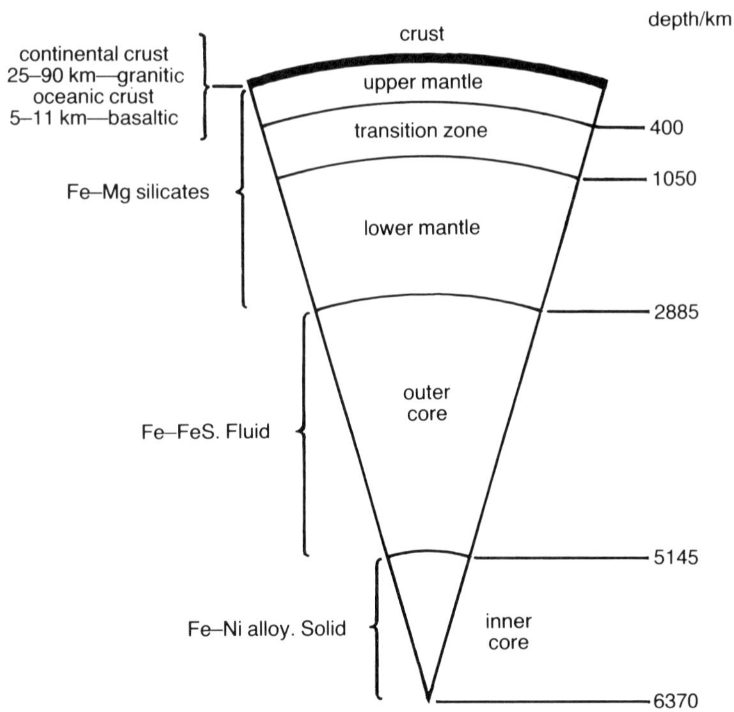

Fig. 2.6 — The internal structure of the Earth.

with increasing pressure and are completed at a depth of about 1000 km. Above the transition zone, the upper mantle is probably quite heterogeneous. This would explain both the chemical differences between ocean island basalts and midocean ridge basalts, which are derived by partial melting of different parts of the mantle, and the different physical properties of these basalt source zones revealed by seismological studies. Radionuclide studies have been particularly helpful in work on the chemical evolution of the upper mantle. This is discussed in Chapter 10. The lower mantle appears to be more homogeneous than the upper mantle and is thought to consist of Fe–Mg silicate with the ilmenite structure and Fe–Mg oxide as periclase. The base of the mantle is marked by a thin transition, below which there is a sudden decrease in seismic velocities. This is the Gutenberg discontinuity below which the core does not transmit S waves and therefore behaves like a liquid, with a negligible modulus of rigidity. The sudden change from the mantle to the core probably represents a major change in density. From analogies with iron meteorites, which are thought to have come from the cores of small planets in the asteroid belt between Mars and Jupiter, and from consideration of the origin of the Earth's magnetic field, which requires a good electrical conductor in the fluid core, it is almost universally accepted that the material making up the outer part of the core is Fe with some Ni. However, the density of such an alloy is higher than that required for the accepted model of the Earth's interior, and so a lower density material must be present as well. Various possibilities have been suggested for this material, particularly Si, S, K and O. The most favoured at the present time is O, but a mixture of low density components is also possible. At a depth of about 5145 km there is another seismic discontinuity marking the boundary between the outer core and inner core. In the inner core, both the density and the transmission velocities of P waves are greater than in the outer core. S waves are transmitted, but only with a very low velocity. These properties are thought to indicate that the inner core is mainly solid Fe–Ni alloy.

2.5 GEOCHEMICAL CLASSIFICATION OF THE ELEMENTS

The general occurrence of elements within the Earth has led to the grouping together of those with similar characteristics into four main classes: atmophile, lithophile, chalcophile and siderophile.

The atmophile elements are those which principally occur in the Earth's atmosphere. This group includes the noble gases. These elements originate within the solid Earth but, as they are gases at normal temperatures and pressures, they migrate by diffusion and by transport in flowing groundwater to the Earth's surface. N is also strongly atmophile but, as it is reactive, it also occurs within the materials of the solid Earth. The highly reactive gases H_2 and O_2 are clearly also atmophile but, equally, possess the characteristics of other classes. Thus they illustrate the important fact that any element may be readily placed in more than one class.

Lithophile elements are those which occur mainly in the Earth's crust. They are the so-called 'oxygen-loving' elements that combine readily with O and Si to form silicates. Lithophile elements characteristically form ionic bonds. Groups I, II and VII are all lithophile, as are the first four groups of the transition elements. Electronegativities of lithophile cations are less than 1.8.

Chalcophile elements are those which tend to form sulphides. These elements tend to occur in the transition group of the periodic table and in the higher periods of groups III–VI. Most have electronegativities between 1.6 and 2.2, and dominantly form covalent bonds with S. Whether any particular element from this group forms ionic or covalent bonds depends upon the difference in electronegativity between it and S. As S has an electronegativity of 2.5, a maximum difference of 0.9 limits the formation of covalent bonds.

Siderophile elements are those which typically form metallic bonds and so generally occur in metallic form. The electronegativities of this class of elements range between 1.8 and 2.4, most being greater than 2.0.

Although a number of elements can be placed within more than one class, Fe is particularly versatile in its geochemical character and may be considered an element of lithophile, chalcophile and siderophile type, depending upon the temperature, pressure and chemical environment of the system in which it is present. The boundary between elements which are essentially lithophile and those which have chalcophile characteristics can be usefully defined with respect to the properties of Fe. Thus those elements which are more easily oxidised than Fe^{2+} are typically lithophile, while those that are less easily oxidised are chalcophile.

2.6 IONIC CHARGE AND SIZE

The acquisition of electrical charge by the gain or loss of electrons is not the only factor that governs the chemical properties of a particular element; the size of the ion is also an important controlling factor. Small highly charged cations, for example, tend to combine with O^{2-} to form complex anions. Thus, Si^{4+} combines to form SiO_4^{4-}. The bonding in these complex anions is partly ionic and partly covalent.

The ability of large anions to group around small cations is governed by the ratio of the radii of the ions concerned and, in turn, the pattern of grouping so controlled governs the coordination of the cation–anion relationships. For small values of the ratio, the anions are placed at the corners of a triangle in three-fold coordination. As the ratio increases, the pattern changes and the anions are placed at the corners of a tetrahedron (4-fold coordination), octahedron (six-fold), cube (eight-fold) and at the midpoint of cube edges (twelve-fold).

In general, the size of ions increases down through the periods of the periodic table and from group I to group VII for each period. There are, however, variations in this, depending upon whether the ionic state is achieved by gain or loss of electrons.

The size and ionic charge also govern the ability of one element to substitute for another in a particular molecular structure. Fe^{2+} and Mg^{2+}, for example, are often found in ionic substitution for each other. The factors which govern ionic substitution of one element for another are similarity of ionic charge and radius; one ion may only completely replace another when they are of similar charge and if the size difference of their ionic radii does not exceed 15% of the radius of the smaller ion. If the difference in ionic radii is greater than this value, partial substitution may occur provided that the size difference does not affect the coordination number.

The role of charge in ionic substitution does not appear to be particularly important, as many examples of substitution of differently charged ions for each other are known, such as Li^{+} for Mg^{2+}, B^{2+} for K^{+}, etc. However, the difference in charge is limited to only one unit, and the substitutions are usually accompanied by another balancing substitution, such as Al^{3+} for Si^{4+}.

2.7 THE CHONDRITIC EARTH

2.7.1 The accretion of the Earth

The accretion of the Earth from the Primitive Solar Nebula can be considered to have taken place either homogeneously, the Earth having a uniform initial composition with separation of the core from a silicate outer envelope at a later date or, heterogeneously, with the refractory elements accreting first and the more volatile elements being added later. In either case, it is believed that the original composition of the Primitive Solar Nebula can be seen today in the meteorites known as carbonaceous chondrites. Achondritic meteorites (the irons and stony meteorites) appear to have originated by the break-up of small planets with cores and mantles. The chondritic meteorites, however, possess a chemical composition which seems to have a combination of core and mantle characteristics. This type of meteorite could have originated by the accretion of only small parts of the Primitive Solar Nebula before chemical separation of any but the lightest elements occurred, either within the nebula or within the larger accreting bodies. This is borne out by studies which show that, although the term 'chondrite' is derived from the presence of small silicate globules (chondrules) which were once molten, the textures of the chondritic meteorites present clear evidence of solid-state recrystallisation with no sign that mass melting had occurred. Different types of chondritic meteorite are recognised according to their composition. These are designated E (for enstatite), C (for carbonaceous) and O (for ordinary); ordinary meteorites are divided into three groups: H (for high Fe), L (for low Fe) and LL (for low Fe, low metal). The degree of recrystallisation increases from C through O to E types while the proportion of volatile components decreases. This shows that the carbonaceous chondrites have not been heated to beyond 180 °C after their formation. It is therefore likely that they are the least altered of all the material that accreted from the Primitive Solar Nebula. However, the most volatile elements, such as H and He, which must have been present in the Primitive Solar Nebula, will have been lost, even from carbonaceous chondrites. They do not, therefore, have totally the same composition as the Earth at the time of its accretion.

2.7.2 Chemical differentiation within the Earth

The recognition that the interior structure of the Earth consists of three main layers of different chemical composition indicates that differentiation of the original material of the Primitive Solar Nebula has taken place. This may have occurred either prior to the accretion of the Earth, allowing the accretion of the refractory condensates first followed by the less refractory and progressively more volatile material later (the theory of heterogeneous accretion), or by separation of components after accretion (homogeneous accretion).

One problem with the heterogeneous accretion hypothesis is that the first materials to accrete would have had a lower density than those accreting later. The initial highly refractory materials would, therefore, have been followed by a metallic Fe–Ni association, producing a layering which is the reverse of what is actually found. A subsequent phase of inversion is, therefore, required. Nevertheless, while this inversion was taking place, silicates would have been accreting. Such a process is unlikely to lead to the formation of a relatively sharp break between the core and mantle. For this reason, among others, the idea generally accepted for the origination of the Earth is one of homogeneous accretion. In this, the overall composition of each planet was determined by its position within the condensing nebula, the panets closest to the Sun being composed of the most highly refractory elements, while the planets most distant from the Sun are dominantly composed of the volatile elements.

However, primitive meteorites are found to consists of a homogeneous mixture of different components and, by analogy, it is considered that the Earth, too, formed from the accretion of different parts of the condensing nebula. In fact, six components, corresponding to six different types of meteorite or meteorite component, have been identified:

(1) Early condensates,	9.2%.
(2) Fe–Ni metal,	31.1%.
(3) Silicate,	11.4%.
(4) Chondrules,	41.8%.
(5) Sulphide,	5.0%.
(6) Volatiles,	1.5%.

The separation of the core from the homogeneous mixture of the components making up the initial form of the Earth must have taken place when the temperature was extremely high. The probable cause of this high temperature is the heating that would have accompanied the accretion process, the kinetic energy of the accreting particles being converted to thermal energy on impact. As melting of metallic Fe to form the core is difficult to envisage in this model, the Fe must have been in a form with a lower melting point than that of metallic Fe. This may account for why the outer core has physical properties which are not in accord with a pure metallic composition.

By these processes the chalcophile and siderophile elements became concentrated in the core. Subsequently, further separation of chalcophile and siderophile elements into the outer core and inner core, respectively, is believed to have taken place. The segregation of the core within the Earth must have caused severe repositioning of the adjacent residual silicate-rich material, causing it to be displaced upwards as the core material sank. Eventually, complete separation would have occurred, resulting in the formation of distinct layers of core and mantle.

Recycling of mantle silicates to produce the crust dates from at least 3800 Ma ago (the oldest continental crust), although it is certain to have been active earlier than that. How much earlier is, however, uncertain, but some of the earliest known rocks from western Greenland consist in part of water-deposited sediments, indicating that there existed at that time both hydrosphere and atmosphere. The original crustal material could have been of anorthositic composition, formed by melting of mantle peridotite and subsequent differentiation of anorthosite from an ultrabasic residuum. Some partial melting of peridotite to yield basalt magmas is also believed to have occurred. During these processes, the lithophile elements in the mantle are believed to have been subject to chemical differentiation. Incompatible elements (those which are not easily incorporated into the minerals found in peridotite) became concentrated in the crust (Table 2.4) following melting and partial melting of the mantle, leaving behind a depleted mantle. Different degrees of partial melting at different pressures produces basalts of different compositions. However, even though the source region for midocean ridge basalts lies at a shallower level in the mantle than that for ocean island basalts, it appears that the midocean ridge basalts are derived from a source that is depleted in incompatible elements compared with that for the ocean island basalts. It is possible that the undepleted source of the ocean island basalts represents the original composition of the mantle before differentiation processes

Table 2.4 — Average abundance of elements in the Earth's crust

Element	Average abundance (ppm)	Element	Average abundance (ppm)
O	4.64×10^5	Pb	1.25×10
Si	2.82×10^5	B	10
Al	8.1×10^4	Th	8.5
Fe	5.4×10^4	Pr	8
Ca	4.1×10^4	Sm	7
Na	2.4×10^4	Gd	7
Mg	2.3×10^4	Dy	6
K	2.1×10^4	Er	3.5
Ti	5×10^3	Yb	3.5
H	1.4×10^3	Be	3
P	1.1×10^3	Cs	3
Mn	10^3	Hf	3
F	6.5×10^2	U	2.7
Ba	5×10^2	Br	2.5
Sr	3.75×10^2	Sn	2.5
S	3×10^2	Ta	2
C	2.2×10^2	As	1.8
Zr	1.65×10^2	Ge	1.5
Cl	1.3×10^2	Mo	1.5
V	1.1×10^2	Ho	1.5
Cr	10^2	Eu	1.2
Rb	9×10	W	1.2
Ni	7.5×10	Tb	1
Zn	7×10	Tl	8×10^{-1}
Ce	7×10	Lu	6×10^{-1}
Cu	5×10	Tm	5×10^{-1}
Y	3.5×10	Sb	2×10^{-1}
La	3.5×10	I	2×10^{-1}
Nd	3×10	Cd	1.5×10^{-1}
Co	2.2×10	Bi	1.5×10^{-1}
Li	2×10	Ag	7×10^{-2}
N	2×10	In	6×10^{-2}
Sc	2×10	Se	5×10^{-2}
Nb	2×10	Hg	2×10^{-2}
Ga	1.8×10	Au	3×10^{-2}

(After Krauskopf, 1979.)

occurred (Table 2.5). Certainly, this is suggested by the abundances of Rb and Sr in these basalts which are similar to those expected in a chondritic Earth. Unfortunately, other trace elements notably Zr and Y, suggest that the source of the ocean island basalts has actually been subject to chemical enrichment in incompatible elements by fluid migration, rendering the true picture unclear. The nature of some of the evidence upon which the solution of this problem may be based is discussed in Chapter 10.

Although the composition of the continental crust is often simplified as granitic, a wide variety of igneous, metamorphic and sedimentary rock types actually occur. Indeed, sections of oceanic crust within continental crust are known, and even sections of upper mantle material have been identified in dismembered ophiolite–flysch associations. By and large though, the two minerals which tend to characterise continental crust are quartz and feldspar which, of course, are also the main constituents of granite. The source of most granites appears to be the mantle, either by fractional crystallisation to produce residual granitic magmas or by stepped partial melting processes such as peridotite → basalt → andesite → granite. If early differentiation took place by either process, however, incompatible elements would have become concentrated in the granitic magmas. Granites can also form by melting of older continental crust (either metamorphosed sedimentary material or older granites), but here too the granitic magmas produced are enriched in incompatible elements. These processes also cause the strongly lithophile elements to be preferentially concentrated in the continental crust.

The separation of siderophile elements to become concentrated in the outer core, and of lithophile elements to become concentrated in the mantle and crust, with strongly lithophile elements occurring particularly in the continental crust, has produced an abundance distribution of the elements in the crust which is very

Table 2.5 — Main radioactive elements in some silicate materials

Material	K (ppm)	U (ppm)	Th (ppm)
Chondrites	880	0.011	0.043
Carbonaceous chondrites	845	0.011	0.04
Earth crust and mantle	200	0.02	0.08
Mantle	54	0.018	0.055
Whole Earth	130	0.012	0.035

(After Mamyrin and Tolstikhin, 1984.)

different from that of either the carbonaceous chondrites or the Earth as a whole. To illustrate this point clearly, U which was probably one of the least abundant elements in the Primitive Solar Nebula is, because of its large atomic number and lithophile geochemical character, fairly abundant in the Earth's crust. Not only is U strongly lithophile, but it is also an incompatible element and its abundance in the continental crust is very significantly enhanced (Table 2.6).

2.7.3 U in the Earth's core

Although the idea that K may be concentrated in the Earth's core has been seriously considered, until recently the possibility that U may also be present in significant concentrations was not considered reasonable. This is because the U silicates and oxides were not thought to be miscible in the metallic Fe–Ni of which the core is probably largely composed. Recent experimental evidence on the nature of the binary system steel–UO_2, however, shows that this may not be the case (Feber *et al.* 1984). In these experiments, stainless steel and UO_2 were melted together by an arc discharge between a tungsten cathode and an anode crucible until observation of the melt was obscured by U oxide vapour at temperatures greater than 3120 °C. After the molten state had been main-

Table 2.6 — Some radionuclides in the Earth's crust

Nuclide	Ultrabasic (ppm)	Basaltic (ppm)	High Ca granite (ppm)	Low Ca granite (ppm)	Shale (ppm)	Sandstone (ppm)	Limestone (ppm)
^{40}K	0.0047	0.98	2.97	4.96	3.14	1.26	0.32
^{50}V	0.096	0.6	0.21	0.1	0.31	0.048	0.048
^{87}Rb	0.056	8.4	31	47	39	17	0.8
^{138}La	—	0.013	0.04	0.049	0.082	0.03	—
^{142}Ce	—	5.3	9	10	6.5	10	1.27
^{144}Nd	—	4.8	7.9	8.8	5.7	8.8	1.1
^{147}Sm	—	0.79	1.3	1.5	0.96	1.5	0.19
^{148}Sm	—	0.6	0.99	1.1	0.72	1.1	0.15
^{174}Hf	0.0011	0.0036	0.041	0.007	0.005	0.007	0.00054
Th	0.004	4	8.5	17	12	1.7	1.7
U	0.001	1	3	3	3.7	0.45	2.2

(After Klement, 1982.)

tained for 7 min, the crucible was cooled to produce an ingot which was examined by electron probe microanalysis. It was found that, in the molten state, two immiscible liquids had been present, one consisting of UO_2 with 1% steel, and the other of steel with 23% UO_2. Above a certain temperature, therefore, there is total miscibility between UO_2 and steel, even though the temperature may be greater than the boiling point of UO_2 (estimated at just over 4000 °C). The presence of 2% Mn, 10% Ni and 16% Cr in the steel does not appear to influence the results in any major way, similar results having been obtained when stainless steel was melted with UO_2, Mo, Ru, Rh and Pd. Even at temperatures that are only a little above the melting point of UO_2, some 26% of dissolved UO_2 can exist in the stainless steel. Indeed, with C present in the steel, two liquids can exist at 2800 °C, one rich in U and the other rich in UO_{2-x}. In this case, several weight per cent of the U-rich liquid will probably mix with liquid Fe.

Although these experiments were carried out at normal pressures, it has been argued that they are applicable to the Earth's core because the effect of increased pressure is unlikely to be very significant. Even the tendency for phase boundary positions to be displaced to higher temperatures is likely to be offset by the chemical effects of other substances in the core. As the temperature at the core–mantle boundary is estimated to be about 3400 °C and, as there is coexistence of two liquids in the UO_2-steel system between temperatures of 3300 and about 5000 °C with total miscibility above that temperature, it is possible that there is continuous absorption of U from the mantle into the core. Most of this absorption would have taken place during the early stages of core formation but, if the processes which cause the removal of Fe from the mantle to the core are continuing, they would be accompanied by a similar growth in U concentrations in the core.

2.7.4 The evolution of the Earth's atmosphere

It is generally accepted that the present-day composition of the Earth's atmosphere has been brought about by both processes of chemical evolution, related to those acting within the solid Earth, and biological processes. In the earlier discussion of the homogeneous accretion model for the origin of the Earth, with the subsequent melting of Fe to form the core, the reactions involved in the liberation of metallic Fe from the oxides and silicates present in chondritic meteorites were not discussed. To form metallic Fe, reduction reactions must have operated. O_2 liberated by this reaction would have been highly reactive and is believed to have combined with C to form CO_2 which then degassed from the solid Earth to form the main component of the early atmosphere.

As mentioned previously, carbonaceous chondrites contain volatile material, and the accretion of the Earth from the Primitive Solar Nebula is likely to have produced an atmosphere by outgassing of these highly volatile components. However, the present-day atmosphere is strongly depleted in the inert gases Ar, Ne and Kr, relative to their chondritic abundances, and in hydrocarbon gases which are also believed to have been produced in the early atmosphere. These observations have led to the idea that the Earth's original atmosphere was lost early in its history. One possible mechanism that has been invoked to explain this loss also accounts for the fact that the Solar System as a whole is empty of dispersed matter. It is possible that, during the formation of the Solar System by accretion from the Primitive Solar Nebula, development of the Sun progressed until the thermal energy released had reached the critical state for initiating H fusion reactions. At that moment, a massive release of high energy particles would have produced an enormously powerful solar wind which could have stripped the primitive atmosphere from the inner planets and swept the interpanetary space of any matter remaining from the Primitive Solar Nebula.

The formation of the Earth's secondary atmosphere is then believed to have taken place by volcanic outgassing. With the exception of O_2, all the main gases found in the atmosphere occur in volcanic emissions or in emissions from fractures in the ground. H and He released in this way are not retained in the atmosphere because the Earth is too small and its surface temperatures too high; the escape velocity of these gases overcomes the Earth's gravitational attraction and they are lost to space. In addition to volcanic sources, some H is formed by the photolytic dissociation of water by ultraviolet radiation from the Sun. This process also releases O_2, but production of O_2 in this way is limited, part of the O_2 being converted to O_3 which forms a layer in the upper atmosphere. This layer shields the Earth's surface from ultraviolet radiation and reduces further photolytic dissociation. O_2 concentrations in volcanic gases are also extremely low, and it is necessary to look elsewhere for the source of the present-day O_2 content of the atmosphere.

Early in the history of the Earth, evidence of the sedimentary rocks known as 'red beds' is missing from the geological record. Red beds are so named because of their generally red colouration which is due to the presence of haematite as a cement or coating around grains of detrital material. Haematite, which has the composition Fe_2O_3, contains Fe in its high oxidation state (Fe^{3+}). These rocks are considered to indicate O_2-rich palaeoenvironments, with O_2 available in the atmosphere. It appears, therefore, that the O_2 content of the atmosphere was too low to allow the formation of the haematite, and yet many of the major features of the Earth's chemical evolution were at a similar stage to that which we see today. There is, however, one important exception to this: the evolution of biochemical processes. The biological process of photosynthesis is an important producer of atmospheric O_2 at the present time. Algal-type organisms were present 3500 Ma ago to form the structures known as stromatolites, and it is believed that the present-day O_2 level in the atmosphere has been slowly built up by photosynthesising organisms since that time. However, the time needed for significant levels of O_2 to be produced is in dispute. Cloud (1968, 1976) considers that the O_2 content of the atmosphere was very low until about 2000 Ma ago, but Windley *et al.* (1984) review the evidence which suggests that high levels of O_2 were present more than 3000 Ma ago. Despite this conflict, an O_2-free atmosphere early in the Earth's history does explain the occurrence in ancient sedimentary rocks of detrital grains of minerals that are chemically unstable in oxidising environments. Uraninite, which is a major U mineral, is an excellent example of this; detrital uraninite is not common in sedimentary rocks younger than about 2500 Ma but is present in older rocks.

BIBLIOGRAPHY

Ahrens, L. H. (1952) The use of ionization potentials, I, Ionic radii of the elements. *Geochim. Cosmochim. Acta* **2**, 155.

Ahrens, L. H. (1953) The use of ionization potentials, II, Anion affinity and geochemistry. *Geochim. Cosmochim. Acta* **3**, 1.

Ahrens, L. H. (1977) Origin and distribution of the elements. *ZPhys. Chem. Earth* **11**.

Alfvén, H. (1967) On the origin of the Solar System. *Q. J. R. Astr. Soc.* **8**, 215.

Allegre, C. J. and Michard, G. (1974) *Introduction to geochemistry*. Dordrecht, Reidel.

Aller, L. H. (1961) *The abundance of the elements*. New York, Interscience.

Allred, A. L., and Rochov, E. G. (1958) A scale of electronegativity based on electrostatic force. *J. Inorg. Nucl. Chem.* **5**, 264.

Arden, J. W. (1977) Isotopic composition of uranium in chondritic meteorites. *Nature* **284**, 431.

Bailey, D. K. (1982) Mantle metasomatism — continuing chemical change within the Earth. *Nature* **296**, 525.

Bell, K., Blenkinsop, J., Cole, T. J. S., and Menagh, D. P. (1982) Evidence from Sr isotopes for long-lived heterogeneities in the upper mantle. *Nature* **298**, 251.

Berry, L. G., and Mason, B. (1959) *Mineralogy: concepts, descriptions, determinations*. San Francisco, California, Freeman.

Bohr, N. (1913) On the theory of the decrease of velocity of moving electrified particles on passing through matter. *Phil. Mag.* **25**, 10.

Bolt, B. A. (1982) Inside the Earth: evidence from earthquakes. San Francisco, California, Freeman.

Brewer, L. (1951) The equilibrium distribution of the elements in the Earth's gravitational field. *J. Geol.* **95**, 490.

Brown, G. C., and Mussett, A. E. (1981) *The inaccessible Earth*. London, Allen and Unwin.

Brownlow, A. H. (1979) *Geochemistry*. Englewood Cliffs, New Jersey, Prentice-Hall.

Bullen, K. E. (1936) The variation of density and the ellipticities of strata of equal density within the Earth. *Mon. Not. R. Astr. Soc. Geophys. Suppl.* **3**, 395.

Bullen, K. E. (1958) Solidity of the inner core. In Benioff, H., Ewing, M., Howell, B. F., and Press, F. (eds), *Contributions in geophysics in honor of Beno Gutenberg*. New York, Pergamon.

Bullen, K. E. (1963) *An introduction to the theory of seismology*. London, Methuen.

Cameron, A. G. W. (1959) The origin of the elements. *Phys. Chem. Earth* **3**, 199.

Cameron, A. G. W. (1959) A revised table of abundance of the elements. *J. Astrophys.* **129**, 676.

Cartledge, G. H. (1928) Studies on the periodic system, I, The ionic potential as a periodic function; II, The ionic potential and related properties. *J. Am. Chem. Soc.* **50**, 2855.

Cartledge, G. H. (1930) Relation between ionization potential and ionic potential. *J. Am. Chem. Soc.* **52**, 3076.

Cartledge, G. H. (1951) The correlation of thermochemical data by the ionic potential. *J. Phys. Chem.* **55**, 248.

Cloud, P. E. (1968) Atmospheric and hydrospheric evolution in the primitive Earth. *Science* **160**, 729.

Cloud, P. E. (1976) Major features of crustal evolution. *Geol. Soc. S. Afr. Annex* **79**, 1.

Cohen, R. S., and O'Nions, R. K. (1982) Identification of recycled continental material in the mantle from Sr, Nd and Pb isotope investigations. *Earth Planet. Sci. Lett.* **61**, 73.

Dimroth, E., and Kimberley, M. M. (1976) Precambrian atmospheric oxygen: evidence in the sedimentary distributions of carbon, sulphur, uranium and iron. *Can. J. Earth Sci.* **13**, 1161.

Elsasser, W. M. (1963) *Early history of the Earth. Earth science and meteoritics*. New York, Wiley.

Feber, R. C., Wallace, T. C., and Libby, L. M. (1984) *Uranium in the Earth's core, Eos 65,* 785.

Fisher, D. E. (1976) Rare gas clues to the origin of the terrestial atmosphere. In Windley, B. F. (ed.), *The early history of the Earth*. London, Wiley.

Fyfe, W. S. (1954) The problem of bond type. *Am. Mineral.* **39**, 991.

Garland, G. D. (1971) *Introduction to geophysics: mantle, core and crust*. Philadelphia, Pennsylvania, Saunders.

Gaskell, T. F. (1967) *The Earth's mantle*. London, Academic Press.

Gmelins handbooks on inorganic geochemistry. Berlin, Springer.

Goettel, K. A. (1976) Potassium in the Earth's core: evidence and implications. In Strens, R. G. T. (ed.), *The physics and chemsitry of minerals and rocks*. London, Wiley.

Goldschmidt, V. M. (1954) *Geochemistry*. Oxford, Clarendon.

Gutenberg, B. (1959) *Physics of the Earth's interior*. New York. Academic Press.

Haack, U. (1983) On the content and vertical distribution of K, Th and U in the continental crust. *Earth Planet. Sci. Lett.* **62**, 360.

Hamilton, P. J., O'Nions, R. K., Bridgwater, D., and Nutman, A. (1983) Sm–Nd studies of Archaen metasediments and metavolcanics from west Greenland and their implications for the Earth's early history. *Earth Planet. Sci. Lett.* **62**, 263.

Henderson, P. (1983) *Rare earth geochemistry*. Amsterdam, Elsevier.

Herzberg, C. T. (1984) Chemical stratification in the silicate Earth. *Earth. Planet. Sci. Lett.* **67**, 249.

Holland, H. D. (1976) The evolution of seawater. In Windley, B. F. (ed.) *The early history of the Earth*. London, Wiley.

Jahn, B. M., and Nyquist, L. E. (1976) Crust evolution in the early Earth–Moon system: constraints from Rb–Sr studies. In Windley, B. F. (ed.), *The early history of the Earth*. London, Wiley.

Jefferys, H. (1959) *The Earth, its origin, history and physical constitution*. Cambridge, Cambridgeshire, Cambridge University Press.
Kanzanas, D., Schramm, D. N., and Hainebach, K. (1974) A consistent age for the Universe. *Nature* **274**, 672.
Klement, A. W. (1982) Natural sources of environmental radiation. In Klement, A. W. (ed.), *CRC handbook of environmental radiation*. Boca Raton, Florida, CRC.
Krauskopf, K. B. (1979) *Introduction to geochemistry*. Tokyo, McGraw-Hill Kogakusha.
Kurz, M. D., Jenkins, W. J., and Hart, S. R. (1982) Helium isotopic systematics of ocean islands and mantle heterogeneity. *Nature* **297**, 43.
Lambert, R. St. J. (1976) Archaen thermal regimes, crustal and upper mantle temperatures, and a progressive evolutionary model for the Earth. In Windley, B. F. (ed.), *The early history of the Earth*. London, Wiley.
Luck, J. M., Birck, J. L., and Allegre, C. J. (1980) ^{187}Re–^{187}Os systematics in meteorites: early chronology of the Solar System and age of the Galaxy. *Nature* **283**, 256.
Lupton, J. E. (1983) Terrestial inert gases: isotope tracer studies and clues to primordial components in the mantle. *Ann. Rev. Earth Planet. Sci.* **11**, 371.
Mamyrin, B. A., and Tolstikhin, L. N. (1984) *Helium isotopes in nature*. Amsterdam, Elsevier.
Manhes, G., Allegre, C. J., Dupre, B., and Hamelin, B. (1979) Lead–lead systematics, the 'age of the Earth' and chemical evolution of our planet in a new representation space. *Earth Planet. Sci. Lett.* **44**, 91.
Mason, B. (1966) *Principles of geochemistry*. New York, Wiley.
McCammon, C. A., Ringwood, A. E., and Jackson, I. (1983) A model for the formation of the Earth's core. *J. Geophys. Res. Suppl.* **88**, A501.
McElhinny, M. W. (1979) *The Earth: its origin, structure and evolution*. London, Academic Press.
Mohorovicic, A. (1909) Das Beben vom 8.x.1909. *Jahrb. Met. Obs. Zagreb (Agram)* **9**, 1.
Moorbath, S. (1976) Age and isotope constraints for the evolution of Archaen crust. In Windley, B. F. (ed.), *The early history of the Earth*. London, Wiley.
Murthy, U. R. (1976) Composition of the core and the early chemical history of the Earth. In Windley, B. F. (ed.), *The early history of the Earth*. London, Wiley.
O'Connell, R. J., and Fyfe, W. S. (1981) *Evolution of the Earth*. Washington DC, American Geophysical Union.
Open University (1979) Atomic structure, *S*101 *Unit* 11. Milton Keynes, Open University.
Pauling, L. (1954) *General Chemistry*. San Francisco, California, Freeman.
Pauling, L. (1964) *Die natur der chmischen Bindung*. Weinheim, Verlag Chemie.
Poldervaart, A. (1955) Crust of the Earth. *Spec. Paper Geol. Soc. Am.* **62**.
Rankama, K., and Sahama, T. G. (1950) *Geochemistry*. Chicago, Illinois, University of Chicago Press.
Ringwood, A. E. (1966) Chemical evolution of the terrestial planets. *Geochim. Cosmochim. Acta* **30**, 41.
Ringwood, A. E. (1979) *Origin of the Earth and Moon*. New York, Springer.
Ringwood, A. E., and Green, D. H. (1969) Phase transitions. In Pembroke, H. H. (ed.), *The Earth's crust and upper mantle*. *Geophys. Monogr.* **13**. Washington DC, American Geophysical Union.
Robinson, A., and Spooner, E. T. C. (1984) Can the Elliot Lake uraninite-bearing quartz pebble conglomerates be used to place limits on the oxygen content of the early Proterozoic atmosphere? *J. Geol. Soc. Lond.* **141**, 221.
Rose, A. W., Hawkes, H. E., and Webb, J. S. (1979) *Geochemistry in mineral exploration*. London, Academic Press.
Rösler, H. J., and Lange, H. (1972) *Geochemical tables*. Amsterdam, Elsevier.
Schidlowski, M. (1976) Archaen atmosphere and evolution of the terrestial oxygen budget. In Windley, B. F. (ed.), *The early history of the Earth*. London, Wiley.
Seitz, M. G., Burnett, D. S., and Bell, P. M. (1974) U, Th, Pu fractionation in geologic systems: early Pu/U abundance in meteorites. *Carnegie Inst. Washington Year book* **73**, 451.
Shaw, D. M. (1976) Development of the early continental crust. Rearchean, Proterozoic and later eras. In Windley, B. F. (ed.), *The early history of the Earth*. London, Wiley.
Staudacher, T., and Allegre, C. J. (1982) Terrestial xenology. *Earth Planet. Sci. Lett.* **60**, 389.
Sun, S. S. (1982) Chemical composition and origin of the Earth's primitive mantle. *Geochim. Cosmochim. Acta* **46**, 179.
Sutton, J. (1979) Evolution of the continental crust. In Bowie, S. H. U., Fyfe, W. S., Ostle, D., Plant, J., and Simpson, P. R. (eds), *Theoretical and practical aspects of uranium geology*. London, Royal Society, 257.
Tarling, D. H. (1978) *Evolution of the Earth's crust*. London, Academic Press.
Urey, H. C. (1952) *The planets*. New Haven, Connecticut, Yale University Press.
Veizer, J. (1976) $^{87}Sr/^{86}Sr$ evolution of seawater during geologic history and its significance as an index of crustal evolution. In Windley, B. F. (ed.), *The early history of the Earth*. London, Wiley.
Vinogradov, A. P., Surkov, Y. A., and Kirnozov, F. F. (1973) The content of uranium, thorium and potassiunm in the rocks of Venus as measured by *Venera* 8. *Icarus* **20**, 253.
Vollmer, R. (1983) Earth degassing, mantle metasomatism and isotopic evolution of the mantle. *Geology* **11**, 452.
Walker, J. C. G. (1976) Implications for atmospheric evolution of the inhomogeneous accretion model of the origin of the Earth. In Windley, B. F. (ed.), *The early history of the Earth*. London, Wiley.
Wedepohl, K. H. (1969–1984) *Handbook of geochemistry*. Berlin, Springer.
White, W. M., and Hofmann, A. W. (1982) Sr and Nd isotope geochemistry of oceanic basalts and mantle evolution. *Nature* **296**, 821.
Williams, I. S., Compston, W., Williams, I. R., and Myers, J. S. (1983) Ion microprobe identification of 4,100–4,200 Myr-old terrestial zircons. *Nature* **304**, 616.

Wilshire, H. G. (1984) Mantle metasomatism: the REE story. *Geology* **12**, 395.
Windley, B. F., Simpson, P. R., and Muir, M. D. (1984) The role of atmospheric evolution in Precambrian metallogenesis. *Fortschr. Mineral.* **62**, 253.
Zindler, A., Jagoutz, E., and Goldstein, S. (1982) Nd, Sr and Pb isotopic systematics in a three-component mantle: a new perspective. *Nature* **298**, 519.

3

The physics of radioactivity

3.1 NUCLEAR COMPONENTS AND FORCES

As mentioned in Chapter 1, the nucleus of an atom is characterised by the number of protons and neutrons it contains. The atomic number Z (the number of protons in the nucleus; also known as the charge number) defines the atom as being of a particular element. α-particle and β-particle emissions from radionuclides cause changes in the number of protons in the nucleus and transformation to a different element. The initial nuclide is referred to as the 'parent' and the product of transformation is called the 'daughter'. A parent radionuclide may give rise to a radioactive daughter which will itself undergo transformation, and so on until a stable daughter is produced. This chain is known as a 'decay series'. The mass number A (the total number of protons and neutrons) defines the isotope of a particular element. Variation in A for a constant value of Z accounts for the occurrence of nuclides with very similar chemical properties but different physical characteristics (isotopes). Nuclides with the same value of A but different values of Z are called 'isotones'. They have different chemical and physical properties. Where nuclides have the same number N of neutrons but are of unequal charge number they are referred to as 'isobars'. For identification of a particular nuclear species among isobars with a given value of A, the number $T_s=1/2(N-Z)$, which measures the number of neutrons in excess of the number of protons, is very useful.

The standard representation of a particular nuclide has the form

$$^{A}_{Z}X,$$

where X is the chemical symbol used, for example, $^{238}_{92}U$. Although the charge number A defines the particular element and therefore renders the chemical symbol redundant, repetition is accepted in the interests of clarity.

Recognition that the nuclei of atoms may contain a number of positively charged particles led to the realisation that the force which holds them together must exceed the electrostatic repulsion experienced between two similarly charged particles. As the gravitational force involved is far too weak for this, the concept of nuclear forces was born. The observation that large energy transfers accompany radioactive emission was also clear evidence for the existence of large forces within the nucleus of an atom.

Although it is the strong force which holds the nucleus together, it acts only on certain particles, of which the proton and the neutron are the two main examples. It does not, for instance, act on electrons. The strong force is characterised by three specific properties:

(1) It is predominantly attractive, whether between neutron and neutron, between proton and proton or between proton and neutron, and its strength is independent of the particles involved.
(2) It is stronger than the electrostatic force of repulsion between similarly charged particles.
(3) It acts only over a very short distance (about 3×10^{-17} m).

The particles affected by the strong force are capable of strong interactions; collision between a proton and a neutron, for example, produces a particle known as a pion. Pions may be neutral, or positively or negatively charged, and as they have a comparatively large mass they are, like neutrons and protons, hadrons. Cosmic rays (which consist of protons, atomic nuclei and electrons) can produce pions when collision with an O or N nucleus occurs in the upper atmosphere.

The electron, positron and neutrino have much less mass than the hadrons and are grouped together in another family of subatomic particles called 'leptons'. These are not produced during strong interactions, which only involve hadrons, but originate during the decay of radionuclides. They are characterised by 'weak interactions', they do not readily give rise to new particles, although some of the higher energy reactions can result in interaction with protons to produce negatively charged pions and positrons. Positrons may also interact with neutrons to produce neutrinos and protons.

Finally, although γ-radiation can be considered as an electromagnetic wave motion, it can also be visualised as a beam of particles known as 'photons'. The relationship between wave motions and beams of particles is given by

$$\lambda=h/p,$$

where λ is the wavelength of particles with momentum p and h is Planck's constant. When dealing with γ-radiation, the concept of photons as particles involved in the transfer of electromagnetic force is useful. The photon is just one representative of force transfer particles, better known as 'exchange particles' or 'gauge particles'. Interactions that produce photons, such as the collision of a positron and an electron, are electromagnetic in character.

3.2 STABLE AND UNSTABLE NUCLEI

The occurrence of naturally unstable nuclei which undergo spontaneous transformation by radioactive decay is not common. By considering the nuclear structure, certain essential features can be identified which control the formation of unstable nuclei. For example, for atoms in which Z is less than about 17, stable nuclei contain roughly equal numbers of protons and neutrons. However, stable nuclides with Z values greater than about 17 contain more neutrons than protons. The amount by which the number of neutrons in a stable nuclide exceeds the number of protons is known as the 'neutron excess'. The presence of small numbers of excess neutrons does not seriously affect the nucleus as they are bound together by the strong force and, being electrically neutral, are not affected by any electrostatic repulsion. In contrast, protons are unlikely to be in excess (producing neutron deficiency) because electrostatic repulsion would oppose the strong force holding the particles together. However, there is a limit to the number of excess neutrons that a nucleus can contain and still remain stable, and the nuclide which contains 83 protons and 126 neutrons (^{209}Bi) forms the upper limit of generally stable nuclei in the scale of increasing complexity. Beyond this, the neutron excess renders the nucleus unstable and it undergoes a β^--decay. Beyond the charge number of 82, the nucleus also becomes unstable because the electrostatic repulsion between the protons overcomes the effect of the strong force (Tables 3.1, 3.2 and 3.3). However, it should be noted that it is not only the heavy natural nuclei which undergo spontaneous transformation; K, for example, has already been noted as having an unstable isotope (^{40}K).

In general, the nuclear configuration of neutrons and protons also affects the stability of nuclides with charge numbers less than 83. Thus, nuclides with an even number of protons and an even number of neutrons are the most abundant and are generally stable. Conversely, nuclides with odd numbers of protons and neutrons are the least stable. Nuclides of intermediate stability are those in which there are odd numbers of protons and even numbers of neutrons and vice versa (Table 3.4).

The binding energy of a nucleus is the energy required to decompose the nucleus into its constituent protons and neutrons. It is given by $E_B = 1.007\ 825N + 1.008\ 665Z$, where E_B is

Table 3.1 — The ^{232}Th decay series

Nuclide	Half-life	Decay	α energy (MeV)	Amount (%)	β energy (MeV)	Amount (%)
^{232}Th	1.39×10^{10} years	α	4.016	77		
	Spontaneous		3.957	23		
	fission $>10^{21}$ years					
^{228}Ra	5.75 years	β^-			0.039	60
					0.015	40
^{228}Ac	6.13 h	β^-			2.180	10
					1.850	10
					1.700	7
					1.110	53
					0.640	7
					0.450	13
^{228}Th	1.913 years	α	5.421	73		
			5.339	27		
^{224}Ra	3.64 days	α	5.686	95		
			5.449	5		
^{220}Rn	55.6 s	α	6.288	100		
^{216}Po	0.145 s	α	6.779	100		
^{212}Pb	10.64 h	β^-			0.569	12
					0.331	83
					0.154	5
^{212}Bi	60.5 min	α				
		(36%)	6.090	27		
			6.051	70		
			5.769	2		
			5.607	1		
		and				
		β^-			2.270	63
					1.550	10
					0.930	8
					0.670	6
					0.450	8
					0.085	5
^{212}Po	3.04×10^{-7} s	α	8.785	100		
^{208}Tl	3.1 min	β^-			1.796	52
					1.521	22
					1.274	21
					0.640	5

(After Ivanovich, 1982; Lederer and Shirley, 1978.)

the binding energy, N the number of neutrons and Z the charge number. The binding energy per nucleon is called the 'binding fraction' which is obtained for any particular nucleus simply by dividing the binding energy by A. A plot of the binding fraction against A shows

Table 3.2 — The ^{235}U decay series

Nuclide	Half-life	Decay	α energy (MeV)	Amount (%)	β energy (MeV)	Amount (%)
^{235}U	7.13×10^{8} years	α	4.397	57		
	Spontaneous		4.598	5		
	fission 3.5×10^{17} years		4.557	4		
			4.503	1		
			4.416	4		
			4.397	57		
			4.367	18		
			4.345	2		
			4.324	3		
			4.217	6		
^{231}Th	25.64 h	β^-			0.302	52
					0.218	20
					0.138	22
					0.090	6
^{231}Pa	3.43×10^{4} years	α	5.058	11		
			5.031	3		
			5.029	20		
			5.012	25		
			4.985	1		
			4.951	23		
			4.934	3		
			4.852	1		
			4.737	8		
			4.713	1		
			4.681	2		
^{227}Ac	22 years	α				
		(1.38%)	4.951	47		
			4.938	40		
			4.870	6		
			4.853	4		
			4.766	1		
					0.046	100
^{227}Th	18.17 days	α	6.038	25		
		and	6.009	3		
		β^-	5.978	23		
			5.960	3		
			5.867	2		
			5.808	1		
			5.755	21		
			5.713	5		
			5.709	8		
			5.701	4		
			5.693	2		
			5.668	2		
^{223}Fr	21 min	α				
		(<0.01%)	5.340	—		
		and				
		β^-			1.120	100
^{223}Ra	11.68 days	α	5.748	9		
			5.717	54		
			5.608	26		
			5.540	9		
			5.434	2		
^{219}At	0.9 min	α				
		(97%)	6.28	100		
		and				
		β^-			—	—
^{219}Rn	3.92 s	α	6.819	81		
			6.553	11		
			6.425	8		
^{215}Bi	8 min	β^-			—	—
^{215}Po	1.83×10^{-3} s	α	7.386	100		
^{211}Pb	36.1 min	β^-			1.355	92
					0.525	6
					0.251	1
^{211}Bi	2.16 min	α				
		(99.73%)	6.623	84		
			6.279	16		
		and				
		β^-			—	—
^{211}Po	0.52 s	α	7.450	99		
^{207}Tl	4.79 min	β^-			1.431	100

(After Ivanovich, 1982; Lederer and Shirley, 1978.)

which are most stable nuclides. The binding fraction is greatest for $A=57$ and decreases with both larger and smaller values of A. The decrease in the binding fraction with increasing atomic mass above $A=57$ occurs because the increased electrostatic repulsion makes the nuclei less stable. The rapid drop in binding fraction below $A=8$ occurs because each nucleon is attracted by only a few other nucleons; the separation between the nucleons, therefore, increases and their stability consequently decreases. $^{4}_{2}He$ is an exception to this, as its

Table 3.3 — The ^{238}U decay series

Nuclide	Half-life	Decay	α energy (MeV)	Amount (%)	β energy (MeV)	Amount (%)
^{238}U	4.468×10^{9} years	α	4.196	77		
	Spontaneous fission		4.149	23		
	8.19×10^{15} years					
^{234}Th	24.1 days	β^-			0.199	73
					0.104	21
					0.060	5
					0.022	1
^{234m}Pa	1.18 min	β^-			2.290	98
^{234}Pa	6.7 h	β^-			1.190	5
					0.680	19
					0.512	63
					0.280	12
^{234}U	2.48×10^{5} years	α	4.774	72		
	Spontaneous fission		4.723	28		
	2×10^{16} years					
^{230}Th	7.52×10^{4} years	α	4.689	76		
	Spontaneous fission		4.621	24		
	$>1.5\times10^{17}$ years					
^{226}Ra	1602 years	α	4.785	94		
			4.602	6		
			5.490	1006		
^{222}Rn	3.825 days	α				
^{218}Po	3.05 min	α				
		(99.98%)	6.002	100		
		and				
		β^-			0.330	100
^{214}Pb	26.8 min	β^-			1.030	100
^{218}At	2 s	α				
		(99.9%)	6.757	4		
			6.694	90		
			6.654	6		
		and				
		β^-			0.670	100
^{214}Bi	19.7 min	α				
		(0.04%)	5.512	40		
			5.448	54		
			5.268	6		
		and				
		β^-				
					3.260	19
					1.880	9
					1.510	40
					1.020	23
					0.420	9
^{214}Po	1.64×10^{-4} s	α	7.687	100		
^{210}Tl	1.32 min	β^-			2.340	19
					1.870	56
					1.320	25
^{210}Pb	22.3 years	α				
		(<1%)	3.72	—		
		and				
		β^-			0.610	19
					0.015	81
^{210}Bi	5.02 days	α				
		(<1%)	4.687	40		
		and	4.650	60		
		β^-			1.161	100
^{210}Po	138.3 days	α	5.305	100		
^{206}Tl	4.19 min	β^-			1.527	100

(After Ivanovich, 1982; Lederer and Shirley, 1978.)

Table 3.4 — Other naturally occurring radioactive nuclides

Nuclide	Element (%)	Half-life (years)	Decay	α energy (MeV)	Amount (%)	β energy (MeV)	Amount (%)
^{40}K	0.0117	1.31×10^{9}	Electron capture (10.67%) and β^-			1.35	100
^{48}Ca	0.187	$>10^{18}$					
^{50}V	0.25	$>10^{16}$	Electron capture (70%) and β^-				
^{64}Zn	48.6	$>8\times10^{15}$					
^{70}Zn	0.62	$>10^{15}$					
^{76}Ge	7.8	$>5\times10^{21}$					
^{82}Se	9.2	1.4×10^{20}	Double β^-				
^{87}Rb	27.83	4.88×10^{10} (5.0×10^{10})	β^-			0.274	100
^{96}Zr	2.8	$>3\times10^{17}$					
^{100}Mo	9.6	$>3\times10^{17}$					
^{113}Cd	12.2	9.3×10^{15}	β^-				
^{116}Cd	7.5	$>10^{17}$					
^{115}In	95.7	5.1×10^{14}	β^-			0.495	100
^{124}Sn	5.64	>2.4×1017					
^{123}Te	0.89	$>5\times10^{13}$	Electron capture				
^{130}Te	34.5	2×10^{21}	Double β^-				
^{138}La	0.089	1.1×10^{11}	Electron capture (67.9%) and β^-			0.205	100
^{142}Ce	11.1	$>5\times10^{16}$					
^{141}Pr	100	$>2\times10^{16}$					
^{144}Nd	23.8	2.1×10^{15}	α	1.83	100		
^{145}Nd	8.3	$>6\times10^{16}$	α				
^{150}Nd	5.6	$>5\times10^{18}$					
^{147}Sm	15.1	1.06×10^{11}	α	2.233	100		
^{148}Sm	11.3	8×10^{15}	α	1.96	100		
^{149}Sm	13.9	$>10^{16}$					
^{152}Gd	0.2	1.1×10^{14}	α	2.14	100		
^{156}Dy	0.057	$>10^{18}$					
^{165}Ho	100	$>6\times10^{16}$					
^{169}Tm	100	$>5\times10^{16}$					
^{175}Lu	97.39	$>10^{17}$					
^{176}Lu	2.61	3.6×10^{10}	β^-			0.565	100
^{174}Hf	0.16	2×10^{15}	α	2.50	100		
^{180}W	0.13	$>10^{15}$					
^{182}W	26.3	$>2\times10^{17}$					
^{183}W	14.3	$>1.1\times10^{17}$					
^{186}W	28.6	$>6\times10^{15}$					
^{187}Re	62.6	4.3×10^{10}	β^-			0.00265	100
^{192}Os	41.0	$>10^{14}$					
^{190}Pt	0.013	6.9×10^{11}	α	3.18	100		
^{192}Pt	0.78	$>6\times10^{16}$	α	2.75	100		
^{198}Pt	7.2	$>10^{15}$					
^{196}Hg	0.15	$>10^{14}$					
^{209}Bi	100	$>2\times10^{18}$	α				

(After Lederer and Shirley, 1978.)

nucleus is the strongly bound neutron pair–proton pair.

3.3 MAGIC NUMBERS

The value of the separation energy of neutrons and protons can be seen generally to follow a smooth curve for most nuclides of the same character, in terms of odd or even numbers of protons and neutrons. However, there are a number of significant exceptions to this. Beyond certain proton numbers, the so-called 'magic numbers' of 2, 8, 20, 50, 82 and 126, the proton separation energy drops by about 2 MeV. For example, the range is from 8.5 to

10.5 MeV for $_{50}Sn$ as the mass number increases from 114 to 120 but is only from 8 to 9 MeV for $_{52}Te$ as the mass number increases from 122 to 126. The increase in value with increasing mass number is due to the increasing number of neutrons which bind a constant number of protons with increasing strength. The same effect occurs for the neutron separation energy when the number of neutrons exceeds these magic numbers. The most marked change occurs when both the neutron number and the proton number are 2. Here the neutron and proton separation energies fall by about 13 MeV.

Nuclei with magic proton or neutron numbers (or both) are called 'magic nuclei'. They have greater numbers of stable isotopes or stable isotones than their even numbered neighbours. The abundance of these nuclides is also greater than that of their neighbours. These properties can be explained in terms of a shell model for the nucleus analogous to that of the extranuclear electron shell model. On this basis, the magic nuclei are taken to possess the number of particles required to give complete shell occupancy. If the number of protons or neutrons is just one larger than this magic number, it will have a small separation energy.

3.4 RELATIVISTIC MOTION: SOME FUNDAMENTALS

As nuclear emissions move with velocities at or approaching the speed of light (Table 3.5), relativistic effects must be considered. Firstly, it should be noted that electromagnetic radiation transfers energy through space and that, when a stream of photons falls upon a surface, a pressure is exerted as if it were a stream of any other type of particle. The momentum p of a photon is given by $p=h\nu/c$ where h is Planck's constant, ν the velocity and c the velocity of light in a vacuum. The energy of the photon is given by $h\nu$.

Secondly, the mass of any particle, such as a β-particle, will change with its velocity, so that at velocity v its mass m is given by

$$m=\frac{m_0}{(1-v^2/c^2)^{\frac{1}{2}}},$$

where m_0 is the mass when stationary or 'rest mass'. Obviously, for cases in which v is small compared with c, the rest mass and the true mass will be virtually identical.

It thus follows that the relativistic momentum p is given by

$$p=\frac{m_0 v}{(1-v^2/c^2)^{\frac{1}{2}}},$$

and the kinetic energy k by

$$k=\frac{m_0 c^2}{(1-v^2/c^2)^{\frac{1}{2}}}-m_0 c^2 .$$

The total energy E of the particle is the sum

Table 3.5 — Initial velocities of radiations

Radiation	Energy (MeV)	Velocity (m s^{-1})
α-particles	1	7.0×10^6
	4	1.4×10^7
Neutrons and protons	1	1.4×10^7
β-particles	0.1	1.6×10^8
	1	2.8×10^8
X-rays and γ-rays	Any	3×10^8

of the kinetic energy and the rest energy (or self-energy) and is given by

$$E = \frac{m_0c^2}{(1 - v^2/c^2)^{\frac{1}{2}}},$$

where m_0c^2 is the rest mass. Thus,

$$E^2 = m_0c^4 + p^2c^2$$

and, for a particle at rest,

$$E = \pm m_0c^2 \; .$$

The fact that E may be positive or negative indicates that, for every particle in the positive energy state, there exists the potential for the creation of an equal and opposite particle in the negative energy state: the so-called antiparticles.

3.5 THE PHYSICS OF DECAY

3.5.1 Decay modes and probability

Nuclides in which Z is greater than 82 exhibit natural transformation due to spontaneous emission of nuclear particles. For these nuclei to decompose, the energy of the nuclear particles must exceed their binding energy. The emission of 2 neutrons and 2 protons as a single entity (the α-particle) causes A to decrease by 4 and Z by 2. Emission of a nuclear electron (or negatron, β^-) increases Z by 1 and leaves A unaffected. Double-electron emission increases Z by 2. Positron (β^+) emission decreases Z by 1 and also leaves A unaffected. Electron capture has the same effect on A and Z as β^+ emission. γ-radiation, defined by $E=hf$ and $c=\lambda f$ (where E is the energy of the radiation, h Planck's constant, f the frequency of the radiation, c the speed of light and λ the wavelength), normally accompanies the emission of β-particles which result in the formation of excited daughter neuclei. Specific levels of de-excitation energy produce the characteristic γ-ray energy spectrum for any particular daughter.

Analogous to binding energy and binding fraction are the concepts of mass defect and packing fraction. The mass defect is the difference between the mass of the nucleus and the value of A for that nucleus, while the packing fraction is the mass defect per nucleon. The mass m of the nucleus is measured in terms of the atomic mass unit scale which can be based on either $^{12}C=12$ or $^{18}O=18$. Unfortunately, the actual value of m, and therefore of the mass defect, depends upon which of these scales is used. Nevertheless, the form of the variation in packing fraction with A is very similar to that of the reciprocal of the binding fraction. Observation of nuclear reactions is, in fact, a very sensitive means of measuring nuclear masses, as a small difference in mass is reflected in a large, and therefore easily measured, energy difference using the Einstein mass–energy relation, $E=mc^2$.

Although radioactivity essentially involves changes in the nuclei of atoms, the mass–energy balances involved can also produce extranuclear changes, giving differences between the initial and final states of the parent and daughter atoms. For example, if the decay of a parent produces a radioactive daughter in an excited state, the excess energy that can be emitted as γ-rays can also be transferred to an extranuclear electron which is then ejected from the daughter atom. This is the internal conversion process. Interaction between the nucleus and extranuclear electrons also occurs in the process of electron capture, in which an inner orbital atomic electron is captured by the nucleus.

Radioactive decay can take place by emission of either α-particles or β-particles (β^+ or β^-) or by electron capture. In fact, the decay of any particular radionuclide need not be restricted to just one of these. ^{40}K, for example, has a branched decay, one branch occurring as the result of β^--particle emission, while the other arises by electron capture.

Whatever the mode of decay, however, all the transformations of the nucleus are characterised by the decay probability: the probability per unit time of transformation taking place from the initial to the final state of the system. The reciprocal of this probability yields the lifetime of the initial state which is defined as the time required for the population number of the initial state to fall to e^{-1} of its starting value. The emission probability λ can also be expressed in terms of the width τ of the decay level, which is best considered as an uncertainty in the energy of the unstable state and related to the average decay time $\bar{T}=\lambda^{-1}$ via Heisenberg's relationship $\lambda=\tau/h$, where h is Planck's constant. The half-life is the time required for the transformation of one-half of the initial number of nuclei:

$$T_{1/2} = \bar{T} \log_e 2 = 0.693\bar{T}.$$

3.5.2 Radioactive emissions

The α-particle

The α-particle is the nucleus of a He atom, ${}^4_2He$, and could alternatively be represented as a ${}^4_2He^{2+}$ ion. The mass of a ${}^4_2He$ atom on the ${}^{12}C$ scale is 4.002 603 6 atomic mass units (amu) or 6.6456×10^{-27} kg. Subtracting the mass of 2 electrons (1.0975×10^{-3} amu or $1.821\,66\times 10^{-30}$ kg) gives the mass of the α-particle (4.0015 amu or 6.6438×10^{-27} kg). The charge/mass ratio is 4.8225 C kg^{-1}, and its size is about 2.4×10^{-17} m. α-particles can only be emitted from a nucleus of atomic number Z and atomic mass number A, where ${}^A_ZM$ is greater than ${}^{A-2}_{Z-4}M+m_\alpha+2m_e$, where ${}^A_ZM$ is the atomic mass of that nuclide, and m_e and m_α are the masses of an electron and an α-particle respectively. The energy of α-particles is measured in electron-volts (eV), which is the energy required to raise the potential of an electron by 1V. This is equivalent to $1.602\,189\,2\times10^{-19}$J. In practice, this unit is usually too small to use directly, and so the common energy unit is the mega electron volt (MeV) which is 10^6 eV or $1.602\,189\,2\times10^{-13}$J. α-particle energies are, therefore, high when compared with energies of a few electron volts for extranuclear transitions which ionise atoms and give rise to visible radiations, and X-ray transitions which are in the range from 10 eV to 10^2 keV.

As a moving object, an α-particle has an associated wave function the wavelength λ of which is given by

$$\lambda = \frac{2.26\times10^{-15}\,(1+4/W)}{E_0^{\frac{1}{2}}}\ \text{m},$$

where E_0 is the kinetic energy in mega electron volts measured in the laboratory and W is the atomic weight of the scattering nucleus. Because of this relationship, α-particles are almost monoenergetic for any particular emission.

α-decay lifetimes are very dependant on the energy of the emitted particle. For example, ${}^{232}Th$ has a lifetime of the order of 10^{10} years and an α-particle energy of about 4 MeV, while ${}^{212}Po$ has a lifteime of only about 3×10^{-7} s and an α-particle energy of 9 MeV. The reason for this lies in the origin of the α-emission. In nuclides with mass numbers greater than 82, this might appear simply to be due to an increase in electrostatic repulsion with increasing mass number. However, a Coulomb barrier exists which contains the emissions. This electrostatic energy barrier can be simply considered by imagining that an α-particle approaches the nucleus from which it has been emitted. At large distances, the potential energy is purely electrostatic and increases in inverse proportion to decreasing distance. When the distance is of the order of the nuclear radius, or smaller, the interaction is no longer purely electrostatic and at some point the attractive nuclear forces overcome the electrostatic forces and a peak in the potential energy function is produced. The form of the potential barrier is determined by the nuclear radius. As α-particle emission must penetrate this barrier, not all nuclides are radioactive. For most nuclides the potential barrier effectively prevents α-particle emission.

The number n of α-particles emitted per second is given by

$$n = \frac{2^{1/2}\pi^2 h^2 \, e^{-2c}}{M^{3/2} R^3 \, (Zze_1^2/R - E)^{1/2}},$$

where z, M and E are the α-particle charge, mass and energy, respectively, outside the nucleus, and R and Z are the radius and charge of the nucleus, respectively. e_1 is the charge on an electron and h is Planck's constant. c is given by

$$c = \frac{2Zze_1^2}{hv} \{\arccos(x^{1/2}) - [x\,(1-x)]^{1/2}\},$$

where $x = E/B$, B is the barrier height determined by $B = Zze_1^2/R$ and v is the velocity of the α-particle outside the nucleus. As E is less than B, an approximation gives

$$c = \frac{Zze_1^2}{hv} - \frac{2e_1}{h}\,(2ZzMR)^{1/2}.$$

These expressions are equivalent to a relationship involving the lifetime τ of the α-radioactivity:

$$\tau = \frac{\log_e 2}{n} = \tau_0 \, e^{-2c},$$

where τ_0 is the lifetime without the barrier,and e^{-2c} is the transmission coefficient of the Coulomb barrier (Stephens and Hurlimann, 1967).

The value of τ_0 is about 3.3×10^{-21} s, but this ignores any time taken to form the α-particle itself. If this is included, a value of about 10^{-15} s is obtained. The theoretical relationship between lifetime and α-particle energies explains the Geiger–Nuttall empirical rule which gives the range in air of α-particles:

$$\log \tau = -57.5 \log R + a$$

where τ is the half-life and R the range in centimetres. a is a constant for any particular decay series; a has a value of 42.3 for the ^{238}U series, 44.2 for the ^{232}Th series and 46.3 for the ^{235}U series. Therefore,

$$\log \tau = -86.3 \log E + 28.5 + \mathrm{a}$$

gives the energy E relationship.

Because the Coulomb barrier increases with atomic number, α-particles are retained in the nuclei of heavy elements except near variations in the level of binding energy. The largest energy release in α-emission occurs when the resultant nuclei have an atomic number of 82 and a neutron number of 126. These conditions represent a relatively stable configuration but, if more than 126 neutrons are present, the nuclear radius changes abruptly. For elements of lower atomic number than 82, instability in the nucleus occurs when there is a drop in the binding energy level. Thus, $^{147}_{62}$Sm is an α-emitter.

In passing through gases, α-particles cause ionisation by removing electrons (Fig. 3.1). The

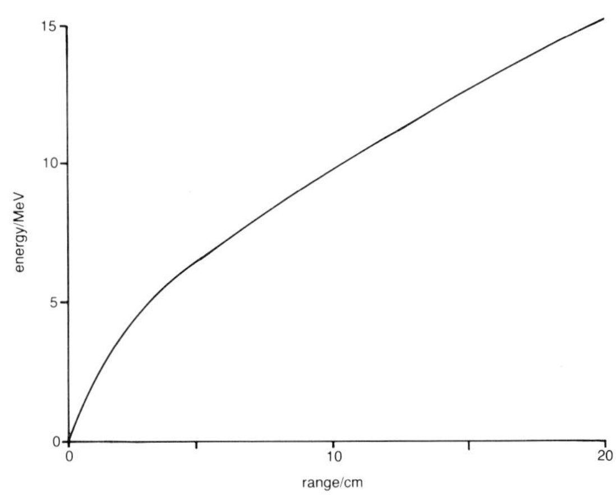

Fig. 3.1 — The energy of α-particles and their range in air at standard temperature and pressure. (Mann *et al.*, 1980, redrawn with permission.)

liberated electrons can, in turn, ionise more of the gas to produce γ-rays. These α-particles do not undergo any deflection from their pathway, unless they approach too closely to an atom, in which case scattering occurs. In air, an α-particle loses between 34 and 35 eV of energy for each ion pair that it creates. Only about 14 eV of this is used to ionise N_2 or O_2, and so about 20 eV is transferred as electron kinetic energy, producing heat or causing excitation of the atoms. Apart from removal of electrons, α-particles also lose energy by collision and scattering. This results in a broadening of the range of α-particle penetration in air known as 'straggling'. Straggling of α-particles with energies below a few mega electron volts also occurs by the acquisition of 1 or 2 electrons, leading to their existence as single charged ions or even as He atoms. In either case, the ability for the α-particle to lose energy decreases, and this affects the range (Fig. 3.2). A useful analogy that has been proposed by Mann *et al.* (1980) to describe the interaction of an α-particle with its incident material is the firing of artillery into a forest of fairly widely spaced trees. The shell could travel in a straight line, tearing off branches and finally be brought to rest through loss of energy. Only if the shell hit the trunk of a tree would it be seriously deflected from its course (Fig. 3.3(a)). Like all decelerating charged particles, decelerating α-particles emit photons. However, as the intensity of this radiation is inversely proportional to the square of the mass of the particles involved, the effect is negligible.

β-particles

If we continue with the analogy in which an α-particle is a shell fired into a forest of widely spaced trees, the β-particle can be considered to be the equivalent of a small-calibre rifle bullet which would ricochet from branch to branch frequently changing direction (Fig. 3.3(b)). β^- decay is only possible if ${}^{A}_{Z}M$ is greater than ${}^{A}_{Z+1}M$ and β^+-decay only if ${}^{A}_{Z}M$ is greater than ${}^{A}_{Z-1}M+2m_e$. The difference of $2m_e$ in β^+-decay is due to the loss of 2 'electrons': a positron from

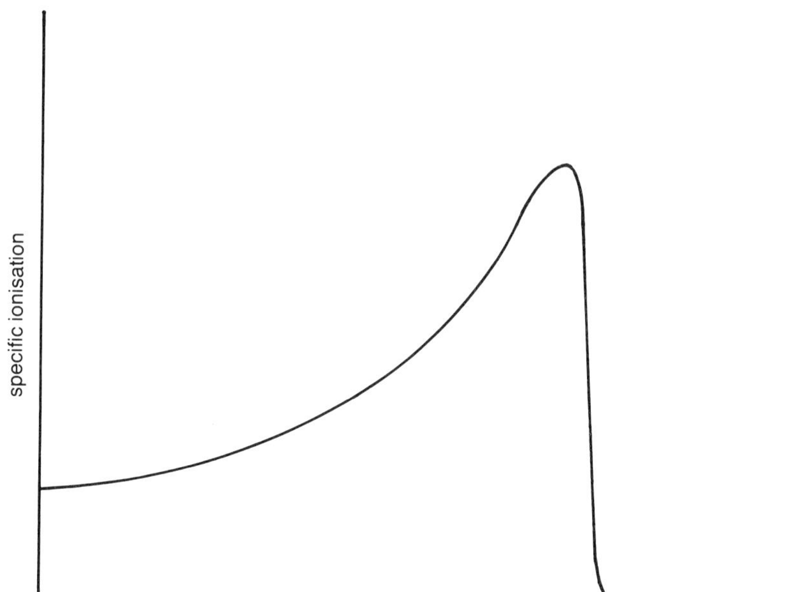

Fig. 3.2 — Variation in specific ionisation with distance travelled by an α-particle.

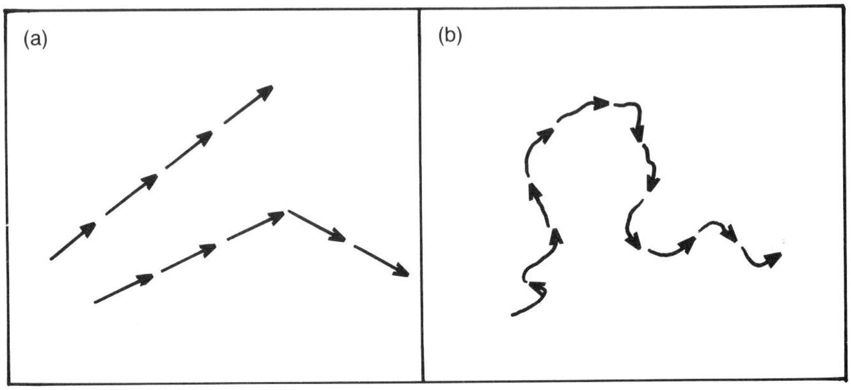

Fig. 3.3 — Diagrammatic representation of (a) α-particle and (b) β-particle tracks. (After Mann *et al.*, 1980.)

the nucleus and an extranuclear electron no longer needed to balance the decreased charge of the nucleus. In β^--decay, the electron emitted from the nucleus can be accepted to balance the extranuclear charge. β-particle energies, like those of α-particles, are usually of the order of a few mega electron volts but, whereas α-particles from a particular source have a well-defined energy, β-particles are emitted with energies that range from (effectively) zero up to a maximum value for that particular source (Fig. 3.4). Indeed, some β-particles may have such high energies that their speed approaches that of light. The explanation of this spectrum of energies is thought to be the loss of energy suffered to various degrees by β-particles on

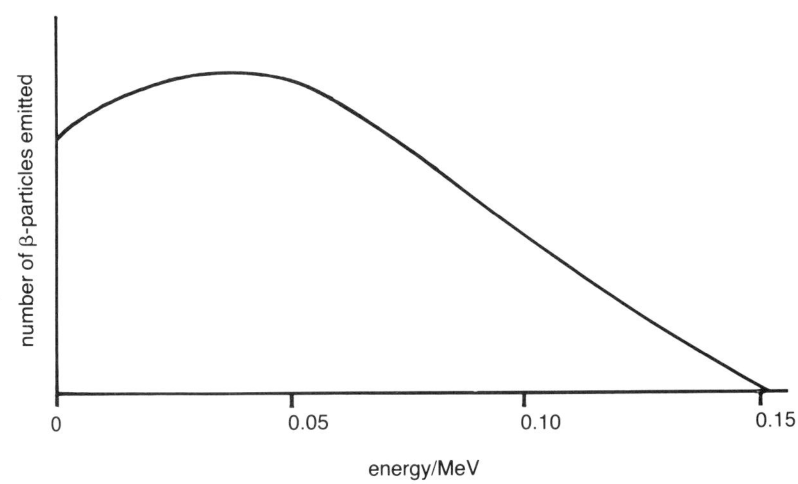

Fig. 3.4 — The β-particle energy spectrum of a simple emitter (^{14}C). (After Mann *et al.*, 1980.)

emission, all the particles originally having the maximum energy value. However, measured rates of heat loss by β^+-emitters correspond to the average energy of the β^+-particles rather than to the maximum energy. Clearly, transfer of energy in an additional way must take place in a β^+-decay. This is the basic theoretical background for the emission of a neutrino with a β^+-particle. In fact, the maximum β^+-particle energy is now recognised as connected via the relationship $E=mc^2$ to the difference between the masses of the parent and daughter nuclei (Fig. 3.5) In 1937, Fermi suggested that the β^--particle is generated within the nucleus as the result of the fact that a neutron changes into a proton, with the additional production of an antineutrino:

$$^1_0n \rightarrow {}^1_1p + \beta^- + \bar{\nu}.$$

The likelihood of production of a β^--particle with a specific energy within the spectrum of energies between zero and its maximum value is thus governed by a probability function which includes the energy carried by the antineutrino, specifically,

$$P\,\mathrm{d}E \propto p\,E(E_{\beta\mathrm{max}} - E)^2\mathrm{d}E,$$

where $P\,\mathrm{d}E$ is the probability of emission in unit time of a β^--particle with energy between E and $E+\mathrm{d}E$, p is the momentum of the β-particle in units of m_0c (where m_0 is its rest mass) and E is the total energy of the β-particle including its rest mass in units of m_0c^2. $E_{\beta\mathrm{max}} - E$ is the energy carried by the antineutrinos. Because this effect slows down β^--particles, it produces an excess of lower energies. In contrast, β^+-particles are accelerated away from the nucleus, producing a deficit of lower energies. The neutrinos emitted with β-particles penetrate all matter practically unhindered, so in practice they have no other bearing on radioactive decay schemes.

The process of electron capture, in which extranuclear electrons from the K, L or M shells are captured by the nucleus, is somewhat similar in effect to β^+-decay but, instead of a

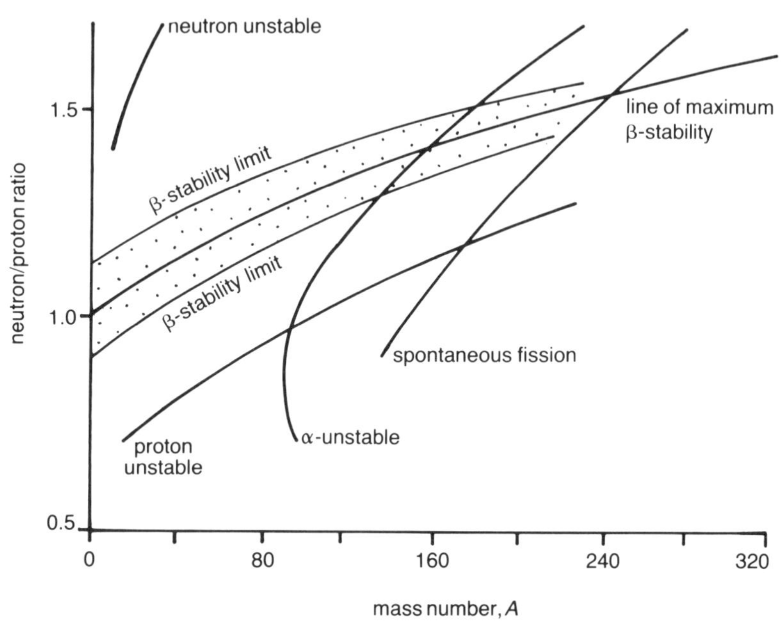

Fig. 3.5 — Nuclear stability limits. The zone of β-stability is shown stippled. (After Segré, 1959.)

neutron decaying to form a proton, a proton is converted to a neutron. Thus, instead of an antineutrino, electron capture is accompanied by neutrino emission:

$$^{1}_{1}p + e^{-} \rightarrow {}^{1}_{0}n + \nu.$$

Electron capture can only occur if $^{A}_{Z}M$ is greater than $_{Z-1}^{A}M$. It is possible for β^{+}-decay to be accompanied by electron capture, in which case the two reactions operate in the ratio of 98.54% electron capture and 1.46% β^{+}-emission.

Although β-particles are affected by extranuclear electrons, in that their pathways through matter change direction, because the mass of a β-particle is some 7000 times less than that of an α-particle, the penetration of β-particles is much greater than that for α-particles of the same energy. In fact, the loss of energy from a 4 MeV β-particle in air at standard temperature and pressure is 2.2×10^{3} eV cm^{-1}, while that for a 4 MeV α-particle is 1.44×10^{6} eV cm^{-1}. With less energetic β-particles, the energy loss is greater; at 10 keV the value is 2.4×10^{4} eV cm^{-1} (Fig. 3.6).

Because the mass of a β-particle is so much less than that of an α-particle, the production of electromagnetic radiation on deceleration is much more marked, and the passage of β-particles through materials is accompanied by X-ray emission. The retardation and stopping of β-particles produce a continuous spectrum of photon energies known as 'braking radiation' or 'bremsstrahlung'. The maximum energy of this radiation is liberated when a β-particle loses all its energy in a collision.

Lifetimes of β-decay transformations range from 10^{-2} s to 10^{11} years and depend upon two factors: firstly, the available energy E according to the relationship E^{-k}, where k is approximately 5 (β-decay is much less sensitive to energy changes than is α-decay); secondly (with high sensitivity), spin and parity changes between the initial and final states. Because the energy available in β-decay is equal to the difference between the atomic masses of the initial and final nuclei, and for β^{+}-emission this mass difference must exceed $2mc^{2}$, two neighbouring isobars ($\Delta T_{s} = 1$) cannot both be stable with respect to β-decay. For large spin change and low energy, the lifetime may, in fact, be so long that the nucleus appears stable. For example, although one of the two $A = 50$ isobars must be unstable, neither exhibits observable β-activity. Other examples are ^{176}Lu (which has a half-life of 3.6×10^{10} years), ^{115}In (with a half-life of 5.1×10^{14} years) and ^{187}Re (with a half-life of 4.3×10^{10} years).

Isobaric pairs with $\Delta T_{s} = 2$ are quite

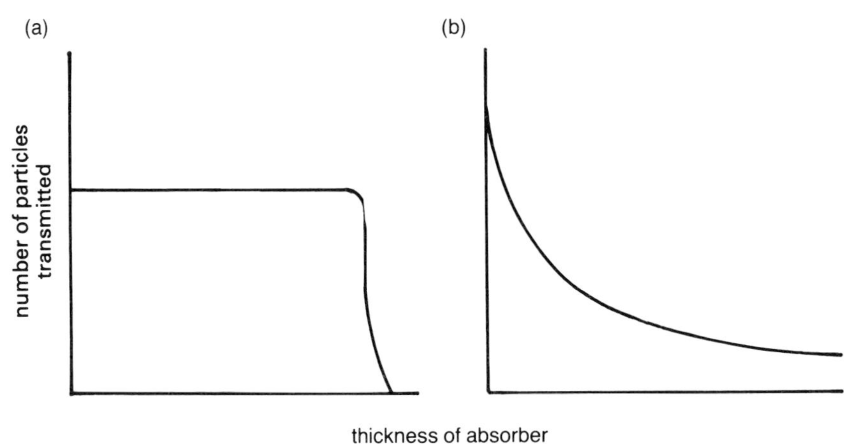

Fig. 3.6 — The penetration of matter by (a) α-particles and (b) β-particles. (After Mann *et al.*, 1980.)

common, and in these cases the higher mass nucleus may decay to the other by either a double-β^+-emission or a double-electron-capture process. The lifetimes for these processes are extremely long (of the order of 10^{20} years for an available energy difference of 5 MeV), but so far the process has not been observed.

γ-radiation

If a stable nucleus is left in an excited state after the emission of an α- or β-particle, or after electron capture, but has not enough energy for the emission of an α-particle, it will lose energy by the emission of electromagnetic radiation of high frequency, γ-radiation, to return to the normal or ground-state energy level. Some nuclear transitions are possible in which the daughter nucleus is not excited and in which no γ-radiation accompanies the nuclear transformation, but these are comparatively rare. Because the physics of the production of a β-particle within the nucleus involves fundamental changes in the nuclear particles, emission of β-particles is more likely to leave the parent nucleus in an excited state than is the case for α-particle emission in which particle transformation is not involved. γ-radiation is therefore much more commonly associated with β-decay than with α-decay systems.

γ-radiation is characterised by its energy and forms a consistent framework with α- and β-particle energies. An excited nucleus might not emit γ-radiation of a single energy to reach the ground state in one step. Instead, emission of one particular energy from a higher to a lower level of excitation may occur, when a second energy can be emitted which may bring the nucleus to the ground state or leave it still in an excited condition. If the latter is the case, a further emission will occur and so on until the ground state energy level is reached. This system is known as a cascade of γ-ray energies and is frequently observed. A problem of terminology arises in assigning the origin of γ-rays to a particular parent because, although it is the excited daughter nucleus that emits the γ-radiation, it is the decay of the parent nucleus that gives rise to the excited daughter nucleus. Therefore the γ-radiation is often incorrectly assigned to the parent in the decay scheme.

γ-ray energies lie in the range of kilo electron volts to mega electron volts. Specifically, γ-rays may be defined as possessing energies between 40 keV and 4 MeV. X-rays, on this basis, have energies between 40 eV and 40 keV. However, this classification is not particularly appropriate, as electromagnetic radiation spanning the X-ray–γ-ray boundary can be produced in the de-excitation of a nucleus (Fig. 3.7). A system which overcomes this difficulty simply defines γ-rays as originating within the nucleus and X-rays as having an extranuclear origin. On this basis, γ-rays can

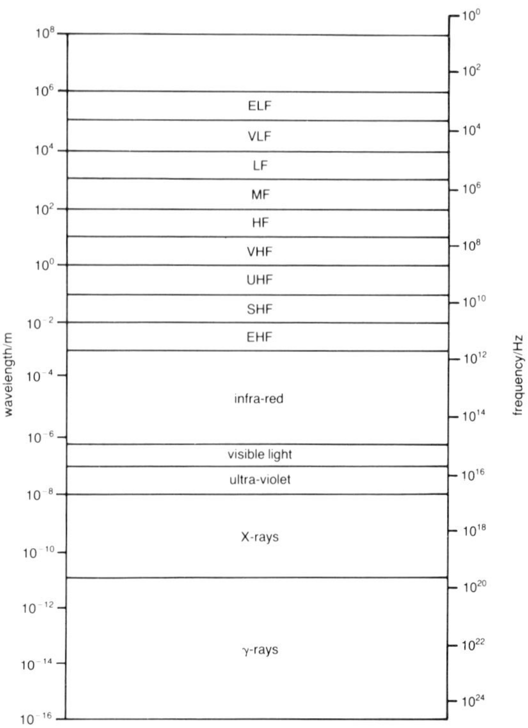

Fig. 3.7 — The electromagnetic spectrum. Frequency bands of radio waves and radar indicated between extremely low and extremely high. V means very, U ultra and S super.

have energies as low as a few kilo electron volts, while X-rays may have energies above 100 keV.

Lifetimes for γ-ray emission are very dependant on changes in spin between the initial and final states. Thus for small spin changes ($\leqslant 2$) the lifetimes may be only 10^{-17}–10^{-16} s, while for larger spin changes (about 4) the lifetime may be several years.

Isomeric nuclei are those in which the final excited level of the nucleus has a long lifetime. Nuclear isomers are typically identified by placing a lower case m after the mass number, as in ^{234m}Pa. They are interesting nuclides in that they are chemically identical with the normal isotope and have the same physical properties except that they are radioactive. They can, therefore, enter into molecular bonding as if they were normal nuclides but cause breaking of the bonds by their radioactivity. The only naturally occurring isomeric nuclide is ^{234m}Pa.

Theoretically, de-excitation by particle emission or γ-ray emission could take place simultaneously, each effect being determined by its own probability function. However, for light elements the probability of particle emission is much greater than that for γ-ray emission, and so γ-radiation is only a minor process. Conversely, for the heavier elements the emission of relatively low energy charged particles from the nucleus is so much inhibited by the potential barrier that γ-radiation is generally the more probable process.

β^--particles emitted from a nucleus normally have a spectrum of energies varying between zero and a maximum value. However, when the energies of β^--particles associated with γ-ray-emitting nuclei are studied, it is found that they often possess discrete energies. In explanation of this phenomenon, it is believed that an excited daughter nucleus can discard some of its energy by transfer to an extranuclear electron The extranuclear electron is then expelled from the atom with kinetic energy equal to the de-excitation transition minus the electron binding energy appropriate to the shell from which the electron is ejected. Thus, instead of a γ-ray, a β^--particle is emitted. This process is known as 'internal conversion' and the β^--particles so produced are called 'conversion electrons'. The relative probability of emission of a γ-ray or a conversion electron can be expressed in terms of a branching ratio called the 'internal conversion coefficient' which is the ratio of conversion electrons to γ-ray transitions.

When a conversion electron is ejected from its shell, X-rays can be emitted if the vacancy so created is filled by another electron. Some of this X-ray energy can, in turn, cause the ejection of an electron from the outer shells of the same atom, if the ionising potential is less than the X-ray energy available. When these outer shell electrons are emitted, they are known as 'Auger electrons' (after their discoverer Pierre Auger). Again, a branching ratio can be established in terms of the probability of either Auger electron emission or X-ray emission; this is called the 'Auger yield' or 'fluorescence yield'. When a γ-ray photon transfers all its energy to a bound electron, which is then emitted, the effect is known as 'photoelectric absorption'. This mainly takes place in the inner electron shells. After emission of the electron, the reorganisation of the atom causes the emission of X-rays and Auger electrons.

The energy liberated following the decrease in mass that accompanies transformation of a parent to a daughter nuclide thus causes a response within the whole atom of the unstable nuclide and not just within its nucleus. This response affects the chemical as well as the physical properties of the atom.

Emission of other heavy particles

When some light elements such as $^{10}_{5}B$, $^{24}_{12}Mg$ or $^{27}_{13}Al$, are bombarded with α-particles, they emit neutrons, as exemplified by the reaction

$$^{24}_{12}Mg + ^{4}_{2}He \rightarrow ^{27}_{14}Si + ^{1}_{0}n.$$

The product then decays further by emission of positrons:

$$^{27}_{14}Si \rightarrow ^{27}_{13}Al + \beta^+.$$

When the α-particle source is removed, neutron emission ceases but, interestingly, the positron emission decays exponentially. This artificial radioactivity was first observed by Joliot and Joliot-Curie in 1934, and it thus became clear that the neutron, because of its uncharged state, could be very effective in producing nuclear transformations simply because it is unaffected by the electrostatic field of the nucleus.

Neutrons can be classified on the basis of their energies into fast, slow or thermal. Fast neutrons have energies greater than 0.1 MeV, slow about 1 eV and thermal about 0.025 eV. The neutrons produced by the nuclear reactions illustrated above are fast neutrons. When fast neutrons collide with heavy nuclei, they do not lose much energy but, if they collide with nuclei of approximately the same mass, they lose energy, slow down and become thermal neutrons. Clearly, the nuclide with a nuclear mass similar to that of a neutron is ^{1}H, with a single proton in the nucleus, H-bearing materials, are, therefore, very effective in slowing neutrons. Slow and thermal neutrons can be captured by the nucleus to render it unstable. This process is accompanied by the emission of γ-radiation. Neutron decay processes are utilised in geological investigations, but their natural occurrence appears mainly to be limited to natural fission reactors (discussed in Chapter 5).

Recently, a new type of natural radioactivity has been detected (Rose and Jones, 1984). In the decay of ^{223}Ra, the emission of particles heavier than α-particles has been observed, resulting in the formation of ^{209}Pb as the decay product. These have been identified as ^{14}C ions, the emission energy of which is 30 MeV. The branching ratio for emission of ^{14}C nuclei relative to α-particles for ^{223}Ra, however, is very small, and so the effect is negligible from a practical viewpoint. Nevertheless, the phenomenon is interesting, as similar decay schemes must occur in other nuclides close to ^{223}Ra, particularly those nuclides close to ^{208}Pb, but the optimum conditions are only likely to be found in artificial decay series and not in natural sources.

Branching ratios

The concept of the branching ratio was considered when dealing with the partitioning of energies in β-decay schemes by internal conversion. In the branched decay of an unstable nucleus, decay need not be restricted to a single type. In fact, it is commonly found that decay takes place by two modes, for example, by electron capture and β^+-particle emission, or by α- and β^--particle emission. The relative probability that each branch is followed can be expressed as a percentage of the total decay. In the branched decay of ^{40}K, for example, 11% decays by electron capture to ^{40}Ar while 89% undergoes β^--particle emission to ^{40}Ca.

When the decay of a parent nucleus leads to the formation of an excited daughter nucleus, the various energy excitation levels produced are also subject to specific probability distributions. Again, the various proportions can be thought of in terms of branching ratios and the occurrence of each particular energy level can be expressed as a percentage.

Conventional representation of decays

In the conventional diagrammatic representation of a decay scheme (Adams and Gasparini, 1970), the parent nuclide is presented as a thick horizontal line at the top of the diagram and the daughter nuclide as a thick horizontal line at the bottom (Fig. 3.8). If the decay is by α-particle or β^+-particle emission or by electron capture, the position of the daughter nuclide is offset to the left, relative to the parent, and if by β^--particle emission the offset is to the right. In some presentations, however, α-particle decays are shown by vertical double lines before offset to the left (Lederer and Shirley, 1978). If the daughter nucleus is present in an excited state, the various excitation energy levels are identified by diagonal lines connected to light horizontal

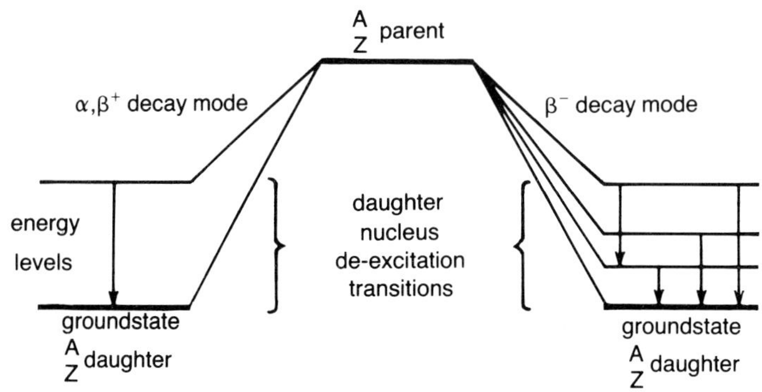

Fig. 3.8 — General radioactive decay scheme.

lines placed, approximately in a linear scale position, in increasing energy order above the position of the daughter nuclide. Vertical arrows, pointing downwards, connect the various energy levels for which γ-ray emissions are observed, and the energies (in mega electron volts) are given for these energy levels and the de-excitation transitions. Thus diagonal lines radiate out from the upper horizontal line representing the parent nuclide to connect with the daughter (to represent the ground state condition), and to the various excitation energy levels. β^--decays, therefore, slope down to the right, while α-, β^+- and electron capture decays slope down to the left. The proportion of the decay following each particular route is given as a percentage for each of the decay modes and for the excitation energy levels. Other data such as half-life values may be also given (Fig. 3.9).

3.5.3 Fission reactions

Nuclear fission is the breakdown of a nucleus into a number of smaller nuclei. Two nuclei of approximately equal mass are generally formed (bipartition), but division into three roughly equal parts (tripartition) is also possible. The

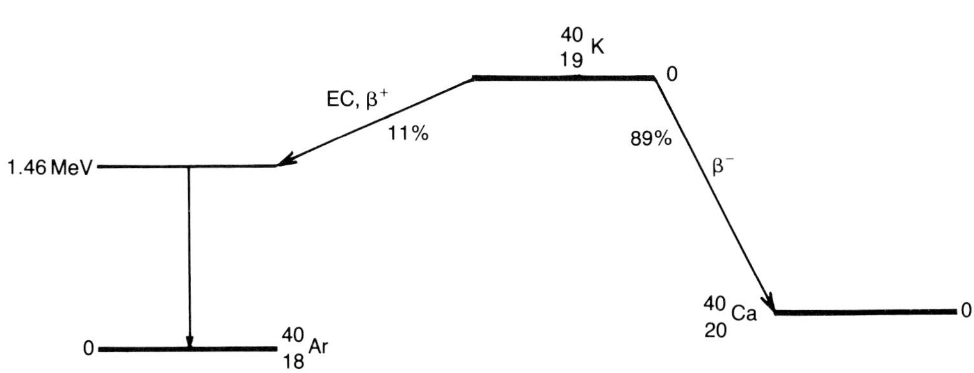

Fig. 3.9a — Simplified decay scheme for ^{40}K. Energies given in MeV. (After Adams and Gasparini, 1970; Lederer and Shirley, 1978).

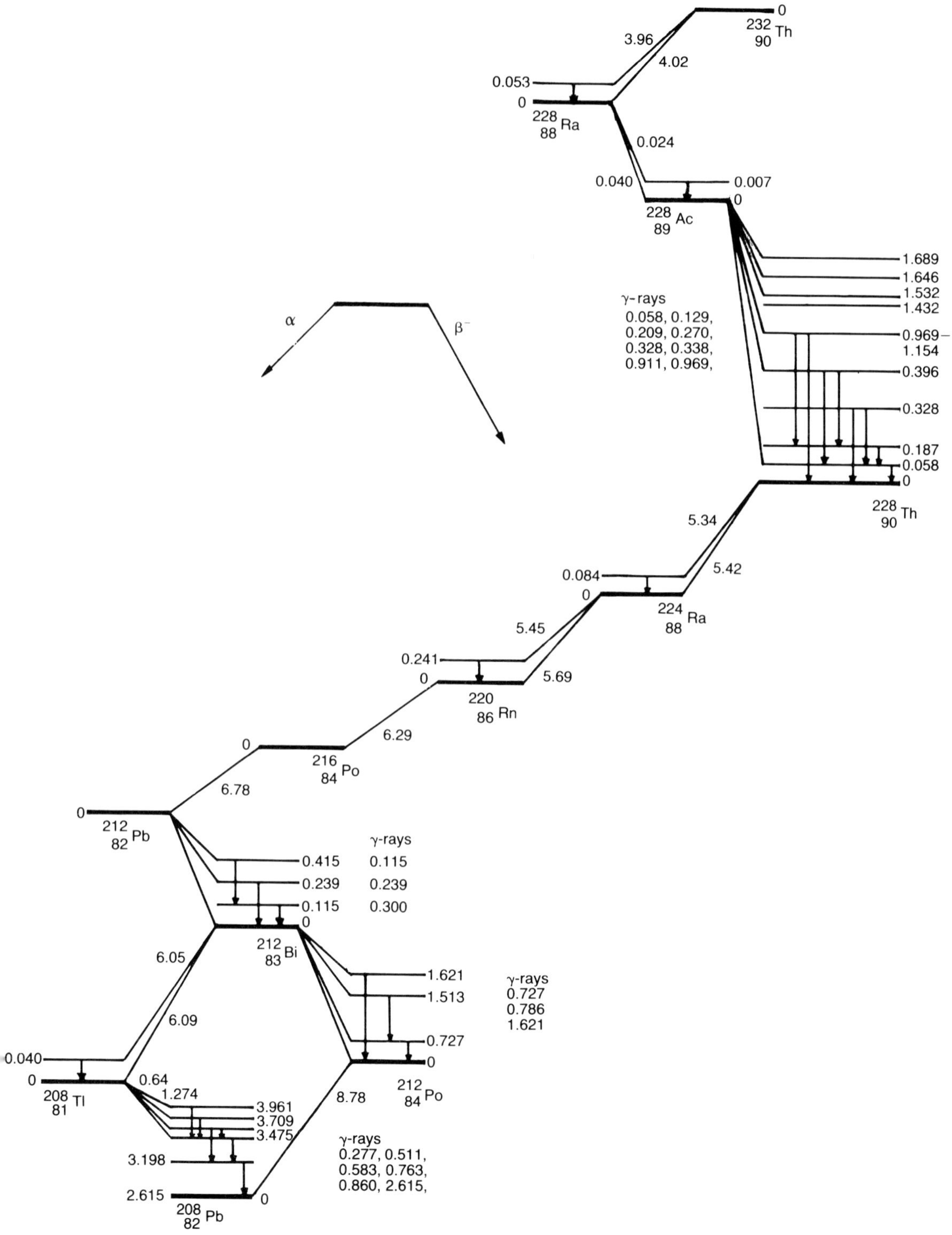

Fig. 3.9b — Simplified decay scheme for ^{232}Th. Energies given in MeV. (After Adams and Gasparini, 1970; Lederer and Shirley, 1978).

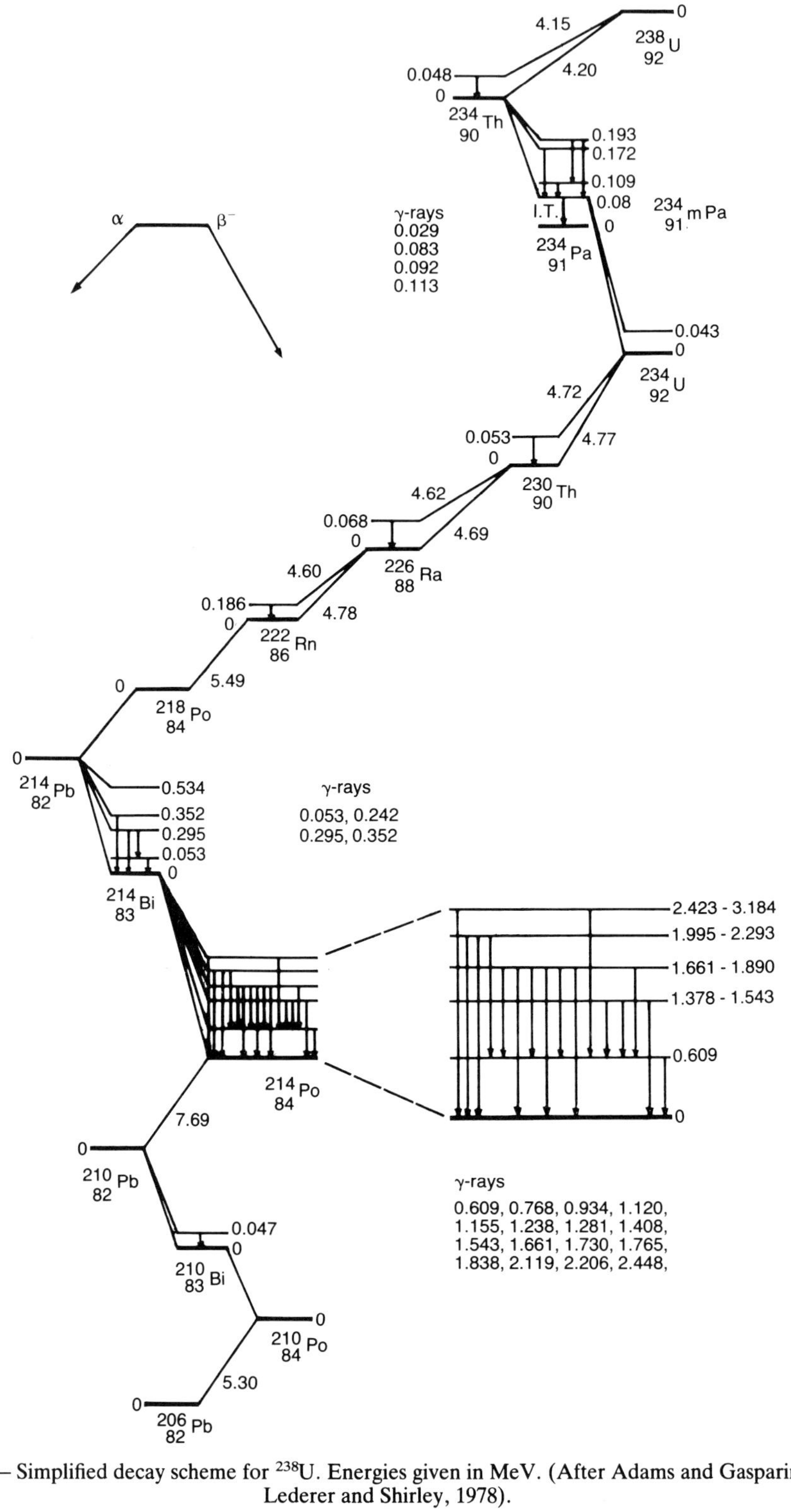

Fig. 3.9c — Simplified decay scheme for ^{238}U. Energies given in MeV. (After Adams and Gasparini, 1970; Lederer and Shirley, 1978).

effect was first observed in 1938 when Hahn and Strassmann found that Ba and La are formed in the transformation produced by neutron bombardment of U. If M_0 is the mass of the original nucleus and M_1 and M_2 the masses of the lighter nuclei resulting from fission, the energy released is given by

$$E=c^2\,[M_0-(M_1+M_2)],$$

and is found to be of the order of 200 MeV.

Consideration of the binding energy relationships above suggests that spontaneous fission should be a common phenomenon, but the only elements to show this effect are U and Th. Moreover, tripartition should be relatively common (and release more energy than bipartition) but, on the contrary, it is rare. Thus, energy arguments alone do not explain the phenomenon.

In the mechanism of fission, the analogy of a charged liquid drop separating into two parts shows that a threshold energy is required to initiate a necking stage after which electrostatic repulsion will become operative. As the binding energy of a nucleus with an even number of neutrons is greater than that with an odd number, it is not surprising that the threshold energy for neutron-induced fission is less in nuclei with odd numbers of neutrons. On this basis, Bohr in 1939 considered the thermal neutron fissionability of U to be due to ^{235}U. Although the presence of the fission energy barrier controls the probability of fission (fission cross-section), it is possible for fission to occur at lower energies. This is because, quantum mechanically, there is a small probability that the nucleus can tunnel through the fission barrier, a process which can even occur in nuclei that are not subject to external excitation, and this appears to be the origin of the spontaneous fission observed in ^{238}U. Most fission thresholds are of the order of 5–6 MeV. Nuclides with odd mass numbers have a much lower rate of spontaneous fission than those with even numbers of protons and neutrons. Spontaneous fission occurs with a half-life of about 10^{16} years in ^{238}U and ^{234}U; in ^{232}Th the half-life is about 10^{21} years and in ^{230}Th it is greater than 10^{17} years.

When the two (or more) fragments of the parent nucleus are formed, they are in a highly excited state because of their proton deficiency. This results in the emission of (prompt) neutrons, followed by γ-rays when the excitation energy has fallen below the neutron binding energy. The resulting fragments (the fission products) are charge deficient, and so further transformation takes place by β-decay or via excited states which also result in the emission of (delayed) neutrons and γ-rays. For every two fission events in the fission of ^{235}U, 5 prompt neutrons are produced.

Although the products of artificially induced fission (Fig. 3.10) are of roughly equal mass (symmetrical fission), in spontaneous fission there is usually a distinct mass difference between the fragments produced (asymmetrical fission). This is also generally true of artificial fission of the heavy elements, and so it appears to be a function of relatively low fission barrier energies. Symmetrical fission becomes established as the excitation energy increases or the mass of the parent nucleus decreases. Although the total amount of kinetic energy released is greater in asymmetrical fission than in symmetrical fission, the number of prompt neutrons emitted by symmetrical fragments is greater than that for asymmetrical fragments. Particularly low numbers of prompt neutrons are emitted from fragments with atomic and neutron numbers close to the magic numbers of 50 and 82 respectively. Between mass numbers 127 and 139, the heavy fragment most likely to be produced has a mass number of about 138–140, independent of the parent source, but the most likely lighter fragment mass number shifts from about 94 to 102 as the mass of the parent nucleus increases.

3.5.4 Fission chain reactions

If sufficient ^{235}U is brought together, a self-sustaining fission chain reaction will be started. Of the 5 neutrons produced by every two fission events, only the equivalent of 2.37 neutrons are

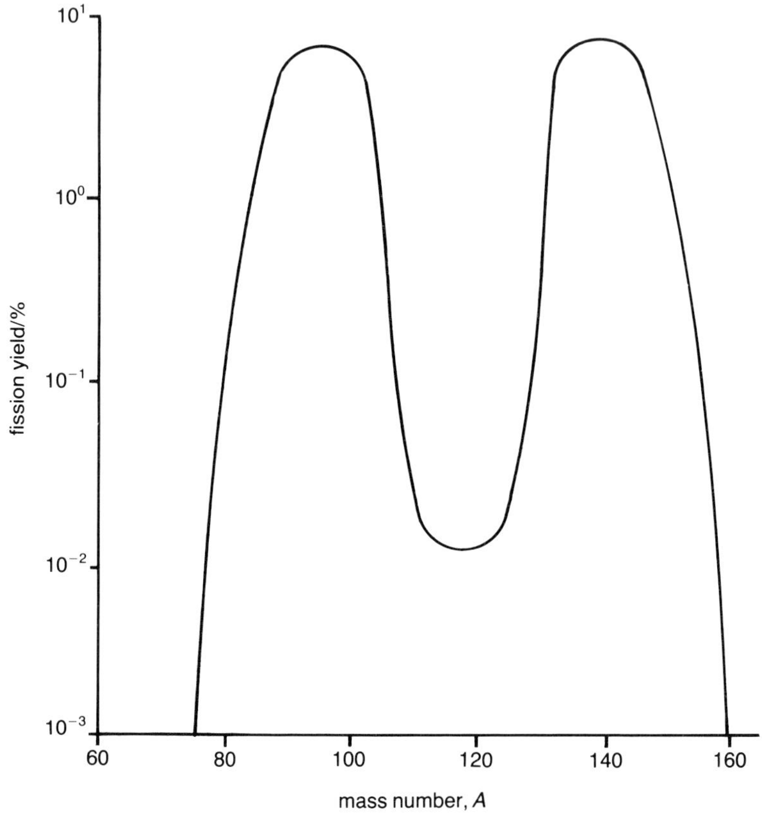

Fig. 3.10 — The symmetrical region fission yield of ^{235}U. (Kathren, 1984, redrawn with permission.)

required to cause more ^{235}U to produce 0.37 nuclei of ^{236}U and two more fissions. With two more fissions another 5 neutrons will be liberated, and the reaction continues. The difference between 2.37 and 5 neutrons can be lost from the system without affecting the result. The actual neutron loss that is likely to be experienced depends upon the mass of fissionable material available. With a smaller mass, a smaller number of fission reactions will be established in each stage of the reaction, and it eventually stops.

The principal naturally occurring isotope of U is ^{238}U. ^{235}U is present as only 1 part in 139. Unless a natural or artificial process concentrates ^{235}U, chain reactions cannot occur. Generally, the presence of ^{238}U in much larger quantities than ^{235}U causes too many neutrons to be lost by capture by ^{238}U. In natural U, for each thermal neutron caught by the nucleus, 1.33 new fast neutrons are produced, but these cause excitation of ^{238}U to produce ^{239}U which decays with a half-life of about 23 min to ^{239}Np which, in turn, decays with a half-life of 2.3 days to ^{239}Pu. In this chain of events, much less than 1 neutron per fission will reach thermal energy and induce more fission. The ^{238}U acts as a moderator of the neutron energies. However, dispersion of a sufficiently large mass of U in a larger mass of a perfect moderator such as He, that absorbs no neutrons at all, does make a chain reaction possible, as fast neutrons are brought to thermal energy levels by collision with the He nuclei and then diffuse until they are captured by U. In this process, it is possible for the 1.33 new neutrons to become available for initiating further fission reactions. The existence of a natural fission chain reaction has been proposed to account for the relative abundance of different nuclides in the Oklo U deposit of Gabon. This is discussed in more detail in Chapter 5.

3.6 SCATTERING, ABSORPTION AND PAIR PRODUCTION

Whatever the type of radiation, when penetration of matter occurs, energy is transferred from the radiation to the atoms making up the material penetrated (Fig. 3.11). Some of the types of energy transfer involving α-particles and β-particles have already been discussed, but other interactions also occur which affect the spectrum of γ-ray energies transmitted. In particular, γ-ray photons may interact with the free or weakly bound outer electrons of atoms and become scattered. The scattered photons suffer energy loss according to both the angle of scattering from their initial path direction and their initial energy. Backscattered photons are those which have been deflected back along their incident pathway. Scattered photons may, of course, be subject to further interaction with outer electrons and thus suffer further energy loss. Scattering can be expressed in terms of an increase in the wavelength of the γ-radiation:

$$\lambda' - \lambda = \frac{h(1 - \cos\theta)}{m_0 c},$$

where θ is the angle of scattering, m_0 the rest mass of the electron, c the velocity of light and h Planck's constant. λ' is the final wavelength and λ the initial wavelength of the radiation. The change in wavelength is thus independent of the initial wavelength of the γ-ray photons. The value of h/m_0c is 2.476×10^{-12} m, and this is known as the 'Compton wavelength'. However, the change in energy expressed in terms of the velocity is dependent upon the initial energy (or frequency) of the incident photon:

$$f' = \frac{f}{(1 + (hf/m_0c^2)\,(1 - \cos\theta)},$$

where f and f' are the frequencies of the incident and final photons. Thus the larger the initial energy hf, the larger is the energy loss hf'. The energy lost by the photon is transferred to the electrons and, if the energy transfer is sufficiently great, ejection of so-called 'Compton electrons' occurs (Fig. 3.12).

Energy loss by Compton scattering will not occur when the photon interactions are with tightly bound electrons as, effectively, the whole mass of the atom acts as the recoil mass. As this is so much larger than m_0, the change in energy is effectively zero, and 'coherent' or 'Rayleigh' scattering occurs. Scattering of X-ray photons in this way is the basis for investigating the structure of crystalline matter by X-ray diffraction.

The absorption of photons follows the exponential relationship

0 $$I_x = I_0\, e^{-\mu x},$$

where I_0 is the incident number of photons and I_x the number remaining after a distance x has been traversed through material with a linear absorption coefficient μ.

As both I_x and I_0 are dimensionless numbers, the customary units of x are the reciprocal of those of μ. x may be given in grams per square centimetre simply by multiplying the linear dimensions of the absorber by its density. This yields the mass absorption coefficient which has the form μ/ρ, where ρ is the density (Fig. 3.13). More fundamentally, an absorption coefficient can be expressed in terms of the number of electrons per square centimetre (electron cross-section μ_e) or atoms per square centimetre (atomic cross-section μ_a):

$$\mu_a = Z\mu_e$$

or

$$\frac{\mu}{\rho} = N\frac{Z}{A}\mu_e,$$

where N is Avogadro's number, and A and Z the mass and atomic numbers of the material. As the cross-sections are small, they are usually expressed in terms of the unit called the barn (10^{-24} cm^2). These relationships apply for electromagnetic radiation and for β-particles,

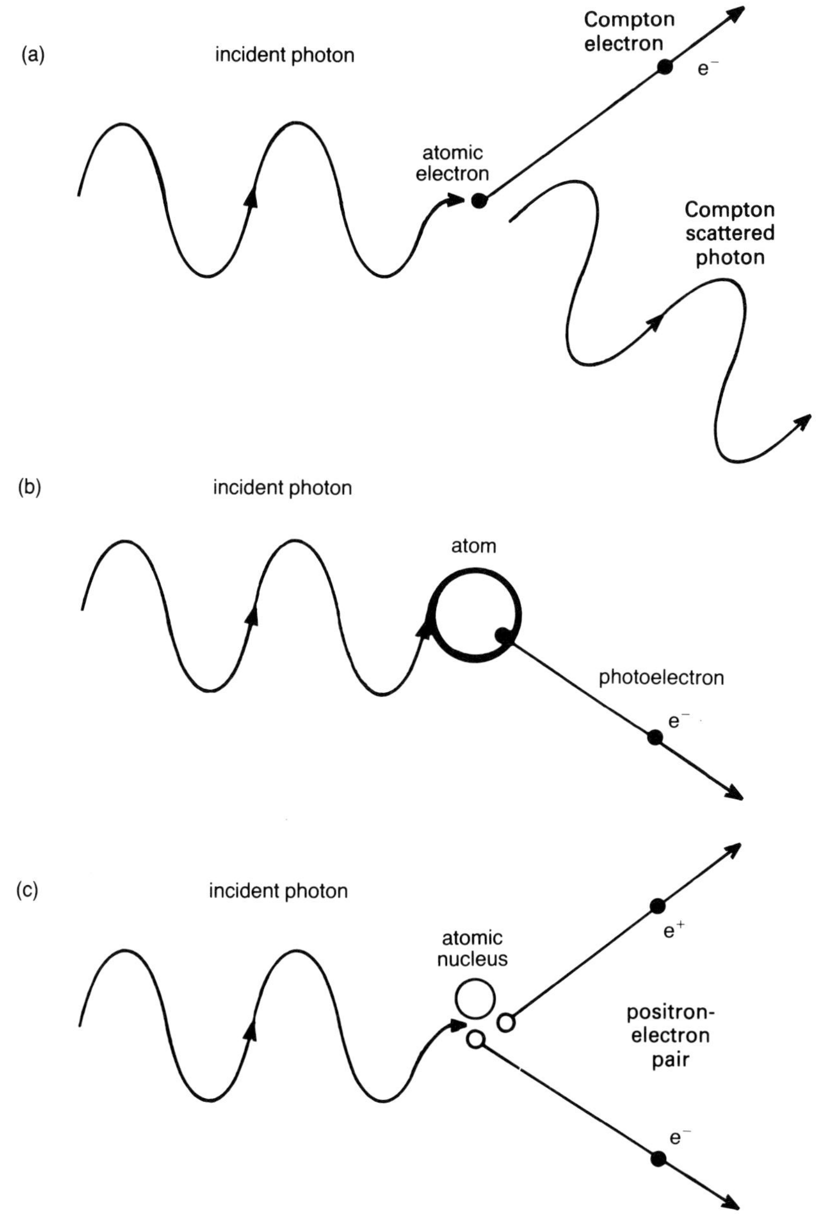

Fig. 3.11 — Diagrammatic representation of (a) Compton scattering, (b) photoeletric interaction and (c) pair production.

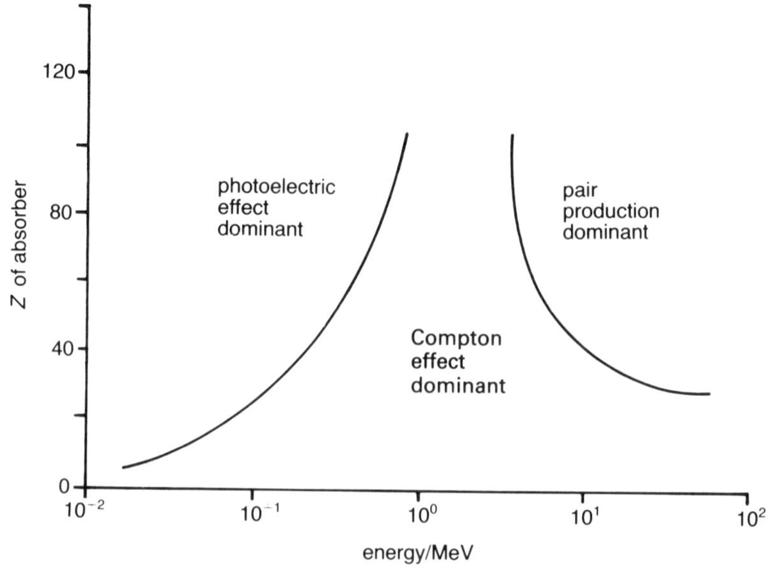

Fig. 3.12 — Relative importance of the three main types of γ-ray interaction. (After Evans, 1955.)

but their numerical values will differ according to the type of radiation concerned. It should be noted, that α-particles do not follow this exponential relationship (Fig. 3.14).

Interaction of photons with the nucleus of an atom may also occur. Coherent scattering is observed for low photon energies but, when the incident photons have a very high energy, scattering from individual nucleons in the nucleus becomes possible. However, it is also possible for the initial γ-ray photon to disappear. Disappearance may occur if the photon energy transfer causes excitation of the absorbing nucleus, a process known as 'resonant absorption', or if it causes the ejection of a nucleon or electron from the nucleus, in which case the effect is called 'direct interaction'.

Resonant absorption will produce different effects depending upon whether the nucleus is within a free atom or bound together in the molecular structure of a solid. For free atoms, the nucleus absorbing a photon will rise to a higher energy state, the energy of which is determined by the recoil energy of the nucleus. Similarly, when decay to a lower energy state occurs, the energies of the emitted photons will depend upon the recoil energy that has been taken by the nucleus. There can thus be no overlapping of absorption and emission energies. In solids, if the bound nuclei have recoil energies greater than the binding energy of the atom in the atomic lattice of the material, a similar process will occur but, when the recoil energy is less than the energy required to displace the nucleus and yet still great enough to cause excitation, there is a finite probability that the atom will go into an excited lattice vibration state. If the recoil energy is less than the lowest vibrational state of the nucleus, the lattice as a whole takes up the recoil. However, the mass of the lattice is so great that little happens, and the term 'recoil free' is used. Even when the recoil energy is great enough to cause excitation, there is a finite probability that the atom remains in the ground state. This probability depends upon the properties of the solid, the temperature and the energy of the incident photons. This is known as the 'recoil-free fraction'. The exis-

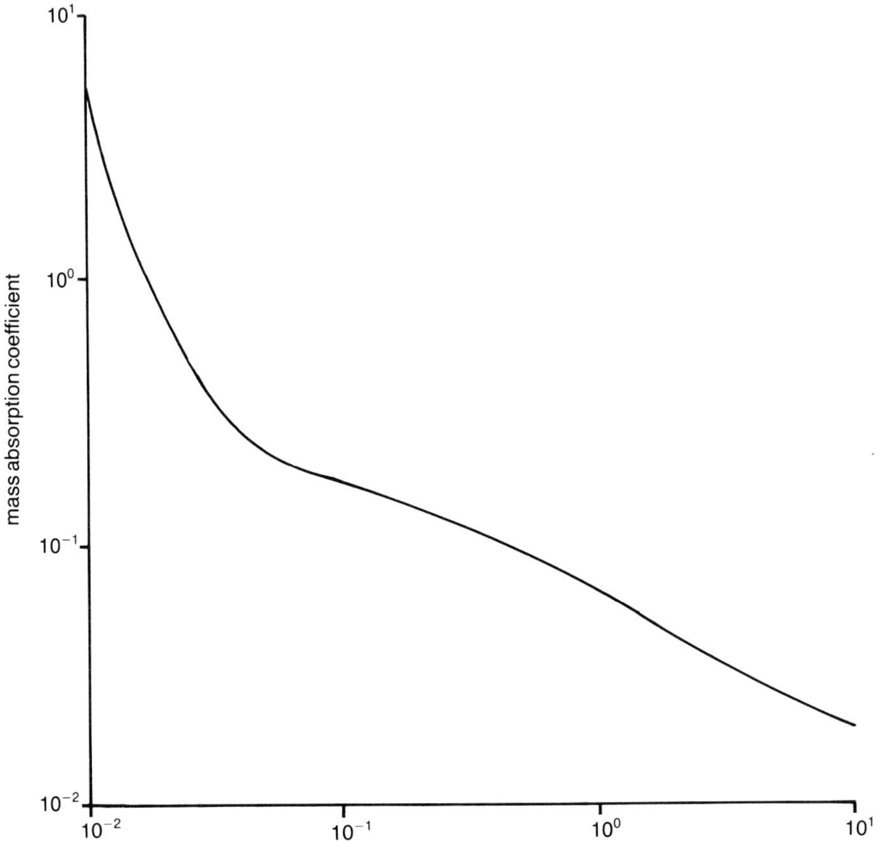

Fig. 3.13 — Total mass absorption coefficient of air. (Adams and Gasparini, 1970, redrawn with permission.)

tence of recoil-free emitters and absorbers of γ-rays have allowed these phenomena (the Mössbauer effect) to be used for the investigation of atomic coordination and oxidation states. This is discussed in more detail in Chapter 9.

Of the various types of nuclear emission, it is only neutron production that can be caused by γ-ray photons. This is because the electrostatic energy barrier is too high to permit emission of charged particles from the nucleus but allows emission of neutrons. However, the threshold values of γ-ray energy for neutron emission are greater than about 5 MeV, which is about twice the level of the most energetic natural γ-rays. The effect cannot, therefore, be caused by natural γ-radiation. Artificial sources of high energy γ-radiation are used to produce neutron emission to detect the presence of elements such as H and Be in which the nuclei have loosely bound neutrons.

The 'disappearance' of γ-rays also takes place in the phenomenon known as pair production, in which photon energy is transferred in the creation of electrons and positrons. This process can best be understood if we follow the reasoning of Dirac in 1923 and recognise that there are two solutions to the relativistic wave equation describing the motion of electrons in

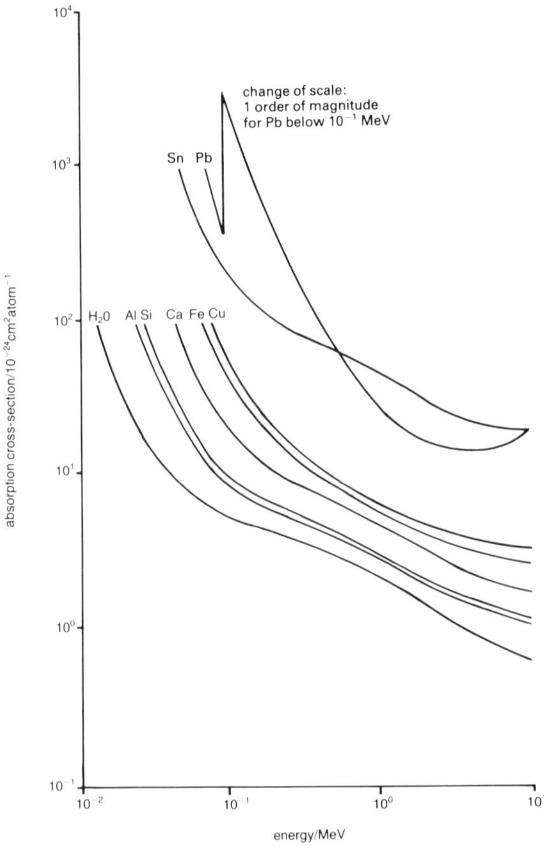

Fig. 3.14 — Typical absorption cross-sections for various materials. (After Adams and Gasparini, 1970.)

electromagnetic fields, corresponding to the positive and negative roots. Thus there are two energy states

$$E \geqslant m_0c^2$$

and

$$E \leqslant -m_0c^2,$$

where m_0 is the rest mass of the electron and c the velocity of light. No electron can exist between these limits, but electrons may cross the divide by means of a quantum jump in energy level if their energy is equal to $2m_0c^2$.

However, another problem then becomes apparent: why do electrons exist which have measurable amounts of energy and so are in the positive energy state? Dirac believed that this could be explained by arguing that 'nearly all' (later modified by Oppenheimer in 1931 to 'all') the negative energy states are occupied. Therefore the excess electrons can only exist in the higher energy state observed. The negative state electrons are unobservable but, when an energy greater than $2m_0c^2$ (1.022 MeV) is transferred to a negative state electron, an ordinary electron is produced. This electron has come from the negative state continuum and so a hole is left in that continuum which appears as a particle with opposite properties of the electron: an antielectron or positron. In order to conserve both energy and momentum, when natural annihilation of an electron and a positron occurs on collision, 2 photons must be produced. An encounter between two γ-rays, each having an energy of at least 0.511 MeV, can therefore lead to the simultaneous creation of an electron and a positron. Moreover, γ-rays with energies greater than 1.022 MeV can raise a negative state electron to the positive state, which would also lead to the creation of electron–positron pairs. This process is observed to be more frequent for γ-ray energies of about 1.022 MeV. However, because both energy and momentum must be conserved in the process, it must take place within the electrostatic field of the nucleus. Once formed, the positron loses energy by interaction with electrons until it is annihilated by combination with an electron, the 2 photons of energy 0.511 MeV being emitted in nearly opposite directions.

A process similar to pair production, but occurring outside the nucleus, is known as 'triplet production'. Here, the interaction of a γ-ray photon with an extranuclear electron causes the production of 2 electrons and a positron. The threshold energy value for this process is twice the level required for pair production.

The absorption coefficient μ for a beam of photons of specific energy passing through material is thus a compound of various processes acting within the atoms of the material. These include photoelectric absorption, Compton scattering and pair production (indicated by the subscripts pa, Cs and pp, respectively). Thus,

$$\mu = \mu_{pa} + \mu_{Cs} + \mu_{pp}.$$

3.7 COSMIC RAYS

Charged particles entering the Earth's atmosphere consist mainly of protons (about 87%) and α-particles (about 12%), although nuclei heavier than He, such as Li, Be, B, C, O and F, are also present (about 1%). Their energies range from a few mega electron volts to 10^{14} MeV. The low energy particles ionise the atmosphere and are stopped, but all other particles undergo collision with the nuclei of the atmospheric gases and produce new particles by a fragmentation process called 'spallation'. Another process in the interaction of these particles with air is the production of neutral and charged pions (π mesons) and, to a lesser extent, nucleons. If these particles have sufficient energy, they produce further particles by secondary collision which, in turn, may undergo further collision to produce more particles in a cascade process (Fig. 3.15). This nuclear cascade propagates down to sea level but can be intercepted by various processes *en route*. Unstable particles will decay to produce particles such as muons, electrons, neutrinos and photons, which do not interact with the nuclei of air; charged particles may lead to the creation of electron–positron pairs. Neutral pions decay very rapidly (lifetime, 2×10^{-16} s) into 2 photons by creation of electron–positron pairs and initiate an electromagnetic cascade which forms the 'soft' component of cosmic radiation at sea level. Charged particles have a rather longer lifetime (2.55×10^{-8} s), but even for these particles the likelihood of being involved in a nuclear interaction is small. Only if the lifetime of the particle is extended by travelling at speeds approaching that of light so that relativistic time dilation is appreciable, will the chance of nuclear interaction be significant. On decay, charged particles produce muons. These muons contribute the 'hard' component of cosmic radiation at sea level but decay with a lifetime of 2.2×10^{-6} s into an electron (or positron) and 2 neutrinos. Because of relativistic time dilation, they only have a low probability of decaying before reaching the Earth's surface, especially if their path is nearly vertical. Decay probabilities increase at shallower angles of incidence because of the longer path length travelled (Fig. 3.16). At sea level, about 80% of the charged particles consist of nucleons and muons and about 20% of electrons.

During nuclear spallation, particles are produced which may be either fragments of the collision nucleus or newly created by the collision. Below about 256 MeV, which is the energy required for pion production, the effect of nucleon–nucleus collision is to cause particles to be ejected from the target nucleus in what is known as the 'evaporation process' as, statistically, a small proportion of the particles of the target nucleus will acquire most of the energy of the collision. Particles produced in this way are nucleons and the nuclei of isotopes of He and H, in particular, ^{2}H and ^{3}H. This process is not very sensitive to ^{1}the energies involved (except that, above 256 MeV, pions are created), as even at high energies these stable nuclear configurations are expelled from the target nucleus before any fragmentation occurs.

The neutrons that are involved in these processes lose energy by repeated collision and finally become captured by N to produce ^{14}C (half-life, 5730 years) which is used in the dating of relatively recent organic matter (less than about 50 000 years old). This method is based on the assumptions that cosmic ray activity has been constant over that time interval and that ^{14}C is only taken into organisms while they are

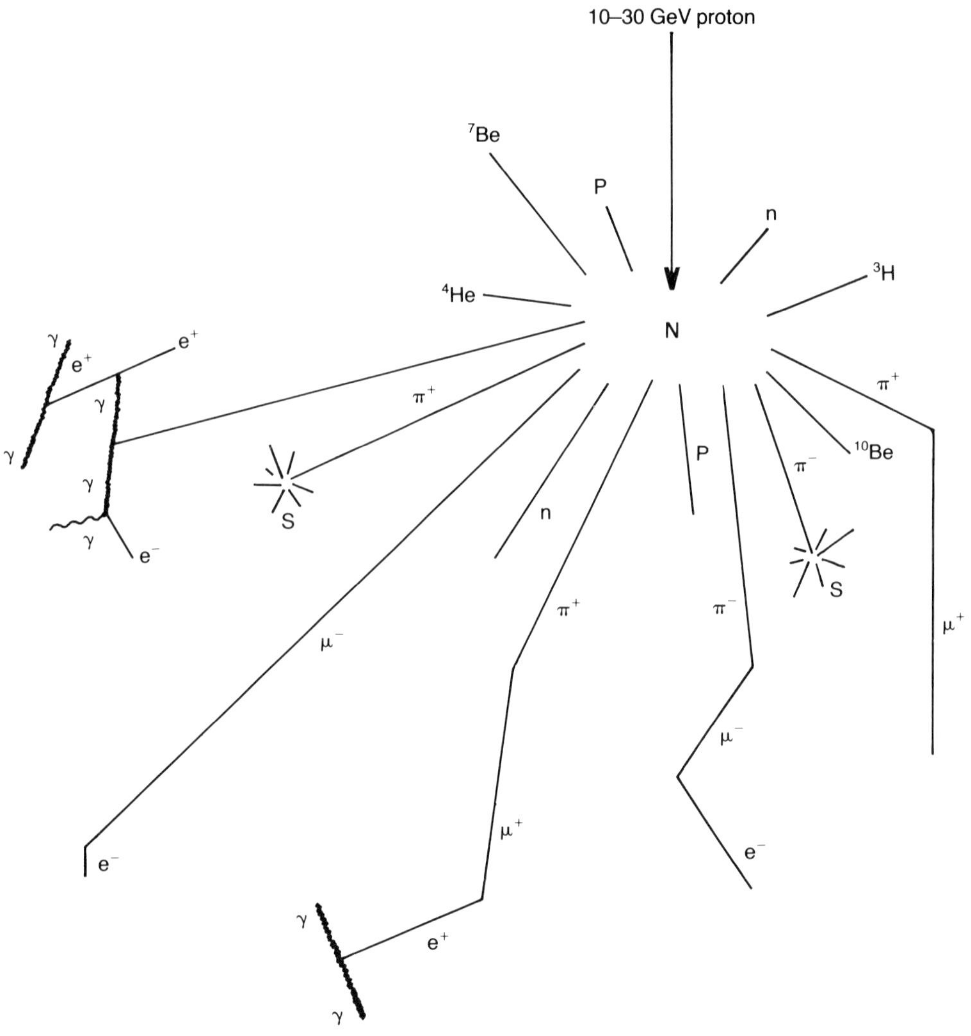

Fig. 3.15 — Diagrammatic representation of the results of the interaction of a cosmic ray proton with a N atom in the atmosphere: p, proton; n, neutron; e^-, electron; e^+, positron; γ, photon; π, pion; μ, muon; s, spallation. (Redrawn with permission from Whicker and Schultz (1982). Copyright CRC Press, Inc., Boca Raton, FL.)

alive. This is discussed in more detail in Chapter 10.

Also produced by interaction with atmospheric N are ^{3}H, ^{3}He, ^{7}Be and ^{10}Be, although all these nuclides can also be formed by interaction with O (as, indeed, can ^{14}C). Interaction with Ar produces ^{26}Al, ^{36}Cl, ^{38}Cl, ^{39}Cl, ^{32}Si, ^{37}Ar, ^{39}Ar, ^{22}Na, ^{24}Na, ^{35}S, ^{38}S, ^{33}P and ^{32}P, while ^{81}Kr is produced by interaction with Kr. With the exception of ^{3}He, all these nuclides are unstable (Table 3.6).

3.8 UNITS

Unfortunately, a wide range of units has been used for measuring activity, absorbed dose and dose equivalent in the past, all of which still appear in some of the literature. This is

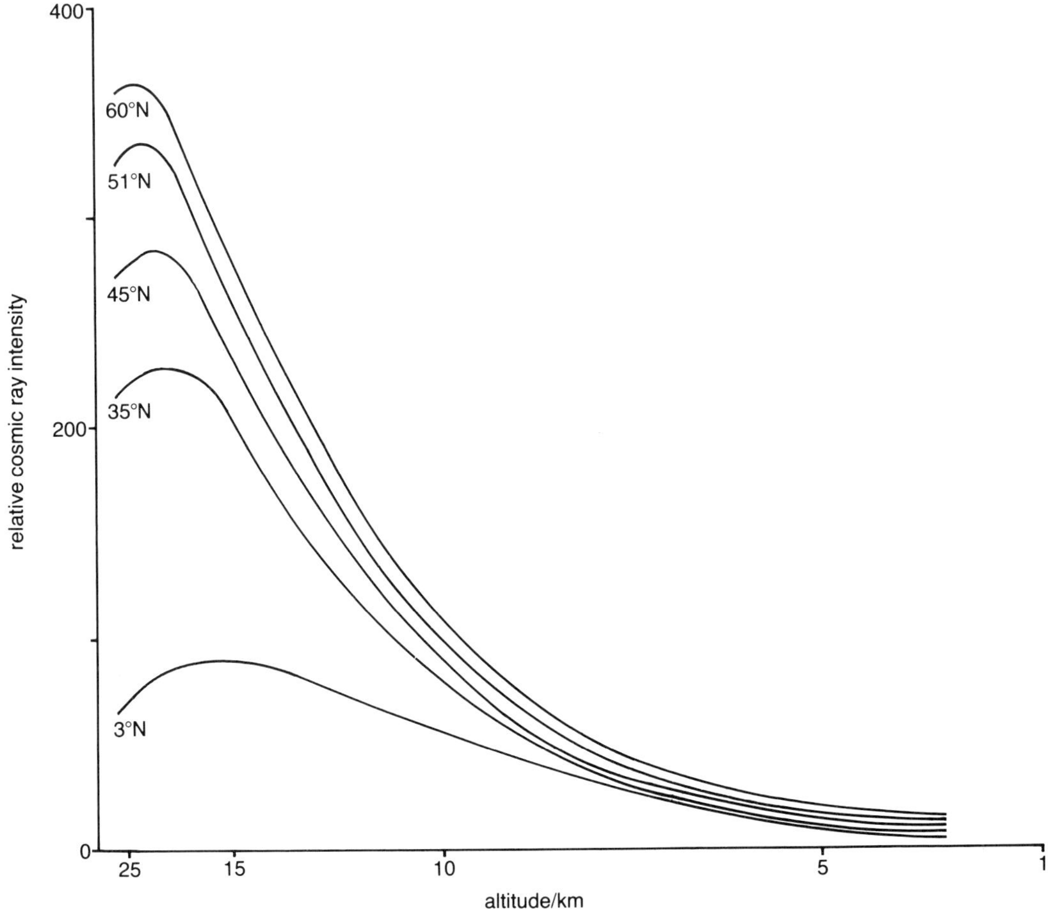

Fig. 3.16 — The effects of altitude and latitude on cosmic ray intensity. (After Kathren, 1984.)

especially true of the measurement of relative values when the absolute level is not considered important.

For activity (the expression for the presence of a radionuclide based on measurement of the rate at which it decays), the following units have been used.

Count	Usually expressed as a rate (for example, counts per minute, counts min^{-1} or cpm), which is the response of a particular radiation detection instrument to the radiation flux. The flux may be α-particles, β-particles or γ-rays, or any two of these in combination. Counts due to cosmic rays and X-rays are also usually included.
Curie (Ci)	1Ci is the activity of 1 g of Ra, which is 3.7×10^{10} atomic disintegrations per second.
Röntgen (R)	1R is the amount of radiation required to produce 1 electrostatic unit (esu) of charge in 1 cm^3 or 0.001 293 g of air at standard temperature and pressure. Alternatively, it is the radiation that will produce 2.083×10^9 ion pairs in air

Table 3.6 — Main radionuclides produced by cosmic rays

Nuclide	Half-life	Production rate (atoms cm^{-2} s^{-1})	Air activity (Bq m^{-3})
^{3}H	12.33 years	0.25	0.167
^{7}Be	53.3 days	0.08	0.017
^{10}Be	1.6×10^{6} years	0.05	10^{-7}
^{14}C	5730 years	2.5	0.067
^{22}Na	2.6 years	8.6×10^{-5}	1.7×10^{-6}
^{25}Al	7.18 s	1.4×10^{-5}	—
^{32}Si	650 years	1.6×10^{-4}	3.3×10^{-8}
^{32}P	14.28 days	8.1×10^{-4}	3.3×10^{-4}
^{33}P	25.3 days	5.8×10^{-4}	2.5×10^{-4}
^{35}S	87.4 days	1.4×10^{-3}	2.5×10^{-4}
^{36}Cl	3.1×10^{5} years	1.1×10^{-3}	5×10^{-10}
^{39}Cl	56 min	1.6×10^{-3}	—
^{39}Ar	269 years	5.6×10^{-3}	—
^{81}Kr	2.1×10^{5} years	10^{-6}	—

(After Lederer and Shirley, 1978; Kathren, 1984.)

(from the charge on 1 electron which is 4.8×10^{-10} esu). This is a measure of the ionising effect of γ- and X-rays and is usually expressed as a rate (for example, microröntgens per hour or μR h^{-1}).

In the Système International (SI), which is the modern system of units based upon metre, kilogram and second for length, mass and time, respectively, the unit for the activity of a radionuclide is the becquerel (Bq), which is equal to one radioactive transition per second. The relationship between the becquerel and the curie is simply 1Ci=3.7×10^{10} Bq.

The effect of radiation on the human body is defined in terms of the 'absorbed dose', which is defined by the energy that is transferred to unit mass of matter such as tissue. The unit used here is the gray (Gy); 1 Gy corresponds to an energy transfer of 1 J kg^{-1}. Absorbed dose was formerly expressed in terms of a unit called the rad. The relationship between the old and new units is simply 1 rad=0.01 Gy.

Different types of radiation, however, will have different biological effects, even though they may transfer the same amount of energy. 1 Gy of α-radiation, for example, is likely to produce a much more harmful effect than 1 Gy of β-radiation. This is because an α-particle is rapidly stopped by tissue and so transfers its energy over a short range. The 'dose equivalent' is an attempt to produce a unit which places all ionising radiations on a scale of the damage they are likely to cause to tissue. The unit used is the sievert (Sv), which is simply obtained by taking the absorbed dose and multiplying it by a factor which is determined by the relative seriousness of the tissue damage likely to be caused by the radiation type involved. For γ-rays, X-rays and β-particles this factor has a value of unity, and so the gray and the sievert are numerically equal. For α-particles, however, the factor is 20, in other words α-particles are assessed as twenty times more likely to cause tissue damage than are γ-rays, X-rays or β-particles. Previously, the unit of dose equivalent was the rem. The relationship between the rem and the sievert is the same as that between the rad and the gray, as no other dimensional factor is involved, that is 1 rem=0.01 Sv. Finally, it must be noted that the risk of serious hereditary defect or fatal cancer arising from exposure to radiation varies with the tissue or organ affected. Table 3.7 gives the fractional contribution of a number of susceptible parts of the human body. Multiplying the dose equivalent to an organ by the appropriate fraction and summing the results yields the effective dose equivalent (also measured in sieverts).

Table 3.7 — Risk weighting factors

Tissue or organ	Fraction
Testes and ovaries	0.25
Breast	0.15
Red bone marrow	0.12
Lung	0.12
Thyroid	0.03
Bone surfaces	0.03
Remainder of body	0.30

(National Radiological Protection Board, 1981, reproduced with permission.)

BIBLIOGRAPHY

Adams, J. A. S. and Gasparini, P. (1970) *Gamma ray spectrometry of rocks*. Amsterdam, Elsevier.

Ahrens, L. H. (1957) Studies on the relative abundance of isotopes. *Geochim. Cosmochim. Acta* **11,** 1.

Ahrens, L. H. (1957) Some features on the relation between mass defect and mass number. *Trans. R. Soc. S. Afr.* **35,** 109.

Allison, S. K. and Warshaw, S. D. (1953) Passage of heavy particles through matter. *Rev. Mod. Phys.* **25,** 779.

Auger, P. (1926) Sur l'effet photelectric compose. *J. Phys. (Paris)* **6,** 205.

Beard, D. B. (1957) The atomic nucleus. *Am. Sci.* **45,** 333.

Bethe, H. A. and Ashkin, J. (1953) Passage of radiations through matter. In Sergré, E. (ed.), *Experimental nuclear physics*. New York, Wiley.

Blackett, P. M. and Occhialini, G. P. S. (1933) Some photographs of the tracks of penetrating radiation. *Proc. R. Soc. Lond. A.* **139,** 699.

Brink, D. M. (1965) *Nuclear forces*. Oxford, Pergamon.

Cohen, B. L., and Halliday, D. (1967) Nuclear reactions. In Condon, E. U. and Odishaw, H. (eds), *Handbook of physics*. New York, McGraw-Hill.

Cork, J. M. (1957) *Radioactivity and nuclear physics*. New York, Van Nostrand.

Des Marais, D. J. (1983) Light element geochemistry and spallogenesis in lunar rocks. *Geochim. Cosmochim. Acta* **47,** 1769.

Eisenbud, L., Garvey, G. T., and Wigner, E. P. (1967) General principles of nuclear structure. In Condon, E. U. and Odshaw, H. (eds), *Handbook of physics*. New York, McGraw-Hill.

Evans, R. D. (1955) *The atomic nucleus*. New York, McGraw-Hill.

Fermi, E., Almaldi, E., D'Agostino, D., Pontecorvo, B., Rasetti, F., and Segré, E. (1935) Artificial radioactivity produced by neutron bombardment. *Proc. R. Soc. Lond. A.* **149,** 522.

Geiger, H., and Nuttall, J. M. (1911) The ranges of α particles from various radioactive substances and a relation between range and period of transformations. *Phil Mag.* **22,** 613.

Geiger, H., and Nuttall, J. M. (1912) The ranges of α particles from uranium. *Phil. Mag.* **23,** 439.

Goldman, D. T., and Muehlhause, C. O. (1967) Neutron physics. In Condon, E. U. and Odishaw, H. (eds), *Handbook of physics*. New York, McGraw-Hill.

Hayward, R. W. (1967) Nuclear electromagnetic radiation. In Condon, E. U. and Odishaw, H. (eds), *Handbook of physics*. New York, McGraw-Hill.

Hill, R. D. (1967) Physics of strongly interacting particles. In Condon, E. U. and Odishaw, H. (eds), *Handbook of physics*. New York, McGraw-Hill.

Hooper, V. D. (1964) *Cosmic radiation and high energy interactions*. London, Logos.

Hubbell, J. H., Veigele, W. J., Briggs, E. A., Brown, R. T., Cromer, D. T., and Howerton, R. J. (1975) Atomic form factors, incoherent scattering functions and photon scattering cross sections. *J. Phys. Chem. Ref. Data* **4,** 471.

Hyde, E. A. (1971) *The nuclear properties of the heavy elements*. New York, Dover Publications.

Hyde, E. K., Perlman, I., and Seaborg, G. T. (1971) *Detailed radioactive properties*. New York, Dover Publications.

Ivanovich, M. (1982) The phenomenon of radioactivity. In Ivanovich M. and Harmon, R. S. (eds), *Uranium series disequilibrium: applications to environmental problems*. Oxford, Oxford University Press.

Johnson, W. H. and Nier, A. O., (1967) Nuclear masses. In Condon, E. U. and Odishaw, H. (eds), *Handbook of physics*. New York, McGraw-Hill.

Kathren, R. L. (1984) *Radioactivity in the environment*. Chur, Harwood.

Lederer, C. M., and Shirley, V. S. (1978) *Table of isotopes*. New York, Wiley.

Mann, W. B., Ayres, R. L., and Garfinkel, S. B. (1980) *Radioactivity and its measurement*. Oxford, Pergamon.

McCantly, I. E. (1970) *Nuclear reactions*. Oxford, Pergamon.

McKay, H. A. C. (1971) *Principles of radiochemistry.* London, Butterworths.

Michel, R., Brinkmann, G., and Stuck, R. (1982) Solar cosmic-ray produced radionuclides in meteorities. *Earth Planet. Sci. Lett.* **59,** 33.

Moon, P. B., and Tillman, J. R. (1936) Neutrons of thermal energies. *Proc. R. Soc. Lond. A.* **153,** 476.

Moszkowski, S. A. (1965) Theory of multiple radiation. In Siegbahn, K. (ed.), *Alpha, beta and gamma ray spectroscopy.* Amsterdam, North-Holland.

Mott, N. F. (1930) The collision between two electrons. *Proc. R. Soc. Lond. A.* **126,** 259.

National Radiological Protection Board (1981) *Living with radiation.* London, HMSO.

Oliphant, M. L. E., Kempton, A. R., and Rutherford, E. (1935) Some nuclear transformations of beryllium and boron and the masses of the light elements. *Proc. R. Soc. Lond. A.* **150,** 241.

Pal, Y. (1967) Cosmic rays and their interactions. In Condon, E. U. and Odishaw, H. (eds), *Handbook of physics.* New York, McGraw-Hill.

Rankama, K. (1963) *Progress in isotope geology.* New York, Interscience.

Reyss, J. L., Yokoyama, Y., and Guichard, F. (1981) Production cross sections of ^{26}Al, ^{22}Na, ^{7}Be from argon and of ^{10}Be, ^{7}Be from nitrogen: implications for production rates of ^{26}Al and ^{10}Be in the atmosphere. *Earth Planet. Sci. Lett.* **53,** 203.

Rose, H. J., and Jones, G. A. (1984) A new kind of natural radioactivity. *Nature* **307,** 245.

Rose, M. E. and Nilsson J. (1967) Weak interaction processes. In Condon, E. U. and Odishaw, H. (eds), *Handbook of physics.* New York, McGraw-Hill.

Rutherford, E. (1919) Collision of α particles with light atoms, IV, An anomalous effect in nitrogen. *Phil. Mag.* **37,** 581.

Segré, E. (1959) *Experimental nuclear physics,* Vol. 3. New York, Wiley.

Seigbahn, K. (1955) *β- and γ-ray spectroscopy.* Amsterdam, North–Holland.

Smith, C. J. and Stokes, A. (1972) *The principles of atomic and nuclear physics.* London, Arnold.

Stephens, W. E., and Hurlimann, T. (1967) Alpha particles and alpha radioactivity. In Condon, E. U. and Odishaw, H. (eds), *Handbook of physics.* New York, McGraw-Hill.

Thiel, K., and Herr, W. (1976) The ^{238}U spontaneous fission decay constant re-determined by fission tracks. *Earth Planet. Sci. Lett.* **30,** 50.

Turnbull, R. M. (1979) *The structure of matter: an introduction to atomic, nuclear and particle physics.* Glasgow, Blackie.

Wheeler, J. A., and Schroder, I. G. (1967) Fission. In Condon, E. U. and Odishaw, H. (eds), *Handbook of physics.* New York, McGraw-Hill.

Whicker, F. W., and Schultz, V. (1982) *Radioecology: nuclear energy and the environment,* Vol. 1. Boca Rouge, Florida, CRC.

4

The Mineralogical Expression of Radioactivity

4.1 METAMICTISATION

The crystalline structure typical of most minerals is not always present in radioactive minerals and, although they may show good external crystal form, they appear amorphous when examined by X-ray diffraction (and by optical methods of investigation). This internal disorder is known as the 'metamict' state. While not all radioactive minerals are metamict, all metamict minerals show some degree of radioactivity, and there appears to be a causal connection between the metamict state and the presence of radionuclides.

The conventional explanation for the origin of the metamict state is that the internal order of an originally crystalline form has been destroyed by α-particle bombardment from radionuclides within the structure. The degree of metamictisation depends upon the number of α-particle sources in the mineral, the strength of bonding, the openness of the structure and the chemical stability of the mineral. α-particles may cause damage in crystal lattices in three ways:

(1) High velocity α-particles may cause ionisation of atoms along their path.
(2) Low velocity α-particles may collide with nuclei of atoms in the mineral.
(3) Heat can be liberated to produce very high temperatures (10^4 °C) for very short periods of time (10^{-11} s) along the path of the α-particle, leading to thermal destruction of the structure.

The last two are considered to be the most important mechanisms for producing radiation damage (Ewing, 1975).

Collision of α-particles with atoms in a mineral causes dislocations and 'Frenkel defects' (the displacement of an atom from its lattice position to a position within the interstices of the lattice). High temperatures result in the melting and resolidification of the mineral as glass, the formation of finely crystalline oxides, the reorganisation of the lattice, or even the self-annealing of dislocations and Frenkel defects. The radiation damage produced in a crystal is a combination of these effects, the relative importance of each being determined by the α-particle energies involved and the strength of the bonds in the mineral. It is also possible that in some minerals the degree of metamictisation is controlled by the ability, effectively, to self-anneal. However, it appears that the nature of the lattice structure, particularly bond strength, exercises the most important control on the amount of damage that is initially produced, irrespective of the intensity of α-activity. $ThSiO_4$, for example, occurs in both a tetragonal form (thorite) and a monoclinic form (huttonite) and, while thorite is almost always found in the metamict state, huttonite is usually normally crystalline. It has been argued that a minimum dosage of 10^{17} α-particles cm^{-2} is needed to allow metamictisation to begin.

In contrast with ideas which relate the metamict state to radiation damage, it has been suggested that, at least in Nb and Ta oxides, metamicitisation is not due to radioactivity but is a product of their complex chemical compositions (Graham and Thornber, 1974). Indeed, investigation of a suite of columbites and ixiolites has shown considerable variation in crystallinity without any observable radioactivity. Moreover, the metamict state has been artificially induced in minerals by means of a shock wave. However, the definition of the metamict state as 'lattice disorder and imperfection in extremely well-crystallised natural minerals which may be annealed out by moderate heating', which was used in these studies, would largely exclude many minerals usually taken to be metamict (on the basis of the lack of internal ordering but with good external structure), as the recrystallisation of many of these minerals by heating does not always restore them to their original structure. Indeed, exact structural restoration appears to be the exception in naturally occurring metamict minerals. This is because metamictisation is often accompanied by a change in chemical composition, particularly an increase in the non-essential water content and the oxidation of Fe^{2+} and U^{4+}.

If these secondary effects are excluded and the structure itself is considered, it seems that the apparent lack of crystalline structure depends upon the scale of domain investigated. Optically amorphous materials may show X-ray diffraction patterns, while Mössbauer spectra from X-ray-amorphous metamict minerals only show line broadening. Therefore, small-scale crystalline domains are present even in minerals which appear to be thoroughly metamict. The mechanism of isomorphous substitution of one element for another in a mineral, which may lead to the existence of many different structural configurations, each with about the same stability and each favoured by a particular composition, is an interesting alternative to radiation damage as a means of producing the metamict state, as compositional variation will cause structural variation within a crystal. Thus the metamict state may be the result of segregation into small domains of different structures and/or compositions.

The presence of radioactive elements may aid in the disproportionation of elements, as the high temperatures produced by radiation may lead to the precipitation in submicroscopic regions of a phase which is different from the original form of the metamict crystal. As these phases would be characteristic of high temperatures, they are unlikely to be in equilibrium with the host crystal. It is therefore possible that unmixing of isomorphous series on the microscopic and submicroscopic scales is more important in creating the metamict state than has previously been supposed.

The presence of well-ordered microdomains within metamict minerals (which would result from this mechanism) accords better with the observed Mössbauer spectra than the usually accepted concept of a three-dimensional random network of coordination polyhedra (glass). However, these two possibilities come together at very small domain sizes, as glasses are also at least partly crystallised on the smallest scale. Interestingly, most metamict minerals are of Pre-Cambrian age, which suggests either that a long time is needed for the production of the metamict state or that, if radiation is either the cause or the catalyst of the change in state, the processes were more effective in Pre-Cambrian times.

4.2 FRACTURING

Radioactive minerals often display profound internal fracturing. They are also commonly found at the centres of groups of fractures that radiate in all directions through the surrounding rock for distances of up to about 1 m (Fig. 4.1). The radioactive minerals concerned are usually metamict, and the formation of the fractures is normally attributed to the increase in stress arising from a volume expansion on metamictisation. This volume expansion arises both from the decrease in density that accompanies the destruction of the ordered lattice structure and from the increase in the non-essential water content of the mineral. Minerals which contain a very high proportion of U (principally uraninite) and are maintained at a slightly higher temperature than surrounding minerals by their radioactivity, may undergo a volume increase by thermal expansion. Uraninite may also expand as a result of auto-oxidation, in which some U^{4+} is converted to U^{6+}. This change in the oxidation state of U causes its coordination number to change, and there is a consequent decrease in the density of the mineral. In pyrochlore, a volume increase of 28–34% has been recorded which, apart from producing brittle fractures, also resulted in the development of plastic deformation and displacements in adjacent biotite (Rankama, 1963).

In quartz around allanite in pegmatites, shattering has been observed which is independent of the position of grain boundaries and of variation in cohesion. Strain rates must, therefore, have been so rapid that shattering was

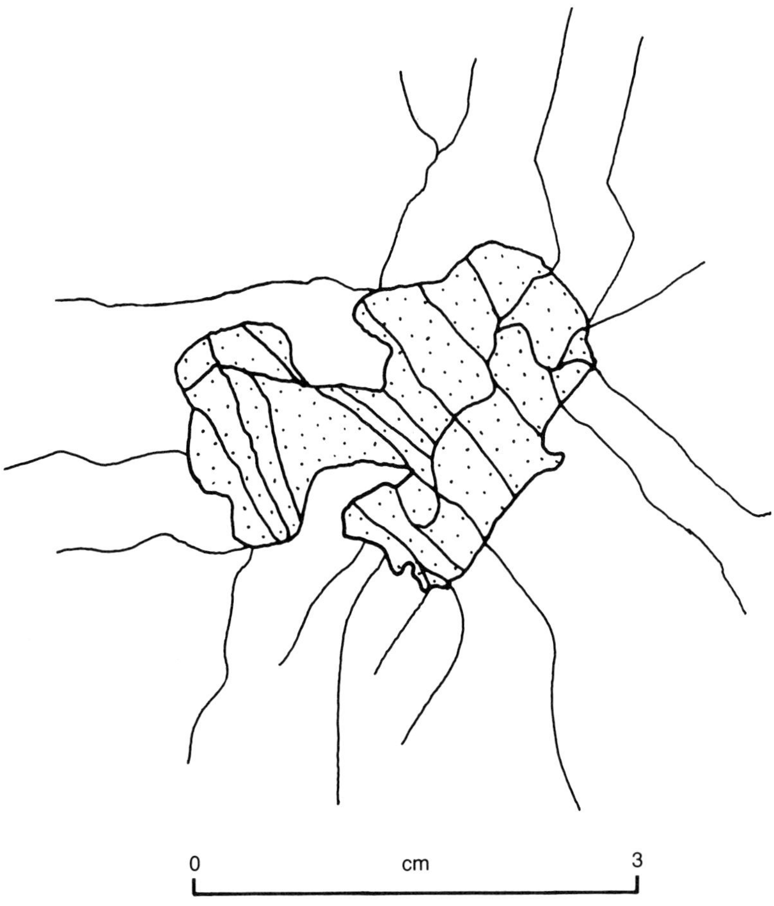

Fig. 4.1 — External fractures around allanite (stippled) in plagioclase. The allanite shows internal fractures. (Heinrich, 1958, redrawn with permission.)

effectively instantaneous. It appears that, in this case, stress built up over a period of time until it reached the short-term strength of the quartz, when explosive shattering occurred. This implies that volume increase within radioactive minerals is progressive. Such explosive shattering of the surrounding rock would also allow the development of internal cracks.

4.3 DISCOLOURATION

In many cases, the presence of radioactive minerals is revealed by the dark red shells in feldspar, calcite, etc., which commonly surround a radioactive mineral. The red colouration is often caused by the presence of haematite along cleavage planes and fractures. The haematite may have been produced during auto-oxidation of Fe^{2+}, which then became available for transport by circulating fluids exploiting the fractures produced in the radioactive minerals and surrounding rock.

The red colouration of feldspar around radioactive minerals is not always the product of haematite precipitation, however. In some feldspars, the red colouration is lost when the mineral is heated to about 500 °C. This observation suggests that the dark colour is formed by a mechanism that is similar to that which causes special colouration in radioactive minerals themselves. Certainly, the colours of some minerals are probably produced by weak radiation applied over long periods of time. These radiogenic colours can be recognised by their disappearance on heating, usually to 200–300 °C.

The colour of a mineral is produced by absorption of some wavelengths from the spectrum making up white light. The colour observed is determined by the wavelength of the reflected (or transmitted) light. Absorption of electromagnetic radiation (including light) requires an energy transfer to the atoms present in the mineral, and this is taken up by labile electrons. The absorption therefore depends upon the structure both of the atoms and of the crystal. Even in an individual crystal, different wavelengths may be absorbed when plane-polarised light is transmitted in different directions through the crystal; this property is known as 'pleochroism'. Structural damage by radiation in a crystal can alter the characteristics of the absorption and thus change the colour of the crystal. In general, this is the way in which colour centres (known as 'F centres' from the German for colour, Farbe) are produced. In detail, the Varley mechanism, in which multiple ionisation of a negative ion is followed by ejection of a positive ion and then by the interaction of an electron and a negative ion vacancy, is most commonly used to explain the effect.

Colour may also be directly related to composition, elements with incomplete electron subshells usually producing distinct colouration. Ti, V, Cr, Mn, Fe, Ni, Co and Cu are particularly effective colouring agents, the effect often being intensified by the presence of an element in two valency states. Thus, Fe-bearing minerals which contain only Fe^{2+} are usually pale coloured, while those in which both Fe^{3+} and Fe^{2+} are present are typified by dark colours. Ions or groups of ions which produce characteristic colours are known as 'chromophores'. Their presence, even in minute amounts, can have a strong colouring effect on a mineral. Radiation can change the state of ionisation of the chromophores, producing a wide variety of colours under different circumstances. Oxidation of Fe^{2+} to Fe^{3+}, particularly, accompanies irradiation. The pale red colour of many feldspars is due to the presence of submicroscopic haematite produced by exsolution of Fe which substitutes for Al in the mineral. Darkening of the red colouration then occurs when radiation oxidises more Fe^{2+} to Fe^{3+}.

A number of other radiation-related colour changes are often observed. The deep blue, purple or dark violet colouration of fluorite is due to the creation of F^- ion gaps by α-particles. Changes in halite, which is usually colourless but can become strongly coloured, are due to the presence of chromophores in proximity to α-particle or β^--particle sources (such as ^{40}K), probably within the halite itself. Deep blue halite is associated with the production of Cl^- vacancies within the structure. Interestingly, sylvite, which has a high ^{40}K content, is colourless, even though it often contains a higher proportion of bivalent chromophores than coloured halite; there seems to be a saturation level beyond which the effect is not observed in the presence of weak radiation. Diamonds in contact with radioactive minerals can develop a green outer zone. Finally, milky quartz appears to form by the action of radiation on ordinary, clear quartz.

4.4 RADIOHALOES

These are the most easily recognised mineralogical features attributable to the presence of radionuclides. They were first described by Michel Levy in 1882, well before Becquerel discovered radioactivity, and were related to radioactivity by Joly in 1907. The haloes consist of a pleochroic, light to dark brown zone within a mineral such as biotite, with an included radioactive mineral at the centre. A wide variety of minerals have been reported to contain pleochroic haloes. Although it has been suggested that β-particles may contribute to their formation, it appears that pleochroic haloes will only form if α-particle emitters are present.

The intensity of radiation required to produce pleochroic haloes varies from mineral to mineral. In biotite, for example, haloes are seen to have developed around relatively weak sources such as zircon and monazite. This is also true of chlorite, hornblende and fluorite. In minerals such as chloritoid, garnet and sphene, however, strong sources such as uraninite and thorianite are required. Where such strong sources are present the radiation causes metamictisation of the surrounding mineral, and so the haloes are more correctly called metamict haloes.

As discussed in Chapter 3, the penetration range of α-particles in minerals is governed by the rate of loss of energy due to ionisation. The range may be expressed by the Geiger–Nuttall rule

$$\log R = A \log \lambda + B,$$

where R and λ are, respectively, the range and decay constants for a particular α-particle emission and A and B are constants. The value of A is almost the same for the ^{232}Th, ^{235}U and ^{238}U decay series.

Geiger also considered that, as the energy of an α-particle is given up, the specific ionisation I caused is proportional to the energy absorbed:

$$I \propto -\frac{\mathrm{d}}{\mathrm{d}x}\left(\frac{MV}{2}\right)^2,$$

where M is the mass, V is the velocity of the α-particle and x is the distance. The distance that an α-particle will travel before it ceases to ionise is given by

$$R - x = \frac{V^3}{a},$$

where R is the overall range as before, V is the velocity and a is a constant. Thus

$$I \propto (R - x)^{\frac{1}{3}}.$$

If I is plotted against x, as I increases, a steep curve is obtained up to the value of R at which it drops to zero. However, straggling of α-particles produces a broadening of this curve as x increases, so that the peak is not so sharp, and the drop in value at R is not immediately to zero. The real result is known as the 'Bragg curve'. Although straggling causes a non-instantaneous fall-off of the specific ionisation, the decrease is still very marked and the range of an α-particle can be determined very precisely. In this way it has been shown that the range of α-particles in air at standard temperature and pressure increases from less than 1 cm at about 1 MeV, to about 2 cm at 4 MeV, 11 cm at 10 MeV and 20 cm at 15 MeV.

Monoenergetic α-particles can only be produced from an infinitely thin source, the thickness of which is effectively zero; otherwise, degradation of α-particle energies will occur by ionisation within the source itself. Such sources, of course, do not exist in natural systems, and so the range of an α-particle of given energy can only be considered as a maximum value.

The penetration of α-particles into biotite ranges from 12.7 μm for an energy of 4.2 MeV to 41.8 μm for an energy of 8.8 MeV (Tables 4.1 and 4.2).

The ionisations produced by α-particles do not normally destroy the crystalline lattice of minerals. However, in some minerals, biotite for example, reordering is consequent upon the oxidation of Fe^{2+} to Fe^{3+} and its subsequent precipitation as Fe_2O_3. In biotite, haematite becomes concentrated within the basal cleavage planes. In has been suggested that somewhere of the order of 10^4 α-particles cm^{-2} represents a threshold value for halo production in biotite, with saturation at a level of 5×10^{15} α-particles cm^{-2}. Where the α-sources are particularly intense, it has been observed that gradual fading of the halo begins after about 5×10^{17} α-particles cm^{-2}, which is beyond the threshold for the development of metamictisation (Rankama, 1963).

Extinction haloes are formed where highly radioactive minerals are included in albite. In some of these haloes, the optical axes show a rotation from their normal positions. When viewed under crossed polars, such haloes go into extinction in a different position from that of the main albite crystal. It has been suggested that, in the formation of these haloes, α-particle radiation has caused either O displacement from Si–Al–O tetrahedra, or the rearrangement of large cations in the structure. This effect is only possible in monoclinic and triclinic crystals, and in the feldspar group it is restricted to triclinic crystals. As birefringence is unaltered in these haloes, it is unlikely that they are caused by stress effects on the crystal lattice produced by thermal expansion of the radioactive mineral.

Calibration of halo widths using α-particle

Table 4.1 — α-particle penetration

Nuclide	Energy (MeV)	Penetration (μm)
^{238}U	4.20 (77%), 4.15 (23%)	12.7
^{234}U	4.77 (72%), 4.72 (28%)	15.3
^{232}Th	4.02 (77%), 3.96 (23%)	12.4
^{230}Th	4.69 (76%), 4.62 (24%)	15.3
^{228}Th	5.42 (73%), 5.34 (27%)	20.0
^{226}Ra	4.79 (94%), 4.60 (6%)	15.3
^{222}Rn	5.49 (100%)	19.2
^{220}Rn	6.29 (100%)	23.9
^{219}Rn	6.82 (81%), 6.55 (11%), 6.42 (8%)	27.2
^{218}Po	6.00 (100%)	23.0
^{216}Po	6.78 (100%)	27.8
^{214}Po	7.69 (100%)	34.2
^{212}Po	8.78 (100%)	41.8
^{210}Po	5.31 (100%)	19.2
^{212}Bi	6.09 (27%), 6.05 (70%)	23.9
^{211}Bi	6.62 (84%), 6.28 (16%)	27.2

(After Rankama, 1954.)

Table 4.2 — Approximate limiting ranges

Energy (MeV)	Range (μm)
4–5	15.5
5–6	20
6–7	28
7–8	34
8–9	42

(After Rankama, 1954.)

irradiations of various energies, total dose and dose rate has attributed the formation of haloes in fluorite to α-particles emitted from ^{210}Po, ^{214}Po and ^{218}Po, all of which are members of the ^{238}U decay chain. The half-lives of these nuclides are 138.3 days, 1.64×10^{-45} and 3.05 min, respectively, and so these haloes could only have been formed by the maintenance of the decay series over long periods of time.

Although most radiohaloes have a width of less than 20 μm (Fig. 4.2), giant pleochroic haloes have been described in biotites from Madagascar (Gentry, 1973, 1978). As with ordinary haloes, these haloes are centred on radioactive inclusions. Surrounding the inclusion, the biotite appears to be unaltered. A bleached annular region then extends between about 25 μm and 60 μm from the edge of the inclusion. It has been suggested that these haloes may have been produced by α-particles of very high energy. Such energies would only be found in the transuranic or superheavy elements. However, none of the artificially produced transuranic elements emits α-particles which are more energetic than those of the ^{238}U or ^{232}Th decay series, the highest energies being about 8.5 MeV. Alternatively, and more probably, the brown colour of biotite may be caused by charge transfer affecting Fe^{3+} rather than Fe^{2+}. The bleached zone may then form by the reduction of Fe^{3+} to Fe^{2+} at slightly elevated temperatures by H produced by radiolysis of OH^- ions from the biotite lattice. These, in turn, are liberated by α-particles emitted from the inclusions. The size of the giant haloes from Madagascar is therefore considered to be due to a particularly high

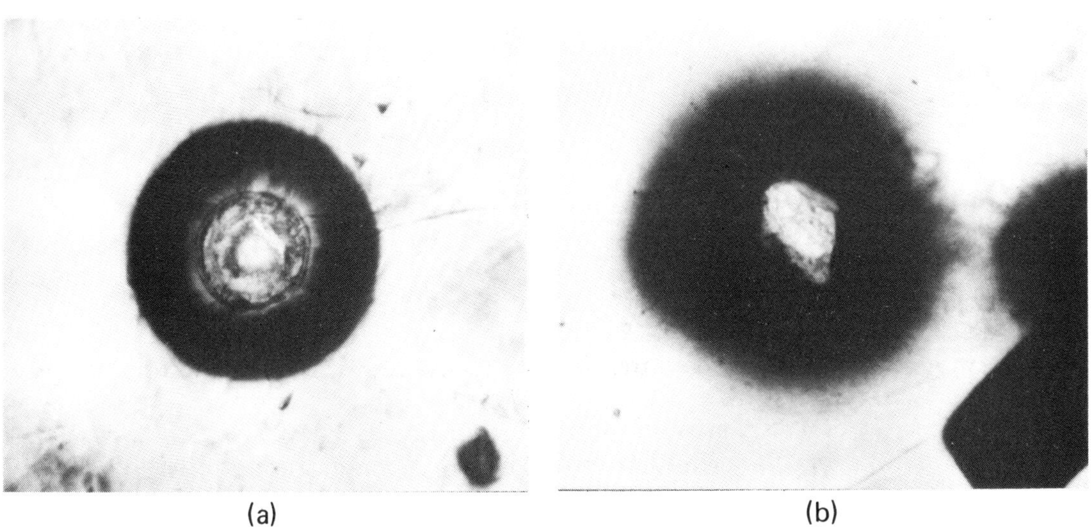

Fig. 4.2 — Photomicrographs of radiohaloes in biotite from the Carnmenellis granite: (a) around a uraninite crystal; (b) around a zircon crystal. The width of the field of view in each case is 1.5 mm. (Source: A. V. Bromley.)

diffusion rate of H in these particular minerals (Vance, 1978).

4.5 FISSION TRACKS

Some of the mechanisms of energy transfer that accompany the penetration of matter by radioactive emissions have already been considered when dealing with metamictisation and the formation of radiohaloes and colour centres. In more general terms, when the displacement of a single atom from the lattice is not involved, the transfer of energy may take place in three main ways (Chadderton, 1965).

(1) Transfer of vibrational energy; effectively, the sudden delivery of a quantity of heat to a small part of a lattice, which is gradually dispersed over a larger volume (the thermal spike).
(2) Displacement of a number of atoms from the lattice so that a series of Frenkel defects are produced, all related to the same initial cause; in particular, the displaced atoms move out and away from the centre of disruption (the displacement spike).
(3) Development of a thermal spike is accompanied by the production of a radial compressive stress; under this stress it is possible for plastic flow of the material to occur (the plastic spike).

When fast heavy charged particles are produced in the spontaneous fission of radionuclides (principally ^{238}U) and are brought to rest when they penetrate a crystal, intense damage results which produces a fission track. By analogy with the three energy transfer mechanisms discussed above, this process can be said to result from a 'fission spike'. The fission spike is in fact an analogue of thermal, displacement and plastic spikes, and the role of each of these energy transfer mechanisms varies along the path of a fission fragment through a crystal. In the early part of the pathway, the fragment loses at least part of its energy in exciting electrons and in the creation of a thermal spike but, as electrons are captured by the fragment, this gradually decreases in intensity. Displacement spikes become important as the charge of the fragment decreases towards zero, when knock-on effects producing Frenkel defects and plastic spikes also become important. The end of the fission track may be considered to be marked by just one extra large displacement spike, the size of which is determined by the difference between the mass of the lattice atoms and the mass of the fragment. When a fission fragment has finally lost its kinetic energy, it comes to rest merely as an atom of impurity within the damaged lattice.

The fact that single fission fragments produce tracks is most clearly seen in crystals in which electron excitation easily leads to damage of the lattice. In these circumstances, it can be seen that the character of the damage depends upon the energy of the fission fragment; at high energies the tracks are continuous but, as energy is lost the tracks become discontinous, the distance separating each section decreasing progressively. This has been interpreted as showing that the thermal spike is responsible for the continuous part of the track, while segments in the discontinuous section are produced by displacement cascades due to knock-on effects and, possibly, displacement spikes.

Even those fission tracks which are produced in especially sensitive artificial materials by radioactive minerals subjected to neutron bombardment in a nuclear reactor, are very small (Fig. 4.3). Under normal conditions, natural fission tracks are only visible using the very high magnifications possible with an electron microscope. Typically, track lengths may be of the order of 10^{-2} μm. However, they can be made visible with the aid of an ordinary optical microscope if etched with a suitable reducing solution, such as HNO_3 or NaOH, because the damaged parts of the crystal are more soluble than the unaffected areas. Among the minerals in which fission tracks have been examined in this way are mica, apatite, zicron, sphene and epidote. Also susceptible to fission track formation is natural glass. The

(a)

(b)

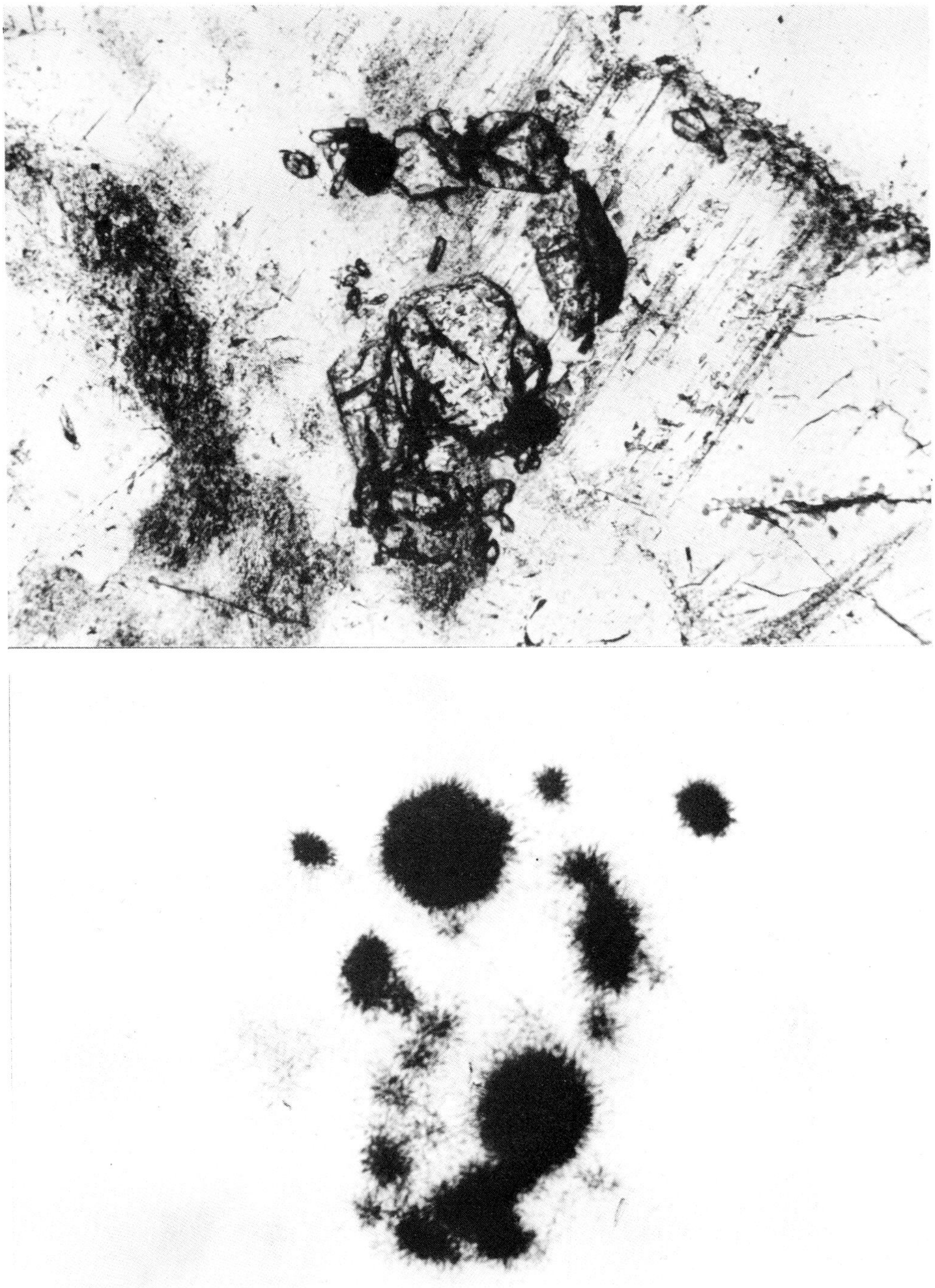

Fig. 4.3 — Use of fission track registration in radioactive mineral location: (a) photomicrograph of biotite granite from Carnmenellis, showing monazite, zircon and apatite accessory mineral clusters: (b) fission tracks in Lexan polycarbonate film. Irradiation of 2×10^{15} neutrons m^{-2}. The section has a width of 1.5 mm. (Source: N. L. Jefferies.)

development of tracks can be used as the basis of a method for determining the presence of ^{238}U. If the abundance of fissionable U and the characteristics of the host mineral are known, measurement of the abundance of fission tracks also provides a means of obtaining the age of the material. The use of fission tracks as a method of U detection is also considered in Chapter 9, while the dating aspects of fission track investigations are discussed in more detail in Chapter 10.

Unfortunately, fission tracks are not permanent. The length of time for which a fission track remains suitable for examination depends upon the nature of the mineral in which the track has been produced and the temperature to which the mineral has been raised. The annealing of fission tracks can also be caused by other factors, such as shock waves, while the presence of water affects the annealing rate of glass. These factors, however, are less important than temperature. The annealing time t for specific reduction in track density is given by

$$\log t = \log A + \frac{v}{2.303kT},$$

where A is a constant which varies with the mineral species, v is the activation energy, k is Boltzmann's constant and T is the temperature in kelvin.

This is a straight-line equation for log t against T^{-1} which has been used to show that complete annealing of tracks in sphene takes place in about 100 years at 500 °C but requires about 10^7 years at 400 °C. For apatite, however, temperatures of only about 200 °C will achieve complete annealing after 100 years, but temperatures of 150 °C will only give total annealing after about 10^7 years. Limiting temperatures for any annealing to occur may be assessed by standardising at 10^6 years, giving 50 °C for apatite and 250 °C for sphene. Fission tracks that develop within a mineral as it cools after solidification from a magma will experience annealing of older tracks as new tracks form. The total number of tracks formed for a particular mineralogical setting is determined by the time that has elapsed since the conditions for approximatley 50% retention of tracks was reached (Faure, 1977).

4.6 α-TRACKS AND α-RECOIL TRACKS

The passage of α-particles through materials causes disruption similar to that produced by fission particles. However, because α-particles have much less energy, their path length is considerably shorter than that of fission fragments. The main use of these tracks is in autoradiography, in which tracks may be etched to lengths of 100 μm or more in highly sensitive detector materials (Fig. 4.4 and 4.5). Although most dielectric materials will record tracks of high energy charged particles, optimum conditions for track formation require a suitable rate of energy loss. Different materials react to particles of different masses. With the exception of muscovite, the materials used for α-autoradiography are all artificial. The study of the occurrence of α-tracks within radioactive minerals requires further development.

α-recoil tracks are produced when the emission of an energetic α-particle from a nucleus causes a large recoil energy to be given to the nucleus. The nucleus then results in damage to the atomic lattice of the mineral involved, producing a disintegration track. As before, etching causes the tracks to become enlarged because of preferential solution and allows them to be viewed easily using an ordinary optical microscope. α-recoil tracks have been observed in muscovite and phlogopite, in which they take the form of pits about 10^{-2} μm deep (Huang and Walker, 1967). Further development of this technique is also needed.

4.7 THERMOLUMINESCENCE

Many minerals, when heated to less than their incandescent temperatures, exhibit the phenomenon of glowing in the dark. This

(a)

(b)

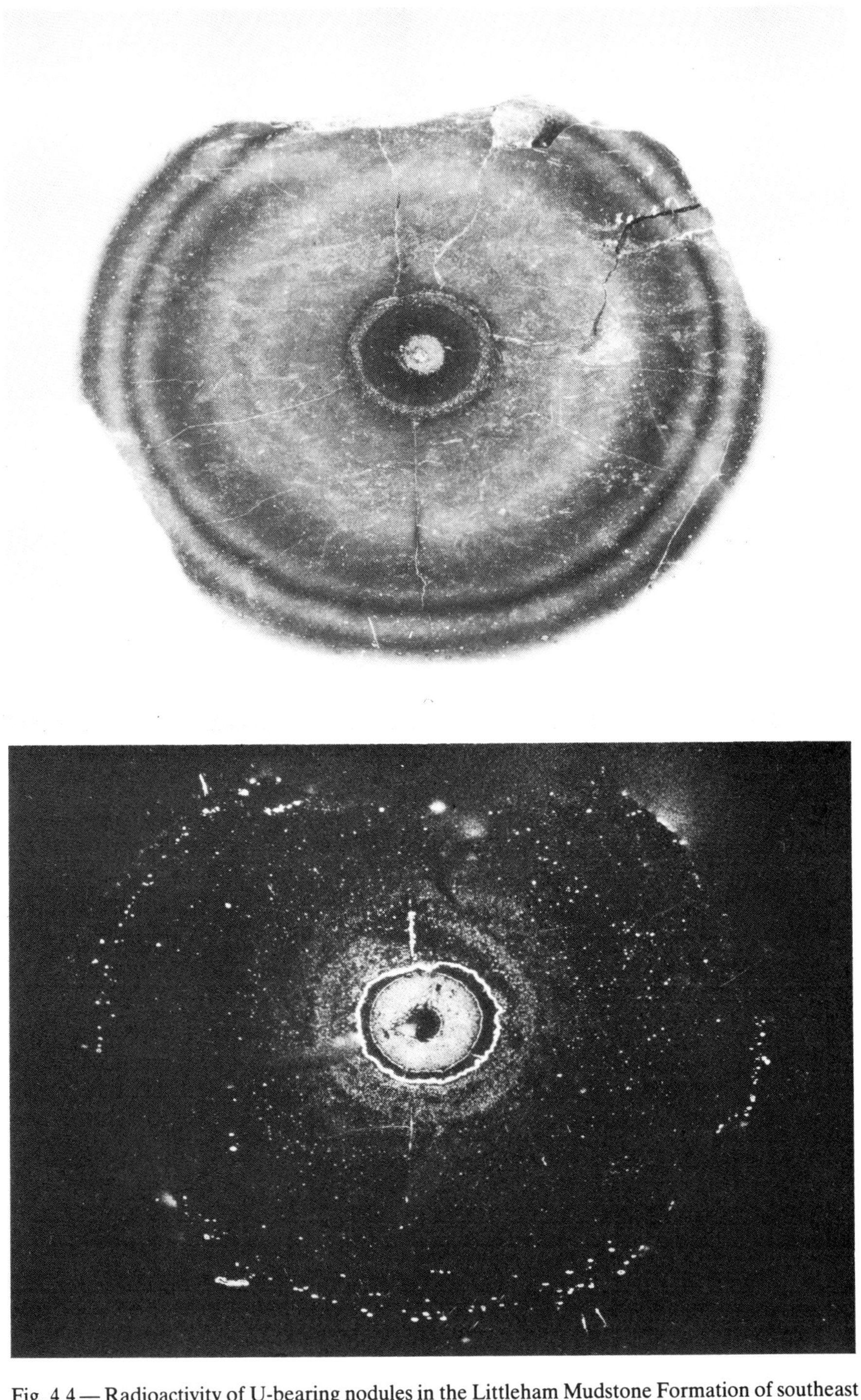

Fig. 4.4 — Radioactivity of U-bearing nodules in the Littleham Mudstone Formation of southeast Devon: (a) section of nodule showing concentric bands: the specimen has a width of 11 cm: (b) contact autoradiograph on nuclear track plastic CR39, the distribution of the U is clearly shown. (Source: N. L. Jefferies.)

(a)

(b)

Fig. 4.5 — Use of α-particle autoradiography for radioactive mineral location: (a) photomicrograph of an intergranular crystal of torbernite in the Dartmoor granite: (b) autoradiograph from a 6 week exposure on nuclear track plastic CR39. The torbenite crystal has a diameter of approximately 50 μm. (Source: M.J. Heath.)

effect, now known as thermoluminescence, was independently discovered by two parties of prospectors in Ontario who placed some radioactive mineral species on a hot camp stove at night. The emission of light is caused by the thermal release of electrons previously trapped in lattice imperfections within the crystal structure at metastable high energy levels. The released electrons are transferred to normal lattice positions and their extra energy is emitted as photons with energies in the visible light range. Temperatures up to about 450 °C usually cause thermoluminescence. Quantitative studies generally follow the practice of measuring the light intensity as a function of the ambient temperature to produce a thermoluminescence curve or 'glow curve'. These curves typically show that several maxima occur, each of which represents a particular energy and a particular trap site in the lattice. Electrons are placed in these traps either during crystallisation at imperfections, distortions or vacancies in the structure or after crystallisation as a result of the transfer of thermal, mechanical or radiation energy to the electrons. Radioactivity, in particular, appears to be very effective in placing electrons in these traps, and most radioactive minerals, or minerals that have been close to a radioactive source, display thermoluminescence.

The distribution of the maxima in the glow curve reflects the characteristics of the crystal involved, any impurities within it and the amount of radiation it has received. It has been demonstrated that the thermoluminescence of radioactive minerals is mainly due to energy transferred to the electrons by γ-radiation. However, when electrons are transferred to trap sites by γ-radiation originating in decay series with α-particle emission, all available sites soon become filled if the α-particle source is intense, and no further increase in thermoluminescence is theoretically possible unless the passage of the α-particles can actually create new traps to supplement those already in the host lattice. When the lattice becomes severely disrupted, however, the metastable trap sites are destroyed and thermoluminescence decreases. It appears, therefore, that there is a limiting α-particle flux rate that cannot be exceeded if thermoluminescence is to become a property of the irradiated material.

Once heated, the thermoluminescent character of the mineral is lost. Use of this property has been attempted for dating geological materials, for geothermometry and for correlation purposes. Unfortunately, the validity of the results of some of these applications is uncertain. Claims which suggest that, if the radiation flux can be accurately determined, the thermoluminescence can be used to indicate the age of a sample have been countered by the argument that thermoluminescence is only caused by the presence of impurities, and so no mineralogical or other useful data can be obtained. Despite this uncertainty, the use of thermoluminescence has been proposed as the basis of a method of U exploration (Charlet *et al.*, 1982).

4.8 PHOTOLUMINESCENCE

With the exception of incandescence produced by heating, the emission of visible light by a mineral is known as 'luminescence'. Special types of luminescence also occur in which photons outside the visible light spectrum are emitted. Emission of light usually occurs when a mineral is irradiated with photons of greater energy than visible light, typically ultraviolet radiation. Indeed, a fundamental property of luminescence is that the wavelength of the emitted light is normally longer than the wavelength of the exciting radiation (Stoke's law). If the luminescence only occurs when irradiation is taking place, the phenomenon is known as 'fluorescence'. When light continues to be emitted after irradiation has stopped, the phenomenon is called 'phosphorescence'. 'Infrared luminescence' occurs when irradiation by visible light causes the emission of infrared energies. However, these simple definitions are not necessarily the best. For example, it is found in practice that fluorescent materials may continue to emit light for short periods of time after the exciting radiation has ceased and yet not long enough to be true phosphorescence; this phenomenon is known as 'slow fluorescence' or 'delayed fluorescence'. True fluorescence, then, is defined on the basis of cessation of the emission of visible light within a certain period of time from the end of excitation. The actual limit used is a decrease in the intensity of emitted light to e^{-1} of its original intensity within a period of 10^{-8} s. Decay times for slow fluorescence and phosphorescence range from 10^{-3} s to many days.

While these different types of luminescence appear to be very similar, they are produced by entirely different mechanisms. Like thermoluminescence, photoluminescence can be considered in terms of the existence of both the ground state and metastable higher energy states of electrons in an atom. Absorption of energy on irradiation causes electrons to move from ground state levels to higher energy levels. In the case of fluorescence, de-excitation then occurs within 10^{-8} s, normally to energy levels above the ground state, so that wavelengths greater than the incident radiations are emitted. Further de-excitation to the ground state results in further emissions of energies lower than the incident radiation.

However, if an atom in an excited state absorbs the incident energy and de-excitation then occurs to the ground state, the wavelength of the emitted light will be shorter than that of the incident light ('anti-Stoke's radiation'). The same effect can be produced if, at the excited level, further excitation occurs (say, through collision), with return directly to the ground state.

Slow fluorescence arises when the transition is to a metastable energy level (as in thermoluminescence), followed by de-excitation to lower energy levels with the emission of radiation. Phosphorescence is related to electrons in metastable energy levels in a similar way. Delayed fluorescence is believed to occur when two atoms in the metastable state collide and produce a non-metastable configuration which radiates in returning to the ground state.

The decay time for fluorescence is thus the lifetime of the higher energy state, while for slow fluorescence and phosphorescence it is the lifetime of the metastable state. Indeed, in all physical respects, phosphorescence is identical with thermoluminescence.

In order for luminescence to occur, it appears necessary that lattice disturbances are present in the mineral. These may be the result either of lattice defects or of the presence of impurities that act as activators. Impurities are usually caused by the presence of minor amounts of foreign elements substituting for one of the main elements in the mineral. Thus, fluorite (CaF_2), the mineral which is generally considered to be synonymous with fluorescence and from which the name was derived, is fluorescent because of substitution of some Ca by rare earth elements. In willemite it is Mn substituting for Zn, and in scheelite Pb for Ca and Mo for W. In scheelite the fluorescence changes from white to yellow as Mo replaces W.

Many U-bearing minerals are fluorescent. Indeed, several types of fluorescence have been recognised (Heinrich, 1958).

(1) Uranyl type which is typical of phosphates, arsenates, some carbonates and some sulphates; Ca, Ba and Na–Ca compositions are strongly fluorescent, Ca–Mg and Mg compositions are less strongly to weakly fluorescent, and Cu, Fe, Pb and Bi compositions are only weakly or non-fluorescent. Two sub-types are recognised:

    (a) Autunite type which displays strong to moderate yellow–green fluorescence.
    (b) Schroeckingerite type which displays strong to moderate whitish green fluorescence.

(2) Uranate type which is orange or red fluorescence exhibited by becquerelite, soddyite, etc.

Th-bearing minerals may also show weak fluorescence.

The fluorescence of minerals varies with the wavelength of incident radiation. It is normally investigated at two wavelengths of ultraviolet light: a short wavelength ranging between 2.3×10^{-7} and 3.1×10^{-7} m, and a long wavelength between 3.1×10^{-7} and 4.0×10^{-7} m. The variability of the fluorescence of radioactive minerals is illustrated by huttonite which fluoresces pinkish white under short-wavelength ultraviolet radiation but is not fluorescent at all at the long wavelength.

BIBLIOGRAPHY

Bass, M. N. (1957) Effect of gamma irradiation on physical properties of minerals. *Am. Mineral.* **42,** 100.

Berman, J. (1953) Nature of crystallisation of metamict minerals. *Bull. Geol. Soc. Am.* **64,** 1395.

Berman, J. (1955) Identification of metamict minerals by X-ray diffraction. *Am. Mineral.* **40,** 805.

Bertel, E., and Mark, T. D. (1983) Fission tracks in minerals: annealing kinetics, track structure and age correction. *Phys. Chem. Minerals* **9,** 197.

Bertel, E., Mark, T. D., and Pahl, M. (1977) A new method for the measurement of the mean etchable fission track length and of extremely high fission track densities in minerals. *Nucl. Track Detection* **1,** 123.

Blazek, M. C. (1979) Classification of pseudomorphs. *Rocks Minerals* **54,** 194.

Chadderton, L. T. (1965) *Radiation damage in crystals.* London, Methuen.

Charlet, J. M., Dupuis, C., Quinif, Y., Bouko, P., and Lair, P. (1982) Natural thermoluminescence of quartz and feldspars, application of a new method in uranium exploration. In *Uranium exploration methods.* Paris, Organisation for Economic Co-operation and Development.

Daliens, M., Delhal, J., and Tarte, P. (1977) Metamictization and U–Pb systematics — a study by infrared absorption spectrometry of Precambrian zircons. *Earth Planet. Sci. Lett.* **33,** 331.

Dobretsov, N. L., Dobretsova, I. L., Sobolev, V. S., and Mali, V. I. (1968) Shock compression of $ZrSiO_4$ and metamict decay. *Dokl. Acad. Sci. USSR Earth Sci.* **182,** 134.

Dollase, W. A. (1971) Refinement of the crystal structures of epidote, allanite and hancockite. *Am. Mineral.* **56,** 447.

Ellesworth, H. V. (1932) Rare element minerals of Canada. *Geol. Surv. Can. Econ. Geol. Ser.* **11.**

Ewing, R. C. (1975) The crystal chemistry of complex niobium and tantalum oxides, IV, The metamict state: discussion. *Am. Mineral.* **60,** 728.

Ewing, R. C. (1975) Alteration of metamict Rare Earth, AB_2O_6-type Nb–Ta–Ti oxides. *Geochim. Cosmochim. Acta* **398,** 521.

Ewing, R. C. (1976) Metamict mineral alteration: an implication for radioactive waste disposal. *Science* **192,** 1336.

Ewing, R. C., and Haaker, R. F., (1980) The metamict state: implications for radiation damage in crystalline waste forms. *Nucl. Chem. Waste Management* **1,** 51.

Ewing, R. C., and Krumhansl, J. L. (1974) Natural gamma ray spectra of euxinite, polycrase and aeschynite. *Can. Mineral* **12,** 357.

Faul, H. (1954) *Nuclear geology.* New York, Wiley.

Faure, G. (1977) *Principles of isotope geology.* New York, Wiley.

Fleischer, R. L. (1979) Where do nuclear tracks lead? *Am. Sci.* **67,** 194.

Fleischer, R. L. (1980) Isotopic disequilibrium of uranium: alpha-recoil damage and preferential solution effects. *Science* **207,** 979.

Fleischer, R. L., Price, P. B., and Walker, R. M. (1975) *Nuclear tracks in solids: principles and applications.* Berkely, California, University of California Press.

Gentry, R. V. (1967) Extinct radioactivity and the discovery of a new pleochroic halo. *Nature* **213,** 487.

Gentry, R. V. (1973) Radioactive haloes. *Ann. Rev. Nucl. Sci.* **23,** 347.

Gentry, R. V. (1974) Radiohaloes in a radiochronological and cosmological perspective. *Science* **184,** 62.

Gentry, R. V. (1978) Are any unusual radiohaloes evidence for SHE? *Ann. Rev. Nucl. Sci.* **28,** 123.

Gentry, R. V., Christie, W. H., Smith, D. H., Emery, J. F., Reynolds, S. A., Walker, R., Cristy, S. S., and Gentry, P. A. (1976) Radiohaloes in coalified wood: new evidence relating to the time of uranium introduction and coalification. *Science* **194,** 315.

Gentry, R. V., Hulett, L. D., Cristy, S. S., McLaughlin, J. F., McHugh, J. A., and Bayard, M. (1974) 'Spectacle' array of ^{210}Po halo radiocentres in biotite: a nuclear geophysical enigma. *Nature* **252,** 564.

Graham, J., and Thornber, M. R. (1974) The crystal chemistry of complex niobium and tantalum oxides, IV, the metamict state. *Am. Mineral.* **59,** 1047.

Graham, J., and Thornber, M. R. (1975) The crystal chemistry of complex niobium and tantalum oxides, IV, the metamict state: reply. *Am. Mineral.* **60,** 734.

Haak, U. (1977) The closing temperature for fission track retention in minerals. *Am. J. Sci.* **277,** 459.

Harwood, J. J., Hausner, H. H., Morse, J. G., and Rauch, W. G. (1958) *The effects of radiation on materials.* New York, Reinhold.

Heath, M. J. (1982) Uranium in the Dartmoor granite: geochemical and radiogeological investigations. *Ph.D. Thesis.* University of Exeter.

Heinrich, E. W. (1958) *Mineralogy and geology of radioactive raw materials.* New York, McGraw-Hill.

Holmes, A. (1926) Estimates of geologic time with special reference to thorium and uranium haloes. *Phil. Mag.* **1,** 1055.

Huang, W. H., and Walker, R. M. (1967) Fossil alpha-particle recoil tracks: a new method of age determination. *Science* **155,** 1103.

Hutcheon, I. D., and Macdougal., J. D. (1976) Particle track studies. In Wenk, H. R., Champness, P. E., Christie, J. M., Cowley, J. M., Heuer, A. H.,

Thomas, G., and Tighe, N. J. (eds), *Electron microscopy in mineralogy*. Berlin, Springer.
International Atomic Energy Agency (1962) *Radiation damage in solids*. Vienna, International Atomic Energy Agency.
Joly, J. J. (1907) Pleochroic haloes. *Phil. Mag.* **13,** 381.
Joly, J. J. (1909) *Radioactivity and geology*. London, Constable.
Joly, J. J. (1923) Pleochroic haloes of various ages. *Proc. R. Soc. Lond. A.* **102,** 682.
Joly, J. J., and Rutherford, E. (1913) The age of pleochroic haloes. *Phil. Mag.* **25,** 644.
Kulikova, I. M., Barinskiy, R. L., Alexksandrov, V. B., and Proshchenko, Y. G. (1973) X-ray absorption spectra of niobium in some metamict minerals. *Dokl. Acad. Sci. USSR Earth Sci.* **210,** 160.
Marfunin, A. S. (1979) *Spectroscopy, luminescence and radiation centres in minerals*. Berlin, Springer.
McAndrew, J. (1957) Natural radiation damage in albite. *Am. J. Sci.* **255,** 715.
Mitchell, R. S. (1973) Metamict minerals: a review, Part 1. *Mineral. Record* **4,** 177.
Mitchell, R. S. (1973) Metamict minerals: a review, Part 2. *Mineral. Record* **4,** 214.
Naeser, C. W., and Faul, H. (1969) Fission track annealing in apatite and sphene. *J. Geophys. Res.* **74,** 705.
Nerurkar, A. P., De, R., Chakraborty, P. N., and Kaul, I. K., (1979) Studies on thermoluminescence, metamictization and sintering properties of zircon sands. *Mod. Geol.* **7,** 13.
Price, P. B., and Walker, R. M. (1962) Observation of fossil particle tracks in natural micas. *Nature* **196,** 732.
Price, P. B., and Walker, R. M. (1962) Simple method of measuring low uranium concentrations in natural crystals. *Appl. Phys. Lett.* **2,** 23.
Price, P. B., and Walker, R. M. (1963) Fossil tracks of charged particles in mica and the age of minerals. *J. Geophys. Res.* **68,** 4847.
Prizbram, K., and Caffyn, J. E. (1956) *Irradiation colours and luminescence*. London, Pergamon.
Pyatenko, Y. A. (1970) Behaviour of metamict minerals on heating and the general problem of metamictization. *Geochem. Int.* 758.
Rankama, K. (1954) *Isotope geology*. London, Pergamon.
Rankama, K. (1963) *Progress in isotope geology*. New York, Interscience.
Reimer, G. M. (1975) Uranium determination in natural water by the fission track technique. *J. Geochem. Expl.* **4,** 425.
Saini, H. S., Lal, N., and Nagpaul, K. K. (1975) Annealing studies of fission tracks in allanite. *Contrib. Mineral. Petrol.* **52,** 143.
Seal, M., Vance, E. R., and Demayo, B. (1981) Optical spectra of giant radiohaloes in Madagascan biotite. *Am. Mineral.* **66,** 358.
Vance, E. R. (1978) Possible mechanism for the formation of bleached, giant haloes in Madagascan mica. In Lodhi, M. A. K. (ed.), *Superheavy elements*. New York, Pergamon.
Vance, E. R., and Anderson, B. W. (1972) Study of metamict Ceylon zircons. *Mineral. Mag.* **38,** 605.
Vance, E. R., and Routcliffe, P. (1976) Heat treatment of some metamict allanites. *Mineral Mag.* **40,** 521.

5

Neutrons in the Earth

5.1 NEUTRON PRODUCTION

As discussed in Chapter 3, primary cosmic ray particles entering the Earth's atmosphere interact with the nuclei of air to produce a variety of secondary particles, including neutrons, either by disintegration or by evaporation processes. Additionally, some solar neutrons are present in the Earth's atmosphere. The evaporation process produces neutrons with energies of about 1 MeV, but direct interactions give higher energies in the range from 1 MeV to 1 GeV. Since the neutrons rapidly lose energy by collision and become captured in the formation of ^{14}C, equilibrium between neutron production and neutron capture is attained near the top of the atmosphere. This equilibrium region is known as the 'free atmosphere', and the neutron energy spectrum within it is independent of altitude.

Close to the Earth's surface, however, these conditions do not apply, and there is variation in both the energy spectrum and the spatial distribution of neutrons. For example, the slow neutron flux over solid ground is 3.1 times greater than that over water. Interestingly, over a water surface, the low energy neutrons are more abundant than in the free atmosphere, but the high energy neutrons are attenuated by about 20% with respect to the free atmosphere. This indicates the progressive conversion of high energy neutrons to low energy neutrons with descent through the atmosphere. Below a water surface, the top 25–30 cm is a zone of rapid decrease in thermal neutron intensity, but thereafter the intensity decreases more slowly with depth.

To investigate neutron characteristics with depth in rock, the neutron flux at 44°N (latitude is important as cosmic rays are affected by the Earth's magnetic field) has been measured in a four-storey garage with concrete floors and low ceilings, on the basis that this simulates various depths within the Earth (Yamashita *et al.*, 1966). On the roof of the building, the thermal neutron intensity was found to be 1.07×10^{-3} neutrons cm^{-2} s^{-1}, count rates decreasing exponentially with depth through the building. Because of interaction with the concrete floors of the garage, the flux intensity of thermal neutrons was found to increase to 1.20×10^{-2} neutrons cm^{-2} s^{-1} at a depth equivalent to 700 g cm^2. Fast neutron intensities for energies between 1 and 10 MeV were found to be 1.7×10^{-3} neutrons cm^{-2} s^{-1} on the roof, increasing to 1.7×10^{-2} neutrons cm^{-2} s^{-1} at 700 g cm^2.

The interaction of cosmic rays with the atmosphere is not the only way in which cosmic rays can produce neutrons. Cosmic ray interaction with water and rock is clearly also possible. Neutron production rates in water typically vary from about 1.85×10^{-5} neutrons g^{-1} s^{-1} at sea level to 1.28×10^{-4} neutrons g^{-1} s^{-1} at a height of 3000 m, while the thermal neutron flux in the ground increases rapidly with shallow depth, being twice the surface value at a depth equivalent of about 10 g cm^{-2}. Similarly, cosmic ray interactions with the ground lead to an annual production of 800–1000 neutrons g^{-1} at sea level.

Generally, neutron production below the Earth's surface decreases rapidly with depth as the cosmic rays become absorbed. Below the zone of cosmic ray influence, neutron production takes place by α-particle interactions with the nuclei of light elements, such as O, Na, Mg, Al and Si, In fact, it is only elements for which Z is less than 16 that can theoretically produce neutrons by this means. Indeed, the only nuclides which are of importance are Al and Na (which have only one naturally occurring isotope), but the less abundant isotopes of O, Mg and Si, F, Be and B could be important in rocks in which these elements are concentrated (Feige *et al.*, 1968). The presence of target nuclei, however, is not the only control over neutron production. α-particles have a spectrum of energies which results in a spectrum of neutron production values. Over a decrease in α-particle energy from 8 to 4 MeV, neutron yields decrease by several orders of magnitude. Because the spectra of α-particle energies for the ^{238}U, ^{235}U and ^{232}Th decay series differ from one another, the neutron production rate varies with the U and Th content of rocks. Thus the composition of the rock is important in terms of the occurrence of both source nuclides and target nuclides. It is also important in that it controls the density of the material. The typical annual neutron production value for sand is about 0.835 neutrons (ppm U)$^{-1}$ and 0.38 neutrons (ppm Th)$^{-1}$. For limestone, the values are 0.64 neutrons (ppm U)$^{-1}$ and 0. 285 neutrons (ppm Th)$^{-1}$, and for granite about 1.57 neutrons (ppm U)$^{-1}$ and 0.7 neutrons (ppm Th)$^{-1}$. Particularly strong sources of neutrons can occur where U and Th occur in the same mineral as light elements such as F, Be and B. The relative neutron yield for F, for example, is over an order of magnitude greater than that for Al. It has also been suggested (Rankama, 1954) that γ-radiation can cause the production of neutrons from Be in the transition

$$^{9}_{4}Be \rightarrow {}^{8}_{4}Be + {}^{1}_{0}n,$$

with ^{8}Be decaying by α-particle emission which could lead to further neutron production from ^{9}Be, although the existence of ^{8}Be has not been proven.

Finally, it is worth noting that, although these neutron production rates are very low when compared with neutron production by cosmic rays, they are significant in zones not penetrated by cosmic rays.

Another method by which neutrons are produced in the ground is the spontaneous fission of nuclides with high atomic numbers. The most important of the nuclides which undergo spontaneous fission is ^{238}U, which has a spontaneous fission half-life of 8.19×10^{9} Ma, 2.2 neutrons being produced per fission event. The annual production by natural U attributable to ^{238}U is therefore 476 200 neutrons g^{-1}. ^{235}U also undergoes spontaneous fission, but the fission half-life is much greater (3.5×10^{11} Ma); although this produces 2.5 neutrons per fission event, the overall annual contribution to total production by ^{235}U is only 200 neutrons g^{-1}. Spontaneous fission of ^{234}U, which has a half-life of 2×10^{10} Ma, also contributes a tiny fraction of neutron production by natural U. The total number of neutrons emitted spontaneously by natural U is 0.016 neutrons g^{-1} s^{-1} or, in a year, approximately 500 000 neutrons g^{-1}. ^{232}Th, with a spontaneous fission half-life of greater than 10^{15} Ma, does not contribute any extra production by spontaneous fission.

5.2 NEUTRON-INDUCED TRANSFORMATIONS

Betweeen 20 and 35 % of the fission events of U in naturally occurring pitchblende ore is induced by neutrons. Although this range may appear quite large, it is actually very limited, bearing in mind the many factors that are operating. Neutron-induced fission in pitchblende is controlled, for example, by its age, as this affects both the U content (which decreases with time by decay) and the $^{235}U/^{238}U$ ratio (because these isotopes have different half-lives). The presence of other nuclides which could act as neutron absorbers also has a profound effect.

^{235}U undergoes fission by the action of thermal neutrons, but fast neutrons are needed to cause fission of ^{238}U. Any neutron, therefore, is likely to cause fission of ^{235}U, and the neutrons released by ^{235}U can, in turn, induce reactions in ^{238}U. The most likely of these to occur is

$$^{238}_{92}U + ^{1}_{0}n \rightarrow ^{239}_{92}U.$$

^{239}U then decays by β^--emission with a half-life of 23.5 min to ^{239}Np which, in turn, decays by β^--emission with a half-life of 2.35 days to the α-emitting ^{239}Pu. The decay process then continues as a chain (the ^{235}U decay series) until the stable ^{207}Pb end product is reached.

The high energy neutrons released in the spontaneous fission of ^{235}U can also cause the reaction

$$^{238}_{92}U + ^{1}_{0}n \rightarrow ^{237}_{92}U + 2^{1}_{0}n.$$

^{237}U decays by β^--emission with a half-life of 6.75 days to form ^{237}Np at the start of another decay series which ends in the stable product ^{209}Bi. ^{237}Np has a half-life of approximately 2.14×10^6 years; it also decays by spontaneous fission with a half-life of greater than 10^{12} Ma.

The simplest neutron-induced reaction is the formation of deuterium from hydrogen:

$$^{1}_{1}H + ^{1}_{0}n \rightarrow ^{2}_{1}H.$$

This occurs as the result of neutron absorption in all H-bearing materials, as discussed in Chapter 3. Another important nuclide to be produced in this way is tritium, by neutron-induced fission of $^{6}_{3}Li$, which is accompanied by the emission of α-particles:

$$^{6}_{3}Li + ^{1}_{0}n \rightarrow ^{3}_{1}H + ^{4}_{2}He.$$

The activity of $^{3}_{1}H$, produced at a constant rate P by neutron irradiation over time t, is given by

$$\lambda_3 N = P(1-e^{-\lambda t}),$$

where N is the number of $^{3}_{1}H$ atoms present after time t and λ_3 is the decay constant for $^{3}_{1}H$.

The production rate P is given by

$$P = Fn,$$

where n is the neutron production in the rock and F the probability that the thermal neutrons are absorbed by Li rather than other elements present:

$$F = \frac{\sigma_{Li} x_{Li}}{\sum_{1}^{92} \sigma_i x_i},$$

where σ_i and σ_{Li} are the neutron absorption cross-section values for all elements present and for Li, respectively, and x_i and x_{Li} are the abundances of all elements present and for Li, respectively (Zito *et al.*, 1980). A typical value of F for granites in which the Li concentration is 40 ppm is 0.034. Combined with the typical annual neutron production for a granite of 51 neutrons cm^{-3}, this gives an equilibrium value of 31 atoms of $^{3}_{1}H$ for every cubic centimetre of rock (Burgess *et al.*, 1982).

5.3 OKLO: A NATURAL FISSION REACTOR

In 1972, analysis of a U sample from Oklo in Gabon found that the proportion of ^{235}U was only 0.7171% instead of its normal abundance in naturally occurring U of 0.7202%. This difference is very significant and required explanation. Further investigation showed that supplies of ore depleted in ^{235}U had only begun in 1970. By mid-1972, about 700 000 kg of U were involved, with a total ^{235}U deficiency of about 700 kg. This is of more than academic interest, for ^{235}U is the main fuel material for nuclear reactors. Studies of individual samples also showed that some sections of the deposit contained only 0.44% of ^{235}U.

Although isotope fractionation is commonly observed in the geological environment, natural processes by which large quantities of ^{235}U are separated from ^{238}U are unknown. This separation is carried out artificially in U-enrichment plants but requires advanced technology. Moreover, the Oklo deposit showed interesting abundances of ^{235}U fission products, and it was concluded that the ^{235}U in the deposit had been depleted as a result of natural fission reactions. Interestingly, the sites of fission reactions were found to be localised within the deposit as a whole, six individual reactor zones being recognised (Fig. 5.1). An excellent review is given by Cowan (1976). It is the presence of the fission products together with the ^{235}U depletion that gives unequivocal proof that Nature herself provided the first nuclear fission reactor on Earth, rather than the nuclear physicists working with Enrico Fermi in 1942!

Among the fission products found at Oklo are isotopes of La, Ce, Pr, Nd, Eu, Sa, Gd, Y, Zr, Ru, Rh, Pd, Nb, Ag, Mo, I, Kr and Xe, but only about half the 30 or so fission product nuclides that would have formed are present. Metals with a valency of 1 or 2 are relatively soluble in water and have been leached away. Thus, Rb, Cs, Sr, Ba and Ca would originally have been present as well. However, because ^{90}Zr is not depleted, it follows that its parent ^{90}Sr, which decays with a half-life of about 30 years, was not removed from the deposit within that period of time. Migration of fission products in the Oklo system is discussed further in Chapter 12.

Because ^{142}Nd is not a direct product of fission of ^{235}U, or of the decay of fission products, its presence in the reactor zone ores can be used to indicate the original abundance of Nd in the ore and to determine the extent of the reactor activity by calculating the abundances of those isotopes of Nd which formed as a result of ^{235}U fission. However,

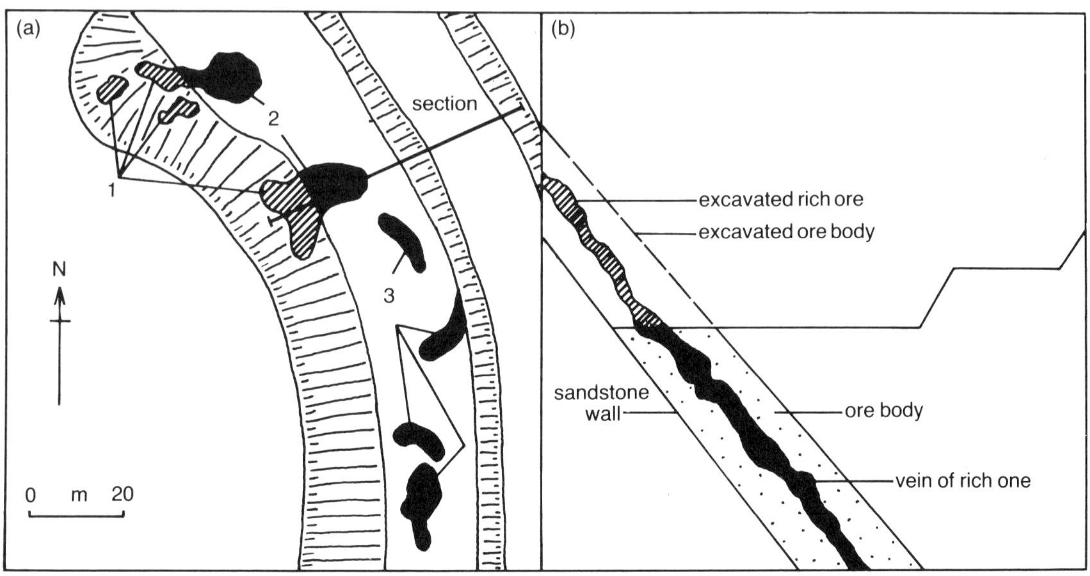

Fig. 5.1 — The Oklo reactors: (a) plan showing the positions of the reactor sites (solid, at and below ground level; diagonal shading, where excavated) and line of cross-section; hachures, quarry sides; 1, reactor sites above ground level projected to ground level; 2, reactor sites at gound level; 3, reactor sites below ground level projected to ground level); (b) representative cross-section. (After Cowan, 1976.)

for complete determination of the system, correction must also be made for conversion by neutron capture of ^{143}Nd to ^{144}Nd and of ^{145}Nd to ^{146}Nd, as during reactor operation the neutron flux would have been very high. When these corrections for natural background level and for neutron capture are made, the abundance ratios for the six Nd isotopes between ^{142}Nd and ^{150}Nd correspond precisely to those measured experimentally in modern reactors which involve reactions of the type believed to have operated at Oklo (Cowan, 1976). Confirmation of the reactor system is also provided by the presence in the reactor zone of ^{232}Th, which must have been produced by the decay of ^{236}U formed by neutron enrichment of ^{235}U. ^{232}Th is almost absent elsewhere in the deposit. Also present only in the reactor zones is ^{213}Bi, produced by the decay of ^{237}U formed by neutron reactions involving ^{238}U.

For any type of continuous nuclear fission reaction to be possible, a sustained flux of neutrons of reasonable intensity is required. On average, the fission of ^{235}U results in the prompt emission of 2.5 neutrons. One of these must be absorbed to produce fission in another nucleus, but the remaining 1.5 neutrons must be allowed to escape or be absorbed elsewhere. A chain reaction in U containing 0.72% ^{235}U is only possible under very special conditions in which ^{2}H is present as a moderator, as it is in the Canadian natural U (CANDU) reactors, or in the British MAGNOX reactors where the moderator is graphite. The alternative to a ^{2}H or graphite moderator is the presence of the^{1}U and a moderator in a specific geometric lattice. However, neither a ^{2}H moderator nor specific geometry and size ^{1}factors are likely to be encountered in nature, especially in six separate zones of one ore body. A reactor system is more likely to develop naturally by bringing a relatively pure mass of U together in such a way that its size is large compared with the distance that a neutron travels before it is captured (Cowan, 1976). If a suitable moderator isotope is present, the size requirement of the deposit is reduced. Water saturating the ore or the presence of graphite could provide a suitable moderator medium. Even given these circumstances, no natural reactor could operate today with U only containing 0.72% ^{235}U, as the neutron flux produced would be too low. However, as ^{235}U has a shorter half-life than ^{238}U, the ^{235}U content of a U ore would have been greater in the geological past. Thus, at the time of the origin of the Earth, ^{235}U accounted for about 17% of total U. The minimum abundance of ^{235}U for the operation of a natural reactor under plausible conditions is 1%, implying that such reactors could have operated until about 400 Ma ago. The Oklo deposit is in fact much older than 400 Ma, having a primary age of 1780 Ma (Lancelot *et al.*, 1975). At that time the ^{235}U content would have been about 3.2%.

Another constraint on the initiation of a reactor system is the abundance of U in the ore. This must average at least 10%, the ore having to be deposited in a zone at least 50 cm thick. Smaller quantities of U do not produce enough neutrons to sustain the reaction, and smaller dimensions allows too many neutrons to escape. Moreover, if the high energy neutrons are to be slowed to thermal energies by water acting as the moderator, about 6% water by weight is required for a reactor about 2000 Ma old. This last requirement is, however, not very stringent. In the sedimentary environment of the Oklo deposit, the rocks were probably originally saturated with groundwater, and so the system was overmoderated. When the reactor started, the heat produced would have caused evaporation and loss of some of this water and eventually the optimum conditions would have been obtained. Finally, the reactor system could only operate if those elements which strongly absorb neutrons (such as Li, B and the rare earths) were present in only minor quantities. Again, this is not a particularly stringent condition as large concentrations of these elements are comparatively rare at Oklo.

In general, the power generated by the reactor was controlled by the local environment. As this varied during the course of the reaction, so did the rate at which the reaction operated. In fact, the Oklo reactor appears to have operated at only very modest power levels.

The history of the rector system is likely to have been controlled not only by the consumption of the ^{235}U fuel and associated nuclear reactions but also by external changes in the groundwater system and in the shape of the ore body. Deposition of new ore material during or after reactor operation is another possibility that must be taken into consideration. The fact that changes occurred with time is shown because the calculated depletion in ^{235}U obtained from measurements of the ratio of ^{143}Nd to ^{144}Nd is about 40% greater than the observed depletion. As Cowan expressed it in 1976, 'knowing the amount of ^{235}U present initially and the amount remaining, we find that the amount consumed was greater than the difference'.

Among the possible causes of this, the most likely appears to be the presence of an additional nuclear reaction. The excess ^{235}U was probably produced by conversion of ^{238}U to ^{239}U as a result of neutron capture. The decay of ^{239}U then proceeded to ^{239}Np and ^{239}Pu, with ^{239}Pu decaying by α-particle emission to ^{235}U. This is a process which also takes place in modern artificial nuclear reactors. As a result, ^{235}U not only is consumed as a fuel but also is created from the ^{238}U present (Fig. 5.2).

Potentially, natural nuclear reactor sites could be found in many geological environments older than about 400 Ma but, to date, only the Oklo reactors have been recognised. The details of the geological setting of Oklo therefore not only are of interest to that specific area but also could provide a useful guideline to the identification of other reactor systems throughout the world. The Oklo deposit is situated about 500 km east–southeast of Libreville and about 100 km northwest of Franceville. It is located within Proterozoic sedimentary rocks overlying Archaean basement which was the source of the U. Erosion of these basement rocks led to the accumulation of U in solution. This was then transported into a deltaic environment where reducing conditions caused its precipitation in bituminous shales which are interbedded within the dominantly sandstone succession. The deposit originally contained about 0.5% U. Fault block displacement of the basement subsequently tilted the succession, about 5° to the east. Fractures were also created which allowed groundwater circulation through the U-rich horizons and redistribution of U to produce zones of high

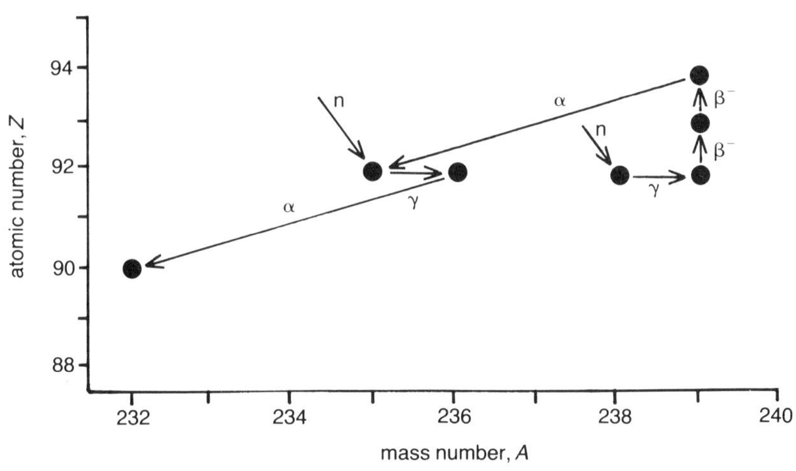

Fig. 5.2 — Neutron-induced non-fission transformations of ^{235}U and ^{238}U and the conversion of ^{238}U to ^{235}U at Oklo. (From 'A natural fission reactor', G. A. Cowan. Copyright 1976 by Scientific American, Inc. All rights reserved, redrawn with permission.)

concentration within the deposit. In these zones, it was possible for the chain reaction to be initiated when enough U had accumulated, that is, when concentrations exceeded 10%. Later disturbance of the system appears to have been produced by exclusion of oxygenated groundwater, possibly by deep burial. Cowan (1976) states that only within the last 4 Ma has the deposit been close to the Earth's surface again, but deoxygenated groundwaters can be indicative of shallow levels where sulphate-reducing bacteria are active.

The total energy released by the reactor zones has been estimated at 15 000 MW years (4.75×10^{17} J), representing the consumption of about 6100 kg of ^{235}U. About two-thirds of this fuel was primary ^{235}U, the remaining one-third resulting mainly from conversion of ^{238}U. There was also a small contribution from neutron-induced fission of ^{238}U and ^{239}Pu. The contribution from ^{239}Pu indicates that the reactor system persisted for longer than the 24 400 year half-life of that nuclide. It is considered that the life of the reactor was between 100 ka and 1 Ma. A similar estimate of the life of the reactor is obtained from calculation of the thermal output of the system. Because the reactor would not have continued to operate if moderator water was lost through boiling, it can be argued that the power output was limited to about 10–100 kW (10–100 kJ s^{-1}), indicating a life of about 150 ka, even at the maximum of the power range. It is interesting to note that throughout this period the reactor zones remained effectively closed to external disturbance, despite the presence of water-filled fractures (Fig. 5.3). Clearly, this stability is one of the main factors that allowed

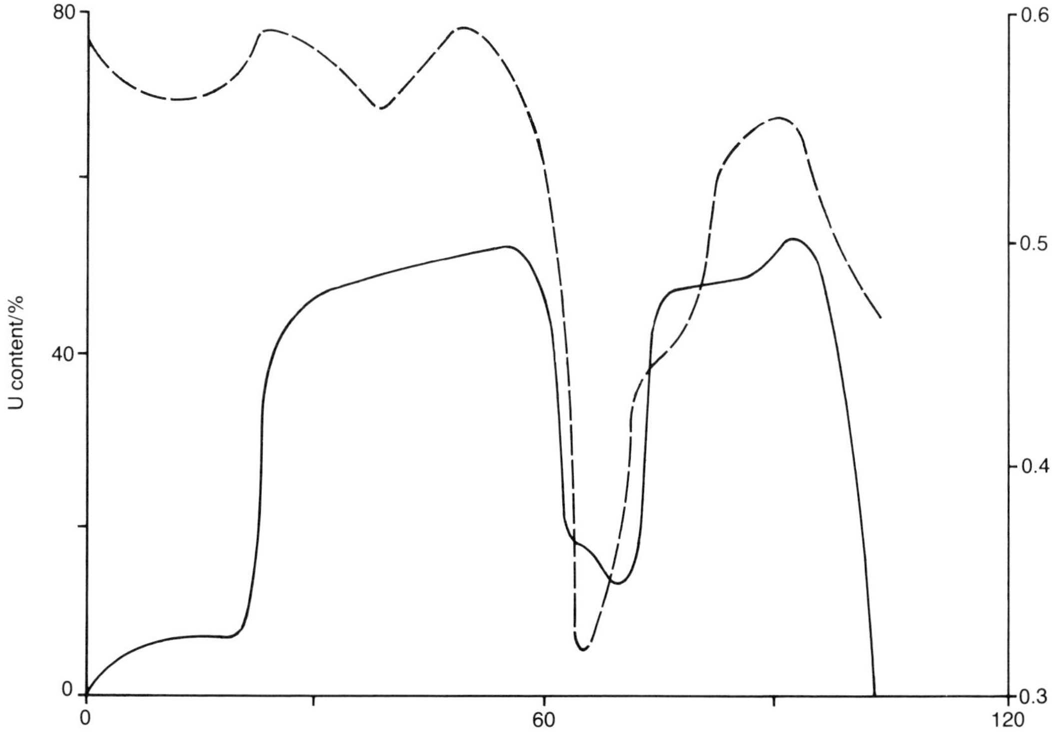

Fig. 5.3 — Variation in U content (solid line) and ^{235}U fraction (broken line) within the Oklo ore body. The zone of low values marks increased reactor efficiency in the vicinity of a water-filled crack. (From 'A natural fission reactor', G. A. Cowan. Copyright 1976 by Scientific American, Inc. All rights reserved, redrawn with permission.)

the reactor to continue, once the initial requirements for the establishment of a chain reaction had been satisfied.

The recognition of the existence of natural fission reactors at Oklo is considered to be significant to the modern problem of the disposal of artificial fission products in radioactive waste. This topic is considered more fully in Chapter 12, but it is interesting to note here that study of the Oklo system indicates that during its operation the only fission products that became released into the environment as a whole were ^{85}Kr and ^{137}Cs, with a little ^{90}Sr. Pu, which is one of the most worrying waste products of modern reactors, appears to have been effectively confined in the Oklo system, probably by the slaty rocks that occur within the sedimentary sequence. Nevertheless, migration of some elements which would normally be considered immobile is shown by the distribution of their daughter products. For example, loss of Sn is indicated by a marked decrease in the abundance of ^{126}Te compared with normal ^{235}U fission (Fig. 5.4).

The fact that U ores can posses $^{235}U/^{238}U$ ratios which deviate from normal also means that checks need to be run on the ores that are being worked at any time. Clearly, great security problems could be created if large quantities of ^{235}U appear to go missing during final enrichment. Variations of a few hundredths of a per cent are now known to be fairly common. For example, the ores from the Colorado Plateau are depleted in ^{235}U by about 0.03%. Although this could be the result of an Oklo-type reaction system, such minor changes can be more easily explained by isotope fractionation within the environment of ore formation. However, although not yet found, the probable existence of other natural reactor systems is considered to be high. There is even the possibility that some have been mined out without being noticed.

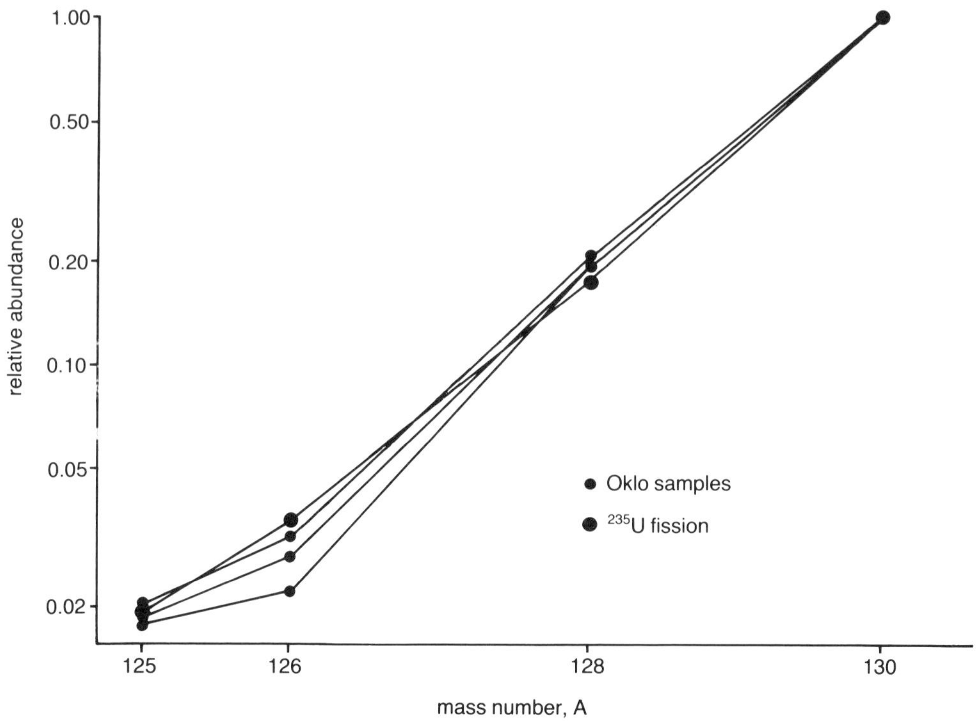

Fig. 5.4 — Comparison of isotopic abundances of Te produced from fission of ^{235}U and the distribution found within three samples from Oklo. (De Laeter *et al.*, 1980, redrawn with permission.)

Finally, speculation allows that, rather than the chance of initiating a reactor system decreasing over geological time by depletion of ^{235}U, it might have increased because of increasing neutron capture by ^{238}U. Thus, if a natural reactor was able to form as late as 800 Ma ago (when the relative abundance of ^{235}U was about 1%), it might have become a breeder reactor in which the ^{238}U component formed the main source of the fuel. The ^{235}U consumed in the reactor would have been replaced by the decay of ^{239}Pu formed from ^{238}U, and the whole zone would then show enrichment in ^{235}U. Various deposits in the Congo are of the appropriate age and, interestingly, appear to be slightly enriched in ^{235}U (Cowan, 1976).

BIBLIOGRAPHY

Bagge, E., and Skorka, S. (1962) Der ubergangeseffekt der ultrastrahlungs-neutronen an der grenzflache luft-wasser. *Z. Phys.* **152**, 34.

Bethe, H. A., Korff, S. A., and Placzek, G. (1940) On the interpretation of neutron measurements in cosmic radiation. *Phys. Rev.* **57**, 573.

Bodu, R., Bouzigues, H., Moin, N., and Pfiffelman, J. P. (1971) Sur l'existence d'anomalies isotopiques rencontrées l'uranium du Gabon. *C. R. Acad. Sci.* **275**, 1731.

Boudin, G., Blain, C., Hagemann, R., Kremer, M., Lucas, M., Merlivat, L., Molina, R., Nieff, G., Prost Marechal, F., Regnaud, F., and Roth, E. (1973) Quelques données nouvelles sur les réactions nucléaires en chaine que se sont produites dans le gisement d'Oklo. *C. R. Acad. Sci.* **275**, 2291.

Brookins, D. G. (1978) Eh–pH diagrams for elements from Z=40 to Z=52: application to the Oklo natural reactor. *Chem. Geol.* **23**, 324.

Brookins, D. G. (1978) Oklo reactor reanalyzed. *Geotimes* **23**, 26.

Brookins, D. G. (1978) Retention of transuranic and actinide elements and bismuth at the Oklo natural reactor, Gabon: application of Eh–pH diagrams. *Chem. Geol.* **23**, 309.

Brookins, D. G. (1979) Thermodynamic considerations underlying the migration of radionuclides in geomedia: Oklo and other examples. In McCarthy, G. M. (ed.), *The scientific basis for nuclear waste management,* Vol. 1. New York, Plenum.

Brookins, D. G. (1981) Alkali and alkaline earth studies at Oklo. In Moore, J. G. (ed.), *The scientific basis for nuclear waste management,* Vol. 3. New York, Plenum.

Brookins, D. G. (1983) Migration and retention of elements at the Oklo natural reactor. *Environ. Geol.* **4**, 201.

Brookins, D. G., Lee, M. J., Mukhopadhyay, B., and Bolivar, S. R. (1976) Search for fission produced Rb, Sr, Cs and Ba at Oklo. In *The Oklo phenomenon, IAEA Symposium* **204**, Vienna, International Atomic Energy Agency.

Brown, W. W. (1954) Cosmic-ray nuclear interactions in gases. *Phys. Rev.* **93**, 528,

Burgess, W. G., Edmunds, W. M., Andrews, J. N., Kay, R. L. F., and Lee, D. J. (1982) The origin and circulation of groundwater in the Carnmenellis granite: the hydrogeochemical evidence. *Rep. Inst. Geol. Sci. Lond.*

Chuukolyukov, Y. A., and Min, D. V. (1977) Fission chains as a characteristic of Precambrian uranium deposits. *Geochem. Int.* **14**, 483.

Cowan, G. A. (1976) A natural fission reactor. *Sci. Am.* **235**, 36.

Davis, W. O. (1950) Energy and distribution of cosmic-ray neutrons. *Phys. Rev.* **80**, 150.

De Laeter, J. R., Rosman, K. J. R., and Smith, C. L. (1980) The Oklo natural reactor: cumulative fission yields and retentivity of the symmetric mass region fission products. *Earth Planet. Sci. Lett.* **50**, 238.

Draganic, I. G., Draganic, Z. D., and Altiparmakov, D. (1983) Natural nuclear reactors and ionizing radiation in the Precambrian. *Precambrian Res.* **20**, 283.

Drozd, R. J., Hohenberg, C. M., and Morgan, C. J. (1974) Properties of the Oklo reactor from plutonium and uranium fission xenon. *International Meeting on Geochronology, Cosmochronology and Isotope Geology,* Paris.

Drozd, R. J., Hohenberg, C. M., and Morgan, C. J. (1974) Heavy rare gases from Rabbit Lake (Canada) and the Oklo mine (Gabon): natural spontaneous chain reactions in old uranium deposits. *Earth Planet. Sci. Lett.* **23**, 28.

Edge, R. D. (1959) The altitude dependence of atmospheric cosmic-ray neutrons and the slow neutron density below a water surface. *Nucl. Phys.* **12**, 182.

Feige, Y., Oltman, B. C., and Kastner, J. (1968) Production rates of neutrons in soils due to natural radioactivity. *J. Geophys. Res.* **73**, 3135.

Fermi, E. (1947) Elementary theory of the chain-reacting pile. *Science* **105**, 27.

Hendrick, L. D., and Edge, R. D. (1966) Cosmic ray neutrons near the Earth. *Phys. Rev.* 145, 1023.

Holliger, P., and Devillers, C. (1981) Contribution à l'étude de la témperature dans les réacteurs fossiles d'Oklo (Gabon) par la mesure du rapport isotopique du lutétium. *Earth Planet. Sci. Lett.* **52**, 76.

Holliger, P., Devillers, C., and Retali, G. (1978) Evaluation des témperatures neutroniques dans les zones réaction d'Oklo par l'étude des rapports isotopiques. In *Natural fission reactors, IAEA Tech. Commun.* **119**. Vienna, International Atomic Energy Agency.

International Atomic Energy Agency (1975) *The Oklo phenomenon, IAEA Symposium* **204,** Vienna, International Atomic Energy Agency.

International Atomic Energy Agency (1978) *Natural fission reactors, IAEA Tech. Commun.* **119**, Vienna, International Atomic Energy Agency.

Kuroda, P. K. (1960) Nuclear fission in the early history of the Earth. *Nature* **187**, 36.

Lancelot, J. R., Vitrac, A., and Allegre, C. J. (1975) The Oklo natural reactor: age and evolution studies by U–Pb and Rb–Sr systematics. *Earth Planet. Sci. Lett.* **25**, 189.

Lederer, C. M., and Shirley, V. S. (1978) *Table of isotopes*. New York, Wiley.

Lingenfelter, R. E., and Flamm, E. J. (1964) Solar neutrons and the Earth's radiation belts. *Science* **144**, 292.

Maeck, W. J., Apt, K. E., and Cowan G. A. (1978) The measurement of ruthenium in uranium ores and ^{238}U spontaneous fission yield. In *Natural fission reactors, IAEA Tech. Commun.* **119**, Vienna, International Atomic Energy Agency.

Naudet, R. (1974) Summary report on the Oklo phenomenon. *Fr. Cent. Energie Atom. Rep. Bull. Inf. Sci. Technol.* **193**, 7.

Naudet, R. (1978) Les reacteurs d'Oklo: cinq ans d'exploration du site. In *Natural fission reactors, IAEA Tech. Commun.* **119**, Vienna, International Atomic Energy Agency.

Newkirk, L. L. (1963) Calculation of low-energy neutron flux in the atmosphere by the Sn method. *J. Geophys. Res.* **68**, 1825.

Northcliffe, L. C., and Schilling, R. F. (1970) Ranges and stopping-power tables for heavy ions. In Way, K. (ed.), *Nuclear data tables*. New York, Academic Press.

Rankama, K. (1954) *Isotope geology*. London, Pergamon.

Segré, E. (1965) *Nuclei and particles*. New York, Benjamin.

Simpson, J. A. (1951) Neutrons produced in the atmosphere by cosmic radiation. *Phys. Rev.* **83**, 1175.

Simpson, J. A., and Uretz, R. B. (1953) Cosmic ray neutron production in elements as a function of latitude and altitude. *Phys. Rev.* **90**, 44.

Swetnick, M. J. (1954) Cosmic-ray neutrons in water at an altitude of 10 600 feet. *Phys. Rev.* **95**, 793.

Wetherill, G., and Inghram, M. G. (1953) *Spontaneous fission in uranium and thorium*. Washington, DC, National Research Council.

Yamashita, M., Stephens, L. D., and Patterson, H. W. (1966) Cosmic ray produced neutrons at ground level: neutron production rate and flux distribution. *J. Geophys. Res.* **71**, 3817.

Yayanos, A. A. (1983) Thermal neutrons could be a cause of biological extinctions. *Nature* **303**, 797.

Zito, R., Donahue, D. J., Davis, S. N., Bentley, H. W., and Fritz, P. (1980) Possible subsurface production of carbon-14. *Geophys. Res. Lett.* **7**, 235.

6

The geochemistry of uranium and thorium

6.1 URANIUM

6.1.1 Properties

U is a member of the actinide group of elements and, because of its position in the periodic table and electronegativity, shows strongly lithophile characteristics. It most commonly occurs as oxides, hydroxides, phosphates, carbonates, sulphates, arsenates, vanadates, molybdenates and silicates and tends to have affinities with hydrocarbon complexes. Ionic bonding is normally developed, but covalent and semi-metallic structures also occur. Although U can exist in four valency states, U^{3+}, U^{4+}, U^{5+} and U^{6+}, in natural materials it is generally present as either U^{4+} (the uranous form) or U^{6+} (the uranyl form). Although in the oxidation of U^{4+} to U^{6+} the intermediate state U^{5+} is formed, this is very rapidly oxidised and may only be preserved in restricted environments with a low redox potential. In strongly oxidising conditions, U^{6+} is the stable form (Fig. 6.1).

In terms of temperature, the geochemical behaviour of U is controlled by (1) high temperature processes and (2) low temperature processes. This division reflects the water content of the system, as high temperature environments are usually characterised by a lower water content than low temperature environments (Table 6.1). It is interesting to note that uraninite, the form of UO_2 characteristic of high temperature environments, displays a wide variety of substitution of U by Th and the rare earth elements, but pitchblende, the low temperature form of UO_2, usually contains much lower concentrations of these elements.

6.1.2 U in igneous rocks

In comagmatic suites of igneous rocks, U appears to follow closely the usual differentiation sequence, the highest concentrations normally occurring in the youngest, most potassic and most silica-rich members (Kimberley, 1978). The high temperature metamorphism and partial melting of sialic crust that leads to the formation of granites also appears to mobilise U, which becomes concentrated in pegmatites. There is often an associated increase in Th, Zr, Hf, rare earth elements, Nb and Ta. Among less silicic rocks, some kimberlites and lamprophyres may be enriched in U. As high U levels are often accompanied by a high alkali content, alkali syenites and carbonatites are often rich in U. A detailed review is given by Boyle (1982).

In high temperature processes, U may crystallise directly from silicate magmas (as uraninite, for example). Other accessory minerals, such as allanite, monazite, apatite, sphene, zircon, xenotime, thorite, thorianite, euxenite, pyrochlore and brannerite, often

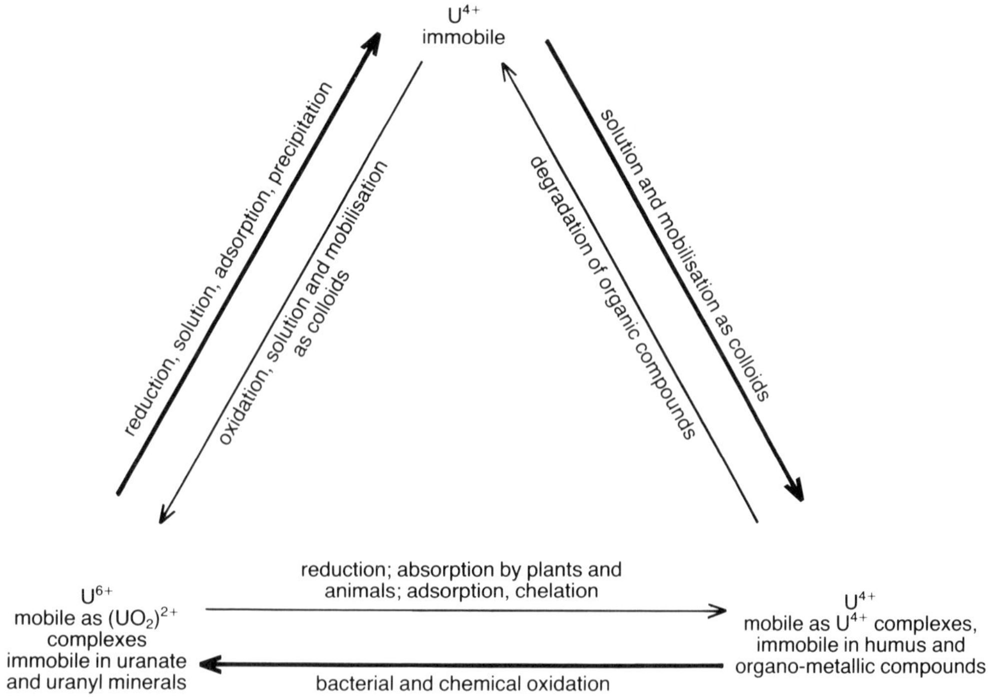

Fig. 6.1 — Simplified geochemical cycle of U. (After Boyle, 1982.)

intimately associated with biotite, may also contain U. Granitic rocks rich in U (and Th) are sometimes referred to as 'hot granites' on account of their high rates of heat production. These granites normally contain more than 15 ppm U and may sometimes have concentrations as great as 120 ppm. They are usually post-orogenic granites which were emplaced at high crustal level and typically display a porphyritic texture (southwest England), but some anorogenic granites also have high U contents (Pikes Peak, Colorado). In granite pegmatites associated with some high grade metamorphic terrains (usually near the contacts with granitic zones), the U content may be as high as 800 ppm, but in other areas they are barren (northern Scotland). The principal radioactive minerals in granite are uraninite, thorite, uranothorite, betafite, monazite, sphene, pyrochlore, allanite, epidote, yttrofluorite, be present. Nepheline syenite pegmatites also contain appreciable levels of U and contain Th-, rare earth element- and Zr-bearing minerals. A high U content is commonly found in other peralkaline igneous rocks. Like uraniferous graites, these are usually post-orogenic instrusions, but they also commonly display zoned or layered structures. The U is usually present in minerals such as steenstrupine, monazite, loparite, thorite and ziron. In carbonatities, the U level may reach 50 ppm, and minerals such as pyrochlore, monazite, perovskite, sphene, ziron and baddeleyite may occur. High levels of rare earth elements are also common. Volcanic rocks are usually U poor because of leaching, but alkali-rich lavas such as rhyolite, trachyte, latite and phonolite may contain 20–25 ppm U.

Unfortunately, the thermodynamic behaviour of U in silicate melts is not well understood xenotime and zircon. In addition, thucolite may

Table 6.1 — Classification of U deposits.

Type	Associations	Minerals	Examples
(1) Petrogenetic			
Igneous	Alkali syenite or carbonatite	Uranothorianite, pyrochlore, betafite, perovskite	Prairie Lake, Ontario, Canada Pocos de Caldas, Brazil Illimaussaq, Greenland
Metamorphic	Pegmatite or skarn	Uraninite, uranothorite, betafite	Bancroft, Ontario, Canada Rössing, Namibia
Placer	Pyritic quartz pebble conglomerate	Uraninite, brannerite	Elliot Lake, Ontario, Canada Witwatersrand, South Africa
(2) Hydrogenetic			
Hydrothermal	Volcanogenic	Uraninite, uranothorite	Rexspar, British Colombia, Canada Ben Lomond, Queensland, Australia
Hydrothermal	Hydrothermal veins	Uraninite, brannerite, thorite, allanite	Bokan Mountain, Alaska, USA Beaverlodge, Saskatchewan, Canada
Syngenetic	Shale, phosphorite evaporite, limestone, calcrete	Pitchblende, carnotite	Ranstad, Sweden Florida, U.S.A.
Epigenetic	Sandstone (tabular, roll front, etc.)	Pitchblende, coffinite, carnotite	Colorado Plateau, USA Wyoming, U.S.A.
Epigenetic	Channel conglomerate (basal type)	Pitchblende, coffinite, autunite	Ningyo Toge, Japan Blizzard, British Columbia, Canada
Epigenetic	Supergene	Gummite, uranophane, carnotite, coffinite	Rössing, Namibia Pocos de Caldas, Brazil (upgrading existing deposits)

(After Kimberley, 1978.)

at the present time, but there is evidence of the cogenetic crystallisation of uraninite, monazite and zircon with biotites in the Hercynian granites of southwest England. It has been argued that U may be concentrated in a gaseous phase evolved from shallow igneous intrusions with deposition of U occurring in brecciated zones (Tilsley, 1978). The common association of uraninite, monazite, fluorite and apatite may indicate that mobilisation of U has occurred as $(UO_2F)^+$ or $(UO_2H_2PO_4)^+$, as well as carbonate and silicate complexes.

6.1.3 U in metamorphic rocks

Apart from some orthogneisses which have inherited high U concentrations in conditions of low U mobility, U in regionally metamorphosed rocks is comparatively rare. Indeed, even where U is found in any abundance it is usually associated with segregations. The question therefore arises whether these rocks were ever once uniformly rich in U. In the high grade gneisses of the Grenville Province of North America, for example, calcite–biotite–apatite segregations contain uraninite, uranothorite, allanite, betafite, zircon and fergusonite. High levels of rare earth elements, Nb, Ta, Zr, Hf and P also occur. Here the deposits probably originated through the reworking of a syngenetic precursor. Similarly, where uraniferous black shales have been metamorphosed, the resulting graphitic schists may show concentrations of uraninite in the graphitic layers.

Other metamorphic rocks in which U may be concentrated are skarns and hornfelses in contact aureols. Uraniferous rocks of this type also probably mainly owe their origin to a syngenetic U-bearing precursor, and so there often appears to have been some stratigraphical control exercised in their development. The U mineralogy of the skarns typically features uraninite, thorite, uranothorite, uranothorianite, pyrochlore and monazite, and the U content of the rock may reach 100 ppm. Although they appear to be mainly reworked syngenetic material, some degree of metasomatic exchange has probably taken place in the formation of these deposits (Boyle, 1982).

In contrast, regional metamorphism is believed to have included U mobilisation in a hydrous phase. Certainly, there is evidence that some U ores have precipitated from high temperature solutions, possibly of the order of 500 °C (definitely within the range 340–350 °C and 700–800 b). This is considered together with other hydrothermal deposits later in this chapter.

Finally, we may note the occurrence of disseminated U deposits associated with albitised zones in granites, gneisses, migmatites and Fe formations (Boyle, 1982). Where granites and granitic gneisses have been affected in this way, they are sometimes known as 'episyenites'. It is generally considered that these deposits have a metasomatic origin. The French episyenites are believed to have formed by dequartzification of granite in an extremely fractured environment (Organisation for Economic Cooperation and Development, 1983).

6.1.4 General characteristics of U in sedimentary rocks

Detrital uraninite occurs in some early Pre-Cambrian rocks, probably indicating a low O_2 content in the atmosphere at the time of their formation. Associated with uraninite may be brannerite. Although possibly having a detrital origin (Andreoli *et al.*, 1985), the brannerite is generally considered to have formed by sorption of U on titaniferous magnetite during diagenesis. Metamorphic reworking has always affected these deposits to a greater or lesser extent.

Black shales and schists which are rich in C are also often rich in syngenetic U, containing up to 400 ppm. Two types have been recognised (Boyle, 1982). These are (1) the kerogen shales, in which uraniferous organic material forms a coating around mineral grains and organic particles, and (2) the sapropel shales, in which the uraniferous organic material occurs as disseminated masses throughout the rock matrix. In neither case has any definite U mineral been identified. U may be present in these shales as a urano-organic complex. Alternatively, the U may simply be sorbed onto the surface of organic and inorganic materials. The U in these rocks, and in phosphatic rocks (another major sedimentary U host), owes its origin to precipitation from hydrous solution.

6.1.5 The mobility of U

Leaching of U

It has been proposed that seven factors are likely to influence the U concentration in natural surface and groundwaters (Langmuir, 1978a). These are as follows

(1) The U content of the source rocks, sediments or soils, and the ease with which U may be leached.
(2) The proximity of the water to the U source.
(3) The degree of hydraulic isolation of the water from dilution by fresher surface or groundwaters.
(4) Climatic effects, seasonal variability and the effects of evapotranspiration.
(5) The pH and oxidation state of the water.
(6) The concentration of suitable complexing agents which can increase the solubility of U, or of metals which will lead to the formation of insoluble U minerals; particularly important are carbonate, phosphate, vanadate, fluoride, sulphate, silicate, Ca^{2+} and K^+ ions.
(7) The presence of highly sorptive materials such as organic matter, oxyhydroxides of Fe^{3+}, Mn and Ti, and clays.

The leachability of U from its source rock clearly depends upon its mineralogical location in the rock. U held in soluble minerals such as uraninite, or in leachable intergranular positions, will be easily removed, while any U held in minerals such as zircon, which are resistant to leaching, will remain in the rock. Suitable chemical conditions must also be present if the U is to enter solution. U solubility is greatly favoured if it is converted to its hexavalent uranyl form, UO_2^{2+} (the uranyl ion). The continued solution and transport of U, however, require the presence either of a suitable Eh–pH environment in which UO_2^{2+} is stable or of complexing agents (Fig. 6.2). Maximum solubility will only be achieved if there is a uranyl complex which is stable in the prevailing Eh–pH conditions. Although the rare mobility of U in a peat bog environment has been attributed by Halbach *et al.* (1980) to the formation of a uranyl fulvate complex which is stable at pH 5.8 of the waters, under normal

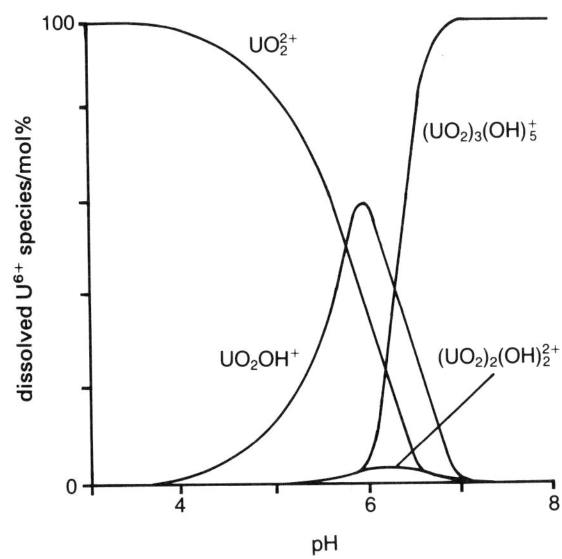

Fig. 6.2 — Distribution of uranyl–hydroxy complexes at 25 °C, for a total dissolved U concentration of 10^{-8} M. (Redrawn, with permission from Langmuir (1978b). Copyright 1978 Pergamon Press.)

circumstances it is the formation of carbonate complexes that most commonly increases U solubility. In groundwater, a rise in P_{CO_2} from atmospheric levels to 100 b increases the solubility by a factor of more than 10^3 for Eh values above −0.05 V. Phosphate can also increase the solubility of uraninite by a factor of 10^3 for Eh values greater than −0.05 V when present in groundwaters at a typical 0.1 ppm. The stability field for U^{4+} in solution is very restricted.

It should also be mentioned that the ideal composition of pitchblende and uraninite, UO_2, is in fact never attained as some of the U^{4+} is always oxidised to U^{6+}. This is probably due to auto-oxidation. In this process the decay of ^{238}U to ^{222}Rn releases free atomic O in the lattice of the mineral, which then reacts with the uraninite or pitchblende. The higher internal temperature maintained by radioactive decay may also aid the process of auto-oxidation.

Hydrogenetic deposits

Chemical or biological precipitation of U from aqueous solutions or sorption of U onto organic or inorganic particles account for most situations in which U concentrations are significantly greater than the average crustal abundance. Deposits from aqueous solution may occur within any type of host rock, provided that a suitable permeability is present, and may occur at any time during the history of the host rock. Thus, in sedimentary rocks, U may be syn-depositional, diagenetic or post-lithification. In igneous rocks, the aqueous solutions may be of syngenetic origin and may be associated with the late stages of crystallisation; connate water in the host rocks may participate in late-magmatic processes; meteoric waters are likely to be involved in groundwater circulation systems at any time in the post-magmatic history. Metamorphic processes may result in the liberation of water from hydrous minerals but, as in other host rocks, the availability of a suitable permeability can allow the passage of aqueous solutions at any time. In any of these environments, the temperature at which the system operates can have a very wide range from the ambient temperature at the ground surface to levels of the order of 500 °C. Strictly speaking, all these processes may be referred to as hydrothermal, simply because they all involve warm water.

The behaviour of U in aqueous solution at relatively low temperatures has been very thoroughly studied (Fig. 6.3). It has been shown by Langmuir (1978b) that the reaction

$$U^{4+} + 2H_2O = UO_2^{2+} + 4H^+ + 2e^-,$$

which has a redox potential of 0.273±0.005 V at 25 °C, is particularly important. In environments which are O_2 deficient, and in which pH values are less than 7, U^{4+} is much less stable than U^{6+}, while U^{5+} (as UO_2^+) has a fairly large stability field. In such environments, U^{4+} species tend to form unstable precipitates of uraninite or coffinite. The greater stability of U^{6+} species and the existence of soluble U^{5+} species provide U with mobility under a wide range of natural conditions. This mobility, moreover, is enhanced by the tendency of U to form complex ions with carbonate, hydroxide, phosphate, fluoride, sulphate and possibly silicate ions. Organic complexes are also important. Uranous complexes are more stable than uranyl complexes under different Eh–pH conditions.

Uranous complexes

Uranous hydroxide complexes have a low mobility at most pH values in natural water and occur at concentrations of less than 0.05 ppb. As the pH rises above 7, the formation of $U(OH)_5^-$ allows this U concentration to increase. At other pH values, the behaviour of the complex varies according to the availability of suitable ligands. In reducing conditions with pH less than 4, uranous fluoride complexes greatly enhance the solubility of U. Fluoride levels need not be high for these complexes to form, and the process can operate in most natural waters. However, if precipitation of fluorite occurs, the reduced mobility of U^{4+} leads to the

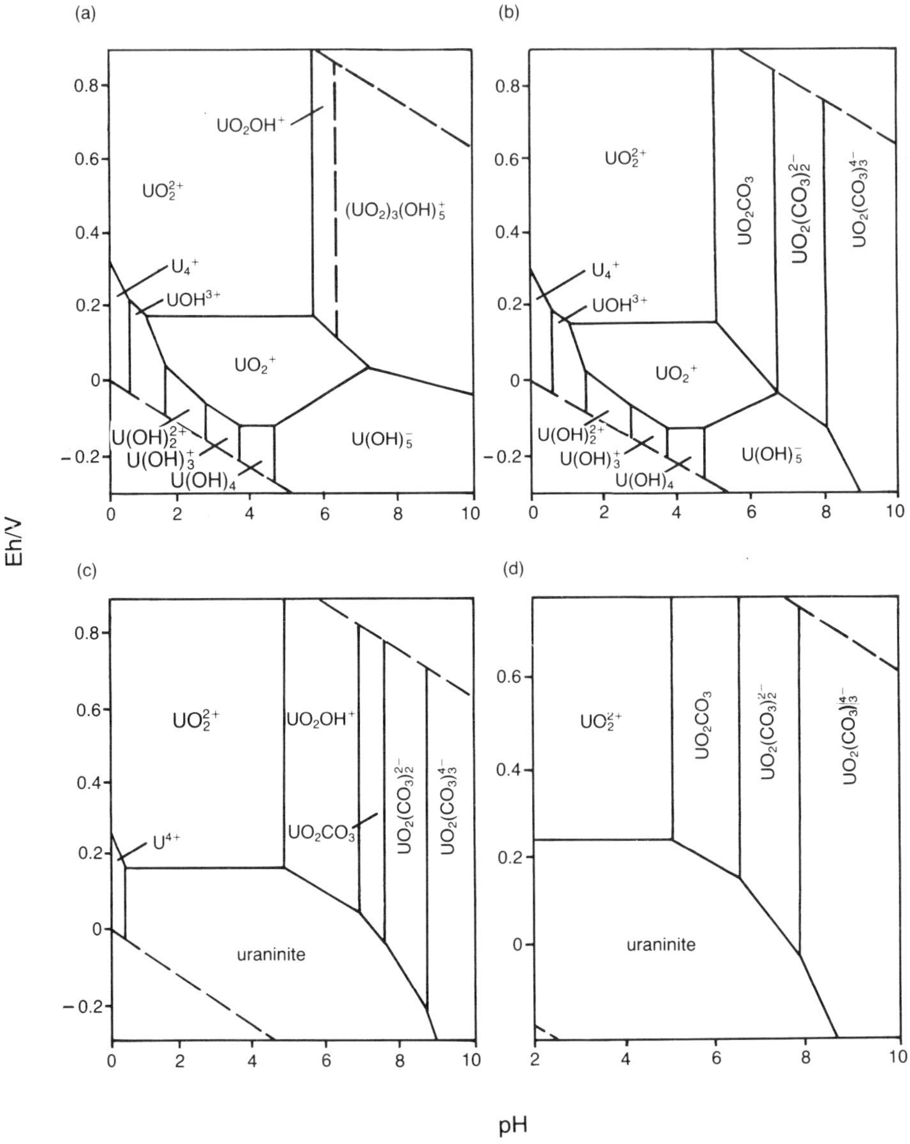

Fig. 6.3 — Stability diagram for U species: (a) open system, $P_{CO_2} = 10^{1.5}$Pa, 25 °C, total dissolved $U = 10^{-6}$ M;(b) closed system, $P_{CO_2} = 10^3$Pa, 25 °C, total dissolved $U = 10^{-6}$ M;(c) open system, $P_{CO_2} = 10^{1.5}$Pa, 25 °C, total dissolved $U = 10^{-8}$ M;(d) closed system, $P_{CO_2} = 10^3$Pa, 25 °C, total dissolved $U = 10^{-6}$ M. (Bouda, 1985, redrawn with permission.)

precipitation of uraninite or coffinite. Uranous chloride, sulphate and phosphate complexes have little effect on the solubility of U at normal concentrations of these ligands in groundwaters at any pH value. Thus the formation of fluoride complexes at pH less than about 3, and hydroxyl complexes at pH greater than about 8, is the only way of significantly increasing uranous concentrations. The solubility of UO_2 reaches a maximum at about 260 °C at a pressure of 750 b and an O_2 fugacity buffered by equilibrium between UO_2 and hydrous or anhydrous UO_3 (Kimberley, 1978). It is interesting to note that, in pure water, uraninite is about eight times as soluble at 100 °C as it is at 25 °C.

Uranyl complexes

The most important uranyl hydroxyl complexes are $UO_2(OH)^+$, $(UO_2)_2(OH)_2^{2+}$ and $(UO_2)_3(OH)_5^+$. The latter two complexes become more important as the dissolved U concentration increases, but the effect of increasing temperature is to increase the importance of UO_2^{2+} and $UO_2(OH)^+$. Very important in determining the solubility of U, however, are the uranyl complexes with fluoride, phosphate, carbonate and sulphate. The uranyl ion itself is only stable under conditions of low pH, being unstable when the pH exceeds 5. Uranyl fluoride complexes are also stable in this pH range as is the uranyl sulphate complex. In groundwater with sulphate levels of 100 ppm, however, UO_2SO_4 may be a significant species up to pH 7. Around pH 6, the silica complex $UO_2SiO(OH)_3^+$ is important (Fig. 6.4).

The most important complexes, however, are the uranyl phosphate and uranyl carbonate complexes. $UO_2(HPO_4)_2^{2-}$ is so stable under normal groundwater concentrations of 0.1 ppm phosphate that it predominates over all other complexes from pH 4 to pH 10. However, in the presence of a P_{CO_2} of 10^{-2} b at 25 °C, uranyl carbonate complexes dominate the solubility of U down to pH 5. As P_{CO_2} falls, this dominance becomes restricted to higher pH levels; so at atmospheric P_{CO_2} ($10^{-3.5}$ b) the limit is pH 5.8. At higher temperatures, the solubility of the uranyl carbonate complexes decreases so that at 100 °C and a P_{CO_2} of 10^{-2} b they are minor species at all pH values. If U is present in solution as a uranyl carbonate complex, increasing temperature is likely to induce the precipitation of uraninite.

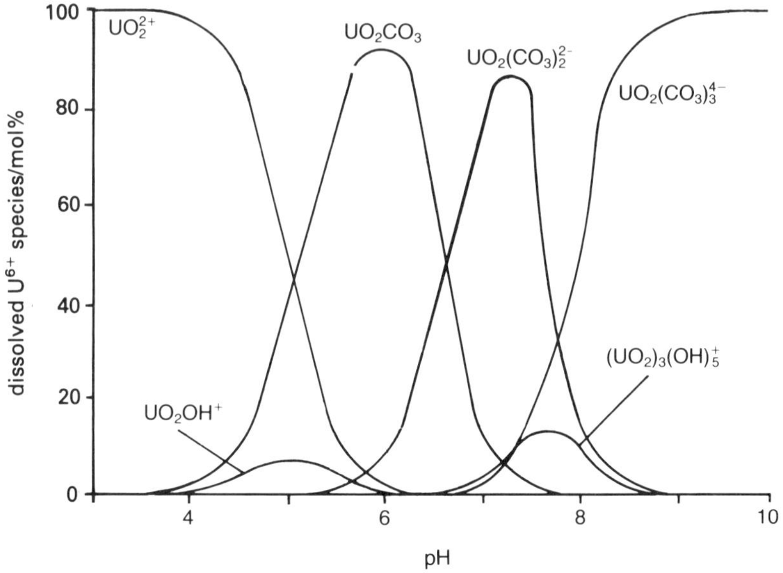

Fig. 6.4 — Distribution of uranyl–hydroxy and carbonate complexes at 25 °C with $P_{CO_2} = 10^3$ Pa and a total U concentration of 10^{-8} M. (Redrawn, with permission, from Langmuir (1978b.) Copyright 1978 Pergamon Press.)

Precipitation of uranous minerals

Uraninite and coffinite are the most common of the uranous minerals. In some Japanese deposits the uranous phosphate ningyoite is found in sufficient quantities to be an ore mineral. Brannerite and davidite also contain U mainly in the uranous state. Uraninite can precipitate from waters which contain only very small amounts of U in solution, over a wide range of pH values at moderate to low redox potential. Reduction of mobile uranyl species, with the subsequent precipitation of uraninite or coffinite, may reflect changes in the oxidation state of more abundant elements in solution, particularly Fe, S or C. Oxidation of HS^- to SO_4^{2-} or Fe^{2+} to Fe^{3+}, for example, would be capable of reducing U^{6+} to U^{4+} and thus lead to the precipitation of UO_2. In this context, the association of uraninite and haematite (ferric oxide) is of interest. In sedimentary U deposits, these two minerals are mutually exclusive but, in vein-type U deposits, they occur together. This shows that at relatively low temperatures the formation of haematite, which takes place by dehydration of ferric oxyhydroxide precipitates, occurs under conditions where the O_2 content is sufficient to keep U in solution. At higher temperatures, however, direct precipitation of haematite can occur, and this leads to a reduction in O and the precipitation of uraninite together with haematite.

At relatively low temperatures, the precipitation of U from aqueous solution is initially as an amorphous mass of oxyhydroxide. It is probably from this precipitate that pitchblende forms by crystallisation and dehydration. Precipitation may occur directly or, because it has colloidal properties, it may be coagulated as a gel.

Colloidal uranous oxyhydroxide has a positive charge, and its formation is thus often accompanied by the simultaneous precipitation of anions and negatively charged colloids, oxides of Si, V and Mn, humic colloids, etc. The botryoidal form of pitchblende and its syneresis cracks are probaly due to the dehydration process (Boyle, 1982). Interaction with SiO_2 in solution (where the SiO_2 concentration is greater than 60 ppm) gives rise to coffinite.

Because of its large size and high charge density, U^{4+} is strongly attracted by a large number of inorganic and organic materials. This leads to the formation of uranous hydrocarbon assemblages of the amorphous gel-like type, to U-bearing Fe and Mn oxides and to U on clay minerals.

Precipitation of uranyl minerals

Uranyl minerals are much more varied in their composition than uranous minerals. They generally originate by alteration of the uranous form in nearby minerals or in solution. The assemblage of mineral species found in any one locality appears to depend strongly upon the presence of other minerals in solution and of precipitating ligands such as vanadate, phosphate, silicate and arsenate. Carnotite and tyuyamunite are the least soluble uranyl minerals, even where V concentrations are low; autunite can be formed when phosphate is present and V practically absent; uranophane is present only in SiO_2-rich environments which are deficient in vanadate and phosphate.

The uranyl ion, UO_2^{2+}, forms a large number of soluble complexes, the occurrence of each depending upon the pH of the system (Fig. 6.5). Those of most interest are

$$UO_2H_2VO_4^-,\quad UO_2HVO_4^{2-}\quad UO_2F^+,\quad UO_2Cl^+\quad UO_2NO_4^+,\quad UO_2SO_4,\quad UO_2(SO_4)_2^{2-},\quad UO_2H_2PO_4^+,\quad UO_2H_3PO_3^{2+},\quad UO_2(HPO_4)_2^{2-},\quad UO_2(CO_3)_2^{2-} \text{ and } UO_2(CO_3)_3^{4-}.$$

In general, the chloride, fluoride and sulphate complexes are stable under low pH conditions, phosphate complexes at neutral pH, and the hydroxide and carbonate complexes under quite a wide range of acid and alkaline conditions (pH 3–10). Among the U–organic complexes which are readily formed are uranyl citrate and tartrate, and uranyl–humic complexes (Boyle, 1982).

The most important uranyl minerals, carnotite, tyuyamunite, autunite and urano-

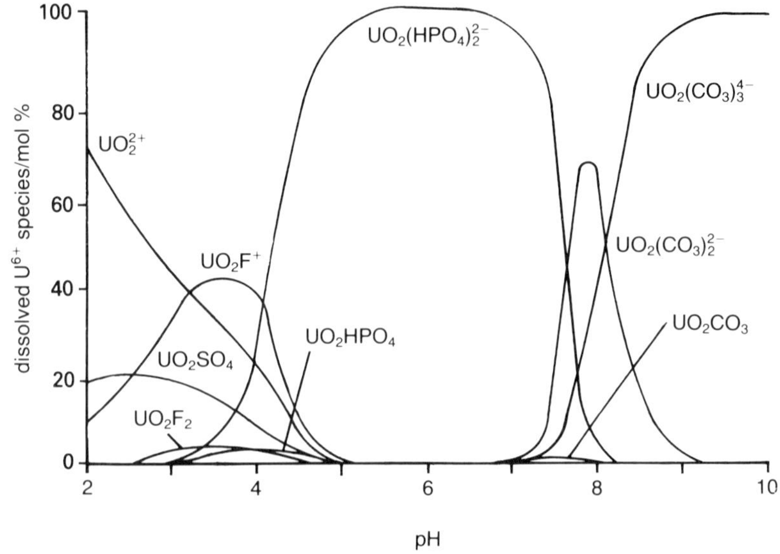

Fig. 6.5 — Distribution of uranyl complexes in the presence of other ions at 25 °C with a total U concentration of 10^{-8} M, $\Sigma F = 0.3$ ppm, $\Sigma Cl = 10$ ppm, $\Sigma SO_4 = 100$ ppm, $\Sigma PO_4 = 0.1$ ppm and $\Sigma SiO_2 = 30$ ppm. (Redrawn, with permission, from Tripathi (1979). Copyright 1979 Pergamon Press).

phane, are least soluble in the pH range 5.0–8.5. Autunite and uranophane are at least 10^2 times more soluble than carnotite and tyuyamunite and rarely precipitate from groundwater except near U-rich source rocks. In fact, carnotite and tyuyamunite are so insoluble that groundwater with only 1–2 ppb can be considered saturated. Solubilities increase above pH 8.0–8.5 by the formation of uranyl carbonate complexes, and below pH 5 by formation of UO_2^{2+}, uranyl phosphate, uranyl fluoride or other uranyl cations.

While the presence of organic acids in solution may aid U mobility, the usual role of organic matter is to remove U from solution. Experimental work has established that carbonaceous material is able to extract U from aqueous solution very efficiently (Moore, 1954). Sub-bituminous coal is most effective, but lignite, peat and canneloid coal are also extremely efficient. The process of extraction leads to the formation of metallo-organic complexes and has been found to be irreversible. As with uranous ions, adsorption of uranyl ions onto organic and inorganic materials is aided by their large size and high charge density. The pH range for maximum adsorption of U by organic matter is 3.5–6.0. Inorganic matter appears to provide better adsorption conditions in the pH range 4.5–7.5. Unfortunately, little is known about the mechanisms involved in the removal of U from solution by organic complexing. It is generally thought, however, that an initial adsorption stage soon gives way to ion exchange in which U becomes bonded through O to carbonyl groups or that U may replace H in carboxyl or other groups (Boyle, 1982).

Among the inorganic materials noted for adsorption of U are oxyhydroxides of Fe, Ti and Mn, phosphate minerals and clay minerals. Adsorption onto Fe oxyhydroxides is particularly effective in the pH range 5–8 but decreases in the presence of CO_2 above pH 8. The enrichment factor for this adsorption process (the ratio of U in the sorbant to U in solution) is of the order of 10^6. In the presence of complexing agents such as carbonate, phosphate, sulphate or fluoride, sorption may be inhibited. U may also be removed from solution, sometimes with Pb and Zn, by coprecipitation with Fe and Mn oxides.

The adsorption of U onto various clay minerals has been investigated experimentally (Borovec, 1981). The results show that uranyl ions and their hydroxo complexes are effectively adsorbed by kaolinite, illite and montmorillonite suspended in solutions at pH 6 and 20 °C. However, impermeable layers of clay do not permit penetration of U-bearing solutions, and the surface area available for sorption is thus restricted. Adsorption of U by clays is of minor importance when compared with adsorption by organic matter and oxyhydroxides of Fe and Mn.

Once formed, hydrous uranyl oxide is very reactive and combines with a wide variety of cations, particularly Cu, Ba, Pb, Bi, Ca, K and Na. These form minerals such as vandenbergite, billietite, clarkeite, curite, etc. Reaction with complex ions such as phosphate, molybdate, vanadate, sulphate, arsenate, etc., gives rise to minerals such as torbernite, autunite, carnotite, zeunerite, zippeite, uranophane and schroeckingerite.

6.1.6 Vein mineralisation

Two main types of U-bearing vein mineralisation have been recognised by Boyle (1982):

(1) Simple, in which the mineral assemblage is, in approximately paragenetic sequence, carbonate (dolomite or calcite), brannerite, chlorite and pitchblende; some later sulphides, arsenides and native metals may occur in small amounts.
(2) Complex, in which the non-sulphide paragenetic sequence is followed by quartz, carbonate and pitchblende with abundant sulphides and native metals.

The difference between simple and complex veins probably reflects differences in the history of remobilisation and in the composition of available source rocks. The common association of U with carbonate suggests that carbonate complexes are the main transporting agent for U. This is in agreement with the thermodynamic data.

In the majority of vein deposits, the water is of meteoric origin, although connate water has clearly played a role is some deposits in sedimentary host rocks. Magmatic waters are often associated with strata-bound or complex sheeted vein systems associated with intrusions. In Canada, vein-type deposits are sub-divided into unconformity-controlled veins, which are closely associated with stable peneplaned surfaces, and classical veins, which exhibit strong structural control. Unconformity-type deposits have been explained by some researchers as mineralisation arising from magmatic fluids. However, the mineralogical and textural characteristics of U deposits seen today may not reflect their original form, and a sedimentary origin with subsequent recrystallisation during deep burial is one alternative explanation. As U is extremely mobile under various chemical and physical environmental conditions, reworking or recrystallisation of deposits is to be expected at any time after their formation. In fact, it has even been suggested by Barbier (1974) that the vein-type mineralisation of the Hercynian province of western Europe may be due to the removal of U held in uraninite-bearing granites during their exposure to Permian weathering, and its subsequent redistribution by meteoric waters. The currently favoured explanation for the unconformity-controlled deposits is somwhat akin to this, in that U in circulating shallow groundwaters with a high Eh is thought to meet low Eh groundwaters issuing from the basement (Sibbald, 1985).

It should be noted that, in any system of flowing groundwater, precipitation of minerals is just as likely by descending as by ascending water. Leaching of U from the Hercynian granites of southwest England accompanied the kaolinisation process which led to the formation of china clay. This is considered to be a product of groundwater circulation in which the thermal draw-down of meteoric water carrying CO_2 caused alteration extending to depths of

2000 m. The dissolved and transported U then appears to have been precipitated in the zones of upwelling (Ball *et al.*, 1979).

6.1.7 Sediment-hosted mineralisation

Sandstone type

The likelihood of repeated U mobilisation noted for vein-type deposits is even greater when dealing with sandstone deposits. Apart from the possible detrital accumulation of uraninite under anoxic atmospheric conditions, these U deposits are usually found at or near oxidation–reduction interfaces in the sedimentary succession. This most commonly arises from the introduction of oxygenated groundwaters, generally under subareal (that is, atmospheric) conditions, which leads to leaching of U and other metals, including V and Mo, from the succession. In some cases, V may also be brought in from elsewhere. Local reducing conditions, if present, may be swamped initially by these oxidising groundwaters. Thus, dark-coloured sandstones can become reddened through the *in situ* oxidation of Fe-bearing minerals and the formation of ferric oxyhydroxides, which are the precursor to haematite. Carbonaceous material in the succession may be destroyed, and pyrite oxidised. However, as the oxidation potential of the groundwaters progressively decreases, a limiting stage is reached when local reducing environments lead to the precipitation of U and other metals from solution. Where organic debris occurs within the succession, local deposition of uraninite or coffinite may occur; even urano-organic complexes may develop. Such organic debris is probably syngenetic with the deposition of the sediments. However, it is also possible for humates to have separated within a sedimentary succession at any time after deposition, and these accumulations will also form precipitation sites for U. Successive precipitation stages from circulating groundwater can lead to the formation of banded or nodular mineralisation around organic centres.

If migrating oxygenated groundwaters carrying U meet O_2-deficient groundwaters, U mineralisation may occur in the reaction–mixing zones (Sanford, 1982). It is in this way that roll-front ore bodies are thought to be formed, the location of the deposit being determined by the position of the boundary between the two groundwater flow regimes as in unconformity-controlled deposits. Again, source rocks are needed from which U can be leached by the oxygenated groundwaters, and these source rocks are likely to supply a range of other metals. The occurrence of U-bearing volcanic sequences or altered granites may be necessary for this to occur. The form of a roll-front deposit is very complex. In plan view it is of a digitate pattern and, although in cross-section a C-shaped roll-front is considered typical (Fig. 6.6), in practice the shape is often irregular or biased to one limb. Boundaries between rocks of different permeabilities can also be the sites for the deposition of U; the contacts between sandstones and mudstones, for example, may be mineralised, while complex depositional configurations may result from permeability variations within sandstones. Because groundwater conditions may change after their formation, these deposits may be found wholly within either the oxidising or the reducing environment. Further mobilisation may then occur.

Shales

The occurrence of high levels of U in association with a wide variety of other metals has long been recognised in Cu-rich shales such as the German Kupferschiefer. The U occurs mainly as uraninite, and the rock as a whole normally contains between 40 and 50 ppm. The origin of the U-bearing metal assemblage is generally considered to be the discharge of hydrothermal solutions into a marine basin deficient in O_2. Quiet, still and probably deep waters are thus indicated. Precipitation probably occurred when the relatively oxygenated hydrothermal fluids encountered these reducing conditions.

Bituminous shales may owe part of their U concentration to post-depositional precipi-

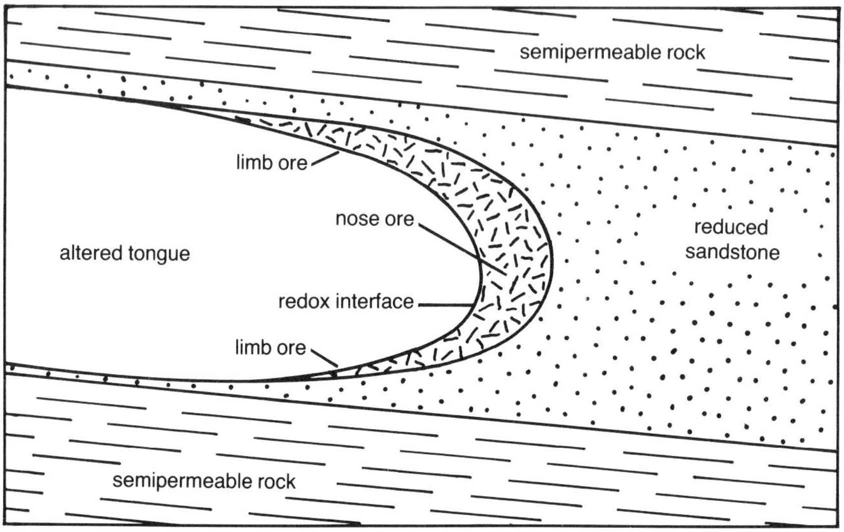

Fig. 6.6 — Simplified cross-section of a roll-front-type U deposit (Goldhaber *et al.*, 1978, redrawn with permission).

tation from migrating groundwater, but most is considered to be syngenetic. Black shales normally contain more than 2% organic C, with organic-rich layers or leaves often present. Because the black colouration is mainly due to disseminated cryptocrystalline pyrite, the U content may or may not be enhanced (Bell, 1978). The shales are usually of marine origin but they can form in a wide range of environments. About 3–7 ppm U is incorporated into detrital material, mainly clays and resistate heavy minerals, but the remainder is either adsorbed onto organic compounds and then combined into U–organic complexes or has been directly precipitated, possibly as colloidal uraninite within the sediments or at the sediment–water interface. The reducing effect of H_2S is probably influential in this process. With U, many other heavy metals are present; the U content may be up to 250 ppm, V up to 1000 ppm, Mn 1000 ppm, and Cr, Ni, Cu, Zn, Zr, Mo, Ba and Pb may also be present.

The euxinic basin, with deep anoxic waters below a surface layer of oxygenated water, is the generally preferred model for the development of marine black shales, but it has been pointed out that a less-well-studied but probably more wisespread euxinic environment occurs on the outer edges of some continental slopes where there is high biological productivity. In these circumstances, anoxic conditions form between depths of about 100 and 1500 m. The U in deep water muds of the Black Sea is derived by removal of U from solution. Apart from the sorption process discussed above, living organisms may play a part in this process. The conditions suitable for uraniferous black shale formation are given by Bell (1978) as the following.

(1) Very slow sedimentation.
(2) High biological productivity.
(3) Anoxic bottom water conditions.
(4) Presence of H_2S.
(5) Biological fixation processes.
(6) Maintenance of U supply in solution.

It is interesting to note that these conditions conform generally to those required for the formation of petroleum source rocks.

Because the U in black shales can be easily

remobilised by biogenetic processes, the interesting suggestion has been made by Bell (1978) that U metallogenic provinces may not form directly by reworking of U in igneous rocks but might involve an intermediate stage in which U enters a black shale. Liberation of U can occur at low grades of organic metamorphism by defluidisation. In general, the U concentration decreases as the grade of metamorphism increases.

Quartz pebble conglomerates

The occurrence of uraninite in conglomerates is well documented from the Elliot Lake deposits of Ontario and the Witwatersrand deposit of South Africa. As noted earlier in this chapter, these deposits are of early Proterozoic age and have been taken to indicate the presence of low O_2 concentrations in the Earth's atmosphere at that time. The presence of an O_2-poor atmosphere is also taken to account for the occurrence of detrital pyrite in association with the uraninite. Other radioactive minerals which often occur in this way are brannerite and monazite. The U content of typical deposits is of the order of 1000 ppm, and there is usually a comparable Th abundance. Au can also be found in economic concentrations.

Although some interpretations invoke a hydrothermal origin for these U deposits, the consensus view at the moment is that the uraninite in the conglomerates is simply due to sedimentation of detrital grains. However, whether they truly indicate low atmospheric O_2 concentrations has been doubted. Robinson and Spooner (1984a), for example, note that detrital uraninite and pyrite occur in the recent fluvial sediments of the River Indus. Moreover, in the Elliot Lake area, the deposits lie above and within a palaeosol in which the contemporaneous formation of both oxidised and reduced profiles occurred (Kimberley *et al.*, 1984). Similar controversy has arisen in the case of the Witwatersand deposit.

Fluvial conditions generally appear to have been present during the deposition of these assemblages. In some cases it has been argued that glacial working of older rocks produced U-rich till which was subsequently reworked before sedimentation of the conglomerates. For the Witwatersrand deposit, pre-concentration is believed to have been produced by hydrothermal activity in the older rocks (Andreoli *et al.*, 1985).

The idea that hydrothermal solutions were responsible for the presence of the uraninite has probably arisen because of the extensive remobilisation of U that is often seen in these deposits. Indeed, in some cases, such as the Paukkajanvaara deposit of Finland, there is evidence of secondary enrichment of U by hydrothermal activity (Boyle, 1982). Clearly, with such ancient deposits, there are often a number of occasions when migration of U could have occurred during metamorphic episodes. The preservation of uraninite in coarse-grained clastic rocks at shallow depth in the Earth's crust would also be aided by reduced permeability arising from metamorphic recrystallisation: access by younger oxygenated groundwaters has not led to the destruction of these deposits.

Phosphorites

Many sedimentary phosphatic formations contain U in small quantities, but some have abundances which can be of the order of 0.01–0.1%. Indeed, the large volume often occupied by these formations taken together with high U make them a major U resource. Of the sedimentary phosphate-bearing materials which are of interest, it is the phosphorites (phosphate rock) which are most important, although some bone beds and phosphatised limestones can also be significant. Both marine and lacustrine phosphorites are recognised as U resource rocks. In Morocco, for example, approximately 6×10^9 kg of U are believed to be present in phosphatic deposits, at an average concentration of about 100 ppm.

Marine phosphorites are mainly composed of carbonate–fluorapatite which is generally present in either the cryptocrystalline form known as collophane or the microcrystalline variety francolite. In a few cases, however, Al and Ca–Al phosphates such as wavellite, crandallite or turquoise are important. Although some of the U in these phosphorites may have been derived from circulating groundwaters, most is considered to be syngenetic. Primary U minerals are only very rarely present, so that the U mainly occurs within the phosphates. However, some U is usually found associated with organic matter in the phosphorites. Typically, this organic matter also contains high levels of Ag, Mo, V, Pb, Zn and Cr. In addition, Pb, As and the rare earth elements occur with U by substituting within the phosphates. Compared with other U-rich deposits, phosphorites usually contain less than 50 ppm Th. This possibly reflects the low values of Th^{4+} found in ordinary natural aqueous solutions. An antipathetic relationship between U and Ca in these phosphorites has been recognised by Heinrich (1958). It is assumed that U^{4+} occurs by substitution for Ca^{2+} in carbonate–fluorapatite and crandallite, and for Cu^{2+} in turquoise. However, U^{6+} is also certainly always present (Avital *et al.*, 1983).

Lacustrine phosphorites contain U abundances which equal or exceed those found in marine phosphorites. Apart from their environment of formation, the features of these rocks are somewhat similar to those of marine phosphorites, but in some cases the U occurs by substitution in the carbonate–hydroxylapatite dahllite (Boyle, 1982).

While phosphatic fossil bone and tooth materials which contain U are widespread in many sedimentary phosphorites and thus contribute to the U content of the rocks as a whole, on occasion fossil bone may be so abundant that it comprises the major component. Consequently, the bone bed itself becomes a U resource. In fossil bones and teeth the U is mainly found substituted for Ca in the carbonate–fluorapatite. Although some of this U may have been derived during life, or soon after death, by adsorption from the aqueous environment in which the animals lived, post-depositional introduction of U by circulating groundwater is also likely to have occurred. Although bone beds can have many origins, such strata are often developed at horizons where changes occur in the sedimentary environment: for instance from marine to fresh water or vice versa. In such examples, the abundance of bone material has been ascribed to catastrophic decease of organisms. Because these rocks can contain abundant diagenetic pyrite, the presence of low Eh conditions has also been inferred. Where circulating groundwater has caused considerable U enrichment (not only in bone beds but also in phosphorites), then U phosphates such as autunite and torbernite can be found.

Phosphatised limestones are not usually rich in U but, in some circumstances where phosphatisation was accompanied by high levels of U in solution, then accumulations of U-rich carbonate–fluorapatite can be formed.

Disseminated deposits

The most important economic U resource which occurs as disseminated mineralisation within sedimentary rocks is found at Olympic Dam (Roxby Downs), South Australia. This deposit is estimated to contain about 2×10^{12} kg of mineralised rock with an average U content of about 0.06%. The U-bearing minerals are distributed within the matrix of a sequence, over 1 km thick, of unmetamorphosed Proterozoic breccias. The U mineralisation is also associated with Cu (average grade, 1.6%) and Au (average grade, 0.6 ppm).

Two types of Cu sulphide mineralisation occur at Roxby Downs, of different ages. The older mineralisation is of strata-bound character while the younger is transgressive (Roberts and Hudson, 1983). U minerals are present within both types and were formed both with and after the Cu sulphides. The main U mineral is uraninite, but coffinite and brannerite are also found. In addition, some bastnaesite is present. The older mineralisation is considered to be syndepositional or syndiagenetic and related to local volcanism in an arid subaerial environment with infilling of fault-controlled basins, while the younger type probably represents deposition from migrating fluids following favourable structural zones.

Supergene precipitates

In arid or semiarid climates, U may be precipitated at or near the ground surface in lacustrine sediments, in playa deposits or in soils. The U is found with calcretes, phoscretes, etc., as carnotite, tyuyamunite, soddyite and uranocircite, and is probably derived by leaching of nearby U-bearing source rocks. Arid conditions also result in high levels of Ca, K, Ba, V, PO_4, CO_3 and SO_4. The other requirements appear to be a suitable porous medium and active evaporation.

6.1.8 Supergene alteration processes

Supergene processes can lead to depletion or enrichment of pre-existing U deposits. The following factors have been identified by Boyle (1982) as influencing these processes.

(1) Type of deposit and its mineralogy (for example, carbonates and sulphates can aid the production of soluble complexes, while the formation of limonite from Fe-rich muds, or of clays from silicates, may cause removal of U from solution).
(2) The lithology of the host rocks (in a similar way to the control exerted by the deposit itself).
(3) Degree of fracturing of the deposit and its host (to permit the ingress of water).
(4) Physiography (strong relief aiding vigorous circulation of oxygenated groundwaters).
(5) Climate (precipitation of U tends to be favoured in arid and semiarid environments).
(6) The position of the water table (controlling the boundary between well-oxygenated and poorly oxygenated conditions).
(7) Radiolysis of supergene water to yield H which can act as a strong reducing agent.
(8) Organic agents (some bacteria aid the solution of U while others may contribute to its precipitation, according to the environment).
(9) Time (clearly, the length of time over which the processes operate is an important control over the extent of the alteration).
(10) Glaciation (Pleistocene glaciations may have removed the higher levels of supergene alteration, leaving only the lower levels; alternatively, glacial deposits may protect U deposits showing fully developed supergene alteration).

Interaction of these various factors makes each U deposit unique, but in general terms the same sequence of alteration can be seen in a wide variety of circumstances. These are best illustrated by reference to a vein-type U deposit with a primary pitchblende–uraninite–coffinite mineral assemblage.

At the top of the section, the leached zone or 'gossan' is the result of mobilisation of soluble ions and alteration under strongly oxidising conditions. Limonite, clay minerals, opal, jarosite, carbonates and sulphates may be present together with thorogummite and, to a lesser extent, gummite. There may be a concentration of resistate minerals. The term 'thorogummite' is used here in the sense of Boyle (1982) to include all secondary oxides and silicates, while 'gummite' includes uranyl oxides, hydroxides, phosphates, molybdates, vanadates, arsenates, sulphates, selenates, silicates and carbonates. Although U may be leached, it is possible for Ra to be left behind to form minerals such as radiobarite or radiocalcite.

Below the gossan, and generally above the water table, the oxidised enriched zone may show the same assemblage of minerals as in the gossan, but the deposition of uranyl compounds is noteworthy. Almost any of the oxides, hydroxides, phosphates, etc., of uranyl type may be formed, depending on the detailed chemistry of the system.

Above the primary ore deposit but below the zone of oxidised enrichment occurs a zone of reduced enrichment where secondary U minerals are precipitated. Here, secondary uraninite or pitchblende ('uranium black'), coffinite and uranophane may be formed; in Mo-rich deposits there may be umohoite, moluranite, sedovite and wulfenite. Considerable enhancement of the U concentration compared with that of the primary ore body can occur in this zone.

The level of the water table is normally taken to mark the boundary between the zones of oxidised and reduced enrichment. However, if downward groundwater flow is concentrated in a fracture system of higher permeability than the surrounding rocks, descending oxygenated water can penetrate well below the general level of the surrounding water table if a forced circulation system (thermal or athermal) is operating. The effects of this deeper penetration of oxygenated groundwaters in the vicinity of the mineralisation are significant when dealing with the supergene alteration of sulphide deposits. Spontaneous electrical potentials, which are most noticeable in the alteration zone of Cu and As sulphide veins, often show detailed anomalies which suggest the presence of a small lateral polarisation superimposed upon the main vertical polarisation. Although U deposits do not produce spontaneous electrical potentials in their own right, the presence of associated Cu mineralisation, with the development of Cu oxides, hydroxides and carbonates in the zone of oxidised enrichment, and of secondary sulphides in the zone of reduced enrichment, may give rise to this effect.

6.1.9 U in organisms

The affinity of U for organic complexes discussed above suggests that living organisms are likely to act as concentrators of U and, indeed, it is generally found that, where U enrichment occurs in the geological environment, the biosphere shows higher U levels than normal. Concentrations between 5 and 85 times greater than normal have been reported from land plants. In water plants the concentrations may be increased a thousand-fold compared with the surrounding water, while it has been reported that bacteria growing on uraniferous bituminous material and some filamentous fungi growing on black shales may have more than 10% U (Capus and Munier, 1978).

The ability of organisms to take up U to these levels shows an interesting tolerance for, although U in trace amounts might be a necessary nutrient and might even have a stimulating effect on the growth of some plants, large concentrations are usually toxic. U occurs in plant tissue as U–protein complexes which are largely present in cell walls, but in animals U becomes combined with phosphate and carbonate in bone.

Most work has been carried out on the uptake of U by plants. Here, there is great variation between one plant species and another. In Britain, those species which show annual growth which dies away (such as grasses) do not usually concentrate U, even in uraniferous ground, but, where successive growth stages are added to produce woody material (such as in heather or gorse), concentration is more likely to occur. The use of biogeochemical prospecting methods is discussed in Chapter 9.

Even in plants that concentrate U, the average amount for the whole plant is usually less than 0.05 ppm. This is because U is only concentrated in certain parts of the plant, for example, in pine needles, second-year wood, roots, etc. Although some coal and lignite deposits have high U contents, the bulk of the U present owes its origin to epigenetic enrichment (principally adsorption) rather than to its incorporation in the organic matter of which the rock was originally composed.

The U content of some petroliferous rocks and bitumens is also high, while the radiogenic gases Rn and He are often present in anomalously large amounts in petroleum reservoirs, and Ra often occurs in oil field brines (Chapter 11). The U in these deposits is thought to be either in colloidal form or within U–hydrocarbon complexes. As for coal and lignite, it is unlikely that the U content of the organisms that were the source of the petroleum was sufficiently high to account for all the U present, and so again an epigenetic origin is involved. Where oxygenated groundwaters penetrate petroleum reservoirs, severe alteration of the petroleum takes place and U can be mobilised.

6.2 THORIUM

6.2.1 Properties

Like U, Th is a member of the actinide series, and because of its position in the periodic table and electronegativity it also shows strong lithophile characteristics. Indeed, because of the mutual substitution of U and Th, particularly in high temperature processes, most of the preceding account of the geochemistry of U applies to a greater or lesser extent to Th. There are differences, however. Th occurs in only one oxidation state in ordinary geochemical environments, that is, as Th^{4+}, and is generally very insoluble in water. Nevertheless it shows strong colloidal characteristics, and hydroxide and silicate colloids may form. In its chemical properties, Th resembles Zr, Hf and some of the rare earth elements (particularly Ce). Some Th is, therefore, to be found in minerals of both Zr and U, and replaces Ce, Y and other lanthanides in most rare earth minerals.

Because of its large size and ionic charge, Th^{4+} has a strong tendency to form complex ions with chloride, fluoride, nitrate, sulphate, carbonate and hydroxide. $ThCl^{3+}$, $ThCl_2^{2+}$, $ThCl_3^{+}$ and $ThCl_4$ are possible, as are

ThF^{3+}, ThF_2^{2+}, ThF_3^{+}, ThF_6^{2-}, $Th(NO_3)_2^{2+}$, $ThSO_4^{2+}$, $Th(SO_4)_2^{2+}$, $Th(SO_4)_2$, $Th(HSO_4.SO_4)^{+}$, $ThO(SO_4)_2^{2+}$, $Th(CO_3)_3^{2-}$, $Th(CO_3)_5^{6-}$, $Th(CO_3)_4(OH)_2^{6-}$, $Th(OH)_4$, $Th(OH)^{3+}$ and $Th(OH)_2^{2+}$.

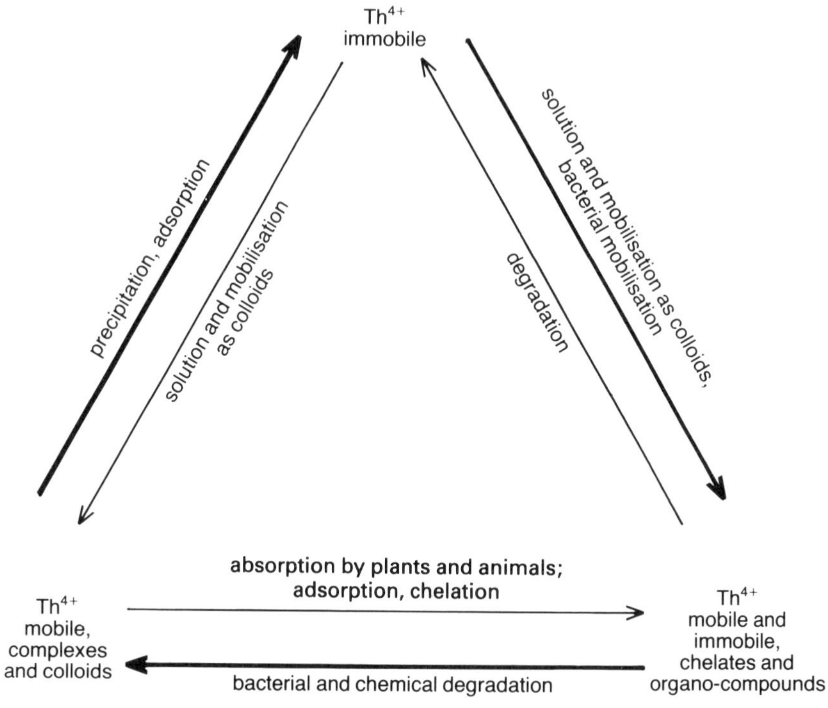

Fig. 6.7 — Simplified geochemical cycle of Th. (After Boyle, 1982).

Silicate, phosphate and organic complexes are also possible, all of which are capable of rendering Th mobile under certain pH and Eh conditions (Fig. 6.7).

6.2.2 Th in igneous rocks

In most igneous rocks, Th is usually present in accessory minerals such as monazite, thorite, zircon, sphene and allanite. Some may also occur as an impurity in fluorite, apatite, haematite and mica. It is also possible for Th to be adsorbed onto clays and onto Fe oxides and hydroxides. The Th content generally increases with the SiO_2 content and closely follows U during differentiation or partial melting. Th increases with the alkali content in comagmatic suites, the K/Th ratio remaining fairly constant at approximately 3×10^3 in a large variety of igneous rocks. Some alkali-rich igneous rocks are especially rich in Th. However, the increase in U with both SiO_2 and alkali content is usually more marked than the increase in Th. By and large, Th/U ratios are usually about 3.5–4.0 in igneous rocks, and Th contents range from about 1 ppm in basic rocks to about 30 ppm in granites.

During post-magmatic tourmalinisation and greisenisation of the Hercynian granites of southwest England, Th and U, present in the accessory minerals monazite and uraninite, were not mobilised to any extent. Where mobilisation did occur, reprecipitation took place almost immediately within the system.

6.2.3 Th in metamorphic rocks

Concentrations of Th in metamorphic rocks are highly variable and reflect the composition of the rock prior to metamorphism, the conditions of metamorphism and any metasomatic processes which may have taken place. In general, however, there appears to be a decrease in the Th content with increasing grade of metamorphism. In particular, where partial melting and mobilisation of granite pegmatites takes place, Th is concentrated into the liquid phase. Where sandstones containing accumulations of heavy minerals have been metamorphosed, bands of high Th content are present. A decrease in the Th/U ratio is sometimes observed with increasing grades of metamorphism, suggesting that Th becomes more easily mobilised than U under conditions of high pressure and elevated temperature, but this is not found everywhere and, indeed, increases in the Th/U ratio have also been recorded.

6.2.4 Th in sedimentary rocks

In sandstones, the Th content is usually fairly low (about 1 ppm) except where concentrations of heavy minerals occur. Monazite sands, for example, or those in which zircon or sphene are concentrated, contain very large quantities of Th. Arkoses have an average of about 5 ppm Th, greywackes about 7 ppm and orthoquartzites up to 9 ppm. Limestones generally have less than 2 ppm Th. Ordinary shales of grey–green colouration have a Th content of about 12 ppm, but in black shales there is a very wide range of Th concentrations (up to 100 ppm). In shales, Th may be present adsorbed onto clay minerals.

6.3 THE HISTORY OF URANIUM MINERALISATION

The evolution of U deposits within the Earth's crust provides a remarkable indication of changes in environmental conditions during the history of the Earth. Although U is often regarded as a ubiquitous element, it requires particular geological circumstances in order to accumulate and form mineral deposits in its own right. Perhaps the most important general control is the oxidation characteristics of the environment.

U mineralisation has occurred in a number of well-defined stages which appear to be related to the progressive evolution of O_2 in the atmosphere. U deposits range in age from late Archaean to Recent, the deposits of each stage being characterised by different types or combinations of types of mineralisation. The main divisions recognised in the following account have been identified and described in detail by Toens (1981).

The Archaean and early Proterozoic

No U deposits older than 2800 Ma have yet been discovered, but the U content of granites of this age is often relatively high (28 ppm). It is believed that during this period the Earth's crust was being enriched in U to a varying extent but that, because oxygenated groundwaters were absent, U concentrations were generally so low that mobilisation and concentration into ore deposits could not occur. Only where granites became eroded did some local concentration of U within the crust take place.

Although the Archaean, by definition, has a younger age limit of 2500 Ma, it is convenient here to include any assemblage of U mineral deposits which are older than about 2200 Ma. During this period, detrital U deposits formed in association with pyritic quartz–pebble conglomerates. Typical examples are the Dominion Group and Witwatersrand Supergroup deposits of South Africa, the Elliot Lake deposits of Canada and the Jacobina deposit of Brazil (Table 6.2).

As noted earlier in this chapter, taken at face value these sedimentary uraniferous sequences appear to have formed before the development of an oxidising atmosphere. In the absence of oxidising conditions, U would have been present as U^{4+} and, therefore, insoluble under normal conditions. As a result, U minerals, principally uraninite, eroded from the crustal rocks were deposited mechanically along unconformities and developed an association with conglomerates or with algal material within sedimentary sequences.

The sedimentary rocks that predate the formation of economically significant Archaean–early Proterozoic U deposits represent several phases of peneplanation and sedimentation during which some pre-concentration of U minerals probably occurred. Reworking of older sedimentary rocks finally yielded the ore deposits. Although the consensus view at present is that an anoxic atmosphere was necessary for fluvial transport of the detrital uraninite which constitutes the major part of these placer deposits, to add to the doubts raised about the Elliot Lake area, it has been suggested that sedimentary U-bearing deposits occur in association with Lower Proterozoic red beds in the Great Slave Lake area of Canada (Stanworth and Badham, 1984), and that the Witwatersrand deposit also contains U precipitated from solution in reducing environments (Simpson and Bowles, 1977). In both cases, U^{6+} would have been present, and this implies the existence of an oxidising atmosphere at that time. However, it has been shown that the Witwatersrand U has undergone later mobilisation during metamorphism, so that the presence of secondary U minerals does not necessarily indicate that an oxidising atmosphere was present at the time of sedimentation. Moreover, even during sedimentation, some U precipitated from solution is to be expected, as mobilisation of U could have occurred through the action of organic acids. This U could then have been later precipitated by microbial activity (Dexter-Dyer *et al.*, 1984). Nevertheless, the occurrence of red beds about 2400 Ma old is difficult to reconcile with the concept of an O_2-poor atmosphere at that time.

This period may be taken to represent the first major stage of U mineralisation (Fig. 6.8).

2200–2000 Ma

This is marked by the almost total absence of U ore deposits. At about 2200 Ma, the actions of photosynthesis and photodissociation of water in the atmosphere by ultraviolet light are thought to have caused O_2 levels in the atmosphere to rise. Mechanical transport of U^{4+} thus became impossible as oxidation to U^{6+} allowed solution of U. Deposition would then have occurred only where reducing conditions were encountered. However, this period may be regarded as a transitional phase in which O_2 levels in the atmosphere were progressively rising. With the passage of time, environments became more similar to those of the present

Table 6.2 — U resources

Country	Resource (t U)	Locality	Age
Algeria	34 000	Abankor	500 Ma
Argentina	8 000	Cordoba	Tertiary
	20 000	Mendoza–Sierra Pintada	Permian–Triassic
Australia	282 000	Jabiluka	1800–1700 Ma
	1 000 000	Roxby Downs	Proterozoic
	30 000	Mary Kathleen–Radium Hill	1730–1550 Ma
	40 000	Yeelirree	Tertiary
Austria	2 000	Forstau	Lower Permian
Bolivia	500	Cotaje, etc	25–12 Ma
Botswana	400	Mokobaesi	Tertiary
Brazil	13 000	Jacobina	2500–2200 Ma
	121 000	Itataia, etc.	Pre-Cambrian
	79 000	Lagoa Real, etc.	2100–1800 Ma
	7 000	Parana Basin	Permian
	23 000	Pocos de Caldas	80–60 Ma
Canada	578 000	Elliot Lake	2600–2500 Ma
	385 000	Athabasca	1800–1600 Ma
Central African Republic	18 000	Bakouma	Early Tertiary
Chile	8 000	National Cu and P_2O_5 (byproduct)	Cretaceous–Tertiary
Egypt	5 000	El Eridya	Late Pre-Cambrian
Finland	3 000	National	Lower Proterozoic
France	50 000	Massif Central	325–280 Ma
	23 000	Lodeve	Permian
Gabon	33 000	Oklo, Okelobondo, etc.	2000 Ma
Germany	13 000	Black Forest	280 Ma
India	57 000	Singhbhum	1600 Ma
Italy	3 000	Central Alps	Carboniferous
Japan	8 000	Ningyo Toge–Tono	Tertiary
South Korea	4 000	Ockcheon	Pre-Cambrian
Madagascar	2 000	Fort Dauphin	500 Ma
Mexico	8 000	Las Margaritas	Tertiary
Namibia	138 000	Rössing	500 Ma
	50 000	Namib Desert	Tertiary
Niger	213 000	Arlit	Carboniferous to Cretaceous
Philippines	300	Larah	270 Ma
Portugal	11 000	Viseu–Guarda	Hercynian
Somalia	5 000	Alio Ghelle	500 Ma
	5 000	Mudugh	Miocene
South Africa	73 000	Dominion Reef	2800–2600 Ma
	412 000	Witwatersrand	2600–2400 Ma
	3 000	Phalaborwa	1950 Ma
	42 000	Karoo	Permo-Triassic
	1 000	North West Cape	Tertiary
Spain	25 000	Cuidad Rodrigo	Pre-Cambrian or Cambrian
	3 500	Mazarete	Lower Triassic
Sweden	5 000	Pleutajokk	1800–1700 Ma
	75 000	Ranstad	Upper Cambrian
Turkey	4 000	Salihli–Köprübasi	Neogene
UK	7 000	Cornwall	Hercynian–Tertiary
USA	320 000	Grants Belt	210–110 Ma
	8 500	Schwartzwalder	Tertiary
	350 000	Wyoming–Texas	Eocene
Yugoslavia	27 000	Zirovski	Permian
Zaire	30 000	Shinkolobwe	500 Ma
Zambia	8 000	Kariba	Permo-Triassic

(After Toens, 1981; Organisation for Economic Cooperation and Development, 1983; J. Hunter.)

day, and important red bed sedimentary successions containing haematite appeared. Because of the solubility of U, radioactive placer deposits formed after this time are generally Th rich.

Despite the absence of marked concentrations of the element, U source rocks were still being formed during this period; the Bushveld granites are enriched in U and Th to such an extent that in some places they approach ore grade themselves, while some alkaline and carbonatite complexes, such as a Phalaborwa, yield U as a byproduct of Cu mining.

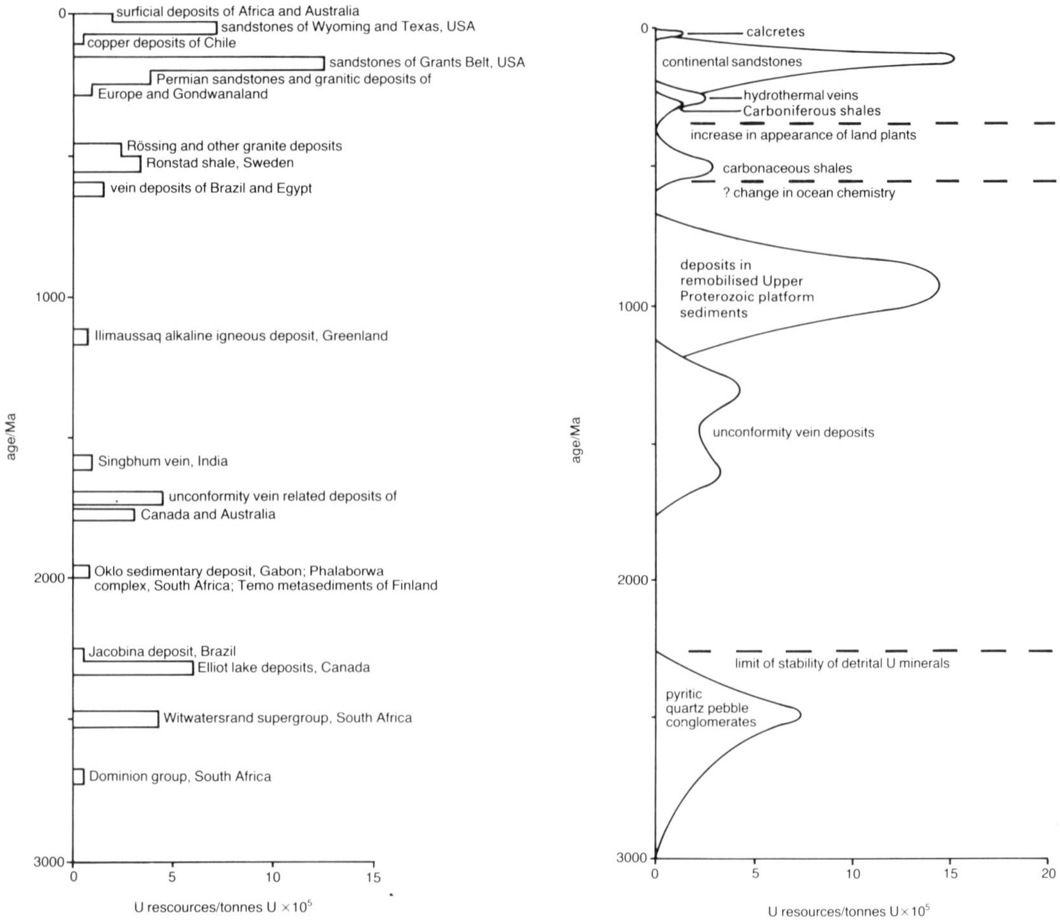

Fig. 6.8 — The world's major U deposits and the evolution of U resources. (After Toens, 1981; J. Hunter.)

2000–1600 Ma

This period saw the formation of some of the world's largest and highest grade U deposits. They occur quite often in tabular and lens-like forms, in unconformity-controlled vein-like bodies and vein associations, in which the main U minerals are uraninite and pitchblende. Examples are the Cluff Lake, Rabbit Lake and Key Lake deposits of Canada, the Ranger, Jabiluka and Nabarlek deposits of Australia, and the Mounana–Franceville deposits of Gabon.

About 2000 Ma ago, the O_2 concentration in the atmosphere had probably risen to such a level that weathering of the older cratonic areas led to a sudden release of U (Robertson *et al.*, 1978). However, below a shallow oxidised zone, reducing conditions still persisted in the lithosphere, and the released U rapidly became deposited. U transported in solution into marine or fresh water basins also became precipitated in poorly oxygenated deeper waters. However, there does not appear to have been any sedimentary preconcentration of U (Sibbald, 1985; Needham and Ewers, 1985). These deposits are associated with the development of mobile belts, and the U deposits now occur in low grade metamorphic rocks. In Australia, the Pine Creek Geosyncline provides an excellent example.

Within the mobile belts, the formation of sedimentary and volcanic rocks and the intrusion and hydrothermal reactivation of U-enriched granites with subsequent tectonic activity caused remobilisation of U. Within an individual deposit there may, therefore, be a wide range of mineralisation ages.

1600–700 Ma

This is usually considered to be a period during which few U ore deposits formed (Toens, 1981), but recent reassessment suggests that this view is incorrect and that this period could be more important in the extent of its U mineralisation than either of the previous two periods. It is now recognised that, during this period, U deposits or protore deposits formed in remobilised Upper Proterozoic platform sedimentary rocks. Examples are the Roxby Downs deposit of South Australia, the Bancroft deposit in Ontario and possibly the Workman Creek deposit of Arizona (although in this case the main mineralisation may be much younger). Minor occurrences of U are found in association with pegmatites in the Namaqua–Natal mobile belt of South Africa, and large deposits of low grade occur in Namaqualand. Alkali igneous complexes of this period at Pilanesberg in Bophuthatswana and in Greenland contain large potential U reserves, but of low grade.

With the mineralisation of the period 2000–1600 Ma, this may be taken to mark the second major stage of U mineralisation.

700–485 Ma

During late Pre-Cambrian and early Cambrian times, a number of U ore deposits were formed in various parts of the world. Disseminated U occurs in granite at Rössing in Namibia and in veins in granite at El Eridya in Egypt, and both disseminated and vein U mineralisation occur at Cordoba in Argentina; vein deposits in schists are found at Abankor in Algeria. Other deposits also occur at Lagoa Real and Itataia in Brazil (Barretto, 1985), at Shinkolobwe in Zaire, and at Alio Ghelle in Somalia. Reworking of older basement has generally been invoked to explain these deposits. At Rössing, for example, it is believed that a proto-sedimentary U ore was remobilised to form U-rich alaskites. An alternative view, however, is that primary concentration of U occurred by partial melting of basement with U becoming concentrated in granitic melts by fractional crystallisation. Finally, boiling of alaskitic magma could have led to the deposition of U in association with pegmatitic granites.

U deposits also occur in association with shales and slates of this period. Examples are at Ranstad in Sweden, Ockcheon in Korea and Cuidad Rodrigo in Spain. In general, these associations are more characteristic of the next stage of U mineralisation and can be considered to have transitional characteristics. The genesis of shale and slate deposits is best considered with the next period.

485–300 Ma

Although there are no U deposits of economic significance of this age, low grade sedimentary deposits were formed in abundance between Cambrian and Carboniferous times. These are the very widespread U-rich black organic shales, examples of which are the Devonian Chattanooga shale of North America, the Carboniferous (Kolm) shales of Norway, Sweden and Estonia, and the Carboniferous shales of Hippolyte in France.

The widespread production of uraniferous shales during this period probably owes much to the development of deep anoxic basins, possibly indicating high eustatic sea level. A global rise in sea level can probably be related to vigorous mantle convention and sea floor spreading, with associated swelling of midocean ridge systems causing the displacement of ocean water. In contrast, periods of low vigour convection and slow rates of sea floor spreading are marked by low global sea level. It has also been argued that high global sea levels lead to climatic equability and that this favours high production rates of marine organisms. It is also possible that, at times of more active mantle convection, the processes leading to depletion of the mantle in U would also be most vigorous. There thus developed a combination of circumstances ideally suited to the formation of these U deposits: firstly, extrusion of U-rich volcanic rocks acting as a source of U in solution; secondly, a high sea level, producing deep marine basins with high organic production rates which, in turn, resulted in organic-rich sediments which produced the reducing environment necessary for the precipitation of U. Whether or not the abundant development of marine organisms with hard skeletons that occurred at the start of Cambrian time reflects a major change in ocean chemistry which might have aided this process is difficult to say.

Certainly, some changes did take place in the chemistry of the oceans, reflecting the progressive increase in atmospheric O_2 levels. The early ocean would have contained acid gases such as CO_2, HCl, and SO_2 in abundance and, although neutralisation would have occurred by reaction with silicate minerals, under a CO_2-rich atmosphere the supply of volcanic acid gases would have maintained the pH below 7.0. Dissolved material would then have included greater concentrations of Ca^{2+} and Mg^{2+} than at the present day, and Fe^{2+} would have been in abundance. Concentrations of rare metals that are precipitated today by reaction with CO_3^{2-} and SO_4^{2-} would also have been present. However, as atmospheric O_2 increased, the pH would also have slowly increased, and precipitation of Ca and Mg carbonates would have occurred. This could account for the extreme development of dolomite in the later part of the Pre-Cambrian. The start of Cambrian time may well mark a critical point in the evolution of the hydrosphere.

300–65 Ma

This period is characterised by the main appearance of uraniferous continental sandstones. The development of land plants during Carboniferous time produced sufficient organic material to allow the development of reducing conditions in terrestrial environments. U weathered from continental rocks and transported in solution was, therefore, very effectively removed by precipitation in these environments. Interestingly, U ore deposits of this type occur at about 50° on either side of the palaeoequator (in contrast with the major oil deposits which were formed at low latitudes) and are therefore closely related to the development of contemporaneous coal fields. While vegetation is necessary as a source of organic matter, a complete vegetation cover will tend to retain U and to inhibit its transport and concentration. Semiarid environments, therefore, appear to have been most suitable for the formation of uraniferous sandstones, especially where granite terrains were being eroded or where contemporaneous silicic volcanism was taking place.

The deposition of U was generally controlled by the presence of reducing environments, where U^{6+} transported in solution became reduced to U^{4+} to be precipitated as uraninite, pitchblende or coffinite. Typically, the reducing environments were produced by organic material contained within the sediments. It is possible that precipitation was aided by the action of sulphate-reducing anaerobic bacteria. U deposits of this type are often associated with concentrations of other elements such as Cu, As, Pb, Zn, V, Mn and P. Apart from precipitation around organic-rich centres, U may also be precipitated from solution at boundaries of contrasting permeability, and where groundwaters of different Eh and/or pH meet.

Examples of U deposits occurring within semiarid sedimentary successions are present throughout the world, but most are found in the Permo-Triassic basins of Gondwanaland. The joining of the southern continents during the Palaeozoic to form Gondwanaland may have caused a retardation in the rate of sea floor spreading towards the end of the Carboniferous, leading to subsidence of ocean ridges and a fall in global sea level. Alternatively, low spreading rates and a decrease in sea level may have resulted from changes in mantle convection. Whatever the cause, a low global sea level is believed to lead to climatic extremes. Both the development of the Carboniferous–Permian glaciation of Gondwanaland and the formation of U deposits may have been the result of these processes for, at the end of this glaciation, climatic amelioration led to the development of impressive sedimentary basins, some land-locked and others open to the sea, often with abundant plant life surrounding them on shore. U deposits of this type are represented in the Karoo Basin of South Africa where U, mainly in coffinite, is associated with organic matter in pods or lenses and in the Transvaal where U is associated with coal seams in the Ecca Group. Similarly, in Niger, the Agadez Basin contains a sedimentary succession like that of the Gondwana basins although not considered as part of the Gondwana sequence. This area represents one of the most important U occurrences in Africa (with Witwatersrand and Rössing) and contains the largest continental sandstone deposits in the world outside the USA. Syngenetic U deposits in sandstones associated with coal seams in the Parana Basin in Brazil, the Argentine deposits of Mendoza Province and U associated with shallow-water marine deposits in the Lake Frome Embayment of Australia are of similar type but different ages.

In the northern hemisphere, continental sandstone U deposits are found in North America and Europe where they are best developed in arkoses under the control of local lithology and hydrology. Important U reserves occur in the Colorado Plateau area of the USA, where precipitation on organic debris and amorphous humates and deposits of roll-front type are well exemplified. In Europe too, sandstones of Permo-triassic age contain U deposits. Only non-marine facies contain this mineralisation, organic matter having acted as precipitation centres.

The association of U deposits with organic centres in continental sandstones is of interest as it indicates the presence of strongly contrasting geological environments. Very commonly, the continental sandstone sequence as a whole shows distinct red colouration. Indeed, the Permo-Triassic sequence of Britain is known as the 'New Red Sandstone' (to distinguish it from a similar facies of Devonian age which is known as the 'Old Red Sandstone'). The colouration of these red rocks is due to the presence of a haematite coating around the detrital grains. This coating has been shown to be of secondary origin, being developed by the diagenetic alteration of Fe-bearing minerals. In this process, Fe^{2+} was dissolved and transported until oxidation caused the precipitation of ferric oxyhydroxide which dehydrated to form haematite. Reducing conditions favour solution of Fe while oxidising environments result in its precipitation (Fig. 6.9). The solution of U, conversely, is favoured by oxidising conditions and precipitation occurs on reduction. As a result of this, in areas where U deposits occur within red bed sequences, the U concentrations are surrounded by a zone which has not been reddened. These haloes have been referred to as 'reduction haloes' or 'bleached haloes' (Fig. 6.10). The former term is incorrect because the Fe content of the haloes is significantly lower than in their red host, and so the colour difference does not reflect merely a change in the oxidation state of Fe. It has, therefore, been argued that the Fe must be both reduced and removed to form these haloes, and hence the term 'bleached haloes'. It is possible that H_2S, originating in the decay of organic matter by anaerobic bacteria, effected the reduction, with solution occurring in highly carbonated water, although the Fe could have entered solution directly in the form of organic complexes. These ideas, however, all presume that haematite was originally present in the haloes. An alternative view is that the haloes represent zones where the diagenetic haematisation shown by the main bulk of the red beds was inhibited. In other words, the haloes essentially represent primary sedimentary material. Interestingly, detailed studies of the haloes in the mudstones of the

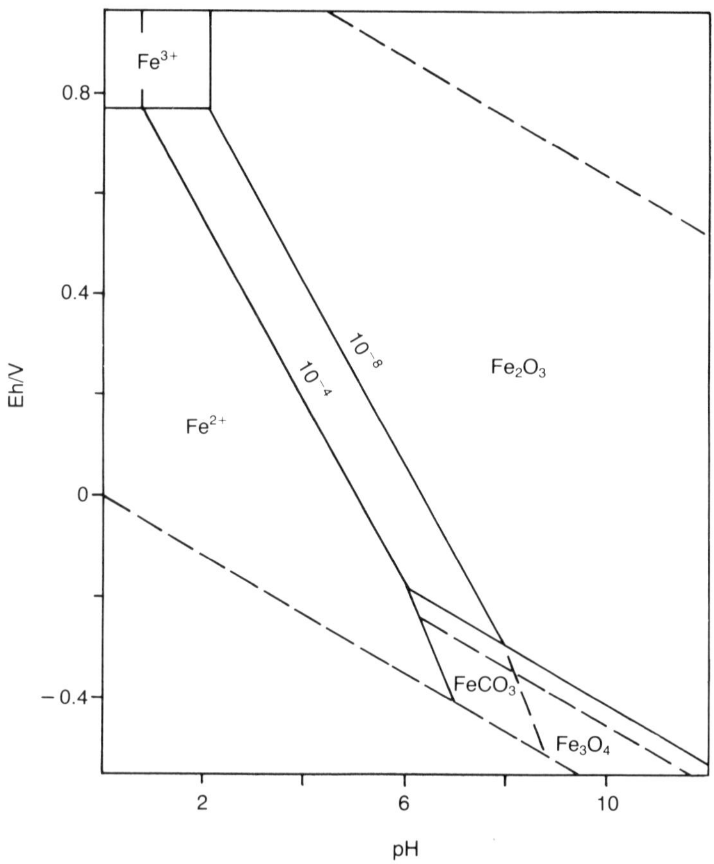

Fig. 6.9 — Stability diagram for Fe species at 25 °C for an open system with $P_{CO_2} = 10^{1.5}$ Pa. The contour lines show the total dissolved Fe activity. (Bouda, 1985, redrawn with permission.)

Permo-Triassic red beds of southwest England have shown that the change from reducing to oxidising environments does not occur at the boundaries between the haloes and the red host but lies entirely within the zones occupied by the haloes (Durrance *et al.*, 1978). Moreover, the majority of the haloes are of spherical shape, despite the mudstones having geotechnical properties characteristic of consolidation that would have given at least 20% vertical shortening. This implies that the diagenetic haematisation of the mudstones occurred after consolidation and that, as the consolidation characteristics of the mudstones were produced by a superincumbent load of about 1000 m, the haematisation did not take place for a considerable period of time after deposition of the mudstones. Indeed, as the mudstones in question occur at the top of the Permian succession, and as 1000 m of cover encompasses the overlying Triassic beds, the haematisation may be interpreted as developing with the onset of a humid climate at the start of the Jurassic Period.

The nodular U mineralisation at the centre of the haloes of southwest England is associated with V, Ni, Cu, Co, As and Ag. The

(a)

(b)

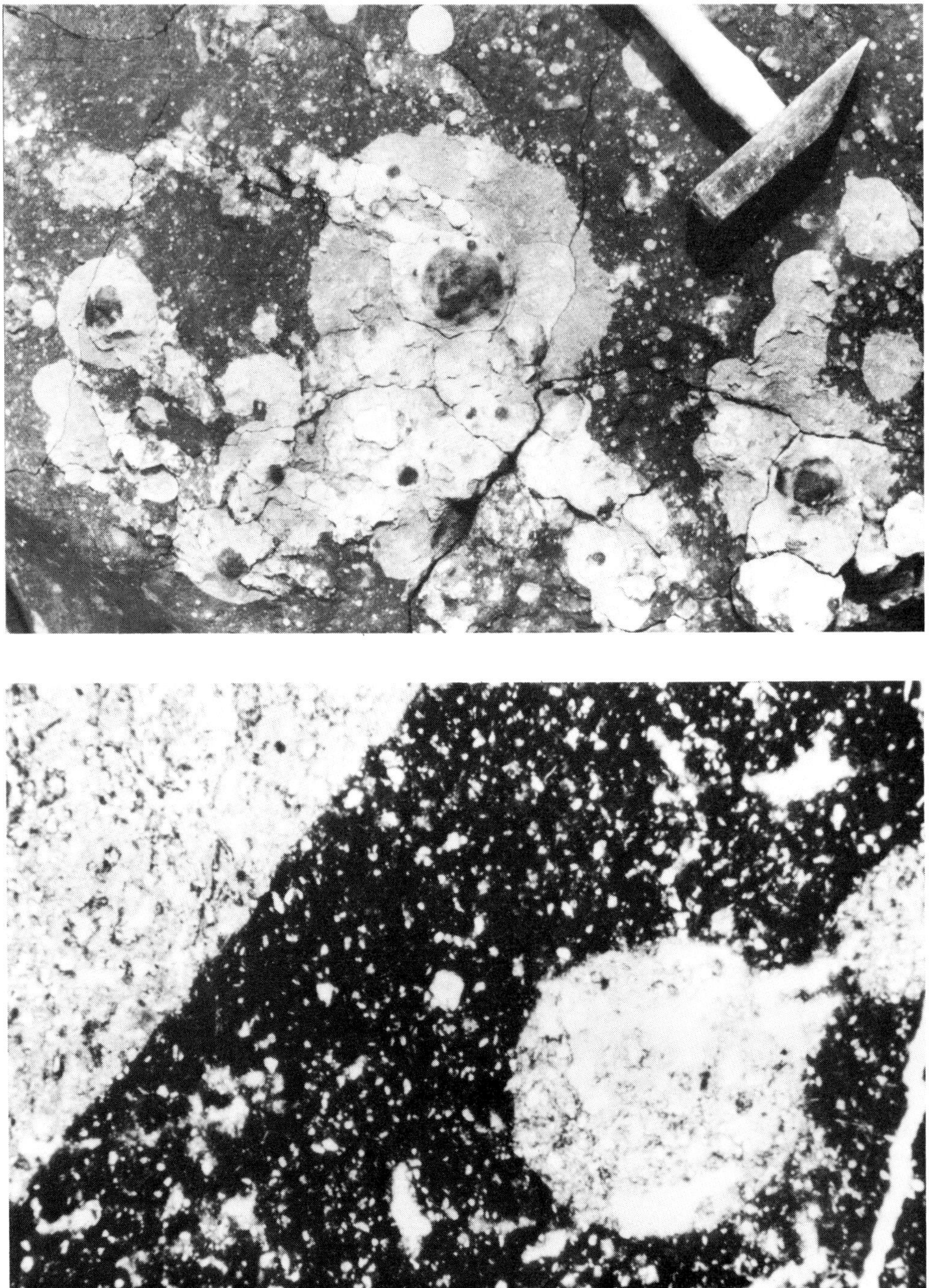

Fig. 6.10 — U-bearing nodules with pale green haloes in red mudstone of the Littleham Mudstone Formation, southeast Devon: (a) cluster of nodules within bedding; (b) the sharp junction between the pale green and red mudstones. The photomicrograph shows a field about 3 mm wide. (Durrance, 1980, reproduced with permission.)

U mineralisation itself is late, coating grains of opaque minerals and occurring as particles within nodules which take a number of forms (Perutz, 1939), ranging from very irregular (finned) varieties to compact spherical varieties. The finned form is considered to have resulted from growth within soft sediments, fins being formed in response to hydrostatic pressure which causes radial displacement into weak partings of lower pressure in the enclosing sediment. These nodules are surrounded by highly irregular haloes. In contrast, the spherical nodules clearly post-date compaction. The mineralogical associations within the nodules indicate precipitation at centres of organic debris. The picture thus presented is one of syngenetic or early epigenetic V–Ni–Cu–Co–As–Ag mineralisation, initially occurring around unconsolidated organic debris, but continuing through to post-compaction. These precipitation sites then remained areas of reducing conditions for a long time, inhibiting haematisation of the host rocks in their vicinity and causing later precipitation of U from solution. However, the age of the U mineralisation cannot be determined, as groundwater carrying U in solution could have passed through the system at any time. This illustrates very well one of the problems of U ore deposits in continental sandstones; even where deposits around organic debris are concerned, the age of the U mineralisation is often unclear.

Hydrothermal circulation through U source rocks can also lead to the formation of hydrothermal vein deposits. U source rocks are of widely varying lithology, but the most suitable appears to be uraniferous granite in which the U is held in easily leachable accessory minerals, principally uraninite, although the leachability of U is governed to some degree by the amount of Th present in the uraninite. Truly 'fertile' granites are those which not only contain high levels of U but also have low Th contents.

Recent investigations into the control of metallisation in British granites by Simpson *et al.* (1977, 1979) have compared the Hercynian Province of southwest England (which is metallised) with the Caledonian Province of Scotland (which is not metallised). These studies have concluded that the major control lies in the nature of the crust into which the granites were emplaced rather than in the primary differences in granite magmatism, although the Caledonian Province is characterised by a more refractory U-bearing accessory mineral suite (Basham *et al.*, 1982). For metallisation to occur, it seems that suitable granite magma must be emplaced in an adequately hydrous environment. However, the hydrothermal vein metallisation of southwest England post-dates the emplacement of the granites by some 20–30 Ma. Furthermore, the associated north Devon ore field lies some 50 km away from the nearest granite. It seems, therefore, that metallisation in this area is related to post-orogenic tension, crustal rifting and basalt magmatism during which vigorous hydrothermal circulation occurred in major fractures.

Primary migration of U may be considered to be the removal of U from its primary mineral, which will depend upon the chemistry of the groundwater, and the chemical, mineralogical and physical properties of the uraniferous minerals and their host. For effective leaching of the main primary U mineral, uraninite, a good intergranular permeability is required. Leaching is aided if the permeability of the minerals themselves has been enhanced by metamictisation. Suitable pH and Eh conditions in the groundwater are also required.

In the Carnmenellis granite of southwest England, it has been demonstrated that, even though leaching of U can be marked, reprecipitation takes place in the immediate vicinity of the primary minerals unless the U forms highly soluble complex ions. Thus, in granite affected by post-magmatic tourmalinisation and chloritisation, U can be considered to have been immobile within the rock as a whole. Post-magmatic kaolinisation, however, was accompanied by considerable loss of U from the altered granite.

Secondary migration is the transport of U

away from its source. This requires a suitable groundwater chemistry. Again with reference to the granites of southwest England, during the formation of the china clay deposits, suitable groundwater chemistry allowed the breakdown of feldspar to form kaolinite, and the transport of U away in solution. Here the complexing agent was probably HCO^-.

Sedimentary lithologies, too, may only act as good U source rocks if the element is present in an easily leachable form. Sandstones are usually acceptable source rocks provided that their U is not contained within Th-rich resistate minerals. Black shales also often contain appreciable quantities of U with low Th levels and may, therefore, form good source rocks. However, it is here worth noting that, under conditions of high grade regional metamorphism, most potential source rocks will become depleted in U as it becomes mobilised within granitic melts.

The U-bearing hydrothermal veins of southwest England show evidence of reactivation on a number of occasions throughout Mesozoic and Cenozoic times. This repeated mineralisation is a characteristic feature of a number of other U deposits. The main control over the development of effective groundwater circulation systems is the presence of a suitable fracture permeability. Provided that fractures remain open or become reopened by tectonic activity, even low geothermal gradients can induce convective movement. If the tectonic history of an area is complex, as in southwest England, episodes of hydrothermal mineralisation are found to be associated with each reactivation of the fracture system.

Finally, we should note that the Cretaceous Period saw the main development of uraniferous phosphorites.

The period 700–65 Ma may be considered to be the third main stage of U mineralisation.

65–0 Ma

Towards the end of the Carboniferous, the proliferation of terrestrial plant life began to reach such proportions that, despite the presence of an O_2-rich atmosphere, widespread reducing conditions developed in areas of organic accumulation to allow U deposits to form. As we have seen, these conditions persisted throughout the Mesozoic. Not surprisingly, therefore, in places where they continued into the Cenozoic, similar deposits were formed (for example, in Wyoming and Texas). In general, however, the Cenozoic is noted for the presence of surfical U deposits unlike those encountered earlier in the Earth's history. These deposits have not been subjected to deep burial and show calcification to a greater or lesser degree. The most important deposits of this type are in the Namib Desert in Namibia and the Yilgaru Province of Australia.

Fluvial uraniferous sediments of valley-fill type are also valuable deposits from this period. It appears that at the time of their formation the climate must have been more humid than at present, so that in the Namib Desert, for example, the sediments consist of arkosic breccias, conglomerates, sands and clays deposited under flash flood conditions. The upper horizons have been cemented with calcite, dolomite or gypsum and, in some places, with carnotite. However, the main occurrences of carnotite are in the lower part of the sequence (usually at the water table) where it is present as coatings, vugh-fills, concretions and veins. It is clear that the formation of carnotite was controlled by climatic factors and run-off characteristics. For example, at Yeelirree, control was exercised by constrictions in the valley floor forcing groundwater near to the surface. In addition, the geochemical nature of the ground (in particular the presence of K and V) is important. It appears that tectonically stable shield areas with lithologies containing an adequate source of U (and K and V), situated in hot semiarid climatic belts within 30°N or 30°S of the equator, and which have internal drainage systems with little or no flow to major rivers or the sea, provide optimum conditions. The preservation of these deposits is due to present-day dry climatic conditions which prevent the flushing-out of the carnotite.

The Bakouma deposit of the Central African Republic is an unusual variation on this theme. Here, U occurs in fluorapatite within sands and clays cemented by limestone and phosphate which occupy solution pits within limestone. Another variation on the theme is the occurrence of U in deposits of marsh and bog environments with highly reducing conditions. Typically, these sediments consist of diatomaceous earth, fluvial sands and even peat.

The continuation of continental sandstone-type U deposits and the development of hydrothermal circulation systems in Tertiary just as in Mesozoic times, and the continued production of specific environments with low redox potentials due to organic accumulations, have resulted in the fact that U deposits within this period are more extensive than previously experienced in the Earth's history. The formation during this period of surficial U deposits is unique in the geological record, but there is no obvious reason why similar deposits could not have formed in previous periods. Probably, older deposits have simply failed to be preserved.

This period may be taken as marking the fourth main stage in the formation of U deposits.

Summary

Thus, changes in the character of U deposits since early in the Earth's history can be seen to be very largely dependent upon the evolution of the Earth's atmosphere and climate, and the evolution of life. Detrital uraninite could only have formed in large amounts in the Archaean and early Proterozoic when probably atmospheric O_2 levels were low; the evolution of marine life and the development of a transitional O_2-enriched atmosphere allowed the unconformity and vein-type deposits of the early–middle Proterozoic to form; organic-rich sediments allowed the formation of uraniferous black shales in the Phanerozoic, and terrestrial plants provided reducing environments for U precipitation in the continental deposits of the Mesozoic and Cenozoic; present-day dry climates have preserved calcrete-type deposits or valley-fill deposits which also accumulated under conditions of active evaporation, and dry climates have also allowed preservation of secondary U minerals. The preservation of secondary U minerals within a dry climate certainly upgraded the deposit at the Rössing mine in Namibia, which has been one of the world's main sources of U (Toens, 1981).

6.4 THE URANIUM AND THORIUM DECAY SERIES IN THE HYDROSPHERE

6.4.1 The hydrological cycle

For the purposes of this account, the hydrosphere may be considered to include rivers and lakes, groundwater and the seas and oceans, each of which are linked within the hydrological cycle. Rivers form the main pathway by which the products of weathering of land surfaces are transported to the sea. However, the time during which water is actually in contact with rock is usually much greater for groundwaters than for rivers. Consequently, groundwaters usually have much higher concentrations of dissolved material than rivers (Table 6.3). The discharge of groundwater into surface streams can, therefore, dramatically affect the chemistry of a river. The chemistry of the sea can be seen as a balance between the input of material (by coastal erosion, by rivers and by the wind) and the output of material (by sedimentation, by precipitation or by organic activity), although some immediate return may arise indirectly from output mechanisms. In studying the chemical balance of the seas, knowledge of the behaviour of the U and Th decay series can be very informative. These systems will be considered, firstly, in

Table 6.3 — Average concentrations of U and Ra in natural waters

	U $(g\,l^{-1})$	Ra $(g\,l^{-1})$
Surface waters		
Seas and oceans	3×10^{-6}	10^{-13}
Rivers	6×10^{-7}	2×10^{-13}
Lakes	8×10^{-6}	10^{-12}
Actively circulating groundwaters		
Sedimentary rocks	5×10^{-6}	2×10^{-12}
Igneous rocks	7×10^{-6}	2×10^{-12}
Uranium deposits	6×10^{-4}	8×10^{-11}
Groundwaters with restricted circulation		
Sedimentary rocks	2×10^{-7}	3×10^{-10}
Igneous rocks	4×10^{-6}	4×10^{-12}
Uranium deposits	8×10^{-6}	6×10^{-11}

(After Brookins, 1984.)

river water and, secondly, in groundwater.

6.4.2 Rivers

U

U can be present in river water over a wide range of concentrations between less than 0.02 $\mu g\ l^{-1}$ (which is the practical detection limit) and more than 30 $\mu g\ l^{-1}$, the average concentration being taken as 0.6 $\mu g\ l^{-1}$. The U content of a river depends upon the U concentration and its mineralogical form in the catchment area and the chemistry of the river as a whole, particularly regarding the presence of ligands with which U may form soluble complexes under the prevailing Eh–pH conditions (Scott, 1982). U concentrations vary considerably in river water, both spatially and temporally. The temporal variation may be seasonal, representing changes in rainfall and run-off, evaporation and transpiration. Unfortunately, the effect of these processes on U levels is not consistent. For example, in a particular area at a particular time, increased rainfall may simply reduce U concentrations by dilution, while in another area, or in the same area at a different time, such an increase may result in more effective leaching and/or transport of U and lead to an enhancement of U concentrations. Indeed, with seasonal rainfall variation, both effects may be observed; for example, early rain after a dry period may increase the U concentration while later rain merely dilutes the system. In humid areas, day-to-day variations in response to meteorological changes can be equal to those resulting from seasonal effects. Indeed, the overall pattern of change within a single small drainage basin can be such that different sites show completely different reaction characteristics in terms of the water chemistry following apparently the same change in meteorological conditions. This variation is probably caused by differences of aspect and microclimate and differences in the nature of the underlying lithologies. Generally, however, the changes brought about by seasonal and shorter-term meteorological variations are usually less than an order of magnitude and thus tend to be accepted as background 'noise', which is always present. Unfortunately, even sampling of an area after a protracted period of constant meteorological conditions will not ensure freedom from this noise, as the effects produced by meteorological variation may well persist for many months in the drainage system.

One of the main factors controlling the U concentration of river water appears to be the availability of HCO_3^-, probably because of the importance of the uranyl carbonate ion in the solution chemistry of U. However, there is also usually a clear link between high U concentrations and high total dissolved solids, suggesting that there is a general control on U concentration by the intensity of weathering. Some doubt thus exists about the control exerted by HCO$^-$, and it may simply be the case that, because HC^{3}O$^-$ is the major anion present in most river waters^3, it correlates with high total dissolved solids. As yet, no clear identification of the major factor has emerged.

Fractionation of ^{234}U from ^{238}U during chemical weathering is generally reflected in the ^{234}U/^{238}U activity ratios obtained from river waters and, indeed, from groundwaters. The average ^{234}U/^{238}U activity ratio in rivers is about 1.2–1.3 but can be greater than 2. Ratios less than unity are rarely, if ever, recorded. Variation in ^{234}U/^{238}U activity ratios has been used to determine mixing ratios for waters of different origins (Fig. 6.11).

In stream sediments, U may be found in detrital material, in which it is usually incorporated into resistate minerals, or it may be adsorbed directly from river water onto clay minerals or organic debris. U may also be coprecipitated with Fe and Mn in some circumstances. Occasionally, U may be removed from river water to sediments directly if reducing bottom conditions exist. Variations in U concentration along the length of a river thus reflect the balance between input of U and its abstraction by sediments and organic material. Equilibrium conditions, however, will only rarely be established, as there is much short-term variation in stream sediment movement, water flow and organic activity.

It is of interest to note that more than 90% of the U transported by rivers to the ocean is in the form of particulate material. The average U content of suspended material in rivers is estimated at 3 ppm, which is slightly higher than the average abundance of U in crustal rocks, because of adsorption effects on suspended clay and organic particles, which form a greater proportion of the average stream sediment than of average crustal rock.

In contrast with the high ^{234}U/^{238}U activity ratios of river waters, river sediments typically show values of less than or close to unity, usually between 0.9 and 1.0, with an average of about 0.95. However, it is possible for activity ratios to be greater than unity in special circumstances, and a value of 1.8 has been reported. Values greater than unity are interpreted as being a result of the fact that the low U activity ratio of inorganic detritus is overwhelmed by the higher ratio of U removed from solution. In general, the occurrence of high ^{234}U/^{238}U

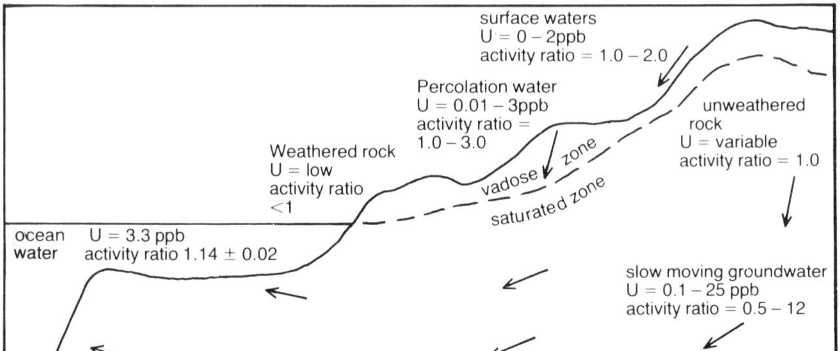

Fig. 6.11 — Variation in the ^{234}U/^{238}U activity ratio and U content within the hydrological cycle. (After Gascoyne, 1982.)

activity ratios in stream sediments may be correlated with high organic content, either originating as detrital particles in the catchment area or actually produced within the rivers themselves. Adsorption of U onto clay particles can also account for higher than normal activity ratios. It has also been suggested that high activity ratios may correlate with high levels of $CaCO_3$ in the sediment. Differences generally exist between suspended sediment and bottom sediment, with lower U concentrations and lower activity ratios in the bottom sediments. This probably reflects the different compositions of the two sediment loads which, in turn, are determined by different grain size characteristics. Suspended sediment is likely to contain more organic debris and clay particles than is the bottom load.

Th

Th is also usually more abundant in the suspended load than in bottom sediments (Scott, 1982). This is interesting because the chemical behaviour of Th in natural waters is quite different from that of U. As discussed in section 6.2, Th is essentially insoluble in normal surface waters, and so Th transport is within particulate matter rather than in solution. The average abundance of Th in river sediments is about 14 ppm. Although ^{234}Th, ^{230}Th and ^{228}Th may also be present, ^{232}Th is the major contributor to the total Th content. The greater part of the ^{232}Th is present within detrital minerals, but the larger values obtained from the suspended sediment suggest that some direct adsorption of Th may occur. ^{230}Th/^{234}U activity ratios are normally greater than unity in both bottom sediment and suspended sediment. The disequilibrium probably arises because of preferential leaching of U during chemical weathering. The addition of ^{230}Th is less likely, simply because Th is much less soluble than U. Interestingly, there is generally a correlation between Th/U mass ratios and ^{230}Th/^{234}U activity ratios in both suspended sediment and bottom sediment. This implies that ^{232}Th must be mobilised to some extent during weathering processes to give a correlation with the ^{238}U decay series.

^{228}Th forms after the decay of ^{228}Ra in the ^{232}Th decay series. Because Ra can usually enter solution in normal waters easily, ^{228}Th/^{232}Th activity ratios in river sediments are typically less than unity, lying in the range 0.8–0.9. In river water, the ^{228}Th/^{232}Th activity ratio can be greater than unity, reflecting the growth of ^{228}Th from ^{228}Ra in solution.

Pa

The behaviour of Pa is very similar to that of Th, as it too is very insoluble in most natural surface waters, but little information is currently available. It is believed that the ^{231}Pa daughter of ^{235}U is mainly present in the inorganic detrital grains of the sediment. Certainly, the ^{230}Th/^{231}Pa activity ratios obtained lie between 20 and 25, which is close to the equilibrium ^{238}U/^{235}U activity ratio of 21.7, suggesting that no chemical isotopic fractionation of this element takes place (Scott, 1982).

Ra

Ra is one of the most interesting elements in river waters. Two isotopes, ^{226}Ra and ^{228}Ra, have been extensively studied. The other isotopes, ^{223}Ra and ^{224}Ra, are of little importance. Rivers form a very important transport medium for this element. In river waters, the activity of ^{226}Ra ranges from less than 0.01 dpm l^{-1} (disintegration min^{-1} l^{-1}) to over 0.3 dpm l^{-1} and is typically about 0.1 dpm l^{-1}. There is no correlation with HCO_3^- or total dissolved solids, neither is there any correlation between the activities of ^{226}Ra and ^{238}U. Therefore, Ra does not simply follow U, and the Ra content of river water appears to be controlled by a different mechanism to that which controls U. Ra is probably carried in solution, concentrations of the element reflecting the composition of source rocks in the catchment area and, in particular, the contribution of groundwater to surface flow; U concentration, in contrast, are mainly controlled by the U content of stream sediment, particularly suspended sediment, and its organic and clay content. Although the solubility of some Ra salts is quite low, because the absolute abundance of Ra is very low there is no theoretical barrier to its mobility.

^{228}Ra is usually found to be greatly in excess of its parent ^{232}Th, a feature attributed to its ability to diffuse out of stream sediments into river water. Like ^{226}Ra, the activity for ^{228}Ra has a very wide range, from about 0.01 dpm l^{-1} to over 0.3 dpm l^{-1}, with an average of about 0.1 dpm l^{-1} (Scott, 1982). The similar ranges and average activity values of the two Ra isotopes are surprising, bearing in mind the different natural abundances, solubilities and distributions of their parent isotopes in crustal rocks. These differences are reflected in the ^{228}Ra/^{226}Ra activity ratio, which is very variable.

Pb

^{210}Pb in river waters may originate in two different ways: by decay of ^{226}Ra in solution, or by decay of atmospheric ^{222}Rn transported into waters or leached from soils and vegetation following wet or dry fall-out. The concentration of ^{210}Pb in water is usually very low except where the pH is less than 4, such as where there is drainage from mine workings. In normal river water, the ^{210}Pb activity is, in fact, less than 0.0001 dpml^{-1}, while in acid waters it may reach 0.14 dpml^{-1} (Scott, 1982). ^{210}Pb is normally associated with river sediments, either adsorbed onto organic particles or clay minerals or coprecipitated with Fe and Mn oxides. The ^{210}Pb may thus occur both in the suspended load and in the bottom sediments.

6.4.3 Groundwater

Stability relationships

This is perhaps the most important environment in which U series disequilibrium is developed. Differences in the solubility of the various members of the U and Th decay series are manifested under a wide range of Eh and pH conditions. U contents range between 0.001 and 1000 μg l^{-1}, and ^{234}U/^{238}U activity ratios between about 0.5 and several hundreds. The dissolved residence time in oxidised water is about 10^8 years for ^{234}U, but ^{232}Th has a dissolved residence time under similar conditions of only 100 years. In groundwater, the mechanisms of diffusion and α-recoil are most obvious in causing preferential uptake of some members of the decay series. In particular, the uptake of Rn is very significant. This may be by diffusion across boundaries, or by recoil diffusion of the higher members of the decay series, as is the case for ^{234}Th in which the low energy α-particle emitted by the parent ^{238}U is sufficient to cause a recoil displacement of 2×10^{-8} m in crystalline materials. For higher energy α-particles, the recoil displacements are proportionally larger. This partition produces the low ^{234}U/^{238}U activity ratios typically found in sediments and the high ^{234}U/^{238}U activity ratios found in most groundwaters (Fig. 6.12). The fact that simple differential leaching is not the mechanism involved in ^{234}U/^{238}U disequilibrium is indicated by the chemical similarity of these isotopes and because the intervening members of the decay series (^{234}Th and ^{234}Pa) are both relatively insoluble and have short half-lives (Osmond and Cowart, 1982).

The half-life of a radionuclide also affects the abundance in which it is likely to be found in groundwaters, unless there is continuous replenishment from adjacent rocks. However, the rate of replenishment and the rate of decay may not be the same and, indeed, the replenishment rate may vary with time.

The geochemical range of a radionuclide, which is a measure of its mobility in the hydrosphere, is determined by both its solubility and its half-life (Fig. 6.13). Thus, ^{238}U and ^{235}U, which have very long half-lives, have very high ranges; with decreasing half-life, the range decreases through ^{234}U, ^{226}Ra, ^{228}Ra, ^{223}Ra and ^{222}Rn to ^{224}Ra, although all are reasonably soluble. The short half-life, relatively insoluble radionuclides, such as ^{234}Pa, ^{231}Th, ^{227}Th, ^{234}Th, ^{210}Po and ^{228}Th, have very short ranges. As a consequence, Th, Pa and Po are rarely

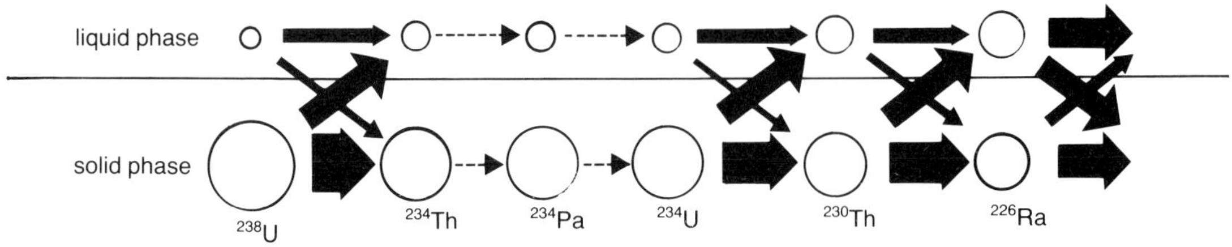

Fig. 6.12 — Origin of ^{238}U series disequilibrium in groundwater by recoil. The sizes of nuclide representation and solid decay arrows indicate relative α-particle activity. The broken arrows represent β-particle decays. (Osmond and Cowart, 1982, redrawn with permission.)

detected in solution. Elements which have intermediate ranges and which are found in some circumstances in solution, are Pb, Bi and Ac.

The concept of range must be used with some caution, as natural systems are complex and the solubility of any nuclide will vary with changing Eh and pH. Also, the rate of supply of a nuclide is controlled by its growth rate relative to the rate of water flow through the rock. Any variation in the concentration of the parent nuclide along the flow path will also affect the supply of a nuclide, especially if the contact time of water with a zone of enhanced parent concentration is shorter than the half-life of the daughter (Fig. 6.14). Within a small well-defined zone of U and Th enrichment in rock adjacent to flowing water, if the flow rate is such

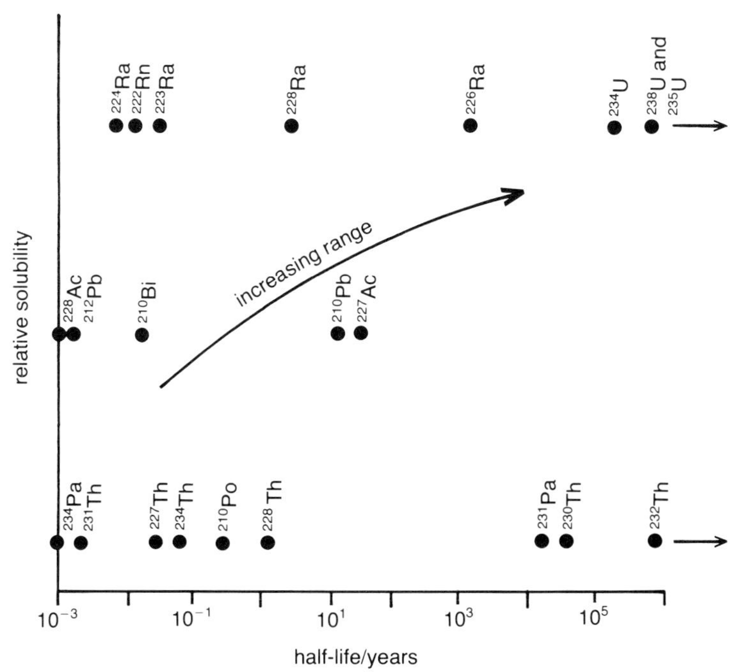

Fig. 6.13 — Range of various radionuclides as a function of half-life and solubility. (Osmond and Cowart, 1982, redrawn with permission.)

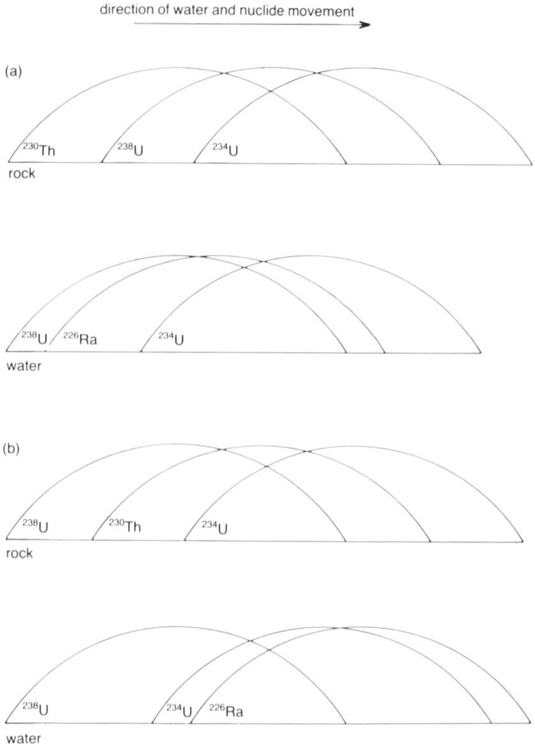

Fig. 6.14 — ^{238}U decay series disequilibrium in rock and water systems across a U-front: (a) fast-moving system; (b) slow-moving system. (Osmond and Cowart, 1982, redrawn with permission.)

that the exposure time in the zone is about equal to the half-life of ^{228}Ra (that is, much less than the half-life of ^{226}Ra and much greater than the half-lives of ^{223}Ra and ^{224}Ra), the concentrations of ^{223}Ra and ^{224}Ra rise and fall very rapidly close to the limits of the enriched zone. ^{228}Ra, however, shows a progressive increase towards the downstream limit of the zone and then decreases. ^{226}Ra shows an overall increase downstream, although the rate of increase is slightly greater over the width of the zone than it is beyond it (Osmond and Cowart, 1982).

It has been recognised that, when describing the solubility of radionuclides, problems arise when concentration is expressed in activity units (such as disintegrations per minute per litre, picocuries per litre or becquerels per litre). Because the mass concentrations of two isotopes of equal activities differ in proportion to their half-lives, the ratio of the mass concentrations of ^{238}U (with a long half-life) to ^{234}U (with a short half-life) in groundwater with an activity ratio of 1.0 is almost 20 000:1. Thus, when measured in terms of specific radioactivity, ^{238}U encounters solubility product limits sooner than ^{234}U. Similarly, although ^{232}Th appears to have a very low abundance compared with that of ^{228}Th and is rarely found in solution at measurable levels, the much shorter half-life of ^{228}Th gives it a level of detectability (in terms of mass) that is about 10^{-10} that of ^{232}Th. Because of this, Osmond and Cowart (1982) consider that some elements which are normally thought to be insoluble may, in fact, only be insoluble with respect to apparently more abundant shorter-lived radionuclides. Indeed, the inverse relationship of solubility and half-life that is usually found in waters in which ^{234}Th > ^{228}Th > ^{230}Th > ^{232}Th is unlikely to be coincidental. Moreover, factors which affect the concentration of isotopes in solution, such as the formation of complex ions, coprecipitation and adsorption onto organic particles or clay minerals, all differ according to the actual concentration present, so that the true solubility of isotopes is very difficult to judge. Nevertheless, systems assuming very simple conditions can be modelled mathematically; for example,

$$A_{\mathrm{w}}^{\mathrm{d}} = \frac{3L\rho A_{\mathrm{R}}^{\mathrm{P}}}{4r\phi}$$

where $A_{\mathrm{w}}^{\mathrm{d}}$ is the steady-state activity in groundwater of a daughter radionuclide formed by the decay of parent of activity $A_{\mathrm{R}}^{\mathrm{P}}$, which is in a rock matrix of bulk density ρ made up of grains of average radius r, and with a fractional porosity ϕ. L is the recoil distance of an intermediate daughter. The values of L, ρ and r are expressed in cgs units. Where the steady state is not obtained, as when the water flow rate is high compared with the half-life of the daughter, some intermediate activity I^{d} is produced which lies between an initial state ${}^{w}A^{d}$ and the steady state A^{d} . This is controlled by^{wi} the decay constant (λ_{d}^{wf}) for the daughter and the duration of the rock interaction (t), such that

$$I_{w}^{d} = A_{wi}^{d} + (A_{wf}^{d} - A_{wi}^{d})\ (1 - e^{-\lambda_{d}t})$$

or

$$\Delta I_{w}^{d} = \Delta A_{wf}^{d}\ (1 - e^{-\lambda_{d}t}),$$

where the prefix Δ indicates the change in activity, that occurs if the water flows from a zone of lower radioactivity to one of higher radioactivity in the rock matrix. Alternatively, if flow is from higher to lower radioactivity,

$$\Delta I_{w}^{d} = \Delta A_{wf}^{d}\ (e^{-\lambda_{d}t}).$$

Aquifers may be classified on the basis of their radionuclide composition into the steady state, augmenting and decaying types (Osmond and Cowart, 1982). In a steady-state aquifer, water flows very slowly through a matrix with a low and uniform U distribution. Such an aquifer has a very high $^{234}U/^{238}U$ activity ratio. However, if the aquifer becomes deeply buried, re-equilibration at high temperatures may actually produce low $^{234}U/^{238}U$ activity ratios. Short-lived radionuclides, such as ^{226}Ra, ^{224}Ra and ^{210}Pb, may generally be considered to be present in steady-state relationship with each other in many aquifers, but redistribution by flowing water results in spatial variation of steady-state values within an aquifer.

An augmented aquifer develops where the radioactivity of the rock matrix is high and yet the rock is resistant to solution. Typical augmented aquifers occur in fractured granite. The ^{238}U content of the water is moderate, and the concentration of ^{234}U increases with time so that the $^{234}U/^{238}U$ activity ratio also increases and may reach 10 or more. Other members of the ^{238}U decay series also increase in concentration with time. Concentrations of the shorter-lived members increase most rapidly, and $^{224}Ra/^{226}Ra$ activity ratios of several hundreds may develop.

A decaying aquifer is one in which groundwater moves from high to low radioactivity areas and becomes progressively depleted in the shorter-lived radionuclides. This is characteristic of many deep aquifers in which recharge waters pass through enriched zones produced by weathering before passing down into the main aquifer, or where descending O_2-rich waters meet a reducing groundwater environment.

In flowing groundwater, the relative rates of dispersion of the nuclides of the ^{238}U decay series generally depend upon the speed of flow. In fast-moving waters, ^{234}U moves ahead of ^{226}Ra and this nuclide in turn moves ahead of ^{238}U while, in the solid matrix, dispersion causes ^{234}U to move ahead of ^{238}U, and ^{238}U ahead of ^{230}Th. In slow-moving waters, ^{226}Ra moves ahead of ^{234}U, and this nuclide moves ahead of ^{238}U while, in the rock matrix, ^{234}U moves ahead of ^{230}Th and this in turn moves ahead of ^{238}U.

Hydrothermal systems

Hydrothermal groundwater systems are worth special mention. In active geothermal areas, the groundwaters are characteristically enriched in Rn and Ra. For this reason, it has been suggested that hot springs indicate U accumulations at depth. However, most geothermal waters are deficient in U and have low $^{234}U/^{238}U$ activity ratios which are unlike the high values that would be expected of groundwaters influenced by large U deposits at depth.

Interestingly, the U concentration of hot springs is often lower than that of nearby cold springs. The different temperatures indicate that different groundwater flow systems are involved: shallow flow for the cold springs and deep circulation for the thermal springs, with different water–rock interactions in each. Circulation rates, in particular, are likely to be different, with rapid movement possible under the thermal head present in geothermal areas. The higher groundwater temperatures in geothermal areas will also affect the nature of water–rock interactions. Thus, even within a single geothermal system, concentrations of ^{238}U and members of its decay series tend to be highest in low temperature waters rich in $CaCO_3$, and relatively low in high temperature systems in which SiO_2 predominates. The ability of ^{238}U to be mobilised may, therefore, be restricted to HCO^--rich waters. Thus, direct association of geothermal systems with U deposits is not necessary. Even in high $^{222}Rn/^{226}Ra$ activity ratio springs, the level of activity found can generally be attributed to the presence of the ^{238}U decay series in normal crustal abundances. Thus, although the excess ^{222}Rn found in many thermal springs could be supplied by a few kilograms of U at depth, this need not be concentrated in one place. If the flow path is sufficiently long, an equivalent mass distributed through a large volume of rock will give the same result.

The mass transfer of thermal waters within rock is facilitated both by the intergranular permeability of highly porous rocks and by the presence of fractures within rocks of high or low porosity. Fluid flow in fissured rocks differs in three important respects from the more pervasive percolation that takes place in intact porous rocks (Garg and Kassoy, 1981):

(1) Fracture-induced permeability is usually much greater than the permeability of interconnected pores. In fact, the matrix permeability of rock only becomes important in the absence of continuous fractures or where fracture apertures are less than 10 μm.
(2) Fracture permeability is usually anisotropic.
(3) Fracture porosity (and hence permeability) is much more sensitive to fluid pressures and rock stresses than is matrix porosity.

The spacing of fractures is an important factor when describing flow through fractured rocks. If the fracture spacing is close, in terms of total linear dimensions of the rock mass, processes which are dominantly fracture controlled on a small scale may be better explained in terms of pervasive flow when considered on a large scale.

The tendency of a body of water to undergo convective flow in a rock layer defined by its conductivity and permeability is determined by the location of that body of water in the rock layer. If a porous medium of thickness h, permeability k and saturated thermal conductivity k_m is saturated with water of density ρ_0, thermal expansion coefficient γ, specific heat c and kinetic viscosity η under the acceleration g, due to gravity, and if heating occurs in one section of the base of the unit so that the temperature is enhanced from T to $T + \Delta T$, a change in the water density of $\Delta\rho = \rho_0 \, \Delta T \, \gamma$ occurs (Elder, 1981). The system can be described using the Rayleigh number R_n which is given by

$$R_n = \frac{g \, \Delta\rho \, kh}{\rho_0 k_m \eta}.$$

At appropriate values of R_n, a circulation system will develop in which cold water moves horizontally into the heated region and rises rapidly as a vertical plume of hot water. As it rises, the plume spreads very little until it reaches the upper surface of the rock layer, when it spreads out and slows down. Outside this region, the unheated water merely takes up a weak induced motion. The bulbous body of hot water above the heated section is called the 'mushroom'.

When convective systems of the mushroom type lose water through surface discharge, there is a progressive collapse of the mushroom until the system approaches the form of a pipe flow model, which ignores the details of how the heat enters the system and considers only forced convection along fractures.

In fact, one of the commonest, and certainly the simplest, systems is that in which the main mass of water passes only once through the rock. This can be illustrated by a model in which a single vertical conduit connects via a horizontal channel with vertical side channels. If cold water at a temperature T charges the vertical side channels, heating occurs at depth in the horizontal connection channel where the temperature increases by ΔT and the density of the water changes from ρ_0 to ρ_1. The vertical conduit of height h then represents the upwelling region of the geothermal area. Convection is driven by the horizontal pressure gradient resulting from the difference between the cold and hot hydrostatic heads between the conduit and side channels. If the topographic head in the recharge channels is zero, the effective head due to heating can be established. For example, in the Wairakei geothermal field in New Zealand, where the temperature is about 250 °C and h about 5 km, the equivalent head produced is as great as 1 km. If in this type of pipe flow model the energy input is large or the resistance to discharge along the pipe is high, the energy cannot be transferred by water flow alone, and a change to the gaseous state occurs. Conversely, at low values of thermal input and low resistance to flow, the water remains in the liquid state.

In established geothermal fields, a permeability of 0.1 darcy is more than adequate to permit all observed hydrothermal phenomena, and the contribution of intergranular permeability is negligible. Interesting calculations have been made for a system in which the rock is represented as a cubic array of blocks of dimension D_1, with narrow spaces of width D_2 separating them. If $D_1 = 1$ m and $D_2 = 0.1$ mm, a permeability of 0.2 darcy results but, if $D_1 = 100$ m and $D_2 = 0.5$ mm, a permeability of 0.2 darcy is still produced, as it is if $D_1 = 1000$ m and $D_2 = 1$ mm (Elder, 1981). It is thus easy to see that even occasional fractures (which need not even be large) give a very significant permeability to the rock mass as a whole.

It has been shown that, in many geothermal areas in which water movement occurs along fracture zones, heat will be lost to the adjacent rocks. If a hydraulic connection exists between the upflow zone and the adjacent rocks, convective charging of the adjacent rocks may occur. Even when an impermeable barrier exists between the two regions, heat transfer will occur by conduction. This relationship has been investigated in terms of a vertical barrier separating a conduit from an adjacent aquifer, and it has been found that, even in this case, conductive heat flow can be sufficient to generate convection in the aquifer and cause discharge of thermal springs.

Fault-controlled charging of a geothermal reservoir has been investigated using a model in which a fracture zone extends downwards through an impermeable surface layer and an interbedded series of permeable and impermeable layers to a fractured basement rock. In this model, it is found that, under the influence of the thermal head, hot water rises through the fault as before, but the presence of the impermeable surface layer suppresses vertical transport, and water is pushed out of the fault zone into the permeable members of the interbedded series. Horizontal flow in these permeable layers then becomes dominant to distances away from the fault zone of between five and ten fault depths.

Tracing the origin and mixing of groundwaters

The presence of naturally occurring radionuclides in hydrothermal systems can be utilised both in dating and in tracing the origin of the waters. The dating of groundwaters will be discussed in Chapter 10, but tracing will be considered here. Naturally occurring radionuclides can be used effectively to determine both sources of water and mixing functions. Sea water, for example, has an essentially constant U content of 13.3 $\mu g\,l^{-1}$ and a $^{234}U/^{238}U$ activity ratio of 1.14. Determination of the isotopic composition of a groundwater can therefore indicate whether or not there has been any direct marine influence. More generally, zones of active water movement may have distinct isotopic compositions which allow them to be traced. Isotopic compositions need not remain constant, and the character of any changes observed can be used to interpret the nature of the water–rock interactions involved in the flow regime.

Groundwaters can also be characterised by their $^{228}Ra/^{226}Ra$ activity ratios which are less variable than $^{234}U/^{238}U$ activity ratios. This may in part be because of shorter half-lives of the two Ra isotopes and partly because each is a member of a separate decay series, the chemical behaviour of the ^{232}Th and ^{238}U parents being different.

The mixing histories of groundwaters can be studied by determining the isotopic composition of possible source waters and developing mixing equations. For example, the mixing of water from three different sources in an aquifer studied in south Nevada and California has been described using equations of the form

$$\frac{V_2}{V_1} = \frac{(C_1 - C_3)\dfrac{(X_4 - X_3)}{(C_4 - C_3)} - (X_1 - X_3)}{(C_2 - C_3)\dfrac{(X_4 - X_3)}{(C_4 - C_3)} - (X_2 - X_3)},$$

where V is the relative volume of water, C is the U content and X is the excess (or deficiency) of ^{234}U, that is, $^{234}U/^{238}U - 1$ multiplied by the U content (Osmond and Cowart, 1982). The subscript 1 refers to the final result of mixing, and subscripts 2, 3 and 4 to the three possible sources.

Where simple mixing of two components is concerned, plots of the ^{234}U excess against the ^{238}U content for different access points in the mixing scheme generally display an approximately linear distribution between the compositions of the two sources. Where three components are present as possible sources, the same type of plot can be used, but here the three sources plot as the apices of a triangle, and the products of mixing, or mixing trends, lie within the triangle.

BIBLIOGRAPHY

Adams, S. S., Curtis, H. S., Hafen, P. L., and Salek-Nejad, H. (1978) Interpretation of post-depositional processes related to the formation and destruction of the Jackpile–Paguate uranium deposit, northwest New Mexico, *Econ. Geol.* **73**, 1635.

Aleinikoff, J. N. (1983) U–Th–Pb systematics of zircon inclusions in rock-forming minerals: a study of armoring against isotopic loss using the Sherman granite of Colorado–Wyoming USA. *Contrib. Mineral. Petrol.* **83**, 259.

Ames, L. L., McGarrah, J. E., Walker, B. A., and Salter, P. F. (1983) Uranium and radium sorption on amorphous ferric oxyhydroxide. *Chem. Geol.* **40**, 135.

Amiel, A. J., Miller, D. S., and Friedman, G. M. (1973) Incorporation of uranium in modern corals. *Sedimentology* **20**, 523.

Amov, B. G. (1983) Evolution of uranogenic and thorogenic lead, 1. A dynamic model of continuous isotopic evolution. *Earth Planet. Sci. Lett.* **65**, 61.

Amov, B. G. (1983) Evolution of uranogenic and thorogenic lead 2. Some differences in the variations of the $^{206}Pb/^{204}Pb$ and $^{208}Pb/^{204}Pb$ ratios. *Earth Planet. Sci. Lett.* **65**, 311.

Anderson, R. F. (1982) Concentrations, vertical flux and remineralization of particulate uranium in seawater. *Geochim. Cosmochim. Acta* **46**, 1293.

Andreoli, M. A. G., Robb, L. J., Meyer, M., Ainslie, L. C., and Hart, R. J. (1985) Granitoids of the pre-Witwatersrand basement: clues to the source of uranium placer mineralisations. In *Recognition of uranium provinces*. Vienna, International Atomic Energy Agency.

Anis, A., and Voijin, V. (1979) Some considerations on the uranium content distribution in marine rocks. *Nucl. Tracks* **3**, 83.

Armstrong, F. C. (1981) Genesis of U and Au-bearing quartz pebble conglomerates. *US Geol. Surv. Prof. Paper* **1161**.

Augustithis, S. S. (1983) *Leaching and diffusion in rocks and their weathering products*. Athens, Theophrastus.

Avital, Y., Starinsky, A., and Kolodny, Y. (1963) Uranium geochemistry and fission track mapping of phosphorites, Zeta Field, Israel. *Econ. Geol.* **78**, 121.

Ball, T. K., and Basham, I. R. (1979) Radioactive accessory minerals in granites from southwest England. *Proc. Ussher Soc.* **4**, 437.

Ball, T. K., Basham, I. R., and Michie, U. McL. (1979) Uraniferous vein occurrences of south-west England — paragenesis and genesis. In *Proceedings of the International Atom Energy Agency*. Vienna, *Meeting on the Geology of Vein and Similar Type Uranium Deposits, Lisbon*.

Barakso, J. J. (1979) Geochemical dispersion of uranium in overburden-covered regions. *Can. Min. Metall. Bull.* **72**, 135.

Barbier, M. J. (1974) Continental weathering as a possible origin of vein-type uranium deposits. *Mineral Deposita* **9**, 271.

Barnes, H. L. (1979) *Geochemistry of hydrothermal ore deposits*. New York, Wiley.

Barretto, P. (1985) Identification of uranium provinces in Brazil. In *Recognition of uranium provinces*. Vienna, International Atomic Energy Agency.

Basham, I. R., Ball, T. K., Beddoe-Stephens, B., and Michie, U. McL. (1982) Uranium-bearing accessory minerals and granite fertility, I, Methods of identification and evaluation; II, Studies of granites from the

British Isles. In *Uranium exploration methods*. Vienna, International Atomic Energy Agency.

Beeson, R. (1980) The recognition of hydrothermal processes in the trace element geochemistry of uranium and thorium enriched tuffaceous rocks, northwest Iran. *J. Geochem. Expl.* **13**, 9.

Bell, R. T. (1978) Uranium in black shales — a review. In Kimberley, M. M. (ed.), *Uranium deposits, their mineralogy and origin, Mineral. Assoc. Can. Short Course Handbook* **3**.

Bloch, S. (1980) Some factors controlling the concentration of uranium in the world ocean. *Geochim. Cosmochim. Acta* **44**, 373.

Borovec, Z. (1981) The adsorption of uranyl species by fine clay. *Chem. Geol.* **32**, 45.

Borovec, Z., Kribek, B., and Tolar, V. (1979). Sorption of uranyl by humic acids. *Chem. Geol.* **27**, 39.

Bouda, S. (1985) Mössbauer effect and geochemical study of solution-mineralization in surficial environment of south west England. *Ph.D. Thesis*. University of Exeter.

Bowie, S. H. U. (1977) Uranium distribution and availability. In Jones, M. J. (ed.), *Geology, mining and extractive processing of uranium*. London, Institution of Mining and Metallurgy.

Bowie, S. H. U. (1978) Global distribution of uranium ores and potential UK deposits. In Duff, P. McL. D. (ed.), *Geological aspects of uranium in the environment, Geol. Soc. Lond. Misc. Paper* **7**.

Bowie, S. H. U. (1979) Mode of occurrence and distribution of uranium deposits. In Bowie, S. H. U., Fyfe, W. S., Ostle, D., Plant, J. and Simpson, P. R. (eds), *Theoretical and practical aspects of uranium geology*. London, Royal Society.

Boyle, R. W. (1982) *Geochemical prospecting for uranium and thorium deposits*. Amsterdam, Elsevier.

Brookins, D. G. (1982) Geochemistry of clay minerals for uranium exploration in the Grants mineral belt, New Mexico. *Mineral. Deposita* **17**, 37.

Brookins, D. G. (1984) *Geochemical aspects of radioactive waste disposal*. New York, Springer.

Burnett, W. C., and Veeth, H. H. (1977) Uranium series disequilibrium studies in phosphatic nodules from the west coast of South America. *Geochim. Cosmochim. Acta* **41**, 755.

Campbell, J. A. (1977) Short papers of the USGS uranium–thorium symposium. *US Geol. Surv. Circ.* **753**.

Cannon, H. L. (1960) The development of botanical methods of prospecting for uranium on the Colorado Plateau. *US Geol. Surv. Bull.* **1085A**.

Capus, G., and Munier, C. (1978) Mise en evidence de fortes concentrations d'uranium dans les corps microbiens de micro-organismes actuels. *C. R. Acad. Sci.* **287**, 191.

Carlisle, D., Merifield, P. M., Orme, A. R., Kohl, M. S., and Kolker, O. (1978) The distribution of calcretes and gypcretes in southwestern United States and their uranium favourability, based on a study of deposits in Western Australia and South West Africa (Namibia). *Rep.* **GJBX 29–78**. US Department of Energy.

Cathcart, J. B. (1978) Uranium in phosphatic rock. *US Geol. Surv. Prof. Paper* **988**.

Cheney, E. S. (1981) The hunt for giant uranium deposits. *Am. Sci.* **69**, 34.

Chung, Y. C. (1974) Radium-226 and Ra–Ba relationships in Antarctic and Pacific waters. *Earth Planet. Sci. Lett.* **23**, 125.

Cochran, J. K. (1982) The oceanic chemistry of the U and Th series nuclides. In Ivanovich, M., and Harmon, R. S. (eds), *Uranium series disequilibrium: applications to environmental problems*. Oxford, Oxford University Press.

Cowan, G. A., and Adler, H. H. (1976) The variability of the natural abundance of ^{235}U. *Geochim. Cosmochim. Acta* **40**, 1487.

Dahlkamp, F. J. (1978) Classification of uranium rocks. *Mineral. Deposita* **13**, 83.

Darnley, A. G. (1982) "Hot" granites: some general remarks. In Maurice, Y. T. (ed.), *Uranium in granites, Paper Geol. Surv. Can.*

De Ment, J., and Dake, H. C. (1948) *Handbook of uranium minerals*. Portland, Oregon. Minerologist Publishing Company .

Derry, D. R. (1980) Uranium deposits through time. In Strangway, D. W. (ed.), *The continental crust and its mineral deposits, Geol. Assoc. Can. Spec. Paper* **20**.

Dexter-Dyer, B., Kretzschmar, M., and Krumbein, W. E. (1984) Possible microbial pathways in the formation of Precambrian ore deposits. *J. Geol. Soc.* **141**, 251.

Diouly-Osso, P. and Chauvet, R. J. (1979) Les gisements d'uranium de la region de Franceville (Gabon). In *Uranium deposits in Africa: geology and Exploration*. Vienna, International Atomic Energy Agency.

Dostal, J., and Capedri, S. (1978) Uranium in metamorphic rocks. *Contrib. Mineral. Perol.* **66**, 409.

Duncan, J. M., Witherspoon, P. A., Mitchell, J. K., Watkins, D. J., Hardcastle, J. H., and Chen, J. C. (1972) Seepage and groundwater effects associated with explosive cratering. *Rep.* **TE-72.2.** University of California.

Durrance, E. M. (1980) A possible example of the survival of bacteria colonies for a long period of geological time. *Spec. Sci. Technol.* **3**, 265.

Durrance, E. M., Bromley, A. V., Bristow, C. M., Heath, M. J., and Penman, J. M. (1982) Hydrothermal circulation and post-magmatic changes in granites of southwest England. *Proc. Ussher Soc.* **5**, 304.

Durrance, E. M., Meads, R. E., Ballard, R. R. B., and Walsh, J. N. (1978) Oxidation state of iron in the Littleham Mudstone Formation of the New Red Sandstone Series (Permian–Triassic) of southeast Devon, England. *Bull. Geol. Soc. Am.* **89**, 1231.

Dyck, W. (1978) The mobility and concentration of uranium and its decay products in temperate surficial environments. In Kimberley, M. M. (ed.), *Uranium deposits, their mineralogy and origin, Mineral. Assoc. Can. Short Course Handbook* **3**.

Dyck, W. (1980) Uranium, radon, helium and other trace elements and gases in well waters of parts of the St. Lawrence lowlands. *J. Geochem. Expl.* **13**, 27.

Dyck, W., and McCorkell, R. H. (1983) a study of uranium-rich reduction spheroids in sandstones from Pugwash Harbour, Nova Scotia. *Can. J. Earth Sci.* **20**, 1738.

Dypvik, H., and Erikson, D. O. (1983) Natural radioactivity of clastic sediments and the contributions of U, Th and K. *J. Petrol. Geol.* **5**, 409.

Elder, J. (1981) *Geothermal systems*. London, Academic Press.

Elsinger, R. J., and Moore, W. S. (1983) ^{224}Ra, ^{228}Ra and ^{226}Ra in Winyah Bay and Delaware Bay. *Earth Planet. Sci. Lett.* **64**, 430.

Evans, A. M. (1982) *Metallization associated with acid magmatism*. Chichester, West Sussex, Wiley.

Fritz, P., and Fontes, J. C. (1980) *Handbook of environmental isotope geochemistry,* Vol. 1, *The terrestrial environment*. Amsterdam, Elsevier.

Fyfe, W. S. (1979) The geochemical cycle of uranium. In Bowie, S. H. U., Fyfe, W. S., Ostle, D., Plant, J., and Simpson, P. R. (eds), *Theoretical and practical aspects of uranium geology*. London, Royal Society.

Fyfe, W. S., Price, N. J., and Thompson, A. B. (1978) *Fluids in the Earth's crust*. Amsterdam, Elsevier.

Gallagher, M. J., Michie, U. McL., Smith, R. T., and Haynes, L. (1971) New evidence of uranium and other mineralisation in Scotland. *Trans. Inst. Min. Metall. B,* **80**, 150.

Garg, S. K., and Kassoy, D. R. (1981) Convective heat and mass transfer. In Ryback, L., and Muffler, L. J. P. (eds), *Geothermal systems*. Chichester, West Sussex, Wiley.

Gascoyne, M. (1982) Geochemistry of the actinides and their daughters. In Ivanovich, M., and Harmon, R. S. (eds), *Uranium series disequilibrium: Applications to environmental problems*. Oxford, Oxford University Press.

Giblin, A. M., Batts, B. D., and Swaine, D. J. (1981) Laboratory simulation studies of uranium mobility in natural waters. *Geochim. Cosmochim. Acta* **45**, 699.

Gmelin's handbooks on inorganic chemistry. Various volumes on uranium and thorium. Berlin, Springer.

Goldhaber, M. B., Reynolds, R. L., and Rye, R. O. (1978) Origin of a south Texas roll-type uranium deposit. *Econ. Geol.* **73**, 1677.

Gorgoni, C., Martinelli, G., and Sighinolfi, G. P. (1982) Radon distribution in groundwater of the Po sedimentary basin (Italy). *Chem. Geol.* **35**, 297.

Govett, G. J. S. (1982) *Rock geochemistry in mineral exploration*. Oxford, Elsevier.

Goyal, K. P. (1978) Heat and mass transfer in a saturated porous medium with application to geothermal reservoirs. *Ph.D Thesis*. University of Colorado.

Goyal, K. P., and Kassoy, D. R. (1977) A fault controlled model of the Mesa anomaly. *Proceedings of the Third Workshop — Geothermal Resources Engineering*. 209. Stanford, California, Stanford University Press.

Grainger, L. (1958) *Uranium and thorium*. London, Newnes.

Green, R. D. (1982) Uraninite and its secondary uranium-bearing minerals. *Rocks Minerals* **57**, 246.

Haji-Vassiliou, A. (1974) Uranium mineralisation — uraninite. *Mineral. Record* **5**, 79.

Halbach, P., Von Borstel, D., and Gunderman, K. D. (1980) The uptake of uranium by organic substances in a peat bog environment on a granite. *Chem. Geol.* **29**, 117.

Hallbauer, D. K. (1975) The plant origin of the Witwatersrand carbon. *Minerals. Sci. Eng.* **7**, 111.

Hamilton, E. I. (1975) The abundance and distribution of uranium in some oceanic, continental utramafic inclusions and host basalts. *Chem. Geol.* **16**, 221.

Harmon, R. S., and Rosholt, J. N. (1982) Igneous rocks. In Ivanovich, M., and Harmon, R. S. (eds), *Uranium series disequilibrium: applications to environmental problems*. Oxford, Oxford University Press.

Harrison, R. K. (1975) Concretionary concentrations of the rarer elements in Permo-Triassic red beds of south-west England. *Bull. Geol. Survey Gt. Brit.* **52**, 1.

Hart, S. R., and Staudigel (1982) The control of alkalies and uranium in seawater by ocean crust alteration. *Earth Planet. Sci. Lett.* **58**, 202.

Heier, K. S. (1979) The movement of uranium during higher grade metamorphic processes. In Bowie, S. H. U., Fyfe, W. S., Ostle, D., Plant, J., and Simpson, P. R. (eds), *Theoretical and practical aspects of uranium geology*. London, Royal Society.

Hein, J. R., Koski, R. A., and Morgenson, L. A. (1982) Uranium and thorium enrichment in rocks from the base of the DSDP hole 465A, Hess Rise, central North Pacific. *Chem. Geol.* **36**, 237.

Heinrich, E. W. (1958) *Mineralogy and geology of radioactive raw materials*. New York, McGraw-Hill.

Hoeve, J., and Sibbald, T. I. I. (1978) On the genesis of Rabbit Lake and other unconformity-type uranium deposits in northern Saskatchewan, Canada. *Econ. Geol.* **73**, 1450.

Holland, H. D., and Schidlowski, M. (1982) *Mineral deposits and the evolution of the biosphere*. New York, Springer.

Hussain, N., and Krishnaswami S. (1982) The behaviour of short-lived radiogenic lead isotopes (^{214}Pb and ^{212}Pb) in groundwaters and laboratory leaching experiments. *Earth Planet. Sci. Lett.* **58**, 430.

International Atomic Energy Agency (1979) *Uranium deposits in Africa, geology and exploration*. Vienna, International Atomic Energy Agency.

International Atomic Energy Agency (1982) *Vein-type and similar uranium deposits in rocks younger than the Proterozoic*. Vienna, International Atomic Energy Agency.

Kassoy, D. R., and Zebib, A. (1978) Convection fluid dynamics in a model of a fault zone in the Earth's crust. *J. Fluid Mech.* **88**, 769.

Kimball, B. A. (1980) geochemistry of spring water, south-eastern Uinta Basin, Utah and Colorado. *US Geol. Surv. Water Supply Paper* **2074**.

Kimberley, M. M. (1978) Origin of stratiform uranium deposits in sandstone, conglomerate and pyroclastic rocks. In Kimberley, M. M. (ed.), *Uranium deposits, their mineralogy and origin, Mineral Assoc. Can. Short Course Handbook* **3**.

Kimberley, M. M., Grandstaff, D. E., and Tanaka, R. T. (1984) Topographic control on Precambrian weathering in the Elliot Lake uranium district, Canada. *J. Geol. Soc.* **141**, 229.

King, P. T., Michel J., and Moore, W. S. (1982) Ground water geochemistry of ^{228}Ra, ^{226}Ra and ^{222}Rn. *Geochim. Cosmoschim. Acta* **46**, 1173.

Kitano, Y. (1975) *Geochemistry of water*. Stroudsberg, Dowden, Hutchinson and Ross.

Lalou, C. (1982) Sediments and sedimentation processes. In Ivanovich, M., and Harmon, R. S. (eds), *Uranium series disequilibrium: applications to environmental problems*. Oxford, Oxford University Press.

Lambert, R. St. J., Chamberlain, V. E., and Holland, J. G. (1976) The geochemistry of Archaen rocks. In Windley, B. F. (ed.), *The early history of the Earth*. London, Wiley.

Langford, F. F. (1978) Mobility and concentration of uranium in arid surficial environments. In Kimberley, M. M. (ed.), *Uranium deposits, their mineralogy and origin, Mineral Assoc. Can. Short Course Handbook* **3**.

Langmuir, D. (1978a) Uranium solution — mineral equilibria at low temperatures with application to sedimentary ore deposits. In Kimberley, M. M. (ed.), *Uranium deposits, their mineraloggy and origin, Mineral. Assoc. Can. Short Course Handbook* **3**.

Langmuir, D. (1978b) Uranium solution — mineral equilibria at low temperatures with applications to sedimentary ore deposits. *Geochim. Cosmochim. Acta* **42**, 547.

Lepp, H. (1975) *Geochemistry of iron*. Stroudsberg, Dowden, Hutchinson and Ross.

Lerman, A. (1978) *Lakes: chemistry, geology, physics*. New York, Springer.

Lerman, A. (1979) *Geochemical processes: water and sediment environments*. New York, Wiley.

Leventhal, J. S. (1981) Pyrolysis gas chromatography — mass spectrometry to characterize organic matter and its relationship to uranium content of Appalachian Devonian black shales. *Geochim. Cosmochim. Acta* **45**, 883.

Lewis, D. M. (1977) The use of ^{210}Pb as a heavy metal tracer in the Susquehanna river system. *Geochim. Cosmochim. Acta* **41**, 1557.

Liebenberg, W. R. (1955) The occurrence and origin of gold and radioactive materials in the Witwatersrand system, the Dominion Reef, the Ventersdorp Contact Reef and the Black Reef. *Trans. Geol. Soc. S. Afr.* **58**, 101.

Louis, C. (1970) Water flows in fissured rock and their effects on the stability of rock massifs *Rep.* **UCRL TRANS 10469**. Lawrence Livermore Laboratory, University of California. (English translation (1967). *Dissertation*. University of Karlsruhe.)

Mangini, A., Sonntag, C., Bertsche, G., and Muller, E., (1979) Evidence for a higher natural uranium content in world rivers. *Nature* **278**, 337.

Mann, A. W., and Deutscher, R. L. (1978) Genesis principles for the precipitation of carnotite in calcrete drainages in Western Australia. *Econ. Geol.* **73**, 1724.

Matthess, G. (1982) *The properties of groundwater*. New York, Wiley.

Maynard, J. B. (1983) *Geochemistry of sedimentary ore deposits*. New York, Springer.

McLennan, S. M., and Taylor, S. R. (1980) Th and U in sedimentary rocks: crustal evolution and sedimentary recycling. *Nature* **285**, 621.

Meneghel, L. (1981) A model of the processes forming sedimentary-hosted uranium deposits. *Econ. Geol.* **76**, 727.

Michie, U. McL., and Cooper, D. C. (1979) Uranium in the Old Red Sandstone of Orkney. *Rep. Inst. Geol. Sci. Lond.* **78–16**.

Miller, W. R., Wanty, R. B., and McHugh, J. B. (1984) Application of mineral solution equilibria to geochemical exploration for sandstone-hosted uranium deposits in two basins in central Utah. *Econ. Geol.* **79**, 266.

Mitchell, A. H. G. (1981) *Mineral deposits and global tectonic settings*. London, Academic Press.

Mitchell, W. S., and Aumento, F. (1974) A geochemical comparison of minerals of oceanic and continental utramafic origin. *J. Geophys. Res.* **79**, 5529.

Moore, F., and Swart, P. K. (1979) The uranium content of ore minerals from St. Michael's Mount Cornwall. *Proc. Ussher Soc.* **4**, 432.

Moore, G. W. (1954) Extraction of uranium from aqueous solution by coal and some other materials. *Econ. Geol.* **49**, 652.

Moore, W. S. (1984) Mechanism of transport of U–Th series radioisotopes from solids into ground water. *Geochim. Cosmoschim. Acta* **48**, 395.

Morton, R. D. (1978) The identification of uraniferous minerals. In Kimberley, M. M. (ed.), *Uranium deposits, their mineralogy and origin, Mineral. Assoc. Can. Short Course Handbook* **3**.

Nash, J. T., Granger, H. C., and Adams, S. S. (1981) Geology and concepts of genesis of important types of uranium deposits. In Skinner, B. J. (ed.), *Econ. Geol. Seventy-Fifth Anniversary Volume (1905–1980)*.

Needham, R. S. (1979) A review of recent studies in the Pine Creek geosyncline with special reference to uranium. *Aust. Inst. Min. Metall.* **169**.

Needham, R. S., and Ewers, G. R. (1985) Pine Creek Geosyncline. In *Recognition of uranium provinces*. Vienna, International Atomic Energy Agency.

Nozaki, Y., and Horibe, Y., 1983. Alpha-emitting thorium isotopes in northwest Pacific deep waters. *Earth Planet. Sci. Lett.* **65**, 39.

Organisation for Economic Cooperation and Development (1983) *Uranium: resources, production and demand*. Paris, Organisation for Economic Cooperation and Development.

Osmond, J. K., and Cowart, J. B. (1976) The theory and uses of natural uranium isotopic variations of hydrology. *A. Energy Rev.* **14**, 621.

Osmond, J. K., and Cowart, J. B. (1982) Groundwater. In Ivanovich, M., and Harmon, R. S. (eds), *Uranium series disequilibrium: applications to environmental problems*. Oxford, Oxford University Press.

Pagenkopf, G. K. (1978) *Introduction to natural water chemistry*. New York, Dekker.

Perutz, M. (1939) Radioactive nodules from Devonshire, England. *Mineral. Petrog. Mitt.* **51**, 141.

Plant, J., Brown, G. C., and Simpson, P. R. (1982) Signature of metalliferous granites in the British Isles. In Maurice, Y. T. (ed.), *Uranium in granites, Paper Geol. Surv. Can.*

Plant, J., Ostle, D., and Miller, J. M. (1983) Natural levels of uranium in Britain — economic and environmental significance. *Inst. Geol. Sci. Lond. Rep.* **83-1**, 32.

Quidwai, H., and Jensen, M. L. (1979) Methodology and exploration for sandstone-type uranium deposits. *Mineral. Deposita* **14**, 137.

Reynolds, R. L., and Goldhaber, M. B. (1983) Iron disulphide minerals and the genesis of roll-type uranium deposits. *Econ. Geol.* **78**, 105.

Rich, R. A., Holland, H. D., and Petersen, U. (1977) *Hydrothermal uranium deposits.* Amsterdam, Elsevier.

Rimsaite, J. (1982) The leaching of radionuclides and other ions during alteration and replacement of accessory minerals in radioactive rocks. *Paper Geol. Surv. Can.* **82-1B**, 253.

Roberts, D. E., and Hudson, G. R. T. (1983) The Olympic Dam copper–uranium-gold deposit, Roxby Downs, South Australia. *Econ. Geol.* **78**, 799.

Robertson, D. S., Tilsey, J. E., and Hogg, G. M. (1978) The time-bound character of uranium deposits. *Econ. Geol.* **73**, 1409.

Robinson, A.; and Spooner, E. T. C. (1984a) Can the Elliot Lake uraninite-bearing quartz pebble conglomerates be used to place limits on the oxygen content of the early Proterozoic atmosphere? *J. Geol. Soc.* **141**, 221.

Robinson, A., and Spooner, E. T. C. (1984b) Post-depositional modification of uraninite-bearing quartz pebble conglomerates from the Quirke ore zone, Elliot Lake, Ontario. *Econ. Geol.* **79**, 297.

Roe, K. K., Burnett, W. C., Kim, K. H., and Beers, M. J. (1982) Excess protactinium in phosphatic nodules from a coastal upwelling zone. *Earth Planet. Sci. Lett.* **60**, 39.

Rosholt, J. N. (1982) Mobilization and weathering. In Ivanovich, M., and Harmon, R. S. (eds), *Uranium series disequilibrium: applications to environmental problems.* Oxford, Oxford University Press.

Ruzicka, V. (1982) Some relationships between granitic plutons and the distribution of uranium deposits. In Maurice, Y. T. (ed.), *Uranium in granites, Paper Geol. Surv. Can.*

Rye, D. M., and Roy, R. F. (1978) The distribution of thorium, uranium and potassium in Archaean granites from northeastern Minnesota. *Am. J. Sci.* **278**, 354.

Sanford, R. F. (1982) Preliminary model of regional Mesozoic groundwater flow and uranium deposition in the Colorado Plateau. *Geology* **10**, 348.

Sarkar, S. C. (1982) Uranium (nickel, cobalt, molybdenum) mineralisation along the Singhbhum copper belt, India, and the problem of ore genesis. *Mineral. Deposita* **17**, 257.

Sassano, G. P., Fritz, P., and Morton, R. D. (1972) Paragenesis and isotopic compositions of some gangue minerals from the uranium deposits of Eldorado, Saskatchewan. *Can. J. Earth Sci.* **9**, 141.

Schwarcz, H. P., Gascoyne, M., and Ford, D. C. (1982) Uranium series disequilibrium studies in granitic rocks. *Chem. Geol.* **36**, 87.

Scott, M. R. (1982) The chemistry of U and Th series nuclides in rivers. In Ivanovich, M., and Harmon, R. S. (eds), *Uranium series disequilibrium: applications to environmental problems.* Oxford, Oxford University Press.

Shirvington, P. J. (1983) Fixation of radionuclides in the ^{238}U decay series in the vicinity of mineralized zones, 1, The Austatom uranium prospect Northern Territories, Australia. *Geochim. Cosmochim. Acta* **47**, 403.

Sibbald, T. I. (1985) Geology and genesis of the Athabasca Basin uranium deposits. In *Recognition of uranium provinces.* Vienna, International Atomic Energy Agency.

Simpson, P. R., and Bowles, J. F. W. (1977) Uranium mineralization of the Witwatersrand and Dominion Reef systems. *Phil. Trans. R. Soc. Lond. A,* **286**, 527.

Simpson, P. R., Brown, G. C., Plant, J., and Ostle, D. (1979) Uranium mineralisation and granite magmatism in the British Isles. In Bowie, S. H. U., Fyfe, W. S., Ostle, D., Plant, J., and Simpson, P. R. (eds), *Theoretical and practical aspects of uranium geology.* London, Royal Society.

Simpson, P. R., Plant, J., and Brown, G. C. (1982) *Uranium '81.* London, Mineralogical Society.

Simpson, P. R., Plant, J., and Cope, M. J. (1977) Uranium abundance and distribution in some granites from northern Scotland and south west England as indicators of uranium provinces. In Jones, M. J. (ed.), *Geology, mining and extractive processing of uranium.* London. Institution of Mining and Metallurgy.

Speer, J. A., Solberg, T. N., and Becker, S. W. (1981) Petrography of the uranium-bearing minerals of the Liberty Hill pluton, South Carolina: Phase assemblages and migration of uranium in granitoid rocks. *Econ. Geol.* **76**, 2162.

Stanworth, C. W., and Badham, J. P. N . (1984) Lower Proterozoic red beds, evaporites and secondary sedimentary uranium deposits from the East Arm, Great Slave Lake, Canada. *J. Geol. Soc.* **141**, 235.

Steacy, H. R., and Kaiman, S. (1978) Uranium minerals in Canada: their description, identification and field guide. In Kimberley, M. M. (ed.), *Uranium deposits, their mineralogy and origin, Mineral. Assoc. Can. Short Course Handbook* **3**.

Stuckless, J. S., and Miesch, A. T. (1981) Petrogenetic modelling of a potential uranium source rock, Granite Mountains, Wyoming. *US Geol. Surv. Prof. Paper* **1225**.

Stuckless, J. S., and Troeng, B. (1984) Uranium mineralisation in response to regional metamorphism at Lilljuthatten, Sweden. *Econ. Geol.* **79**, 509.

Szabo, B. J., and Rosholt, J. N. (1982) Surficial continental sediments. In Ivanovich, M., and Harmon, R. S. (eds), *Uranium series disequilibrium: applications to environmental problems.* Oxford, Oxford University Press.

Szalay, A. (1974) Accumulation of uranium and other micrometals in coal and organic shales and the role of humic acids in these geochemical enrichments. *Ark. Mineral. Geol.* **5**, 23.

Tatsch, J. H. (1976) *Uranium deposits: origin, evolution and present characteristics.* Sudbury, Tatsch.

Thiel, K., Saager, R., and Muff, R. (1979) Distribution of uranium in early Precambrian gold bearing conglomerates of the Kaapvaal craton, South Africa: a review of a case study for the application of fission track micromapping of uranium. *Minerals. Sci. Eng.* **11**, 225.

Tilsley, J. E. (1978) Uranium mineralisation in shallow intrusive environments. In Kimberley, M. M. (ed.), *Uranium deposits, their mineralogy and origin, Mineral. Assoc. Can. Short Course Handbook* **3**.

Toens, P. D. (1981) Uranium provinces and their time-bound characteristics. *Trans. Geol. Soc. S. Afr.* **84**, 293.

Tremblay, L. P. (1982) Geology of the uranium deposits related to the sub-Athabasca unconformity, Saskatchewan. *Paper Geol. Surv. Can.* **81-20**.

Tripathi, V. S. (1979) Comments on 'Uranium solution — mineral equilibria at low temperatures with applications to sedimentary ore deposits'. *Geochim. Cosmochim. Acta* **43**, 1989.

Turcotte, D. L., Ribando, R. J., and Torrance, K. E. (1977) Numerical solution of two-temperature thermal convection in a permeable layer with application to the Steamboat Springs thermal system, Nevada. In Heacock, J. G. (ed.), *The earth's crust, Geophys. Monogr.* **20,** Washington, DC, American Geophysical Union.

Turekian, K. K., and Bertine, K. K. (1971) Deposition of molybdenum and uranium along the major ocean ridge systems. *Nature* **229**, 250.

Turekian, K. K., Nozaki, Y., and Benninger, L. K. (1977) Geochemistry of atmospheric radon and radon products. In Donath, F. A., Stehli, F. G., and Wetherill, G. W. (eds), *Annual review of Earth and planetary sciences,* Vol. 5. Palo Alto, Annual Reviews.

Turner, B. R. (1985) Uranium mineralisation in the Karoo Basin, South Africa. *Econ. Geol.* **80**, 256.

Vassiliou, A. H. (1980) The form of occurrence of uranium in deposits associated with organic matter. *Econ. Geol.* **75**, 609.

Walton, N. R. G. (1982) A detailed hydrogeochemical study of groundwaters from the Triassic sandstone aquifer of south-west England. *Rep. Inst. Geol. Sci. Lond.* **81-5.**

Watson, J. V. (1976) Mineralisation in Archaen provinces. In Windley, B. F. (ed.), *The early history of the Earth.* London, Wiley.

Watson, J. V., and Plant, J. (1979) Regional geochemistry of uranium as a guide to deposit formation. In Bowie, S. H. U., Fyfe, W. S., Ostle, D., Plant, J., and Simpson, P. R. (eds), *Theoretical and practical aspects of uranium geology.* London, Royal Society.

Watters, R. A., and Sagala, F. P. (1979) Geochemical reconnaissance for uranium and base metals using heavy mineral separates in central and southern Sumatra. In Watterson, J. R., and Theobald, P. K. (eds), *Geochemical exploration, 1978.* Ontario, Association of Exploration Geochemists.

Wittke, W. (1973) General report on the symposium 'Percolation through fissured rock'. *Bull. Int. Assoc. Eng. Geol.* **7**, 3.

Wolf, K. H. (1981) *Handbook of strata-bound and stratiform ore deposits,* Vol. 8. Amsterdam, Elsevier.

Wollenberg, H. A. (1973) Fission track radiography of uranium and thorium in radioactive minerals. In Jones, M. J. (ed.), *Geochemical exploration, 1972.* London, Institution of Mining and Metallurgy.

Zhmodik, S. M., Mironov, A. G., and Nemiroovskaya, N. A. (1982) Uranium distribution in iron hydroxides from weathered mantle as shown by neutron-fragment radiography. *Dokl. Akad. Sci. USSR Earth Sci.* **250**, 219.

7

Radioactivity, heat generation and hydrothermal circulation

7.1 SOURCES OF HEAT IN THE EARTH

7.1.1 Primordial and continuing sources

An essential advance towards understanding the geophysical behaviour of the Earth was the recognition that rocks are to a greater or lesser extent radioactive and act as heat generation systems. To consider this mechanism of heat production in isolation from other heat sources, however, would fail to indicate the relative importance of radiogenic heat in present-day processes. Since the time of the Earth's accretion, a number of heat production mechanisms have operated, including the accretion process itself, adiabatic compression, exothermic recrystallisation, tidal forces and radioactivity (Bott, 1982). Radiogenic heat may be subdivided into the heat produced by the decay of short-lived isotopes during and soon after the accretion of the Earth (one of the primordial sources) and the heat produced by long-lived radionuclides (a continuing source).

7.1.2 Primordial sources

Accretion

The accretion mechanism, in which collision of planetesimals under gravitational attraction resulted in coalescence to form larger bodies in the Primitive Solar Nebula, was accompanied by the conversion of much of the kinetic energy of the planetesimals to heat energy (Kaula, 1979). Moreover, as the Earth grew in size, its gravitational attraction increased and the kinetic energy of the impacting planetesimals also increased. Thus, during the accretion process the Earth must have experienced a large rise in temperature. The total energy liberated is estimated at about 2×10^{32} J, and the Earth's internal temperature may have risen from about 25 °C when the Earth had a radius of about 1000 km to about 4000 °C by the time that the radius had increased to about 5000 km. The temperature of the Earth's surface would essentially have remained low as a result of radiation heat loss. Because the amount of energy available for conversion depends upon the size of the impacting body, more heat is available from a few large impacts than from many small impacts. For example, a planetesimal large enough to produce an impact structure 1000 km in diameter would liberate about 3×10^{27} J (Grieve, 1980). Thus, if the Earth received large impacts occasionally throughout its history, local severe disruptions to the thermal state would have occurred. Indeed, the cumulative impact energy imparted by bodies after the main accretion had been completed is estimated at about 7×10^{28} J or roughly 10^4 J kg^{-1}. For example, about 10^5 planetesimals which produced impact structures with diameters greater than 1 km are likely to have hit the Earth between 4600 and 3900 Ma ago.

Core formation

Another probable major source of internal heat early in the Earth's history is related to the formation of the core. However, whether or not this mechanism actually contributed heat to the Earth as a whole depends upon when the core was formed. There are two views on this, as discussed in Chapter 2. In the first, the Earth is considered to have accreted from chondritic material that remained uniform throughout the accretion process (homogeneous accretion). The heat from impacting particles would have resulted in melting and the separation and sinking of Fe, Ni and S to form the core. The second view, in contrast, is that the more refractory elements separated from the Primitive Solar Nebula before the Earth accreted and that the material which formed the core represents the first stage of the formation of the Earth from these condensates, the less refractory components accreting at a later stage (heterogeneous accretion). If the latter view is accepted, the origin of the core would have played only a minor part in the thermal history of the Earth.

The present-day consensus, however, favours the first of these views. This is because, although Pb isotope ratios (controlled by the decay of U) show that the Earth's crust first became a closed chemical system about 4550 Ma ago, the age of meteorites is about 4600 Ma. Now, Pb differs from lithophile U in having siderophile and chalcophile affinities, so that separation of the core from a homogeneous Earth would have removed large quantities of Pb from the mantle but have left U behind. This would not affect the U/Pb ratio of the bulk Earth, but the U/Pb ratio of the mantle would have been significantly increased. The age of initiation of the U–Pb systems in the crust may, therefore, have been the time of core formation, suggesting that this occurred within about 50 Ma of the formation of the Earth. Further, if core segregation is to have taken place within an initially homogeneous Earth, temperatures must have been high enough for total melting of all but a thin crustal layer. Only during and soon after the accretion process itself were temperatures high enough for this to be possible. Of course, if U was abstracted from the mantle with Pb during core formation (as suggested in Chapter 2), this line of reasoning is invalid. Whatever the mechanism and timing, the maximum energy liberated by the separation of the core from a homogeneous Earth is estimated at about 10^{31} J. If this energy was available, it would have been sufficient to raise the temperature of the whole Earth by about 2200 °C, provided that core formation was instantaneous. A lesser, but nevertheless large, increase would have accompanied core formation extended over a longer period of time.

Adiabatic compression

Adiabatic compression must have been another important source of heat early in the Earth's history. Heat would have been liberated as internal parts of the Earth became progressively compressed by burial to increasing depth during accretionary growth. This would, therefore, have been an accelerating process that continued throughout the growth of the Earth. The heat-generating effect of adiabatic compression must have been greatest at the centre of the Earth, where pressure was greatest, and must have caused an increase in temperature of the order of 900 °C. The effect, however, decreased progressively towards the surface. It should thus be noted that the outer core is unlikely to have remained liquid without the presence of other sources of heat, although we do not know how much superadiabatic heat there is in the outer core, or even whether the outer core is undergoing thermally induced convection.

The adiabatic temperature gradient in the Earth is generally taken to be about 0.25 °C km^{-1}. If true, and the mantle is assumed not to be undergoing thermal convection, this would give a rise in temperature of about 750 °C over the depth range of the mantle. However, convection is probably taking place and the relative importance of the various factors that control the temperature gradient in the mantle is very uncertain.

Extinct radionuclides

Short-lived radionuclides could have liberated sufficient heat to affect significantly the temperature of the Earth for about 5–15 Ma after the end of the nucleosynthesis. The principal nuclides responsible were ^{26}Al, ^{36}Cl, ^{60}Fe, ^{244}Pu and ^{129}I. Of these, ^{26}Al, with a half-life of about 0.73 Ma, was by far the most important and would have been a significant heat source for about 10 Ma. As the half-lives of ^{36}Cl and ^{60}Fe are only about 0.3 Ma, these nuclides can only have been effective heat sources for about 4 Ma. An estimate of the heat produced by ^{26}Al has been obtained from the well-studied Allende meteorite, in which ^{26}Mg, the daughter of ^{26}Al, is found concentrated in inclusions that are thought to be highly refractory early condensates of the Primitive Solar Nebula. Because the ratios of ^{26}Mg to other isotopes in the Allende meteorite are greater than the nucleosynthetic ratios, the origin of the excess ^{26}Mg is believed to be the decay of ^{26}Al. Thus the original $^{26}Al/^{27}Al$ ratio from the Allende meteorite can be determined and used to place an upper limit on the ^{26}Al concentration in the Earth at the time of accretion. This indicates a maximum heat production of 0.17 μW kg^{-1}, which would have been sufficient to cause complete melting of the Earth. However, whether or not this mechanism was effective in the early history of the Earth depends upon the time interval between nucleosynthesis and the Earth's accretion: if there was only a delay of 10 or 20 Ma, the heat from this source would certainly have been sigificant; if, however, it was of the order of 100 Ma, it would have produced only limited effects. Certainly, the evidence that the age of meteorites is close to that for the accretion of the Earth supports the view that heat liberated by the decay of short-lived radionuclides was extremely important in the early history of the Earth.

7.1.3 Continuing sources

Tidal forces

Because of the gravitational attraction of the Moon and Sun, the solid Earth experiences tidal interactions, just as the oceans. These interactions act as a brake on the rotation of the Earth and thus progressively cause a lengthening of the day. Soon after formation, the Earth possibly had a day length of less than 10 h, perhaps even only 2–4 h, but the day length has subsequently increased to the 24 h of the present. As a result of this slowing, loss of rotational energy must have occurred by conversion to other forms of energy: partly within the solid Earth by tidal action, partly in the oceans (particularly shallow seas) as a result of tides and partly to increase the Moon's orbital energy and to increase its orbital radius from the Earth (by about 2 cm a year). Solid Earth tide conversion results in the liberation of heat. If all the rotational energy were converted in this way, temperatures within the Earth would significantly increase, but the actual proportion of energy liberated as internal heat is unknown. One estimate, however, is an annual release of about 3×10^{19} J out of a total annual availability of tidal energy of 3.3×10^{21} J. This is therefore an important continuing source of heat in the Earth. Indeed, the effects of tidal heating have recently been clearly revealed in studies of Io, the innermost satellite of Jupiter. Io is a comparatively large satellite and is trapped in a complex orbital path between Jupiter and Europa. Because of the orbital character of these satellites Io is subject to extreme tidal interactions. The conversion to heat of this large amount of tidal energy means that Io is in a state of continuous volcanic activity (Gore, 1980).

Because the Moon is gradually receding from the Earth, in the past, when the Moon was closer, tidal effects in the Earth were greater than today. The amount of heat liberated by the tidal effect would, therefore, have been greater when the Earth and Moon were closer together. Unfortunately, little is known about the formation of the Earth–Moon system but, if the Moon was captured by the Earth's gravitational field, the tidal effect of capture would have added a considerable amount of energy to both bodies, perhaps as much as 10^{30} J.

Long-lived radionuclides

Although all radionuclides generate heat as they decay, only those which are relatively abundant and have half-lives which are comparable with the age of the Earth can be considered to have maintained a significant level of heat production throughout the Earth's history (Rybach, 1976). Of the naturally occurring radionuclides, only ^{40}K, ^{232}Th, ^{235}U and ^{238}U fall into this category. Of these, ^{235}U generates most heat per unit mass at 0.56 mW kg^{-1}. ^{238}U generates 0.096 mW kg^{-1}, ^{40}K produces 0.028 mW kg^{-1} and ^{232}Th produces 0.026 mW kg^{-1}. However, the importance of a particular radionuclide as a heat producer depends upon its isotopic abundance. ^{235}U, for example, accounts for only 0.71% of total U, while ^{40}K, similarly, accounts for an even smaller proportion of total K. Nevertheless, the abundance of K in the Earth is so great that ^{40}K is an important heat source. Thus, at the present time, K contributes about 15% of total radiogenic heat, Th contributes 42% and U (mainly ^{238}U) contributes 43%. In absolute terms, the annual heat production is 8.8×10^2 J kg^{-1} for ^{40}K, 17.7×10^3 J kg^{-1} for ^{235}U, 3.0×10^3 J kg^{-1} for ^{238}U and 8.2×10^2 J kg^{-1} for ^{232}Th. At the present-day isotopic ratios the annual heat production from natural K is 6.2×10^{-3} J kg^{-1}, and that from natural U is 3.1×10^3 J kg^{-1}. The heat production value for total Th is the same as that for ^{232}Th which is the only naturally occurring non-radiogenic representative of Th.

In the radioactive decay process, a proportion of the mass of each decaying nuclide is converted to energy. Most of this energy is in the form of the kinetic energy of the emitted particles and of γ-rays. For β-decays, however, some of the energy is carried by neutrinos. All the decay energy is absorbed within the Earth and converted to heat, except that of the neutrinos which do not readily interact with matter. Therefore, they generally pass through the Earth and their energy is lost to space. Determination of heat production by each particular radionuclide must, therefore, take into account the loss of neutrino energy by β-particle emitters. Direct determination of neutrino energy is possible in the decay of ^{40}K but, because β-particle emission in the ^{232}Th, ^{235}U and ^{238}U decay series accounts for only a small fraction of the total decay energy, a general relationship is used in which, on average, neutrinos represent 0.6667 of the decay energy of a β-particle decay. An empirical relation for the energy of emitted α-particles is

$$\log_e T = AK^{-1/2} - B$$

where T is the half-life of the α-emitter, K is the kinetic energy of the α-particle, and A and B are constants. In the U and Th series, about 90% of the energy is provided by α-particles.

Because the abundance of even the long-lived radionuclides has decreased significantly since nucleosynthesis, the heat production at the time of the Earth's formation must have been significantly greater than today (Fig. 7.1). Thus, in absolute terms, 4000 Ma ago the heat produced annually from K was 9×10^{-1} J kg^{-1}, while 1×10^3 J kg^{-1} was produced by Th and 11.8×10^3 J kg^{-1} by U. Moreover, the relative abundance of the heating-producing radionuclides was such that ^{235}U and ^{238}U provided about half the total and 40 K about 40%. The relative position of ^{232}Th has thus changed because, with its much longer half-life, its heat

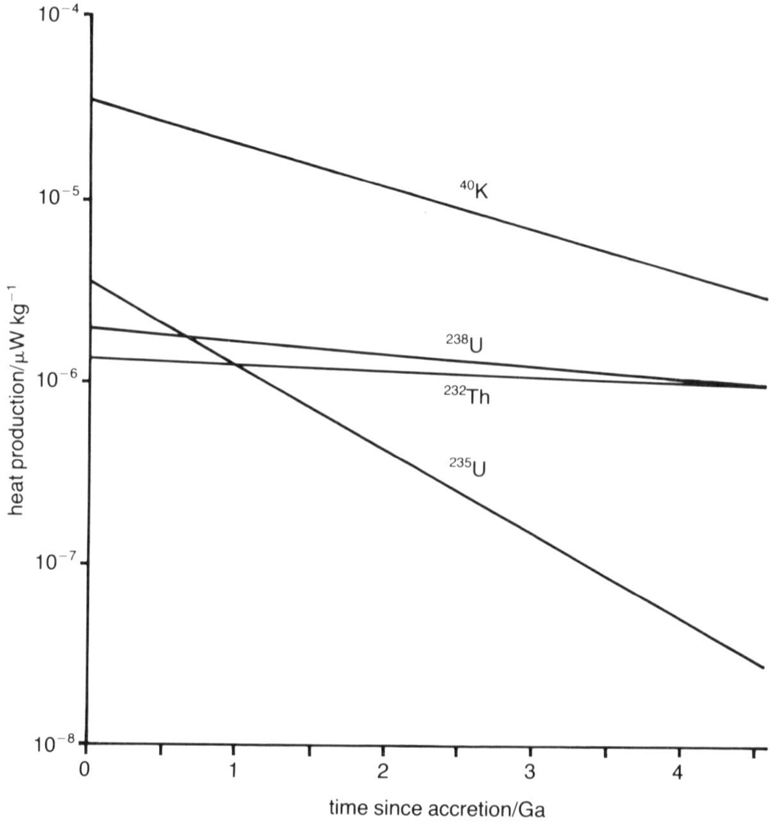

Fig. 7.1 — Variation in radiogenic heat production due to ^{40}K, ^{232}Th, ^{235}U and ^{238}U through geological time for the Earth (assuming a chondritic composition). (Redrawn from Open University (1981); copyright 1981 The Open University Press.)

production has decreased much less than that of 40 K ^{235}U and ^{238}U.

7.2 THE IMPORTANCE OF RADIOGENIC HEAT

Radiogenic heat production in the Earth is controlled by the abundance of the main radionuclides in geological materials as discussed in Chapters 2 and 6. To obtain an estimate of radiogenic heating of the whole Earth, estimates must first be made of the radionuclide composition of the core as well as of the mantle and crust. In crustal rocks, a typical granite produces annually about 5×10^{-3} J kg^{-1} due to ^{40}K, 8×10^{-2} J kg^{-1} due to ^{232}Th and 10^{-2} J kg^{-1} due to U; this gives a total of about 2.5×10^{-2} J kg^{-1}. Granites with high U and Th contents will exceed this value by a factor of 2 or 3.

The average annual heat production in the upper continental crust is also about 2.5×10^{-2} J kg^{-1}, but the lower continental crust is a comparatively poor heat producer at about 1.25×10^{-2} J kg^{-1}, so that the average value for the continental crust as a whole is about 1.8×10^{-2} J kg^{-1}. It should be noted, however, that, although the heat production of the lower continental crust is very difficult to determine, it appears to be laterally extremely variable. In comparison with the continental crust, oceanic crust is a poor producer of radiogenic heat. For example, typical gabbro and basalt produce annually only about 5×10^{-3} J kg^{-1}, or one-fifth of the radiogenic heat produced by average upper continental crust. Peridotite, which may represent mantle material, has an even lower annual heat production capability of about 8×10^{-5} J kg^{-1}.

On the assumption of an average chondritic composition of 0.01 ppm ^{238}U, 7×10^{-5} ppm ^{235}U, 0.04 ppm ^{232}Th and 0.1 ppm ^{40}K, the bulk Earth has an annual heat production of about 1.5×10^{-4} J kg^{-1}. Now the heat production in the core can be estimated if the heat production in the mantle and crust is calculated and the mass of each of the various layers is known. In fact, the mass of the upper continental crust is about 8×10^{21} kg, as is the mass of the lower continental crust, and so the annual heat production is about 2×10^{20} J and 10^{20} J respectively. The oceanic crust has an estimated mass of 7×10^{21} kg with an annual heat production of about 3.5×10^{19} J and the mantle about 4×10^{24} kg with an annual heat production of about 3.2×10^{20} J. The upper and lower continental crust each form 0.14% of the whole Earth, the oceanic crust 0.12% and the mantle 68.19%. Summing the heat produced by each of these layers in a year and dividing by the mass of the Earth gives a value of about 1.1×10^{-4} J kg^{-1}, that is, less than that for the bulk Earth based upon a chondritic composition. This therefore suggests that the arguments which favour the presence of significant levels of K, U or even Th in the core may be valid.

Concerning the possibility that K may be present in the core, it should be noted that there is good evidence from crustal and upper mantle rocks that the Earth is depleted in K relative to the chondritic composition. If, however, the Earth is indeed assumed to have a chondritic composition, the K that is not present in the crust and upper mantle could be concentrated in the core. Some of this K will be ^{40}K, and the possibility that the core could make a significant contribution to radiogenic heating of the Earth is thus supported. The fact that K might occur in the core is also suggested by the presence of djerfisherite ($K_6Mg(Fe,Cu)_{24}S_{26}Cl$) in some iron meteorites. This occurs in association with troilite (FeS), which typically contains K up to a few per cent. Moreover, at high temperatures and pressures, K undergoes electron transfer between the s and d subshells and becomes chalcophile in character. The separation of some K during core formation is, therefore, plausible. The arguments which suggest the presence of U in the core were outlined in Chapter 2.

Now the bulk Earth heat production value for a chondritic composition is close to that for the present-day annual heat loss from the Earth of about 10^{21} J, suggesting that there is an approximate balance between heat production and heat loss. Not only does this appear to lend support to the idea that the Earth has a chondritic composition but also to the fact that the present-day heat flow from the Earth can be mainly attributed to the decay of long-lived radionuclides. Nevertheless, despite these conclusions, there are a number of uncertainties which relate both to the possible heat contribution from long-lived radionuclides in the core and to the possible continuing contribution of primary heat to the observed heat loss.

Firstly, knowledge of the distribution of the heat-producing radionuclides in the Earth is, in fact, very sketchy, and estimates of the contributions made by the mantle and the lower continental crust are subject to large errors.

Secondly, because rocks have low thermal conductivity values, the present-day observed heat loss from the Earth represents heat production that occurred some time in the geological past. However, as noted above, the abundances of the main heat-producing radionuclides have decreased with time, and so equating present-day heat production with present-day heat loss is invalid. In fact, heat production in the past was greater than at present, and thus it may be concluded that the coincidence of values is probably due to chance.

Thirdly, the contribution of primordial heat

sources to present-day heat loss from the Earth is very uncertain. Estimates range as high as 50% of the total heat flux.

Fourthly, if K or U is, indeed, present in the Earth's core, the thermal consequences at the surface of the Earth need only be considered if there is transfer of heat across the core–mantle boundary. This, in turn, depends upon the nature of the boundary itself and the origin of the motions in the outer core which are thought to generate the Earth's magnetic field. These, however, are the subject of much debate and, until further evidence becomes available, the contribution from the core must be viewed as uncertain.

Finally, the removal of U to the core from the mantle, in the manner described in Chapter 2, will result in a decreased capability for radiogenic heat production in the lower mantle. Because Th may behave simiarly to U, it is possible that enrichment of the core in Th has also occurred, with subsequent depletion of the lower mantle. The loss of both U and Th may, therefore, mean that the lower mantle makes a smaller contribution to the heat production of the Earth than supposed previously.

7.3 HEAT FLOW FROM THE EARTH

The mean heat flow from the Earth is about 80 mW m^{-2}. To determine heat flow it is necessary to measure the increase in temperature with depth in the Earth and the thermal conductivity of the rocks over which that temperature gradient is developed. Heat flow may be determined for both continental and oceanic areas, but the procedures involved are somewhat different.

On land, a borehole is needed which extends to a sufficient depth below the surface that the influence of such effects as recent climatic change and seasonal temperature variations no longer can be considered as possible disturbances. Usually this means a depth of about 100 m. Even at this depth the temperature gradient may, however, be influenced by circulating groundwater, so that false heat flow values result from calculations simply based upon the thermal conductivity of the rock. Also, it is clear that some time must pass before heat generated during drilling and the effects of the circulating drilling mud have died away, and thermal equilibrium has been re-established. Because boreholes are expensive to drill, those made especially for the purpose of heat flow studies are comparatively few, so that boreholes drilled for other purposes, such as oil or gas exploration, are often used. For this reason, the available heat flow data in continental areas are somewhat sporadic and do not equally represent all different geological terrains. Thermal conductivity measurements must be carried out on individual rock samples obtained from the borehole core, and uncertainty therefore arises about the representative nature of the selected samples for any lithological unit as a whole. Finally, even with a moderately deep borehole, disturbance of vertical temperature gradients can arise by lateral heat flow through the sides of hills. In continental areas the mean heat flow is about 60 mW m^{-2}.

Heat flow determination in oceanic areas is much simpler than on land, and the results are subject to fewer uncertainties in their interpretation. Temperature gradients are usually measured by means of a gravity-driven probe carrying thermistors at different points along its length. The probe is designed to penetrate the top 20 m or so of the soft sediment forming the ocean floor and is hollow, so that a core of this sediment is obtained. It is this core which is the sample used for measurement of thermal conductivity; a heated needle is inserted into the core and the increase in temperature with time noted. Because of the ease of measurement of both temperature gradient and thermal conductivity, heat flow values for oceanic areas are much more common and more representative of different environments than continental values. Nevertheless some uncertainties are present as a result of, for example, sea-floor topography or the rapid deposition or erosion of the sediment. The mean heat flow for oceanic areas is about 95 mW m^{-2}.

7.3.1 Continental heat flow

In general, heat flow values from continental crust vary inversely with the age of the last major episode of metamorphism or magmatism (Vitorello and Pollack, 1980). This relationship is the basis for the recognition, on a large scale, of heat flow provinces (Fig. 7.2). In rocks younger than about 1000 Ma, heat flow decreases rapidly with increasing age. The rate of decrease is lower for rocks between 1000 and 1500 Ma, however, and for older heat flow provinces the decrease is very gentle; provinces older than 1500 Ma can effectively be considered to have more or less constant heat flow values (Sclater *et al.*, 1980). Even though the rate of decrease in heat flow in young provinces is comparatively rapid, it is significantly less than would be expected if loss of heat had occurred solely by conduction from a source consisting only of heat-generating long-lived radionuclides.

For any single heat flow province, the heat A generated by radioactivity can be calculated for different rock types at various sites and compared with the measured heat flow q at those sites. Where the contribution from metamorphic or magmatic heat is small, the relationship between A and q is found to be expressed as a straight-line equation:

$$q = bA + q^*,$$

where q^* represents the heat flow originating

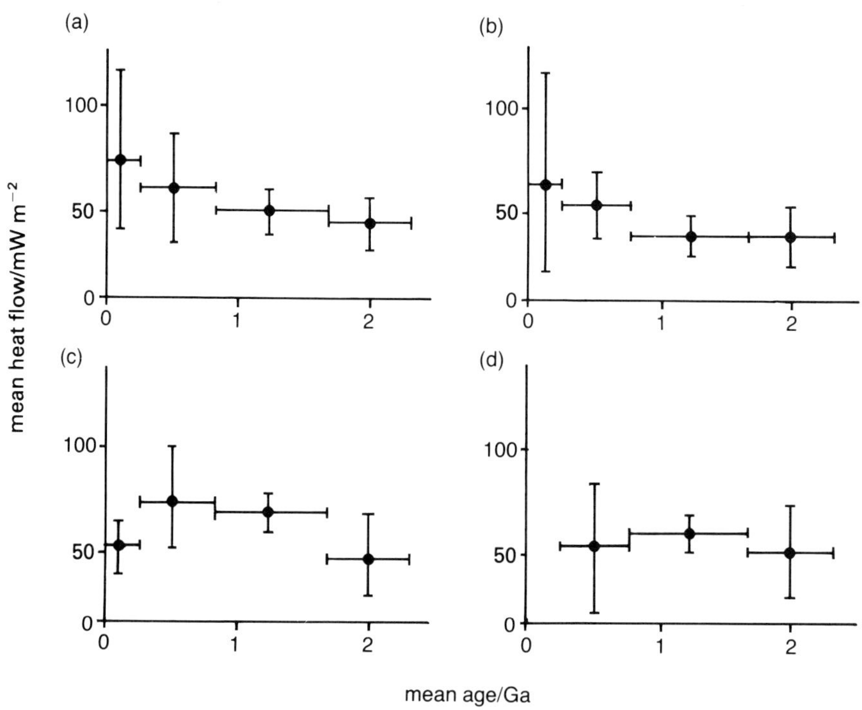

Fig. 7.2 — Heat flow values (and standard deviations) for continental areas for the age groups 0–250 Ma, 250–800 Ma, 800–1700 Ma and >1700 Ma: (a) North America; (b) Eurasia; (c) Australia; (d) Africa. (After Sclater *et al.*, 1980.)

lower in the crust and mantle (the 'reduced heat flow'), and b is a function of the thickness of that part of the continental crust which acts as the radiogenic heat source. This has been taken to imply that heat generation occurs uniformly throughout the source region of the upper crust, but this is not the case. In fact, the same relationship would be obtained if the radionuclides were distributed in such a way that heat generation decreased exponentially with depth (Fig. 7.3). This model is more realistic, bearing in mind the generally lithophile character of the radionuclides involved, and their likely depletion in the lower crust through extraction by partial melting and the generation of granitic magma (Jaupart *et al.*, 1981). Moreover, the physical characteristic represented by factor b is very unclear. For example, it may be related in various ways to non-conductive heat flow and the presence of microfractures permitting convective heat transfer.

The relationship between q^* and heat flow already noted, indicates that the reduced heat flow is almost constant, contributing about 60% of observed values. Possibly this value (which is about 50 mW m^{-2}) represents an equilibrium value that is reached by continents when the rate of decrease in heat flow with time has fallen to zero. In fact, it is now recognised that continental heat flow is made up of three separate components (Fig. 7.4). Firstly, there is the background heat from deep sources in the Earth; secondly, there is the heat released from the last major magmatic or metamorphic event; thirdly, there is the heat generated by the long-lived radionuclides concentrated in the upper crust. The change in value of these sources with time is most noticeable where the decay of the heat of formation (metamorphism or magmatism) is involved; effectively, this is totally lost within about 1000 Ma. Deep background heat is more or less constant today for all provinces that formed within the last 2500 Ma, although some decrease consequent upon the decay of radionuclides in the mantle or core will have taken place. The radiogenic heat

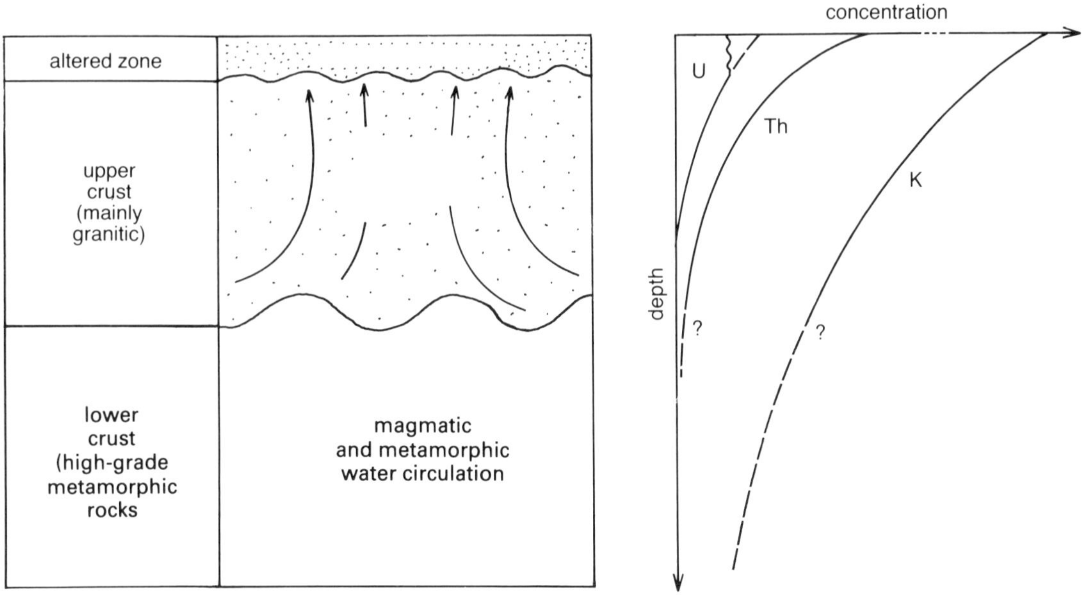

Fig. 7.3 — Processes controlling the distribution of radionuclides in the continental crust and the possible variation in concentration of K, Th and U with depth. (Jaupart *et al.*, 1981, redrawn with permission.)

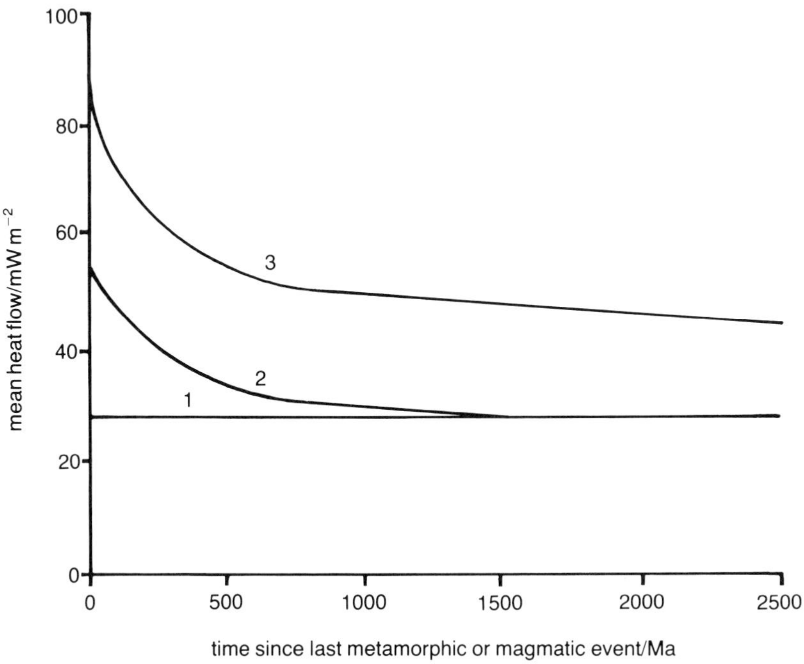

Fig. 7.4 — The summed components of continental heat flow and their variation with age of crust: 1, background heat of deep origin; 2, heat of formation; 3, radiogenic heat. (After Vitorello and Pollack, 1980.)

source in the upper crust will also have shown a gradual decrease as the result of decay of the long-lived radionuclides.

7.3.2 Oceanic heat flow

As the 'basaltic' crust in oceanic areas has a low abundance of radionuclides, only two of the heat sources found in continental areas are present: the deep background heat and the heat of formation. This does not mean that heat flow values are lower in oceanic than in continental areas; in fact, the average heat flow in oceanic areas is significantly greater. The high heat flow values at or near constructive plate margins are the main reasons for this. The thermal conductivity of the rocks making up the oceanic crust is rather similar to that of continental crust, and so differences are unlikely to arise through this mechanism. Now, heat of formation appears to be lost in a manner related to $t^{-\frac{1}{2}}$, where t is the age of crust less than about 100 Ma. After the heat of formation has been lost, the heat flow in oceanic areas is about 50 mW m^{-2}, which may be taken as an equilibrium value for oceanic crust (Fig. 7.5). Interestingly, the decrease in heat flow with increasing distance from a constructive margin is controlled by the thickening of the oceanic lithosphere. Thickening of oceanic lithosphere is this way has been seen by Hager (1978) as a possible mechanism for producing plate motion.

If the contribution from radiogenic heat is considered to be negligible in oceanic areas, it follows that, with similar equilibrium conditions of about 50 mW m^{-2} for both continental and oceanic areas and some radiogenic heat originating in the lower continental crust, the heat supply from the mantle beneath the

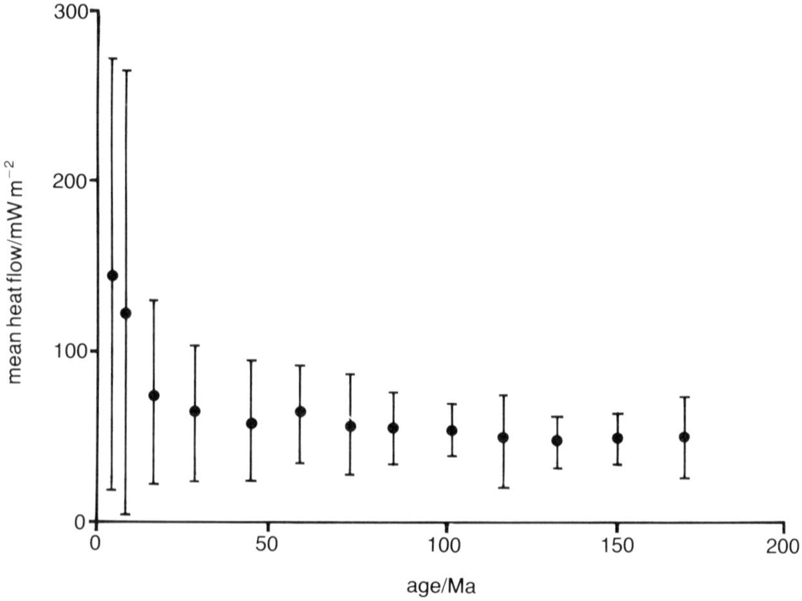

Fig. 7.5 — Average heat flow values (and standard deviations) for ocean basins as a function of age of crust. (After Sclater *et al.*, 1980.)

continental crust must be less than that beneath oceanic crust that has reached equilibrium. Major lateral differences in the thermal character of the mantle are, therefore, present.

A model which accommodates this conclusion is one in which the main heat-producing radionuclides in the mantle beneath the continents have, through differentiation, been concentrated into the continental crust, leaving the mantle depleted in radionuclides and, consequently, a poor source of radiogenic heat. In contrast, the mantle beneath the oceans has not suffered so much vertical differentiation and is a more significant source of radiogenic heat. However, below a depth of about 400 km this situation is unlikely to be true and, within the transition zone above the lower mantle, conditions are probably much more laterally uniform.

An alternative model involves the idea that mantle convection gives an equilibrium heat flow for oceanic areas which, by coincidence is equal to that of the equilibrium heat flow for old continental areas. Certainly, mantle convection is probably taking place, and different scales of cell and rates of movement may occur beneath oceanic and continental areas which could account for the different mantle heat supply noted above.

7.4 HYDROTHERMAL SYSTEMS AND GEOTHERMAL ENERGY

Hydrothermal groundwater circulation at active ocean ridges (constructive plate margins) is well recognised. In continental areas in which there is young igneous activity or even active volcanoes, shallow hydrothermal circulation systems are also present. Such areas have great importance as sources of geothermal energy but, whereas these hydrothermal systems may ultimately owe their origin to radiogenic heat production in the Earth, this connection is now rather indirect and they are not considered in this chapter.

7.4.1 Hot dry rock systems

Geothermal energy is potentially available for recovery from any part of the Earth's crust that is at a high temperature, simply by virtue of the increase in temperature with depth. In most circumstances, however, technical problems mean that heat extraction from great depth is not practical with today's technology. Nevertheless, where the continental crust is rich in the main heat-producing radionuclides, high temperatures are found at shallower depths than normal, and exploitation is feasible. Such areas are usually granitic and their age is not important; magmatic heat of formation could have long since been lost. Because the rocks involved are generally of low porosity, and thus have a low water content, they are referred to as a 'hot dry rock' geothermal resource. These hot dry rock areas are currently receiving a great deal of attention throughout the world as a possible energy alternative to fossil fuels.

Before dealing with the characteristics of such systems, it is worth noting that the term 'hot dry rock' is not necessarily a completely accurate description. The use of this term dates from before detailed *in situ* investigations of permeability and porosity were carried out in granite. Some of the results of these investigations have shown that, even at depths of several hundred metres, a granite body can possess fracture porosity and permeability which are similar to those found at the surface and that, far from being dry, the granite at depth contains a significant volume of water held within fractures. Other results, however, have shown granite to be relatively free of fractures below the top few tens of metres and, therefore, apparently truly dry. The conflicting results obtained from single granite intrusions probably reflect the occurrence of fracturing within granites on different scales. Major fractures with high permeability and large aperture are probably widely separated, but between these are minor fractures with smaller aperture and lower permeability; within these, yet more minor fractures may occur on various scales. The dimensions of these fracture separations are probably determined by the stress history of each individual igneous body. Where intensive work has been carried out in the Carnmenellis granite of Cornwall, minor low permeability fractures occur with a separation of about 10 m; more important fractures with moderate permeability occur with separations of about 100 m, and major fracture systems appear to occur with separations of several kilometres. Although individual fractures can be recognised, it is more convenient to think in terms of fracture zones rather than discrete fractures. The recognition of both high and low permeability areas within single granite bodies is probably simply due to the different locations of study areas within this multiscale system (Heath and Durrance, 1985).

The general principle behind exploitation of hot dry rock geothermal energy is to drill at least two boreholes to a depth where the ambient temperature is high enough to warrant the expense of establishing an extraction system; perhaps somewhere of the order of 200 °C at a depth of about 5 km in most high heat production granites (Batchelor and Pearson, 1979). The boreholes are drilled to curve with depth until at the extraction level they are quite shallowly inclined. An artificial fracture system is then generated between and around the boreholes using explosives, hydraulic fracturing or other techniques. This fracture system acts as a heat exchanger, cold water being pumped down one borehole to circulate through the hot granite to be extracted from the other borehole as hot water or steam. Because of the relatively low thermal conductivity of granite, a large surface area is needed within a large volume of granite for a sufficiently large heat exchange to take place. If the energy extracted is to compete in cost-effectiveness with electricity generated using fossil fuels, it has been estimated that eight fractures, 300 m in radius, would be needed in a reservoir operating at a temperature of about 200–250 °C. The economic lifetime of each heat extraction system of this type would be about 20 years.

7.4.2 Hydrothermal circulation in high heat production granites

Water contained within fractures in high heat production granites is likely to be set into convective circulation. Quantitative studies of such convection systems have been made using the Conway granite, New Hampshire, as a model. These studies have shown that, in plutons of this type, with radiogenic heat production values of about 4×10^{-7} W m^{-3} (which contrasts with the low heat production character of the host rocks in the area), active hydrothermal circulation is possible provided that the overall permeability of the granite is more than 0.05 m darcy. For high flow rates to become established the permeability must be in the order of 0.5 m darcy or more (Fehn *et al.*, 1978).

Convective flow regimes are not necessarily confined to the pluton itself, however, but can impinge on the envelope of host rocks (Fig. 7.6). If the host has a higher permeability than the pluton, an enhanced flow rate within both host and granite may develop. Topographic control is important in determining the geographical location of sites of draw-down (recharge) and upwell (discharge) within the convective system. The overall diameter of the convection cells can be extremely variable but typically ranges between about 5 and 20 km except where the system becomes locked onto major structural discontinuities. However, the aspect ratio of cells can change with time and even reversal of flow is possible.

In the Carnmenellis granite of Cornwall, where fractures with effective hydraulic apertures of 50 μm have a separation of about 10 m, and some major fractures with effective hydraulic apertures of 200 μm occur

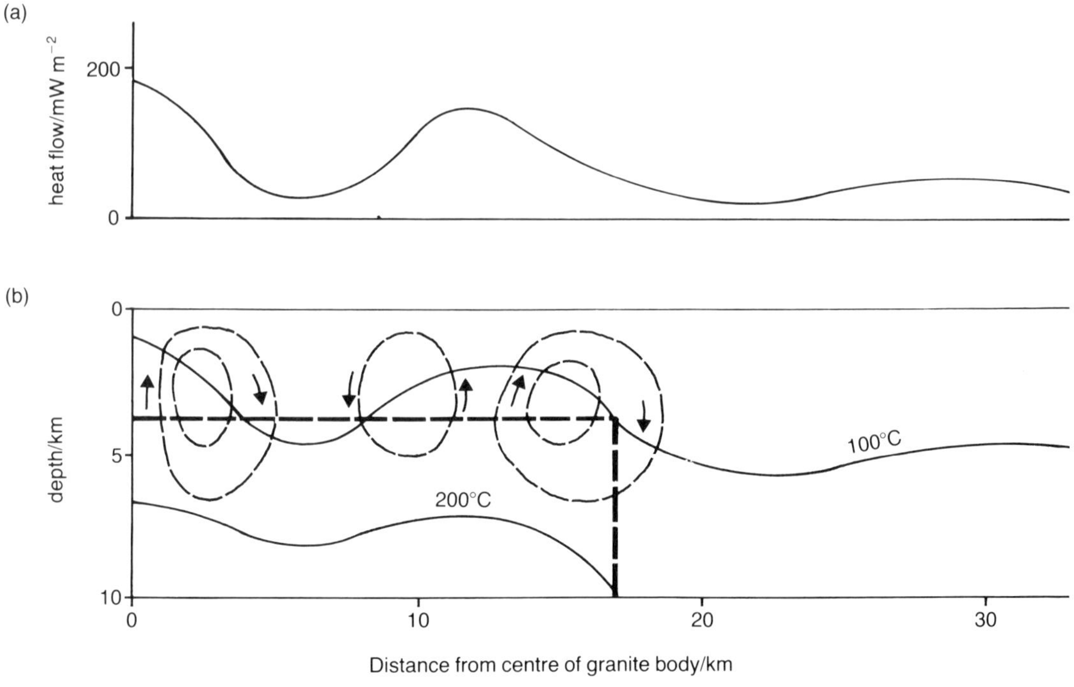

Fig. 7.6 — Model of convective circulation of groundwater (light broken lines) in an abnormally radioactive granite pluton: (a) heat flow; (b) temperature distribution and groundwater movement. The origin of each diagram is the centre of the pluton and the heavy broken lines mark its upper and outer limits. (After Fehn *et al.*, 1978.)

with separations of about 100 m, the overall permeability of the granite is of the order of 20 m darcy, the greater part of this being contributed by the more widely separated major fractures. This granite is very similar to the Conway granite in the contrast which its radiogenic heat production makes with the surrounding host rocks, and active hydrothermal circulation is taking place. In its geological and mineralogical characteristics and its structural setting, the Carnmenellis granite is very similar to the other late Carboniferous plutons making up the Cornubian granite batholith of southwest England. Active hydrothermal circulation systems are, therefore, probably present throughout the areas of both exposed and subsurface granite (Edmunds *et al.*, 1984), especially as the area of high heat flow in southwest England is largely coincident with the subsurface extent of the batholith (Fig. 7.7).

One point of specific detail shown by the distribution of heat flow values needs to be considered. While the highest values are located within the granite plutons, as would be expected, the distribution is not even, and a wide range of values is observed, with high heat flow near the margins of the plutons as well as in their central sections (Fig. 7.8). This observation generally supports the conclusion that active hydrothermal circulation cells are present. The variation in heat flow from high values near the centre of a pluton to lower values with distance from the centre and then rising again as the margin is approached is characteristic of the Conway-type hydrothermal circulation model. Considerable local variation in heat flow may, therefore, be expected in the same way that hydrothermal circulation at active ocean ridges results in extreme variation in heat flow over short distances.

Mapping of these hydrothermal systems is extremely difficult because of their sluggish nature, but the use of naturally occurring radionuclides as tracers of these systems has proved very informative. The approach most extensively used is mapping of dissolved ^{222}Rn concentrations in surface waters. The basis of this technique, discussed in detail in Chapter 9, is that the loss of ^{222}Rn from the ground is principally controlled by fractures. For ^{222}Rn to enter surface waters in large amounts, an effective transport mechanism is needed, and this is most readily accomplished by discharge of groundwater. If the discharging groundwater has been part of a deep-circulating hydrothermal system, it is likely to contain a large amount of ^{222}Rn unsupported by its parent, ^{226}Ra. Within an area of relatively uniform geology, variation in the pattern of ^{222}Rn activity can indicate the overall nature of the groundwater circulation system (Mogro-Campero and Fleischer, 1977).

It should be realised, however, that the concentration of ^{222}Rn in surface waters is extremely variable, and many factors can act to change the equilibrium value with time. Rainfall, for example, may cause gentle run-off, transporting ^{222}Rn into streams and leading to increased concentrations of the gas. However, heavy rainfall can cause both dilution and increased stream flow. The latter, in turn, causes increased turbulence and an increase in natural degassing of the water. Moreover, sample sites are likely to vary considerably in their overall environmental setting. Even though care may be taken to avoid areas of turbulence, and checks can be run for major disturbances, small-scale effects are unlikely to be recognised and cannot be allowed for even if identified.

If large areas are to be surveyed, samples are likely to be collected under various meteorological conditions, and even changes in air pressure and temperature will cause variation in the level of dissolved ^{222}Rn in stream waters. For this reason, interpretation of ^{222}Rn survey data may not be carried out directly in terms of ^{222}Rn activities. Instead, significance can be attached only to high values by using exclusion isolines in the graphical treatment of results. These lines are constructed by working outwards from the area of highest value and are drawn to include all readings greater than the value of the isoline.

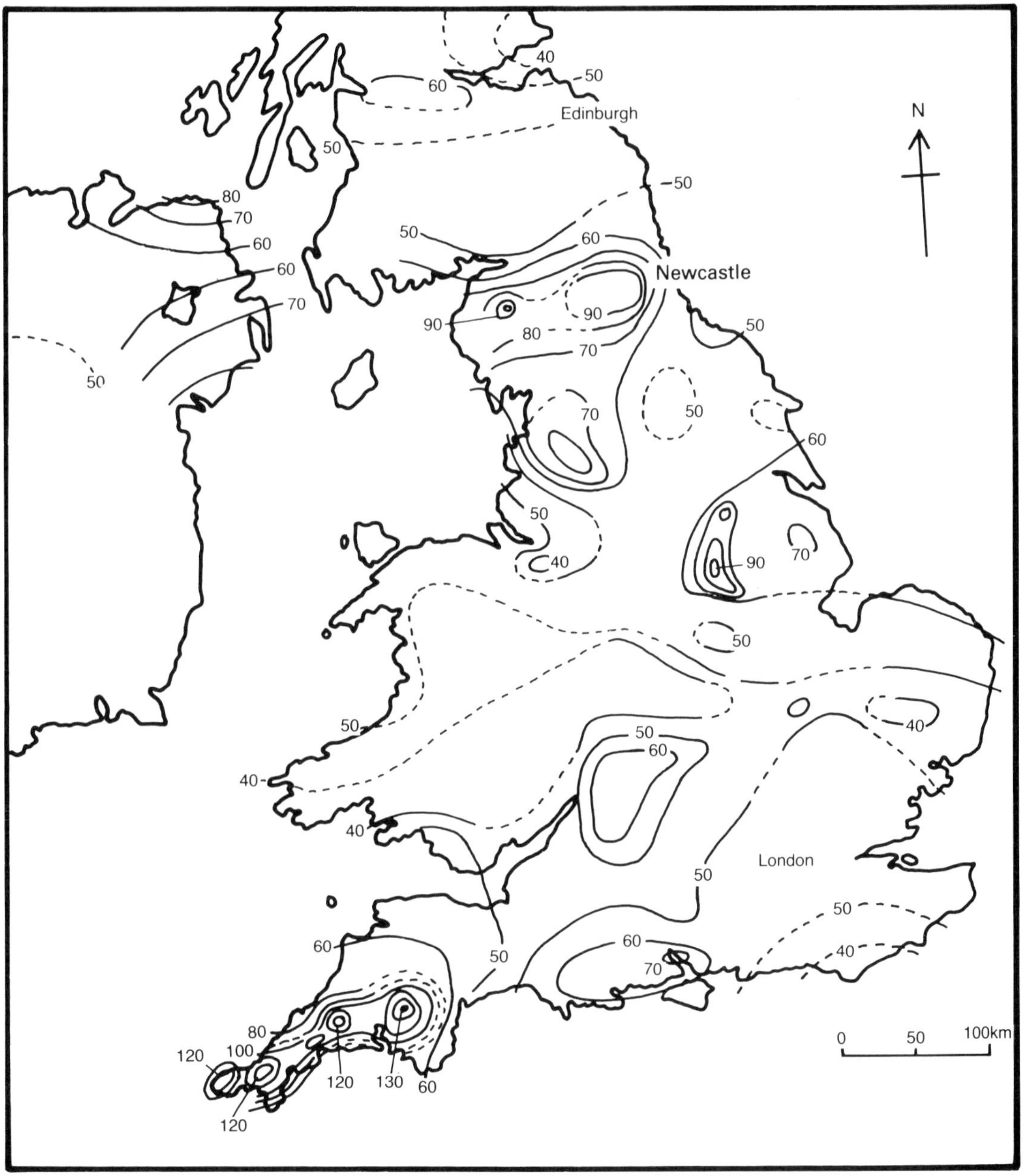

Fig. 7.7 — Heat flow ($m Wm^{-2}$) in the southern part of United Kingdom. The broken lines indicate uncertainty. (Downing and Gray, 1986, redrawn with permission.)

Clearly, the values of exclusion isolines decrease outwards until all values outside the outermost line merely reflect background levels. The pattern of ^{222}Rn exclusion isolines thereby obtained may be interpreted in terms of ascending and descending limbs of hydrothermal convection cells (Fig. 7.9).

Confirmation of the use of ^{222}Rn in surface

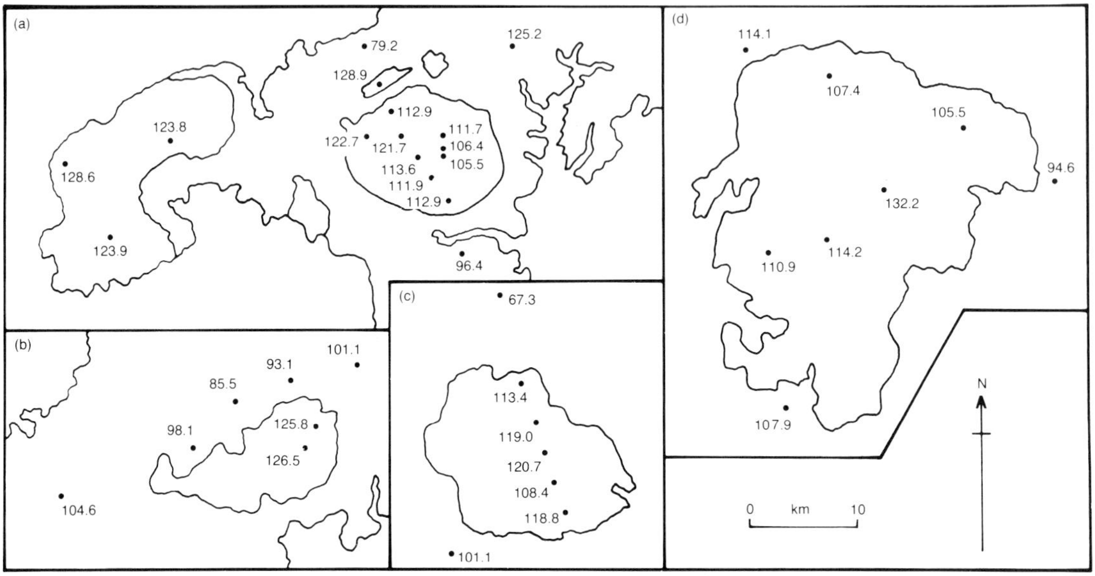

Fig. 7.8 — Heat flow in southwest England (mW m^{-2}), showing the outlines of the plutons: (a) Land's End and Carnmenellis granites; (b) St. Austell granite; (c) Bodmin Moor granite; (d) Dartmoor granite. (After Wheildon *et al.*, 1980.)

waters as an indicator of hydrothermal circulation in the Carnmenellis granite, Cornwall, has recently been obtained from mapping the varition in ^{4}He and ^{222}Rn concentrations in streams (Gregory and Durrance, 1986). Like ^{222}Rn, ^{4}He is a radiogenic gas which is preferentially carried to the surface of the Earth by upward-moving groundwater. In this area the ^{4}He and ^{222}Rn levels in stream water show a close direct correlation with heat flow. This is illustrated in Fig. 7.10. Recently, too, Fehn (1985) has modelled hydrothermal circulation in the granites of southwest England which closely matches the results of Gregory and Durrance.

Finally, it is important to note that, because ^{3}He in the Earth is mainly of primordial origin while ^{4}He is mainly radiogenic, ^{3}He/^{4}He ratios have been used to identify local variations in the contribution of deep background heat to the total heat flow in an area (Hooker *et al.*, 1985). The argument used is that, where there is a higher than normal heat flow from the mantle, then crustal levels of ^{3}He, as shown by groundwaters, will be enhanced. In the case of the Carnmenellis granite, Hilton *et al.* (1985) have found such extremely low ^{3}He/^{4}He ratios in groundwater that the abnormally high heat flow in this area is now proved to be due to the highly radioactive nature of the granite. Indeed, the role of radiogenic heat in controlling post-magmatic geochemical and mineralogical processes within and around the granites of southwest England is profound; radiogenic heat alone has probably been sufficient to give such active hydrothermal circulation that metallisation occurred. An account of the methods used in Rn and He surveys is given in Chapter 9.

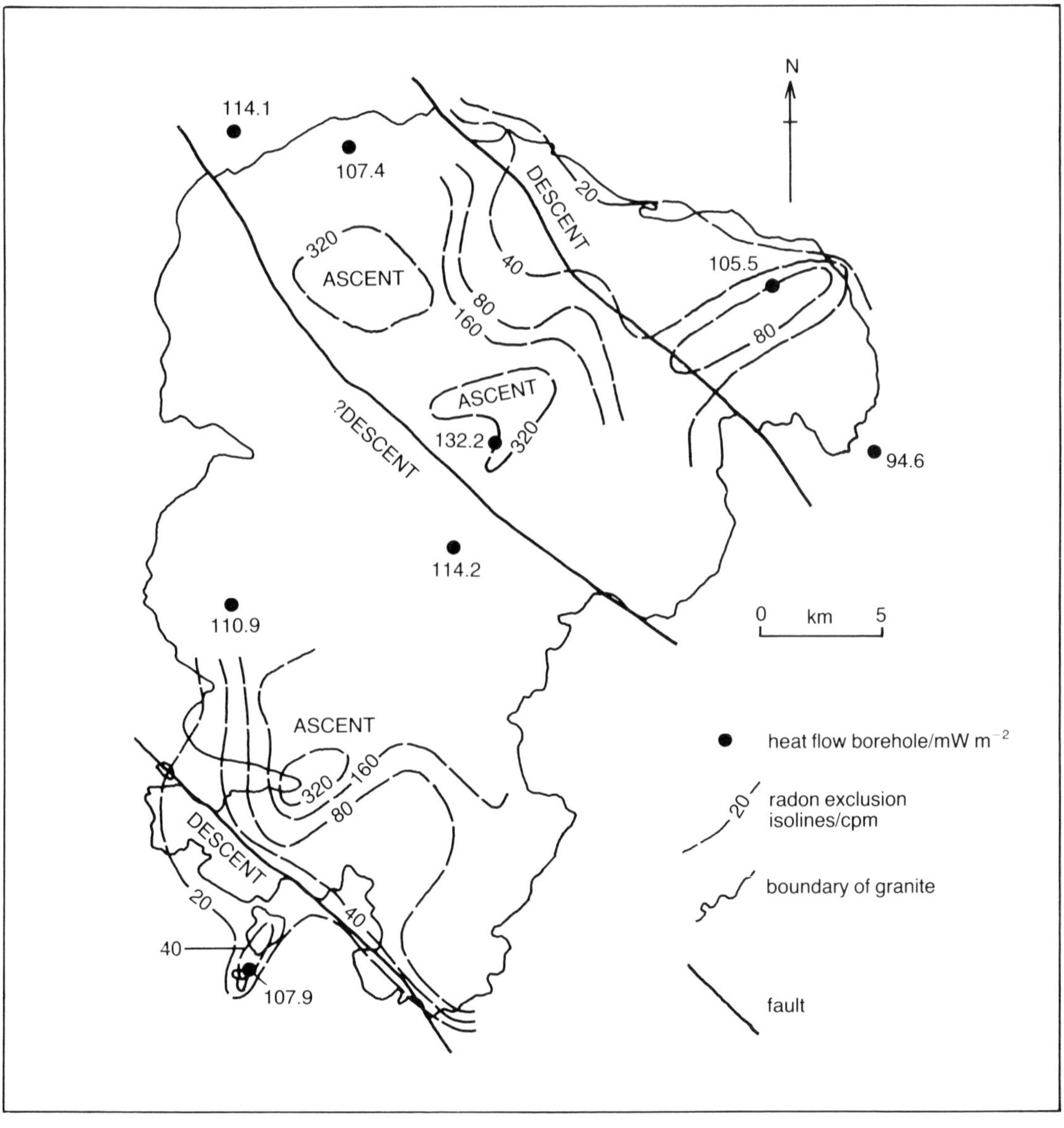

Fig. 7.9 — ^{222}Rn distribution in the stream waters of Dartmoor and possible positions of hydrothermal convective circulation systems in relation to heat flow. 1 cpm (count min^{-1}) is equivalent to 0.3 Bq l^{-1}. (Durrance and Heath, 1985, redrawn with permission.)

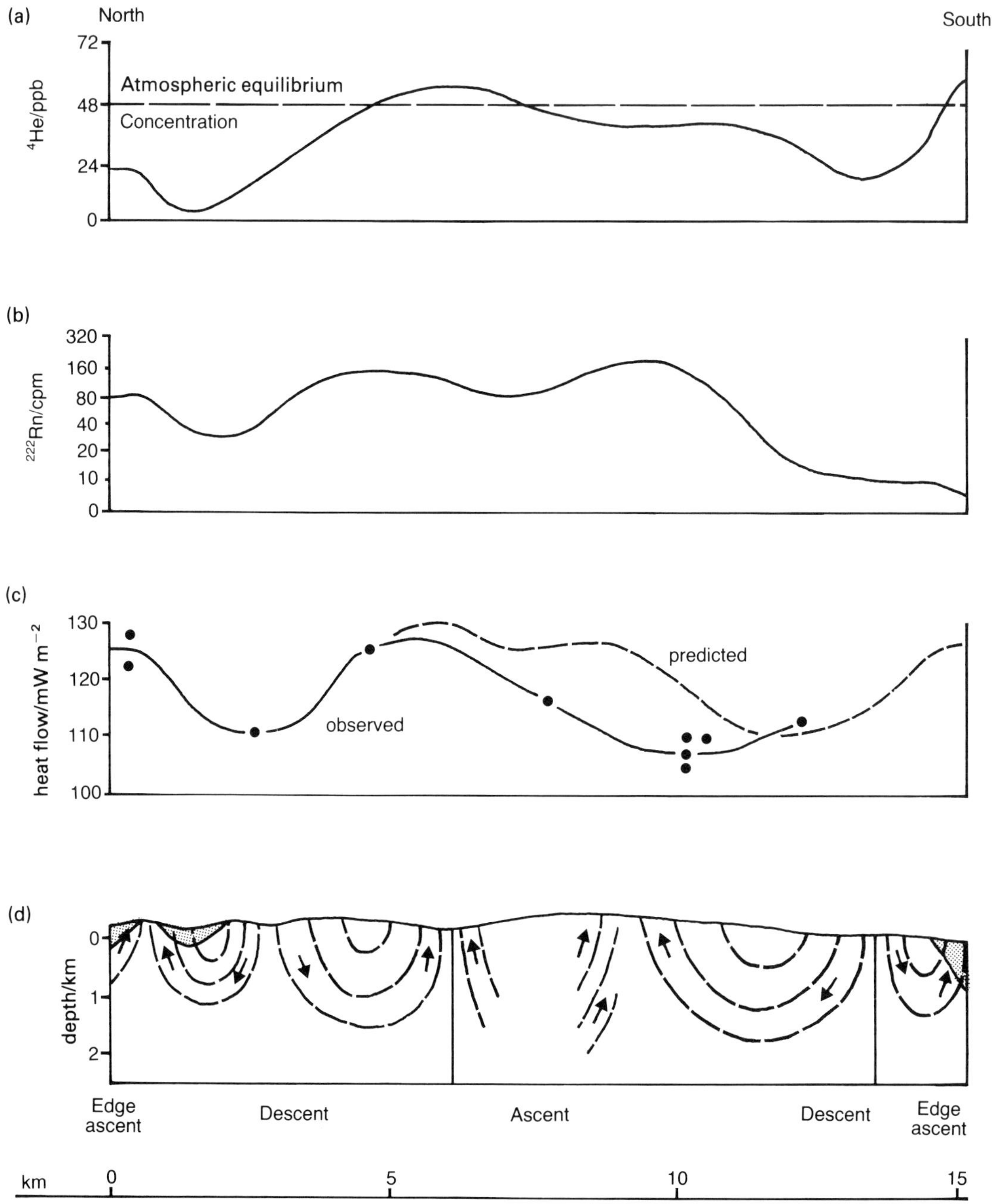

Fig. 7.10 — North–south profiles across the Carnmenellis granite, Cornwall: (a) ^{4}He concentrations in stream water; (b) ^{222}Rn activities in stream water where 1 cpm (count min^{-1}) is equivalent to 0.3 Bq l^{-1}; (c) heat flow; (d) interpreted hydrothermal circulation system in relation to main fractures and the granite margin (host rock stippled). (After Gregory and Durrance, 1986.)

BIBLIOGRAPHY

Armstead, H. C. H. (1978) *Geothermal energy: its past, present and future contributions to the energy needs of Man*. London, Halstead.

Batchelor, A. S., and Pearson, C. M. (1979) Preliminary studies of dry rock geothermal exploitation in southwest England. *Trans. Inst. Min. Metall. B* **88,** 51.

Bott, M. H. P. (1982) *The interior of the Earth: its structure, composition and evolution*. London, Arnold.

Bowen, R. (1979) *Geothermal resources*. London, Applied Science.

Cathles, L. M. (1981) Fluid flow and genesis of hydrothermal ore deposits. In Skinner, B. J. (ed.), *Econ. Geol. Seventy-Fifth Anniversary Volume (1905–1980)*.

Chung, Y. C., and Craig, H. (1972) Excess-radon and temperature profiles from the eastern equatorial Pacific. *Earth Planet. Sci. Lett.* **14,** 55.

Clarke, S. P., and Turekian, K. K. (1979) Thermal constraints on the distribution of long-lived radioactive elements in the Earth. In Bowie, S. H. U., Fyfe, W. S., Ostle, D., Plant, J., and Simpson, P. R. (eds), *Theoretical and practical aspects of uranium geology*. London, Royal Society.

D'Amore, F., Ferrara, G. C., Nuti, S., and Sabroux, J. C. (1976) Variation in radon 222 content and its implications in a geothermal field. In Augustithis, S. S. (ed.), *Proceedings of the International Congress on Thermal Waters, Geothermal Energy and Vulcanism of the Mediterranean Area 1, Athens*.

Downing, R. A., and Gray, D. A. (1986) Geothermal Energy: The potential in the United Kingdom. *British Geological Survey*. London, H.M.S.O.

Durrance, E. M., Bromley, A. V., Bristow, C. M., Heath, M. J., and Penman, J. (1982) Hydrothermal circulation and post-magmatic changes in granites of southwest England. *Proc. Ussher Soc.* **5,** 304.

Durrance, E. M., and Heath, M. J. (1985) Thermal groundwater movement and radionuclide transport in south west England. *Mineral Mag.* **49,** 289.

Edmunds, W. M., Andrews, J. N., Burgess, W. G., Kay, R. L. F., and Lee, D. J. (1984) The evolution of saline and thermal groundwaters in the Carnmenellis granite. *Mineral Mag.* **48,** 407.

Elder, J. (1981) *Geothermal systems*. London, Academic Press.

Fehn, U. (1985) Post-magmatic convection related to high heat production granites of southwest England: a theoretical study. In *High heat production granites, hydrothermal circulation and ore genesis*. London, Institution of Mining and Metallurgy.

Fehn, U., Cathles, L. M., and Holland, H. D. (1978) Hydrothermal convection and uranium deposits in abnormally radioactive plutons. *Econ. Geol.* **73,** 1556.

Glikson, A. Y. and Lambert, I. B., (1976) Vertical zonation and petrogenesis of the early Precambrian crust in Western Australia. *Tectonophysics* **30,** 55.

Gore, R. (1980) What Voyager saw: Jupiter's dazzling realm. *National Geographic* **157,** 2.

Gregory, R. G., and Durrance, E. M. (1986) Helium, radon and hydrothermal circulation associated with the Carnmenellis radiothermal granite of southwest England. *J. Geophys. Res.* In press.

Grieve, R. A. F. (1980) Impact bombardment and its role in proto-continental growth on the early Earth. *Precambrian Research,* **10,** 217.

Gupta, H. K. (1980) *Geothermal resources*. Amsterdam, Elsevier.

Haenal, R. (1980) *Atlas of subsurface temperatures in the European Community*. Brussels, Commission of the European Communities.

Hager, B. H. (1978) Oceanic plate motions driven by lithospheric thickening and subducted slabs. *Nature* **276,** 156.

Heath, M. J. and Durrance, E. M., (1985) Radionuclide migration in fractured rock: hydrological investigations at an experimental site in the Carnmenellis granite, Cornwall. *CEC Rep.* **EUR 9668 EN.** Luxembourg, Commission of the European Communities.

Hilton, D. R., Oxburgh, E. R., and O'Nions, R. K. (1985) Fluid flow through a high heat flow granite: constraints imposed by He and Rn data. In *High heat production granites, hydrothermal circulation and ore genesis*. London, Institution of Mining and Metallurgy.

Hooker, P. J., O'Nions, R. K., and Oxburgh, E. R. (1985) Helium isotopes in North Sea gas fields and the Rhine Rift. *Nature* **318,** 273.

Jaupart, C., Mann, J. R. and Simmons, G., (1982) A detailed study of the distribution of heat flow and radioactivity in New Hampshire (USA). *Earth Planet. Sci. Let.* **59,** 267.

Jaupart, C., Sclater, J. G., and Simmons, G. (1981) Heat flow studies: constraints on the distribution of uranium, thorium and potassium in the continental crust. *Earth Planet. Sci. Lett.* **52,** 328.

Kaula, W. M. (1979) Thermal evolution of Earth and Moon growing by planetesimal impact. *J. Geophys. Res.* **84,** 999.

Langseth, M. G. and Von Herzen, R. P. (1976) Heat flow through the floor of the world oceans. In Maxwell, A. E. (ed.), *The sea,* Vol. 4, New York, Interscience.

Mogro-Campero, A. and Fleischer, R. L., (1977) Subterrestial fluid convection: a hypothesis for long-distance migration of radon within the Earth. *Earth Planet. Sci. Lett.* **34,** 321.

O'Nions, R. K., and Oxburgh, E. R. (1983) Heat and helium in the Earth. *Nature* **306,** 429.

Open University (1981) The Earth's origin and evolution. S237 Block 7, Milton Keynes, Open University.

Parsons, B., and Sclater, J. G. (1977) An analysis of the variation of ocean floor bathymetry and heat flow with age. *J. Geophys. Res.* **82,** 803.

Pollack, H. N. (1982) The heat flow from the continents. *Ann. Rev. Earth Planet Sci.* **10,** 459.

Pollack, H. N., and Chapman, D. S. (1977) On the regional variation of heat flow, geotherms and lithospheric thickness. *Tectonophysics* **38,** 279.

Rao, R. U. M., Rao, G. V., and Reddy, G. K. (1982) Age dependence of continental heat flow — fantasy and facts. *Earth Planet. Sci. Lett.* **59,** 288.

Rinehart, J. S. (1980) *Geysers and geothermal energy*. New York, Springer.

Roberts, A. A., Friedman, I., Donovan, T. J., and Denton, E. H. (1975) Helium survey, a possible technique for locating geothermal reservoirs. *Geophys. Res. Lett.* **2,** 209.

Roy, R. F., Blackwell, D. D., and Birch, F. (1968) Heat generation of plutonic rocks and continental heat flow provinces. *Earth Planet. Sci. Lett.* **5,** 1.

Rybach, L. (1976) Radioactive heat production; a physical property determined by the chemistry of rocks. In Strens, R. G. J. (ed.), *The physics and chemistry of minerals and rocks*. London, Wiley.

Rybach, L., and Buntebarth, G. (1981) Heat generating radioelements in granitic magmas. *J. Volcanol. Geotherm. Res.* **10,** 395.

Rybach, L., and Muffler, L. J. P. (1981) *Geothermal systems*. Chichester, West Sussex, Wiley.

Sclater, J. G., Jaupart, C., and Galson, D. (1980) Heat flow through oceanic and continental crust and the heat loss of the Earth. *Rev. Geophys. Space Phys.* **18,** 269.

Tatsch, J. H. (1976) *Geothermal deposits: origin, evolution and present characteristics*. Sudbury, Tatsch.

Torgersen, T. and Jenkins, W. J., (1982) Helium isotopes in geothermal systems: Iceland, The Geysers, Raft River and Steamboat Springs. *Geochim. Cosmochim. Acta* **46,** 739.

Torgersen, T., Lupton, J. E., Sheppard, D. S., and Giggenbach, W. F. (1982) Helium isotope variations in the thermal areas of New Zealand. *J. Volcanol. Geotherm. Res.* **12,** 283.

Verhoogan, J. (1980) *Energetics of the Earth.* Washington, DC, National Academy of Sciences.

Vitorello, I., and Pollack, H. N. (1980) On the variation of continental heat flow with age and the thermal evolution of continents. *J. Geophys. Res.* **85,** 983.

Wheildon, J., Francis, M. F., Ellis, J. R. L., and Thomas-Betts, A. (1980) Explanation and interpretation of the S.W. England geothermal anomaly. *Proceedings of the Seminar on Geothermal Energy*. Strasbourg, Commission of the European Communities.

White, D. E. (1981) Active geothermal systems and hydrothermal ore deposits. In Skinner, B. J. (ed.), *Econ. Geol. Seventy-Fifth Anniversary Volume (1905–1980).*

Young, H. W., and Lewis, R. E. (1982) Hydrology and geochemistry of thermal ground water in southwestern Idaho and north central Nevada. *US Geol. Surv. Prof. Paper* **1044J.**

8

Detection and measurement of radioactivity

8.1 RADIATION DETECTORS

8.1.1 General description

The detection of charged particles or high frequency electromagnetic radiation (photons) is based on the interaction of these emissions with a detector material, in the same way that light is detected in the eye by the interaction of electromagnetic radiation of visible frequencies with the retina. This analogy can be extended to the second stage in the process because, just as the retina converts the received light energy into an electrical signal which is passed to the brain, in the detection of radioactivity the incident radiation interacts with the detector material to generate a signal. Just as the brain can assess both the frequency of visible light (colour) and its amplitude (intensity), artificial detection systems should indicate both frequency (or energy) and intensity. Moreover, detectors should be capable of distinguishing types of charged particles (α, β, etc.) and, when dealing with nuclear reactions which proceed at statistically fixed rates (even though individual events occur at random), of determining the precise time interval between events. They should also be able to isolate time-coincident events from a background of non-coincident data.

A wide variety of materials can be used to detect ionising radiation. Gases have been commonly employed, as have some solids and liquids; gases ionise when exposed to radiation, some semiconducting crystals may release electrons, other materials may be susceptible to radiation damage, and yet others show significant increases in temperature. Any detection system, however, can be defined in terms of five response characteristics:

(1) Intrinsic efficiency, which is the proportion of emissions reaching the detector that are actually detected.
(2) Total efficiency, which is the ratio of the detected emissions to the source emissions.
(3) Geometry, which is the fraction of the 4π solid angle around the detector from which emissions can actually be received.
(4) Range, which is the frequency or intensity spectrum to which the detector is limited.
(5) Resolution, which is the ability to distinguish between emissions of similar intensity and fequency.

Some of these characteristics are more applicable to photon detection systems, while others are of greater importance in the detection of charged particles.

8.1.2 Gas ionisation

Ionisation chambers

Ionising radiations may be detected qualitatively using a gold leaf inside a gas-filled chamber. When the leaves are charged, they diverge but, when ionising radiations are present, discharge of the electrostatic field occurs and the leaves collapse. The length of time taken for the leaves to collapse gives some indication of the strength of the radiation. The Lauristen electroscope is based on the same principle (Mann *et al.*, 1980).

The more commonly used detection system, however, consists of a chamber, the walls of which are transparent to ionising radiations, filled with an inert gas into which project two insulated electrodes, one acting as the anode and the other as the cathode for an applied voltage. When radiation passes through the gas, energy is transferred, causing atomic or molecular excitation which results in ionisation of the gas. Positive–negative ion pairs, such as molecular positive ions and electrons, are usually formed but in the so-called electronegative gases, such as O_2, the electrons may become attached to neutral molecules to form negative ions. The latter is not specifically utilised in radiation detectors although, if the gas contained within the chamber is air, some negative ions will, of course, be produced. In the absence of an applied voltage, recombination of ions and electrons takes place but, when a potential difference exists between the electrodes, the positive ions and the electrons are accelerated towards the cathode and anode, respectively, where they produce an output signal. Typically, the walls of the chamber form the cathode and a wire passes through the centre of the chamber as the anode. Linking these electrodes is a DC voltage via a current-measuring device. If there is a constant source of ionising radiation, the characteristics of the ionisation current depend upon the magnitude of the applied voltage. It should also be noted that the speed at which the electrons travel through the dectector will be about three orders of magnitude greater than that of the positive ions, because of the more frequent collision of the latter with the gas molecules.

As voltage increases, the ionisation current also increases, but this gradually levels off until the current is independent of the applied voltage. At very high voltages, however, the ionisation current again increases with increase in voltage (Fig. 8.1). This variation occurs because, in the region of low voltage, recombination of ions occurs before they reach the electrodes but, as the applied voltage increases, progressively more and more ions are collected until a saturation point or plateau is reached. The voltage required to produce this saturation current depends upon the rate at which ionisation occurs in the chamber, and the current itself represents the integration of a large number of different ionisation events. As each ion pair transfers 1.602×10^{-19} °C of charge, 6.25×10^{6} ion pairs in 1 s will produce a current of 10^{-12} A. Approximately 34 eV is required to produce one ion pair in air, whether the radiation consists of α-particles, β-particles or photons.

When the ionisation current is large, it can be measured directly. However, for small currents, the charge is usually collected on a capacitor circuit. Alternatively, a pulse ionisation chamber can be employed. The latter can also be used to determine the rate of ionising events and the energy distribution of charged heavy particles. Pulse ionisation chambers consist essentially of electrodes, which have large areas compared with their separation, across which is applied a high voltage (but not high enough to cause discharge through the gas). As the radiation passes through the gas, ionisation takes place, and a suitable resistance–capacitance circuit in parallel with the electrodes records an output voltage pulse. The resistance and capacitance

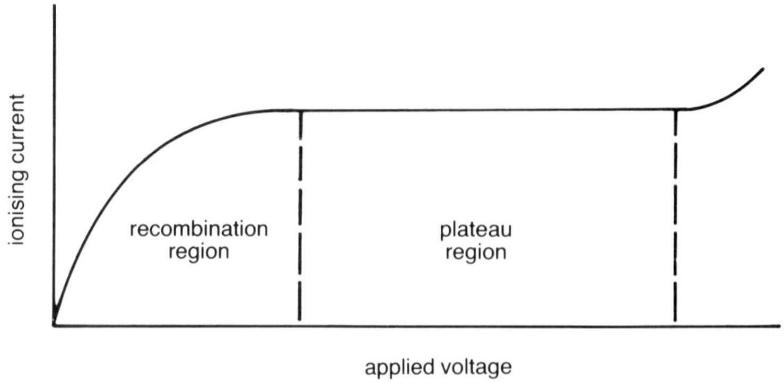

Fig. 8.1 — Variation in ionisation current with applied voltage for a typical ionisation chamber.

values can be chosen so that the voltage pulse returns rapidly to zero, allowing further pulses to be recorded. Individual pulses may be compared on the basis of, for example, pulse shape, which depends upon the distribution of ionisation events within the whole geometry of the detector.

Although ionisation chambers can be designed to give full 4π geometry, in which case they are highly sensitive, they are little used for geological purposes. They are more commonly employed as dosimeters and for detecting α-particles and fission fragments. As α-detectors, they have been employed for monitoring Rn in mines.

In purely electropositive gases, beyond the plateau region the electrons acquire enough energy to produce secondary ionisation, causing the ionisation current to increase again with an increase in the applied voltage. Here, each electron of a given energy produces a certain number of secondary electrons and an avalanche effect (the 'Townsend avalanche') results. Consequently, the number of electrons collected at the anode is greater than the number produced by the radiation. With increasing voltage, the size of the avalanche finally reaches such a level that electrons of different energies begin to interact, and the ionisation current is no longer proportional to the initial ionisation. The effect of secondary ionisation is enhanced if the anode is in the form of a fine wire. Two types of detector make use of the characteristics of ionisation at high voltages: proportional counters and Geiger–Müller counters. Proportional counters work in the region where gas amplification occurs (that is, where secondary ionisation is produced), but at voltage levels at which the size of the output is proportional to the initial ionisation. Geiger–Müller counters, however, operate at voltages beyond the level at which the output becomes independent of the initial ionisation.

Proportional counters

These are normally operated in pulsed mode. The gas contained is purely electropositive or is a mixture of polyatomic gases and an electropositive diatomic or monatomic gas. In the latter case, the amount of polyatomic gas is small and performs a quenching function; when low excitation and ionisation potentials are selected, it prevents the emission of photoelectrons from the cathode.

Pure gas counters

One of the gases typically used in these detectors is Ar. The cathode is normally the enclosing case of the gas chamber and the anode is a very fine central wire. A simple cylindrical arrangement is normally used, but some employ combination systems to give 4π geometry. The cathode is normally steel, Cu or Al, but it can be glass or graphite. The anode is usually made of stainless steel. It is important that the anode has a uniform cross-section throughout its length. This is because the system depends upon the linear gas amplification that takes place close to the anode; here, the number of ion pairs formed is directly proportional to the number of electrons produced by ionisation. It therefore follows that the position of the ionisation event within the detector does not affect the character of the generated pulse; the number of electrons produced is the only factor that controls the pulse height via a proportionality M:

$$M = \frac{m}{1 - mP},$$

where m is the average number of ion pairs formed for each electron entering the region near the anode and P is the probability that a photoelectron will be produced from the counter walls for each such event (because of the interaction of photons produced by the inelastic collisions of the accelerating electrons). In a proportional counter, mP is much less than unity. If n electrons are produced in the track of the ionising radiation, N electrons will reach the anode:

$$N = \frac{nm}{1 - mP}.$$

Thus the total number N of electrons reaching the anode at any particular applied voltage is proportional to the number m ion pairs formed, and $M = N/n$. However, as m relates only to the average number of ion pairs formed per electron, it is subject to statistical variation, and the output pulse height will vary statistically from one ionisation event to another (Mann *et al.*, 1980). Apart from photoelectrons produced by photon interaction, when positive ions are neutralised at the cathode, photoelectrons are also generated. This gives rise to the phenomenon known as 'after-pulsing'. However, if the counter uses a moderate applied voltage, this effect is very minor.

Mixed gas counters

If the gas mixture has a lower ionisation potential than the energies of the photons liberated in the inelastic collisions of the accelerating electrons, this too will produce photoelectrons. The effect of after-pulsing is reduced, however, by the presence of a gas with a low ionisation potential. The most common gas mixture used is 90% Ar and 10% CH_4, in which the CH_4 forms the low ionisation potential component. Ionised Ar can thus acquire electrons from the CH_4; de-excitation then occurs by disassociation when it is electrically neutralised near the cathode, instead of by emission. This greatly increases the stability of proportional counters. Ch_4 can be replaced by N_2, CO_2 or any other electronegative gas, while Ar can be replaced by Kr or Xe. Each gas has its own particular properties. Ar, for example, is best for detection of X-rays, while Xe and Kr have better photoelectric efficiency for low energy γ-rays. Apart from the difference in the gas content, this type of proportional counter is similar to those employing a single gas.

In a proportional counter the electrons are collected in a very short time. It is, therefore, possible for the counter to recover and be ready to record another ionisation event within a microsecond or so. Thus for count rates of up to about 2000 counts $\min^{-1}$, the efficiency of a proportional counter is independent of the count rate. For higher count rates, however, the 'dead-time' effect becomes quite noticeable and the efficiency of the detector decreases. The dead time is simply the time that the system requires to process a detected event; if another event occurs within this time, it will not be correctly processed. Essentially, there are two types of dead time. Extendable dead time relates to those systems in which the occurrence of an event within the dead time of a previous event causes an extension of that dead time.

Thus no other event can be processed until time $\tau+\Delta t$ has elapsed, where τ is the dead time of the system and Δt the time that passed between the first and second events. Non-extendable dead time relates to systems in which the occurrence of the second event within the dead time of the first event does not alter the response of the system; that is, there is no interaction between the system and the second event. Corrections for dead-time losses can only be applied easily for the non-extendable type but, in practice, any system usually contains aspects of both extendable and non-extendable dead time, and modification may be needed to make the non-extendable component dominant.

The linearity of the response of a counter to different count rates can be checked by inserting an increasing number of metal foils between the window of the detector and a collimated monoenergetic radiation source. The logarithm of the recorded count rate is then plotted against the total thickness of the metal foil absorber. Such a check is important because manufacturers' dead-time values may be purely instrumental; that is, they are based upon the assumption that the radiation is evenly distributed in time. In fact, because radioactive decay is a statistical phenomenon in time, this condition is not always satisfied, and episodes of high radiation may cause the actual dead time to be somewhat greater than indicated.

Another problem that can be encountered is the 'end effect' of the detector. Unless the gas amplification within the detector is collimated into the middle part of the anode wire, abnormal electrical fields at the ends of the wire can affect the character of the gas amplification. This can be partly offset by making the wire very long with respect to its radius. Alternatively, correction is possible by dividing the wire into unequal lengths by a glass bead, and substracting the recordings obtained from one part from those obtained from the other.

If the passage through the gas of a photon of moderate energy is considered, it is apparent that, after ionisation due to photoelectric absorption, the excited gas atom may emit a lower energy photon. A pulse height corresponding to either the full energy of the initial photon or this initial energy minus the energy of the escaping photon may therefore be recorded. As a result, two peaks will be found in an energy spectrum: the full energy peak and the lower energy escape peak. The emitted photons can come from the K or L electron shells, but the L photon escape peak cannot usually be resolved, and so only the K escape peak is observed. The magnitude of the escape peak depends upon the nature of the gas within the chamber, the pressure under which the gas has been contained and the size of the detection chamber. In general, however, Ar usually gives only a minor escape peak, while Kr produces a large escape peak. The size of the escape peak in any system can be reduced by increasing the size of the counter and by increasing the pressure of the gas.

Increasing the size of the counter and the pressure of the gas will also reduce the effect of decreased efficiency that results from backscattering of photons by the wall of the counter. In this process, backscattered low energy photons can have so little energy that they are unresolved from the photoelectric peak. Overestimation of the photoelectric peak can thus occur. Because the Compton backscattering cross-section decreases with increasing photon energy, this effect is most important for X-rays and low energy γ-rays. The resolution by proportional counters of closely spaced energy peaks is very good, typically in the 40–180 keV range (Adams and Gasparini, 1970).

Geiger–Müller counters

If the applied voltage across an ionisation chamber is increased beyond the proportional region, the number of electrons collected at the anode becomes independent of the number of electron–ion pairs formed by the initial ionising event (Fig. 8.2). In this 'Geiger region', gas amplification becomes of greater importance and increases the initial ionisation by many orders of magnitude. Here, the factor mP, which was much less than unity in proportional

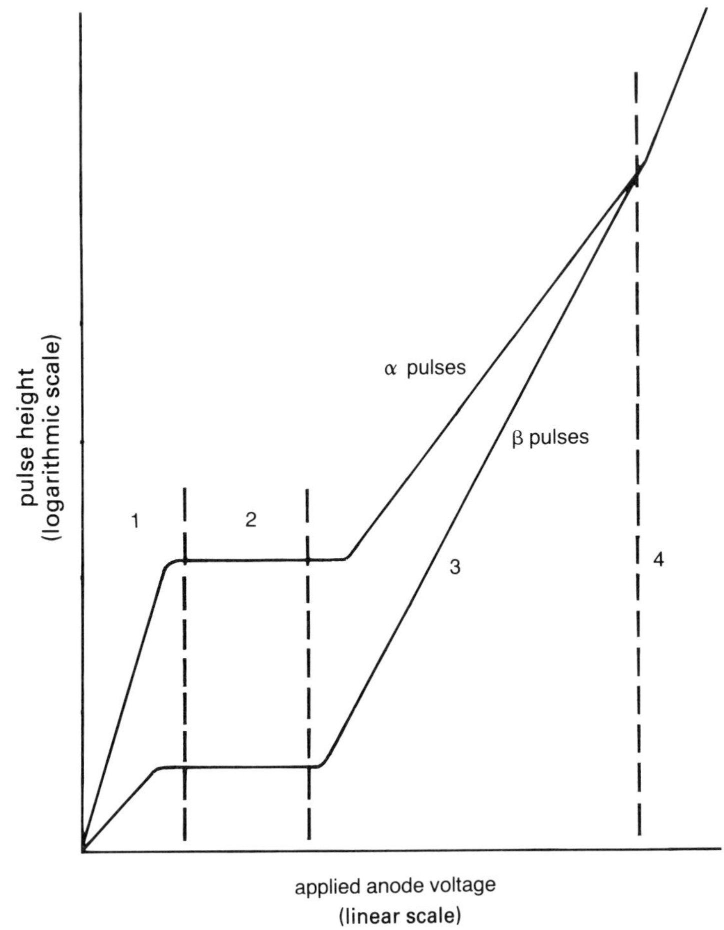

Fig. 8.2 — Variation in pulse height with applied voltage in a gas ionisation detector: 1, recombination region; 2, ion chamber region; 3, proportional region; 4, Geiger-Müller region.

counters, is increased as the multiplication in the avalanche near the anode increases. As mP approaches unity, many ion pairs and photoelectrons are generated and an electrical discharge takes place. Under these circumstances, it is necessary to use a quenching gas in the chamber to prevent the development of multiple pulses or continuous discharge.

The typical configuration of a Geiger–Müller counter follows the cylindrical arrangement already discussed. Any metal can be used for the cathode, but the anode is usually W. The gas is again normally Ar, but He or Ne is also possible. Organic gases such as ethyl alcohol can be used as the quenching agent but, because they suffer rapid decomposition, the life of detectors in which they are used is rather limited. Halogen gases such as Br or Cl also have good quenching characteristics and have the advantage that the life of the detector is longer. They also allow higher voltages to be employed, and they give a steadier response both to background and to high count rates. As noted in the general introduction to gas ionisation, in the Geiger region the response to any given radiation will increase as the applied voltage is increased until a plateau level of output is reached. In Geiger–Müller detectors, the applied voltage is chosen to lie in the middle of this plateau (usually between 800 and 1500 V). However, the actual voltage chosen depends upon the quenching agent employed, and Br-quenched detectors have been designed which operate at only 350 V. The gas pressure is usually about 10^5 Pa or slightly less.

Although α-particles, β-particles and γ-rays will produce gas ionisation, the ionising effect due to β-particles is much smaller than that produced by the other two main types of radiation. In practice, however, α-particles are absorbed by the chamber walls and are not detected. The absorption of γ-radiation by the walls varies with the energy of the incident radiation and, for all typical metals used, the intrinsic efficiency of the detector increases with the photon energy. The detector response, therefore, also depends upon the energy of the incident radiation. Moreover, this is not the only factor that influences intrinsic efficiency, and the opacity of the walls to the electrons liberated in the chamber is also of significance. Bi, for example, tends to give a high intrinsic efficiency (certainly higher than Al) because of the high photoelectric cross-section provided by high atomic number elements.

Because of the response characteristics of a Geiger–Müller detector, the output cannot be used directly for energy analysis of the incident radiation. Only if differential absorption techniques are used can some idea of the nature of the radiation be obtained. These detectors are, therefore, of very limited use, although they do have some positive features. Because Geiger–Müller detectors are based upon a gas-filled tube and a high voltage source, their construction is both simple and inexpensive. The output response is large and so requires no amplification, again aiding the simplicity and cheapness of their design. For these reasons, Geiger–Müller detectors are still employed where routine measurements of activity are needed at low cost, and where detailed information on the nature of the radiation is not wanted. As such, they are a popular geophysical tool for the prospector. Finally, it is worth noting here that liquids and solids are, in fact, better absorbers of penetrating radiation than gases. Liquid–gas ionisation chambers have consequently been developed for more specific uses.

8.1.3 Scintillation detectors

Scintillation

Although solids and liquids may respond to radiation by becoming ionised, this type of detector utilises another property, possessed by materials known as 'phosphors', to behave as transducers and to emit light when struck by charged particles or γ-radiation ('scintillation'). This method was originally developed as the spintharoscope which was used for the measurement of very low activities. It involved the observation and counting under a microscope of individual light flashes from activated ZnS under α-particle bombardment (Adams and Gasparini, 1970).

A variety of materials scintillate in this way: some gases, chiefly the rare gases, emit light by electron de-excitation after they have been subjected to ionising radiation; some organic substances, whether in solid, liquid or gaseous form, emit light by de-excitation of loosely bound electrons in complex molecules; activated crystalline solids such as NaI and ZnS behave as semiconductors; finally, if very energetic particles travel faster than light through a material, light is emitted in a forward direction, relative to the direction of motion of the particle (the latter giving rise to systems known as Čerenkov counters).

In this section, only crystalline solid scintillation systems will be discussed. Use of the scintillation properties of activated ZnS for detecting α-particles has already been mentioned. Similar properties for anthracene make it suitable for the detection of β-particles, while Tl-activated NaI is suitable for detecting γ-rays. The suitability of different phosphors for detecting different types of radiation is related to their density; the density of NaI is much greater than that of anthracene. However, the mechanisms responsible for producing scintillations in organic and inorganic phosphors are different. As noted above, scintillation in organic phosphors arises by de-excitation of loosely bound electrons in complex ions but, to understand why inorganic crystals behave as phosphors, the nature of electrical conductivity in a semiconductor must be considered.

Electrical conductivity in semiconductors

In individual atoms, the electrons occupy well-defined energy levels but, when atoms are bound together in crystals, the mutual interaction between outer shell electrons causes their energies to merge into an energy band called the valence band. Electrons having energies within the valence band can move through the crystal by exchanging with electrons in adjacent atoms, but they remain bound within the atomic framework. If, however, electrons within the valence band receive additional energy, they are able to move freely and enter the conduction energy band (Fig. 8.3). The energy gap between the valence and conduction bands controls the electrical conductivity of the crystal. Conductors such as Cu have valence and conduction bands which are overlapping or are part of a continuum. In these conductors, electrons can move under an applied potential difference. As the electrons move, however, interaction with bound electrons causes lattice vibrations to occur. Different materials react with different degrees of lattice vibration and this accounts for their different electrical resistivities.

Where the valence and conduction bands are separated by an energy gap, some electrons will occur in the valence band and some in the conduction band, such that

$$\frac{n_c}{n_v} = e^{-E/kT}$$

where n_c is the number of electrons in the conduction band, n_v is the number in the valence band, E is the energy gap between the valence and conduction bands, k is Boltzmann's constant and T the temperature in kelvin (Mann *et al.*, 1980). In an insulator, E is a few electron-volts and thus significantly greater than kT at normal temperatures; so very few electrons enter the conduction band and the material has a very high resistivity. Semiconductors, however, have an energy gap of only about 1 eV, and there is a significant number of electrons in the conduction band, although these are numerically small compared with the electrons in the valence band. The variation in the resistivity of a semiconductor with temperature is therefore very marked.

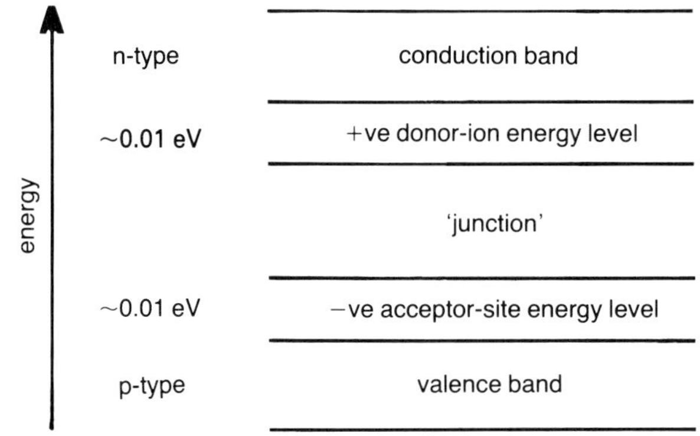

Fig. 8.3 — Semiconductor energy levels.

This will be developed further in the discussion of semiconductor detectors.

This rather simple picture of electrical conductivity is, however, complicated by the fact that most crystals contain impurities and defects in their structure. These can form electronic traps which have energy levels intermediate between the valence and conduction bands of the pure crystal. If such electron traps have energies just a little lower than that of the conduction band, some electrons which have entered the conduction band following excitation may return to the electron trap or enter the valence band when de-excited. De-excitation causes the electrons in the trap to return to the valence band, but this may only occur in response to thermal excitation or some other process such as those which give rise to fluorescence and phosphorescence.

Crystal detectors

Perhaps the most widely used phosphor in radiation detection is NaI. At the temperature of liquid N (77 K), pure NaI crystals are extremely efficient scintillators. However, as the temperature increases, their response deteriorates, and at room temperature they are only about 10% as efficient, the remaining energy being lost in phonon production (vibration) and heat. Furthermore, for maximum efficiency, pure NaI crystals are limited to a thickness of about 1.3 cm because of strong self-absorption of their fluorescent radiation, even at liquid N temperature.

Adding about 0.2% Tl to a melt of NaI, enables crystals of Tl-activated NaI (NaI(Tl)) to be grown which have energy levels below the conduction band. This is due to the presence of Tl halide complexes at sites within the NaI crystal which act as artificial luminescence centres of the type described in Chapter 4. Transfer of electrons to the luminescence centres followed by de-excitation gives a broad band of emitted wavelengths. Furthermore, as NaI(Tl) is transparent to these wavelengths, there is no constraint on the usable thickness of the crystal. NaI(Tl) crystals are, therefore, widely employed as scintillators. The overall scintillation effect is governed by the amount of Tl added; as more is added, the scintillation response passes from the NaI to the luminescence centres. Pure NaI responds by emitting light with a wavelength of about 300 nm, while NaI(Tl) responds at 430 nm. The efficiency of a NaI(Tl) crystal at 300 K is intermediate between that of a pure NaI crystal at 77 K and one at 300 K.

The type of interaction that takes place between a γ-ray and a NaI(Tl) crystal depends upon the energy of the γ-ray. Low energy γ-rays liberate extranuclear electrons by photoelectric transfer from one of the Na, I or Tl atoms. The excited atom then emits an X-ray. Intermediate energy γ-rays initially produce Compton electrons through multiple scattering, but eventually photoelectric interactions occur; again an X-ray is emitted by the excited atom. High energy γ-rays (energy greater than 1.022 MeV) first produce a positron–electron pair, but annihilation of the positron then produces 2 photons of energy 0.511 MeV which, in turn, interact in the same way as low and intermediate energy γ-rays. Again, Compton electrons and photoelectrons are liberated which then interact with the valence electrons in the crystal lattice, exciting these into the conduction band. Finally, the lattice de-excites with the emission of radiation with a wavelength peak at about 430 nm.

Only a fraction of the γ-radiation energy received by a NaI(Tl) crystal, however, is re-emitted in this way, the remainder being taken up in lattice vibrations and heating of the crystal. Different crystals possess different responses, and the intrinsic efficiency varies from less than 1% to as much as 40%. At normal temperatures, the average intrinsic efficiency of NaI(Tl) crystals is about 10%.

The emitted signal from the scintillator (especially from low energy sources) is, therefore, usually very weak and requires amplification by means of a photomultiplier system. Here, both the optical efficiency of the phototube which collects the photon signal and the photoelectric efficiency of its conversion to an electrical signal are important considerations. Typically, optical efficiencies are high for commercial phototubes (usually around 90%), but photoelectric efficiencies are much lower (perhaps only 10% or so). On leaving the cathode of the phototube, the electrons are directed to a multiplier dynode which liberates more electrons than it receives. A chain of multiplier dynodes enhances the signal to a level sufficient for recording. Multiplications of the order of 10^6–10^8 are often employed.

One important factor which bears upon the suitability of a phosphor as a scintillation detector is the decay time constant, which is the length of time for which the photon signal persists. Clearly, this has to be long enough to allow normal recording systems to operate. The decay time constant is defined as the time taken for the signal intensity to decrease to e^{-1} of its peak value. For NaI(Tl) crystals, it is usually of the order of 10^{-6} s.

NaI(Tl) scintillation detectors have three general drawbacks. Firstly, the crystals are hygroscopic; that is, they absorb water from the atmosphere and must, therefore, be kept inside a sealed container. Secondly, the crystals are rather susceptible to shock and can be easily broken. Finally, some K usually substitutes for Na in the crystal structure; there is, therefore, an internal source of γ-radiation which acts as a background that can give rise to problems when high accuracy is required for analytical purposes.

At the present time, the main alternative to NaI(Tl) is CsI(Tl), although high efficiency detectors using Bi–Ge compounds have been developed which appear to have a very promising future. CsI(Tl) crystals are not hygroscopic and are less easily damaged than NaI(Tl). They are also unlikely to be contaminated with K and have a higher intrinsic efficiency than NaI(Tl), but they give poorer resolution of closely spaced energies.

With regard to another problem, it should be noted that the output from the phototube is temperature dependent. The dark current rate (the temperature-dependent background signal) is given by

$$n_e = aT^2 \, e^{-b/T}$$

where n_e is the rate of dark current electron emission, T the temperature in kelvin, a the photocathode area, and b a constant which depends upon the photoelectric material used (Adams and Gasparini, 1970). The dark current output can thus be minimised by operating the system at a low temperature and can be maintained at a constant rate by thermostatic temperature control.

Finally, it is important to appreciate the scale of these various effects. For example, if a 3 keV source is detected with 10% efficiency, 300 eV of emitted energy is obtained. The wavelength of this emitted energy will be about 430 nm; so the photons will have an energy of 3 eV and the output will be only 100 photons. If the phototube has an optical efficiency of 90%, this is reduced to 90 photons, while a 10% photoelectric efficiency reduces this further to only 9 photoelectrons. Thus, only 1 photoelectron is emitted for every 0.3 keV of energy received by the detector (Mann *et al.*, 1980)!

Liquid detectors

Liquid detectors are scintillation counters in which the radioactive material to be analysed is dissolved in an organic liquid which also carries organic flourescent agents in solution. Typical solvents are toluene, xylene and dioxane, with a fluorescent agent such as *p*-terphenyl at a concentration of about 0.5–1%. Radiation energy causes ionisation of the solvent and excitation of the fluorescent agent. De-excitation of the agent then results in emission of violet–ultraviolet light. The method is used for the measurement of α-particle and β-particle emissions, especially for low energy β-particles. It is therefore suitable for those applications which require measurement of ^3H and ^{14}C activities (Chapter 10).

Coincidence counting

This approach is particularly suitable for low energy emissions when discrimination between background and true signal is required. Two detectors are used with separate photomultipliers, the output from which is passed to a coincidence analyser which only allows those outputs which occur within a very short time period of each other to be recorded. Thus, irregular background effects are considerably reduced. If a radionuclide emits two different types of radiation, coincidence counting systems can employ different detector types: β-detectors and γ-detectors, for example. However, the different detectors will have different operating efficiencies, and correction must be made for this. If the efficiency of the β-detector is E_b and that of the γ-detector is E_g, and if N_0 is the true number of disintegration events, and N_b and N_g the number of recorded β-emissions and γ-emissions, respectively, such that $N_b=N_0E_b$ and $N_g=N_0E_g$, the number N_c of coincidence events is given by $N_c=N_0E_bE_g$ and the true number of disintegrations by $N_0=N_bN_g/N_c$. Further complications arise, however, because other differences also exist between the two detectors: different dead times, backgrounds, sensitivities to other radiations, and so forth. These must be considered before true activity can be determined.

These problems can, of course, be largely overcome if the two detectors measure the same type of emission but of different energies: γ–γ detection, for example. However, it is important to realise that, generally, this is only possible where the γ-emissions take place as part of a cascade. Under other special circumstances, γ–γ coincidence counting has been successfully used for the detection of very small concentrations of ^{22}Na and ^{26}Al. In these examples, decay is by β^+-particle emission upon annihilation of which two coincident γ-rays are produced which are emitted at about 180° to each other. In addition, a higher energy γ-ray accompanies each β^+-particle emission. Detectors placed at opposite ends of a source may then be linked in a coincidence mode to distinguish events due to these nuclides from the general background.

8.1.4 Semiconductor detectors

General description

As discussed earlier, the resistivity of a crystal is determined by the energy gap between the valence band and the conduction band. Examples at the extreme range of this energy gap are graphite (zero) and diamond (6 eV); graphite is a good electrical conductor while diamond has a resistivity in the order of $10^6\,\Omega$m. Within this range lies the class of materials known as semiconductors, in which the resistivity is inversely related to temperature. The energy gap is usually greater than 3 eV. At 0 K, all defect-free semiconductors are perfect insulators.

As before, when ionising radiation encounters a semiconductor, electrons in the valence band are excited and enter the conduction band, or highly excited bands above the conduction band, leaving a positive hole in the valence band. Highly excited electrons exist for only about 10^{-12} s before they de-excite to the conduction band. In this process, the radiation emitted causes more electron–hole pairs to be formed. As a result, the valence band is characterised by positive holes while the conduction band has a negative charge. Therefore, if two electrodes are placed on opposite faces of the crystal semiconductor and a voltage is applied, the electrons (and holes) move to the appropriate electrode and the induced charge can be measured.

This simple system is, however, usually disrupted by the presence of impurities and imperfections in the crystal lattice. These give rise to electron traps at energy levels between those of the valence band and the conduction band. Both absorption and emission of electrons are possible from these sites; where impurity atoms are responsible, they are known as ‘acceptors’ and ‘donors’, respectively. Donors provide sites where loosely bound electrons can be excited to the conduction band, even by thermal energies at room temperature.

These arise where a higher valency impurity is introduced into the structure. In contrast, acceptor sites are created where a lower valency impurity is present. These sites accept electrons from the valence band, leading to the formation of positive holes. Again, thermal energies are sufficient to accomplish this.

Semiconductors in which the presence of donor atoms increases conductivity by emission of electrons (negative charge carriers) are known as 'n-type' semiconductors, while the presence of acceptor atoms (holes which effectively act as positive charge carriers) they are give rise to 'p-type' semiconductors. It should be noted, however, that in each case it is the dominant type of charge carrier that is the basis of the classification. The other type will also be present as a minority carrier.

Semiconductor properties may be modified by artificially introducing impurity atoms into growing crystals. This process, which is known as 'doping', allows the efficiency with which a semiconductor will detect ionising radiation to be greatly enhanced.

To have good detection characteristics, a semiconductor should have the following properties (Adams and Gasparini, 1970):

(1) Within the crystal, there should normally be very few charge carriers, either electrons in the conduction band or holes in the valence band.
(2) There should be few trapping centres which hold electrons or holes.
(3) Recombination of electrons and holes should not occur during the detection process.
(4) Electrons and holes must have high mobility so that charge collection is rapid.
(5) For good resolution, the production of an electron–hole pair should require low average energies.

Because semiconductors are not good insulators, the application of an electrical potential to collect the charge carriers often results in a leakage current. The solution to this difficulty in Si and Ge, two common semiconductors used for radiation detection, has been to remove the charge carriers from the material as a whole by the use of rectifying contacts (Fig. 8.4). This can be achieved by diffusing high concentrations of the donor impurity into a p-type semiconductor, or of acceptor atoms into an n-type semiconductor, to make a p–n junction rectifying diode. A p–n junction can also be made by putting a surface of an n-type semiconductor on a p-type layer and then coating with a pure metal (such as Ag or Au). The latter method is used in the construction of surface barrier detectors which are used for α-particle spectrometry, but it produces a sensitive volume that is too small to be used for γ-ray spectrometry.

The sensitive volume can, however, be increased by purification of the semiconductor to produce very high quality crystals, but so far this has only proved possible with Ge. More commonly, the volume is increased by compensation procedures in which electron donor ions (usually Li) are drifted into p-type crystals where they form complexes with the acceptor impurities, thereby neutralising the acceptor atoms. This method is particularly suitable for Ge crystals, as the high mobility of Li ions in Ge allows large Ge(Li) detectors to be made. Because of its high atomic number, Ge is a good material for the total absorption of γ-rays. Conductivity is reduced even further by cooling to liquid N temperature (77 K).

Ge and Si are generally the most satisfactory materials for semiconductor detectors, as they also have suitable mobilities for the charge carriers. The latter can be increased further by cooling to liquid N temperature. To prevent redistribution and precipitation of Li within the detectors, they must be stored at temperatures below 222 K. Li-drifted detectors are also known as p–i–n detectors. Today, this procedure is still used for the manufacture of Si(Li) X-ray detectors, but it is clear that, if very pure crystals can be grown, there is no need for Li drifting. In fact, in the last few years the development of crystal growth techniques has

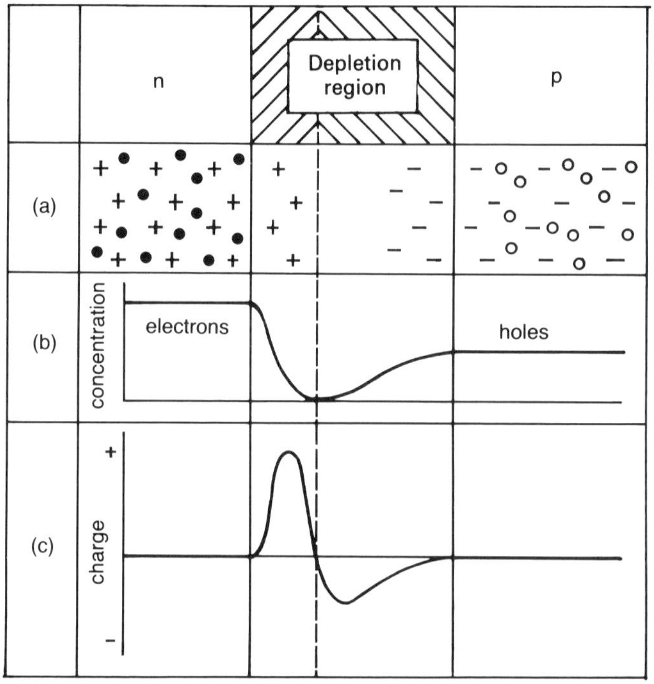

Fig. 8.4 — Diagrammatic representation of a p–n junction showing (a) the distribution of acceptors (−), donors (+), electrons (●) and holes (○), (b) electron and hole concentrations and (c) charge distribution. (After Mann *et al.*, 1980.)

progressed to the point that Ge crystals of sufficient purity are now grown routinely, and Ge(Li) detectors are no longer produced in substantial quantities. Modern γ-ray detectors still need cooling to liquid N temperature in order to reduce the electrical noise to an acceptable level, but there is no danger of destruction of the detector if the temperature should rise to room levels.

Two detector designs are available: planar and coaxial. Planar detectors consist of a flat surface, usually disc shaped, while coaxial detectors have cylindrical geometry.

Charged particle detectors

Most charged particle detectors used for analytical purposes are surface barrier diodes. In practice, charged particle detection is much simpler than photon detection because the particles interact with the material of the detector to produce direct ionisation, and the path of the particle is more clearly defined. The efficiency with which a particle is detected is, therefore, almost 100% as far as the interaction is concerned and, for standard geometries, the overall detection can be calculated using the solid angle alone. For example, a small detector of circular cross-section placed at a large distance x from a point source (where x is greater than three times the detector diameter) subtends a solid angle of A/x^2, where A is the cross-sectional area of the detector. At smaller values of x, the solid angle is $2(1-\cos\theta)$, where θ is the maximum angle subtended by the detector.

As a charged particle penetrates matter, it loses energy along its path length. The energy loss per unit path length (the stopping power) determines the total penetration range of the particle in the detector (Fig. 8.5). Path length and range, however, are not necessarily the same. Only if the path length is straight, as it normally is for heavy charged particles such as protons, will they be equal. For electrons, however, the path length is much greater than the range. This is because electrons lose energy by collision with other electrons, and this produces angular deviations from their original direction. The energy transferred from the incident electrons cause ionisation and radiative collisions. Ionisation is the main process for electrons up to a few megaelectronvolts, but radiative energy losses increase with increasing energy and are roughly equal to ionisation losses at about 9 MeV, although this, of course, varies with the detector material.

It is possible to calculate the ranges of electrons of different energies in different materials, but statistical variation about these values is considerable. Range is expressed in milligrams per square centimetre (thickness multiplied by density), which is also known as the 'surface density'. For thin foils, this is calculated by dividing the weight by the area.

Heavy charged particles lose energy in a straight line while undergoing collision with electrons. The theoretical loss of energy E is given by

$$\frac{\mathrm{d}E}{\mathrm{d}x} = -\frac{4\pi Z_1 Z_2 e^2 N}{mv^2} \log_e\left(\frac{2mv^2}{I}\right) ,$$

where m is the mass of an electron with charge e, Z_1 is the atomic number of the moving particle, Z_2 is the atomic number of the stopping material, I is the average ionisation poten

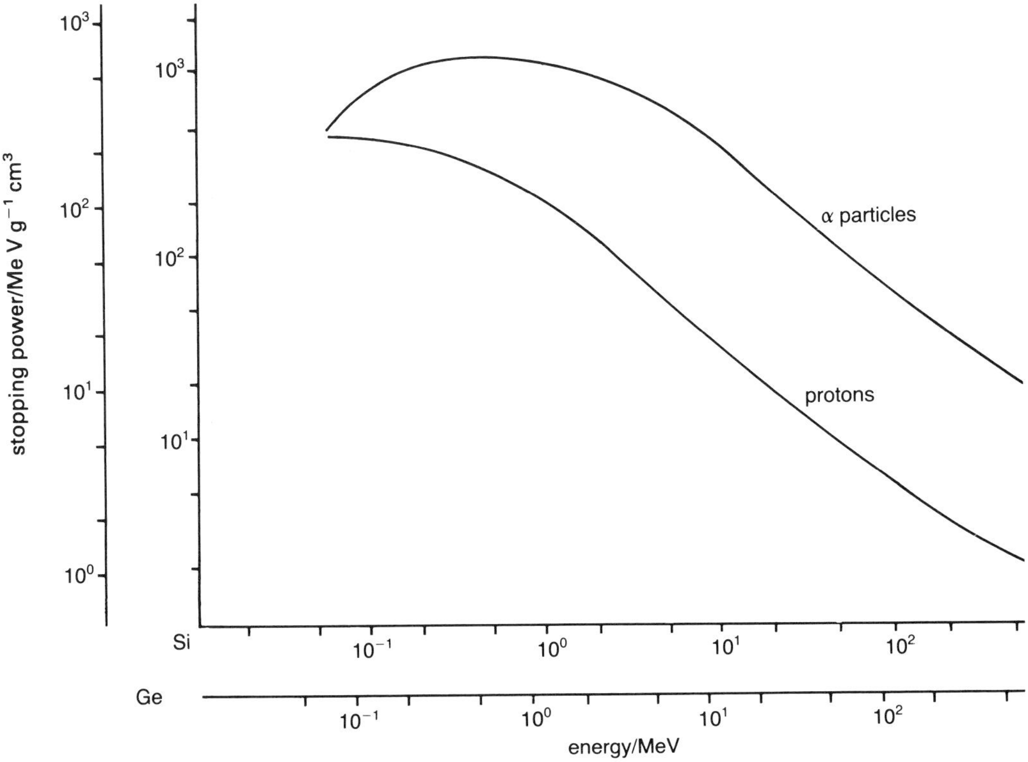

Fig. 8.5 — Stopping power as a function of energy of protons and α-particles in Si and Ge. (After EG&G Ortec, 1983

tial of the stopping atoms and v is the ion valency. As the logarithmic term only changes slowly with energy,

$$\frac{dE}{dx} \propto -\frac{Z_1}{E}$$

and

$$R = \int_{E_{max}}^{0} \frac{dE}{dE/dx},$$

where R is the range (Canberra, 1984).

Unfortunately, higher Z ions do not behave as if they were simple multiples of protons, since they are not completely stripped of electrons. Indeed, the amount of electron stripping actually present is difficult to calculate, and experimental measurements are needed.

Si surface barrier detectors consist of a p–i–n diode in which a depletion region is formed by applying a bias voltage. The resulting electric field collects the electron–hole pairs produced by the charged particles. The resistivity of the Si must be high enough to allow a sufficiently large depletion region at moderate bias voltages. Typically, the diode has a very thin (40 $\mu g\,cm^{-2}$) evaporated Au contact on the etched Si surface wafer, which forms the front rectifying contact, while the rear ohmic contact is composed of 40 $\mu g\,cm^{-2}$ of evaporated Al. When a reverse bias voltage (positive polarity on the Al contact) is applied, an electric field is created. Depending upon the voltage applied, the electric field can be made to decay to zero over the thickness of the detector (totally depleted), over only part of the thickness (partially depleted), or in such a way that it still has a significant value over the whole thickness (over-depleted). Over-depleted systems are produced when the applied bias voltage is larger than the minimum necessary for total depletion. An electronic field larger than zero exists, therefore, immediately under the Al back contact. Because the free carriers are swept out of the region of the electronic field in the detector, this region is called the 'depletion region'. Depletion depths can be calculated from knowledge of the detector resistivity. Commercial detectors are generally available with depletion depths of 10^{-3}–10^{-4} m. Their cost is approximately proportional to the depletion depth. This depth is an important detector parameter because in energy spectrometry, for example, each output pulse must be generated with an amplitude proportional to the energy of the charged particle. Therefore, the sensitive depth of the detector must be sufficient to absorb completely all the particle energy. Where spectrometry of heavy charged particles is involved, very large electric fields are needed to ensure complete charge collection.

Detector characteristics are also specified in terms of surface area. This too is important because it affects both sensitivity and signal resolution. For measurement of very low intensity radiation, or when accurate count rates are required, a large surface area is desirable. However, because detector capacitance and junction noise are also directly proportional to the area, smaller detectors give better resolution. There always needs to be some compromise between sensitivity and resolution. Resolution is also better for thin detectors. Typically, α-particle resolutions of 15–35 keV and β-particle resolutions of 7–30 keV are available in commercial detectors, but for α-particles the ultimate resolution (with no significant contribution from noise) is probably less than 10 keV. β-particle resolution is nearly always determined solely by the broadening of energy peaks by electronic noise. Areas of 25–450 mm^2 are available in commercial detectors.

For α-particle spectrometry, even the thinnest detectors are suitable but, in thin detectors, only very low energy electrons are collected. Protons have intermediate ranges.

Because the range of particles is very limited within the detectors, it is clear that, if there is any variation in the thickness of totally depleted detectors, considerable variation in the output characteristics can result. For example, a 10 MeV α-particle incident upon a 50 μm detector will lose approximately 5.9 MeV in traversing the detector. The rate of energy loss of the particle on leaving the detector, however, will be about 160 keV μm^{-1}. A variation of 1 μm in the thickness of the detector will, therefore, cause an energy spread of 160 keV, which is many times greater than the resolution of the detector. Precise control over the thickness of detectors is, therefore, essential during their manufacture.

Because the presence of defects in the Si crystal affects the behaviour of the detector, any radiation damage suffered by a detector during use will cause its response characteristics to deteriorate. This is a minor problem when light charged particles are involved, but the detection of, say, fission fragments is severely limited in this way. The particle dose that will lead to significant damage to the detector varies according to the type of particle concerned and is about 10^8 fission fragments cm^{-2}, 10^9 α-particles cm^{-2}, 10^{10} protons cm^{-2}, 10^{12} fast neutrons cm^{-2} and 10^{13} electrons cm^{-2}. Partially depleted detectors are more susceptible to damage than are fully depleted detectors.

Physical damage resulting from shock can also cause change in the output characteristics of a surface barrier detector. These detectors must, therefore, be treated with care. To prevent contamination, the surface of the detector must not be handled in any way. Finally, it should also be noted that, as the temperature increases, the leakage current increases (by a factor of 3 for an increase of 10 °C, giving a noise width increase of about 1.7 keV) until an upper limit is imposed by the ultimate and permanent breakdown of the detector at between 45 and 55 °C.

Neutron detection

Because neutrons are uncharged, they do not cause ionisation when they interact with a detector. Measurement of secondary processes is, therefore, required for their detection. Three different procedures are commonly employed (Canberra, 1984).

(1) Detection of nuclear reaction products following neutron capture.
(2) Detection of recoil protons following elastic scattering.
(3) Detection of β-particles or γ-rays from long-lived nuclides following neutron capture.

Neutron-induced nuclear reactions can result in the emission of α-particles or fission products. Commonly, the detection of thermal neutrons is based on the reaction

$$n + {}^{10}_{5}B \rightarrow \alpha + {}^{7}_{3}Li \ .$$

Here, a gas detector filled with BF_3 is used. The α-particle emitted has an energy of 1.8 MeV and can be easily detected. For low count rates, the Geiger region is used, while an ionisation chamber system is employed for high count rates. An alternative system uses a gas chamber filled with ^{3}He, such that

$$n + {}^{3}He \rightarrow p + {}^{3}H \ .$$

In this case, both the proton and the ^{3}H are detected. Because ^{3}He counters are operated at higher pressures than BF_3 counters, they have higher efficiencies. Fast neutrons may be measured by slowing them to thermal velocities by means of a H-rich material such as water or plastic.

Capture of neutrons results in the emission of fission fragments which can be detected. ^{235}U is used as the fissionable material for thermal neutrons, and ^{238}U or ^{237}Np for fast neutrons. Fission fragments are detected in an ionisation chamber.

Fast neutrons can also be measured by detecting the recoil protons associated with neutron interactions using a surface barrier detector. For a specific neutron energy, there will be a continuum of proton energies up to the energy of the neutron itself, at which there is a sharp cut-off. Recoil protons can also be measured using H gas detectors.

The third method of neutron detection is based on the activation of a long-lived radionuclide, commonly ^{115}In. Neutron bombardment changes this to ^{116}In which decays by β-particle emission with a half-life of 54 min. Since the capture cross-section of the ^{115}In is known, the neutron flux can be calculated from measurements of the β-activity of ^{116}In.

Photon detection

The choice of a detector for a particular application is determined primarily by the energy of the photons to be detected. Commercially available detectors cover a range from less than 1 keV to over 10 MeV. Individual detectors can only be used for specific parts of this range, although some are available which extend from 5 keV to 10 MeV.

Typically, a wide-range detector consists of a reverse electrode coaxial Ge system. This is similar in geometry to other coaxial Ge detectors, but the electrodes are arranged in an opposite mode to that in a conventional coaxial detector; the p-type electrode is on the outside and the n-type is on the inside. The advantages of this arrangement are that it allows reduced window thickness and gives increased resistance to radiation damage. Ion-implanted B p-type electrodes and diffused Li n-type electrodes are typically used. Because the ion-implanted contact is extremely thin, when used in conjunction with a thin window (it is actually embedded into the surface) the energy response can extend to as low as about 15 keV.

Radiation damage in photon detectors is principally due to the creation of hole traps in the Ge by neutrons or charged particles. In most detectors the hole charge carriers are collected by the inner electrode in the coaxial configuration, but in the reverse electrode system they are, of course, collected by the outer electrode. In the coaxial configuration, a much greater proportion of the active detector volume is situated within any given distance of the outside contact than of the inner contact. It follows, therefore, that, on average, holes have less distance to travel if they are attracted to the outside contact than to the inside contact. In general, the effect is to render the detector some ten times more resistant to radiation damage than are conventional coaxial Ge detectors.

The conventional coaxial Ge detector is known alternatively as the 'pure Ge', 'hyperpure Ge' or 'intrinsic Ge' detector. It consists of a cylinder of Ge with an n-type contact on the outer surface and a p-type contact on the surface of an axial well. Typically, the Ge has a net impurity level of about 10^{10} atoms cm^{-3}, and so with moderate reverse bias the entire volume between the electrodes becomes depleted. The n and p electrodes are usually diffused Li and Au surface barrier, respectively. Ion-implanted B may be substituted for the surface barrier contact. The usual energy range is from 50 keV to about 10 MeV. Resolution is about 1 keV at 122 keV and about 2 keV at 1.33 MeV. Similar resolutions are obtained from reverse electrode systems.

Planar Ge detectors are constructed from a circular disc of pure Ge with a diffused Li n-type contact on one side and a surface barrier of ion-implanted p-type contact on the other. Typically, these detectors are grooved on the rear n-type contact in order to reduce leakage current along the surface between the contacts. This type of detector is useful down to energies of 3 keV and up to several hundred kilo-electronvolts. The lower limit requires detector windows which are particularly transparent to photons. These are usually made of Be. Areas range from 25 up to 1000 mm^2, with depths between 5 and 13 mm. Resolution is extremely good, being of the order of 200 eV for most detector sizes at 5.9 keV and about 500 eV at 122 keV.

Special low energy Ge detectors have also been developed which have some important advantages over the normal planar or coaxial detectors. These consist of a circular disc of Ge on which there is an ultrathin implanted p-type outer contact on the front face and the wall. The diffused Li n-type contact occupies only a spot on the rear face, which covers much less than the full area of the disc. Thus the capacitance of the detector is much less than that of a planar detector of similar size. This results in lower preamplifier noise, which is a function of detector capacitance. The detectors are available with active areas of between 5 and 25 cm^2, and with thicknesses of between 5 and 20 mm. Extremely good resolution is provided at low and moderate energies.

8.1.5 Source–detector geometry

The radiation from a point source subtends a solid angle of 4π. Many detectors, however, only accept radiation travelling in a certain direction, as governed by the detector window. For a point source of radiation and the flat end window of a cylindrical, placed at a distance d from the source, the area A of the spherical wave front that the window admits is given by

$$A = 2\pi r^2 \left(1 - \frac{d}{r}\right) ,$$

where r is the distance from the point source to the margin of the detector window. The fraction of the radiation received by the window is thus

$$\frac{A}{4\pi r^2} = \frac{1}{2}\left(1 - \frac{d}{r}\right) = G .$$

G is known as the 'source–detector geometry', 'geometry factor', 'geometrical efficiency' or 'extrinsic detector efficiency'. The ratio d/r defines the half-angle θ of the cone subtended by the detector window, as $d/r = \cos\theta$. When $\theta = \pi/2$, the solid angle is 2π but, when $\theta = \pi$, the solid angle is 4π and the detector completely surrounds the source. For most exploration geophysics a 2π geometry is the best that can be employed.

8.1.6 Background

The response of a detector in the absence of a specific radioactive source is known as its 'background'. This is due to cosmic radiation, radiation from radionuclides within the materials of which the detector is constructed and delayed die-away of the detector after responding to higher activity. One of the characteristic features of Geiger–Müller detectors, for example, is that the background can increase to such an extent after recording high activities that a break in use is needed to restore the detector to a normal background response. The effect is most noticeable when organic quenching gases have been used, and if the detector has been heated. Normally, Geiger–Müller detectors are designed to work at maximum temperatures of about 50–80 °C.

Where the activity of specific samples is to be measured, the background count rate of the detector is usually measured first by substituting an inactive sample under the same geometry and operating conditions. It is also good practice to follow the determinative measurement with another background count in order to minimise any time-dependent background effects. Background can be determined if the detector is heavily shielded by Pb blocks. Note that the background rate is subtracted from any higher count rate after the dead-time correction has been applied to the higher count rate.

8.1.7 Dead-time corrections

For a non-extendable dead-time interval of τ s, and an observed count rate of n counts s^{-1}, the system is insensitive for $n\tau$ counts that is, it is sensitive for $1-n\tau$ counts. The dead-time corrected count rate N is, therefore, given by

$$N = \frac{n}{1 - n\tau} .$$

With a system involving an extendable dead time, it is also necessary to know the statistical distribution of the time intervals between successive events. Even though radioactive decay is a random process and the events arrive in random sequence, the time interval is characterised by a Poisson distribution. This allows determination of the probability that no events occur within the dead-time interval using $e^{-N_i\tau}$, where N_i is the mean input rate. Thus

$$n = N_i e^{-N_i\tau} ,$$

which does not allow solution of N_i for the observed values of n and τ. An approximation for N_i, however, is given by

$$\frac{n}{1 - n - (n\tau)^2/2} ,$$

which allows correction of count rates with accuracies of about 0.1% for extendable dead-time losses of 10% or less.

Non-extendable dead-time systems show count rates which asymptotically approach τ^{-1}, for event rates considerably greater than τ^{-1}, but the system is never paralysed. In contrast, extendable dead-time systems reach a maximum observed count rate for an event rate of τ^{-1} which decreases to zero for event rates greatly in excess of τ^{-1}. This means that, for extendable dead-time systems, an observed count rate could be due either to a real value less than τ^{-1} or to an 'alias signature' from an exceptionally strong source. Checks can be run by decreasing the signal a little (say, by increasing the distance between source and detector) and noting the change in the observed count rate; if this decreases, the first value was correct but, if the count rate increases, there is an alias signature.

8.1.8 Resolution

On average, between 25 and 35 eV of energy is used in the creation of one ion pair in most gas ionisation detectors for different types of radiation. For a NaI(Tl) scintillation crystal detector, about 200 eV is needed per photoelectron received by the cathode of the phototube. A similar energy is needed in liquid scintillation systems. However, only about 3 eV is needed for the production of an electron–hole pair in a semiconductor detector. Thus, for a given quantity of radiation, the number of electrons liberated in a semiconductor detector is much greater than that for a NaI(Tl) detector and, because there is a distribution of energy, the standard deviation of the pulse height of the semiconductor response will be much smaller than that for the NaI(Tl) response.

Resolution of NaI(Tl) detectors is defined in terms of the percentage ratio $\Delta E/E$, where ΔE is the full peak width measured at half the maximum peak height and E is the energy of the 662 keV γ-ray of the decay of ^{137}Cs. For semiconductor detectors, the full width (ΔE) at half-maximum alone is used. The energy employed as a standard for Ge(Li) detectors depends upon the detector size. For sizes greater than about 10 cm, the 122 keV γ-ray from ^{57}Co and the 1.332 MeV γ-ray from ^{60}Co are used as reference energies. Smaller Ge or Si detectors also use the 122 keV γ-ray from ^{57}Co but, in addition, the 5.9 keV γ-ray from ^{55}Fe is used.

Because of the nature of the interaction of the ionising radiation with the detector, a monoenergetic incident beam will become broadened. If it is assumed that this broadening follows a Poisson distribution, the full width at half-maximum is given by $2.35\sigma^2$, where σ is the standard deviation of the peak which is given by

$$\sigma = \left[\frac{\sum_{i=1}^{k} (n_i - \mu)^2}{k}\right]^{\frac{1}{2}}$$

Here n is the number of events recorded in the series between $i = 1$ and $i = k$, and μ is the true mean of the distribution. However, since it is impossible to obtain the value of μ from experimental observations, an estimate of σ has to be obtained by substituting the average number

$\bar{n}$ of decays occurring in k observations. Moreover, since the pulse height is proportional to the number n of charges collected by the detector which, in turn, is proportional to the radiation energy E,

$$\frac{\Delta E}{E} = 2.35(\bar{n})^{-1/2} ,$$

or, if the average energy required to produce an electron–hole pair is $\bar{E}_p$ and E_γ is the energy of the ionising radiation,

$$\Delta E_\gamma = 2.35(E_\gamma \bar{E}_p)^{1/2} .$$

As the variance is given by σ^2, the full width at half-maximum is given by 2.35 times the variance (Mann *et al.*, 1980).

In the whole detector, therefore, the overall variance (taken as the full width at half-maximum, ignoring the constant 2.35) is the sum of the variance for each component (n, charge collection and the electronics). Although the response of detector systems has improved greatly in recent years, the variance on charge collection caused by recombination and electron trapping is still significant. The variance in the electronic processing also cannot be ignored. The major factor in obtaining good resolution, however, is the average number of electrons collected for a given amount of radiation energy. Interestingly, however, experimental work has found that the resolution obtained with pulse ionisation chambers is two or three times better than the values based upon the variance of the distribution. This discrepancy can be explained by the fact that the energy of a charged particle is transferred to a gas molecule in a larger amount of interactions than those which actually produce ionisation. In fact, interactions between charged particles and gas molecules cause excitation of the gas without ionisation. The difference between total and partial ionisation can be expressed in terms of the Fano factor F:

$$F = \frac{\langle (n - \bar{n})^2 \rangle}{\bar{n}} ,$$

in which n is the number of electron–ion pairs produced (which must be a moderately large number, as indicated by the term within the angular brackets). The value of $\bar{n}$ is equal to the energy of the ionising particle divided by the experimentally measured average energy expended in producing an electron–ion pair. If the value of F is unity, the statistical fluctuations must follow a Poisson distribution, but pulse ionisation chambers give results which do not agree with this. This is not surprising, since the probability that a photelectron will interact with a gas moelecule to produce further electron–ion pairs cannot be ignored.

An analogous mechanism can be seen to apply when dealing with semiconductor detectors. If all the energy lost by ionising radiation in a semiconductor were spent breaking covalent bonds in the detector's sensitive volume, no fluctuations would occur in the number of electron–hole pairs produced by ionising radiation of any given energy. At the other extreme, if the energy entering the semiconductor detector that is partitioned between breaking covalent bonds and producing lattice vibrations or phonons were completely uncorrelated, a Poisson distribution would result. In fact, neither of these suppositions agrees with reality. As the incoming ionising radiation gives up energy, a large shower of electrons is created. After many generations, the energy of these electrons approaches the ionisation energy needed to create an electron–hole pair in the semiconductor detector, and so there are several possible competing mechanisms for energy loss. Again, when there are no fluctuations in the number of electron–hole pairs, F is zero while, in the case of a Poisson distribution, F is unity. Since the energy necessary to create electron–hole pairs in semiconductor detectors is much smaller than the energy of most ionising radiation, it can be calculated that F is closer to zero than to one. The Fano factor has been measured for Ge(Li) and Si(Li) detectors, and values of 0.06 and 0.07, respectively, have been obtained, but the true values of F for Ge and Si are still unknown (Zulliger and Aitken, 1970). For all practical purposes, an assumed value for F of 0.1 for both Ge and Si is found to give satisfactory agreement with measured results in most cases.

Thus, better resolution is obtained from semiconductor detectors than from pulsed ionisation chambers, simply because of the larger number of charge carriers produced in the former by the ionising radiation.

Because proportional counters and scintillation detector systems involve more than one statistical distribution in the production of the measured signal, the determination of resolution is more complex. In proportional counters, the production of electron–ion pairs is followed by gas multiplication; in scintillation counters, the light output from the scintillator is converted to electrons at the photocathode of the photomultiplier tube. For practical counters, the resolution is decreased by a factor of about 2 from that of a pulsed ionisation chamber, giving a full width at half-maximum of about 8 keV for 500 kV incident radiation. The resolution of scintillation assemblies is, however, theoretically very complex and depends critically upon the behaviour of the multiplier phototube (Adams and Dams, 1970).

8.2 COUNTING SYSTEMS

8.2.1 Pre-amplifiers

Most detectors can be considered to be capacitors into which a charge is deposited. The detector bias provides an electric field which collects this charge. During charge collection, a small current flows which produces a voltage drop (the 'pulse voltage') across the bias resistor. In the counting circuit, a preamplifier is employed which is separated from the high voltage by another capacitor. The rise time of the pulse is related to the collection time of the charge, while the decay time of the pulse is the resistance–capacitance time constant characteristic of the pre-amplifier itself. Rise times range between a few nanoseconds and a few microseconds, while decay times are usually set at about 50 μs.

Three types of pre-amplifiers are available: charge sensitive, voltage sensitive and current sensitive. The former type is commonly used with most solid-state detectors. In these pre-amplifiers, an ouput voltage pulse is produced which is proportional to the output charge and which is essentially independent of the capacitance of the detector. Si surface barrier detectors, in particular, need a high bias voltage and require pre-amplifiers which are not affected by the capacitance of the detector, although the latter will still affect the noise level.

The pre-amplifier performs three essential functions (Canberra, 1984):

(1) Conversion of charge to a voltage pulse.
(2) Signal amplification.
(3) Pulse shaping.

Pre-amplifiers are normally mounted on or very close to the detector assembly in order to reduce the capacitance of the leads. This minimises both reduction of the effective signal size and degradation of the rise time. Pre-amplifiers also serve for impedance matching between the high impedance of the detector and the low impedance of the connections (by coaxial cable) to the main amplifier which may be located some distance from the detector.

Because the decay time of a pre-amplifier pulse is long, the pulse may not have returned to zero voltage before another pulse occurs. To overcome this problem, the pulse is shortened and only the detector information contained in the pulse rise time is preserved. This is achieved by means of a resistance–capacitance clipping technique in which the pulse is differentiated to remove the slowly varying decay time and then integrated somewhat to reduce the noise (Fig. 8.6). A simple peak (unipolar pulse) of near-Gaussian shape is produced. A second differentiation may then be used to produce a bipolar

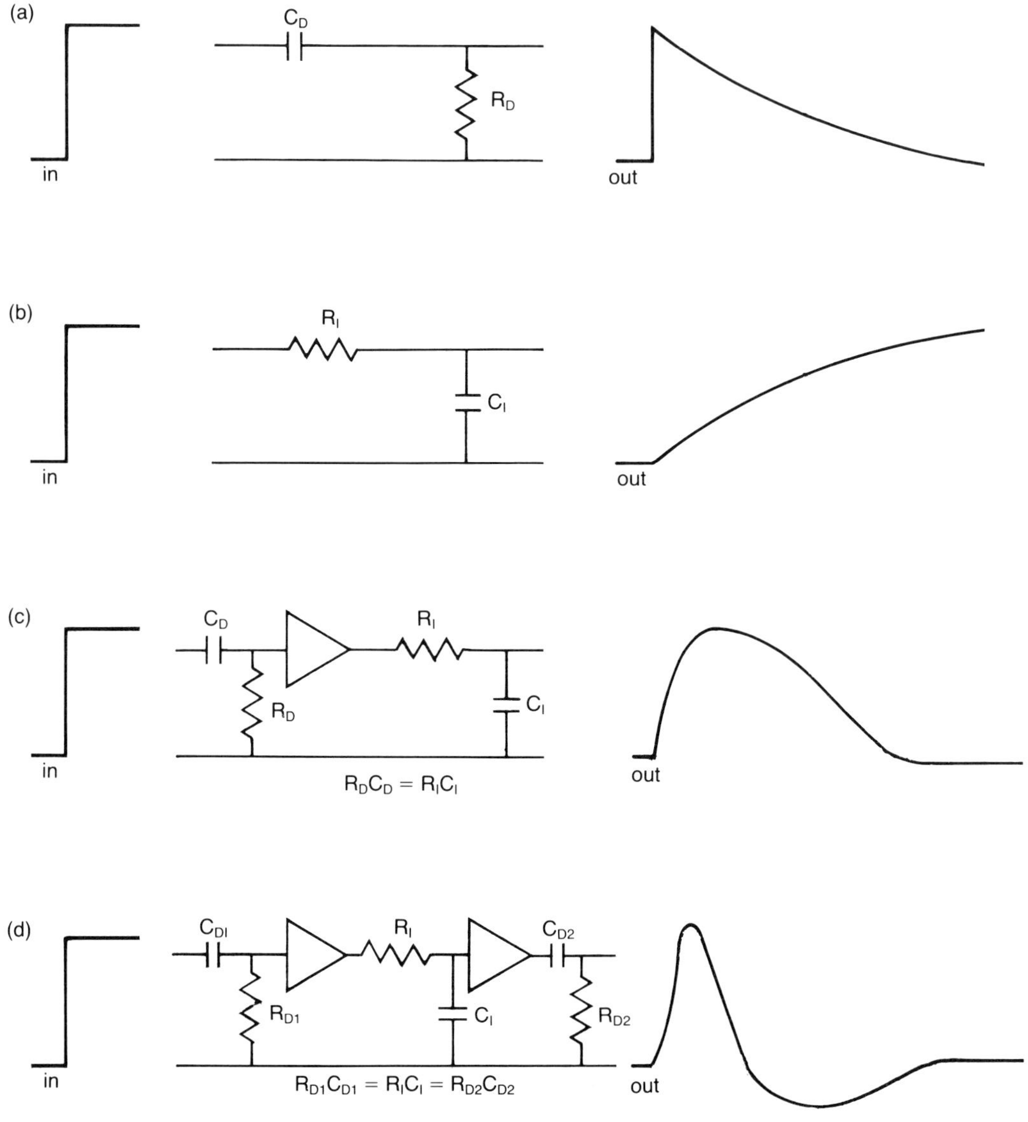

Fig. 8.6 — Resistor (R)–capacitor (C) pulse shaping: (a) CR differentiation; (b) RC integration; (c) CR–RC pulse shaping; (d) doubly differentiated CR–RC–CR shaping. (After EG&G Ortec, 1983.)

pulse. This has nearly equal positive and negative areas; so the net voltage is almost zero and has the advantage that, when the bipolar pulse passes from one stage of the circuit to another through a capacitor, no charge is left on the capacitor between pulses. Simple unipolar pulses, conversely, produce a charge which must be allowed to leak off, and the system must be reset to zero with a baseline restorer. Unipolar pulses are, however, usually subject to less noise, while bipolar pulses have zero crossing points which are more or less independent of time for a wide range of amplitudes. Most general-purpose pre-amplifiers have a resistance–capacitance feedback network which results in a quasi-step function output. For extremely low noise detectors such as Si(Li) and planar Ge, the feedback resistor is eliminated and the output signal becomes a true stay function which builds up in random staircase fashion and is reset by so-called pulsed optical feedback circuitry.

The design of good pre-amplifiers for use with solid-state radiation detectors is a demanding science. A Ge detector, for example, is an extremely high source impedance (of the order of $10^{10}\,\Omega$) with appreciable capacitance ($(1-5)\times10^{-11}$ F), and generates only a very small charge (5.4×10^{-4} C) with the absorption of a 1 MeV particle. The total event lasts for a period of between 1 ns and a few microseconds. In order to amplify signals from such a source without loading it and distorting the signal, the pre-amplifier must have an input stage impedance that is greater than that of the signal source. The use of a field-effect transistor in the input stage provides such a high impedance input while also providing a low noise contribution to the signal. The output signals from the pre-amplifier are usually fed to the main amplifier through low impedance shielded coaxial cables which minimise the chances of picking up unwanted signals which would increase the noise in the system. The pre-amplifier must, therefore, have a low impedance output stage that is capable of high current drive in order to prevent attenuation of the signal. To transform from high to low impedance and give high current drive, a bipolar transistor in an emitter–follower configuration is used as the output driver stage. Such a circuit can be designed so that the output impedance of this stage matches the impedance of the output signal cable as closely as possible.

Current-sensitive pre-amplifiers are not normally used in radiation detection systems because they have low input impedances and so are not compatible with high output impedance detectors. Voltage-sensitive pre-amplifiers are similar to charge-sensitive pre-amplifiers and may be used with proportional counters or cooled semiconductor detectors. However, if the output of the detector is large (as it is, for example, from scintillation detectors and Geiger–Müller detectors), the signal may be too large for charge-sensitive pre-amplifiers, and the voltage-sensitive type must be employed. In this case, however, a very low gain is often used, and the pre-amplifier is mainly for matching the output impedance of the detector to the impedance of the signal cable.

8.2.2 Amplifiers

General characteristics

The main amplifier usually consists of a voltage-sensitive circuit often called a linear amplifier. Although the overall gain that is normally available ranges between about 5 and 2000, amplification itself is initially a minor function. The major role of the amplifier is, in fact, to convert the pre-amplifier output signal into a form suitable for measurement. Many different techniques are available to treat the signal, in order to optimise measurement either of each characteristic of the signal (specialist amplifiers), or in a non-specialised way (general amplifiers).

Good amplifiers should have low noise characteristics. Each amplifier has a noise value, which is the amplitude of the noise pulse divided by the gain. Noise may be the result of random movements of charged particles (thermal noise), discrete charges in non-metallic media (shot noise), hum from the amplifier power supply (internal pick-up noise) and hum from nearby sources of electromagnetic radiation (external pick-up noise). Different amplifier designs are inherently more noisy than others, but all should keep pick-up noise to a minimum.

As each new pulse in the output from the pre-amplifier rides on the tail of the preceding pulse, the first job of the amplifier is to restore a baseline reference level. This is achieved by means of pulse differentiation. A high pass filter is used in which all information from the rapidly rising leading edge of the pulse is passed on, but which clips the slow trailing edge. After differentiation, the signal is integrated to recover its information content. In practice, both differentiation and integration are carried out by filtering selected frequency components from the signal. This means that the frequency characteristics of the filters employed must be selected so as to preseve the integrity of the information and to maintain the reference between events. However, these conditions cannot both be met at the same time. If, in order to provide a rapid return to baseline for high count rate detection, the time constant of the amplifier is short, the signal quality is degraded because too much noise is allowed through the amplifier. Conversely, if, in order to achieve high resolution, the integration time constant is long, the pulse is widened, and baseline shift will occur as events pile up on one another. With resistor–capacitor filters, the point of compromise is reached quite early, but active filtering and the production of a near-Gaussian pulse shape affords dramatic improvement over resistor–capacitor filtering. However, it is important to recognise the basic difference between high count rates and high resolution, and to design counting experiments accordingly.

Pulse shaping is of utmost importance in the design of amplifiers for two reasons: firstly, to prevent pulse overlap and, secondly, to enhance the signal-to-noise ratio. Aspects of pulse shaping have already been considered when dealing with pre-amplifiers. Amplifiers also usually employ near- or semi-Gaussian pulse shaping in which one pulse differentiation stage is followed by several stages of active integrator filtering, all set to the same cut-off frequency (the 'shaping time constant').

In segregating successive energy pulses in any pulse shaping method, baseline stabilisation must also be carefully considered if optimum amplifier performance is to be achieved. Clearly, the reference voltage must remain constant for each pulse if comparisons are to be permitted. Normally, the unipolar output pulse is related to a 0 V DC reference by a baseline restoration circuit. This is essentially a biased diode. When a unipolar pulse, for example, returns to the baseline and goes into undershoot, the diode becomes conducting and shorts the signal nearly to ground potential, effectively cancelling the undershoot. Unfortunately, this process distorts low amplitude pulses, and more complex diode restorers or active restorer circuits are needed. The effectiveness of an amplifier's baseline restorer determines its application to high count rates.

Some methods of pulse shaping have been considered under pre-amplifiers, three other main systems exist (EG&G Ortec, 1983).

(1) Semi-Gaussian pulse shaping.
(2) Gated integrator pulse shaping.
(2) Delay line pulse shaping.

Semi-Gaussian pulse shaping in amplifiers is similar to that in pre-amplifier circuits. Typically used in the bipolar mode with a second differentiation following the active filter, amplifiers produce a negative amplitude between 50 and 80% of the positive amplitude. This results in a reduction of baseline shift at high count rates, but the unipolar pulse remains better for high resolution work.

Gated integrator pulse shaping is usually employed where there is a requirement for high throughput signal processing, as there is in high performance Ge γ-ray detectors. In this process, ballistic deficit effects caused by long charge collection times are eliminated. Moreover, an improvement in throughput by a factor of 4 can be achieved by gated integrator pulse shaping compared with semi-Gaussian shaping. Gated integrator output is obtained by integrating the entire unipolar signal so that the problems associated with charge collection effects are removed, even when short shaping time constants are used. Otherwise, detector current pulses with equivalent charge but different rise times produce different pulse heights when processed by a charge-sensitive pre-amplifier and a semi-Gaussian filter amplifier. This would result in distortion of the spectrum in direct proportion to the pulse amplitude;

distortion being most pronounced with short shaping time constants.

Delay line pulse shaping (Fig. 8.7) can be used to produce an essentially rectangular output pulse from each step function input pulse. Because the pulse immediately returns to the baseline, this method is ideal for preventing overlap, but the method suffers from a poorer signal-to-noise ratio than simple capacitor–resistor, resistor–capacitor or semi-Gaussian shaping. The main characteristic of delay line shaping is that it preserves the fast rise time of an input pulse while clipping the length of the signal. This makes the delay line amplifier very suitable for timing purposes and for the measurement of high count rates. It is possible for a second delay line shaper to follow the first, so that a doubly differentiated delay line shape is obtained. This has almost 100% negative amplitude and so is free of baseline shift problems. However, the circuit is even more susceptible to noise than the single delay line system.

Fast timing amplifiers

When a detector signal passed on from a preamplifier or photomultiplier tube is of sufficient amplitude for recording, no amplification should be employed. The best rise times and least noise will be obtained on time resolution.

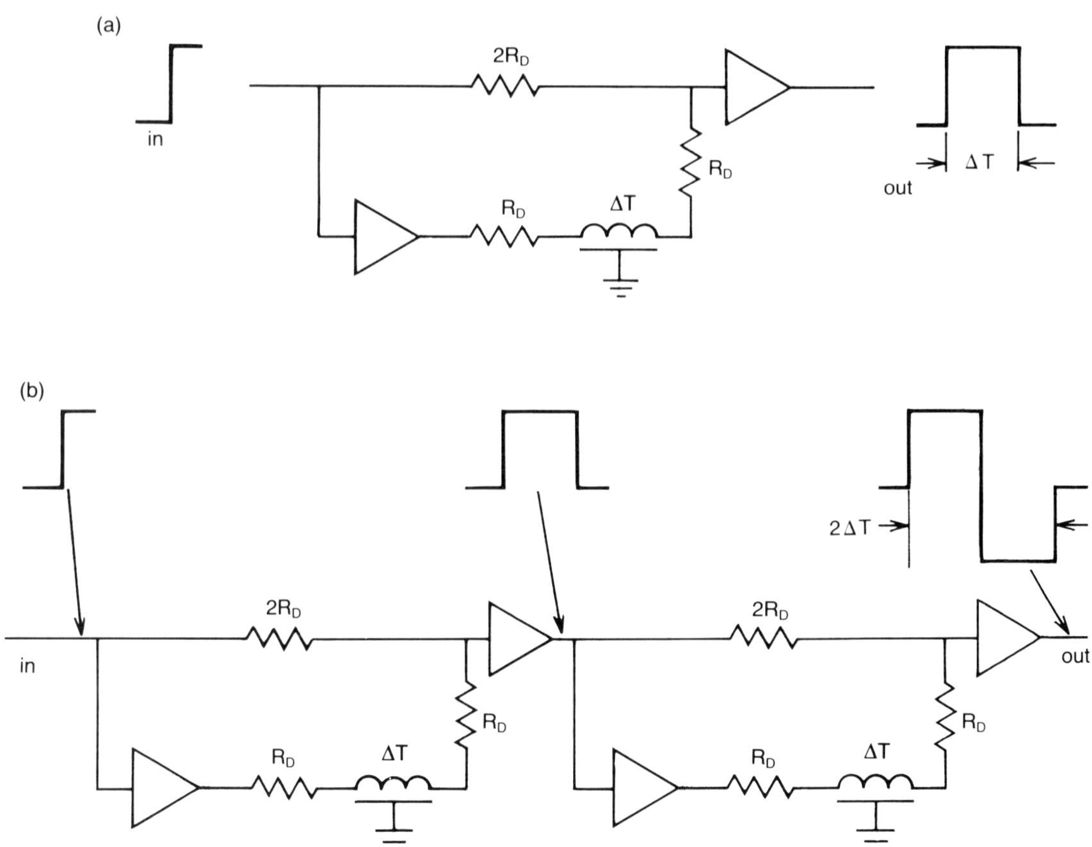

Fig. 8.7 — Delay line pulse shaping: (a) single delay line shaping; (b) double delay line shaping. (After EG&G Ortec, 1983.)

However, when the signal must be amplified or shaped, an amplifier designed for timing should be used. Timing amplifiers are recognised by their wide bandwidth, and the possession of integrating and differentiating time constants in the nanosecond range (spectroscopy amplifiers have these time constants in the microsecond range). The output from timing amplifiers is usually a negative pulse, since most timing circuits are designed to accept negative input signals. Integration times are set as long as possible without disturbing the rise time of the input signal, and differentiation times are set to give rapid baseline return without affecting the pulse amplitude.

Biased amplifiers

When detailed information about a small portion of the total spectrum is required, a biased amplifier can be used to expand the region of interest to cover the full memory of the analyser system. The expansion is carried out by setting the threshold of the biased amplifier to the voltage at the low energy end of the section of interest so that only pulses which reach the bias level are passed. The post-gain of the amplifier is then set so the section of interest is expanded to cover the full input range.

Well-designed amplifiers should have good overload characteristics so that, if a signal greater than the input range of the amplifier is received, it is not unduly paralysed. During the dead time that arises from overload, other signals will be lost, and so accurate measurement becomes impossible. Unfortunately, the dead time produced in this way is variable, and so correction cannot be made. The problem can be overcome to some extent, however, by the use of dead-time circuits which deliberately introduce a dead time that is larger than the largest paralysis time of the amplifier.

Pile-up rejectors

Pile-up rejection circuits are very useful features of amplifiers. Immediately after each signal, they generate an inspection period of equal length to that of the busy period during which the signal is being processed. If a second event occurs within this period, an inhibiting signal is produced which takes the amplifier out of circuit, thus discarding the distorted amplifier output.

8.2.3 Analysers

Pulse height analysers may consist of a simple discriminator, that is, a circuit which sorts pulses according to their amplitudes. In its simplest form (integral operation), a discriminator allows only pulses greater than a specified amplitude to proceed for further analysis. A slight refinement on this system is the differential comparator, which passes a steady signal only when the input amplitude exceeds a set value. This allows a background effect to be taken into account and produces a rectangular output pulse. This pulse can then be passed to a scalar or pulse counter. However, as most data consist of a range of pulse heights, it is normal to use either a single-channel analyser and scaler or a multichannel analyser.

Single-channel analysers

A single-channel analyser uses two discriminator circuits with separately adjustable thresholds, so that it produces an output logic pulse when the input lies between the discriminator levels. With this system, all voltage pulses within a specified range can be selected and counted. Additional single-channel analysers and scalars can be added for additional voltage ranges, but it is rare for more than three ranges to be measured in this way because of the complexity of the required circuits. More commonly, if a more complete voltage spectrum is required, the single-channel analyser discrimininators are set to a narrow range and then stepped through increasingly higher voltages. This can be carried out automatically by means of a single-channel analyser sweep. The voltage difference between the lower and upper level discriminators is often referred to as the 'window'. The scalar (perhaps more commonly known as a decade counter) records the number of pulses received within a specific interval of time.

The technique employed to produce an output logic signal from a single-channel analyser is used as the basis of classification of the different types of analyser (Fig. 8.8). Non-timing analysers produce an output pulse if the input signal lies within the window voltages, the output occurring when the trailing edge of the input signal recrosses the lower level discriminator threshold. In this way, two peaks which are simultaneous on input but have different widths will become separated in time on output, the broad peak being delayed with respect to the narrow peak. In contrast, timing analysers produce an output logic signal that is precisely in time with the input, irrespective of peak width or height. This type of analyser can thus be used where the precise timing of an event is required, such as for coincidence measurement and pulse-shape discrimination. In this system, the unipolar pulse is differentiated to a bipolar signal, and the zero cross-over time (the moment of zero charge as the signal changes from positive to negative or vice versa) marks the start of the output pulse. Either double-delay line-shaped pulses or resistor–capacitor shaped pulses may be used, but the former provide better timing resolution. Because the output signal from a double-delay line-shaping amplifier crosses the baseline at a fixed fraction (effectively 50%) of the charge collected from the detector, zero crossing timing can be considered as timing at a constant fraction of the input signal amplitude. A trailing edge constant-fraction technique can be used with either unipolar or bipolar signals to obtain a time pick-off pulse after the peak time of the signal. For two signals of identical rise times but different amplitudes, which lie within the window voltages, the output signal will occur at a time which is independent of amplitude.

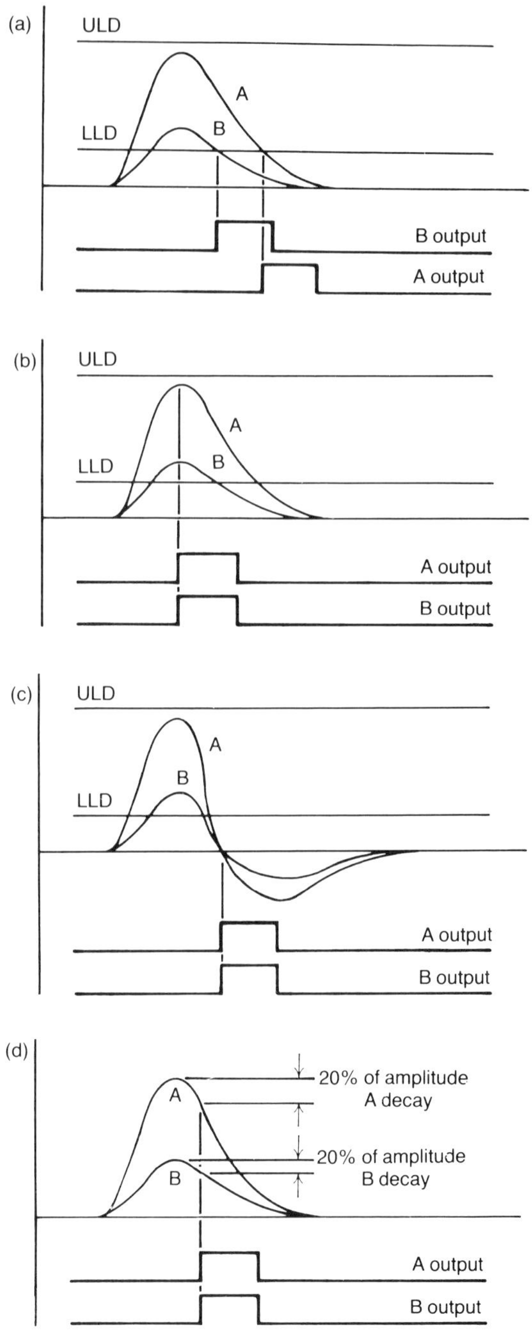

Fig. 8.8 — Single-channel analyser output triggering: (a) non-timing output; (b) peak-sensing output; (c) zero cross-over output; (d) constant-fraction. ULD, upper level discriminator; LLD, lower level discriminator. (After EG&G Ortec, 1983.)

Multichannel analysers

The multichannel analyser is the most convenient way of collecting the entire spectrum at once. It consists essentially of an analogue-to-digital converter, a memory and a display oscillosope. However, in practice, multichannel analyser systems can be more complex, with various built-in output devices, amplifiers, preamplifiers and data analysis functions. Because the multichannel analyser collects pulses in all voltage ranges at once, it provides a major improvement over single-channel analyser systems. In the multichannel analyser, the input pulse is first checked to see whether it is within the selected single-channel analyser range and then it is passed to the analogue-to-digital converter. The pulse is then converted to a number proportional to the pulse voltage. That number is taken to be the address of a memory location and one count is added to the contents of that memory location. After collecting data for some time, the memory contains a list of numbers corresponding to the number of pulses of each discrete voltage. The display reads the memory many times per second and produces a point plot of memory contents against memory locations, which is equivalent to the number of pulses against voltage or energy. Each memory location is a channel, and the number of channels ranges between a minimum of 256 and a maximum of 65 536 (Fig. 8.9). However, analogue-to-digital converters are only available from 256 to 16 384 channels (Table 8.1).

The number of channels available and the resolution of the analyser system varies from one type of detector to another. In general, the higher resolution detectors require more channels and better analogue-to-digital converter resolution to provide a sufficient number of channels for each peak (so that peak areas can be determined). It is common practice, however, always to use a multichannel analyser with a larger memory than that required, so that an unknown spectrum can be obtained in one half of the memory while a reference spectrum is stored in the other half for comparison. NaI(Tl) detectors usually work with a 512 or 1024-channel analyser, Si surface barrier detectors with 256 to 1024 channels, Si(Li) detectors with 512 or 1024 channels, and Ge detectors with between 2048 and 16 384 channels.

The digitising of an analogue input signal uses fast Wilkinson and successive approximation processes in which a capacitor is charged by the input pulse up to the amplitude of the input and then discharged at a constant rate (the ramp). During the discharge time, a crystal-controlled pulse train is counted in a register. When the capacitor charge has returned to the baseline, the number in the register represents the signal pulse height.

The resolution of the analogue-to-digital converter is related to the slope of the ramp and

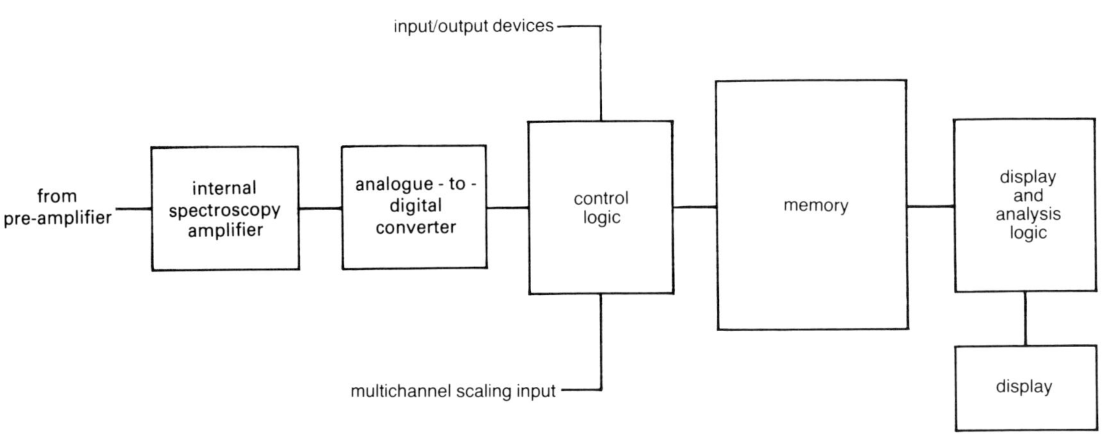

Fig. 8.9 — Diagrammatic representation of a multichannel analyser system. (After Canberra, 1984.)

Table 8.1 — Typical detector and multichannel analyser functions.

Detector type	Function	Resolution	Energy range	Nominal analogue-to-digital conversion gain (channels full scale)	Multichannel memory size (channels)
Si surface barrier	α-particle spectroscopy	15 keV	4 MeV–8 MeV	256–2048	256–4096
Si surface barrier	β-particle spectroscopy	15 keV	20 keV–2 MeV	1024	1024–4096
Si(Li)	X-ray spectroscopy	180 eV	1 keV–50 keV	1024	1024–4096
NaI(Tl)	γ-ray spectroscopy	8%	0.1 MeV–2 MeV	256–1024	256–4096
High purity Ge	γ-ray spectroscopy	1.8 keV	3 keV–10 meV	4096–8192	4096–16384

(After EG&G Ortec, 1983).

the crystal-controlled clock frequency. It is usually referred to as the number of channels but is, in fact, the reciprocal of the number of channels occupied. The slope of the ramp is also considered in the conversion gain of the converter, which is the number of channels per input voltage range (again often referred to simply as 'channels'). Spectroscopy systems typically have resolutions of 2048, 4096, 8192 or 16 384 channels (1, 12, 13 or 14 bits) and clock frequencies of 50–450 MHz. The dead time of the converter is again the time during which it is busy processing one pulse and, therefore, unable to accept another. This can be minimised by using a higher clock frequency which allows a shorter ramp time for the same number of channels. The dead time τ (also known as the conversion time) is typically given by an equation such as

$$\tau = 0.7 + 0.0022(N + X)\,\mu s$$

where N is the converted channel number and X the amount of digital offset. The 0.7 μs fixed time follows the ramp generation, but the total dead time includes the input pulse width and the time taken to get the output to the multichannel analyser memory. Because values of N may be of the order of 1000, even with a zero value of X the dead time is several microseconds. The percentage dead time for a particular system should be indicated on the analyser. Compensation for dead time is obtained by acquiring data for live-time periods only, that is, by gating off the dead time with a clock. True time (or clock time) recording is usually possible with live time stored in channel 0 and true time in channel 1.

Digital offset is the capability of the analyser to subtract a specified channel from the converted channel number before the memory is accessed. This is used frequently in α-spectrometry to remove lower energy ranges and to expand the higher range across the whole memory. It can also be used to collect data at twice the range of a multichannel analyser memory. If, for example, a 2048-channel converter gain is used with a 1024-channel analyser, a digital offset of 1024 will allow converter channels from 1024 to 2047 to be stored in the 1024 channels of data memory.

Another factor which is of importance in converters is linearity, expressed in terms of integral and differential non-linearity. Integral non-linearity is the deviation from a linear relationship of pulse height against channel number. Differential linearity is the variation in channel width from one channel to another. Typical values are about 1 channel in 4000 (0.025%) for integral non-linearity and about 1% for differential non-linearity.

Modern multichannel analyses usually have an energy calibration function incorporated into their processing system which provides a linear calibration scale for the energy axis (energy=$A+Bx$, where x is the channel number A or B are constants). The accuracy of this, however, depends upon the linearity of the converter in the analyser. With high resolution detectors, the errors that arise from non-linearity tend to be unacceptable when dealt with in this way, and a more complex algorithm is applicable (energy = $Ax + Bx + C$, where C is yet another constant). This formula is also better for describing the type of non-linearity found with NaI(Tl) detectors. As lower and upper level discriminators allow rejection of pulses from the converter if they lie outside the window, it is clear that the independent single-channel analyser output based upon these discriminators can be used for multichannel scaling. All that is needed after a pulse height spectrum has been collected is to set the discriminators to bracket any peak of interest.

The zero level of the analogue-to-digital converter (the 'analogue baseline'), which is slightly dependent upon input pulse shape and conversion gain, can usually be adjusted over a small range to allow precise setting of channel 0 to 0 V. Long duration counting should normally employ some type of external stabilisation to keep the zero and gain steady. In a typical stabilising system, a peak in the spectrum is maintained at the same position by changing the converter gain and/or the zero setting. This corrects drift anywhere in the detector, preamplifier, amplifier and analogue-to-digital converter system.

Visual display to provide a graphic representation of the accumulated data is an integral part of all multichannel analysers. This is usually presented in the form of a linear horizontal energy scale and a linear vertical intensity scale. The horizontal scale can be adjusted over the whole range of the memory channels. Logarithmic scales are useful for comparison of data which would otherwise have different linear scales. Horizontal scale expansion is usually provided so that the spectrum can be examined in detail, and different spectra may be overlapped for comparison. Groups of adjacent channels may be defined as a region of interest and intensified on the display for ease of identification. The upper and lower limits of these defined regions may also be used to specify sections for data processing. For example, an intensified region can be the zone between the boundaries of a peak and the average of its end points, which is useful for calculation of the peak area. Moveable cursors are also useful for setting regions of interest, specifying points of scale expansion and indicating special features in the data. Alphanumeric characters can be displayed in combination with the graphics in an interactive way so that qualitative and quantitative results can be output to a graphics plotter.

Data acquisition is usually available in three different data analysis modes (EG&G Ortec, 1983): pulse height analysis, multichannel scaling and list (which can be either a single or multiple input). Data processing can also be carried out.

(1) *Pulse height analysis.* In this mode, the pulses from the detector have amplitudes which are proportional to the energies of the detected radiation. By counting the number of pulses of each height and constructing a histogram, the multichannel analyser records the energy spectrum of the radiation. Multi-input pulse height analysis allows simultaneous collection and storage of the multiple spectra seen by the detector. Pulse height analysis is the traditional operating mode of multichannel analysers and is used for nuclear and X-ray spectroscopy and time spectroscopy. With multiple input from several detectors, pulse height analysis can be handled with a multiplexer–router in which each detector input goes to the same analogue-to digital converter, but for which a different section of the memory is selected. For low to moderate count rates, this provides a simple way to collect data from several samples but, for high count rates, or when each detector must have a different input voltage range, a digital multiplexer must be used or multiple analogue-to-digital converters placed within one multichannel analyser. A slowly varying DC signal may also be sampled and digitised to obtain an amplitude distribution curve or a histogram of the distribution of the input voltage waveform. This is useful in noise source analysis.

(2) *Multichannel scaling.* In the multichannel scaling mode, the individual channels of the memory are analogous to a sequence of counters, each channel counting the data for a pre-determined interval of time (the 'dwell time'). At the end of each dwell time, the counting process is automatically passed to the next channel of the memory. This results in the production of a time histogram of the count rate data, each channel representing a sequential time interval. The dwell time for each channel can be set by an internal clock or by an external channel-advance signal. This mode is useful for applications where analysis of count rate relative to elapsed time is of interest. This mode is, therefore, used for measurement of the decay rate of short-lived radionuclides, Mössbauer analysis and line-scan analysis in electron microprobes.

(3) *List mode.* In the list mode, individual memory channels store sequential energy values, that is, the digitised version of the input analogue voltages. Input events can be stored until the memory is full, at which point the data are dumped and the storage sequence restarted at the beginning of memory. This mode is useful for multiparameter pulse height analysis for time-related spectroscopy and for digitising slowly changing DC signals.

(4) *Data processing.* The incorporation of microprocessors within multichannel analysers allows extensive data analysis to be performed. These are the so-called 'smart' analysers. One aspect of this capability, energy calibration, has already been considered, and the ability of multichannel analyses to determine peak areas has also been mentioned. In fact, smart multichannel analysers can find peaks, compute their centroids and net areas and then output the centroid as energy and the area as intensity. In calculating net areas, the background continuum data are stripped from the calculations. Ratio calculation is possible, allowing the user to determine the unknown amount of one source by calculating the ratio of intensity of a spectral peak of the unknown source to a peak of a known source. Spectrum stripping techniques permit the subtraction of known spectra (such as background or calibration sources) from any measured spectrum. This can be accomplished by simple subtraction or by an iterative process in which mutiples or fractions of the known spectrum are subtracted (fractional stripping). This enables the user to determine the amount of an unknown source and the existence of peaks in the unknown spectrum which do not exist in the known spectrum. Spectrum smoothing allows easier visual judgement and can give improved determination of the end points of peaks. Most modern multichannel analysers allow the user to enter simple algebraic expressions in order to implement corrections to the compound data and to turn the output into directly meaningful values. Such procedures can be incorporated into programming sequences which relate to input, storage, composition and output. For instance, they could allow the operation of automatic sample changers.

In conclusion, it is important to note that equipment manufacturers' specifications are continuously being up-dated and improved. Details of the current position should therefore be obtained by direct enquiry to the respective Technical Information departments.

BIBLIOGRAPHY

Adams, F., and Dams, R. (1970) *Applied gamma-ray spectrometry*. Oxford, Pergamon.

Adams, J. A. S., and Gasparini, P. (1970) *Gamma-ray spectrometry of rocks*. Amsterdam, Elsevier.

Andresen, E. G. (1965) Die totzeitkorrektur. *Kerntechnik* **7**, 79.

Ayres, J. B. (1973) Preamplifiers. In Krugers, J. (ed.), *Instrumentation in applied nuclear chemistry*. New York, Plenum.

Bevington, P. R. (1969) *Data reduction and error analysis for the physical sciences*. New York, McGraw-Hill.

Birks, J. B. (1964) *The theory and practice of scintillation counting*. Oxford, Pergamon.

Bowie, S. H. U., Davis, M., and Ostle, D. (1972) *Uranium prospecting handbook*. London, Institution of Mining and Metallaurgy.

Bransome, E. D. (1970) *The current status of liquid scintillation counting*. New York, Grune and Stratton.

Brown, R. G (1983) *Introductions to random signal and Kalam filtering*. New York, Wiley.

Canberra (1984) *Edition Six catalogue*. Meriden, Connecticut, Canberra Industries Inc.

Chiang, H. H. (1969) *Basic nuclear electronics*. New York, Wiley.

Conner, F. R. (1972) *Signals*. London, Arnold.

Conner, F. R. (1973) *Noise*. London, Arnold.

Cunninghame, J. G. (1975) Methods of detection and measurement of radioactive nuclides. In Coomber, D. I. (ed.), *Radiochemical methods in analysis*. New York, Plenum.

Delaney, C. F. G. (1980) *Electronics for the physicist with applications*. Chichester, West Sussex, Ellis Horwood.

Dixon, W. J., and Massey, F. (1969) *Introduction to statistical analysis*. New York, McGraw-Hill.

EG&G Ortec (1983) Radiation detection: Measurement: Analysis. *Catalogue 1983/84*. Oak Ridge, Tennessee, EG&G Ortec.

England, J. B. A. (1974) *Techniques in nuclear structure physics*. London, Macmillan.

Fano, U. (1947) Ionization yield of radiations, II. Fluctuations of the number of ions. *Phys. Rev.* **72**, 26.

Goulding, F. S., and Pehl, R. H. (1974) Semiconductor radiation detectors. In Cerney, J. (ed.), *Nuclear spectroscopy and reactions*. New York, Academic Press.

Halmstadt, H. V., Enke, C. G., and Crouch, S. R. (1981) *Electronics and instrumentation for scientists*. Reading, Massachusetts, Benjamin.

Hatch, K. F. (1973) Amplifiers. In Krugers, J. (ed.), *Instrumentation in applied nuclear chemistry*. New York, Plenum.

Hurley, P. M. (1956) Direct radiometric measurement by gamma-ray scintillation spectrometer. *Bull. Geol. Soc. Am.* **67**, 395.

Ivanovich, M. (1982) Spectroscopic methods. In Ivanovich, M., and Harmon, R. S. (eds), *Uranium series disequilibrium: applications to environmental problems*. Oxford, Oxford University Press.

Jones, M. H. (1977) A practical introduction to electronic circuits. Cambridge, Cambridgeshire, Cambridge University Press.

Knoll, G. F. (1979) *Radiation detection and measurement*. New York, Wiley.

Lanoo, M., and Bourgoin, J. (1981) *Point defects in semiconductors*. Berlin, Springer.

Laurie, P. (1980) *The micro revolution*. London, Futura.

Leck, J. H. (1967) *Theory of semiconductor junction devices*. Oxford, Pergamon.

Littauer, R. (1965). Pulse electronics. New York, McGraw-Hill.

Loevinger, R., and Berman, M. (1951) Efficiency criteria in radioactivity counting. *Nucleonics* **9**, 26.

Mann, H. M. (1967) Germanium gamma ray spectrometers. *Trans. Nucl. Sci.* **14**, 10.

Mann, W. B., Ayres, R. L., and Garfinkel, S. B. (1980) *Radioactivity and its measurement*. Oxford, Pergamon.

Milam, J. K. (1973) Single-channel analysers. In Krugers, J. (ed.), *Instrumentation in applied nuclear chemistry*. New York, Plenum.

Millman, J. (1979) *Microelectronics: digital and analog circuits and systems*. New York, McGraw-Hill.

Morris, N. M. (1976) *Logic Circuits*. London, McGraw-Hill.

Morrison, R. (1977) *Grounding and shielding techniques in instrumentation*. New York, Wiley.

Mort, J., and Pai, D. M. (1976) *Photoconductivity and related phenomena*. Amsterdam, Elsevier.

Mowle, F. J. (1976) *A systematic approach to digital logic design*. Reading, Massachusets, Addison-Wesley.

National Council on Radiation Protection and Measurement (1978) *A handbook of radioactivity measurements procedures*. NCRP Rep. 58. Washington, DC, National Council on Radiation Protection and Measurement.

Papoulis, A. (1977) *Signal analysis*. New York, McGraw-Hill.

Parratt, L. G. (1961) *Probability and experimental errors in science: an elementary survey*. New York, Wiley.

Pearman, R. A. (1984) *Solid state industrial electronics*. Virginia, Reston.

Price, W. J. (1964) Nuclear radiation detection. New York, McGraw-Hill.

Ross, W. A. (1973) Multichannel analysers. In Krugers, J. (ed.), *Instrumentation in applied nuclear chemistry*. New York, Plenum.

Shklovskii, B. I. (1984) *Electronic properties of doped semiconductors*. Berlin, Springer.

Siegbahn, K. (1965) *Apha, beta and gamma-ray spectrometry*. Amsterdam, North-Holland.

Staub, H. H. (1953) Detection methods. In Segré, E. (ed.), *Experimental nuclear physics*. New York, Wiley.

Stever, H. G. (1942) The discharge mechanisms of fast Geiger-Müller counters from dead time experiments. *Phys. Rev.* **61**, 38.

Steyn, J. J., and Nargolwalla, S. S. (1973) Detectors. In Krugers, J. (ed.), *Instrumentation in applied nuclear chemistry*. New York, Plenum.

Stoneham, T. J. (1984) *Digital logic techniques, principles practice*. Washington, Van Nostrand.

Taylor, J. M. (1963) *Semiconductor particle detectors*. Washington DC, Butterworths.

Zulliger, H. R., and Aitken, D. W. (1970) Fano factor, fact and fallacy. *Trans. Nucl. Sci.* **17**, 187.

9

Exploration methods

9.1 INTRODUCTION TO EXPLORATION TECHNIQUES

The qualitative detection of radionuclides is quite straightforward. The Geiger–Müller counter, which indicates the discharge induced when transformation emissions (principally β-particles and γ-rays) cause ionisation in a discharge tube, was first used for radiological exploration in the early 1900s. Geiger–Müller counters, however, have severe limitations in terms of the quality of the data produced. At low count rates, the background effect may be significant while, at high count rates, under-recording of the true activity is caused by the dead time of the discharge tube. Moreover, no discrimination between different sources of radiation is possible unless variable external shielding is used.

Many modern systems involve the detection specifically of γ-rays by scintillation detectors. The main scintillator used is NaI(Tl) in combination with a photomultiplier tube. In the simplest systems, the total γ-radiation intensity is detected and output directly as counts or as a count rate. More useful, however, are the NaI(Tl) systems which include a single-channel energy analyser with several discriminator settings, allowing a number of energy windows to be pre-selected. These systems permit not only the recording of the total γ-radiation but also the resolution of activity due to specific radionuclides. Even greater information about γ-ray source radionuclides can be obtained by the use of multichannel analysers. These very capable systems are not normally used with NaI(Tl) detectors but with Ge detectors. The drawback with the latter, however, is the need to operate the detector at liquid N temperature, and so for field use a nearby souce of liquid N is needed. Unfortunately, this often makes their use impractical.

One difficulty that has to be faced with the use of any exploration system based upon the detection of radioactive emissions is the statistical time-dependent nature of the radioactive decay process itself. This makes short count periods, which yield only low numbers of counts, very prone to error, with poor repeatability. As the percentage statistical error Y varies inversely with the square root of the number N of counts as $Y=100/N^{1/2}$, there is always a need for compromise between count time and statistical error for any detector system. This compromise can be partly achieved by employing large detectors. As a general guide, it should be noted that, in order to produce results within the normally accepted 5% error, measurement must be based on count numbers greater than about 500.

Single-channel analyser γ-ray spectrometers allow the main naturally occurring radionuclides to be recognised and quantified. ^{238}U and ^{232}Th, for example, are measured by recording the characteristics of daughters ^{214}Bi and ^{208}Tl, respectively. Other members of the decay series can also be used. Rn detection, in particular, is often used as a guide to the presence of U and Th; ^{222}Rn is part of the decay scheme of ^{238}U, while ^{220}Rn is a daughter of ^{232}Th. These isotopes of Rn may be distinguished from one another by their very different half-lives (3.825 days for ^{222}Rn and 55.6 s for ^{220}Rn). Interestingly, while γ-ray spectrometry systems using single-channel analysers quantify U and Th by assuming that the respective decay series are in equilibrium, the theoretical basis of Rn detection is that the decay series are in disequilibrium! For Rn to be detected, it must diffuse out of its parent mineral and be transported to the point of sample collection. Because Rn is gaseous at normal temperatures and pressures, this loss from the parent appears to take place quite commonly. As Rn is the only naturally occurring radioactive gas, its presence and abundance are easily detected by measuring the activity of gas (air) samples taken from the atmosphere, soil, groundwater or surface water. Interpretation of the results of such surveys in terms of the abundance of the parent U and Th is, nevertheless, fraught with difficulty, as discussed later.

Practical γ-ray spectrometry and Rn detection systems for U exploration both suffer from problems related to the fact that they do not measure U directly but take an indirect approach. Geochemical prospecting for U, however, based upon rock, soil or water samples, is a long-established routine method. In terms of the number of samples analysed, the most commonly employed analytical techniques are optical fluorimetry, X-ray fluorescence and neutron activation, which yield good results where the U concentrations are greater than about 1 ppm in rock and soil and greater than about 5 ppb (μg l^{-1}) in water. However, for low concentrations (particularly important when dealing with water analysis), the results obtained may be unreliable. An alternative method, not yet widely available, is laser-induced fluorimetry. This technique has been specifically developed for analysis of U in aqueous solutions and has a sensitivity of 0.005 ppb.

α-particle spectrometry using solid-state detectors is essentially a laboratory-based method. However, α-particle detection in the field is the basis for Rn exploration sytems, as Rn is an α-particle emitter. The Rn flux in the ground, for example, can be measured by recording the number of α-particle tracks formed in nuclear track plastic in a given period of time. As the track lengths produced by α-particles in such plastic are proportional to the energies of the particles, the distribution of these data yield information on the α-particle energy spectrum. However, for recognition of α-particle-emitting radionuclides, the nuclear track plastic must be placed against a flat surface of the material being investigated, and even then the determination of track lengths is almost impossible if many closely spaced α-particle energies are present. The technique is, therefore, essentially confined to laboratory use.

Field detection of Rn by α-track recording is most commonly carried out using the α-cup technique. Although some solid-state α-particle detector–counting systems have been developed, they are very expensive. In the α-cup method, the activity of Rn in the ground is measured over a long period of time by placing a section of α-sensitive plastic within a buried inverted cup. Integration times of hours, days, weeks or even a year are employed in this approach, but it still generally suffers from the same disadvantages as the other Rn detection methods when interpretation of the data is attempted in terms of the distribution of the parent U and Th.

9.2 PORTABLE γ-RAY SPECTROMETRY

9.2.1 The NaI(Tl) single-channel analyser system

Equipment and operation

In practice, this is a system in which four windows selectively record (1) the total count above a certain threshold, (2) the count for the 'Th window', (3) the count for the 'U window' and (4) the count for the 'K window' (Fig. 9.1).

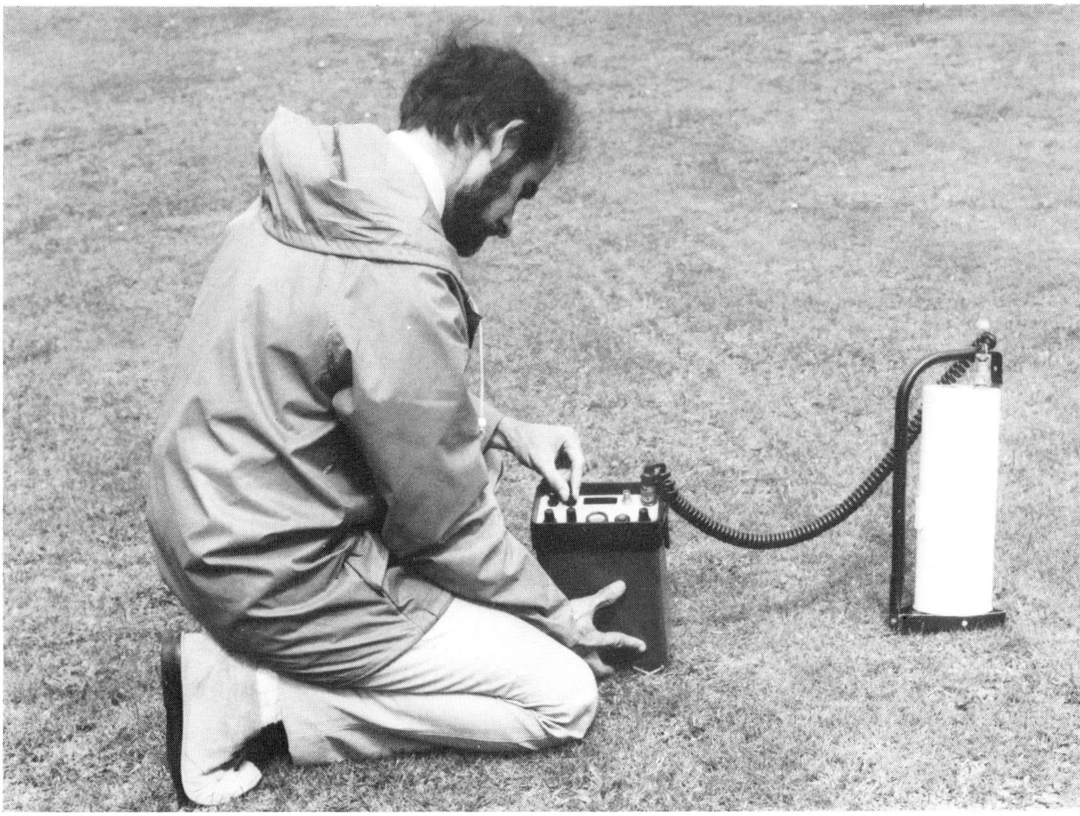

Fig. 9.1 — The Geometrics DISA 400A γ-ray spectrometer. The detector element contains a NaI(Tl) scintillation crystal beneath a photomultiplier assemblage.

This method is principally used as a means of obtaining the *in situ* concentrations of Th, U and K in rocks and soils, together with the total γ-ray activity (Ward, 1981). Three basic requirements must be met, however, if the method is to provide acceptable data: firstly, the sample must be reasonably homogeneous; secondly, the decay series must be in secular equilibrium (if disequilibrium conditions exist, errors will result); thirdly, a consistent operational sampling geometry must be used.

Natural γ-ray activity is overwhelmingly due to the presence of the ^{238}U and ^{232}Th decay series and ^{40}K. Each of these produces a characteristic spectrum of γ-ray energies (Adams and Gasparini, 1970). The presence of several well-defined peaks at various points in the spectra of the ^{238}U and ^{232}Th decay series and the single 1.46 MeV peak for ^{40}K provide the theoretical basis for conventional four-window spectrometry. Above 2.5 MeV, the dominant contribution to the total γ-ray intensity spectrum is the 2.62 MeV peak of the ^{232}Th decay series due to decay of ^{208}Tl (Fig. 9.2). Therefore, if only energy at 2.62 MeV is recorded, and on the assumption that the ^{232}Th decay series is in equilibrium, the count rate obtained may be converted to a Th content by a simple conversion factor k_{Th}. In practice, the recording of a single energy level is not appropriate because spreading of energies occurs. It is, therefore, peak area rather than peak height which is the more reliable guide to the abundance of a particular radionuclide in

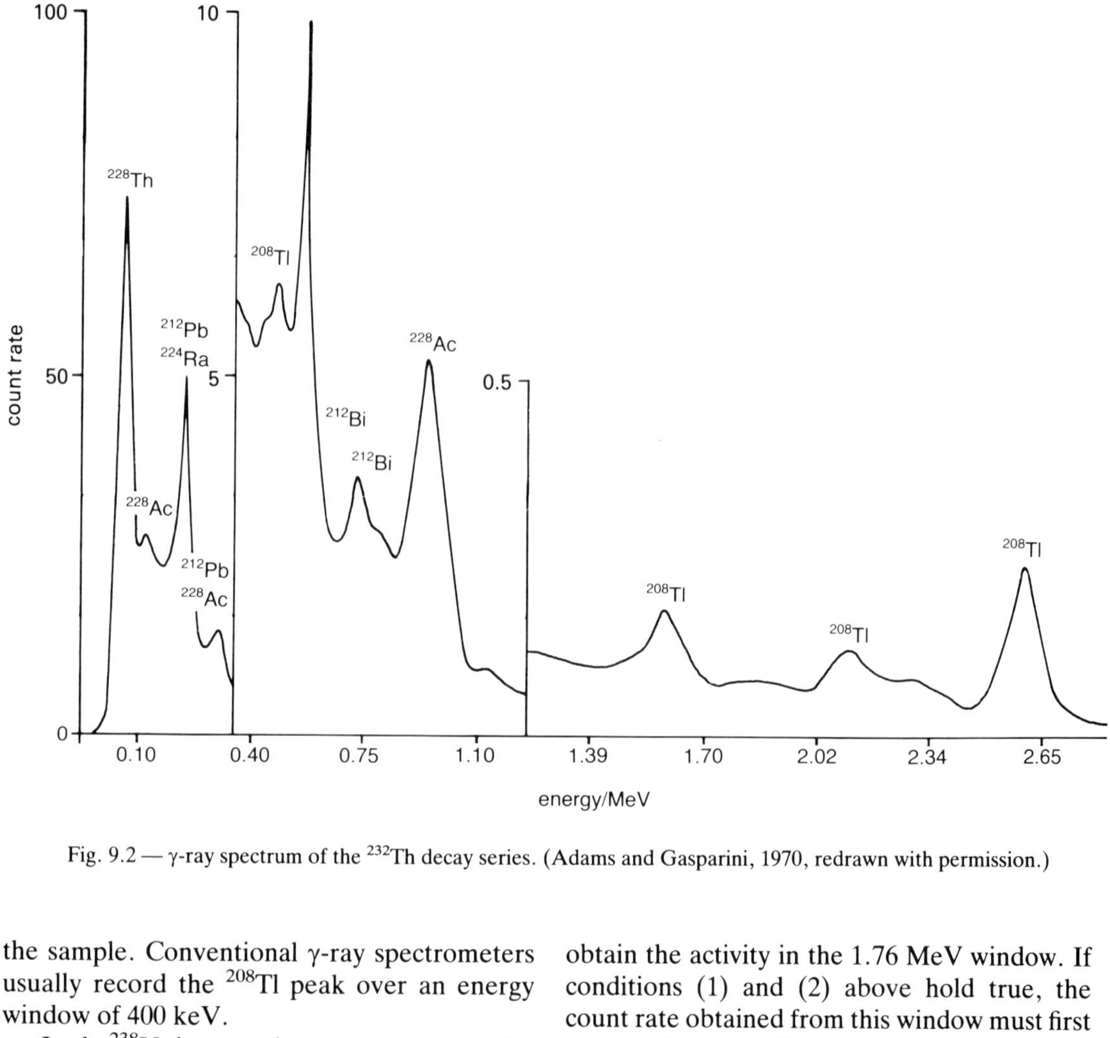

Fig. 9.2 — γ-ray spectrum of the ^{232}Th decay series. (Adams and Gasparini, 1970, redrawn with permission.)

the sample. Conventional γ-ray spectrometers usually record the ^{208}Tl peak over an energy window of 400 keV.

In the ^{238}U decay series γ-ray spectrum, the main contributor is ^{214}Bi, which gives rise to a number of peaks between 0.6 and 1.76 MeV (Fig. 9.3). Above about 1.6 MeV and below about 2.5 MeV, the combined spectra of the ^{238}U and ^{232}Th decay series accounts for the natural γ-ray activity. Measurement of the 1.76 MeV peak will provide data that can be converted to a U content if (1) the ^{238}U decay series is in equilibrium, (2) the Th content of the sample is known or has been previously determined by the spectrometer, and (3) the ^{232}Th decay series is also in equilibrium. An energy width of 200 keV is usually employed to obtain the activity in the 1.76 MeV window. If conditions (1) and (2) above hold true, the count rate obtained from this window must first be split into its 'U' and 'Th' components by application of the appropriate stripping constant S_1 to the activity recorded by the Th window. The count from the U window can then be converted to a U concentration by means of another conversion factor k_U.

In terms of the energy of γ-rays emitted during the decay of ^{40}K, a monoenergetic peak, with some spreading, should be recorded at 1.46 MeV only, but the spectrum obtained in practice is much wider (Fig. 9.4). Note that energies less than 0.4 MeV which originate outside the nucleus are referred to as X-rays. X-ray emission is conveniently covered when dealing with

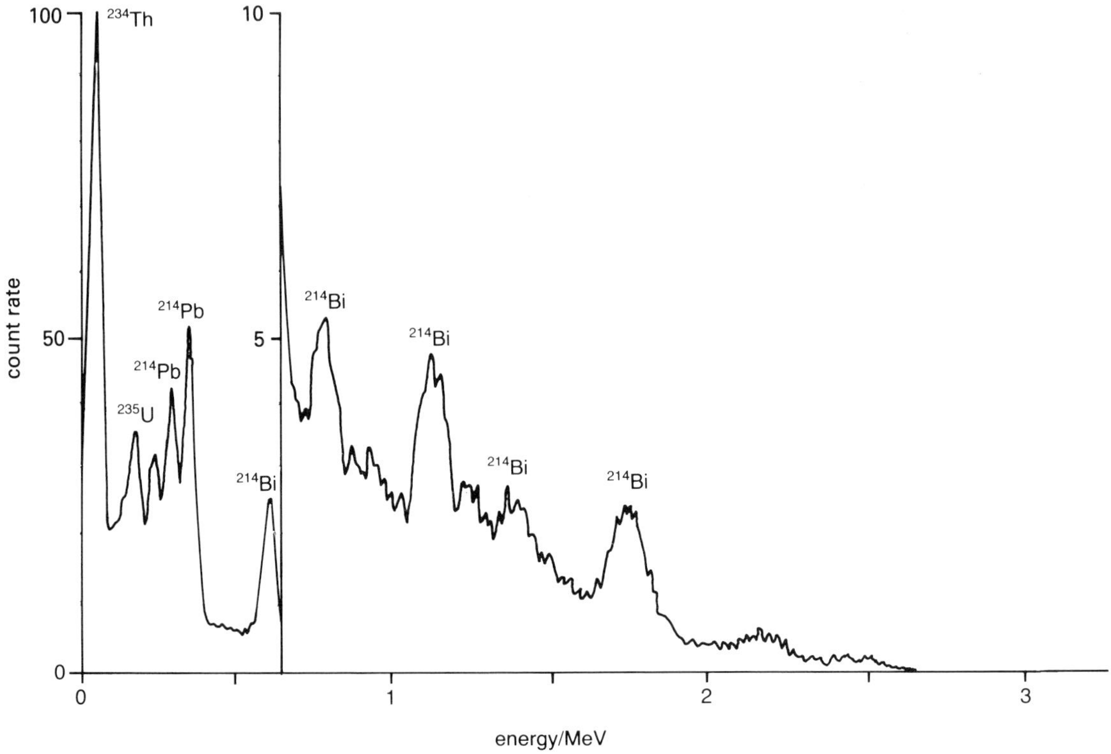

Fig. 9.3 — γ-ray spectrum of the ^{238}U decay series. (Richards and Walraven, 1975, Council for Mineral Technology, redrawn with permission.)

the decay of ^{40}K, although it also occurs in the other decay series.

When a vacancy occurs in an inner electron shell in the atomic structure of an element, it is filled by an electron coming in from an outer shell. The energy difference between the two shells involved is liberated and released as an X-ray photon. Internal conversion phenomena (the internal interaction between electrons and the nuclear field) and orbital electron capture by the nucleus are the main causes of vacancies in the inner shells. However, vacancies can also arise because of the interaction of orbital electrons with γ-rays from outside the atom. The X-rays emitted display a line spectrum of energies characteristic of the element and the orbital shell involved, but X-rays can also display a continuous spectrum. When electrons approach the electrostatic field of a nucleus, the electrostatic interaction produces an acceleration and causes deflection of the electrons from their original pathways with a consequent loss of energy by photon emission. The emitted radiation is continuous over a range of energies because of the statistical nature of the electron energy loss. This radiation is called Bremsstrahlung or braking radiation. Bremsstrahlung is particularly important in the γ-ray spectrum of ^{40}K.

In a practical γ-ray spectrometer, the 1.46 MeV peak of the ^{40}K decay is measured using a 200 keV window. Therefore, by similar argument to that used for Th and U above, the K content of the sample can be determined if the accompanying ^{238}U and ^{232}Th decay series are in secular equilibrium, and the contribution of the ^{238}U and ^{232}Th decay series to the K window can be removed by the use of the

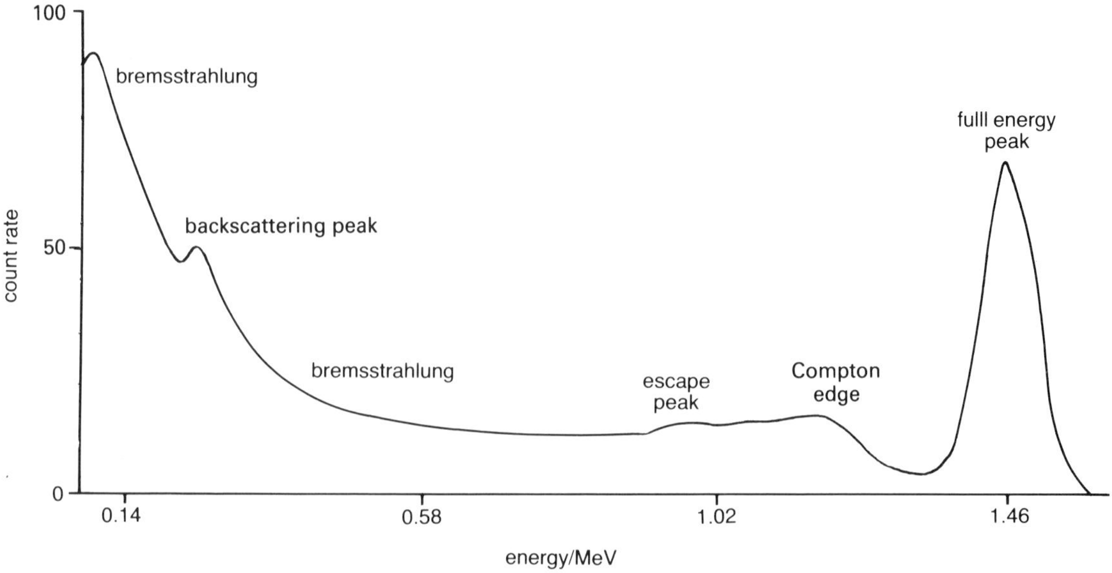

Fig. 9.4 — γ-ray spectrum of ^{40}K. (Adams and Gasparini, 1970, redrawn with permission.)

appropriate stripping constants S_2 and S_3. The conversion factor k_K is then used to convert the count rate to a K content. Note that the isotopic ratio $^{39}K/^{40}K$ is assumed to be constant at 8500. Thus,

$$_{e}Th_{ppm}=Th_{count}k_{Th},$$

$$_{e}U_{ppm}=(U_{count}-Th_{count}\ S_1)\ k_U,$$

$$_{e}K_2O\%=(K_{count}-U_{count}\ S_2-Th_{count}\ S_3)\ k_K,$$

where Th_{count}, U_{count} and K_{count} are the count rates for the appropriate energy windows, and the prefix e is used to indicate an equivalent radiometric composition (Fig. 9.5).

Unfortunately, the count rates for each energy window given by the spectrometer may not lead directly to the required radioelement concentrations. It is often found that a 'background' count rate is also present. True counts are given by

$$Th_{true}=Th_{count}-Th_b,$$

$$U_{true}=U_{count}-U_b,$$

$$K_{true}=K_{count}-K_b,$$

where Th_b, U_b and K_b are the background counts for each energy window. The values of k_{Th}, k_U, k_K, S_1, S_2, S_3, Th_b, U_b and K_b have to be determined for each type of spectometer by reference to known standard samples analysed by conventional chemical methods. Manufacturers usually supply values for the first six of these factors as they relate essentially to the behaviour of a particular instrument, but the background count rates depend upon the specific environment of a survey and must be determined independently. Independent determination of the conversion factors and stripping constants is also advisable.

In conducting a survey, it is important that adequate count times are employed so that statistical error in the counting of radioactive decays is kept to acceptable levels. Moreover, the geometry of the spatial environment at each point should be similar. Normally, it is the Th window which records the lowest count rate and which thus determines the counting period

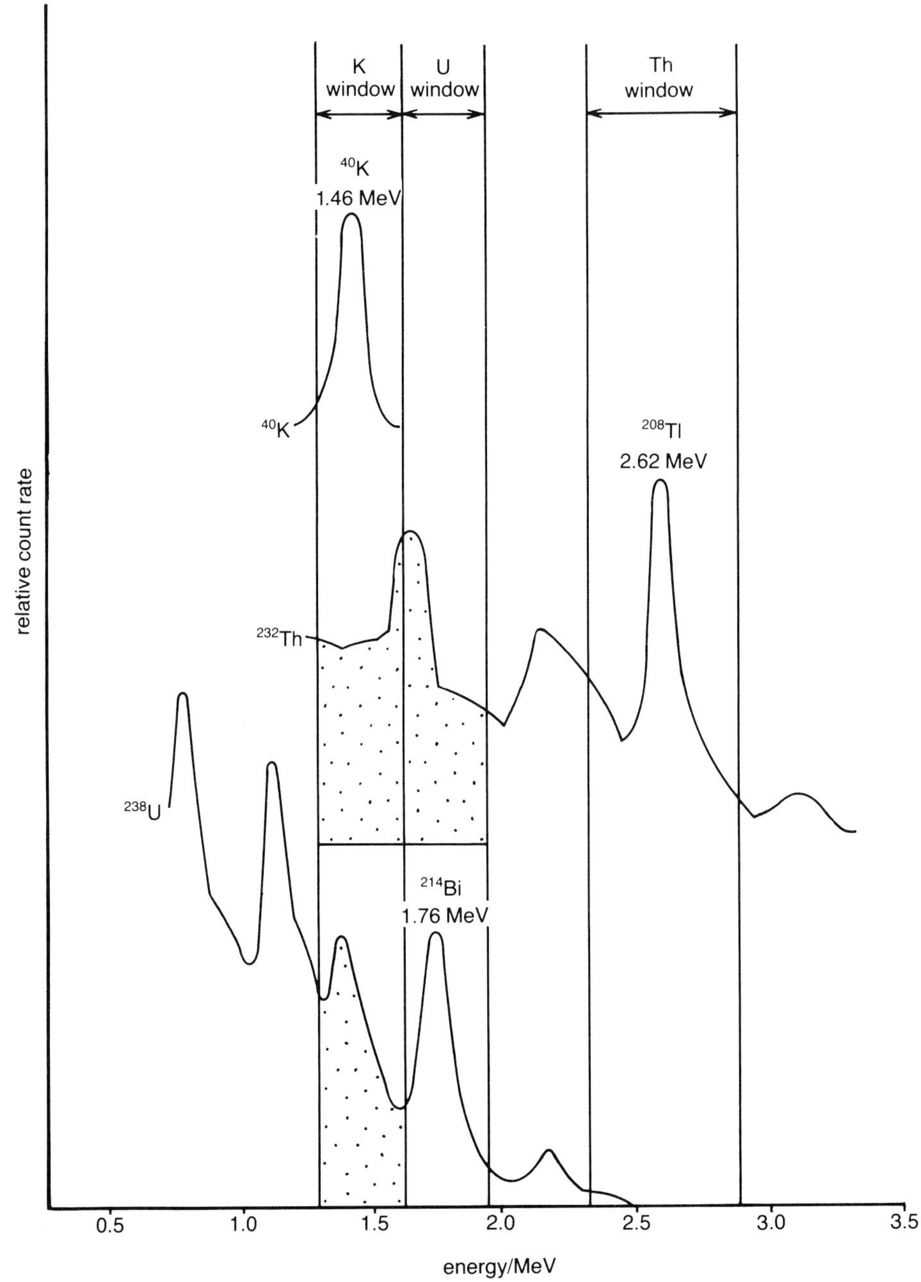

Fig. 9.5 — Energy windows and areas of overlap of the ^{40}K and the ^{232}Th and ^{238}U decay series spectra (stippled) recorded by a single-channel γ-ray spectrometer. (Richards and Walraven, 1975, Council for Mineral Technology, redrawn with permission.)

which must be employed. A minimum of 500 counts in the Th window is generally needed for results of analytical quality. This can necessitate count periods of 5 min or more with a portable spectrometer employing a detector crystal 3 inches (7.62 cm) thick and 3 inches in diameter but is minimised if the detector is placed directly onto the soil or rock. Because the count rates from the K and U windows and the total count are so much higher, the memory for these windows can overrun the decade counter display. In these circumstances, the integrity of the lower decades is maintained, so that the lost decades can be obtained by a short count period used in conjunction with the lower decade values from the longer count period.

The maintenance of spatial 2π geometry (hemispherical geometry with an infinitely extensive planar surface) for each survey point is ideal in field applications. Changes in sample geometry from station to station will produce spurious variations in the resultant data. It is clear, for example, that, if the detector is placed on horizontal ground adjacent to a vertical face rising from that surface, there is approximately 50% more material contributing to the signal than if the vertical face were absent. In fact the situation is more complex than this and is governed also by the distance to the vertical face and by the strength and energy band of the source. Generally, conversion factors and stripping constants are derived for a 2π source–detector configuration.

Depth of investigation

If the detector is placed on a flat surface of a uniform material which extends to infinity in all directions about the detector, γ-rays will be received from a 2π geometry. Within this geometry, the greater the distance which a point source is from the detector, the smaller the recorded intensity will be. The attenuation of γ-rays is controlled by the relationship

$$I = I_0\, e^{-\mu x},$$

where I_0 is the initial intensity and I the intensity after travelling a distance x through material with a linear absorption coefficient μ. The intensity, therefore, decreases exponentially with distance. Conversely, the volume of material acting as the source of γ-rays will increase with distance. However, the volume increase is governed by a geometrical relationship (Fig. 9.6).

The half-thickness, the distance at which the intensity has fallen by 50%, is determined by

$$x_{1/2} = \frac{0.693}{m_\mu \rho}$$

where m_μ is the mass absorption coefficient and ρ the density of the material. m_μ is related to the absorption cross-section a_μ by the relationship

$$m_\mu = \frac{N_a a_\mu}{w},$$

where N_a is Avogadro's number and w the atomic weight of the element concerned. For mixtures of elements or compounds,

$$m_\mu = m_{\mu_1} a_1 + m_{\mu_2} a_2 + m_{\mu_3} a_3 \;\ldots\ldots\; m_{\mu n} a_n,$$

where m_{μ_1}–m_{μ_n} are the mass absorption coefficients of the constituents, and a_1–a_n are the atomic proportions by weight of each of the elements in the mixture or compound.

To determine actual depths of investigation for any soil or rock type, it therefore appears necessary to obtain at least a major element chemical analysis. By calculation on the basis of its major element chemistry, the Mercia Mudstone of East Devon (Tables 9.1–9.3) has $x_{1/2}$ values for γ-rays of 1.46 MeV, 1.76 MeV and 2.62 MeV of 5.8 cm, 6.4 cm and 7.8 cm respectively. Interestingly, Dartmoor granite and pure SiO_2 have half-thickness values that are very simlar to these figures. It appears, therefore, that ordinary lithologies have approximately equal half-thickness values, irrespective of the composition of the rock or soil type and despite variation in density and composition. Clearly,

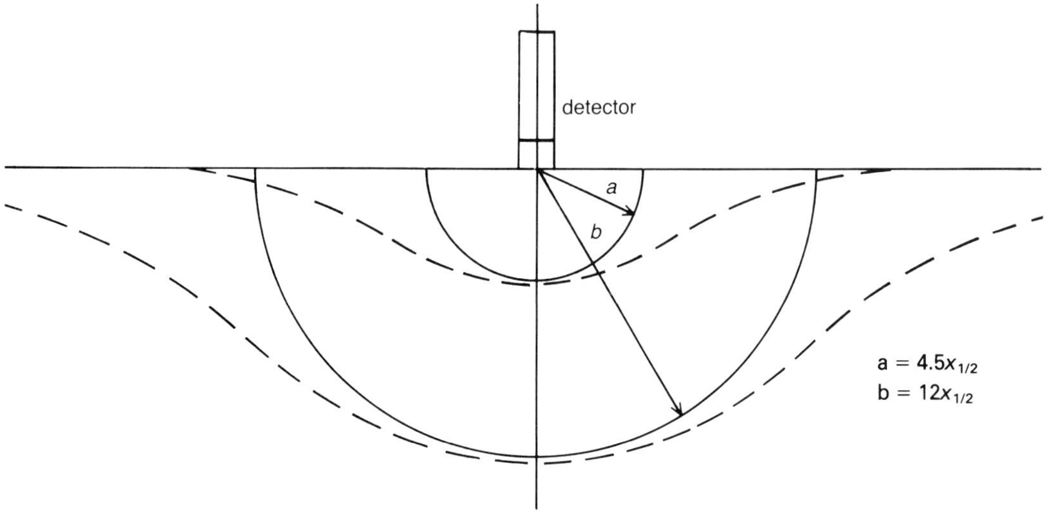

Fig. 9.6 — γ-ray sample geometry. Hemispheres a and b respectively provide 60% and 99% of the sample. The broken lines show the actual form of the source when an uncollimated detector is used. The half-thickness $x_{1/2}$ of an absorber is given by $0.693/m_{\mu}\rho$ where m_{μ} is the mass absorption coefficient and ρ is the density. (After Heath, 1982.)

the combination of density and composition remains reasonably constant for a wide range of geological materials.

For any particular γ-ray energy, 99% of the radiation recorded by a detector placed on flat surface is generated within a distance of about $12x_{1/2}$ from the detector. For the most energetic γ-ray (about 2.62 MeV) this distance is 108 cm, which defines the limit of the zone investigated for the ^{232}Th decay series; the zones for the ^{238}U decay series and for ^{40}K are 88 cm and 80 cm respectively. Where the detector is not shielded from lateral air-borne γ-rays, the volume of rock or soil sampled is not, however, hemispherical in shape because γ-rays emitted near the surface will suffer less attenuation in travelling through the air than through the solid soil or rock. In practice, most energy is received from the soil or rock immediately below the detector. Indeed, some 60% of the received radiation comes from within a radius of only $4.5x_{1/2}$. Effectively, therefore, portable γ-ray survey results only relate to the top 20 cm or so of the soil or rock profile in the immediate vicinity of the detector (Figs. 9.7 and 9.8). Unfortunately, this near-surface environment is subject to weathering which often produces open system chemistry and disequilibrium in the ^{238}U decay series (discussed more fully in Chapter 10). Clearly, direct quantitative analysis by γ-ray spectrometry of this kind has its limitations. The method, however, has the advantages of being relatively cheap and simple to operate, with power to the instrument usually being supplied by replaceable dry cells with a life of several days under normal field use. Interpretation of the results in a comparative fashion is generally reliable.

9.2.2 Portable multichannel analyser systems

Portable systems for use with multichannel analysers have recently become available with the introduction of pure Ge detectors that, unlike older Ge(Li) detectors, do not have to be continuously maintained at liquid N temperature. However, they must still be operated at this low temperature, and so their use is limited by the availability of a liquid N supply. Also, the liquid N cryostat must be able to withstand the rigours of field use (Canberra,

Table 9.1 — Chemical composition of Mercia Mudstone.

Chemical component	Percentage	Element	Atomic proportion (%)
		O	45.69
SiO_2	50.84	Si	26.03
TiO_2	0.79	Ti	0.52
Al_2O_3	15.35	Al	8.92
Fe_2O_3	6.60	Fe	5.07
CaO	6.69	Ca	5.24
MgO	5.37	Mg	3.53
Na_2O	0.30	Na	0.24
K_2O	5.22	K	4.75
Total	91.16		99.99

Table 9.2 — Absorption coefficients for the major elements in the Mercia Mudstone.

Element	Energy level, 1.46 MeV		Energy level, 1.76 MeV		Energy level, 2.62 MeV	
	a_μ (barns cm^{-1})	$m_\mu = \mu/\rho$ (cm^2g^{-1})	a_μ (barns cm^{-1})	$m_\mu = \mu/\rho$ (cm^2g^{-1})	a_μ (barns cm^{-1})	$m_\mu = \mu/\rho$ (cm^2g^{-1})
O	1.385	0.052 134 39	1.260	0.047 429 12	1.020	0.038 395 00
Na	1.925	0.050 428 22	1.740	0.045 581 87	1.420	0.037 199 00
Mg	2.085	0.051 649 18	1.895	0.046 942 54	1.550	0.038 396 27
Al	2.260	0.050 445 27	2.065	0.046 092 69	1.700	0.037 945 56
Si	2.450	0.052 535 66	2.220	0.047 603 74	1.820	0.039 026 49
K	3.325	0.051 211 88	3.030	0.046 668 27	2.510	0.038 659 19
Ca	3.505	0.052 666 97	3.190	0.047 933 70	2.645	0.039 744 40
Ti	3.850	0.048 406 47	3.520	0.044 257 34	2.905	0.036 524 88
Fe	4.580	0.049 390 53	4.160	0.044 861 27	3.515	0.037 905 61

Table 9.3 — The absorption of γ-rays by the Mercia Mudstone.

Element	Atomic proportions × μ/ρ		
	1.46 MeV	1.76 MeV	2.62 MeV
O	0.023 82	0.021 67	0.017 54
Si	0.013 68	0.012 39	0.010 16
Ti	0.000 25	0.000 23	0.000 19
Al	0.004 50	0.004 11	0.003 38
Fe	0.002 50	0.002 27	0.001 92
Ca	0.002 76	0.002 51	0.002 08
Mg	0.001 82	0.001 66	0.001 36
Na	0.000 12	0.000 11	0.000 09
K	0.002 43	0.002 22	0.001 84
μ/ρ	0.051 88	0.047 17	0.038 56
ρ	2.30	2.30	2.30
μ	0.119 324	0.108 491	0.088 688
$x_{1/2}$ (cm)	5.81	6.39	7.82

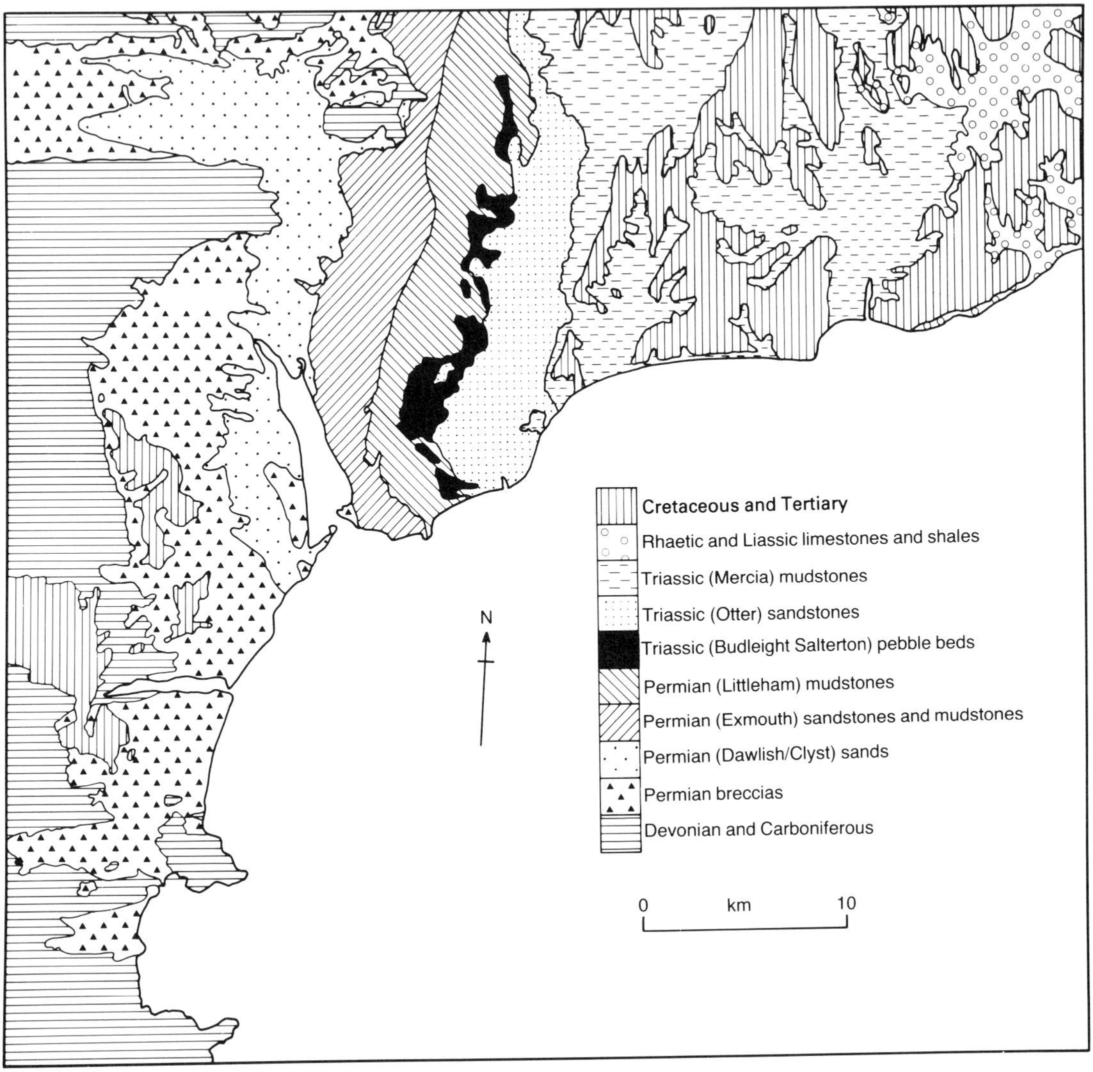

Fig. 9.7 — Geological map of southeast Devon.

1984). This is accommodated by the use of so-called multiattitude cryostats which allow operation of the detector in any orientation without spillage of liquid N, even when the detector is full. The design of the system incorporates two fill-and-vent ports which are arranged so that one is able to vent at all times regardless of the orientation of the Dewar flask. Unlike most single-port Dewars which can only be filled to half their capacity, this type of two-port Dewar can be completely filled, so that the external dimensions of the cryostat are kept to easily managed proportions, with consequently much improved volume–lifetime characteristics. Indeed, since the unit can hold sufficient liquid N to keep the detector cool overnight, there is no need to wait for cool-down to occur each morning that the detector is used. Typical holding times are of the order of 24 h, and the cool-down time is about 2 h, with a

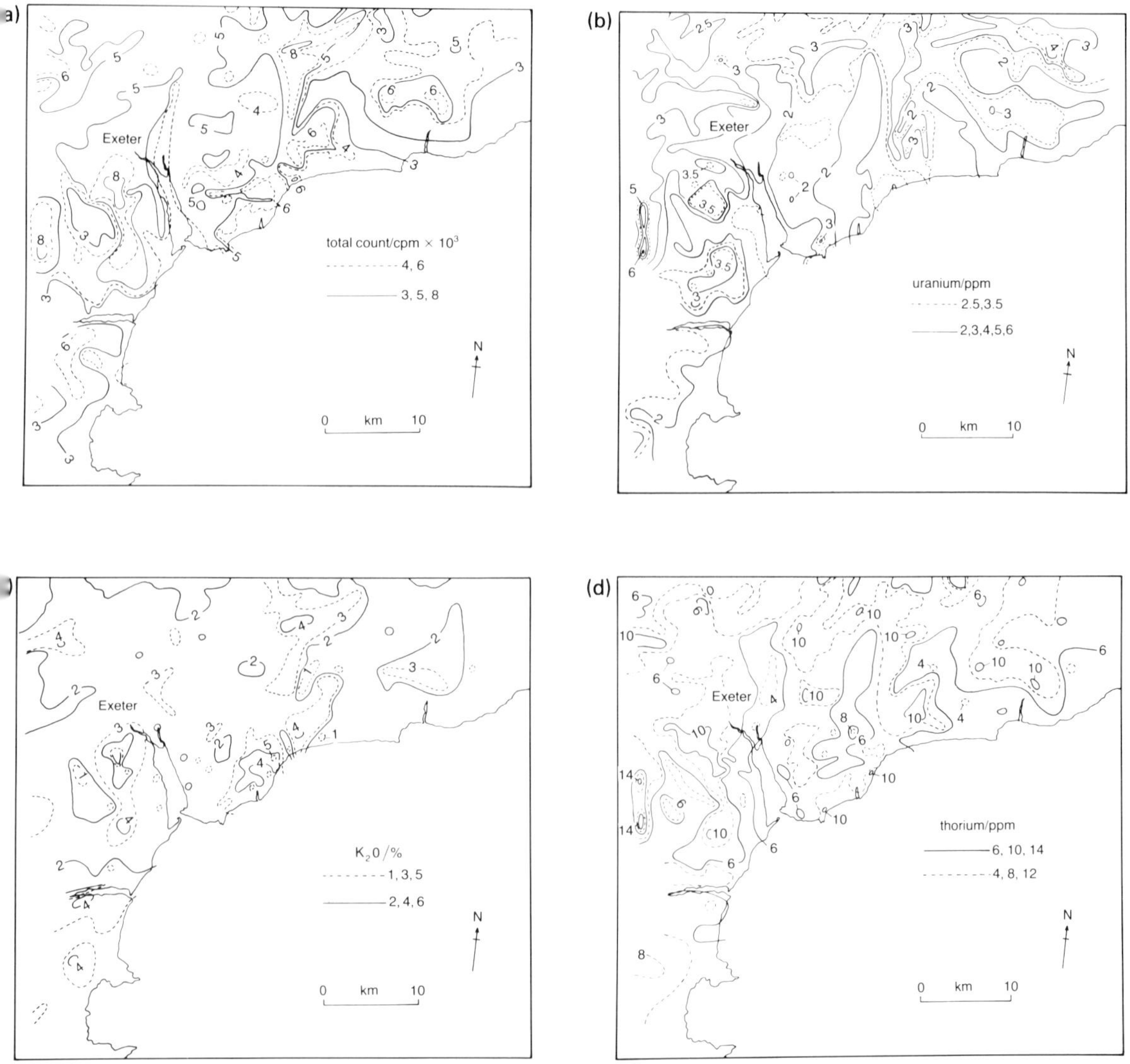

Fig. 9.8 — Results of a γ-ray spectrometer survey of grassland soils in southeast Devon: (a) total count; (b) eU; (c) eK_2O; (d) eTh. (Durrance and Heath, 1985, reproduced with permission.)

liquid N capacity of 1.4 l. The detector itself is usually of the coaxial type, permitting a slimline system which is suitable for shielding when used for low count rates. Pre-amplifiers may be operated at either normal or liquid N temperatures depending upon the detector characteristics; in the latter case the cryostat also serves to cool the pre-amplifier. The typical weight of a practical detector with the cryostat filled with liquid N is of the order of 8 kg, and overall dimensions are about 50×24 cm (Fig. 9.9).

The development of portable multichannel analysers for use with portable Ge detectors was only possible with the introduction of microelectronics. Size and weight are important

Fig. 9.9 — Canberra Series 10 portable multichannel analyser used with a portable Ge detector. (Source: Canberra Industries Inc.)

considerations, and a typical portable multichannel analyser has a mass of about 5.5 kg and measures about 11cm×23 cm×28 cm. Portable multichannel analysers usually have a unipolar-DC restored spectroscopy amplifier with fast and slow shaping that makes them suitable for use with NaI(Tl) as well as Ge detectors. They normally run off any external 12 V DC supply (such as a car battery) but have internal rechargeable batteries sufficient for 8–10 h of continuous operation. They can also run off ordinary mains AC supplies, should their use in the laboratory for long counting periods be necessary. Characteristic features are as follows: a 100 MHz analogue-to-digital converter; a 1–4k memory, with 24-bit capacity and direct memory access to data storage; 1024- or 4096-channel conversion gain; conventional visual display and output to cassette recorder; microprocessor capabilities such as learn–execute and define–use functions. With learn–execute, complete analytical sequences can be defined, stored on cassette tape and field-loaded to perform specific tasks. Define–use is an expansion of this capability which allows up to five user-defined equations to be entered, either manually or via a cassette tape, to enable computations to be carried out. The results may then be displayed or output to cassette (Canberra, 1984). Because of the very recent introduction of this capability, there is as yet little information on its geological applications in the field.

9.3 RADON TECHNIQUES

The tendency of Rn to migrate from its point of generation within U- or Th-bearing minerals, particularly in the presence of moving groundwater, frequently results in the formation of extensive Rn anomalies in soil gas and in surface waters and groundwaters around deposits which are not detectable by γ-ray surveys. This is because γ-rays have such limited ranges in rocks and soils that, unless the top 50 cm or so of the soil is enriched in U or Th, attenuation is so great that impracticable count times would be needed to detect an anomaly. For example, in an area where there is a superficial cover of glacial boulder clay derived from rocks containing normal levels of radionuclides, a U deposit in the underlying rock would be completely masked to γ-ray detectors. In such a situation, the use of Rn techniques has great potental. However, may difficulties are also inherent in the method.

9.3.1 Soil gas methods

Rn production and migration

Direct Rn detection systems are based on the fact that Rn is the only naturally occurring radioactive gas. Each of its three major isotopes decay by α-particle emission: ^{220}Rn (in the ^{232}Th decay series) to ^{216}Po with a half-life of 56 s; ^{219}Rn (in the ^{235}U decay series) to ^{215}Po with a half-life of 4 s; ^{222}Rn (in the ^{238}U decay series) to ^{218}Po with a half-life of 3.825 days. In practice, the low abundance and short half-life of ^{219}Rn allow the presence of this isotope to be ignored in exploration systems. With its comparatively short half-life, the migration of ^{220}Rn from its point of generation is severely restricted, and detectable levels are unlikely to be encountered at any great distance. Despite its longer half-life, even ^{222}Rn requires mechanical transport, and not simply diffusion, to allow high levels away from U sources to develop. The main means of mechanical transport is in solution in flowing groundwater, but above the water table any moving current of air will also transport Rn.

Numerous studies have shown that ^{222}Rn is not supported in solution by its parent ^{226}Ra and that ^{222}Rn has, therefore, migrated directly from its original place of generation. There are three possible steps involved in this (Andrews and Wood, 1972).

(1) Formation and recoil of the Rn atom by disintegration of the parent Ra atom.
(2) Diffusion of newly formed Rn atoms through the mineral grains in which they are generated.
(3) Diffusion and transport through the rock mass.

^{222}Rn atoms have an initial recoil energy of 100 keV, which corresponds approximately to recoil ranges of 3×10^{-6} cm, 10^{-5} cm and $(6–9)\times10^{-3}$ cm in rock, water and air, respectively. Only those ^{222}Rn atoms that are produced in a very thin surface layer of a rock particle adjacent to a pore can, therefore, escape in this way. Having left the mineral, the ^{222}Rn atom may be trapped within the pore or pass through it and enter another mineral. Atoms which remain within the pore constitute the 'direct recoil fraction' of the emanation (Fleischer, 1983). It is possible, however, for atoms which have entered other minerals to diffuse back into the pore where they would have created pockets by vaporisation. This component within the pore is the 'indirect recoil fraction'. The 'emanation coefficient' is the percentage of the ^{222}Rn generated that actually escapes. It is theoretically the sum of the direct and indirect recoil fractions plus that component which diffuses through the source mineral. Measured emanation coefficients for minerals are very variable, even within individuals of the same species. These ranges reflect the varying degree of physical continuity of the individual grains. The emanation coefficient for carnotite, for example, ranges between about 6 and 32%, for sphene it is about 9% and for pitchblende it is between 0.06 and 16.6%. These data on their own are not very meaningful, although they do show that secondary U-bearing minerals tend to have higher coefficients than primary U minerals (uraninite, for example, has an emanation coefficient of 0.6%), probably because of their more open structure. Comparison of measured values with theoretical values also shows that recoil and diffusion alone cannot account for the levels of emanation and that additional factors must, therefore, also be acting. Transport of ^{222}Rn along discontinuities within mineral grains is probably very significant.

Emanation coefficients also vary greatly within and between different rock types. Granite, for example, has been reported to have values between 7 and 33%, granodiorite between 17 and 40% and gneiss between 1 and 14%. In general, however, rocks usually show emanation coefficients of about 5%. Most soils, in contrast, have an emanation coefficient of about 40%, although in sandy soils it may be only 10%. In all materials, the presence of water in the pores enhances both the indirect recoil and the diffusion factors by limiting the recoil range and inhibiting adsorption of ^{222}Rn on the solid walls of the pores. However, emanation coefficients increase with decreasing grain size, as surface area per unit mass increases.

Emanation coefficients also vary with time. Changes in atmospheric pressure and temperature tend to alter the openness of fractures, while the emanation coefficient often decreases with increased rainfall. However, the effect of rainfall is complex and appears to depend upon the recent rainfall history. Surface freezing and a snow cover will also affect the rate of ^{222}Rn loss. At high latitudes, the ^{222}Rn trapped within frozen ground during winter will be released over a short period of time with the spring thaw. ^{222}Rn levels are, therefore, unusually high at this time, and the identification of anomalies is made easier than at other times of the year. This phenomenon has given rise to what has become known as the 'radon season'. Another factor sometimes used to define the movement of ^{222}Rn is the 'effective diffusion coefficient' (Telford, 1983), which is the diffusion coefficient of ^{222}Rn in the pores divided by the porosity of the medium as a whole. The effective diffusion coefficient for ^{222}Rn in most soils lies between 0.004 and 0.04 $cm^2\ s^{-1}$.

Because of the time-dependent nature of the ^{222}Rn flux and the variability due to differences in the physical properties of mineral, rock and soil materials, it has been argued that it is not possible to construct meaningful ^{222}Rn isograds from observations of ^{222}Rn concentrations and that it is only feasible to delineate anomalous areas. Even in anomalously high ^{222}Rn areas, temporal variations of up to a factor of 50 have been recorded over a period as short as 3 weeks in the temperate climate of southern Britain (Ball *et al.*, 1983).

In general, however, the details of ^{222}Rn release and migration tend, in practical terms, to be overwritten by the larger effects due to the presence of fractures within rock and soil. On a large scale, the faulting of rock masses very significantly enhances ^{222}Rn transport, even to the extent that fault traces on the ground surface produce ^{222}Rn concentrations which might be interpreted as due to buried U ore bodies. This is especially true when the general levels of U in the rocks is above average anyway. On a small scale, too, proximity to cracks in the soil results in much higher Rn concentrations than at sites away from such cracks. This variation can be on a scale of a few centimetres up to several metres.

Finally in this list of difficulties, the migration of the immediate parent ^{226}Ra, and of ^{238}U and ^{234}U, must be considered. Each of these nuclides is easily transported in solution in groundwater (as discussed in Chapter 6), so the ^{222}Rn source may be carried some distance from an ore body.

The detection and measurement of Rn are essentially carried out by α-particle recording systems. Soil gas systems may be grouped as either integrating or non-integrating, depending upon the time over which measurements are made. The integrating systems employ measurement times of days or longer, while non-integrating systems, known as emanometers or 'sniffers', use measurement times of no more than a few minutes.

Rn emanometers

Rn emanometers (sniffers) consist typically of a sampling probe, a pump and an α-particle detection and counting arrangement. The sampling probe is very important, as obtaining a sample of soil gas uncontaminated by the atmosphere is by no means simple. The most basic probes consist of a hollow tube which is inserted into a hole produced by augering or driving a stake to a depth of 50 cm or so into the ground. Augering, however, considerably disturbs the ground adjacent to the hole, and the repeated withdrawal of the auger to clear the flights as the hole deepens, together with reinsertion after each clearing, causes the introduction of air from the atmosphere in considerable quantities. In comparison, the driving of a solid stake does not allow the atmosphere to enter the hole and, when the stake is withdrawn prior to insertion of the probe, soil gas is sucked into the hole from the surrounding ground. However, when the stake is driven into the ground, compaction of the soil takes place adjacent to the hole and its normal permeability can be considerably reduced. This is especially true of clay soils in which the compaction can be so great that the ground adjacent to the hole becomes impermeable to the passage of soil gas. Another disadvantage of using a simple stake is the difficulty that can be experienced in removing the stake from the ground!

A refined soil gas probe has been designed by the British Geological Survey to overcome the difficulties of augering or using a solid stake (Fig. 9.10). The probe consists of a hollow outer tube and a solid inner rod which fits snugly inside this tube. Attached to the solid rod is a sliding hammer which, on the downward stroke, hits the top of the outer tube and, on the upward stroke, strikes against a flange at the top of the inner rod. This flange is also struck on the downward stroke of the sliding hammer. The whole arrangement is made of stainless steel for strength and to avoid corrosion and is about 1m long. A very important feature of the arrangement is that the inner rod extends for some 4 cm below the bottom of the outer tube. Also, the end of the rod is pointed, while the end of the outer tube is tapered. A boss on the outer tube near the top has two flat grooves to take the extraction bar. The probe is inserted into the ground simply by pushing the inner rod into the outer tube and then firmly striking down with the sliding hammer. When the outer tube is at a sufficient depth (50 cm minimum), the inner rod is extracted by firmly striking up with the sliding hammer. A cavity is thus created below the base of the hollow outer tube. A gas pumping system can then be placed and sealed in the top of the tube.

Problems exist with this system in that compaction of the ground around the sampled volume still occurs, but at least this is kept to a minimum. Also, in dry or sandy soils the contact between the ground and the steel of the outer tube may not be air tight; when gas is pumped from the tube, atmospheric air may be drawn down around the outside of the probe to contaminate the soil gas sample. Indeed, atmosphere can be drawn into the sample volume even during the withdrawal of the inner rod, and so a certain volume should always be vented to the air to purge the system before any measurements are made. Gross contamination by drawing atmosphere down the side of the tube can be checked by noting the rate at which gas can be pumped from the probe (depending on the soil type) and sometimes by hearing the air 'hissing' down the side of the probe.

Normally, a simple hand pump is used to extract the soil gas from the probe, an adaptor fitted with O-ring seals connecting the pump to the probe. A small filter is often placed between the probe and the pump to prevent any fine particles of soil from being carried through the

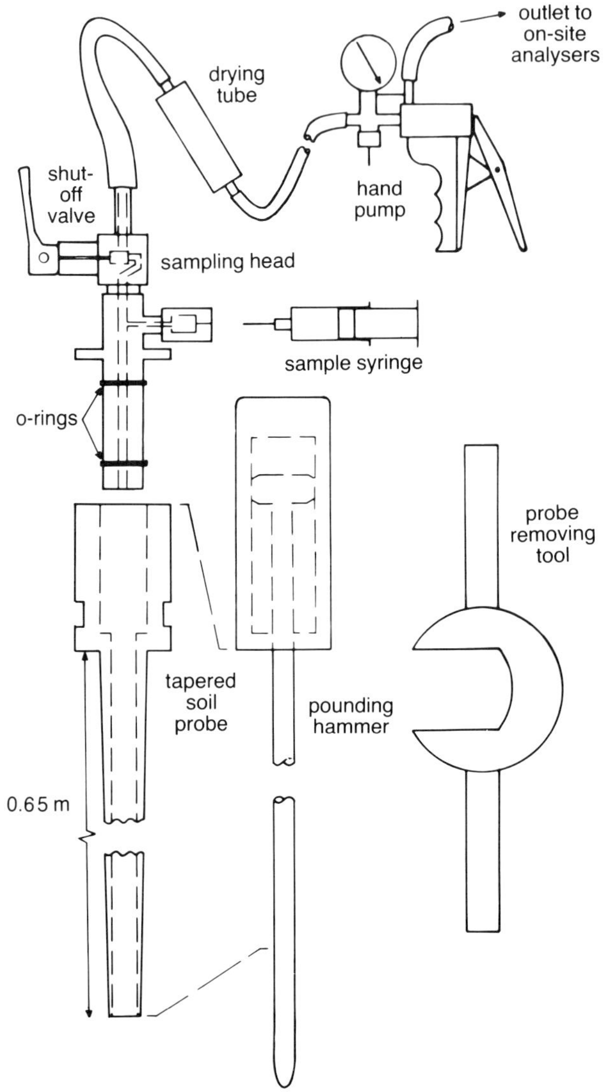

Fig. 9.10 — Soil gas sampling probe. (Gregory and Durrance, 1985a, reproduced with permission.)

system. A dehydration unit (such as a tube filled with self-indicating silica gel) is also usefully inserted between the probe and the pump to prevent moisture from being passed on into the detector arrangement (Fig. 9.11).

In most systems, the soil gas is pumped directly into a scintillation cell consisting of a container the inner surface of which is coated with activated ZnS. The scintillation cell may be filled either while it is within its photomultiplier counting system or externally before being placed into the photomultiplier. Cells may be either sealed themselves, in which case they have a transparent glass window at one end, or, if filled within the photomultiplier, open ended. In each case, some circulation system is needed

Fig. 9.11 — EDA Electronics RD200 with soil gas attachment. An open-ended scintillation cell (shown separately) is contained within the unit.

to draw soil gas into the cell. Closed cells are first evacuated and then the vacuum is simply released by drawing soil gas from the probe (here the pump is not needed for the probe but for the cell). With open-ended cells, a pair of one-way valves (fill and extraction) and a circulating pump venting to the atmosphere are connected to the probe itself.

Because of their different half-lives, ^{220}Rn and ^{222}Rn activities may be distinguished by taking three separate 1 min counts, C_1, C_2 and C_3, on the same sample immediately after collection (Morse, 1976). The total Rn activity is then given simply as the average of these results, while

$$^{222}\text{Rn activity} = 0.87\,C_3 + 0.32\,C_2 - 0.34\,C_1$$

and

$$^{220}\text{Rn activity} = \text{total (Rn} - {}^{222}\text{Rn) activity}$$

A variation on this theme is to pump the soil gas over an adsorbing surface of charcoal. The Rn is trapped on the charcoal until it is released by heating and can then be passed to the detection–counting arrangement. This modification permits samples to be collected *en masse* and then taken back to the laboratory or field base for analysis, but a correction for ^{222}Rn decay between the times of sample collection and analysis must be applied. The other techniques require analysis on site but no correction for delay is needed.

Integrating systems

The main problem with all the procedures considered so far is that, because the sample is collected over a very short time, it may not be truly representative of the site. In the simplest approach, it can be argued that repeating the same procedure at each site will yield results which can be compared with one another and which will allow anomalous areas to be delineated, but this will only be true if all other natural factors remain constant. The time-dependent variations can be minimised, however, by covering the survey area very rapidly so that atmospheric conditions remain constant, but the real problem lies in the fact that soil types are likely to vary from one site to another. Even within one soil type, there are likely to be subtle changes in texture over small distances. These changes in soil conditions mean that the ability of the gas to enter the sampled volume differs considerably from one site to another, a close-textured clayey soil tending to produce lower Rn counts than an open-textured sandy soil of the same true Rn content. Attempts to overcome this difficulty have been made by trying to measure the equilibrium Rn concentration. In this approach, soil gas is pumped from the sample volume using a series of increasing (usually logarithmic) delay times. Sample 1, for example, could be collected immediately, sample 2 after 1 min, sample 3 after 5 min, sample 4 after 10 min, and so forth. As the sequence progresses, the count rate obtained approaches a steady value for the site. In sandy soils, this may be reached very quickly, but clayey soils can require several hours to achieve equilibrium.

Other methods of soil gas sampling also incorporate a time-integration effect. Indeed, the earliest techniques were of this kind (Dyck, 1972). Essentially, the sample is collected in an inverted open-ended container made of disposable material (old oil cans have been used) which is pushed into the ground to below the organic-rich soil layers. The side and top are then covered with leaves or organic matter. After the container has been left for a sufficient length of time (hours or longer), a hole is punched in the top and the gas is pumped either into a sealed container for transport to the detector or directly to a detector–counter system. However, setting out the collection cans is time consuming and, with the development of more reliable methods in recent years, this approach is rarely used today.

The newer methods of integrated monitoring include the α-particle meter. This is a solid-state detector which incorporates counting and display units. It is in the form of a cylindrical probe some 35 cm long and 5 cm in diameter with a decade counter display at one end. It has a mass of about 1 kg. To measure the Rn concentration in the soil, the α-particle detector is planted with the display end uppermost and is left for a period of time after which the display is recorded. The elapsed time is also given by the meter. Typical integration times are 1–2 days. For survey of a particular area, about 20 meters need to be distributed over part of the site, and the counts recorded. The meters are then moved to another part of the site, the counts recorded, moved again, and so on until the whole area has been covered. Because the time taken for each part of the survey is long, there is the danger that medium-term time-dependent variations become apparent. One way of reducing this effect is to use random site distribution over the survey area at each stage rather than to cover the whole site in systematic subsections. The recorded data from the whole site may then be processed to remove any random variations by a smoothing programme. Perhaps the main disadvantage of this method is the cost. Each meter is expensive and, when planted, is usually unsupervised. Even in remote areas, loss and damage can be sustained; in populated areas with inquisitive inhabitants, such dangers are greatly increased.

Other integrating systems which do not involve the burial of expensive equipment include the α-particle detection card and track etching methods. The first of these systems uses a thin collecting membrane mounted on a small card, with two Si semiconductor detectors attached, one on each side. Again, the card is buried in the soil. After about 12 h, the card is recovered and the electrical charge distribution on it is measured by a portable card reader. In the track etching method, instead of a solid-state semiconductor detector, a strip of cellulose nitrate film is taped inside the bottom of an inverted cup (Pedersen *et al.*, 1980). α-particles from Rn trapped within the cup produce radiation damage in the film which can be later etched, allowing the α-tracks to be counted. The cups are usually buried for periods of about 3 weeks, but sometimes for much longer periods, before the films are recovered. The method is thus cheap in terms of capital outlay, but processing of the film and counting is more time consuming (and, therefore, costly) than is the case with the other methods.

In any integrating system, the α-activity of both ^{220}Rn and ^{222}Rn will be recorded. Discrimination between these two isotopes is, of course, important where identification of U or Th sources is required. Various techniques have been adopted to achieve this in the different types of integrating system. For example, in the track etching method, a membrane can be inserted to delay the passage of the soil gas into the detector area long enough to cause substantial attenuation of the short half-life (55.6 s) isotope, ^{220}Rn; the longer half-life of ^{222}Rn (3.825 days) means that it is virtually unaffected by this delaying membrane. The approach used in the card system is simply to take a second recording a few hours after the main count without rejuvenating the card in the meantime.

Of all these integrating methods, the track etching system is probably the most versatile. The low capital cost means that a large number of detector cups can be planted over the survey area, each being subject to the same fluctuating conditions. The system also permits very long integration times to be used where count rates are low, or if significant levels of short-term variations are likely to be present in the Rn concentrations. Certainly, integration times of 6 months or a year can eliminate a great deal of the time-dependent variation present in data collected over short periods. However, even where long-term integration methods are used, the effect of spatial noise is not reduced. Changes in the distribution of Rn concentrations in relation to soil type and fracturing will still be present, although here, too, processing of the data by smoothing programmes can be very useful.

9.3.2 Water gas methods

Rn in surface waters and groundwaters

A different approach to Rn surveys is to sample water gas rather than soil gas. The concentrations of dissolved gases in surface waters or groundwaters are the result of spatial and temporal integration processes. As the availability of groundwater samples over a survey area is usually limited, most areal surveys of this type concentrate on surface waters. However, the concentration of ^{222}Rn in surface waters is usually an order of magnitude less than that found in groundwaters and, although ^{222}Rn anomalies in surface waters may be caused by U deposits, high ^{222}Rn concentrations may also be caused by the influx of groundwater. Indeed, the interaction between groundwaters and surface waters is probably such that the location of U mineralisation by this method is very difficult, especially as high ^{222}Rn concentrations may develop in groundwaters within aquifers containing only the normal crustal abundance of U. When this method is applied to U exploration, therefore, not only must the ^{222}Rn concentration at a particular site be recorded, but also information is required on the groundwater–surface water relationship at the site. In fact, ^{222}Rn concentrations in surface waters are often more useful in studying the hydrogeology of an area than as a method of U exploration. For example, some early work in the Wasatch

Mountains of Utah found that high ^{222}Rn levels in small streams were associated with the influx of groundwater and that identification of the lithological units which form the major aquifers of the district was possible using this technique. Interestingly, none of the high ^{222}Rn concentrations was supported by ^{226}Ra, indicating that migration from its source of ^{222}Rn itself had taken place. However, in other areas it has been found that the ^{222}Rn content of stream waters can be accounted for solely by the presence of ^{226}Ra in stream sediments. There appears, therefore, to be conflicting evidence for the origin of ^{222}Rn in stream waters but, in fact, this is not the case. Because of the relatively short half-life of ^{222}Rn (the very short half-life of ^{220}Rn effectively renders stream water methods impractical for this isotope) and its loss by aeration of water due to turbulence, the source of high ^{222}Rn concentrations must be local. In some cases, this will be ^{226}Ra in stream sediments, but in others it will be the discharge of groundwater. In any particular environment, either of these sources may be dominant but, in practice, it is usually found that the latter is the most important.

Other factors which control ^{222}Rn levels in stream waters include the volume of flow, the degree of turbulence, the water temperature, the pH of the water and the presence of U, either in solution or in the stream sediments. The possibility of considerable time-dependent variation in the Rn concentration at a particular site is, therefore, obvious and is, indeed, often found to be the case. However, results from Canada have been reported which seem to contradict this (Dyck, 1972). In a study of the relative magnitude of the various factors which lead to time-dependent variation, such as temperature, water depth, alkalinity, pH, size of the water body, etc., it was found that changes in these environmental factors may cancel one another. The major factors affecting ^{222}Rn in surface waters were found to be the depth of water, temperature and alkalinity. The controls exerted by distance from the stream sediment and temperature are obvious, but the role of alkalinity is less direct. In fact, it is related to the presence in stream sediments of calcite which is derived from U-bearing mineralisation in the district. Therefore, alkaline (carbonate-rich) waters are associated with calcite-rich sediments and hence with U. Moreover, carbonate-rich waters also tend to be relatively rich in dissolved U through the formation of complex ions. Turbulence was also considered to be an important factor, and the pH was found to be negatively correlated with ^{222}Rn, probably because of the discharge of groundwater into the streams.

As a general guide to the presence of U mineralisation, ^{222}Rn in groundwater may be a useful technique, but the problems caused by addition of ^{222}Rn to stream waters by influx of groundwater and by degassing effects make such studies unsuitable for precise determination of U distribution. A further limiting factor is the permeability of the rocks; this has a major influence on the migration of ^{222}Rn. It has been suggested that the long-distance migration of ^{222}Rn is only effectively carried out by hydrothermal convective circulation of groundwater (Mogro-Campero and Fleischer, 1977).

Sample collection

In the survey of a large area, the problems associated with time-dependent variations when water samples are collected over a long period of time can be minimised by the collection of samples *en masse* in as short a time as possible, and their return to the laboratory or field station for analysis, rather than carrying out analyses in the field. Techniques of sample collection vary, but glass bottles are normally used with thin metal screw caps. The bottles should be thoroughly rinsed in the water to be sampled before the sample is collected, taking care not to disturb the underlying stream sediment at the sample site; it is good practice to perform this rinsing a few metres downstream of the sample point itself. When the sample is collected, the bottle should be filled slowly to avoid degassing of the water during collection.

The bottle should be completely filled so that no air bubble is present when the cap has been screwed into place. This is best achieved by fitting the cap under water and then inverting the bottle to check for the presence of an air bubble. If the bottle is not completely filled, there is a risk of degassing of the sample by agitation during transport. It should be realised that the samples are often at a lower temperture than their surroundings, so that some expansion of the samples may occur as they warm up. The thin metal of the caps can usually accommodate such expansion. Any porous sealing disc or ring within the caps should be removed to avoid trapping ^{222}Rn, leaving the seal as a metal-to-glass contact.

At each site, the time of collection should be noted so that a correction for the decay of ^{222}Rn between collection and analysis can be applied. Water temperature and pH should also be recorded to aid interpretation.

Analysis and calibration

Analytical systems vary in their specific details but, as with the measurement of ^{222}Rn in soil gas, all operate on the basis of α-particle detection. Here, however, it is first necessary to liberate the dissolved ^{222}Rn from the water sample. Total liberation is difficult to achieve but, fortunately, is not required, provided that the procedures used are uniform and result in the liberation of a constant proportion. In one of the most commonly used methods, degassing is simply achieved by bubbling air through the water sample. Typically, a 130 ml sample is poured into a glass bubbler tube, taking care that degassing is kept to a minimum by ensuring a slow smooth flow, and the tube stoppered. A hand pump is then used to evacuate a sealed scintillation cell to a standard vacuum. Air is drawn through the sample as the vacuum is released, liberating ^{222}Rn and transporting it into the cell. A regulator valve between the bubbler tube and the scintillation cell ensures a standard degassing time of 3 min. A drying compound is also usually present to prevent moisture from being carried into the scintillation cell (Fig. 9.12).

After the ^{222}Rn has been passed to the scintillation cell, the cell is transferred to a detector–photomultiplier system for counting the α-particle activity. Previously, of course, any background count associated with the cell itself must have been determined. Once in the detector, the cell is left for 5 min to allow equilibration before a 5 min count is carried out. The results of the 5 min count are usually converted to counts per minute. When the count is completed, the degassing apparatus and the scintillation cell must be thoroughly flushed through with fresh air to remove the ^{222}Rn and any daughter products. Several flushings are usually needed. If the count rate has been high (of the order of several hundred counts per minute), the background produced may be so great, even after many flushings, that the cell remains inoperative for some time. Cells should, therefore, be numbered and used sequentially until the work is completed, putting to one side cells which have unacceptable background counts. If the analyst has some idea of which samples are expected to give high values and which will give low values, it is sensible to analyse the low activity samples first.

There is no fixed time limit before which samples must be analysed but, clearly, with a half-life of less than 4 days, too long a delay between collection and analysis will increase the percentage analytical error. Ideally, samples should be analysed within 24 h of collection. Because the solubility of ^{222}Rn is temperature dependent, analysis should be carried out under reasonably uniform temperature conditions. Cells should be checked from time to time to ensure that they do not leak. The build-up of daughter nuclides can be checked by recording the background levels of activity of a cell throughout its lifetime.

Although the adoption of standard procedures of this type allows comparison of the results from different samples and the recognition of anomalous and background ^{222}Rn concentrations, standardisation of the analytical

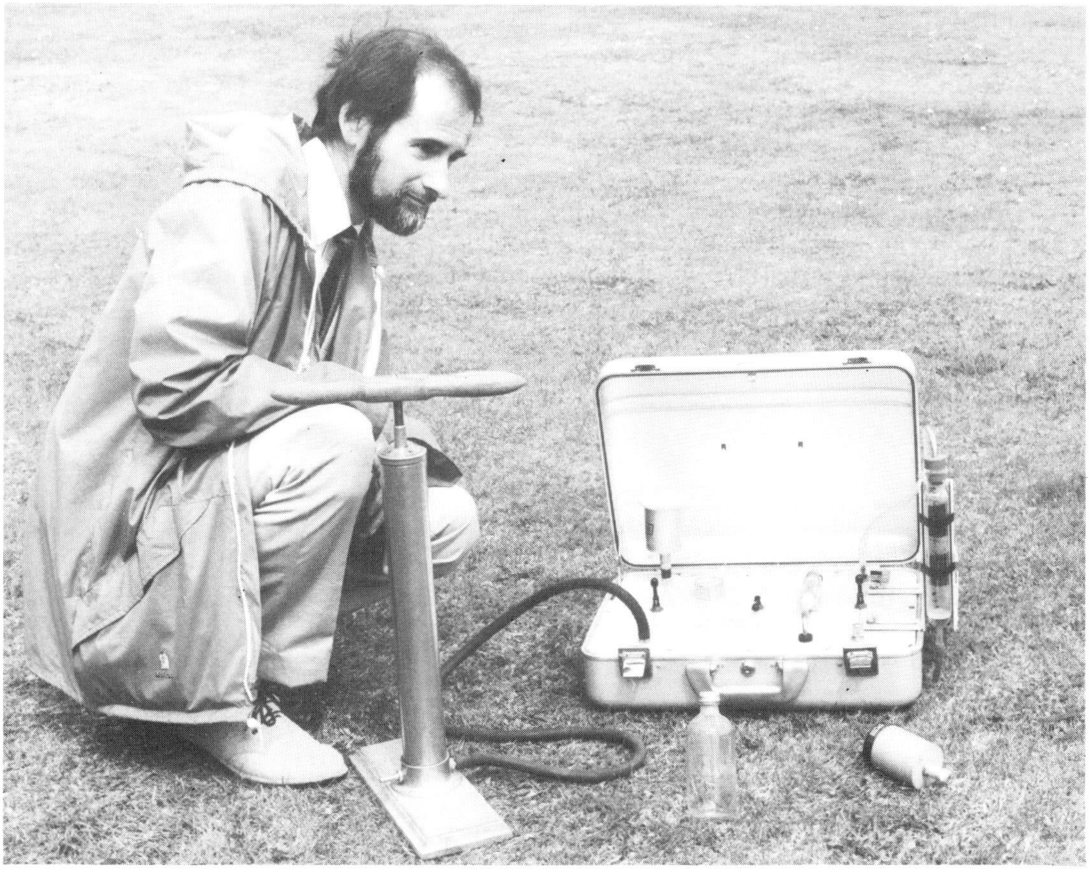

Fig. 9.12 —EDA Electronics RDU200 water-degassing unit. Mounted on the degassing unit is a sealed scintillation cell (the glass window is covered with a protective plastic cap). A vacuum pump is attached at the lower left. After degassing of the water sample the scintillation cell is placed in the RD200 (Fig. 9.11) for counting.

system is necessary for determination of ^{222}Rn activities. The activity of ^{222}Rn in water samples is not only of interest for correlation with data obtained by other procedures but is also of importance in determining the ^{222}Rn content of drinking water, as high levels of ingested ^{222}Rn are detrimental to health. This is considered further in Chapter 12.

The standardisation procedure is quite straightforward. Solutions containing ^{226}Ra of known activity may be obtained from radiochemical centres. These solutions are made up to a large volume (about 1 l) with 0.5 M HCl, and about 5 ml of this, in turn, is made up to another 1 l with the same acid. This second solution forms a batch of high activity from which other more dilute solutions can be made up. After preparation, the solutions must be left undisturbed while the growth of ^{222}Rn approaches equilibrium with ^{226}Ra. Theoretically, 99% equilibrium will be achieved within 25.4 days. The solutions can then be analysed using the same procedure as that employed for water samples. By analysing a number of solutions of different activity, a correlation graph can be constructed which links count rate to activity. This is usually a linear relationship over the range of activities encountered in natural waters, and a typical correlation may be of the order of 1 count min^{-1} to 10 pCi l^{-1}.

Although the unit is now obsolete, the 'pCi l^{-1}' is still often used in the presentation of activity data. The modern equivalent is the 'Bq l^{-1}', where 1 pCi l^{-1}=0.37 Bq l^{-1}.

Interpretation

As discussed earlier in this chapter, interpretation of ^{222}Rn activity data is fraught with difficulty. Qualitative assessments can be made but are of limited use. Quantitative interpretations generally give unreliable results because of the complex nature of the factors affecting ^{222}Rn concentrations. To overcome these limitations to some extent, a semiquantitative interpretation method has been developed for ^{222}Rn activity distributions based upon surface water surveys (Durrance, 1978). This method works on the basis that, although there are many factors which can act to reduce the ^{222}Rn activity of a sample, only the influx of ^{222}Rn, or its immediate parent, will give rise to high activities. If an area of uniformly high ^{222}Rn input is considered, the results of an activity survey in which sample points have an irregular distribution will show an apparently random pattern of high and low values, the high values being the true indicators and the low values merely being the result of factors such as turbulent degassing immediately above the sample point. It may be inferred, therefore, that only the high values have significance. Instead of constructing contour lines of equal value as a first stage in the interpretation of these data, a better approach is simply to draw exclusion isolines. These are lines of a given activity value constructed so that they enclose all data points that have values equal to or greater than the value of the isoline. Areas of consistently high activity within such a plot may then be taken as geologically meaningful (Fig. 9.13).

Although surveys of ^{222}Rn in stream water have been carried out in areas of known U mineralisation, it is interesting that, when the results are interpretated using the exclusion isoline method, rather than showing the distribution of U deposits, the ^{222}Rn distribution pattern usually reflects the major structural alignment of the area, particularly fracture zones. It appears, therefore, that ^{222}Rn exclusion isolines can be used to give a good indication of groundwater movement by fracture flow. Moreover, where hydrothermal circulation of groundwater is taking place, the distribution of ascending and descending limbs of convection cells can be delineated using the ^{222}Rn distribution (Durrance and Heath, 1985). Obviously, however, the level of ^{222}Rn activity that is recorded will depend upon both the effectiveness of water transport and the U concentrations within the rocks of the area. Resolution of the effects of these two interacting variables can only be made using other data on U distribution derived by γ-ray spectrometry or geochemical methods.

9.4 HELIUM METHODS

9.4.1 He production and migration

There are two naturally occurring isotopes of He: ^{4}He (the most abundant) and ^{3}He. These isotopes may be either primordial or radiogenic in origin. Most ^{4}He is radiogenic, formed by the α-decay of naturally occurring radionuclides in the Earth. As an α-particle is a He nucleus, it can, in fact, be considered to be a He^{2+} ion. Acquisition of extranuclear electrons produces He atoms. The main sources of ^{4}He are the ^{238}U and ^{232}Th decay series: eight He atoms are formed during the decay of ^{238}U to ^{206}Pb, and six atoms during the decay of ^{232}Th to ^{208}Pb. Because of their relatively low abundance, other naturally occurring α-particle emitters can effectively be considered to make a negligible contribution today. However, during the very early history of the Earth, decay of short-lived superheavy nuclides may well have made a substantial contribution to the initial ^{4}He concentration. Because it is an inert gas, ^{4}He generated within the Earth does not become incorporated into stable compounds and so migrates to the atmosphere. However, with its low escape velocity, ^{4}He is not retained even within the atmosphere and is continuously lost to space. The atmospheric concentration,

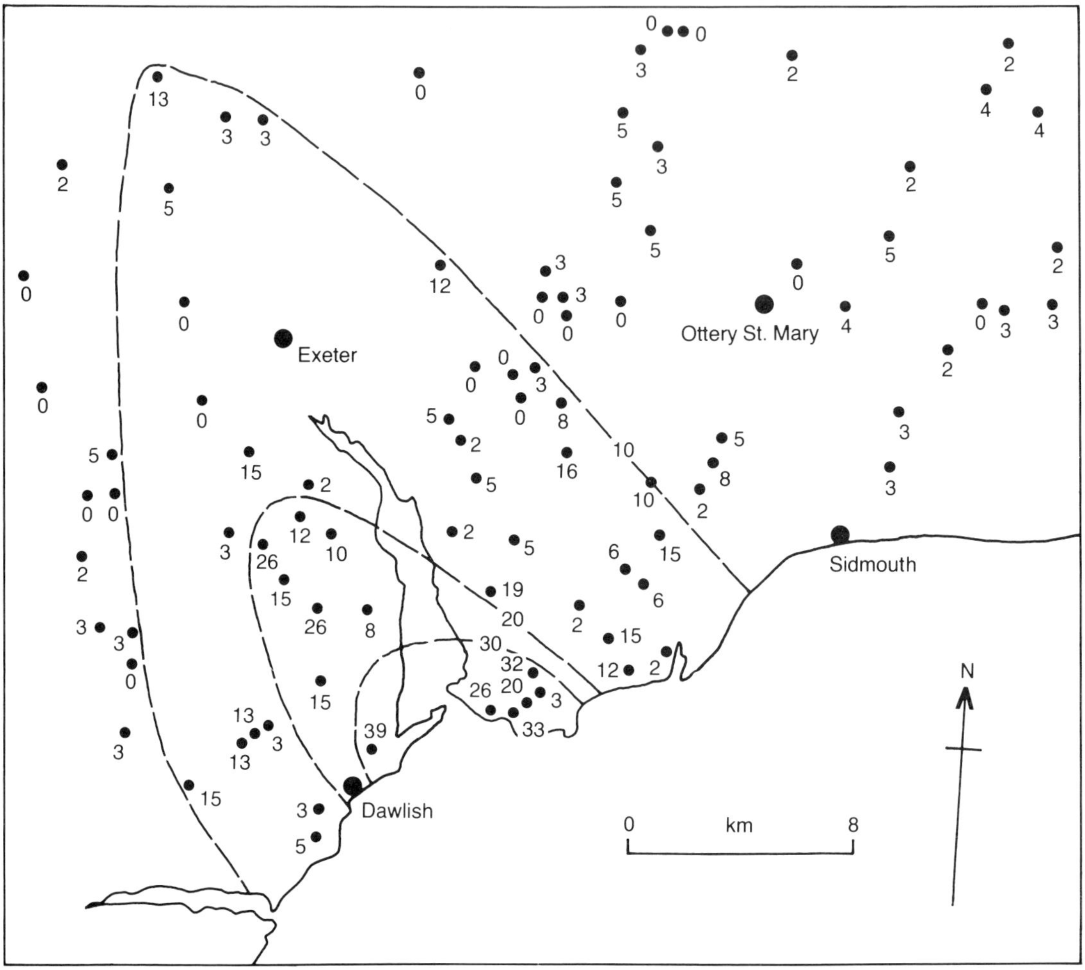

Fig. 9.13 — ^{222}Rn distribution shown by exclusion isolines (counts per minute) in the stream waters of southeast Devon. 1 cpm (count min^{-1}) is equivalent to 0.3 Bq l^{-1}. The northwest–southwest trend of the isolines reflects the direction of Tertiary tear faults in the area. (Durrance, 1978, reproduced with permission.)

therefore, represents an equilibrium between production and loss.

On the assumption of average U and Th abundances of 4 ppm and 12 ppm, respectively, the annual production of ^4He in the crust today can be calculated as 8.2×10^{-13} cm^3 g^{-1} (Dyck, 1976). Because of the greater abundance of U and Th when the Earth first developed a continental crust, the production about 4000 Ma ago would have been of the order of 31×10^{-13} cm^3 g^{-1}. On this basis, the ^4He content of average continental crust should be about 9×10^{-3} cm^3 g^{-1}, but considerably less than this is actually found.

The low values arise because of the generally easy loss of He by diffusion out of the minerals in which it is formed and, especially if it enters fractures, the great mobility with which it migrates to the atmosphere. In detail, however, the loss of He by diffusion depends to a large extent on the nature of the material concerned. The controls are similar to those governing Rn diffusion, but the rate of loss is much greater than that of Rn (Fig. 9.14).

On average, mantle-derived igneous minerals (in nodules in kimberlites) have He contents of only up to 2×10^{-5} cm^3 g^{-1}, while minerals in crustal igneous rocks normally contain less than 1×10^{-9} cm^3 g^{-1} of the gas. Sedimentary rocks at shallow levels in the crust, however, can have very high He concentrations, and values of a few hundred volume parts per million are normal. Where the concentrations reach several volume per cent, they may even be exploited as commercial sources of He. Interestingly, very high He levels are usually associated with oil or gas reservoirs. The He content of the atmosphere is generally about 5.24 ppm; this is fairly constant because of the rapid mixing by the wind and by diffusion, despite significant variation in the degassing rate of the Earth from one place to another. The He content of sea water is about 0.04 ppm (Mamyrin and Tolstikhin, 1984).

The isotopic ^{3}He/^{4}He ratio is generally very low. In the atmosphere, its value is 1.4×10^{-6}. In deep oceanic water over constructive margins, however, not only is the total He content higher than normal but also the ^{43}He/ He ratio may be two orders of magnitude greater than the atmospheric value. The additional He flux in these areas must have a ^{3}He/^{4}He ratio of 1.6×10^{-5} to account for the observed difference, and it has been suggested that this is of mantle origin. Interestingly, mantle-derived igneous rocks in oceanic areas also have a ^{3}He/^{4}He ratio of about 10^{-5}, and so the high sea water ratio in these areas may simply result from direct degassing of submarine volcanic sources. The evolution of the ^{3}He/^{4}He ratio from a theoretical value of 10^{-4} at the time of the Earth's origin to today's value of 10^{-5} is consistent with the production of radiogenic ^{4}He.

Although ^{3}He is mainly of primordial origin, it should be noted that there is a significant radiogenic component derived from the decay of ^{3}H which is produced from Li by the following reactions:

$$^6\text{Li(n)}\rightarrow{}^3\text{H}+{}^4\text{He},$$

$$^7\text{Li}(\alpha)\rightarrow{}^3\text{H}+2\,{}^4\text{He},$$

which also result in the formation of ^{4}He. Neutron production in the Earth was discussed

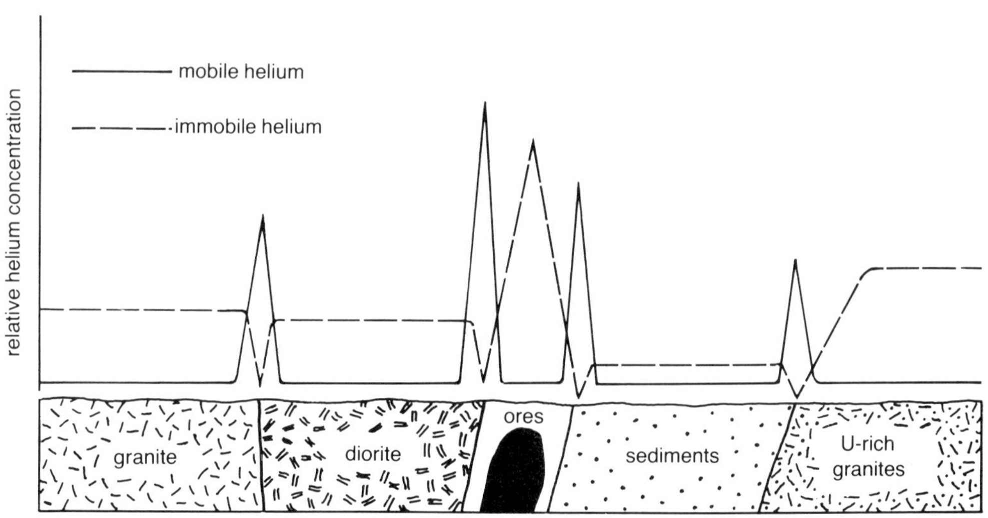

Fig. 9.14 — ^{4}He production over different rock types (immobile ^{4}He) contrasted with ^{4}He anomalies related to fracture and lithological boundaries (mobile ^{4}He). (Eremeev *et al.*, 1973, redrawn with permission.)

in Chapter 5. Because of the lithophile character of Li, ^{238}U (a source of neutrons and α-particles) and ^{232}Th (a source of α-particles), ^{3}H production is greatest in continental crust. ^{3}H has a half-life of 12.35 years, decaying by β^{-}-particle emission to ^{3}He:

$$^{3}H \rightarrow {}^{3}He + \beta^{-}.$$

In the atmosphere, ^{3}H is also generated by cosmic ray bombardment, and a significant proportion of ^{3}He is produced by this means. He abundance values and $^{3}He/^{4}He$ ratios can thus yield valuable information on the evolution of the crust and mantle and, in terms of mantle out-gassing, give an indication of the intensity of currently active processes (Hilton *et al.*, 1985).

Because of the production of ^{4}He during the decay of U, ^{4}He surveys may give useful information on the location of possible economically valuable U deposits (Butt and Gole, 1984). In fact, 1 g of ^{238}U and its decay products in equilibrium will annually produce 1.17×10^{-7} ml of ^{4}He (at standard temperature and pressure). ^{4}He is thus similar to ^{222}Rn in its generation. It is also similar in that migration by diffusion or by mechanical transport will cause a dispersion halo around the source. Distortion of this halo in the direction of easy transport will, in turn, be just as apparent for ^{4}He as it is for ^{222}Rn. Unlike ^{222}Rn, however, ^{4}He is stable, and an anomalous concentration can occur at some distance from the source. Some explorationists view this as an advantage of ^{4}He over ^{222}Rn, in that the concentration of ^{4}He will continue to build up as long as an intact trap is present. Certainly, if the trap is located close to the source, very strong ^{4}He anomalies may develop which can be readily related to the U deposit. Nevertheless, the high diffusivity of ^{4}He means that it is likely to encounter a free-flowing transport path before trapping, and there need be no readily observable spatial association between source and anomaly. Therefore, the problem of interpreting the results of ^{4}He surveys may be even more profound than those associated with ^{222}Rn.

9.4.2 Sample collection and problems of interperetation

In ^{4}He surveys, sample collection may be from soil gas or from groundwaters or surface waters. Soil gas ^{4}He is usually collected in a similar way to ^{222}Rn, in that a probe is hammered into the ground, a central rod withdrawn and the top sealed. Access to the hollow rod is, in this case, by means of a rubber septum. At each site, the hollow probe is first purged by withdrawing and discarding a gas sample of slightly greater volume than that of the probe itself. The gas within the probe is then left to equilibrate for some pre-determined time (usually a few minutes) before a sample is extracted using a glass or plastic syringe with the needle inserted through the septum. Because the ^{4}He in the sample can be lost or contaminated by diffusion of atmospheric ^{4}He through the walls of the syringe, analysis must be carried out very soon after sample collection (within 24 h) to avoid significant reduction in the size of any anomaly present. Here, it is important to note that soil ^{4}He values may be either greater or less than atmospheric concentrations, and variations in anomalies of either sign can be significant in interpreting migration pathways in different environments.

Where analysis of samples cannot take place soon after collection (in surveys of remote areas, for example), one approach to reducing diffusion losses has been to contain soil gas samples within small-diameter Cu tubes. These may be used in two ways. In the first method, a Cu tube merely replaces the normal body of the syringe and is used in combination with the soil probe as discussed above. Long-term storage of the sample is achieved simply by sealing the ends of the tube by pressure contact. Alternatively, a tube may be sealed at the top and inserted vertically into the ground for direct sampling. In this case, the lower part of the tube is perforated so that soil gas may diffuse easily into it and become trapped. The tube is left in the ground for an appropriate length of time (which may be several days) and then removed and the unperforated upper section sealed. Sealed tubes may be stored for a week or more before analysis. The other way of overcoming the problems caused by delay between sample collection and analysis is, of course, to have a mobile analytical system.

A completely different method of soil gas sampling involves the collection of actual soil samples. A sample is obtained by augering to a depth of about 1 m, and the soil (plus, of course, some air) placed into a container which is then sealed. After sufficient time has elapsed for equilibration, the free gas in the container is analysed. The ^{4}He content must then be calculated from known parameters of soil mass, density, porosity, moisture content, free gas volume and concentrations of other gases.

The results obtained following any of these sampling procedures generally tend to be somewhat ambiguous. Even in areas of known U mineralisation, the soil gas concentrations are only of the order of 5.4 ppm, which is only a few per cent greater than the atmospheric concentration. Not surprisingly, therefore, in some cases, U mineralisation has been found not to produce any significant He anomaly while, in other cases, although a He anomaly may be produced, it is displaced laterally from the U occurrence. Another problem that has been encountered is that, if surveys are repeated at different times of the year (under different meteorological conditions), the results obtained may be very different. In examples where He anomalies have been found in summer, they have not been found in winter and vice versa. It is clear that, in some areas, summer aridity which has led to the drying and cracking of the soil has caused rapid loss of the soil gas, penetration of atmospheric air and the destruction of the He anomaly. Moreover, in dry soil conditions it is actually very difficult to ensure that soil gas probes are not in direct contact with the atmosphere via such cracks. Conversely, in winter, loss of anomaly may occur if conditions are very wet and cause He to be flushed out of the soil. The optimum conditions thus appear to exist where a surface soil layer is sufficiently damp to act as a coherent cover; that is, neither so dry that it cracks nor so wet that it washes the He out of practical sampling depths. Interestingly, it has also been noted that strong wind may cause the destruction of He anomalies. However, frost conditions can provide a suitable barrier to He exchange with the atmosphere, and surveys carried out while the surface is frozen may be more successful than those at other times.

Becaue of the high rate of loss of He by diffusion, and the impracticality of *in situ* measurement systems for the gas, there is no satisfactory approach to the long-term monitoring of He in soil gas. It is not possible, therefore, to smooth out variable meteorological effects over periods of several months or a year, in ways that are comparable with the integrating methods of ^{222}Rn collection.

As with all other exploration techniques, it must be realised that the absence of an anomaly does not imply the absence of a ^{4}He source. Also, surveys conducted over a long period, during which there are variable meteorological conditions, may show variations in the results which are due to those changes rather than to geological factors. Obtaining reliable soil gas samples is especially difficult and, unless the surface layers of the soil have good coherence, there will possibly be some direct linkage between the sampled point and the atmosphere. This will vary from site to site and can again cause variations in the results which do not reflect any fundamental geological change. Except where samples are collected over a short period of time and interpreted in a relative or comparative way, it is difficult to correct for such problems. Perhaps only by obtaining samples from a depth of 4–5 m can these time-dependent effects be eliminated. Inserting soil gas probes and collecting soil samples from this depth, however, are both time consuming and costly and may, indeed, be impractical in some conditions.

In unfractured ground, a He halo will extend upwards and outwards from the source, and direct interpretation of the pattern of He distribution in the soil is possible. However, if a fracture passes close to the source, the He anomaly will be found along the fracture trace. Indeed, because the fracture is unlikely to be an open parallel-walled slot, the migrating He will be directed along specific channels. As a result, localised He anomalies may develop, which could be misinterpreted in terms of source position (Eremeev *et al.*, 1973).

Although He surveys of a reconnaissance type have been carried out, and anomalous areas defined which have been suggested as indicating potential U deposits, the uncertainties in interpretation of soil gas distributions make any expensive follow-up hazardous. He soil gas surveys, like those for Rn, are more safely considered as a means of fracture mapping. However, the interesting suggestion has been made that combined ^{4}H–^{222}Rn or ^{4}He–CO_2+O_2 soil gas surveys may well differentiate between barren fracture systems and those intersecting U-bearing deposits. In this approach it is argued that, even though channel flow will always cause localised anomalies, the limited migration of ^{222}Rn may allow background effects to be assessed. If this is possible, changes in the absolute ^{222}Rn and ^{4}He concentrations or in $^{222}Rn/^{4}He$ ratios could indicate nearby U mineralisation. Similarly, CO_2+O_2 anomalies may develop over mineralised fractures (Fig. 9.15).

Groundwaters or surface waters may be sampled for He in preference to soil gas. In most circumstances, the He content of surface waters is close to 0.044 ppm at standard temperature and pressure, but it can be greater in samples collected from great depth in the oceans or lakes. In shallow surface waters, the greatest concentration is of the order of 0.15 ppm. Groundwaters, however, can contain considerable concentrations of He and, where these groundwaters discharge into surface streams, the He level of the latter is enhanced. In groundwaters, the He concentration may exceed 1 ppm.

Sampling of groundwaters for He is difficult. If samples are brought from depth, care must be taken to ensure that degassing of the He does not occur with the release of pressure as the sample is brought up to the surface. A variety of suitable self-sealing sampling systems have been developed which are designed to prevent He loss and to avoid mixing of waters or the introduction of air.

In water samples, the He itself is obtained by degassing in a manner similar to that described for Rn, but with the gas passing into an ordinary container from which it may be sampled directly via a septum and syringe. Clearly, if there is any delay between sample collection and analysis, water samples must be protected from loss of He by diffusion and, if transport of samples is involved, by agitated degassing. Here too, sealed Cu tubes may be used, with care taken to ensure that no air bubble is trapped inside.

An alternative approach to degassing water samples is to equilibrate the samples with a known volume of atmospheric air and then to calculate the He content of the water from the He concentration in that volume. Yet another method has been simply to suspend ping-pong balls in the water and allow sufficient time for diffusion of the He through their walls to produce equilibrium with the air inside. The air in the balls may then be sampled simply by inserting the needle of a syringe through the wall.

In lake water, the He concentration generally increases with depth. This is probably due to agitation of the surface by wave action and consequent degassing. At depth, however, lakes appear to collect He released from the lake floor over periods of several months and may, therefore, provide a suitable means of carrying out integrated regional surveys. In cold winter conditions, when lakes have an ice cover, He concentrations become enhanced, even at shallow depths.

Rivers and streams present the same problems of degassing of He as were discussed

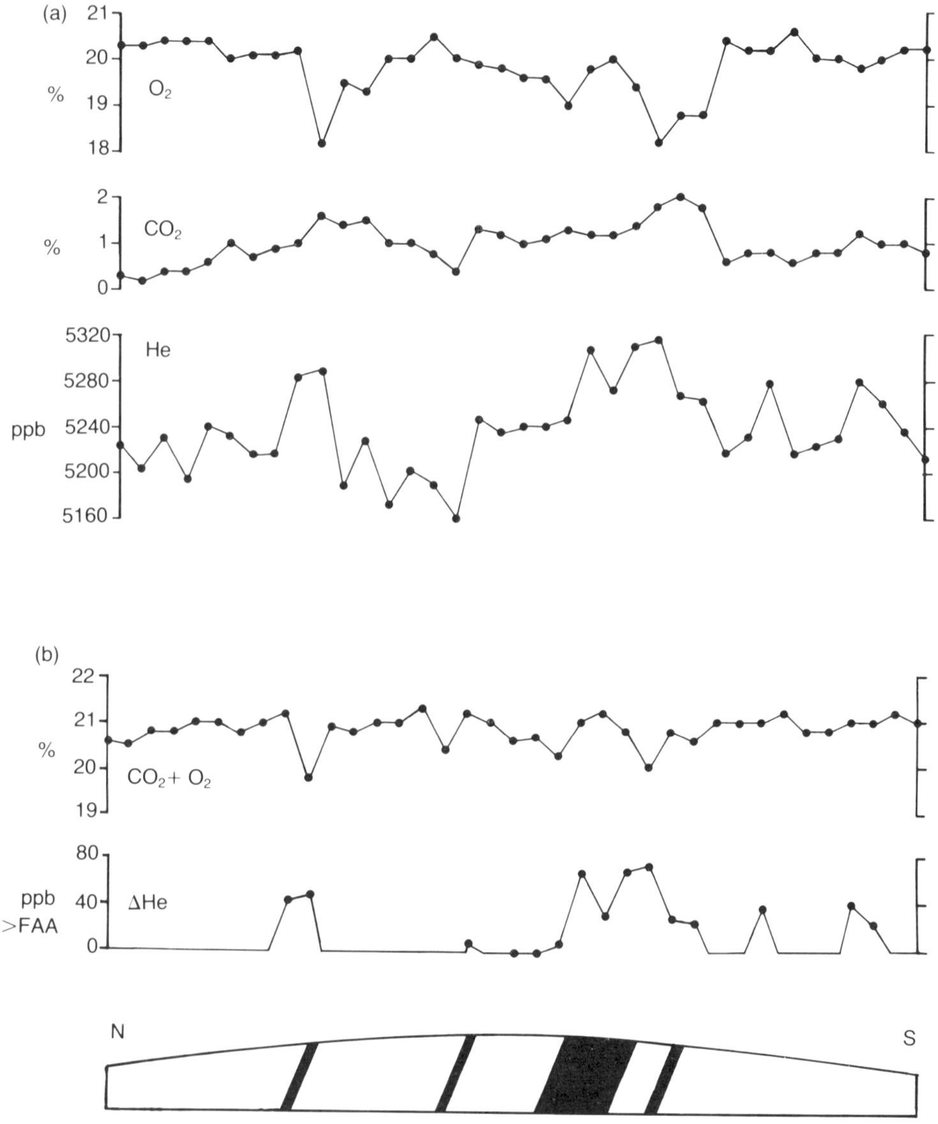

Fig. 9.15 — Soil gas results over stratiform mineralisation on the northern flank of the Dartmoor granite: (a) raw data for O_2, CO_2 and ^4He; (b) processed data for CO_2+O_2 and ^4He anomalies greater than atmospheric level (5240 ppb). The mineralised zones are shown solid. (After Gregory and Durrance, 1985b.)

for Rn, with the added factor of more rapid loss due to diffusion. Generally, He values above atmospheric levels are only likely to be found where groundwaters discharge, and mapping of He, like Rn, in surface waters can be considered an effective means of investigating groundwater flow regimes. High concentrations of He in groundwater can also be interpreted in terms of the structural characteristics of an area in the same way as soil gas values. Thus the presence of major fractures may allow He to migrate from depth and then to become trapped in shallow groundwaters. The direct association of high He concentrations with U mineralisation seems, from results to date, to be very tenuous.

Generally, groundwater samples are collected from boreholes, which usually show a distinct increase in He concentrations with increasing depth. Part of the reason for this probably lies in the dilution of near-surface groundwater by fresher meteoric water. In aquifers in sedimentary formations, the He content usually increases in the down-dip direction. This has been interpreted in some areas as indicating an increasing age of the groundwater, the He having equilibrated with the adjacent rock matrix.

There have been a number of attempts to calculate the age of groundwaters in sedimentary rock aquifers from their ^{4}He content. These are based upon the production rate of ^{4}He from the known U and Th content of the matrix, the porosity and the ^{4}He release rate of the rock (Fig. 9.16). Generally, where cross-correlation has been possible, the ages determined have been found to be considerably greater than those obtained by other methods. This is usually explained as being due to enhancement of ^{4}He in the groundwater by leakage through faults and fractures from other parts of the sedimentary succession. The implied assumption that groundwaters represent closed systems is, therefore, incorrect.

In thermal waters, ^{4}He contents are often very high indeed. This partly reflects the long residence time usually associated with such waters but is also related to the occurrence of major fracture sysems and may even partly reflect above-average U and Th levels. Where thermal waters from depth rise along fractures which intersect aquifers at shallow levels, charging of the aquifer with thermal water can result in high ^{4}He concentrations which extend for considerable distances away from the feeder.

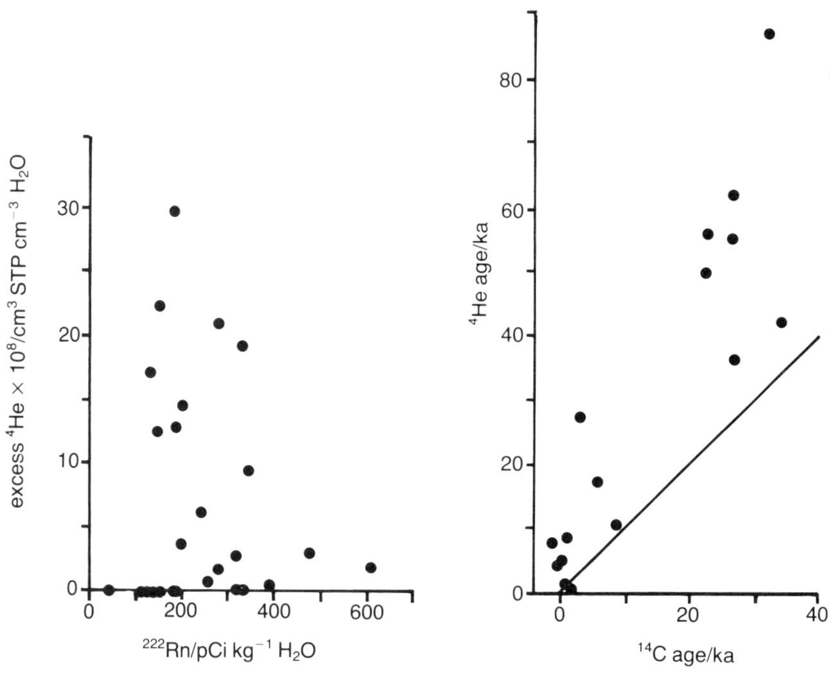

Fig. 9.16 — Excess ^{4}He correlated with ^{222}Rn and excess ^{4}He age correlated with ^{14}C age for groundwaters from central England. (After Andrews and Lee, 1979; Gascoyne, 1982.)

Thus, ^{4}He surveys of groundwater are potentially useful for obtaining information which will help to identify and locate areas of active hydrothermal circulation.

Although not a source environment for ^{4}He, oil and gas reservoirs are usually rich in He and can even be commercial sources of the gas. As noted before, in some areas the He concentration can reach several per cent. This association probably simply reflects the fact that both He and petroleum naturally migrate upwards through the crust, with accumulations occurring wherever that upward movement is impeded. Thus, conditions which favour the development of petroleum reservoirs also serve as He traps. The average He content of such reservoirs varies with the geological setting. Reservoirs on ancient platforms contain about 1250 ppm He, while in young platforms the average He content is around 230 ppm. In Tertiary–Recent fold belts, the value is only 20 ppm. However, these averages mask considerable ranges of values, and some overlap is present between the young and ancient platforms. Because He will usually leak to some degree from petroleum reservoirs, He surveys can be used to locate oil and gas traps.

9.4.3 Analysis

He analysis is carried out using a mass spectrometer (Reimer, 1976). For ^{3}He/^{4}He ratios, the instrument must be of high quality in order to separate these close charge/mass ratios. Unfortunately, such instruments do not give absolute abundances with great accuracy. Measurement of ^{4}He for survey work requires good calibration in terms of absolute concentration and of sensitivity and precision. Modified leak detectors are usually employed for this purpose. These are commercially available instruments which consist of a mass spectrometer specifically tuned to ^{4}He. They are used in industry to test for leaks in sealed pressure vessels. The vessel is filled with ^{4}He under pressure, and the leak detector then 'sniffs' for any ^{4}He leakage to the surrounding atmosphere. Such instruments require modification to the inlet system so that they can operate under either constant-volume or constant-pressure conditions. Individual workers often express preference for either of these, although in practice there appears to be little advantage of one over the other.

One of the most common analysers in use today is based upon a Du Pont leak detector and utilises a constant-pressure inlet (Fig. 9.17). In this system, about 5 ml of the sample is injected through a septum into a piston chamber which serves as the reservoir to the analyser. Before entering the spectrometer, however, the sample is passed through a trap consisting of activated charcoal in a U-tube immersed in a liquid N bath. The trap serves to remove all components in the sample except H, Ne and, of course, He. Thus the He in the sample is pre-concentrated. The sample is then pumped to the mass spectrometer (Fig. 9.18). A variable leak valve is adjusted to give a suitable flow rate which is selected according to the desired recording period; the optimum flow rate and recording period must be determined experimentally.

Generally, each sample for analysis is ratioed to water-saturated laboratory air before and after each measurement. This is necessary because the samples themselves are generally water-saturated. Duplicate samples are always analysed to reduce errors. If considerable differences emerge from these duplicates, resampling and new analyses may be required, or the results discarded. Overall calibration of the system is carried out using commercially available ^{4}He standards. These are suitable for calibration over the entire working range required. The detection limit is less than 1 ppm ^{4}He and the sensitivity of the system is about ± 10 ppb. Precision is also about ± 10 ppb.

9.5 RADIOGENIC GASES AND EARTHQUAKE PREDICTION

One very interesting application of radiogenic gas surveys is in areas of earthquake activity; it is possible that monitoring of Rn and He may allow successful prediction of earthquakes

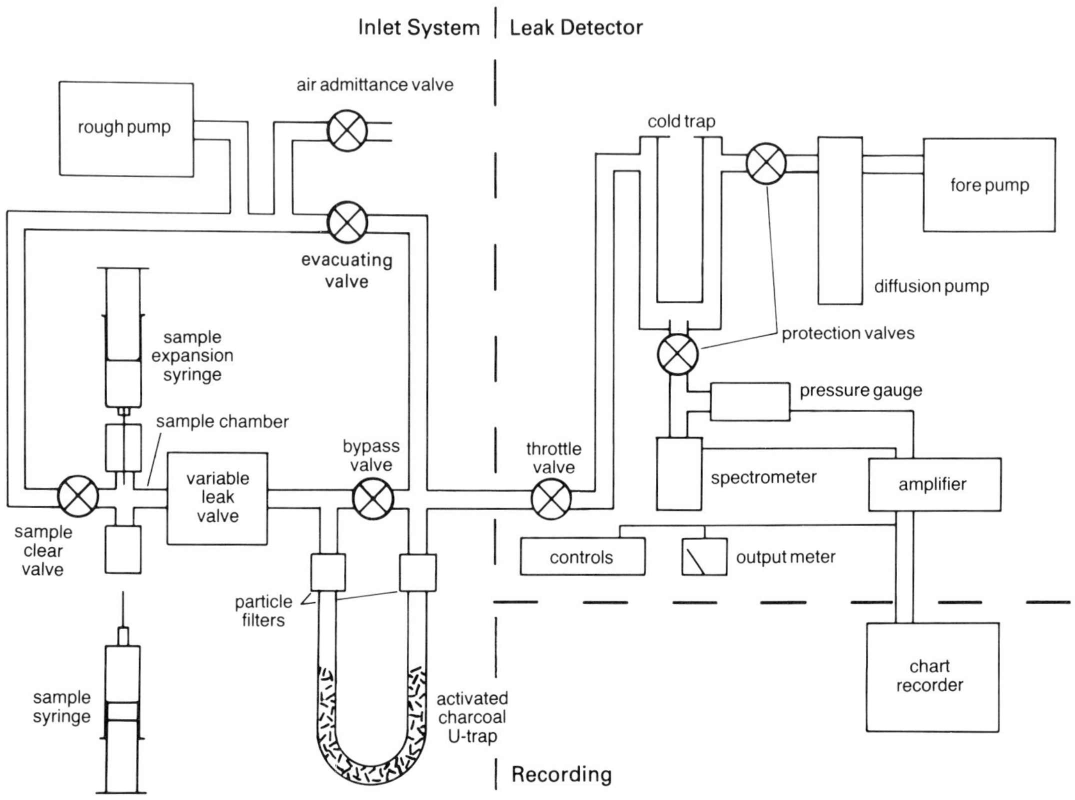

Fig. 9.17 — Diagrammatic representation of ^{4}He analyser. (Gregory and Durrance, 1985a, reproduced with permission.)

(Reimer, 1979). To understand the rationale, it must be appreciated that, although an earthquake is a short-term phenomenon, it is the climax of long-term events. In the precursor phase of the fault movement that generates an earthquake, microfracturing of the ground takes place as stress builds up. The rock affected may be considered to contain a large number of randomly orientated cracks (pores), of which those with a critical orientation will respond to low applied stress by propagating as their walls fail; greater stress is needed for propagation of pores of other orientations. The intensity of microfracturing thus increases as the rate of stress increase grows; it can be envisaged as an initial, slow and relatively quite roll on a drum which progressively increases in volume and speed as a crescendo until a climax is reached (the earthquake event) when resistance to fracture of the whole rock mass has been overcome. The main movement will, of course, cause much cracking of the ground as well, and any history of fore-shocks and after-shocks will also give rise to a complex system of crack-forming activity.

When cracks are propagating in the pre-cursor phase, the overall permeability of the ground will be increased and will aid the transport of He and Rn through the rock, both by providing additional permeable pathways and by extending the volume of cracks. This second factor increases the grain–void contact area and enhances the release of radiogenic gases from their parent minerals. Thus,

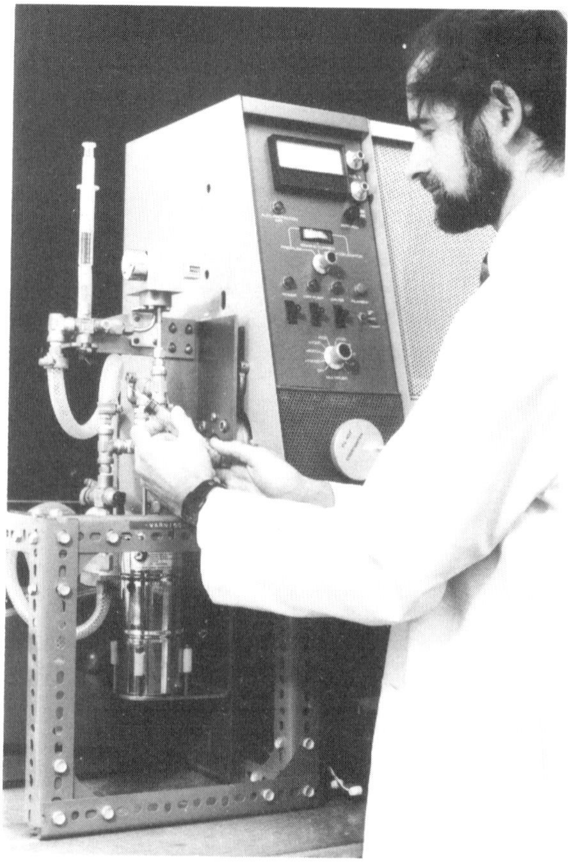

Fig. 9.18 — ^{4}He mass spectrometer based upon a Du Pont 120SSA leak detector.

degassing of the rock as a whole will become progressively more effective as microfracturing proceeds. This can be monitored by observing the changes in the concentrations of these gases in groundwaters or in the soil (Bashorin, 1980). Generally, however, groundwaters have been studied in preference to soil gases because they are less susceptible to variations brought about by short-term meteorological changes. Sources of groundwater that can be sampled include wells, boreholes and springs; the deeper the groundwater source, the smaller is the variation due to surface changes (such as rainfall) and the smaller is the interference with the seismically produced gas signal.

^{222}Rn has received more attention than ^{4}He in these studies because its detection and measurement are comparatively simple. In fact, methods can be used for ^{222}Rn which require no great scientific training or the use of sophisticated equipment and so have the potential for widespread use in the developing countries. Some of the earliest results using ^{222}Rn have been reported in connection with the Tashkent earthquake of 1966, where the ^{222}Rn content in a well water showed a fairly steady rise over the first 5 years of that decade, and rapidly increased 1 year before the earthquake itself (Fig. 9.19). A sudden acceleration in the rate of increase in ^{222}Rn appears, therefore, to be a good warning indicator. After the main shock, ^{222}Rn concentrations showed

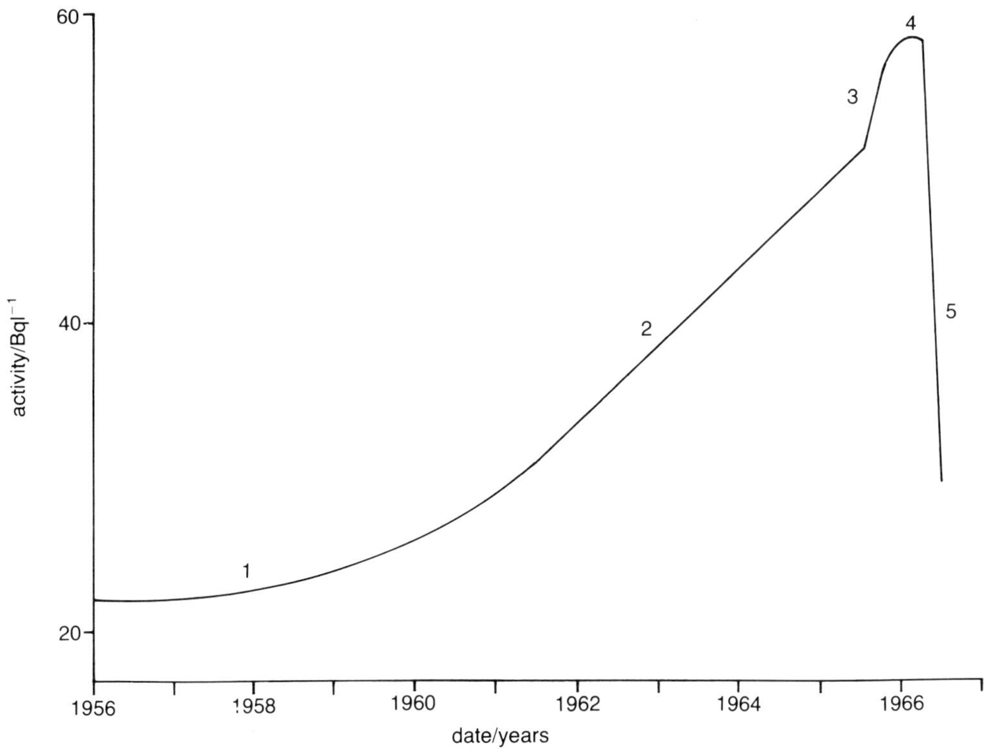

Fig. 9.19 — Variation in ^{222}Rn content of well water in the period leading up to the Tashkent earthquake of 1966; 1, steady state; 2, gradual increase due to microfracturing; 3, rapid acceleration of microfracturing; 4, new equilibrium achieved after intense microfracturing; 5, variation due to earthquake.

rapid rise and fall as new equilibrium conditions in the ground became established. The ^{222}Rn concentration at the time of the earthquake was about three times the previous level. On this occasion, ^{4}He levels increased twelve-fold, but in the Tashkent earthquake of 1975 the ^{4}He content only increased five-fold.

Clearly, the size of a ^{222}Rn signal will in part depend upon the presence of parent U-bearing minerals in the rocks affected by the microfracturing. If there are none of these, no radiogenic gases can be generated, regardless of the intensity of the fracturing. Even if U-bearing minerals are present in deeply buried rock units, the limited half-life of ^{222}Rn means that transport rates would have to be very high to allow significant quantities to reach the near-surface environment. Similarly, the very short half-life of ^{220}Rn makes the presence of Th-bearing minerals no help. However, the quantity of U-bearing minerals needed to produce a recognisable signal for earthquake prediction purposes does not have to be great; normal crustal abundances of U are usually adequate.

The requirement for near-surface source rocks is not as stringent for stable ^{4}He. Even if the ^{4}He source is at great depth, its upward transport will be increased by the microfracturing. This may explain why the amplitude of the ^{222}Rn signal is usually less than that of the ^{4}He signal produced in the earthquake precursor phase. It should be noted that this effect will also be observed if the microfracturing allows deeply seated groundwaters to mix progressively with shallow groundwaters.

The distance between the epicentre of the earthquake and the site at which a radiogenic gas signal can be recorded appears to be very variable and ranges to over 250 km. The size of the signal changes with distance from the epicentre, but is also affected by the magnitude of the earthquake. It has also been observed that, the greater the length of time for which the increase in gas concentration is maintained, the greater is the magnitude of the eventual earthquake. Much work, however, remains to be carried out on investigating these effects in a quantitative manner, and the theoretical basis needs refinement. In some cases, for example, a decrease in ^{4}He before an earthquake has been reported. Furthermore, lateral migration of radiogenic gases within the fracture system may cause complex redistribution to occur. This approach to earthquake prediction may also be applicable to the prediction of volcanic eruptions.

9.6 GEOCHEMICAL METHODS

The use of geochemical methods in the search for radioactive minerals may seem surprising, since they usually show radiation anomalies which are easily detected by the methods already considered (Grimbert, 1972). Furthermore, geochemical methods require the collection, preparation and analysis of samples, all of which are time consuming and expensive. However, the presence of only a thin cover of non-radioactive soil or rock may completely mask a radioactive source, making it undetectable in γ-ray surveys, while ^{222}Rn methods can produce results which are difficult to interpret on their own. Moreover, geochemical methods provide a direct approach in that they can concentrate on the element of interest (usually U) rather than on a decay product. Also, if early results indicate that other elements are also of interest, further analyses of samples can be made without the need to resurvey the ground.

Geochemical methods are applicable to surveys of rock, soil, groundwaters and surface waters, stream sediment and even biological material (Boyle, 1982). Where possible, samples are usually collected on a grid pattern, the dimensions of which are determined by the type of survey being undertaken; preliminary or reconnaissance investigations cover a large area and employ a widely spaced grid, while final detailed surveys may use a grid of only a few metres and cover a very small area. Deviation from a grid pattern may have to be accepted in areas where true grid sampling is impossible; for example, exposure of rocks may be so poor that only an irregular sampling programme is possible, taking advantage of all natural and man-made outcrops; where stream waters or sediments are sampled, the sampling pattern may be of dendritic form, controlled by the location of the main streams and their tributaries within a drainage basin; biogeochemical surveys may be restricted by the distribution of species of interest.

9.6.1 Water surveys

Water samples are usually collected in glass or polypropylene containers and are not usually treated in any way if analysis is to be carried out within a short period of time. If, however, there is any delay between collection and analysis which might allow growth of algal material within the sample, leading to the removal of, say, U from solution, acidification of the water is desirable. Some geochemists argue that acidification is good practice anyway, to avoid absorption by or adsorption onto the walls of the container, but this procedure prevents measurement of certain sample properties, such as alkalinity, specific conductivity, etc. The acid employed must also be selected with regard to any effect that it might have on the analytical procedure to be used. If water samples are very turbid, filtering is likely to be necessary. Suspended solids may be removed by millipore or nucleopore filters, with or without surfactant; the water and filtered material are then analysed separately. Although some geochemists favour field filtration, others consider that, if filtering is needed, it should be carried out in the laboratory.

When exploration for U deposits is carried

out in areas in which stream waters have very low U contents, a procedure may be used at each sample site to enhance the U concentration by means of cation–anion exchange resins (Parslow and Dwairi, 1977). In this technique, a standard large volume of water is passed through (or over) the resin, and the resin is then analysed. Many other elements besides U can, of course, also be determined by analysis of the resin. Finally, at each water sample site, the temperature, Eh, pH and aquatic biological activity should be noted. Also, for large-scale survey work extending over a long period of time, the monitoring of those factors which cause temporal variations in the water chemistry is required, so that results from one part of the survey area may be compared with those from other parts.

In this context, of particular importance are temperature and rainfall, but the effects of these are not usually straightforward. Temperature changes may lead to freezing or melting of water, thereby changing the flow of streams; rainfall, as well as affecting the flow volume, may wash large amounts of particulate matter into streams and displace water previously in contact with soil and rock. The changes thus produced in the chemistry of the streams can be very complex. Even over one dry–wet–dry cycle, the concentrations of different elements at a single site can vary significantly. Furthermore, the pattern of change varies from one site to another, even though the sites all lie within a single drainage basin and apparently experience the same meteorological conditions. In practice, the situation is considerably more involved because, even within small drainage basins, micrometeorological variations will occur because of changes in altitude and aspect. Although attempts to allow for such changes may be partly successful, as a general rule it is advisable to carry out the sampling programme as quickly as possible, drainage basin by drainage basin, and to try to plan the programme so that the work takes place during the season of most settled (yet acceptable) climatic conditions. Even then, the recent meteorological history of the area may have a profound effect.

9.6.2 Stream sediment surveys

Stream sediment surveys require more treatment than water surveys, both in terms of the collection of the samples and in the interpretation of the resulting data. The major components of stream sediments are as follows:

(1) Detrital rock and mineral fragments produced by the physical breakdown of rocks within the catchment.
(2) Secondary detrital material produced by chemical weathering of those rocks.
(3) Material precipitated from solution under favourable Eh–pH conditions.
(4) Exchangeable elements adsorbed onto detrital and precipitated components.
(5) Organic matter including detrital and *in situ* organic debris.

Because of the variable nature of these interacting components, the relationship between the compositions of stream sediment and bedrock can be very largely obscured. Stream sediment anomalies may, therefore, be displaced over large distances, and the effect of sorting of grain sizes, precipitation and adsorption may cause local variations not related to the character of the bedrock. In particular, precipitation of Fe and Mn oxides with the coprecipitation of trace elements may occur. This is especially likely to be encountered where acid groundwaters experience a change in pH on entering streams, or at the confluence of streams with different pH values. Moreover, the mineralogical location of elements in the bedrock will influence their abundance and behaviour in stream sediments; for example, those elements held in resistate accessory minerals will tend to occur in the detrital fraction, while those originally held in soluble phases may be removed completely. Subsequently, soluble phases are likely to be precipitated or adsorbed by oxides or organic matter and become incorporated into the sediment. The dispersion of such mobile elements may be extensive. The loose bonding of mobile elements will have facilitated their initial leaching, and so their presence may actually provide a good indication of nearby mineralisation. Because both Fe and Mn oxides scavenge stream waters for metal ions, they may be sampled specifically to aid in the location of mineral deposits. The concentration of a trace metal in these exchangeable sites readily equilibrates with its concentration in the stream water; these materials act like natural ion exchange resins, giving results which are proportional to the stream water chemistry.

The highest concentrations of trace metals are found in the finest sediment fraction, where the heavy minerals tend to be concentrated. However, the best indication of mineralisation is provided by intermediate grain sizes, ranging between the 80 and 200 μm mesh sizes for sieved samples; most stream sediment surveys actually involve analysis of the <80 μm or <100 μm mesh fractions. Samples are usually collected with a plastic scoop, taking care to avoid large pebble-sized material, twigs, etc., and put in a polythene bag. They then have to be dried (a sandbath is often used) and disaggregated. The sample can then be sieved and the appropriate size fration separated for analysis.

9.6.3 Soil surveys

Sampling

Selecting the horizon or horizons from which to collect soil samples is an important consideration. Different soil types develop in different ways and have different components contributing to their composition. Moreover, soils may develop directly on bedrock or they may be formed on overlying drift material, such as glacial deposits, alluvium, periglacial deposits, wind-blown sand, loess, etc. Clearly, where the soil is developed on material derived from elsewhere, the soil composition will not reflect the composition of the underlying bedrock.

The distribution of U and Th in soils is generally very complex. However, there is often a tendency for Th to be enriched in the B or illuvial horizon (the layer often composed of a mixture of inorganic fragments and organic matter, in which material leached from higher in the soil profile is precipitated), because of its association with ferric oxyhydroxides, while U occurs principally in the A or eluvial horizon (the layer rich in organic matter, which is often leached). The C horizon (formed by the breakdown of the underlying rock) tends to have U and Th contents similar to those of the coherent rock below, although in some cases downward-percolating water has produced U enrichment. In general, it is usually best to sample the B horizon but, in the delineation of soil anomalies, samples from either the A or the C horizon can often be found to give the best results. Unless other factors dictate the horizon to be sampled, it is, therefore, good practice to precede the main survey by a pilot survey to determine which horizon in the area of interest gives the best results. With drift deposits, only material collected close to rockhead is satisfactory.

The U content of soils is usually about 1 ppm but can be as high as 8 ppm. Th contents are usually about 5 ppm but can be as high as 50 ppm. The occurrence of both these elements within resistate minerals is possible, Th more commonly than U. U also occurs as organic complexes and adsorbed onto clays and oxides. In fact, the general distribution of U in soils is very similar to that in stream sediments. Sieving of samples to obtain the optimum grain size for delineating anomalies should, therefore, be carried out after drying. Soil samples are best obtained from the side-walls of pits dug specifically for the purpose. This, however, is a time-consuming and, therefore, costly exercise. Consequently, samples are normally obtained by augering (hand or power) or by overburden drilling. With these procedures, care must be taken to ensure that the samples are not contaminated by debris from higher in the hole. When samples have to be obtained from just above rockhead in areas of thick glacial deposits, the efficiency of surveys can be improved (thereby minimising cost) by carrying out initial shallow seismic or electrical resistivity surveys to determine overburden thickness (Boyle, 1982). These will show the form of the bedrock surface and can indicate the presence of buried valleys which might contain placer deposits. Even the structure of the bedrock, including fracture zones which might contain hypogene deposits, could be revealed. Detailed information about the overburden thickness will also allow an effective costing of the drilling programme to be made.

Interpretation of the results of soil surveys can be extremely difficult, depending upon the type of terrain concerned. It should always be borne in mind that transport of elements takes place in soils. This is mainly the result of movement of ground and surface waters, as already discussed, but is partly by downhill migration of solid soil material. In some circumstances, it is also possible for solid soil particles to be moved over considerable distances by the wind. These effects will vary from one soil type to another and according to meteorological circumstances. Groundwater movement is greater through open-textured sandy soils than through compact clayey soils, leading to a much greater degree of downhill displacement, and even to such efficient removal of U that any potential U anomaly is washed-out of near-surface horizons. Displacement of Th is less marked than that of U, but significant movement may be encountered. The assumption that a soil profile has developed entirely *in situ* is, therefore, not completely valid.

U removed in solution from one soil location may be precipitated or adsorbed elsewhere, particularly if soils with a high organic content are encountered, in which case U enrichment is likely to take place. Displacement of an anomaly may occur, therefore, simply because of variation in the soil type. Such migration also takes place within the soil profile itself, and so the distribution of elements with depth is not constant. U and Th concentrations may be highest in the A horizon, just because of its greater concentration of organic matter. However, the B horizon may have the highest levels if there has been adsorption onto or coprecipitation with Fe and Mn oxides and/or adsorption onto clay minerals. Variation in the B horizon composition will, therefore, also influence the position of any anomaly. It is very difficult to correct quantitatively for these effects. All that is possible is to record the soil characteristics and topographical setting of each site. As with surveys using water samples, a record of the meteorological conditions during the survey period should be kept. These data should then be integrated with the results of the soil survey when attempting, at least, to distinguish between false anomalies and those which truly indicate bedrock conditions. The problems caused by wind-blow dispersion of soil particles, both modern and associated with Pleistocene glaciation, are impossible to overcome. These loess deposits, where thick, are easily recognised, but in many areas the contribution of thin pockets of loess to pedological variation has only fairly recently been recognised as an important process.

Although most of the difficulties caused by migration of material can be reduced or eliminated if soil samples are taken as close to the bedrock as possible, with loess-derived soils even this is unsatisfactory. All that can be done is to recognise the presence of such soils from their textural and mineralogical features and to disregard any data from those sites.

While most soil survey work is carried out by chemical analysis, studies of clasts and mineralogy are also extremely important. The presence of rock fragments in soils can give a good indication of bedrock characteristics, and mapping of clast type and texture can provide information on directions of movement and the derivation of soil material. Examination of heavy and light mineral separates can also help to determine the distribution of elements within the soil. In the heavy assemblage, quantitative analysis of resistate mineral concentrates can be used to indicate bedrock variation. Generally, only B and C horizon samples should be used for this. The light mineral assemblage can be studied for secondary minerals and for the gangue minerals that are likely to be associated with a radioactive ore deposit. In particular, vein quartz which is haematised or darkened by radioactivity can be a useful indicator. Indeed, recognition of radioactive damage in any mineralogical component may prove a valuable indicator of nearby radioactive ore deposits.

Fertilizers

The influence of agricultural activity on the soil profile and composition are of great concern in soil surveys. Agricultural use of fertilizers is widespread on both pastoral and arable land. Even where soils show no deficiencies, maintenance dressing is often carried out to prevent gradual loss of productivity. Both organic and inorganic fertilizers are used. The most commonly used organic fertilizer is manure in semi-liquid form, but liquid slurry is also employed.

Inorganic fertilizers are principally used to provide an easy and rapid (but not cheap) means of maintaining or enhancing the level of N, P and K in the soil. Because the cost of these fertilizers is relatively high, recommended application rates are published which deal with different types of soil and different crops (Ministry of Agriculture, Fisheries and Food, 1981). Recommendations for N, however, are not based upon soil characteristics but upon the requirements of the crops to be grown. Inorganic fertilizers are normally applied as 'compound fertilizer' containing N, P and K in specific amounts; they are, therefore, simply known as 'NPK' fertilizers. Their formulation is given as the percentage ratio of the active constituents. For example, a 50 kg bag of 15:15:20 NPK contains 7.5 kg N, 7.5 kg P and 10 kg K (the N is in the form of NH_4NO_3, the P is present as P_2O_5 and the K as KCl). The remainder of the material is a clay base (usually kaolinite) which allows the fertilizer to take the form of solid granules which break down in the rain.

Straight fertilizers may contain any of the three active ingredients on their own. NH_4NO_3 accounts for 17 kg (34%) of each 50 kg bag of nitrate fertilizer, triple superphosphate contains 22.5 kg (45%) P_2O_5, and muriate of potash contains 30 kg of KCl (60%) in each 50 kg bag. Compound fertilizers are generally more popular with farmers because it is less expensive to apply one dressing of a compound fertilizer than three dressings of straight fertilizer. The use of compound fertilizers, however, requires that farmers balance their requirements year by year according to the formulations which are available. Thus the application of an excess of P one year may be cancelled by the use of a fertilizer containing less P the following year. Generally, P and K are not continuously undersupplied.

In the UK, P and K indices, defined by the P and K content of the soil, are provided by the Ministry of Agriculture, Fisheries and Food, and by fertilizer manufacturers, after conducting soil tests. The scales range from 0 (deficient) to 9 (excess). For all arable crops, it is usually recommended that the soil is maintained at indices 2 and 3, and grassland at index 1. Less than maintenance dressing is recommended where the indices are greater than 3, to allow savings in fertilizer. The recommendations stress the need for farmers to use fertilizers efficiently, not only because of their cost but also to prevent nutrient imbalance in the soil and excessive leaching of nutrients into drainage water. Interestingly, it appears that the net cost-application ratio for the UK is relatively insensitive to changes in the price of either the fertilizer or the crop produced. Current recommended application rates have, therefore, remained stable for some time and are likely to do so in the future.

Because it is the gross effect that the use of fertilizer may have on the radionuclide content of soils that is of interest, only the worst possible case will be considered; that is, for index 0, which will require maximum rates of application. It is worth pointing out here that the use of inorganic fertilizers will smooth the differences in natural soil conditions. Thus, where a soil is naturally low in K, for example, it will be artificially enhanced, while those soils which have a natural abundance of K will be unaffected.

For all newly sown grassland, the recommended application rates for P_2O_5 and K_2O are 100 kg ha^{-1} and 125 kg ha^{-1}, respectively. If the grass is undersown with cereal, in addition to the cereal fertilizer rate, an extra 50 kg ha^{-1} of P_2O_5 and 30 kg ha^{-1} of K_2O is recommended for application immediately before the cereal harvest. Established grassland used for grazing has a recommended application rate of 60 kg ha^{-1} per year of both P_2O_5 and K_2O; for conserved grassland which may be cut several times in a year, the application rates vary between 50 and 100 kg ha^{-1} of P_2O_5 per cut and between 80 and 120 kg ha^{-1} of K_2O per cut. Generally, there is unlikely to be more than two cuts in any one year. For all cereal crops, an application rate of 75 kg ha^{-1} per year is recommended for both P_2O_5 and K_2O. This application rate is also recommended for oil seed rape and for maize, whether grown for forage or for grain. Kale, forage rape and stubble turnips require 75 kg ha^{-1} of P_2O_5 per year and 100 kg ha^{-1} of K_2O per year. Fertilizer requirements for potatoes are high at 350 kg ha^{-1} per year for both P_2O_5 and K_2O, but sugar beet has a recommended application rate of 100 kg ha^{-1} of P_2O_5 per year and 200 kg ha^{-1} of K_2O per year. Mangels, fodder beet and swedes have recommended rates of 150 kg ha^{-1} of P_2O_5 and 125 kg ha^{-1} of K_2O per year.

Where farmyard manure is applied, the inorganic fertilizer rates are reduced by 20 kg of P_2O_5 per 10 t of manure and by 40 kg of K_2O per 10 t of manure. Use of undiluted slurry allows reduction at the rate of 1 kg of P_2O_5 per 1000 l of cow slurry, 2 kg for pig slurry and 5.5 kg for poultry slurry, and at the rate of 4.5 kg of K_2O per 1000 l of cow slurry, 2.7 kg for pig slurry and 5.5 kg for poultry slurry.

From these values, it is possible to determine that the likely maximum rate of application of P_2O_5 is about 100 kg ha^{-1} per year, and about 150 kg ha^{-1} per year of K_2O, if root crop areas are excluded. For root crops, these values should be multiplied by 3 to get an approximate application rate.

The effect on the radionuclide content of adding K to the soil is obvious; ^{40}K will be present in both organic and inorganic fertilizers. This, of course, is of interest to normal radiometric surveys as well as to geochemical soil surveys. If the effective source zone for ground-based γ-ray spectrometry is taken as 10 cm in thickness, and if the fertilizer is distributed uniformly within this zone, the annual increase in K_2O is only in the order of 0.0075%. Thus, continued application at the maximum rate, with no loss of K through removal in run-off or in harvested crops, would need to continue for 133 years for the K_2O content of this layer of soil to be increased by 1%. Generally, only if spillage of fertilizer has occurred to give a large mass over a small area, will there be any significant effect as far as K_2O is concerned. Similarly, it can be argued that it would take approximately 200 years of maximum application to increase the P_2O_5 content of the soil by 1%, again assuming that none is removed.

The fact that both K_2O and P_2O_5 are lost is, however, clear when surface water chemistry in an agricultural area and the natural requirements of the plants are considered. To get a better idea of the effect of using fertilizers, the level of maintenance dressing should be taken into account. This ranges between about 40 kg ha^{-1} of both P_2O_5 and K_2O for grasses, and about 200 kg ha^{-1} of P_2O_5 and 100 kg ha^{-1} of K_2O per year for potatoes. Thus the net annual gain in P_2O_5 and K_2O for grassland and cereals is, at most, only about 70 kg ha^{-1} of P_2O_5 and 120 kg ha^{-1} of K_2O. Therefore, although the application of fertilizers over several hundred years will have a noticeable effect on the level of ^{40}K in the soil, at present such effects can be ignored, except in unusual circumstances, especially if only grassland areas are sampled.

P_2O_5 itself is not a source of radioactivity in soils. However, the ability of phosphate-rich rocks to remove U from solution has already been mentioned in Chapter 6. U occurs in many phosphatic sedimentary formations of a wide range of geological ages and geographical distributions. The U is mainly of syngenetic origin, but may be locally redistributed to cause variation in concentration within a particular formation. The association between U and P_2O_5 appears to occur both in marine and non-marine sedimentary rocks.

The P_2O_5 content of inorganic fertilizers is derived from crushed phosphatic rock originating in many different parts of the world (Dee *et al.*, 1957). Concentration of the P_2O_5 in triple superphosphate fertilizer is then achieved by processing with H_2SO_4. The fact that high U levels have been reported in phosphatic fertilizers is, therefore, not surprising (Table 9.4).

Table 9.4 — U content of some inorganic fertilizers.

Specification (N:P:K)	uranium (ppm)
17:17:17	26
20: 8:14	11
8:20:16	51
8:20:16	48
0:25:25	39
20:10:10	20
25: 5: 5	4
0:24:24	35

Analyses of different types of phosphatic fertilizer used in southern England have shown that, for 20% P_2O_5 fertilizer (a typical representative of a compound fertilizer), the U content is usually about 40 ppm. Thus, for a 100% P_2O_5 content, the U level is 200 ppm. Although there are many uncertainties, this appears to be a reasonable value, considering the range of U concentrations found in phosphatic rocks. However, it must be borne in mind that, while some processes of phosphate extraction may remove some of the U from the soil, other processes may preferentially concentrate U (and indeed Ra). Therefore, at the annual gain rate of 70 kg ha^{-1} of P_2O_5 discussed previously, the likely increase in U in soils in southern England will typically be about 14 kg ha^{-1}. With normal soil density, taking an effective thickness of 10 cm as before, this is an increase of 0.007 ppm U per year.

At maximum rates of application, a period of about 150 years would thus be needed to increase the U content of the soil by 1 ppm, but at normal application rates the period would be two or three times greater. Again, except in unusual circumstances such as spillage of fertilizer at a sample site or the use of a fertilizer extraordinarily rich in U, the effect of inorganic fertilizer on soil U can generally be ignored.

The use of inorganic fertilizers on U in stream water should also be considered. Phosphate in groundwaters and surface waters in East Devon (an area of intensive agriculture), for example, is principally due to dissolved fertilizer; a small input is directly from rainfall, but there is no geological control. The P content of the surface waters is about 0.3 ppm. If this has all been derived by solution of phosphatic fertilizers containing 200 ppm U, an addition of 0.06 ppb U will also have occurred, provided that all the U behaves similarly to P_2O_5, goes into solution and stays there (that is, if U is not liberated where P_2O_5 is fixed and if direct precipitation, coprecipitation, adsorption and organic take-up of U have not occurred). Clearly, there are great uncertainties in following U and P_2O_5 between use of the fertilizer and the sampled stream water but, if these order-of-magnitude calculations are correct, then this level of increase can usually be ignored. Even at the highest P level found in East Devon (1.14 ppm), the U addition is at most only about 0.23 ppb. Thus, while variation in background U levels can be expected to result from the use of inorganic fertilizers, this is of minor importance unless the survey is concerned with detecting anomalies of very low amplitude.

Possible exceptions could arise, however, where the contents of a bag of fertilizer, or an open bag with some fertilizer left in it, has been dumped in the stream above the sample point. Fortunately, the first situation is likely to be rare, bearing in mind the cost of fertilizer and most farmers' interest in keeping water courses pollution free, but it is not uncommon to find old fertlizer bags in stream courses in agricultural areas. Usually, however, such discarded bags are empty. Experiments carried out with old fertilizer bags still containing some fertilizer, in streams of different acidity, have shown that only low pH streams give a degree of leaching which significantly alters the U content of those streams. Therefore, it appears that this source of difficulty need only be considered in areas of intensive agriculture where the streams naturally have low pH values.

Fall-out

Another problem in the interpretation of soil anomalies is the presence of fall-out produced by atmospheric testing of nuclear weapons or, near nuclear establishments, the more localised effects of controlled or accidental discharge of radionuclides to the environment. The latter are, however, generally well known from the detailed monitoring of the areas surrounding nuclear establishments. The effect of fall-out is much wider in its geographical extent and can possibly interfere with survey results from areas at very large distances from the test sites themselves. Major accidents at nuclear power plants, such as that which occurred at Chernobyl in 1986, can also affect a wide geographical area.

Clearly, the effects of radioactive fall-out were a major problem during the time of intense atmospheric testing of nuclear weapons in the 1950s and early 1960s. During fission of ^{238}U, ^{235}U and ^{239}Pu, the nucleus splits into two fragments of unequal (but similar) mass numbers. These fission products are unstable, having a high excess of neutrons, and decay by β-particle emission in a series of short decay chains. The products can be classified as either short-lived (those with half-lives less than a few days), intermediate-lived (with half-lives between a few days and a few years) and long-lived (with half-lives longer than a few years). In addition to these fission products, neutron activation effects after a nuclear explosion produce a whole series of low mass number radionuclides with half-lives of less than 6 years.

Fall-out was a serious problem in the interpretation of the results of γ-ray surveys carried out before 1970, even for areas remote from test sites, because of the effective dispersion of fission products by the global circulation system in the upper atmosphere. Deposition of fall-out can occur principally in two ways: dry deposition, which produces an even distribution of fall-out onto the leaves of grass, shrubs and trees; rainfall (or wet) deposition, which purges the surface soil layer and vegetation of activity and transfers it to the subsurface soil and, by run-off, to ditches and streams. Initially, however, rainfall may be very effective in transferring radioactive particles from the atmosphere to the ground. On the basis of their different half-lives, fall-out nuclides can also be considered as either 'new fall-out' or 'old fall-out'. New fall-out occurs within a period of about 4 weeks of a large atmospheric nuclear test and is characterised by the presence of ^{140}La, while old fall-out is characterised by ^{144}Ce, ^{102}Rn, ^{106}Rh, ^{95}Zr, ^{95}Ni and ^{137}Cs.

The effects of fall-out were noted in the results of air-borne γ-ray surveys of southern England during the late 1950s. Higher activity on southwesterly slopes was considered to be due to the influence of the prevailing south-westerly wind on the distribution of fall-out, and localised points of higher activity were identified as haystacks, dry fall-out having been concentrated into one point on the ground. Virtually all the problems related to fall-out in the interpretation of U and Th exploration data disappear within a year or so of the cessation of atmospheric release.

Soil transport

Long-lived fall-out nuclides can be used in the study of soil transport. The only fall-out nuclide with a sufficiently long half-life for this is ^{137}Cs (30.1 years). This nuclide will remain in the environment for a considerable period and can be used as a tracer of short-term erosion and sedimentation within river catchments. This is very important in the assessment of potential river exploitation schemes. Because ^{137}Cs is strongly adsorbed onto clay minerals and organic matter, its initial distribution is controlled by the intensity of fall-out, the proportion of clay in the soil and its ion-exchange capacity, and the abundance of organic matter (McHenry and Ritchie, 1977). In an area in which these factors can be assumed to be reasonably uniform, redistribution of soil particles can be traced by measuring the ^{137}Cs activity of samples collected throughout the area.

Temporal variation in the supply of ^{137}Cs must be taken into consideration when it is used as a spatial tracer. The first appearance of significant amounts of ^{137}Cs did not occur until 1955–1956, thereafter rapidly building up to a maximum in 1963–64. There was, however, a slight decrease in the rate of deposition from 1959 to 1962. During the second half of the 1960s and through to the late 1970s, a number of fluctuations occurred as a result of Chinese and French atmospheric nuclear tests (Fig. 9.20). The uptake of ^{137}Cs by the soil is very rapid and, in any soil sequence, two temporal markers are present: one marks the first appearance of ^{137}Cs and the other the maximum activity. These are used as the basis for establishing rates of movement.

Samples of soil or sediment for these studies have to be collected very carefully. Generally, it is only the upper 20 cm or so of the profile that is

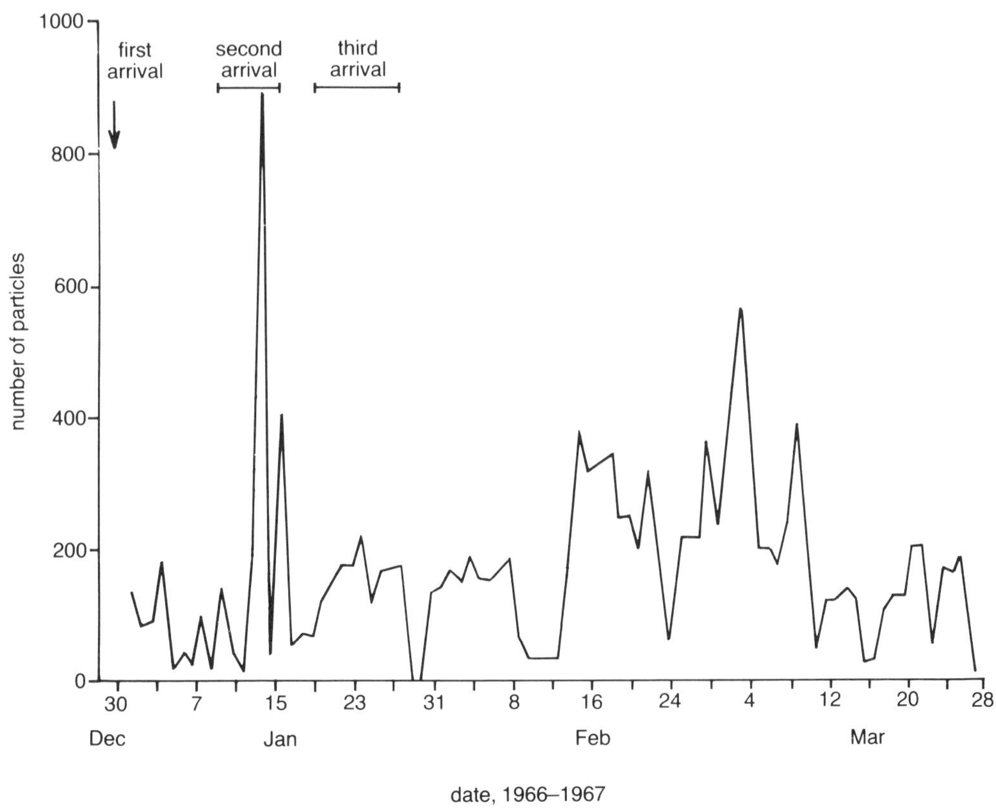

Fig. 9.20 — Daily variation in $10^{-6} - 3\times10^{-5}$ m fall-out particles in 580 m^3 of ground level air in Japan following a Chinese atmospheric nuclear test on 28 December 1966. (After Yoshikawa *et al.*, 1968.)

of interest. Unfortunately, this material is the most unconsolidated, and it is difficult to avoid compression of soil, which in any one section may show different degrees of compaction. Other difficulties can result from suction effects during the removal of the coring tube from the soil or sediment. Furthermore, core must be stored at its *in situ* temperature in order to prevent gas formation and expansion of the core. A procedure whereby eight cores are collected at each site and then compounded in 10 cm increments is recommended in order to reduce errors introduced by coring (Ritchie and McHenry, 1978).

^{137}Cs is measured using γ-ray spectrometry. The ^{137}Cs peak has an energy of 0.662 MeV. The only interference is from a 0.666 MeV peak of ^{214}Bi, which is corrected by measurement of the 0.609 MeV ^{214}Bi peak and applying a proportional correction to the sum of the 0.662 MeV and 0.666 MeV peaks. The resulting peak area is proportional to the concentration of ^{137}Cs and may be calibrated by the use of an appropriate standard. Multielement analysis is also possible for the Th and U decay series and ^{40}K.

Once deposited as fall-out, the movement of ^{137}Cs through the soil profile is controlled by the chemical composition of the soil and its physical properties. Generally, ^{137}Cs resides in the top 10 cm of undisturbed soils, the concentration decreasing exponentially with depth. Where the soil is particularly permeable and free of adsorption sites, however, detectable ^{137}Cs concentrations may extend to a depth of 30 cm. Where there is a high clay mineral content in the soil, more ^{137}Cs is retained near the surface. Consequently, neither activities nor concentrations of ^{137}Cs can be used to indicate the age or mobility of soils, as the variations may merely reflect clay mineral contents. Instead, ^{137}Cs concentration is usually presented as a fraction of the clay mineral content. Where organic matter is also present in abundance, fixing of ^{137}Cs will occur, trapping it in the extreme top of the soil profile. Variations in the amount of organic matter may then affect the ability of ^{137}Cs to descend to lower soil levels.

The general ^{137}Cs distribution model for soils recognises, essentially, that sites at which there is little or no erosion have high concentrations of ^{137}Cs in the upper soil layer. The total input of ^{137}Cs in the area can be determined from such sites, taking into account the loss due to radioactive decay and land use, concentrations being greater in forested areas than in open grassland. Sites showing some erosion on uncultivated hilltops and valley sides display a truncated ^{137}Cs profile. This is produced simply by the removal of the upper layers of soil, as part of a single-stage or multistage erosional process, followed by the development of a new soil profile. Clearly, the lower (old) level will show the tail of the original ^{137}Cs profile, while the upper level develops a new profile in accordance with the prevailing ^{137}Cs fall-out and the incorporation of residual ^{137}Cs. Where sites are in cultivated areas, there is usually a relatively even distribution of ^{137}Cs within the cultivated section, but the total amount of ^{137}Cs is generally less than in uncultivated soils. The variation in total ^{137}Cs has been taken as showing the effect of loss of ^{137}Cs adsorption sites and thus may be a function of soil erodability. Finally, at sites where there is active deposition of sediment, ^{137}Cs concentrations may show marked variability, both laterally and with depth, depending upon the history of sediment deposition.

Because there is a clear relationship between the loss of radionuclides strongly adsorbed in soils and the loss of the soil itself, it is possible to estimate soil loss from ^{137}Cs data. In general, the relationship between radionuclide loss and soil loss has the form

$$Y = A\left(\frac{x}{b}\right)^n,$$

where Y is the ^{137}Cs loss, x is the soil loss and A, b and n are empirically determined values. This relationship has been refined for the situation where sheet erosion occurs, such that

$$Y=1.6x^{0.68},$$

where Y is the ^{137}Cs loss expressed as a percentage of the input and x is the soil erosion in tonnes per hectare.

The pattern of change in soil distribution within even small cultivated areas can thus now be quantified; previously, only subjective assessment of erosion and deposition was possible. The method used involves, firstly, measurement of ^{137}Cs concentrations in undisturbed uneroded soil with a forest cover (to establish the base level of ^{137}Cs) and, secondly, the measurement of the soil profiles along traverses over the cultivated area to determine the depth of mixing of ^{137}Cs in the soil by the various cultivation methods used in that area. If the depth of accumulated ^{137}Cs is greater than the depth of disturbance, recent deposition of the soil has occured. If, however, the situation is reversed, soil erosion has taken place. Obtaining data over the whole area gives the pattern of soil movement, while repeating the surveys at successively later dates allows quantification of the rates of deposition and erosion (Campbell *et al.*, 1982).

9.6.4 Biological surveys

It has long been recognised that certain plants preferentially take up some trace elements from the soil and produce concentrations in their tissues greatly in excess of those in the soil. Furthermore, where the abundance of one of these elements in the soil is high, concentrations in the plant are also high. This is the basis of biogeochemical exploration, in which collection of specific organisms (plant and animal) and direct analysis for U and Th, or indirect analysis for other indicator elements, can be used to identify areas of anomalous concentration which may lead to the discovery of economic deposits.

A variation on this theme, known as geobotany, is based on the observation that some plant species only grow, or will grow particularly vigorously, where the soil contains a particular trace element in abudance, while other species are inhibited by that element. Thus, hidden mineral deposits may be detected without the need for chemical analysis. Just as plant species are affected in their growth, animals too can show marked physiological variation if they live in areas of unusual soil chemistry. Observation of these characteristics is the basis of geozoology; again, hidden mineralisation may thus be identified. In the case of radionuclides, it might be thought that the primary cause of these variations is radioactivity, but that is not the case. Levels of activity, even in plants and animals which concentrate U and Th, are usually very low. Moreover, many stable nuclides have been found to produce similar effects. Instead, the main control appears to be exercised by the toxicity of the elements and the nutritional requirements of the organisms (Gough *et al.*, 1979).

Geobotanical and geozoological surveys are, in principle, much easier to carry out than biogeochemical surveys. However, identification of species and the recognition of unusual physiological features require expert assessment. Such observations cannot often be made on a casual basis, although it is well known that a number of valuable discoveries of metalliferous mineralisation have been made by noting changes in vegetation. Except in the most obvious cases, expert assessment is also needed to distinguish variations caused by soil chemistry from those which result from other factors, such as micrometeorology, inbreeding, etc. For these reasons, conventional geobotanical and geozoological surveys are rarely carried out on a systematic basis in unknown ground. Where the methods have proved successful, there has usually been a very clear indication of unusual conditions.

One aspect of geobotanical investigations that is rapidly developing, however, is the study of variations in the electromagnetic spectral response of plants produced by changes in their physiological characteristics. Certainly the frequencies characteristic of healthy plants may be quite different from those where toxic trace elements have affected growth. These variations can be detected by satellite remote sensing methods, and so it is possible to survey large areas on a reconnaissance basis very easily. It is already apparent that the discovery of economic mineral deposits using this approach is feasible and can lead to spectacular success. Even so, detailed follow-up surveys on the ground are needed to confirm the indication and to provide information about the nature of the geochemical anomaly.

Interpretation of the results of biological surveys is, therefore, usually based in one way or another on the use of chemical analysis. In normal biogeochemical surveys the organisms most commonly used are plants (Brooks, 1979), because of their widespread occurrence, but animals may be employed in special circumstances. Shellfish such as mussels, for example, have a wide distribution in near-shore marine environments and can be used to determine variation in trace element concentrations in sea water (Polikarpov, 1966).

Biogeochemical surveys are subject to many difficulties, especially where their application to U and Th exploration is concerned. This is because the uptake of U and Th varies not only between species but also between the different tissues of the plant (roots, branches, leaves, etc.) or animal (liver, kidney, etc.), and with the age of individuals, the amount of transpiration by plants, the nature of the ground in which plants grow (pH, the presence of complexing ions, leachable or resistate minerals, meteorological conditions, etc.), and the detailed food chain of animals.

The case of plants which concentrate U and Th is particularly interesting in terms of whether these elements are necessary to their survival. Because of its widespread occurrence in small quantities in organisms, it is generally believed that U is a nutrient in many species, but Th does not appear to be essential in any way. Even the requirement for U is marginal, and its absence does not appear to restrict the growth of individuals or the occurrence of any species. There are thus no known plants which act as specific indicators of U or Th. Nevertheless, where U and Th concentrations are high in plants, there is often an accompanying suite of enriched elements such as Cu, Pb, Zn, rare earth elements, Se, S, Ti, V, Co and Ni, and these may well be critical in determining the presence or absence of particular species.

Sampling of plants for chemical analysis is usually carried out on the basis of a closely-spaced grid, provided that the occurrence of the species of interest is reasonably uniform. Where large plants are sampled, material should be collected from all parts of the plant. For example, if the leaves or first- or second-year twigs of trees are wanted, they should be collected from all around the tree and not just from one small part. This is to ensure that very local variations which affect different parts of the tree (such as aspect) do not give avoidable errors in the results. If roots are collected, great care is needed to ensure that soil particles are not trapped in the material to be analysed. Similarly, with plants which have a low growth profile, such as lichens and mosses, it is important that none of the underlying rock or soil is sampled together with the plant. After collection, the plant material is usually ashed, and the ash analysed by conventional chemical methods.

In southwest England, the vegetation on the U-rich granite of Dartmoor illustrates well the variation in uptake of U that occurs. For example, grasses contain little U compared with woody plants such as heather and gorse. In the woody plants, it is found that the U is mainly concentrated in the older wood; that is, in at least second-year growth: first-year twigs and leaves show markedly lower concentrations. A single year is probably insufficient to allow concentrations to build up. The very low levels of U found in grasses, therefore, appear to reflect the annual nature of their growth, in contrast with that of perennial heather and gorse. In other circumstances, it is found that pine needles are most effective in concentrating U.

Because of uncertainties in the response of plants to variations in soil chemistry, it is often recommended that, before surveying unknown ground, trials should be run over known mineralisation in a similar area, so that the response of different species to mineralised and barren ground can be determined for the type of mineralisation sought. Moreover, the best tissues to sample of the species showing greatest sensitivity to mineralisation can also be determined.

Much effort, therefore, is needed before exploration actually commences and, although an optimum choice may be made, there is no guarantee that similar conditions will be found elsewhere, even in close proximity to the trials. Clearly, changes in groundwater chemistry and flow regime, or in micrometeorological environment, can undermine the basis of the method. It can also be argued that, with highly sensitive analytical methods available, biogeochemical surveys do not show any features that would be missing from the results of a soil survey and that the additional complexity involved in sampling plants is unnecessary. Furthermore, this is yet one step more removed from the source of the elements of interest and the mineral deposit itself.

9.6.5 Rock surveys

Sample collection

Rock samples for chemical analysis may be collected on the basis of a rectangular grid, at regular points along a traverse, or on an irregular basis, depending on the nature of the survey and the availability of sample sites. Where exposures have to be used, regular sampling is usually impossible; drilling, in contrast, normally permits access to samples at any point on a traverse or grid. Whole rock samples or mineral separates may then be analysed, either for U and Th specifically or for other indicator elements. Analysis is commonly carried out for all major and most minor elements as a routine practice.

A major disadvantage of simple sampling systems is that, where radioactive minerals are concentrated into narrow zones, this mineralisation may be missed. It is, therefore, good practice not only to obtain samples from 'typical' rock sites but also to sample joint surfaces, fracture infill, shear zones, haematised zones, nodular concretions, etc. In this way, an indication of both the primary distribution of the elements of interest and their secondary reworking can be obtained. Indeed, secondary effects can serve as a major guide to the location of economic ore reserves.

Because of its great chemical mobility, U is easily leached from weathered rocks. Collection of near-surface samples can, therefore, give unsatisfactory results. Even when apparently fresh unweathered material has been collected, it must be recognised that the results of U analyses may be misleading. Mineralogical examination of samples before chemical analysis is, therefore, important. Signs of alteration in feldspars, micas, etc., should be taken as a cautionary guide. Moreover, leaching of U is not the only process to occur, as precipitation of U may occur elsewhere in the near-surface environment. The U concentrations obtained cannot, therefore, be assumed to be minimum values, as enrichment may have occurred. In interpreting U distributions, the possibility that local anomalies are produced by this mechanism must always be considered. Only where additional information on the entire ^{238}U decay series is available can the situation be quantitatively assessed with certainty.

Mineralogical separation can make a useful contribution to understanding U mobility. Where U is contained entirely within resistate minerals, leaching will be inhibited. Even so, the resistate minerals should be examined for surface pitting and other signs of corrosion. This may only be possible by scanning electron microscopy as, although the corrosion effects may be profound, they are on too small a scale to be recognised optically. Detailed examination of resistate minerals using electron microprobe analysis often shows that even apparently fresh resistate minerals, such as monazite and zircon, display chemical changes from centre to rim which reflect alteration by fluids (Fig. 9.21).

Autoradiography

The spatial distribution of radionuclides within a rock or mineral sample can be determined by autoradiographic techniques, which facilitate the identification of host minerals and the recognition of zones of concentration within minerals (Bowie, 1977). Although ordinary photographic emulsions may be used for this purpose, they are not particularly suitable where resolution of well-defined tracks is required, and their light-sensitivity makes their handling difficult. More usefully, specially formulated nuclear track plastics are employed. A typical example is CR39, a thermoset plastic formed as a cross-linked polymer from the monomer allyl diglycol carbonate. This material possesses a high sensitivity to ionising radiation, combined with a very high degree of isotropy and homogeneity and high mechanical strength (Henshaw *et al.*, 1979). The procedure normally used is to cut a parallel-sided slab of the sample to a thickness of about 10 mm. The surfaces of the slab are polished, and it is mounted between two sheets of Perspex, with the CR39 positioned between the upper surface of the slab and the Perspex. The whole sandwich is then clamped together by means of corner bolts. Care must be taken to ensure that all the surfaces are clean before final assembly. The outline of the sample should be scored onto

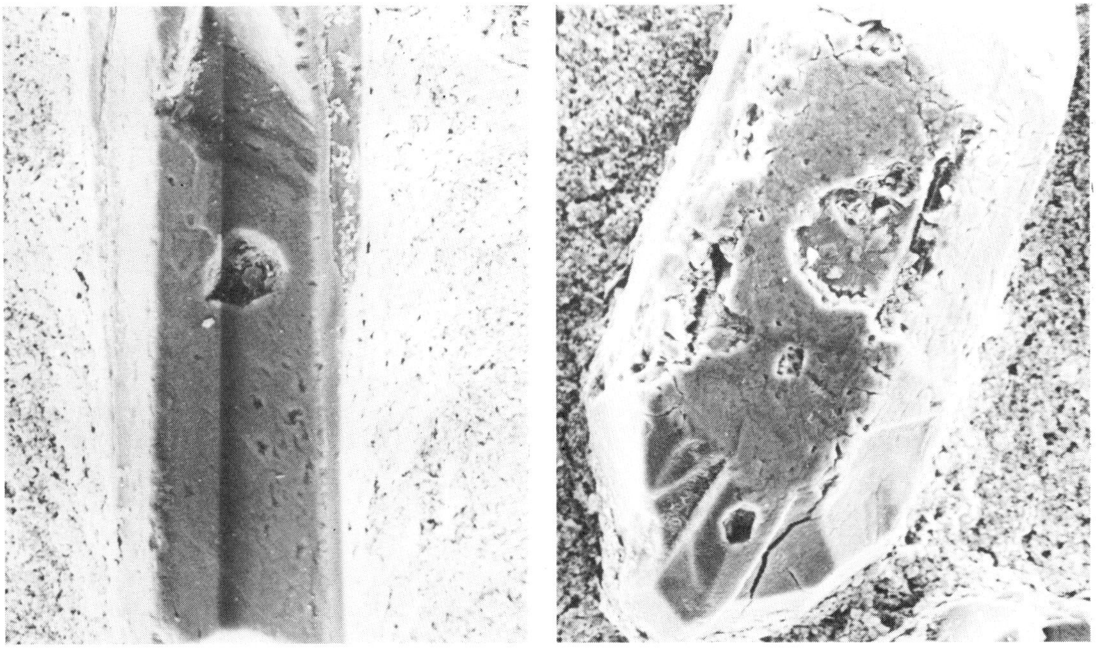

(a) (b)

Fig. 9.21 — Scanning electron photomicrographs showing differential dissolution of zircon crystals in the Dartmoor granite: (a) reasonably fresh zircon; (b) zircon crystal with considerable solution pitting. Disequilibrium in the ^{238}U decay series results from dissolution. In both cases the width of the field of view is 0.5 mm. (Source: S. Bouda.)

the plastic with a sharp point for future reference. The sandwich is then left for an appropriate length of time (a few days to several weeks), depending on the concentration of radionuclides present. At the end of this period, the CR39 is removed and the α-tracks made visible by etching. Illustrations of the use of this procedure are also given in Chapter 4.

Uniform etching of the α-tracks is achieved under specific conditions. Typically, a 6.25 N NaOH solution is used at a temperature of 75°C. Specially constructed stainless steel etching tanks are used, with polythene or polystyrene covers to reduce loss of NaOH by evaporation. Periodic checking of the concentration of the NaOH solution is, nevertheless, recommended to ensure that no change occurs by evaporation during etching. The temperature of the solution must also be maintained within close limits, and so an accurate thermostat control over the heating coil is needed. Etching times are typically a few hours, sufficient to etch the most energetic tracks to their termination point. When etching has been completed, the CR39 is placed in weak acetic acid to neutralise the NaOH and is then thoroughly washed in distilled water to remove any remaining acetic acid or NaOH.

The distribution of α-tracks in the CR39 can then be examined and compared with the distribution of minerals on the surface of the rock slab (Fig. 9.22). The sample is repositioned using the scored outline on the CR39. Examination of the sample may be by reflected light microscopy, or it may be thin-sectioned and examined by transmitted light. Correlation between the CR39 track density and the chemical composition of the sample may also be made using scanning electron microscope and electron probe analysis (Southworth, 1975).

There have been many studies of the mineralogical distribution of radionuclides by this method. For example, it has been shown that, in the granites of southwest England, U occurs mainly as uraninite, with monazite of secondary importance. Previously, U was thought to occur principally in zircon. The differences between resistate zircon and easily leached uraninite has an important bearing on the ability of hydrothermal circulation systems to leach U from the bulk rock and to produce vein systems.

It is possible to measure the length and orientation of a track in order to determine the energy of the α-particle that produced it, provided that a suitable calibration source is available. The real track length L is given by the relationship

$$L = \frac{Z + \Delta}{\sin \theta},$$

where Z is the length of the track measured from vertically above the surface of the plastic, Δ is the depth of plastic removed by etching (which can be determined experimentally) and θ is the angle by which the track is inclined to the normal to the surface of the plastic. The real track length is, of course, proportional to the energy of the α-particle that produced it. Such data have been used in studies of disequilibrium in the ^{238}U decay series in rock samples, but the problems involved in using this method with geological samples are immense, owing to the degradation of the α-particle energies in passing through the surface skin of the sample, the presence of ^{232}Th and the distribution of the radionuclides in such a way that neither simple point sources nor sheet sources exist. Generally, it is unlikely that the time-consuming process of track length determination, in which hundreds or thousands of measurements must be made to obtain statistically acceptable data, will give useful results.

Before leaving the topic of autoradiography, it should be noted that great enhancement of fission activity can be obtained by neutron activation of the sample. This approach, combined with fission track measurement, has been successfully used with short exposure times for the detection of low concentrations of radionuclides (Clarke *et al.*, 1979). An illustration of the results is given in

(a)

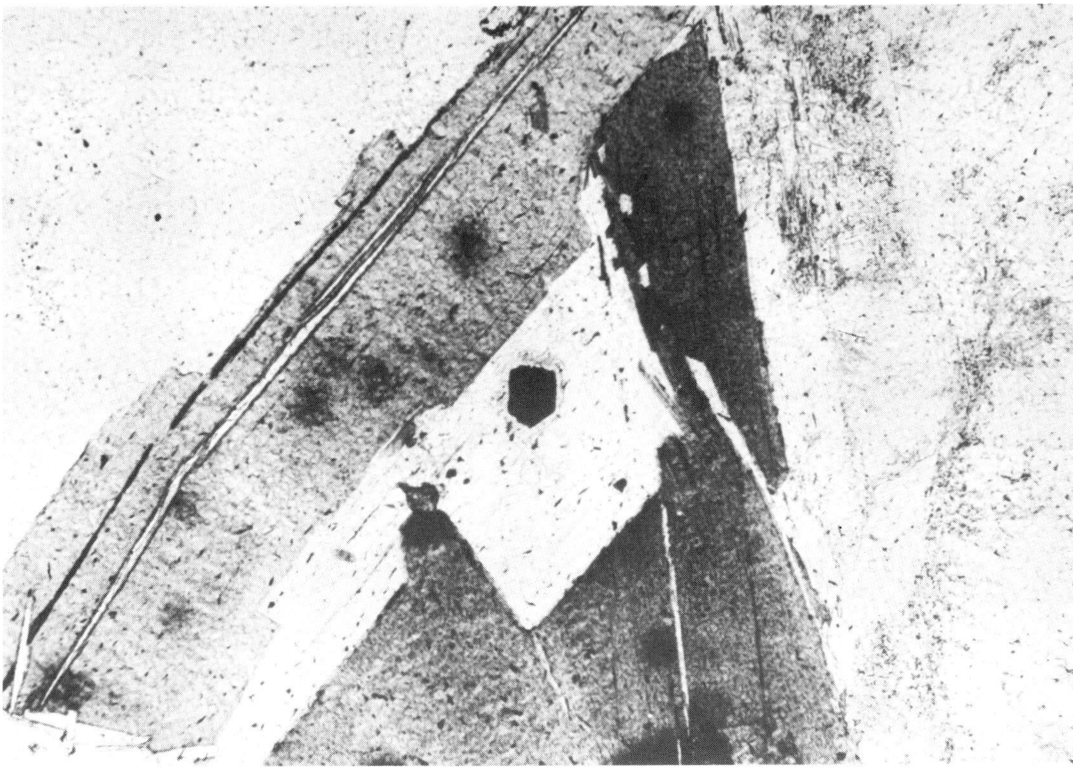

(b)

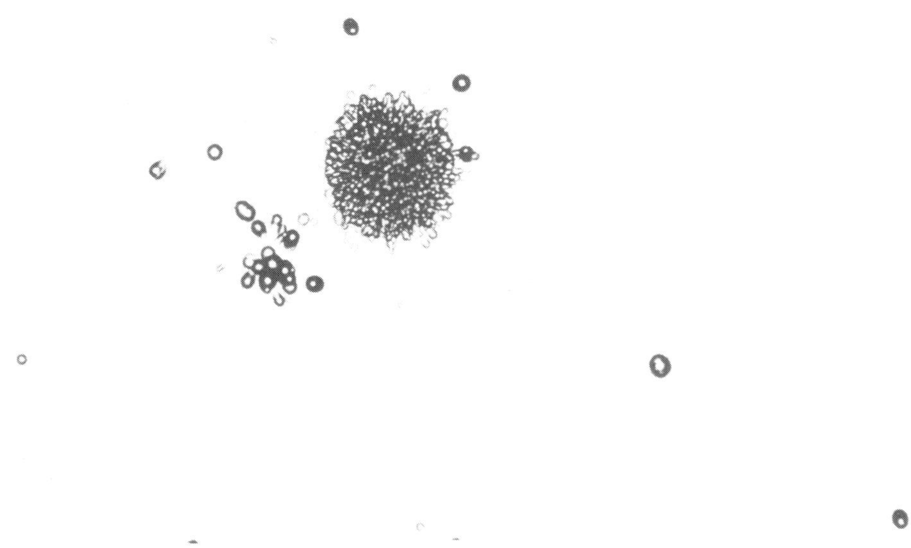

Fig. 9.22 — Use of α-particle autoradiography for radioactive mineral location: (a) photomicrograph of uraninite inclusion within biotite granite from Carnmenellis; (b) autoradiograph with a 2 week exposure on nuclear track plastic CR39. The section has a width of 1.5 mm. (Source: N. L. Jefferies.)

Chapter 4. Neutron activation of samples, however is more commonly employed as a preliminary stage in a method of chemical analysis, that will be considered in more detail later in this chapter.

9.7 GEOCHEMICAL ANALYSIS

Among the many and varied methods of chemical analysis (Jeffery and Hutchinson, 1981), the most common systems used for determination of radionuclide abundances in rocks and soils are X-ray fluorescence spectrometry and neutron activation analysis, while laser-induced fluorimetry is a technique specially developed for the analysis of U in water samples. This account will deal only briefly with X-ray fluorescence and laser-induced fluorimetry, as these methods do not directly concern radioactivity. The production of artificial radionuclides in neutron activation analysis, however, is discussed in more detail.

9.7.1 X-ray fluorescence spectrometry

When atoms are irradiated with a beam of X-rays, they become excited and emit electromagnetic radiation of longer wavelength than that of the incident radiation, a property that forms the basis for X-ray flourescence spectrometry. The first component in the system is, therefore, the X-ray source or 'tube'. When electrons of sufficient energy are made to strike matter, they cause X-rays to be emitted in a continuous spectrum of energies up to that of the incident electrons, independent of the material involved. Within this spectrum, the intensity varies but is greatest at about one-third of the maximum energy. However, the total energy produced is dependent upon the element irradiated and is governed by the relationship

$$E=1.1\times10^{-9}ZV,$$

where Z is the atomic number of the element and V is the accelerating potential of the electrons. Because the energy of the electrons is proportional to V, the energy of the radiation is proportional to ZV^2. In an X-ray tube, not all the energy spectrum is used; selective filtering of low energies takes place at the source anode and at the tube window. The effect of anode filtering varies with the atomic number of the material, and so changes in the energy relationship occur.

In addition to the continuous spectrum, line spectra are produced when the incident electrons have sufficient energy to remove electrons from the inner shells of excited atoms. The X-ray photons emitted when an outer electron falls into the vacancy thus created have an energy that is determined by the atomic structure of the target material. Line spectra are also produced when matter is irradiated with energetic X-rays, but in this case no continuous spectrum is produced. The spectral lines produced have energies that are characteristic of the element irradiated and make up what is known as the 'characteristic spectrum', which is divided into the K, L and M series, depending upon which shell developed a vacancy. Electron transitions of specific energy are only possible for a specific element, and so each element is characterised by a particular group of emissions.

X-ray tubes are available with various anode materials, including Au, Pt, W and Cr (Norrish and Chappell, 1977). This variety is needed because a tube cannot be used for the analysis of small quantities of elements of which the anode is composed, or if the anode line spectrum interferes with measurement, either directly or through Compton scattering. Also, the continuous spectrum produced by the anode is generally more intense than the line spectrum, an effect which increases with increasing the atomic number of the anode material. Thus Au, Pt and W generally provide the best targets for general-purpose X-ray tubes. Where X-ray tubes have Ti, Cr or Ag anodes, they have thin Be windows which allow low energy X-rays to be transmitted. This makes them particularly suitable for analysis of the low atomic number ($Z<20$) elements. Windowless tubes are available for very light ($5<Z<9$) elements. A Mo tube is usually used for analysis of U and Th.

The sample itself must first be homogenised, usually by the preparation of a fused bead or a pressed powder disc. For U and Th analyses, pressed powder discs are normally prepared. Sample preparation consists of grinding in a tungsten carbide mill to less than 50 μm in size. To 5 g of this powder is then added 3 drops of 2% aqueous polyvinyl phyrolidone (PVP). Boric acid powder is added as a bonding agent, and the mixture is then pressed into a circular disc under a load of 3 t for 3 min.

The radiation emitted from the sample is collected and passed to an analysing crystal which disperses the X-rays. Again, a wide range of crystals are available because it is not possible for a single crystal to disperse the whole range of X-ray wavelengths used for analysis. For U and Th analysis, a (200) LiF crystal is usually used. The dispersed X-rays are then further collimated and passed to the detector, where they are analysed and counted. Either NaI(Tl) scintillation detectors or proportional counters are used in these instruments. For both U and Th, the determinations are based upon quantitative analysis of the Lα lines; wavelengths of 0.956 and 0.968×10^{-10} m are used for Th and 0.911 and 0.923×10^{-10} m for U. The maximum sensitivity for these elements is less than 1 ppm (possibly as low as 0.2 ppm for U).

9.7.2 Neutron activation analysis

The other main method of U and Th determination in rocks and soils is neutron activation analysis or, more specifically, instrumental neutron activation analysis. In essence, this technique can be thought of as induced γ-ray spectrometry, in that either non-radioactive nuclides are artificially converted to short-lived γ-ray emitters or the γ-ray emission of naturally occurring radionuclides is increased. The γ-radiation emitted is then detected and measured by the methods outlined later in this chapter when dealing with natural γ-ray spectrometry in the laboratory.

The neutrons used to induce the radioactivity are most commonly supplied by nuclear reactors but may also originate in accelerators or in radioactive sources. In reactors used for this purpose, the thermal neutrons produced have an average energy of 0.025 eV, by energy loss from the original fission (fast) neutron energy of about 1 MeV. The position of the sample within the reactor governs the energy of the neutron flux to which it is exposed. Epithermal neutrons with energies between 0.1 and 10^5 eV may comprise between 1 and 40% of the neutron flux, depending on position, and may also be used for activation purposes. Typical thermal neutron fluxes of about 10^{12} neutrons cm^{-2} s^{-1} are employed.

Neutron capture cross-sections are more-or-less insensitive to neutron energies within the thermal range, but for epithermal neutrons there can be a strong dependence upon energy. The neutron capture cross-section, as well as the neutron flux, clearly affect the rate of production of the new radionuclide. Where a pure epithermal flux is required, the ability of some nuclides to capture thermal neutrons efficiently is exploited. For example, Cd may be used to filter out thermal neutrons in a mixed thermal–epithermal flux. Some elements are best irradiated in a thermal flux, and others in an epithermal flux, depending upon which transition is required for their later analysis.

Although of more limited application, accelerators have been found to be useful adjuncts to nuclear reactors for some specific purposes. The accelerator produces a beam of ^{2}H nuclei (deuterons) with maximum energies between 150 eV and 600 keV and a beam current of 1–5 mA. The deuteron beam is made to react with a ^{3}H-bearing target to produce 14 MeV neutrons in the formation of ^{4}He, or with a ^{2}H-bearing target to produce 3 MeV neutrons in the formation of ^{3}He. Either of these neutron beams may be moderated to produce thermal neutron fluxes of 10^6-10^7 neutrons cm^{-2} s^{-1}.

The equation for the production rate p of the product of neutron capture is given by

$$p = f\sigma N,$$

where f is the neutron flux, σ the cross-section and N the abundance of the target nuclei (Goles, 1977). Production will be offset, however, by decay of the product radionuclide according to the expression

$$N_t = N_0 e^{-\lambda t},$$

where N_0 is the original number of nuclei with decay constant λ and N_t is the number remaining after time t. Combining these expressions gives the relationship which governs the number of product nuclei at any time t during which the processes of formation and decay take place together:

$$N_t = \frac{f\sigma N(1 - e^{-\lambda t})}{\lambda}.$$

However, this is only really valid for neutron irradiation in which flux is constant and the value of $f\sigma$ is reasonably small, in which case the value of N does not decrease significantly during irradiation. These conditions generally apply for most geochemical work but, in some cases, burn-up of the target may occur.

If the change in N_t with t is considered, a linear growth rate is obtained when t is short compared with the half-life of the nuclide produced but, as t increases to several half-lives, the rate of increase in N_t gradually decreases and becomes asymptotic to a value given by $f\sigma N/\lambda$. At this stage, the system is said to have become saturated with induced radioactivity. In the irradiation of a sample for geochemical analysis, some short-lived nuclides will usually reach saturation levels. This is not achieved deliberately, however, because with long irradiation times the incremental activity of the product nuclide decreases exponentially. The cost of long irradiation times can also be high.

Clearly, some time will elapse between the cessation of irradiation and the commencement of counting procedures. During this period, decay of the radionuclides will not be balanced by production. If the time at the end of irradiation is taken as the new starting point, and the time since then taken as t_1, the decay rate is given by

$$\lambda N_0 \, e^{-\lambda t_1}.$$

Generally, however, absolute decay rates are not measured. Instead, the response of the sample is compared with the response of very precisely controlled standards. On this basis, factors which are characteristic of the detection and measurement system are constant. Thus the activity for a particular radionuclide is given by

$$xf\sigma N \, e^{-\lambda t_1} \, (1 - e^{\lambda t}),$$

where x is the proportionality constant for these factors. Both sample and standard may be irradiated in the same flux for the same length of time, and so the comparison depends only upon the different delays before beginning observation, the times taken for observation and the relative abundances of the specific target nuclides.

Detection and counting systems are similar to those used for conventional γ-ray spectrometry as discussed in Chapter 8. Multichannel analysers with 4000 channels are generally adequate for this purpose and are used in conjunction with Ge(Li) or pure Ge detectors. To reduce the effects of backscattered photons, it is best not to shield the system. Only simple counting circuits are needed, but stability of the system is essential.

The design of the specimen holder for use within the reactor is important and varies according to the type of reactor used. Generally, a specimen rack with spaces for housing standards is employed. Several standards are usually included at different positions in the rack. Standards of all elements of interest, in various combinations, should be prepared.

Some carefully chosen elements can also be used as flux monitors for various sections of the neutron energy spectrum. It is also recommended that one or two standard rocks are analysed together with each batch of unknowns. Thus, at least two or three, and perhaps as many as ten, positions in the rack must be kept for standards. Irradiation times also vary with the type of reactor but are typically of the order of 6–12 h. After irradiation, the samples may be left for 2–3 days for some decay to take place, thus making handling of the samples easier (Goles, 1977).

In general, the data refinement and processing techniques employed are similar to those used in conventional γ-ray spectrometry. The use of pile-up rejection is particularly useful in minimising errors associated with high count rates. Another important source of error in interpretation, however, arises when spectral interference occurs in the response of one element owing to the presence of another element. For example, in the analysis of Na in ultrabasic rocks, the reaction used is ^{23}Na (n, γ) ^{24}Na, but this may be interfered with by the reaction ^{24}Mg (n, p) ^{24}Na. Corrections for this can be applied using the known branching ratios if the nuclide producing the interference has a well-defined peak elsewhere in the spectrum.

Abundances can be determined with a precision of about ±5%, although with some elements a precision of ±1% is possible. With careful use of standards, accuracies of the same magnitude should be obtained. Detection limits range from as low as 2 ppb (for Au) to as high as 1% (for Fe), depending upon the nuclear properties of the element. In some special applications, detection limits as low as 0.1 ppb have been obtained.

For the analysis of U and Th, three techniques are available: thermal instrumental neutron activation analysis, epithermal in-strumental neutron activation analysis and delayed neutron emission. The thermal instrumental neutron activation analysis technique utilises a thermal neutron irradiation flux. It is suitable for the determination of Th with moderate sensitivity, and U with poor sensitivity, in most rocks and some minerals. The analysis of U and Th is usually a byproduct of the determination of other elements. The epithermal instrumental neutron activation analysis technique, which, of course, utilises an epithermal neutron flux, is more costly than the thermal instrumental neutron activation analysis technique, but the sensitivity is increased twenty- to forty-fold for Th and fifty- to onehundred-fold for U. Both these elements can be detected down to 0.1 ppm in rocks and minerals, and other elements can be analysed at the same time. The delayed neutron emission technique is based upon the decay by β-particle emission of fission products to excited daughters that are unstable against prompt neutron emission. Neutron sources are thus formed in fissionable materials which decay with half-lives ranging between a few seconds and a few tens of seconds; analysis must, therefore, be carried out at the reactor site. The analysis itself is both rapid and cheap and has a high sensitivity for U, although not so for Th. Another technique which is specific for U is the counting of induced fission tracks for ^{235}U. Useful data can often be obtained by combining several of these approaches.

9.7.3 Laser-induced fluorimetry

This technique is used specifically for the analysis of aqueous solutions for U and involves, essentially, measurement of the characteristic green fluorescence emitted by uranyl compounds under ultraviolet excitation (Robbins, 1978). The excitation is provided by a N laser which gives a 4 ns pulse at a wavelength of 337 nm (Fig. 9.23). In the fluorescence spectrum produced in a natural water sample, there is, in addition to any fluorescence produced by U, an emission band due to the presence of dissolved organic C (Fig. 9.24). This emission, however, is very short lived (4–10 ns), in contrast with that of U which persists for much longer (Robbins and Castledine, 1979).

In practice, the fluorescent signal is measured between 1 and 3 μs after the

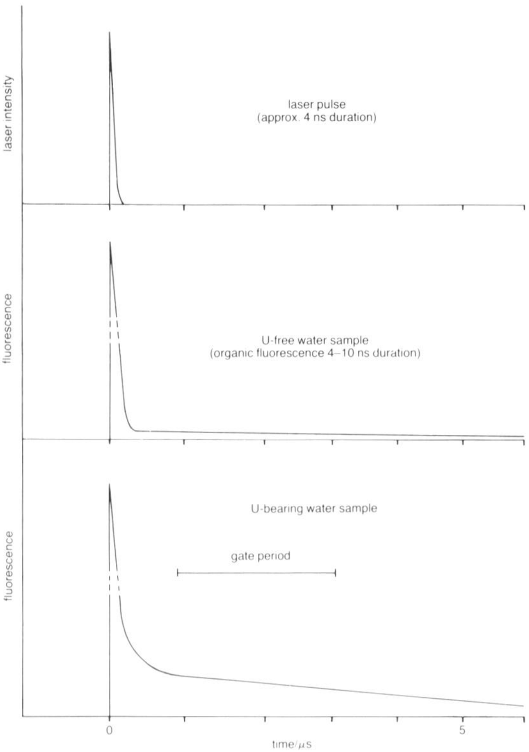

Fig. 9.23 — Fluorescence of water samples in the Scintrex UA3 analyser. (After Scintrex, 1978.)

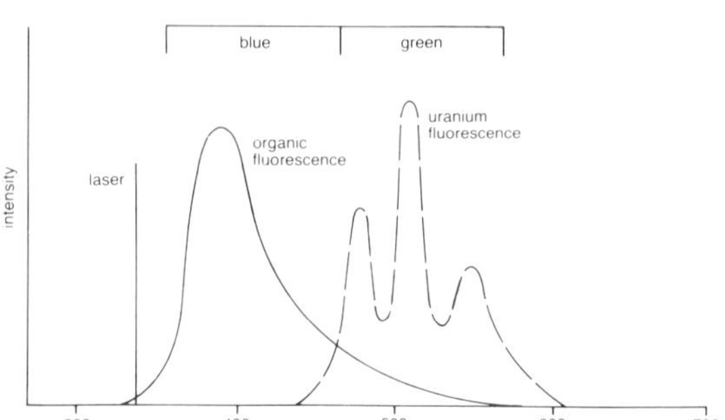

Fig. 9.24 — Persistence of fluorescence of water samples in the Scintrex UA3 analyser. (After Scintrex, 1978.)

laser pulse. Instruments are designed to indicate activity on an arbitrary scale which can be varied to change the range and sensitivity (Fig. 9.25). As U may be present in solution in a variety of chemical forms, each sample is treated by adding 0.8 ml of a sodium pyrophosphate–sodium monophosphate complexing agent ('Fluran'), which converts all the U present to uranyl pyrophosphate.

The analyser is usually calibrated using standard solutions based upon distilled or deionised water. However, natural waters carry many elements in solution, some of which may interfere with the fluorescence of the U in the sample if present in sufficient concentrations. Among the most important interfering agents are Fe^{3+} and humic acids, which act as 'inner filters', absorbing part of the laser pulse, and Mn, Ca and Mg, which act as 'quenching filters', which reduce the lifespan of the uranyl fluorescence. Checks must, therefore, be run for these interference effects using natural waters as the matrix for a series of standard U solutions, the response of which can be compared with that of standards prepared with distilled or deionised water. With care taken in the analytical and calibration procedures, the method is sensitive to 0.05 ppb U in solution (Fig. 9.26).

9.8 MÖSSBAUER SPECTROSCOPY

In the analysis of rocks and minerals, a useful parameter bearing upon the mobility of U is the oxidation state of Fe. FeO and Fe_2O_3 determinations are possible by conventional chemical methods, but these all appear to suffer from inaccuracies due to oxidation of samples during preparation and analysis (Durrance *et al.*, 1978). Even when digestion of samples takes place under an inert atmosphere, oxidation problems still arise, possibly because of recombination as components enter solution. An alternative method, Mössbauer spectroscopy, which does not involve digestion or any complex processing of samples but utilises the nuclear properties of elements held

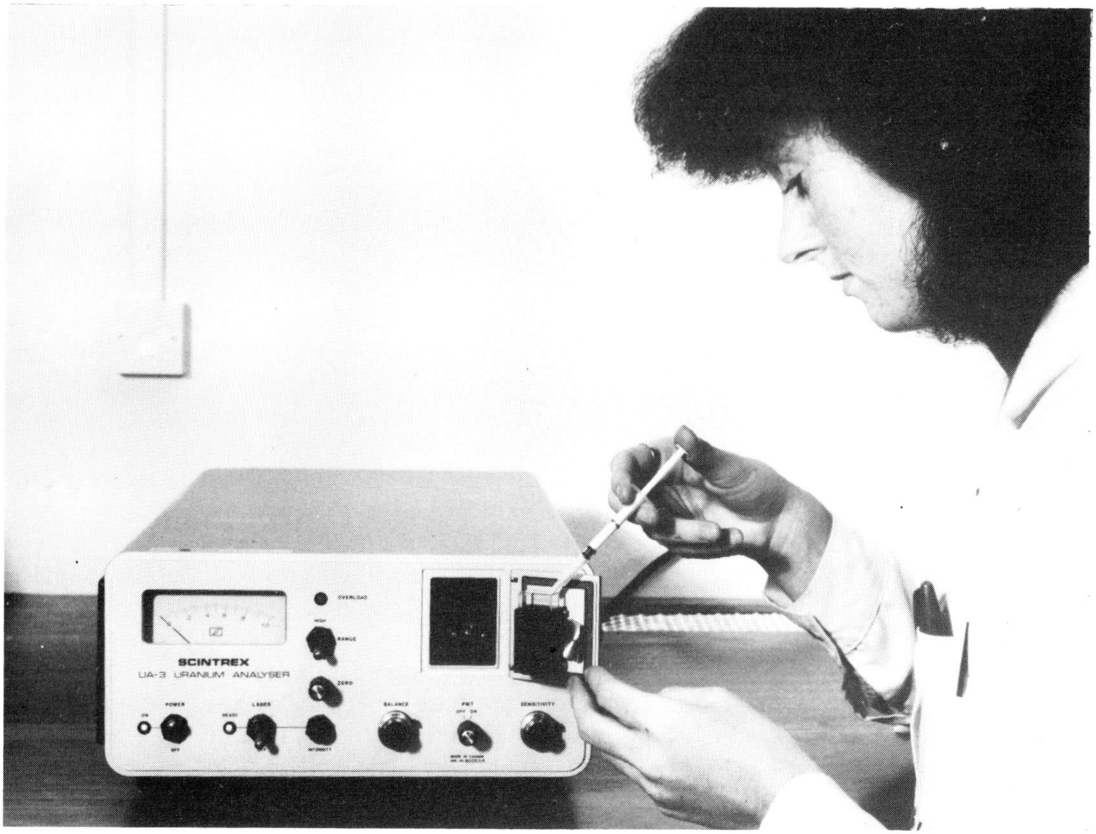

Fig. 9.25 — The Scintrex UA3 uranium analyser.

in different atomic configurations, can be used to determine ferrous/ferric ratios in a wide range of minerals (May, 1971).

The 'Mössbauer effect' is the phenomenon of recoil-less resonant absorption or emission of a γ-ray photon in a solid, and its measurement allows slight variations in the energy of interactions between a nucleus and extranuclear electrons to be distinguished (Gibb, 1976). The analytical system employed consists of a monoenergetic γ-ray source of moderate intensity, a sample holder mounted on a vibrator in line with this source, and a γ-ray detector and counter system of the type described in Chapter 8 (Fig. 9.27).

The energy difference E between the energy E_g of the ground state and the energy E_e of an excited state of the nucleus of an isolated atom of a gas is given by

$$E = E_e - E_g.$$

If a photon is emitted from a nucleus of mass M, moving with velocity V_x in direction x, the total energy above ground state at rest is given by

$$E + \frac{MV_x^2}{2},$$

The emitted γ-ray will have energy E_γ, and the nucleus a new velocity $V_x + v$ due to recoil. By conservation of energy,

$$E + \frac{MV_x^2}{2} = E_\gamma + \frac{M(V_x + v)^2}{2}.$$

The difference ΔE in energy of the nuclear transition is given by

$$\Delta E = E - E_\gamma = \frac{MV_x^2}{2} + MvV_x.$$

or

$$\Delta E = E_R + E_D,$$

where E_R is the recoil kinetic energy and E_D is a Doppler effect energy. The mean kinetic energy $\bar{E}_M$ of a free atom in a gas is given by

$$\bar{E}_M = \frac{M\bar{V}_x^2}{2} \approx \frac{kT}{2},$$

where k is Boltzmann's constant and T is the

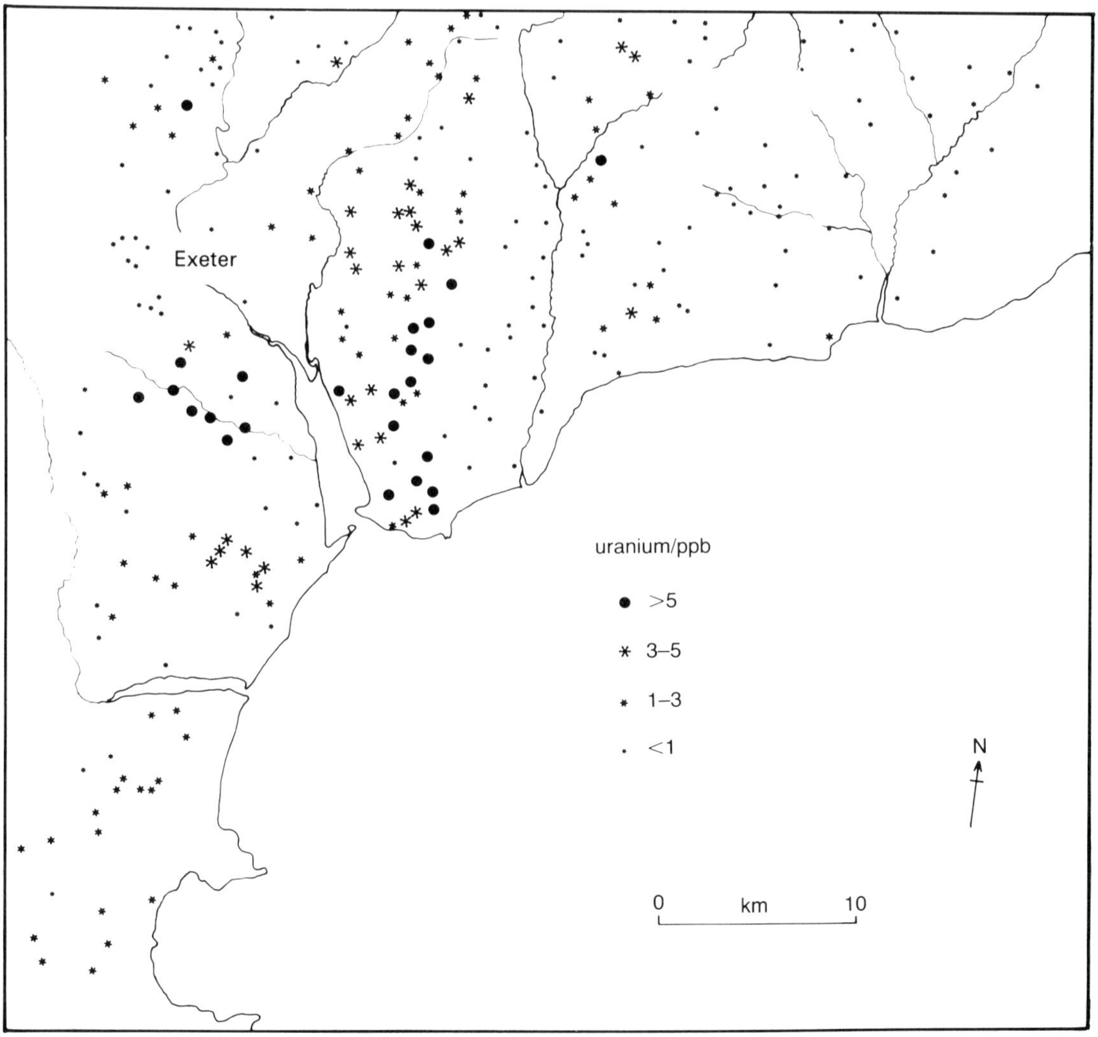

Fig. 9.26 — Uranium in the stream waters of southeast Devon. (Durrance, 1984, reproduced with permission.)

temperature in kelvin. Thus, the mean Doppler effect energy is

$$\bar{E}_D = 2(\bar{E}_M E_R)^{1/2}.$$

Therefore, the γ-ray distribution is displaced by E_R and broadened by twice the geometric mean of the recoil energy and the average thermal energy, with a Gaussian distribution. In terms of γ-ray energy,

$$E_R = \frac{p^2}{2M},$$

where p is the recoil momentum of the atom, which will be equal and opposite to the momentum of the γ-ray photon,

$$p = -p_\gamma = -\left(\frac{E_\gamma}{c}\right),$$

where c is the velocity of light. Hence,

$$E_R = \frac{E_\gamma^2}{2Mc^2}$$

and

$$\bar{E}_D = E_\gamma \left(\frac{2\bar{E}_M}{Mc^2}\right)^{1/2}.$$

In addition to the emission of a γ-ray photon by the nucleus, γ-ray photons can be absorbed. The Gaussian energy distribution for absorbed

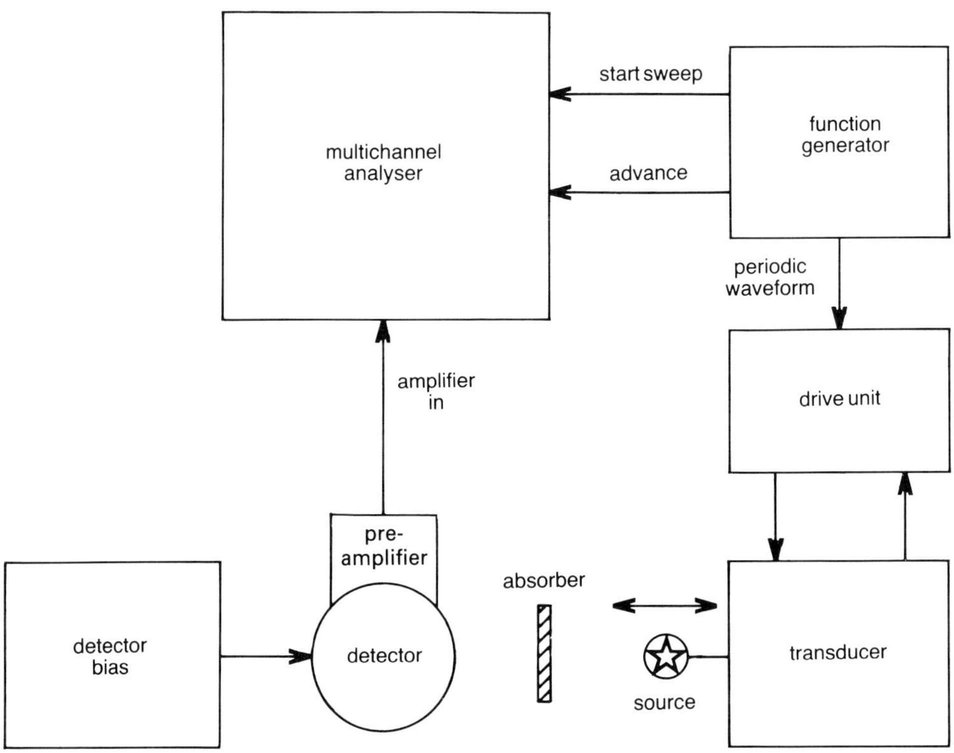

Fig. 9.27 — Diagrammatic representation of the lay-out of a system for Mössbauer spectroscopy. (After Canberra, 1984.)

and emitted photons will be the same, but their mean energies will be $-E_R$ and $+E_R$, respectively. Thus the amount of resonance overlap of these distributions will be very small.

The mean lifetime of the excited state is a very important factor when considering γ-ray energy distribution. The uncertainties in energy and time are related to Planck's constant ($h=2\pi\hbar$) by Heisenberg's uncertainty principle

$$\Delta E \, \Delta t \geqslant \hbar .$$

At ground state, the nuclear level has an infinite lifetime and zero uncertainty in energy. The excited state, however, has a mean life of a microsecond or less, so that there will be a spread of γ-ray energies of width r_s at half-maximum peak height, where

$$r_s \, \tau = \hbar ,$$

and τ is the mean life ($\tau = \log_e 2 t_{1/2}$, where $t_{1/2}$ is the half-life). However, for free atoms, this is many orders of magnitude less than both E_R and $\bar{E}_D$, and so can be neglected. This is not the case where bound atoms are concerned.

In crystalline solids, the chemical binding and lattice energies are very much greater than E_R, being of the order of 1–10 eV. Therefore, if an emitting atom is held in position within such a crystalline solid, the recoiling mass is the mass of the whole crystal and not just that of the atom. Even if the crystalline material is in the form of a fine powder, each particle will contain so many atoms that both E_R and $\bar{E}_D$ are negligible. Although the recoil of the atom as a whole is thus limited, the nucleus of the atom is not rigidly bound in this way but is free to vibrate. This vibrational energy may be transferred to the lattice and increase the vibrational energy of the whole crystal. However, the lattice vibrational energy levels are quantised so that only specific energies can be transferred. If these energies are not present, the whole crystal recoils but, where energy transfer of this type is possible, the Mössbauer effect is produced. The average energy transferred per event is the free-atom recoil energy, and

$$f = 1 - \frac{E_R}{\hbar\omega},$$

where f is the fraction of photons emitted without transfer of recoil energy and $\hbar\omega$ is one photon.

Thus the value of f will, in practice, be determined by the free-atom recoil energy (which is proportional to E_γ^2), the properties of the lattice and the temperature of the crystal. It will be greatest for small γ-ray energies, strongly bound atoms and low temperatures. Alternatively, f may be expressed as

$$f = e^{-E_\gamma^2 \langle x^2 \rangle / (\hbar c)^2}$$

where $\langle x^2 \rangle$ is the component of the mean square vibrational amplitude of the atom in the direction of the emitted γ-ray. This shows that there is an upper limit to the values of E_γ which can be used to observe the Mössbauer effect; the 155 keV line from ^{188}Os is near this limit. This relationship also shows why the effect cannot be observed in either gases or any but the most viscous liquids, for which values of $\langle x^2 \rangle$ are high.

The crystalline character of the γ-ray source, therefore, needs to be taken into account in any study of the Mössbauer effect in a target crystal. Moreover, the nuclear properties of the source must also be considered. Firstly, the source should generate γ-rays with an energy profile approaching the natural Heisenberg linewidth, should not be subject to any significant line broadening and, in general, should have a single emission. Secondly, the Debye temperature of the matrix in which the source is embedded should be sufficiently high to give a substantial recoil-free fraction. Debye temperatures for metals range between −185 °C (for Pb) and 1727°C (for Be). High melting temperature metals are usually used. Thirdly, the source matrix should not contain any of the nuclide which is to be investigated in the target sample; otherwise, self-resonance will occur which will produce an effective source linewidth greater than the Heisenberg natural linewidth. Fourthly, non-resonant scattering within the source should be reduced by careful selection of the overall composition of the matrix or by limiting the depth to which the generating source is diffused into the matrix. Finally, the ground state nuclide should be stable and should have a high natural abundance so that the source radionuclide may be easily produced.

Because the radionuclides which provide the γ-ray source should, ideally, all be in identical chemical environments to ensure satisfactory emission, any defects in the source can lead to line broadening. This is an important consideration because, in the production of Mössbauer source radionuclides, irradiation by neutrons at high flux rates may be employed which cause radiation damage and lattice defects to form. To overcome this difficulty, the source is sometimes annealed after its formation to remove the defects; alternatively, the radionuclide may be removed chemically and then incorporated into a new matrix. It is, of course, also necessary for all the γ-ray-generating nuclides to occupy the same type of chemical site in Mössbauer sources. Manufacturers supply a wide range of matrix materials and source nuclides of different strengths and sizes. Clearly, the properties of any particular source must be suitable for the investigations planned, but the strength of the source should not be so great that handling is difficult. However, the half-lives of typical sources are relatively short and, if the source is to give usable service for a few years, it must be of reasonable strength to begin with; otherwise costly replacement becomes necessary.

If γ-rays from such a source are passed through an absorber containing the same material as the source, resonant absorption will take place with re-emission over a 4π solid angle. The transmission of the γ-rays in the direction of incidence will, therefore, be reduced. The amount of this decrease is affected by the difference in the relative values of E_γ for the source and the absorber. The effective E_γ values can be changed by applying a Doppler effect shift. This is achieved by moving the source and the absorber relative to each other with a specific velocity. When the effective E_γ values match exactly, resonance will be at a maximum and transmission at a minimum. If the velocity is higher or lower, the resonance will decrease until it is effectively zero. Thus to obtain a range of effective E_γ values, a vibrator system is used in the direction of the incident γ-ray beam. A closing velocity between source and absorber (that is, the higher energy side) is conventionally defined as the positive velocity. Maximum velocities range up to about 200 ms^{-1}. Although, theoretically, it does not matter whether it is the source or the absorber that is vibrated, in transmission Mössbauer spectroscopy it is usually the source that is vibrated. This enables the sample, or the environment of the sample (temperature, for example), to be changed easily. However, where the source vibrates, the solid angle that it subtends at the detector will also change during each vibrational cycle. Therefore, if the amplitude of the source movement is large in terms of the source–detector geometry, a correction factor may have to be applied.

A large solid angle in the geometry of the system should generally be avoided anyway, because the Doppler shift $E_\gamma V/c$ is only accurate along the axis of the vibrational motion. Any photon moving at an angle θ to this direction will have an effective Doppler shift of $(E_\gamma V \cos\theta)/c$, and so distortion in the shape of the Mössbauer absorption line will occur. It is necessary to use either a large separation between source and detector or collimation of the γ-rays to avoid this problem.

Generally, a Mössbauer spectrum can be obtained either by counting the transmitted γ-rays over a fixed period of time while the source is moving at a constant velocity and then repeating at different velocities, or by repeated scanning through a range of velocities. In the preceding discussion, reference has generally been made to vibrators which, in fact, provide the mechanism for the second mode, because it is difficult to obtain constant-velocity movement and it is tedious to change the velocity settings. However, the first mode, which can be based upon various systems, such as motion along a screw thread at constant rotation, is useful when study of a very small velocity range not centred on zero velocity is required, and when the spectrometer needs to be calibrated in terms of absolute velocity. In vibrator systems, various types of waveform have been used. The sine wave is fairly straightforward to produce but clearly results in a non-linear velocity distribution. The triangular waveform (symmetrical double ramp), which scans the spectrum with constant acceleration in both directions, is better. Calibration of vibrator systems in terms of velocity is usually carried out by reference to the response of a standard sample material or by direct measurement of the velocities generated by the vibrator itself.

Just as the properties of the source are important in the production of good Mössbauer spectra, consideration must be given to the general characteristics of the sample itself. Firstly, the absorber should not be too thick; otherwise there will be an increase in the measured linewidths which will decrease the resolution of a multiline spectrum or the precision of location of a single line. An excessive absorber thickness will also decrease the transmitted γ-ray intensity and thereby increase the counting time needed to give statistically acceptable results. The sample must, however, be thick enough to allow the resonance to be observed. There is, therefore, an optimum thickness, the value of which depends upon the type of resonant absorber and its abundance in the sample. For example, for the Mössbauer spectrum of Fe, the optimum thickness is given by about 10 mg cm^{-2} of total Fe in the sample. Such information is, however, very limited in its useful application, because it is found that the optimum thickness of the sample also depends upon the size of the particles of which it is composed. Some experimentation with sample thicknesses and particle sizes may be necessary to determine optimum conditions. In geological work, samples are usually in the form of powders which must be contained. Generally, there are few specific requirements for such containment, except that the windows of the container must be transparent to γ-rays and not contain any of the resonant absorber being studied. Organic plastics and Al foil are in most common use.

As the recoil-free fraction of many absorbers is very low at room temperature, experiments are often conducted with the sample at a low temperature so that the Mössbauer effect is enhanced. Also, study of the temperature dependence of hyperfine effects (discussed later) is often of interest. To achieve this, it may be necessary to cool both the source and the absorber but, with good sources, cooling of the stationary sample alone is easier to carry out and may be quite satisfactory. Various types of cryogenic equipment have been designed for this purpose, with suitably stable cryostat systems. Standard coolants used are liquid N (boiling point, 77.3 K), liquid H (boiling point, 20.4 K) and liquid He (boiling point, 4.2 K). Temperatures above 300 K (room temperature) are sometimes required (in studies of Fe, for example), in which case a temperature-controlled heating system (furnace or attached heating coils) is utilised. In all systems of cooling and heating, it must be remembered that the windows must be transparent to γ-rays and be free of the absorber under study.

Detectors used in Mössbauer spectroscopy are similar to the types of γ-ray detector described in Chapter 8. Scintillation detectors are frequently used for γ-ray energies between 50 and 100 keV, but their resolution decreases with decreasing γ-ray energy; so they can only be used where there is low background radiation and no other lines near that of the Mössbauer transition. Scintillation detectors, however, have a high efficiency. Below 40 keV, gas-filled proportional counters give better resolution than scintillation detectors but have a lower efficiency. Ge(Li) and pure Ge detectors give a very high degree of resolution but their sensitivity can be lower than for other systems. They are very useful when there are several γ-rays of similar energy to be separated.

The output from the detector is passed to a multichannel analyser of the type described in Chapter 8. This may be used in either pulse height analysis mode or time (multiscaler) mode. In the pulse height analysis mode, a reference waveform identical with that supplying the velocity drive system is given a direct current shift so that it is always positive. When a γ-ray that has passed through the sample is detected, the waveform is sampled so that a pulse is produced which has a maximum voltage characteristic of the velocity at that instant. The analogue-to-digital converter transforms this pulse into a train of pulses from a constant-frequency clock, so that the number of pulses is proportional to the voltage. These pulses then pass through the channel addresses of the analyser until terminated, when the number stored in the open channel at that position is increased by one. This gives a measure of the counts in various channels, each of which represents a velocity increment. Unfortunately, this system produces a dead-time error, which may be up to 100 μs, during which no other count can be recorded. A fixed dead time must, therefore, be selected which is at least the expected value of the maximum time for storage; otherwise the dead-time error will change from one channel to another, giving apparently higher count rates for the lower channels. Another problem is the linearity and stability of the analogue-to-digital converter. Changes in baseline of 1% can render the interpretation of Mössbauer spectra, which may be a similar order of magnitude, very difficult.

In the time mode, the analogue-to-digital converter is not used, and so the problems associated with pulse height analysis are avoided. Instead, the channel address is advanced sequentially with a dwell time in each channel of, say 50 or 100 μs under the control of a crystal oscillator. While a channel is open, it accepts all input pulses from the detector. The only dead time is during the change from one channel to the next, and this is very short. Baseline constancy is also determined by the crystal oscillator. Again, the system must be linked to the vibrator driver. This type of system is almost always used in preference to pulse height analysis.

The random nature of γ-ray emission requires that, in order to obtain statistically acceptable counts over the whole of the spectrum, long count periods may have to be employed. If there are N counts registered in a particular channel, there is a standard deviation on that count of $N^{-1/2}$. Thus, for standard deviations of less than 0.1%, 10^6 counts are needed, and count times of the order of several days may be required. Of course, with such long count periods, the danger of equipment malfunction during the course of an experiment, or of accidental interference with the system, increases.

As with other types of γ-ray spectral analysis, curve fitting is an integral part of the interpretation procedures in Mössbauer spectroscopy. This is normally carried out directly by feeding the digitised output to a computer. The general problem of curve fitting can be illustrated by considering N digitised values Y_i. The function to be fitted contains a number of variables represented by the vector V, and gives a calculated value at the data point i of $A(V)_i$. The error e is then given by

$$e = Y_i - A(V)_i.$$

The best possible fit to the N data points is such that the overall error is a minimum. The statistics of radioactive counting processes mean that the values of Y_i will have a Poisson distribution with a variance equal to the mean. The quantity $F(V)$ that needs to be a minimum for best fit is therefore given by

$$F(V) = \sum_{i=1}^{N} \frac{[Y_i - A(V)_i]^2}{Y_i}.$$

In effect, this is a χ^2 function with $N-n$ degrees of freedom. In general, a computed fit with 400 degrees of freedom is within 25–75% confidence limits of a χ^2 distribution if $380<F(V)<419$, and within the 5–95% limits if $355<F(V)<448$ (Gibb, 1976).

In comparison with other applications of curve fitting in radioactivity, the χ^2 values accepted for Mössbauer work can be rather high. This is because the Mössbauer spectrum is taken to be made up of one or more Lorentzian curves and the peaks within the spectrum may not be strictly Lorentzian in shape. Also, if the counting rate is greater than about 10^5 counts s^{-1}, dead-time effects will distort the usual Poisson distribution. The decision whether a fit is acceptable or not is often made semisubjectively by the experimentalist in the light of the known parameters under which the experiment was conducted.

The characteristics of the Mössbauer spectrum obtained by these procedures can give valuable information about the crystalline structure of materials and the sites occupied by specific atoms. The features of the spectrum may be considered in terms of three major properties: the isomer shift, quadrupole splitting and magnetic hyperfine interactions (Fig. 9.28).

The isomer shift or, more strictly, the chemical isomer shift is seen in the displacement of the absorption line from a position about zero velocity. As a result, it is sometimes called the 'centre shift'. This effect is the result of a change in the electrostatic interaction between the

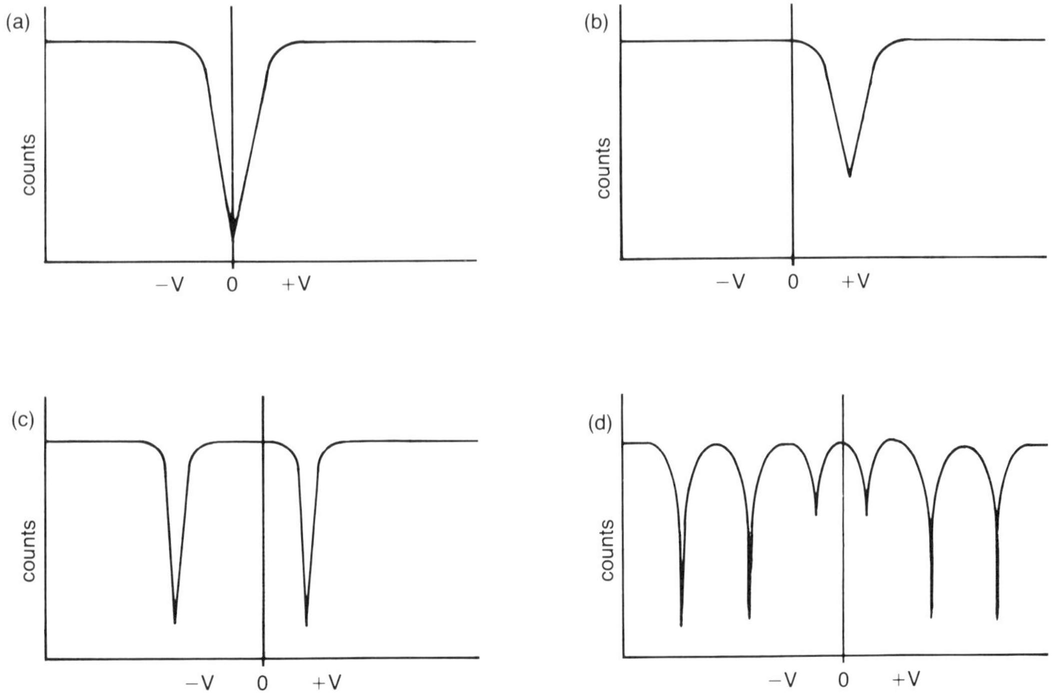

Fig. 9.28 — Characteristic features of Mössbauer spectra: (a) simple absorption; (b) the isomer shift; (c) quadrupole splitting; (d) magnetic splitting.

electronic and nuclear charges in an atom, caused by the change in size of the nucleus when it passes from its ground state to an excited state. If the radius of the nucleus changes by ΔR, there will be a change of electrostatic energy given by

$$\Delta W = \frac{1}{5\varepsilon_0} Ze^2R^2 \frac{\Delta R}{R} |\psi_S(0)|^2,$$

where e is the charge on an electron, Z is the atomic number of the atom, ε_0 is the permittivity of a vacuum, R the radius of the nucleus and $|\psi_S(0)|^2$ is the non-relativistic Schrödinger wave function at $r=0$ (r is an increment of R) representing the s-electron density at the nucleus.

The value of $\Delta R/R$ (typically about 10^{-4}) is characteristic of each transition, and may be of either sign; when positive, the nucleus shrinks on de-excitation. In Mössbauer spectroscopy it is the difference in energy transitions between the source and the absorber which is measured. The isomer shift is given by

$$\delta = \frac{1}{5\varepsilon_0}\left[Ze^2R^2\frac{\Delta R}{R}|\psi_S(0)_{\text{absorber}}|^2|\psi_S(0)_{\text{source}}|^2 \right].$$

This is related to the measured shift in Doppler velocity units by

$$V = \frac{c}{E_\gamma}\,\delta.$$

Because the change in nuclear radius is constant for a particular transition, the nuclear aspects of the isomer shift can be taken as fixed, and the method used to investigate relative changes in electron density. $|\psi_S(0)|^2$ includes contributions from all the s-electrons in the atom but is most sensitive to changes in the outer valence shells.

Magnetic hyperfine interactions arise because the nucleus has a magnetic moment when the spin quantum number is greater than zero. The energy of the nucleus is then governed by interaction with a magnetic field. If I is the spin quantum number, μ the magnetic moment of the nucleus and B the magnetic flux density, the energy levels E_m of the nucleus in the field will be given by

$$E_m = \frac{-\mu B}{I}\, m_2,$$

where m_2 is the magnetic quantum number with possible values I, $I-1$, ... , $-I$. Thus, in effect, the magnetic field splits the energy level into $2I+1$ non-degenerate equi-spaced sublevels with a separation of $\mu B/I$. In Mössbauer spectroscopy, it may be found that a transition occurs from a ground state with spin quantum number I_g and magnetic moment μ_g to an excited state with values I_e and μ_e. In a magnetic field, both states will split according to the above relationship. Transitions may take place between sublevels provided that the rule $\Delta m_2=0$, ± 1, etc., is obeyed. As a result, the Mössbauer spectrum contains a number of resonance lines, but these lines are symmetrical about the centroid. In geological applications of Mössbauer spectroscopy, these effects are normally produced in response to the presence of a magnetic field possessed by a component of the material being analysed. This is known as an 'internal field'.

Electric quadrupole interactions concern the γ-ray transitions that occur between the hyperfine multiplets of the nucleus in its ground state and in an excited state. Only those nuclear states with $I>\frac{1}{2}$ have a nuclear quadrupole moment and show quadrupole hyperfine splitting. This may be caused either by $I_g>\frac{1}{2}$ or $I_e>\frac{1}{2}$ (irrespective of a lower value of I_g). The nuclear quadrupole moment is a measure of the deviation of the nuclear charge from spherical and is given by

$$eQ = \int \rho r^2\,(3\cos^2\theta - 1)\,\mathrm{d}r,$$

where e is the charge on the proton, ρ is the

charge density in the volume element dr at a distance r from the centre of the nucleus and at an angle θ to the spin axis. Q is positive when the nucleus is elongated along the spin axis and negative when flattened in the spin.

The electrostatic potential at the nucleus due to a point charge q at distance r is given by

$$V = \frac{q}{4\pi\,\varepsilon_0\,r},$$

where ε_0 is the permittivity of a vacuum. In a system of Cartesian axes, the three finite principal values of the electrostatic potential may be defined as V_{xx}, V_{yy} and V_{zz}. However,

$$V_{xx} + V_{yy} + V_{zz} = 0,$$

and so it is only necessary to specify two parameters to describe the field gradient (second derivative of V). The asymmetry parameter η, however, is defined in terms of all three principal values of V:

$$\eta = \frac{V_{xx} - V_{yy}}{V_{zz}},$$

such that $|V_{xx}| > |V_{yy}| > |V_{zz}|$ and $0 \leqslant \eta \leqslant 1$, with the z axis then referred to as the major axis, and x the minor axis. In the simple case in which the electric field gradient possesses axial symmetry ($\eta = 0$), the possible energy levels are given by

$$E_q = \frac{e^2qQ}{4I(2I-1)}\,[3I_z^2 - I(I+1)],$$

where the quantum number I_z can take the $2I+1$ values of I, $I-1$, ..., $-I$. When $I=\pm\frac{3}{2}$, for example, there are two energy levels at $+e^2qQ/4$ for $I_z=\pm\frac{3}{2}$, and $-e^2qQ/4$ for $I_z=\pm\frac{1}{2}$. Of course, more complex systems exist, but in Mössbauer spectroscopy the most common example of this is, in fact, very similar and concerns the transition $I_g=\frac{1}{2}\rightarrow I_e=\frac{3}{2}$, in which the ground state is unsplit (at $I_z=\pm\frac{1}{2}$) but the excited state has two levels with $I_z=\pm\frac{3}{2}$ and $\pm\frac{1}{2}$. The separation of these levels is $(e^2qQ/2)(1+\eta^2/3)^{\frac{1}{2}}$, and so the resultant Mössbauer spectrum is a doublet, the separation of the peaks being known as the 'quadrupole splitting'. This allows determination of η and e^2qQ. The second term is known as the 'coupling constant', which is the product of the nuclear constant eQ and the maximum value of the potential. These parameters may be directly related to the valence electrons of the atom concerned. Thus the total value of q for either a completely filled or a half-filled shell is zero, while an excess of electrons along the z axis gives negative values of q and an excess in the xy plane gives positive values. Only p- and d-electrons have any effect, the spherically symmetrical s-electrons giving no electric field gradient. However, it should be noted that some materials, although possessing high symmetry with either filled or half-filled shells, still show quadrupole splitting. This is because there may be charges external to the atom which produce a large electric field gradient at the nucleus. This is known as the 'lattice contribution'. Finally, it should be remembered that both magnetic and quadrupole hyperfine effects result from the directional interaction of the nucleus with its environment; so, when both are present together, the results may be complex, especially if their respective principal axes are not parallel (Gibb, 1976).

Identification of the atomic parameters governing the chemical isomer shift, the magnetic hyperfine interaction and quadrupole splitting can thus all be assessed merely from the positions of the spectral lines, and the contribution to the spectrum of a specific element which may occupy different sites in a single crystal can be determined. However, if a sample which is composed of numerous small, randomly orientated crystals is considered, the intensity of the spectral lines will be governed by the abundance of the particular site occupancies. The relative proportion of different site occupancies can, therefore, be assessed simply by apportioning the various components of the spectrum as

(a)

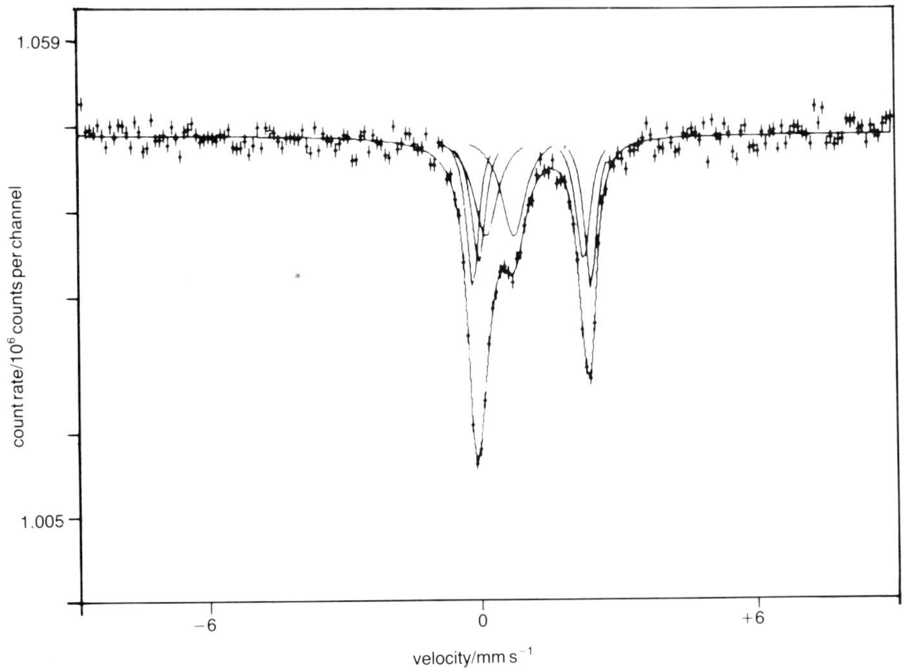

(b)

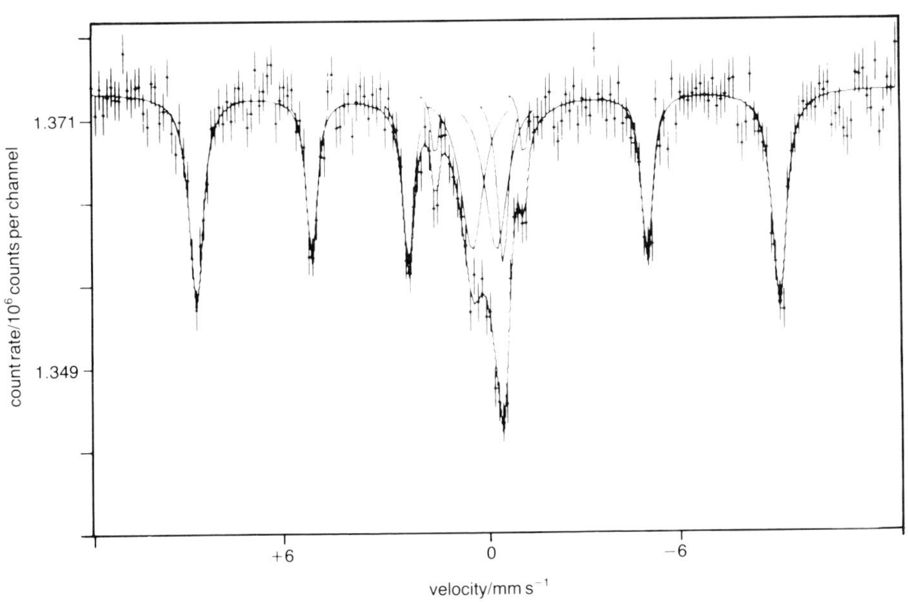

Fig. 9.29 —Mössbauer spectra from the Littleham Mudstone Formation of southeast Devon (ordinates show counts per channel$\times 10^6$; abscissae show velocity in millimetres per second): (a) a green mudstone at −200 °C containing a Fe-bearing clay mineral; (b) a red mudstone at −200 °C containing haematite in addition to the Fe-bearing clay mineral. (Durrance *et al.*, 1978, reproduced with permission.)

percentages of the total. Provided that the recoil-free fractions are equal, comparison is direct. However, equality is not always the case; indeed, any ratio of recoil-free fractions may also change with temperature.

In geological (and, indeed, other) applications, Mössbauer spectroscopy is principally used for investigating the occupancy of lattice sites by Fe ions and can yield Fe^{2+}/Fe^{3+} ratios without involving the digestion of samples (Fig. 9.29). Almost invariably, Fe^{2+}/Fe^{3+} ratios determined by Mössbauer spectroscopy are greater than the chemically determined ratios for the same samples. This implies that great caution is needed in interpreting the chemical data.

9.9 DISEQUILIBRIUM IN THE ^{238}U DECAY SERIES

9.9.1 Disequilibrium characteristics

Before considering how radioactive disequilibrum can be used as an exploration tool for U ore deposits and for monitoring active geological processes, some general discussion of disequilibrium is needed. It will be recalled that a decay chain is in secular equilibrium when there is neither gain nor loss of any parent or daughter nuclide, other than by radioactive decay and growth. Disequilibrium develops in chemically open systems when the different physical and chemical properties of each nuclide permit their fractionation. The ^{238}U series is particularly susceptible to the development of disequilibrium because it contains several chemically distinct nuclides with long half-lives (^{234}U, ^{230}Th and ^{226}Ra). Numerous studies have shown that these nuclides may be separated during weathering and other alteration processes.

The isotopic fractionation of ^{238}U and ^{234}U is particularly important. Disequilibrium between these isotopes was first observed in the secondary mineral schroeckingerite (Cherdyntsev, 1955), an example of which exhibited a $^{234}U/^{238}U$ activity ratio of 3.7, while excess ^{234}U was noted in solutions from leached U minerals. $^{234}U/^{238}U$ activity ratios between 0.56 and 1.3 were also found in other secondary U minerals. However, although the possibility of disequilibrium in the ^{238}U series was recognised, it was initially thought to be unimportant, as it was believed that in most alteration processes equilibrium would be restored faster than the rate of alteration. Loss of ^{222}Rn from the decay series was only recognised as a problem that could arise during sample preparation (despite the use of exploration methods based on ^{222}Rn in soil gas).

When disequilibrium in the ^{238}U decay series was discovered to be a common occurrence, it was realised that the members of the decay series could be classified into a number of groups. Moreover, this can be extended to include the ^{235}U and ^{232}Th decay series (Fig. 9.30). The long-lived nuclides ^{230}Th and ^{226}Ra in the ^{238}U series, and ^{231}Pa in the ^{235}U series, may suffer chemical fractionation and are placed in separate groups, ^{231}Pa with its daughters. ^{222}Rn and its daughters also form a separate group, as this nuclide is gaseous and may be lost from the system. The ^{210}Pb group is headed by a nuclide with a relatively long half-life (22 years). The U group contains all three common U isotopes, although fractionation of ^{234}U and ^{238}U does, of course, occur.

On this basis, a six-fold classification of disequilibrium patterns has been established by Rosholt (1959):

(1) *Daughter product deficiency*. This may be due to very recent introduction of the parent, or to chemical leaching of daughters; the latter process is considered more important, especially where U is fixed by V or phosphate; ^{226}Ra is the most easily leached daughter, while loss of ^{222}Rn does not appear to be a major cause of long-term disequilibrium.

(2) *Time-related daughter product deficiency*. Here, the amount of each daughter present is such that they have all been generated over the same period of time;

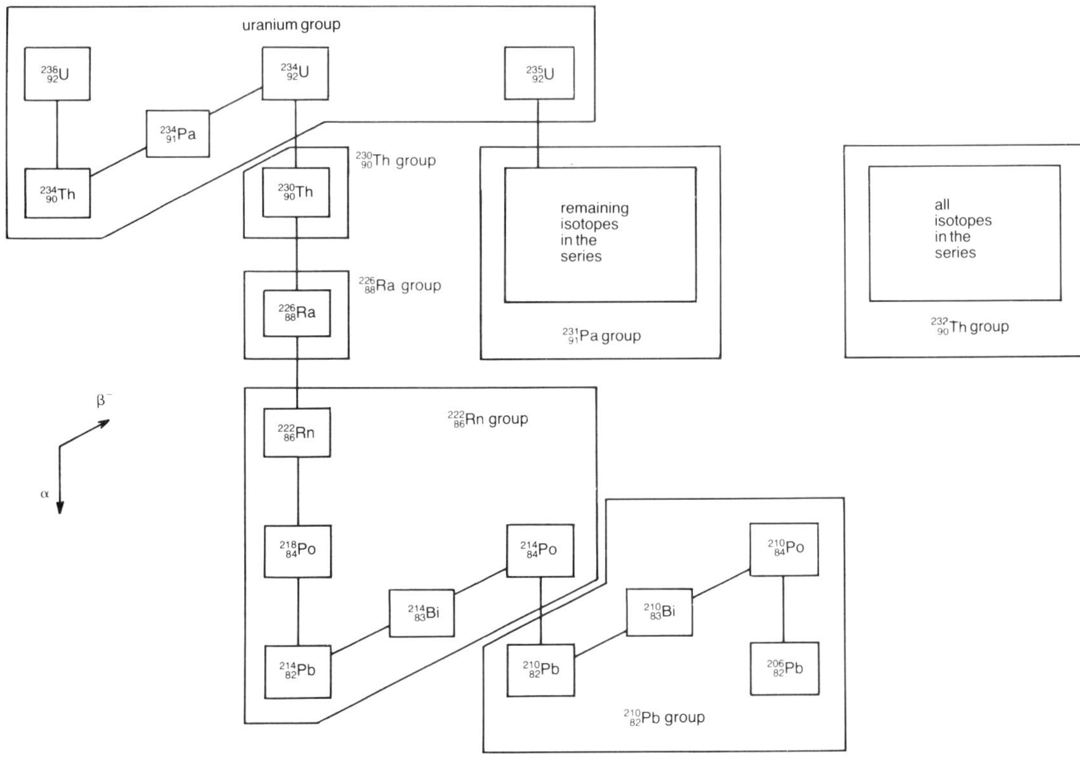

Fig. 9.30 — Classification of the ^{238}U, ^{235}U and ^{232}Th decay series into disequilibrium groups. (After Rosholt, 1959.)

that is, they have all grown following the introduction of ^{238}U but have not had time to reach equilibrium.

(3) *^{230}Th deficiency*. This usually arises through deposition of ^{238}U from a solution from which other U daughters were also deposited but which did not carry the relatively insoluble Th isotope.

(4) *Daughter product excess*. This is usually found in oxidising environments and is produced by leaching of the parent ^{238}U; daughter product addition does not normally occur but is sometimes indicated by a low ^{230}Th content; ^{231}Pa is particularly important in this type of disequilibrium because it is easily hydrolysed and precipitated from solution once mobilised and is thus the least mobile of the U decay products.

(5) *^{226}Ra excess*. In some very acid environments associated with oxidising sulphide deposits, leaching of U and other daughters takes place, but Ra is retained as an insoluble sulphate.

(6) *The presence of Ra isotopes exclusively*. This type of disequilibrium, found in oil field brines and radioactive hot springs and their deposits, results from the coprecipitation of Ra with Ba and Sr sulphates and Fe oxyhydroxides; the low ^{210}Pb content of these precipitates often indicates a very recent origin.

This system of classification closely reflects the disequilibrium observed in the geological environment. In the sandstone-type U deposits of New Mexico, for example, disequilibrium is almost universal. As ^{231}Pa is the least mobile of the nuclides involved, it may be used as an index to which excess or deficiency of other isotopes can be related (Table 9.5). On this basis, there is a consistent deficiency of ^{238}U, ^{234}U, ^{230}Th and ^{236}Ra within the ore bodies, while the adjacent rock has an excess of ^{226}Ra. This indicates that ^{238}U, ^{234}U and ^{230}Th have been removed. ^{226}Ra, although mobilised, was soon redeposited. $^{234}U/^{238}U$ activity ratios of about 1.5 were found in recent marine deposits soon after the widespread existence of disequilibrium was recognised. Later work, however, has revealed that $^{234}U/^{238}U$ activity ratios are very variable in these deposits, ranging between a 40% ^{234}U excess and a 40% ^{234}U deficiency.

In general, ^{234}U is leached preferentially to ^{238}U. This is because the recoil associated with α-particle emission disrupts the chemical bonds holding the isotope. Fractionation is then facilitated by the different chemical properties of the intermediate nuclides, ^{234}Th and ^{234}Pa. It is also suggested that, in all but the most reducing environments, ^{234}U is produced in the hexavalent state and is removed as UO_2^{2+}, changing later to the tetravalent state. Such selective removal of ^{234}U probably accounts for the $^{234}U/^{238}U$ activity ratio of 2.3 obtained from groundwater near Wolverhampton (Hill and Crookall, 1963).

Preferential removal of ^{234}U, of course, leaves the altered rock depleted in this isotope. Thus, in the Conway granite of New Hampshire, there is a 15% ^{234}U deficiency in the near-surface environment, although equilibrium exists at depth (Richardson, 1964). Similarly, in the alteration of U-bearing mineral deposits in Wyoming, extreme ^{234}U deficiency occurs; ^{234}U enrichment in the altered zone near the ore is attributed to deposition from waters which carry ^{234}U away from the ore in solution (Rosholt *et al.*, 1964). In these enriched zones, the decay of ^{234}U has produced high ^{231}Pa and ^{230}Th concentrations.

In the Elliot Lake U mining district of Canada, however, ^{238}U deficiency is commonly developed as a result of leaching, while ^{238}U enrichment occurs at sites of recent deposition. ^{231}Pa excess is found in highly weathered material, probably as a result of the leaching-out of all other members of the decay series. Generally, the degree of disequilibrium decreases with depth. It seems likely, therefore, that rapid changes in the nature of radioactive disequilibrium can occur even over very small areas and over very small changes in depth. Moreover, highly leached zones may be located very close to sections which are virtually unaffected by leaching (Ostrihansky, 1976).

The fractionation of U and Ra in stream waters and sediments is another important aspect of disequilibrium. In the streams of the Bancroft area of Ontario, ^{226}Ra is enriched in clastic sediments relative to ^{238}U, as a result of either the leaching of U or the precipitation or adsorption of Ra. In organic sediments, however, a ^{226}Ra deficiency has been produced by preferential precipitation of U. Ra is found to be enriched in stream waters relative to U, because of the removal of U by organic matter.

In Finnish well waters, $^{234}U/^{238}U$ activity

Table 9.5 — ^{238}U series disequilibrium in ore deposits (nuclide activities).

$^{238}U > ^{231}Pa > ^{230}Th > ^{226}Ra$	Recent leaching of daughters or addition of U parent. Generally the former especially where V or PO_4 complexes present
$^{238}U \gg ^{231}Pa > ^{230}Th > ^{226}Ra$	Recent accumulation of U parent
$^{238}U < ^{231}Pa > ^{230}Th < ^{226}Ra$	Recent deposition of U plus daughters excluding ^{230}Th
$^{238}U \ll ^{231}Pa > ^{230}Th > ^{226}Ra$	U leaching. Surficial environments
$^{238}U < ^{231}Pa > ^{230}Th \ll ^{226}Ra$	Differential leaching of all decay series members. Surface ores
$^{238}U = 0$ $^{231}Pa = 0$ $^{228}Ra < ^{226}Ra$	Addition of Ra. Oil field brines and hot springs

(After Levinson *et al.*, 1982.)

ratios between 0.76 and 4.67 have been reported, with an average ^{234}U excess of 60%. These waters also show $^{238}U/^{226}Ra$ activity ratios between 1 and 20. This indicates a high U mobility in the waters studied.

The excess of ^{226}Ra over ^{238}U in many oil field brines is attributed to the highly reducing nature of the brine environment, which favours Ra mobility but in which U^{4+} is highly insoluble. However, where the brines encounter waters rich in sulphate, the Ra is precipitated as $RaSO_4$.

In the Grenville rocks of Canada, disequilibrium in the ^{238}U decay series extends to considerable depth in boreholes. The most common types of disequilibrium observed are daughter enrichment (commonly ^{226}Ra) and ^{222}Rn loss. Interestingly, the ^{232}Th decay series is not disturbed (Szoghy and Kish, 1978).

The possibility of radioactive disequilibrium must, therefore, be taken into account in the interpretation of radiometric and geochemical exploration data. γ-ray spectrometry in which the 1.76 MeV ^{214}Bi peak is measured, for example, is based upon the assumption of equilibrium in the ^{238}U series, and the development of disequilibrium will render unreliable any U values determined by this method. Conversely, there are a number of U exploration methods which assume disequilibrium in the ^{238}U decay series. For example, if U is being leached from a U-bearing granite source, transported and then precipitated, possibly to ore grade, in a favourable environment, disequilibrium studies can be used to trace these processes and to enable the location of ore deposits to be predicted (Levinson *et al.*, 1982). This is a particularly valuable approach when used in combination with detailed investigations of the mineralogy and geochemistry of the environments in question.

Low $^{238}U/^{206}Pb$ mass ratios coupled with high $^{206}Pb/^{204}Pb$ and $^{207}Pb/^{204}Pb$ mass ratios can be used to indicate that U levels were much higher in the past than at present and that leaching of U must have occurred. If the extent of U leaching indicated by the degree of disequilibrium can be related to the total volume of rock affected, the mass of U mobilised can be estimated. It has also been argued that Th/U mass ratios of 5 or greater in igneous rocks (rather than the normal value of about 3.5) suggest that leaching of U has occurred (Stuckless and Nkomo, 1978). It should be recognised, however, that the character of the U-bearing mineral in the source rock is also important; minerals rich in Th, for example, are usually relatively unaffected by leaching processes.

The variable leaching and fractionation of ^{234}U and ^{238}U in U-bearing minerals are controlled by the location of the U atoms in the crystal lattice. In U ore minerals, the ^{234}U recoil atoms are able to occupy the positions of ^{238}U atoms. When leaching occurs, therefore, both isotopes may enter solution to about the same extent, producing a $^{234}U/^{238}U$ activity ratio in the leachate of about 1.0. If U is only a minor constituent of the mineral, however, the ^{234}U recoil atoms leave their original positions, are not able to find suitable alternative sites and migrate to microfractures from which they can then be leached. This leads to $^{234}U/^{238}U$ activity ratios in the leachate of greater than 1.0. The study of $^{234}U/^{238}U$ activity ratios in groundwaters or surface waters can, therefore, indicate whether the U in solution has been derived from a suitable source rock or from trace occurrences. This distinction is not necessarily given by the absolute quantity of U in solution, which is determined more by the Eh–pH conditions of the water and the presence or absence of suitable complexing ions in solution. Consequently, it is clear that, taken together, the $^{234}U/^{238}U$ activity ratio and total U can greatly enhance the effectiveness of hydrogeochemical exploration. An abrupt increase in the $^{234}U/^{238}U$ activity ratio, combined with a decrease in U concentration, is particularly significant in helping to locate U ore deposits (Cowart and Osmond, 1977).

In contrast with the usefulness of U isotopes, Th is only very locally dispersed in water, except in special geochemical circumstances (in water with a very low pH, for example), and so the use of ^{230}Th is limited.

^{226}Ra is extremely important in disequilibrium studies: firstly, it is the parent of the short-lived nuclides upon which a number of exploration methods are based; secondly, with its mobility and relatively long half-life (1602 years), it is a potential pathfinder for U in its own right. However, the very mobility of Ra means that it can give rise to significant radiometric anomalies which may be incorrectly interpreted as due to U deposits.

^{226}Ra is potentially mobile and can be leached from its parent ^{230}Th to form soluble complexes. Precipitation of Ra takes place by formation of insoluble compounds (such as radiobarite as discussed above), by coprecipitation with Fe and Mn oxides and hydroxides or by adsorption onto Fe and Mn oxides, clay minerals or organic matter. Ra can, therefore, be considered similar in its behaviour to U but, while U and Ra may be leached from a source rock together, Ra may be preferentially left behind (if, for example, water passes through a clay-rich zone which adsorbs the Ra).

Radioactive springs and well waters typically carry Ra in solution without U, giving rise to anomalies which may incorrectly be attributed to U mineralisation. As these anomalies are often associated with hydrothermal circulation systems, it has been suggested that, at elevated temperatures, radium carbonate is soluble while the uranyl carbonate complex is unstable. Thus, Ra can remain in solution but U cannot. In passing through a U-bearing host rock, circulating groundwaters may thus cause dispersion of Ra. Even trace amounts of the ^{238}U decay series in rocks may give rise to significant Ra concentrations in groundwaters by this means, producing false anomalies which might be interpreted erroneously as indicating the presence of economic U deposits elsewhere in the circulation system. Because of the short half-life of ^{222}Rn, it might be suggested that ^{222}Rn anomalies can be reliably interpreted as ^{226}Ra anomalies but, owing to the emanation and transport characteristics of Rn, this is not normally the case. In fast-moving groundwater systems, in particular, separation of ^{222}Rn from its parent can be considerable, highlighting the need for caution in the interpretation of γ-ray spectrometry data based upon ^{214}Bi. False ^{226}Ra anomalies can also give rise to false ^{4}He anomalies, because of the eight α-particles released in the decay of ^{238}U five are formed between ^{226}Ra and ^{206}Pb and all the radionuclides involved have relatively short half-lives.

^{210}Pb and ^{210}Po have also been used in exploration work. The ^{210}Pb activity in a sample indicates the amount of ^{222}Rn that has decayed in it. As ^{210}Pb has a half-life of 22 years, such measurements are unaffected by loss of ^{222}Rn during sample preparation. All intermediate decay products between ^{222}Rn and ^{210}Pb have very short half-lives and the series can be reasonably assumed to be in equilibrium (Levinson *et al.*, 1982). Thus, ^{210}Pb provides a record of ^{222}Rn activity in an environment. Following the decay of ^{222}Rn, ^{210}Pb is deposited in soils or stream sediments, reaching its maximum activity with unsupported ^{226}Ra after about 100 years. Thus ^{210}Pb acts as a natural integrator of the ^{222}Rn flux over this period of time. The ^{210}Pb method is, therefore, considered to be an alternative to the Rn survey methods, which require long integration periods discussed earlier in this chapter. In turn, the decay of ^{210}Pb to ^{210}Bi and then to ^{210}Po allows detection of the ^{210}Pb by measurement of the α-particle activity of ^{210}Po, assuming that the relatively short half-lives between ^{210}Pb and ^{210}Po will ensure equilibrium conditions. Although the method cannot be a substitute for surveys of ^{222}Rn in water, it is considered to be particularly suitable for soil surveys, as ^{210}Pb is very effectively sequestered by organic matter in the A horizon. Sampling of soil below the A horizon is believed to eliminate the surface contributions of ^{210}Pb which are liable to change with the weather and with the seasons. In addition, the use of $^{210}Pb/U$ ratios is a valuable exploration method for U-bearing mineralisation, as it measures the integrated excess ^{222}Rn concentration with respect to the ^{222}Rn emanation factor at each site. However, just as ^{222}Rn migration times are controlled by soil porosity and permeability, so ^{210}Pb and ^{210}Po concentrations will be greater in clayey soils than in sandy soils, all other factors being equal. Therefore, by normalising the ^{210}Pb or ^{210}Po values to the surface area of the soil sample, it is possible to identify anomalous features due to (1) U mineralisation at depth, (2) decreased ^{222}Rn migration in clayey soils and (3) minor U occurrences in the soil samples.

It is also of interest to note here that long-term integration of the ^{222}Rn flux can be recorded by certain plants, such as mosses and lichens. These derive nutrients from atmospheric moisture and have long lifetimes during which they accumulate atmospheric ^{210}Pb formed from the decay of ^{222}Rn. Mapping of ^{210}Pb levels using these plants can allow areas of anomalous U to be identified, but it must not be assumed that anomalies are based solely on atmospheric distributions, as the plants may also show a distribution controlled by the substrate.

Lastly, it should be noted that ^{206}Pb, the stable end member of the ^{238}U decay series, can also be used for exploration purposes as, indeed, can ^{207}Pb, the end member of the ^{235}U decay series. In particular, it has been observed that $^{206}Pb/^{204}Pb$ ratios are less than 18 in the rocks of areas of good ore potential but, where the ore potential is poor, this ratio rises to over 20.

9.9.2 Measurement of U series disequilibrium

The method most often used for the measurement of disequilibrium in the U decay series is α-particle spectrometry, although γ-ray spectrometry is also widely employed. Neutron activation analysis, fission track analysis and α-track analysis are also used, but the first and second of these depend on access to a reactor or other neutron source, and the second and third are of limited sensitivity and accuracy. Mass spectrometry may also be employed.

α particle spectrometry

A general discussion of α-spectrometry was presented in Chapter 8. Here, U decay series measurements will be considered specifically, using arrangements of the type illustrated in Fig. 9.31. It should first be noted that, in some cases, the α-particle energies emitted by different nuclides are very similar, and straightforward resolution of their abundances is not possible. In these circumstances, it is necessary to carry out chemical separation of the elements of interest before α-particle spectrometry is worthwhile. Such techniques, however, can only be successfully applied where there is a strong chemical distinction between the nuclides involved. Also, the extraction efficiency of these chemical techniques varies with the element concerned, and should be known. This information is obtained by adding a 'spike' of known activity of a particular isotope of each element of interest before the separation process begins. The amount of the spike recovered after separation is then measured. Clearly, the spike should not contain isotopes that are also present in the sample, and it must have α-particle energies that are distinguishable from those of the natural radionuclides. Moreover, there must be no chemical fractionation between the spike and other isotopes of the same element in the sample.

The chemical procedures usually involve three stages: (1) pre-concentration or extraction, (2) separation and (3) preparation of a suitable source for α-particle spectrometry (Lally, 1982). Of the members of the ^{238}U decay series, isotopes of U, Th, Pa, Ra, Rn, Pb and Po are usually investigated in this way. These may occur in rocks, soils or waters of various compositions, usually in concentrations at the level of parts per billion or parts per million.

The kind of pre-concentration or extraction procedure adopted depends upon the nature of the sample. For rocks and soils, some type of acid digestion is usually used to obtain a liquid sample. This is then passed through a pre-concentration process such as adsorption onto ion-exchange resins or coprecipitation. Various

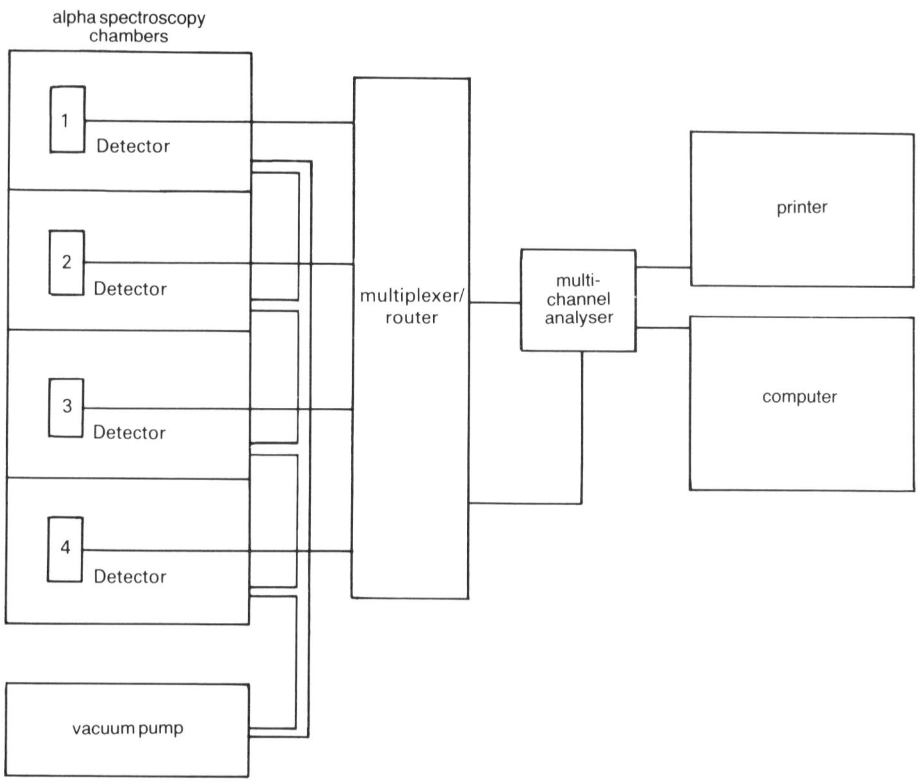

Fig. 9.31 — Diagrammatic representation of a four-detector α-particle spectrometer. (After EG & G Ortec, 1983.)

ion-exchange resins are commercially available for effective adsorptive pre-concentration. For coprecipitation, Fe and rare earth hydroxides are usually used.

Ion-exchange resins are commonly employed in the separation process, usually by the column-separation method. They are essential for separation of U, Th and Pa and may also be used for separation of Ra, Pb and Po. In the column-separation method, the solution is passed through a vertical cylinder filled with the appropriate ion-exchange resin which removes the ion of interest. As the solution passes down the column, equilibrium is established with the resin according to the distribution coefficient K_d, which is defined by

$$K_d = \frac{C_1}{C_2},$$

where C_1 is the amount of the ion (in grams) adsorbed per gram of dry resin and C_2 is the amount of the ion which remains in solution (measured in millilitres) after equilibrium has been established. This method must be used where K_d values are less than 10 or where separation of elements with similar K_d values is required. Under other circumstances an alternative approach, the batch process, may be employed. In this, sufficient resin of the appropriate type is mixed with a known volume of solution. After time for adsorption to occur has elapsed, the resin is separated, and desorption of the adsorbed ions is carried out (Lally, 1982).

The other separation technique, which is used for Ra, Pb and Po, is conventional precipitation. Ra can be precipitated as sulphate but, unfortunately, Pb is coprecipitated at the same time, and further treatment is necessary.

Pb separation involving sulphate, chloride, chromate and sulphide have all been employed. Po is frequently coprecipitated with Te or MnO_2. Finally, He and Rn are separated by pumping off, or by degassing followed by collection in cold traps (as discussed in Chapter 8).

The method selected for the preparation of sources for α-spectrometry depends upon the spectrometric method to be employed. In general, the main requirement is that the source should be as thin as possible and yet uniformly distributed on a flat disc to be placed in a vacuum chamber housing the detector (Fig. 9.32). The necessity of a thin source arises from the need to minimise the effects of self-absorption of the α-particles, which causes degradation of energy and loss of resolution. Thin sources are obtained either by evaporation of an organic complex of the element of interest directly onto a metal planchet or by electrodeposition from a variety of electrolytes. However, as thin sources require lengthy preparation, during which short-lived but important nuclides are lost, alternative thick-source techniques have been proposed for specific purposes. One of these involves the computation of a theoretical equilibrium α-particle energy spectrum for the sample, against which the actual measured spectrum can be compared and the major feature of disequilibrium identified (Killeen and Carmichael, 1976). Source preparation for mass spectrometry also depends upon the particular type of spectrometer to be used but, in general, a purified fraction of the element is evaporated onto a Re or W filament prior to measurement. For ionisation chambers and gas scintillation counting, O_2 and water vapour have to be removed before counting commences.

Because one of the most interesting aspects of U disequilibrium studies is the determination of $^{234}U/^{238}U$ and $^{230}Th/^{238}U$ activity ratios in rocks (as well as determination of ^{238}U and

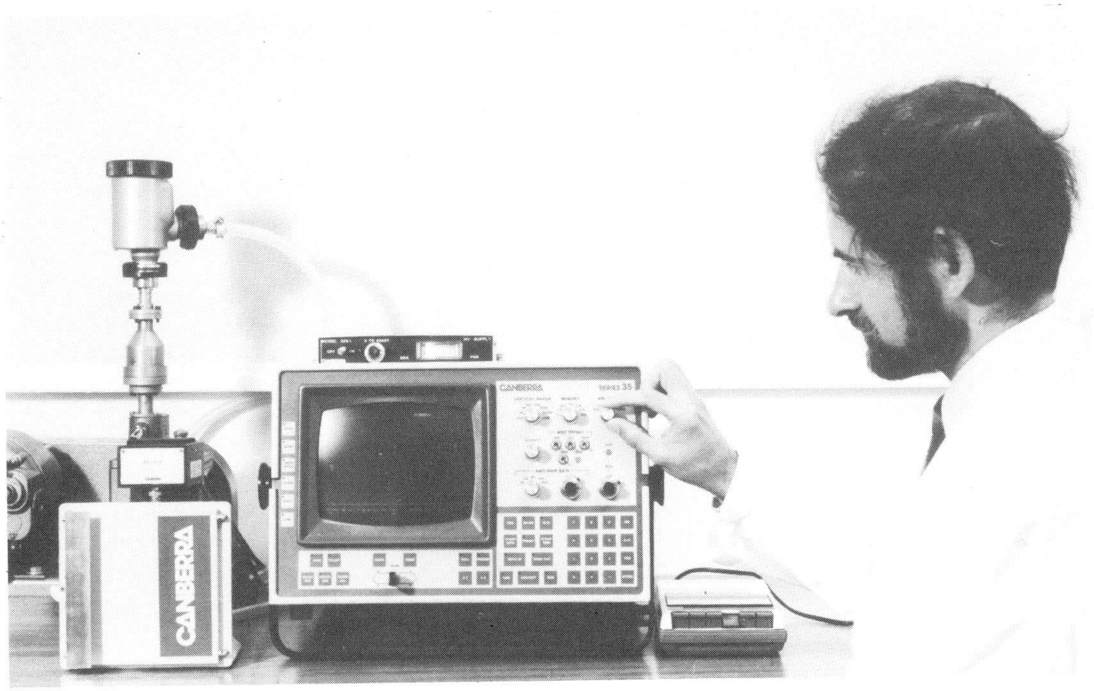

Fig. 9.32 — Single-chamber α-particle spectrometer using a Canberra Series 35 multichannel analyser.

^{232}Th abundances), a brief account of the methods used by Heath (1982) for the separation of U and Th will be given to illustrate the general principles outlined above. Firstly, the rock is reduced to gravel-sized fragments in a jaw crusher, and 100 g of this material is ground to a grain size of less than 50 μm using a tungsten mill. The powder is then carefully dried, and 0.1 g is pressure digested in a polytetrafluoroethylene beaker in a stainless steel bomb at 110 °C for 2 h with 6–7 ml of concentrated HF solution, 0.75 ml of concentrated HCl solution and 0.75 ml of concentrated HNO_3 solution. The acid is then evaporated off in a fume cupboard under an infrared lamp until almost dry, when the residue is taken up in deionised water and transferred to a volumetric flask in which the solution is made up to 100 ml with deionised water. It is then convenient to transfer the solution to a polypropylene container from which a carefully weighed sample is taken for separation. At this stage the spike is added; in this case a precisely weighed amount of a ^{232}U–^{228}Th tracer solution (half-lives, 74 years and 1.91 years, respectively). The activity of the tracer should be approximately equal to that expected from the U content of the sample so that the sample and spike peaks in the α-particle energy spectrum are of similar magnitudes. The activities of the ^{232}U and ^{228}Th in the standard solution must be determined regularly, and the time that has elapsed since the last determination must be recorded, so that the exact activity of these nuclides at the time of counting is known. As ^{228}Th has a short half-life, the effect of ingrowth of this isotope after source preparation needs to be minimised by avoiding delay before counting commences. Similarly, with a half-life of only 3.64 days, ^{224}Ra ingrowth will take place. This isotope will, therefore, contribute to the α-particle energy spectrum, and so its activity at the time of counting must be known. ^{228}Th is generated during decay of ^{232}Th but, with its long half-life, its contribution can be ignored.

After weighing and spiking, the sample is washed into a beaker with 6 M HCl and evaporated to dryness on a hot-plate at 80 °C under an infrared lamp. If the sample contains excess Fe (shown by a yellow colour), it should be washed with 25 ml of 6 M HCl into a separation funnel and an equal amount of an active solvent (such as isobutyl methyl ketone) added. After thorough mixing, the acid is allowed to separate. It is then run off into a beaker and evaporated to dryness.

The next stage is the preparation of the ion-exchange columns. Typically, two resins are used, one in chloride form (such as Amberlite) and one in nitrate form (such as Dowex). The columns are prepared by plugging the bottom with glass wool, filling with resin suspended in deionised water to within 1 cm of the opening of the S-pipe and plugging the top with glass wool. The chloride column is then generated by eluting with 60 ml of 0.1 M HCl followed by 60 ml of 9 M HCl, and the nitrate column by 60 ml of 0.1 M HCl followed by 60 ml of 7 M HCl.

The sample is then dissolved in 9 M HCl and transferred to the chloride exchange column which adsorbs U, Fe^{3+} and Pa, while Th and other cations are passed through the column and collected separately. U, Fe^{3+} and Pa are then flushed from the column by adding 0.1 M HCl. The two solutions, one containing U and the other Th, are then evaporated to dryness, taken up in 7 M HNO_3 and, in turn, added to the nitrate column. Here U and Pa (or Th) are removed while Fe and other cations are passed through. The U and Pa (or Th) are then released by flushing with 0.1 M HCl, and five drops of $HClO_4$ are added to destroy organic matter before evaporating to dryness.

The final step is the preparation of the thin source. The first stage is to dissolve the sample in five to ten drops of 18 M H_2SO_4 and then evaporating to dryness in a furnace at 360 °C for 1 h to remove Cl^-. The residue is then dissolved in $(NH_4)_2SO_4$ solution (1 M for U, and 2 M for Th) and transferred to an electrodeposition cell for about 6 h for Th and 3 h for U, at a current of 0.5 A. About 3 min before electrodeposition ends, five drops of NH_4OH solution are added to raise the pH and to remove the remaining U and Th from solution. The planchets are then rinsed in deionised water and dried. The samples are then ready for transfer to the vacuum chamber containing the α-particle detector and counting can commence as discussed in Chapter 8.

To keep the counting error due to the statistical nature of the radioactive decay process down to a limit of 1%, each peak due to a specific nuclide must contain at least 10^4 counts. To achieve this, it may be necessary to utilise count times of a day or more, depending on the U and Th content of the sample. Generally, it is good practice to run counts regularly on blank sources for the same length of time as that required by the natural sources, so that a check is made of any variation in the background. This is important because, although under conditions of constant geometry it may be thought that background effects remain the same, this is not the case. It is true that the background due to noise in the detector, electronics and cosmic rays will be effectively constant, but the background caused by recoil nuclei, such as daughters of the ^{232}U spike, ^{228}Th, ^{224}Ra and ^{220}Rn, moving across the gap between the source and the detector, is variable and must be known accurately when counting of low activity samples is taking place. Because the background due to embedded recoils will be different for U and Th sources, these should, if possible, be determined on separate dedicated U and Th detectors.

Once the appropriate spectrum has been obtained, it is necessary to determine the peak areas. Peak integration is ideally carried out over about the same number of channels for each of the peaks in the spectrum, such that the limits of the integration are given by those channels in which the count has fallen to zero. However, this is often not possible, as background spread across the whole spectrum, or thick sources combined with poor detector resolution and high activity, may give rise to peaks with long low-energy tails. Under such circumstances, the limits of integration are often set at 1% of the peak's maximum count, but this is only possible where peaks do not overlap. Where a low energy tail encroaches on another peak, it may be necessary to compute the exponential decay of the tail to zero and thus to use a theoretical zero-count channel as an integration limit. Alternatively, it is common practice to use the shape of the low energy tail of an isolated peak in the spectrum to give a simple approximation for peaks where overlap makes direct identification of shape impossible. Detector efficiency is determined using a calibration source.

Provided that sources are counted soon after separation, the U isotopes ^{238}U, ^{235}U, ^{234}U and ^{232}U are easily identified, and the only correction that is applied after integration is for background. However, for Th, the analysis of the spectrum is more complex because of rapid ingrowth of ^{224}Ra under the ^{228}Th peak. True ^{228}Th activity is obtained by taking the 94.5% count rate for the main ^{224}Ra peak and, after allowing for background, using this to calculate the 5.1% component that is present in the ^{228}Th count rate.

Extraction efficiences for U and Th are determined from the ^{232}U and ^{228}Th count rates in their respective spectra using the known detector efficiencies. The extraction efficiencies are given by

$$Y_U(\%) = \frac{(^{232}\mathrm{U})100}{(E/100)W_s(^{232}\mathrm{U})_s},$$

where E is the detector efficiency, W_s is the weight of added spike in grams and $(^{232}\mathrm{U})_s$ is the concentration of ^{232}U in the spike (in disintegrations per minute per gram), and

$$Y_{Th}(\%) = \frac{(^{228}\mathrm{Th})100}{(E/100)W_s(^{232}\mathrm{U})_s R_s},$$

where R_s is the ^{228}Th/^{232}U activity ratio in the spike. (^{232}U) and (^{228}Th) are the integrated count rates of the peaks in their respective spectra (Ivanovich, 1982). Thus the U concentration in a sample of weight W is given by

$$U(\text{ppm}) = \frac{(^{238}\text{U})W_s\,(^{232}\text{U})_s}{0.747W(^{232}\text{U})}$$

and the ^{230}Th/^{234}U activity ratio is given by

$$\frac{^{230}\text{Th}}{^{234}\text{U}} = \left(\frac{^{230}\text{Th}}{^{234}\text{U}}\right)_m \left(\frac{^{232}\text{U}}{^{228}\text{Th}}\right)_m \left(\frac{^{228}\text{Th}}{^{232}\text{U}}\right)_s,$$

where the suffix m refers to measured activity ratios in the respective spectra and the suffix s to the calibrated spike activity ratio.

γ-ray spectrometry

One major advantage that γ-ray spectrometry has over α-particle spectrometry in U series disequilibrium studies is that no chemical separation of elements is required before counting. However, this advantage tends to be more than offset by the lower efficiency of γ-ray detectors, which limits their application to samples with relatively high U contents; otherwise, excessively long count times are involved, with the consequent problems of variable background effects. The capital cost of the equipment is also comparatively high. Moreover, where NaI(Tl) detectors have been used, the only nuclides that have been successfully measured are ^{226}Ra and ^{228}Th; the detection of ^{230}Th, ^{232}Th, ^{234}U and ^{238}U is not possible because of insufficient energy resolution. Nevertheless, with high purity Ge detectors, considerable improvement is possible, allowing, for example, measurement of ^{234}Th, ^{234}Pa and ^{208}Tl; difficulties still occur with ^{234}U, as there is superimposition of its 53.3 keV peak on the 53.2 keV peak from ^{214}Pb. A typical arrangement is shown in Fig. 9.33.

Integrated peak areas are obtained for γ-ray data just as they are for α-particle spectrometry. In this case, however, processing of peak areas to take account of three factors is necessary before true isotopic activity ratios can be obtained. Firstly, a correction may be needed for self-absorption. As the samples measured in γ-ray spectrometry are usually comparatively thick (they can be slices of rock), there is a possibility that significant attenuation of intensities will occur and will vary with the energy of the individual γ-rays. Checks for self-absorption need to be made, therefore, for low energy γ-rays in sections of different thickness (say, 1, 2 and 3 mm) by ratioing their activities to that of a higher energy γ-ray. If the ratio remains constant, self-absorption effects can be ignored. For rock samples, thicknesses of about 2 mm or less usually permit this simplification. The second factor that must be taken into account is the branching ratio related to the

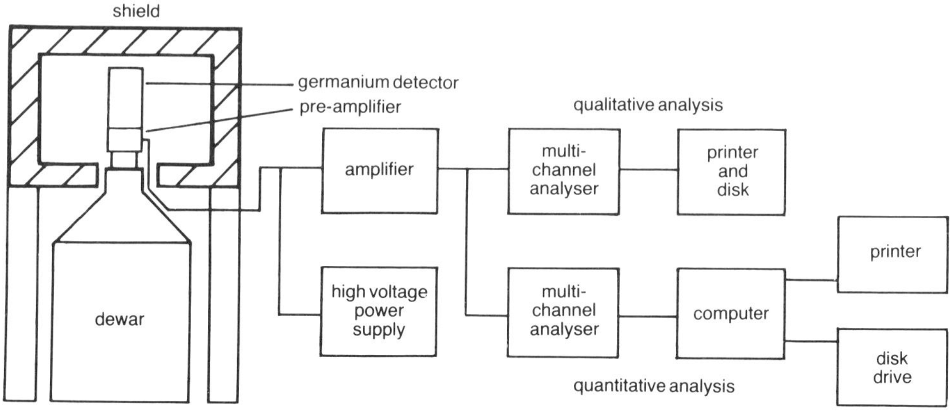

Fig. 9.33 — Diagrammatic representation of a γ-ray spectrometer system. (After EG & G Ortec, 1983.)

particular energy being studied that is, the proportion of the total decay represented by that particular energy. This is obtained from the physics of the decay scheme involved. Finally, correction must also be made for internal conversion using the internal conversion coefficient (the proportion of γ-rays which are actually emitted). Again, this is given by the physics of the decay. Correction for detector efficiency and calibration by means of standard sources has to be applied in γ-ray spectrometry as in α-particle spectrometry.

Errors in α-particle and γ-ray spectrometry

Errors in activity ratios determined by α-particle spectrometry are usually quoted as ± one standard deviation. This is because the most easily quantifiable error is that due to the counting factors mentioned above. As discussed in Chapter 8, for a Poisson distribution this error is inversely related to the square root of the number of counts measured. However, there are other sources of error; these include contamination of reagents by U and Th, problems associated with separation of U and Th during chemical processing, contamination effects in glassware or detectors, non-equilibration of the spike with U and Th isotopes in the sample (which leads to selective loss of either sample or spike nuclides during chemical separation) and possible loss of U and Th from the spike (which leads to selective adsorption of Th onto glassware). These errors can be assessed by repeated analysis. Accuracy can only be checked against standards and the results obtained by other techniques. In addition, problems related to the sample itself can also give rise to errors in interpretation of the results; in age dating, for example, difficulties can be caused by open system conditions, as discussed in Chapter 10, while any contact of samples with organic matter can lead to excess Th or depletion of U.

In γ-ray spectrometry, the main source of error is, again, in counting and is usually presented as the standard deviation of the count rate. The other sources of error lie in peak integration and in the uncertainty of self-absorption effects, but the method is generally less susceptible to analytical errors than α-particle spectrometry.

β-particle spectrometry

β-particle spectrometry is not a common procedure in geological applications but can be used in the determination of ^{228}Ra following extraction of its immediate daughter ^{228}Ac . However, this approach is not very successful because of difficulties in obtaining complete extraction of ^{228}Ac.

β-particle spectrometry itself is not easy. Because β-particles have only a low mass and travel quickly, they produce relatively few ion pairs per unit length of track. As discussed in Chapter 3, β-particle tracks are very erratic, and the β-particle energy spectrum develops a considerable low energy tail. Individual β-particle sources, however, can be quantified by measurement of the disintegration rate. This is carried out by a system rather similar to that used in α-particle spectrometry. The sample, which should be as thin a source as possible, is housed in a vacuum chamber. The detector is of the Li-drifted Si type, and a depletion depth of 5 mm is sufficient to stop β-particles with energies up to 2.5 MeV. A collimator is needed in front of the detector to prevent β-particles from entering from the edges. A resolution of 30 keV can be easily obtained, and improvement is possible by cooling the detector or, for low energies, by using ultrathin sources. Conversion electrons can also be monitored by this system; alternatively, a Si surface barrier detector can be used.

Another application of β-particle spectrometry is in the measurement of ^{210}Pb (half-life, 22.3 years) by measuring the growth of its daughter ^{210}Bi (half-life, 5.02 days). This involves the extraction of ^{210}Pb and ^{210}Bi from the sample and their separation using exchange column techniques. A pure ^{210}Pb source is thus obtained, and the growth of ^{210}Bi within this is monitored over a period of 30–35 days using a β-particle detector. The ^{210}Pb activity can then be determined from count rates recorded after the growth of ^{210}Bi over this length of time. However, the method is by no means simple, as the initial extraction of ^{210}Pb and ^{210}Bi is difficult when the samples have complex chemistry.

Conclusion

While these methods for investigating disequilibrium give very specific results, it is worth noting that a broad idea of ^{238}U decay series disequilibrium can be obtained from field data collected by a NaI(Tl) single-channel spectrometer system. This approach uses the total count in combination with the eK_2O, eU and eTh values obtained from the individual energy windows and assumes that the ^{232}Th decay series is in equilibrium. Variations in the total count about the value found at sites where the ^{238}U decay series is in equilibrium then indicate either gain or loss of daughter nuclides (Fig. 9.34). That is,

$$\text{total count} - (C_1 eK_2O + C_2 eU + C_3 eTh) \gtrless 0,$$

where C_1, C_2 and C_3 must be determined empirically for each particular instrument and survey area (Durrance and Heath, 1985).

The use of detailed γ-ray, α-particle and β-particle spectrometry in conjunction with chemical separation techniques, however, provides a very full picture of the quantitative distribution of nuclides within the ^{238}U decay series, as well as information on the ^{235}U and

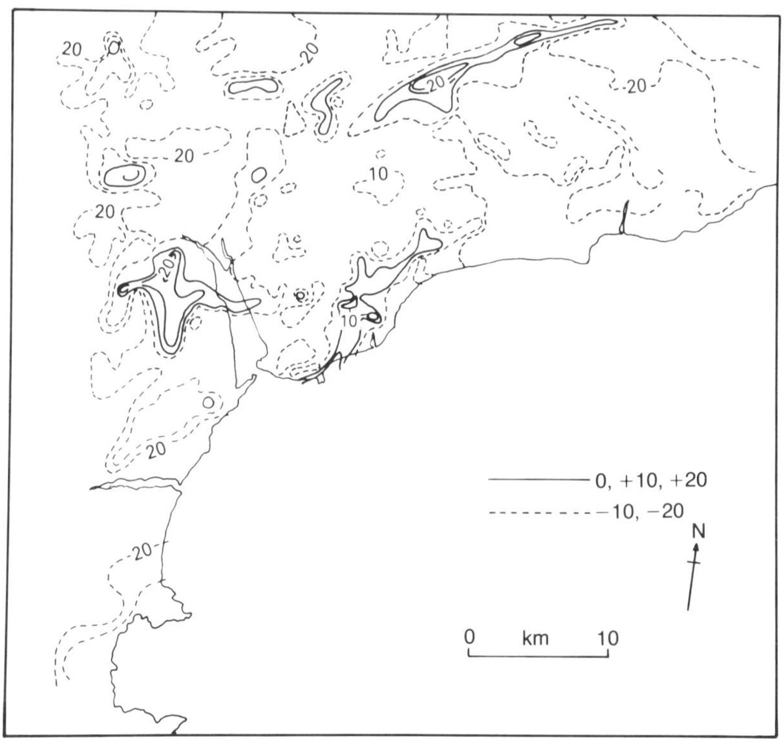

Fig. 9.34 —Disequilibrium in the ^{238}U decay series in grassland soils of southeast Devon, from a NaI(Tl) single-channel analyser system. Values based upon 1% of 5 min total counts and C_1=43.95, C_2=25.11 and C_3=11.45. (Durrance and Heath, 1985, reproduced with permission.)

^{232}Th series. The exact nature of any disequilibrium in these decay chains can, therefore, be determined. Such results can be applied to the investigation of many different processes that have taken place within the Earth during the last million years and, indeed, of processes which are still taking place today.

BIBLIOGRAPHY

Adams, J. A. S., and Gasparini, P. (1970) *Gamma ray spectrometry of rocks*. Amsterdam, Elsevier.

Amirkhanov, K. I., Anokhina, L. K., Chalbov, R. I., and Batyrmurzayev, A. S. (1980) Identification of radiation effects and ages of dike rocks from Mössbauer spectra of iron. *Dokl. Acad. Sci. USSR Earth Sci.* **2155,** 212.

Aller, R. C., and DeMaster, D. J. (1984) Estimates of particle flux and reworking at the deep-sea floor using ^{234}Th/^{238}U disequilibria. *Earth Planet. Sci. Lett.* **67,** 308.

Anderson, R. F., Bacon, M. P., and Brewer, P. G. (1983) Removal of ^{230}Th and ^{231}Pa at ocean margins. *Earth Planet. Sci. Lett.* **66,** 73.

Andrews, J. N., and Kay, R. L. F. (1982) ^{234}U/^{238}U activity ratios of dissolved uranium in groundwaters from a Jurassic aquifer in England. *Earth Planet. Sci. Lett.* **57,** 139.

Andrews, J. N., and Kay, R. L. F. (1983) The U contents and ^{234}U/^{238}U activity ratios of dissolved uranium in groundwaters from some Triassic sandstones in England. *Isotope Geol.* **1,** 101.

Andrews, J. N., and Lee, D. J. (1979) Inert gases in groundwater from the Bunter Sandstone of England as indicators of age and palaeoclimatic trend. *J. Hydrol.* **41,** 233.

Andrews, J. N., and Wood, D. F. (1972) Mechanism of radon release in rock matrices and entry into ground waters. *Trans. Inst. Min. Metall. Lond. B* **81,** 197.

Asikainen, M. (1981) State of disequilibrium between ^{238}U, ^{234}U, ^{226}Ra and ^{222}Rn in groundwater from bedrock. *Geochim. Cosmochim. Acta* **45,** 201.

Atkinson, B. K. (1982) Subcritical crack propagation in rocks: theory, experimental results and applications. *J. Struct. Geol.* **4,** 41.

Bajo, C. (1979) Thorium and uranium determinations in rock solutions by the induced fission track method. *Nucl. Tracks* **3,** 101.

Ball, T. K., Nicholson, R. A., and Peachey, D. (1983) Effect of meteorological variables on certain soil gases used to detect buried ore deposits. *Trans. Inst. Min. Metall. Lond. B* **92,** 183.

Barrios, N., Morel, O., Zlosilo, M., and Schlein, W. (1978) Determination of uranium in aqueous solution by X-ray fluorescence. *X-ray Spectrom.* **7,** 31.

Basham, I. R. (1981) Some applications of autoradiographs in textural analysis of uranium-bearing samples — a discussion. *Econ. Geol.* **76,** 974.

Basham, I. R., and Easterbrook, G. D. (1977) Alpha particle autoradiography of geological specimens by use of cellulose nitrate detectors. *Trans. Inst. Min. Metall. Lond. B* **86,** 96.

Bashorin, V. N. (1980) The helium survey as a method of studying long-lived faults. *Dokl. Acad. Sci. USSR Earth Sci.* **255,** 91.

Beasley, T. M., Carpenter, R., and Jennings, C. D. (1982) Plutonium, ^{241}Am and ^{137}Cs ratios, inventories and vertical profiles in Washington and Oregon continental shelf sediments. *Geochim. Cosmochim. Acta* **46,** 1931.

Belyaev, I., and Sobornov, O. P. (1981) Uranium, thorium and potassium in reference samples. *Geostand. Newslett.* **5,** 109.

Berelson, W. M., Hammond, D. E., and Fuller, C. (1982) Radon 222 as a tacer for mixing in the water column and benthic exchange in the southern California borderland. *Earth Planet. Sci. Lett.* **61,** 41.

Berger, G., Bosch, B., and Marce, A. (1983) Ten years quantitative analysis of tritium in rain, surface water and groundwater in Brie (Seine-et-Marine) France. *Rept. Inst. Geol. Sci. Lond.* **82,** 6.

Bertel, E., Mark, T. D., and Pahl, M. (1977) A new method for the measurement of the mean etchable fission track length and of extremely high fission track densities in minerals. *Nuc. Track Detection* **1,** 123.

Beyth, M., and McIntear, C. (1982) Evaluation of stream sediment data for favourability of uranium mineralisation in Tertiary granite plutons of the Alaska Range. *Mineral Mag.* **46,** 475.

Bibby, D. M., Chaix, R. P., and Andeweg, A. H. (1974) The determination of uranium and thorium by the use of delayed neutrons. *Rep. Natl. Inst. Metall. S. Afr.* **1625.**

Blackburn, J. A. (1970) *Spectral analysis: methods and techniques*. New York, Dekker.

Bowie, S. H. U. (1977) Radiographic techniques. In Zussman, J. (ed.), *Physical methods of determinative mineralogy*. London, Academic Press.

Bowie, S. H. U., Davis, M. and Ostle, D. (1972) *Uranium prospecting handbook*. London, Institution of Mining and Metallurgy.

Boyle, R. W. (1974) Elemental associations in mineral deposits and indicator elements of interest in geochemical prospecting. *Paper Geol. Surv. Can.* **74,** 45.

Boyle, R. W. (1982) *Geochemical prospecting for uranium and thorium deposits*. Oxford, Elsevier.

Bradshaw, P. M. D., Thomson, I., Smee, B. W., and Larsson, J. O. (1974) The application of different analytical extractions and soil profile sampling in exploration geochemistry. *J. Geochem. Expl.* **3,** 209.

Brits, R. J. N. (1979) A routine method for the determination of the ^{234}U/^{238}U ratio in natural water. *Chem. Geol.* **25,** 347.

Brooks, R. R. (1979) Indicator plants for mineral prospecting — a critique. *J. Geochem. Expl.* **12,** 67.

Butt, C. R. M., and Gole, M. J. (1984) Helium emanometry under Australian conditions — preliminary results. *J. Geochem. Expl.* **22,** 359.

Cameron, E. M., and Ballantyne, S. B. (1977) Reconnaissance level geochemical and radiometric exploration data for the vicinity of the Rabbit Lake uranium deposits. *Can. Min. Metall. Bull.* **70,** 76.

Campbell, B. L., Loughran, R. J., and Elliot, G. L. (1982) Cesium-137 as an indicator of geomorphic processes in a drainage basin system. *J. Aust. Geogr. Studies* **20,** 49.

Canberra (1984) *Edition six catalogue,* Meriden ,Connecticut Canberra Industries Inc.

Cassidy, J. (1982) A gamma-radiometric survey of the northern units of the Leinster granite. *Rep. Geol. Surv. Ireland* **RS82/1.**

Cerling, T. E., and Spalding, B. P. (1982) Distribution and relationships of radionuclides to streambed gravels in a small watershed. *Environ. Geol.* **4,** 99.

Cerling, T. E., and Turner, R. R. (1982) Formation of freshwater Fe-Mn coatings on gravel and the behaviour of ^{60}Co, ^{90}Sr and ^{137}Cs in a small watershed. *Geochim. Cosmochim. Acta* **46,** 1333.

Chalov, P. I. (1959) The U234/U238 ratio in some secondary minerals. *Geochemistry* **2,** 203.

Cherdyntsev, V. V. (1955) *Proceedings of the Third Session of Committee for Absolute Age Determinations 1954,* 175.

Cherdyntsev, V. V., Orlov, D. P., Isabaev, E. A., and Ivanov, V. (1961) Uranium isotopes in nature, II, Isotopic composition of uranium minerals. *Geochemistry* **10,** 927.

Clarke, G. J., Gulson, B. L., and Cookson, J. A. (1979) Pb, U, Tl, Hf and Zr distributions in zircons determined by proton microprobe and fission track techniques. *Geochim. Cosmochim. Acta* **43,** 905.

Clarke, W. B., and Kugler, G. (1973) Dissolved helium in groundwaters: a possible method for uranium and thorium prospecting. *Econ. Geol.* **68,** 243.

Coker, W. B., and Jonasson, I. R. (1977) Geochemical techniques for uranium in the Grenville Province of Ontario. *Can. Min. Metall. Bull.* **70,** 67.

Collins, W. (1978) Analysis of airborne spectroradiometric data and the use of Landsat data for mapping hydrothermal alteration. *Geophysics* **43,** 967.

Conaway, J. G., and Killeen, P. G. (1980) Gamma ray spectral logging for uranium. *Can. Min. Metall. Bull.* **73,** 115.

Cowart, J. B., and Osmond, J. K. (1977) Uranium isotopes in groundwater: their use in prospecting for sandstone-type uranium deposits. *J. Geochem. Expl.* **8,** 365.

Cumming, G. L. (1974) Determination of uranium and thorium in meteorites by the delayed neutron method. *Chem. Geol.* **13,** 257.

Dahlbacka, J., Lindblom, P., and Ehlers, C. (1979) An absolute semi-quantitative method to determine U and Th concentrations in geological samples with nuclear emulsions. *Chem. Scr.* **13,** 96.

Dee, T. P., Nunn, R. J., and Sharples, K. (1957) The use of different types of phosphate rock in single and triple superphosphate production. *Proc. Fert. Soc.* **74,** 165.

Del Fiore, G., Peters, J. M., and Roelands, I. (1974) Precise determination of uranium in rocks and minerals by neutron activation analysis using ^{133}I fission products. *Chem. Geol.* **13,** 309.

Dickson, B. L., Clark, G. J., and MacGregor, B. J. (1979) Technique for correcting for overburden effects in ground level radiometric surveys of uranium ore bodies. *Geophysics* **44,** 89.

Dikun, A. V., Korobeynik, V. M., and Yanitskii, I. N. (1976) Some features of the development of helium surveying. *Int. Geol. Rev.* **18,** 98.

Dodd, P. H. (1977) Uranium exploration technology. In Jones, M. J. (ed.), *Geology, mining and extractive processing of uranium.* London, Institution of Mining and Metallurgy.

Dooley, J. R., Granger, H. C., and Rosholt, J. N. (1966) Uranium-234 fractionation in the sandstone-type uranium deposits of the Ambrosia Lake District, New Mexico. *Econ. Geol.* **61,** 1362.

Dooley, J. R., Tatsumoto, M., and Rosholt, J. N. (1964) Radioactive disequilibrium studies of roll features, Shirley Basin, Wyoming. *Econ. Geol.* **59,** 586.

Dostal, J., and Capredi, S. (1975) Partition coefficients of uranium for some rock forming minerals. *Chem. Geol.* **15,** 285.

Duane, M. J., and Williams, C. T. (1980) Some applications of autoradiographs in textural analysis of uranium-bearing samples. *Econ. Geol.* **75,** 766.

Duckworth, H. E. (1960) *Mass spectroscopy.* Cambridge, Cambridgeshire, Cambridge University Press.

Durrance, E. M. (1978) Radon in the stream waters of East Devon. *Proc. Ussher Soc.* **4,** 220.

Durrance, E. M. (1984) Uranium in the New Red Sandstone of south east Devon. *Proc. Ussher Soc.* **6,** 108.

Durrance, E. M., and Heath, M. J. (1985) Thermal groundwater movement and radionuclide transport in south west England. *Mineral Mag.* **49,** 289.

Durrance, E. M., Meads, R. E., Ballard, R. R. B., and Walsh, J. N. (1978) Oxidation state of iron in the Littleham Mudstone Formation of the New Red Sandstone Series (Permian–Triassic) of south-east Devon, England. *Bull. Geol. Soc. Am.* **89,** 1231.

Durrance, E. M., Meads, R. E., Brindley, R. K., and Stark, A. G. W. (1980) Radioactive disequilibrium in uranium-bearing nodules from the New Red Sandstone (Permian–Triassic) of southeast Devon. *Proc. Ussher Soc.* **5,** 81.

Duval, J. S. (1980) Radioactivity method. *Geophysics* **45,** 1690.

Dyck, W. (1972) Radon methods of prospecting in Canada. In Bowie, S. H. U., Davis, M., and Ostle, D. (eds), *Uranium prospecting handbook.* London, Institution of Mining and Metallurgy.

Dyck, W. (1976) Helium in mineral exploration. *J. Geochem. Expl.* **5,** 3.

Dyck, W., Jonasson, I. R., and Liard, R. F. (1976) Uranium prospecting with ^{222}Rn in frozen terrain. *J. Geochem. Expl.* **5,** 115.

Dyck, W., and Pelchat, J. C. (1977) A semi-portable helium analysis facility. *Paper Geol. Surv. Can.* **77,** 85.

Dyck, W., and Tan, B. (1978) Seasonal variations of helium, radon and uranium in lake waters near the Key Lake uranium deposits, Saskatchewan. *J. Geochem. Expl.* **10,** 153.

EG & G Ortec (1983) Radiation detection: measurement: analysis. *Catalogue 1983–1984.* Oak Ridge, Tennessee EG & G Ortec.

Eremeev, A. N., Sokolov, V. A., Sokolov, A. P., and Yanitskii, I. N. (1973) Application of helium surveying to structural mapping and ore deposit forecasting.

In Jones, M. J. (ed.), *Geochemical exploration 1972.* London, Institution of Mining and Metallurgy.

Evans, D. W., Alberts, J. J., and Clark, R. A. (1983) Reversible ion-exchange fixation of cesium-137 leading to mobilization from reservoir sediments. *Geochim. Cosmochim. Acta* **47,** 1041.

Ferguson, J., and Goleby, A. B. (1980) *Uranium in the Pine Creek geosyncline.* Vienna, International Atomic Energy Agency.

Fergusson, R. B., and Price, V. (1976) Hydrogeochemical and stream sediment reconnaissance in the eastern United States. National Uranium Resource Evaluation Program. *J. Geochem. Expl.* **6,** 103.

Fink, C. W. (1984) Chronology of weathered materials and soil age determinations in pedostratigraphic sequences. *Chem. Geol.* **44,** 311.

Fleischer, R. L. (1982) Alpha-recoil damage and solution effects in minerals: uranium isotopic disequilibrium and radon release. *Geochim. Cosmochim. Acta* **46,** 2191.

Fleischer, R. L. (1983) The theory of alpha recoil effects on radon release and isotopic disequilibrium. *Geochim. Cosmochim. Acta* **47,** 779.

Fleischer, R. L., and Likes, R. S. (1979) Integrated radon monitoring by the diffusion barrier technique. *Geophysics* **44,** 1963.

Fletcher, W. K. (1981) *Analytical methods in geochemical prospecting.* Amsterdam, Elsevier.

Gascoyne, M. (1982) Geochemistry of the actinides and their daughters. In Ivanovich, M., and Harmon, R. S. (eds), *Uranium series disequilibrium: applications to environmental problems.* Oxford, Oxford University Press.

Ghosh, P. C., and Bhalia, N. S. (1981) On the behaviour and measurement of thoron (^{220}Rn) in soil. *Indian J. Earth Sci.* **8,** 1.

Gibb, T. C. (1976) *Principles of Mössbauer spectroscopy.* London, Chapman and Hall.

Goles, G. G. (1977) Instrumental methods of neutron activation analysis. In Zussman, J. (ed.), *Physical methods of determinative mineralogy.* London, Academic Press.

Golterman, H. L., Clymo, R. S., and Ohnstad, M. A. M. (1978) *Methods for physical and chemical analysis of fresh waters.* Oxford, Blackwell.

Gough, L. P., Shacklette, H. T., and Case, A. A. (1979) Element concentrations toxic to plants, animals and man. *US Geol. Surv. Bull.* **1466.**

Gough, L. P., and Severson, R. C. (1981) Biogeochemical variability of plants at native and altered sites, San Juan Basin, New Mexico. *US Geol. Surv. Prof. Paper* **1134D.**

Gregory, L. P. (1976) Radon in waters — simplified radiochemical method and results for Christchurch artesian waters. *New Zealand J. Sci.* **19,** 369.

Gregory, R. G., and Durrance, E. M. (1985a) Helium, carbon dioxide and oxygen soil gases: small scale variations over fractured ground. *J. Geochem. Expl.* **24,** 29.

Gregory, R. G., and Durrance, E. M. (1985b) Stratiform copper deposits in Mid Devon: soil gas geochemistry as an exploration method. *Proc. Ussher Soc.* **6,** 258.

Grimbert, A. (1972) Use of geochemical techniques in uranium prospecting. In Bowie, S. H. U., Davis, M., and Ostle, D. (eds), *Uranium prospecting handbook.* London, Institution of Mining and Metallurgy.

Grosz, A. E. (1983) Application of total count aeroradiometric maps to the exploration for heavy-metal deposits in the coastal plain of Virginia. *US Geol. Surv. Prof. Paper* **1263.**

Hamilton-Jones, B. B. (1978) Theory and practice of geochemical prospecting for uranium *Minerals Sci. Eng.* **10,** 182.

Harrison, R. K. (1975) Concretionary concentrations of the rarer elements in Permo-Triassic red beds of southwest England. *Bull. Geol. Surv. Gt. Brit.* **52,** 1.

Heath, M. J. (1982) Uranium in the Dartmoor granite: geochemical and radiogeological investigations in relation to the south west England geothermal anomaly. *Ph.D. Thesis* University of Exeter.

Hendry, G. L. (1975) X-ray fluorescence. In Nicol, A. W. (ed.), *Physicochemical methods of mineral analysis.* New York, Plenum.

Henshaw, D. L., Fews, A. P., and Webster, D. J. (1979) A technique for high sensitivity alpha-autoradiography of bronchial epithelium tissue. *Phys. Med. Biol.* **24,** 1227.

Hill, C. R., and Crookall, J. O. (1963) Natural occurrence of ^{234}U-enriched uranium. *J. Geophys. Res.* **68,** 2358.

Hilton, D. R., Oxburgh E. R. and O'Nions, R. K. (1985) Fluid flow through a high heat flow granite: constraints imposed by He and Rn data. In *High heat production granites, hydrothermal circulation and ore genesis.* London, Institution of Mining and Metallurgy.

Hinkle, M. E., and Dilbert, C. A. (1984) Gases and trace elements in soils at the North Silver Bell deposit, Pima County, Arizona. *J. Geochem. Expl.* **20,** 323.

Hinkle, M. E., and Kilburn, J. E. (1979) The use of vacutainer tubes for collectors of soil samples for helium analysis. *US Geol. Surv. Open-file Rep.* **79–1441.**

Honda, M., Kurita, K., Hamano, Y., and Ozima, M. (1982) Experimental studies of He and Ar degassing during rock fracturing. *Earth Planet Sci. Lett.* **59,** 429.

Hooker, P. S., O'Nions, R. K., and Oxburgh, R. E. (1985) Helium isotopes in North Sea gas fields and the Rhine Rift. *Nature* **318,** 273.

Howarth, O. (1973) *Theory of spectroscopy: an elementary introduction.* London, Nelson.

Hurley, P. M. (1954) The helium age method and the distribution and migration of helium in rocks. In Faul, H. (ed.), *Nuclear geology.* New York, Wiley.

Hussain, N., and Krishnaswami, S. (1980) U-238 series radioactive disequilibrium in groundwater: implications for the origin of excess U-234 and the fate of reactive pollutants. *Geochim. Cosmochim. Acta* **44,** 1287.

Ivanovich, M. (1982) Spectroscopic methods. In Ivanovich, M., and Harmon, R. S. (eds), *Uranium series disequilibrium: application to environmental problems.* Oxford, Oxford University Press.

Jacobs, J. J., and Tolmay, R. T. (1979) The determination of uranium in solids and solutions by X-ray flourescence spectrometry. *Rep. Natl. Inst. Metall. S. Afr.* **2025.**

Janot, C. (1981) New investigation techniques for minerals. *Bull. Min.* **104,** 135.

Jeffery, P. G., and Hutchinson, D. (1981) *Chemical methods of rock analysis.* Oxford, Pergamon.

Jonasson, I. R., and Gleeson, C. F. (1976) On the usefulness of water samples in reconnaissance surveys for uranium in the Yukon Territory. *Paper Geol. Surv. Can.* **76–1C,** 241.

Joshi, S. R., and Durham, R. W. (1976) Determination of ^{210}Pb, ^{226}Ra and ^{137}Cs in sediments. *Chem. Geol.* **18,** 155.

Killeen, G. A., and Carmichael, C. M. (1976) Radioactive disequilibrium determinations, 1, Determination of disequilibrium in uranium ores by alpha spectrometry. *Paper Geol. Surv. Can.* **75–38,** 1.

Kiser, R. W. (1965) *Introduction to mass spectrometry and its applications.* Englewood Cliffs, New Jersey, Prentice-Hall.

Korkisch, J., and Huber, H. (1976) Determination of uranium in minerals and rocks. *Talanta* **23,** 283.

Lally, A. E. (1982) Chemical procedures. In Ivanovich, M., and Harmon, R. S. (eds), *Uranium series disequilibrium: applications to environmental problems.* Oxford, Oxford University Press.

Lambert, G., Bristeau, P., and Poluan, G. (1976) Emission and enrichments of radon daughters from Etna volcano magma. *Geophys. Res. Lett.* **3,** 724.

Lederer, C. M., and Shirley, V. S. (1978) *Table of isotopes.* New York, Wiley.

Levinson, A. A. (1980) *Introduction to exploration geochemistry.* Wilmette, Illinois, Applied Publishers.

Levinson, A. A., and Bland, C. J. (1978) Examples of the variability of disequilibrium and the emanation factor in some uraniferous materials. *Can. J. Earth Sci.* **15,** 1867.

Levinson, A. A., Bland, C. J., and Lively, R. S. (1982) Exploration for U ore deposits. In Ivanovich, M., and Harmon, R. S. (eds), *Uranium series disequilibrium: applications to environmental problems.* Oxford, Oxford University Press.

Linden, A. H. (1977) Method of detecting small or indistinct radioactive sources by airborne gamma ray spectrometry. In Jones, M. J. (ed.), *Geology, mining and extractive processing of uranium.* London, Institution of Mining and Metallurgy.

Lively, R. S., Harmon, R. S., Levinson, A. A., and Bland, C. J. (1979) Disequilibrium in the 238 uranium series in samples from Yeelirree, Western Australia. *J. Geochem. Expl.* **12,** 57.

Mamyrin, B. A., and Tolstikhin, I. N. (1984) *Helium isotopes in nature.* Amsterdam, Elsevier.

Martin, F. S., and Miles, G. L. (1958) *Chemical prospecting for nuclear fuels.* London, Butterworths.

May, L. (1971) *An introduction to Mössbauer spectroscopy.* London, Hilger.

McAllister, W. D. N., and Firman, R. J. (1980) Stream sediment uranium anomalies in Eskdale, Cumbria, England. *Trans. Inst. Min. Metall. B* **89,** 177.

McCall, P. L., Robbins, J. A., and Matisoff, G. (1984) ^{137}Cs and ^{210}Pb transport and geochronologies in urbanized reservoirs with rapidly increasing sedimentation rates. *Chem. Geol.***44,** 33.

McCallan, M. E., O'Leary, B. M., and Rose, C. W. (1980) Redistribution of cesium-137 by erosion and deposition on an Australian soil. *Aust. J. Soil. Sci. Res.* **18,** 119.

McCorkell, R. H. (1983) A fission track method for the determination of uranium in natural waters for geochemical exploration. *Can. Min. Metall. Bull.* **76,** 89.

McCorkell, R. H., and Card, J. W. (1978) The decay products of ^{222}Rn in etched track radon detection. *J. Geochem. Expl.* **10,** 277.

McHenry, J. R., and Ritchie, J. C. (1977) Estimating field erosion losses from fall-out cesium-137 measurements. *Proceedings of the Symposium on Erosion and Solid Matter Transport in Inland Waters, Paris, IAH Symp.* **122,** 26.

McHenry, J. R., and Ritchie, J. C. (1977) Physical and chemical parameters affecting transport of ^{137}Cs in arid watersheds. *Water. Resour. Res.* **13,** 923.

McHenry, J. R., Ritchie, J. C. and Gill, A. C. (1973) Accumulation of cesium-137 in soils and sediments in selected watersheds. *Water. Resour. Res.* **9,** 676.

Megumi, K., and Mamuro, T. (1974) Emanation and exhalation of radon and thoron gases from soil particles. *J. Geophys. Res.* **79,** 3357.

Michie, U. McL., and Cooper, D. C. (1979) Uranium in the Old Red Sandstone of Orkney. *Rep. Inst. Geol. Sci. Lond.* 78-16.

Middleditch, B. S. (1979) *Practical mass spectrometry: a contemporary introduction.* New York, Plenum.

Minatidis, D. G., and Slatt, R. M. (1976) Uranium and copper exploration by nearshore lake sediment geochemistry, Kaipokok region of Labrador. *J. Geochem. Expl.* **5,** 135.

Ministry of Agriculture, Fisheries and Food (UK) (1981) *Lime and fertilizer recommendations* Vol 1, *Arable crops and grassland.* Alnwick, Ministry of Agriculture, Fisheries and Food (UIC).

Mogro-Campero, A., and Fleischer, R. L. (1977) Subterrestial fluid convection: a hypothesis for long distance migration of radon within the Earth. *Earth Planet. Sci. Lett.* **34,** 321.

Morse, R. H. (1971) Comparison of geochemical prospecting methods using radium with those using radon and uranium. In Boyle, R. W., and McGerrigle, J. I. (eds), *Geochemical exploration, Can. Inst. Min. Metall. Spec. Vol.* **11.**

Morse, R. H. (1976) *Radon counters in uranium exploration, IAEA STI Publ.* **434,** Vienna, International Atomic Energy Agency.

Moxham, R. M. (1976) Gamma ray spectrometer measurements of $^{238}U/^{235}U$ in uranium ore from a natural reactor at Oklo, Gabon. *J. Res. US Geol. Survey,* **4,** 589.

Moxham, R. M., and Tanner, A. B. (1977) High resolution gamma ray spectrometry in uranium exploration. *J. Res. US Geol. Survey.* **5,** 783.

Nhyham, J. W., Drennon, B. J., and Crowell, J. M. (1983) Evaluation of an automated assay system to measure soil radionuclides. *J. Soil Sci. Am.* **47,** 486.

Nicholls, G. D., and Wood, M. (1975) Elemental analysis

using mass spectrographic techniques. In Nicol, A. W. (ed.), *Physicochemical methods of mineral analysis.* New York, Plenum.

Norrish, K., and Chappell, B. W. (1977) X-ray fluorescence spectrometry. In Zussman, J. (ed.), *Physical methods in determinative mineralogy.* London, Academic Press.

O'Connor, P. J. (1981) A reconnaissance radiometric survey of the Irish Republic. *Rep. Geol. Surv. Ireland* **81/1**.

Oliver, B. M., Bradley, J. G., and Farrar, H. (1984) Helium concentrations in the Earth's lower atmosphere. *Geochim. Cosmochim. Acta* **48,** 1759.

Olsen, C. E., and Brookins, D. G. (1982) Investigations of uranium hydrogeochemistry — the effect of water treatment on observed uranium concentrations in natural water supplies. *Mod. Geol.* **3,** 115.

Osmond, J. K., and Cowart, J. B. (1982) Groundwater. In Ivanovich, M., and Harmon, R. S. (eds), *Uranium series disequilibrium: applications to environmental problems.* Oxford, Oxford University Press.

Ostrihansky, L. (1976) Radioactive disequilibrium determinations. 2. Radioactive disequilibrium investigations, Elliot Lake area, Ontario. *Paper Geol. Surv. Can.* **75–38,** 19.

Overstreet, W. C., and Marsh, S. P. (1981) Some concepts and techniques in geochemical exploration. In Skinner, B. J. (ed.), *Econ. Geol. Seventy-Fifth Anniversary Volume (1905–1980).*

Parslow, G. R., and Dwairi, I. (1977) Extraction of uranium ions from water: a rapid and convenient method using prepackaged ion exchange resins. *J. Geochem. Expl.* **8,** 541.

Pedersen, C. P., Dunbier, J., and Gingrich, J. E. (1980) Experience with the Track Etch method for uranium exploration in Northern Australia. In Ferguson, J., and Goleby, A. B. (eds), *Uranium in the Pine Creek geosyncline.* Vienna, International Atomic Energy Agency.

Plant, J., Goode, G. C., and Herrington, J. (1976) An instrumental neutron activation method for multi-element geochemical mapping. *J. Geochem. Expl.* **6,** 299.

Plant, J., Jeffery, K., Gill, E., and Fage, C. (1975) The systematic determination of accuracy and precision in geochemical exploration data. *J. Geochem. Expl.* **4,** 467.

Polikarpov, G. G. (1966) *Radioecology of aquatic organisms, the accumulation of biological effect and radioactive substances.* Amsterdam, North-Holland.

Reedman, J. H. (1979) *Techniques in mineral exploration.* London, Applied Science.

Reeves, R. D., and Brooks, R. R. (1978) *Trace element analysis of geological materials.* New York, Wiley.

Reimer, G. M. (1975) Uranium determination in natural water by the fission track technique. *J. Geochem. Expl.* **4,** 425.

Reimer, G. M. (1976) Design and assembly of a portable helium detector for evaluation as a uranium exploration instrument. *US Geol. Surv. Open File Rep.* **76–398.**

Reimer, G. M. (1979) The use of soil-gas helium concentrations for earthquake prediction, studies of factors causing diurnal variation. *US Geol. Surv. Open File Rept.* **79–1623.**

Reimer, G. M., and Carpenter, B. S. (1974) Thorium determinations in glasses using fission track techniques. *Nature* **247,** 101.

Reimer, G. M., Denton, E. H., Friedman, I., and Otton, J. K. (1979) Recent developments in uranium exploration using the US Geological Survey's mobile helium detector. *J. Geochem. Expl.* **11,** 1.

Richards, D. J., and Walraven, F. (1975) Airborne geophysics and ERTS imagery. *Minerals Science and Engineer,* **7,** 234–278.

Richardson, K. A. (1964) Thorium, uranium and potassium in the Conway granite, New Hampshire, USA In Adams, J. A. S., and Lowder, W. M. (eds), *The natural radiation environment.* Chicago, Illinois, University of Chicago Press.

Ritchie, J. C., and McHenry, J. R. (1978) Fall-out cesium-137 in cultivated and non-cultivated north central United States watersheds. *J. Environ. Qual.* **7,** 40.

Robbins, D. A. (1978) Field technique for the measurement of uranium in natural waters. *Bull. Can. Inst. Min. Metall.* **71,** 793.

Robbins, D. A., and Castledine, C. (1979) Laser fluorescence as a basic analytical technique for uranium in the mining industry. *Bull. Can. Inst. Min. Metall.* **72,** 98.

Rogers, A. W. (1967) *Techniques of autoradiography.* Amsterdam, Elsevier.

Rose, A. W., Hawkes, H. E., and Webb, J. S. (1979) *Geochemistry in mineral exploration.* London, Academic Press.

Rose, A. W., and Korner, L. A. (1979) Radon in natural waters as a guide to uranium deposits in Pennsylvania. In Watterson, J. R., and Theobald, P. K. (eds), *Geochemical exploration, 1978.* Ontario, Association of Exploration Geochemists.

Rosholt, J. N. (1959) Natural radioactive disequilibrium of the uranium series. *US Geol. Surv. Bull.* **1084.**

Rosholt, J. N., Doe, B. R., and Tatsumoto, M. (1966) Evaluation of the isotopic composition of uranium and thorium in soil profiles. *Bull. Geol. Soc. Am.* **77,** 987.

Rosholt, J. N., Harshman, E. N., Shields, W. R., and Garner, E. L. (1964) Istopic fractionation of uranium related to roll features in sandstone, Shirley Basin, Wyoming. *Econ. Geol.* **59,** 570.

Scintrex (1978) *UA3 uranium analyser instruction manual.* Toronto, Scintrex.

Scott, R. H., Strasheim, A., and Kokot, M. L. (1976) The determination of uranium in rocks by inductively coupled plasma-optical emission spectrometry. *Anal. Chim. Acta* **82,** 67.

Severne, B. C. (1978) Evaluation of radon systems at Yeelirrie, Western Australia. *J. Geochem. Expl.* **9,** 1.

Simmons, G., and Richter, D. (1976) Microfractures in rocks. In Strens, R. G. J. (ed.), *The physics and chemistry of minerals and rocks.* London, Wiley.

Smart, P. (1981) *Electron microscopy of soils and sediments.* Oxford, Clarendon.

Smith, J. N., and Ellis, K. M. (1982) Transport mechanisms for Pb-210, Cs-137 and Pu fallout radionuclides through fluvial-marine systems. *Geochim. Cosmochim. Acta* **46,** 941.

Smythe, J. R., Thompson, J., and Wolfsberg, K. (1980) Microautoradiographic studies of the sorption of U and Am on natural rock samples. *Radioactive Waste Management* **1,** 34.

Smythe, L. E., and Finlayson, R. J. (1978) *Rapid spectroscopic methods for trace element investigations in natural waters.* Canberra, Australian Government Publishing Service.

Soonawala, N. M., and Telford, W. M. (1980) Movement of radon in overburden. *Geophysics* **45,** 1297.

Southworth, H. N. (1975) Scanning electron microscopy and microanalysis. In Nicol, A. W. (ed.), *Physicochemical methods of mineral analysis.* New York, Plenum.

Stanners, D. A., and Aston, S. R. (1984) The use of reprocessing effluent radionuclides in the geochronology of recent sediments. *Chem. Geol.* **44,** 19.

Stanton, R. E. (1976) *Analytical methods for use in geochemical exploration.* London, Arnold.

Steinnes, E. (1976) Simultaneous determination of uranium, thorium, molybdenum, tungsten, arsenic and antimony in granitic rocks by epithermal neutron activation analysis. *Anal. Chem.* **48,** 1440.

Stuckless, J. S., Millard, H. T., Bunker C. M., Nkomo, I. T., Rosholt, J. N., Buch, C. A., Huffman, C., and Keil, R. L. (1977) A comparison of some analytical techniques for determining uranium, thorium and potassium in granitic rocks. *J. Res. US Geol. Surv.* **5,** 83.

Stuckless, J. S., and Nkomo, I. T. (1978) Uranium-lead isotope systematics in uraniferous alkali-rich granites from the Granite Mountains, Wyoming: Implications for uranium source rocks. *Econ. Geol.* **73,** 427.

Szabo, B. J. (1982) Extreme fractionation of $^{234}U/^{238}U$ and $^{230}Th/^{234}U$ in spring waters, sediments and fossils at Le Pomme de Terre valley, southwestern Missouri. *Geochim. Cosmochim. Acta* **46,** 1675.

Szoghy, I. M., and Kish, L. (1978) Determination of radioactive disequilibrium in uranium-bearing rocks. *Can. J. Earth Sci.* **15,** 33.

Tammenmaa, J., Grasty, R. L., and Peltoniemi, M. (1976) The reduction of statistical noise in airborne radiometric data. *Can. J. Earth Sci.* **13,** 1351.

Telford, W. M. (1983) Radon mapping in the search for uranium. In Fitch, A. A. (ed.), *Developments in geophysical exploration methods* Barking, Essex, Applied Science.

Thurber, D. L. (1962) Anomalous U-234/U-238 in nature. *J. Geophys. Res.* **67,** 4518.

Tikoo, B. N., and Murty, D. S. R. (1980) A rapid method for uranium analysis using UA3 "Scintrex" uranium analyser. *Curr. Sci. India* **49,** 861.

Tolmay, R. T., and Jacobs, J. J. (1980) A rapid method for the determination of uranium in ores and carbonaceous materials by X-ray fluorescence spectrometry. *Rep. Natl. Inst. Min. Metall. S. Afr.* **2034.**

Top, Z., and Clarke, W. B. (1981) Dissolved helium isotopes and tritium in lakes: further results for uranium prospecting in central Labrador. *Econ. Geol.* **76,** 2018.

Torgersen, T. (1980) Controls on pore fluid concentrations of ^{4}He and ^{222}Rn and the calculation of $^{4}He/^{222}Rn$ ages. *J. Geochem. Expl.* **13,** 57.

Vandergraaf, T. T., Abry, D. R. M., and Davis, C. E. (1982) The use of autoradiography in determining the distribution of radionuclides sorbed on thin sections of plutonic rocks from the Canadian Shield. *Chem. Geol.* **36,** 139.

Wallenberg, H. A., and Smith, A. R. (1964) Studies in terrestial gamma radiation. In Adams, J. A. S., and Lowder, W. M. (eds), *The natural radiation environment.* Chicago Illinois, University of Chicago Press.

Walton, N. R. G. (1982) A detailed hydrogeochemical study of groundwaters from the Triassic sandstone aquifer of south-west England. *Rep. Inst. Geol. Sci. Lond.* **81–5.**

Waples, D. (1981) Organic geochemistry for exploration geologists. Minneapolis, Minnesota, Burgess.

Ward, S. H. (1981) Gamma-ray spectrometry in geological mapping and uranium exploration. In Skinner, B. J. (ed.), *Econ. Geol. Seventy-Fifth Anniversary Volume (1905–1980).*

Webb, J. (1980) Total count gamma ray detection and spectrometry. *Aust. Soc. Expl. Geophys.* **11,** 139.

Wilkes, P. G. (1979) Acquisition, processing and interpretation of airborne gamma ray spectrometry data. *Rep. Aust. Bureau Min. Res. Geol. Geophys.* **186.**

Yeates, A. N., Wyatt, B. W., and Tucker, D. H. (1982) Application of gamma-ray spectrometry to prospecting for tin and tungsten granites, particularly within the Lachlan fold belt, New South Wales. *Econ. Geol.* **77,** 1725.

Yliruokanen, I. (1975) Uranium, thorium, lead lanthanoids and yttrium in some plants growing on granitic and radioactive rocks. *Bull. Geol. Soc. Finland* **47,** 71.

Yoshikawa, K., Fujita, A., Murayama, N., Matsunami, T., and Mamuro, T. (1968) Fallout particles in the ground-level air from the Chinese nuclear explosion of December 28, 1966. *J. Geophys. Res.* **73,** 3637.

Zussman, J. (1977) *Physical methods in determinative mineralogy.* London, Academic Press.

10

Radiometric and Isotopic Dating

10.1 HISTORICAL BACKGROUND

The experiments of Rutherford and Soddy reported in 1902 showed that radioactive decay is a statistical process in which the number of atoms that decay in a given period of time is directly proportional to the number of atoms present. It must be borne in mind, however, that the observed statistical process only occurs when a large number of atoms are present; if only a few atoms are present, the relationship need not be obeyed. The operation of life insurance companies provides an analogy to this. For a sufficiently large population, ignoring catastrophic effects (such as war, famine, plague, etc.), life insurance companies can very reasonably predict the survival rate of the population as a whole. The fate of an individual, however, is uncertain. The same is true of radioactive decay processes.

In 1906, Rutherford proposed that the age of U-bearing minerals could be determined by measuring the amount of He that had accumulated in those minerals, assuming that all the He present had formed radiogenically and that none had been lost. It was realised that the second of these assumptions was unlikely to be true, even in the more dense and compact minerals, but that the method would still yield a minimum age. Measurements were carried out on a sample of fergusonite and another unspecified mineral from Glastonbury, Connecticut, and an age of 500 Ma was obtained.

The problems of He loss from samples led Boltwood in 1907 to use Pb, the product of U decay, for dating purposes, as it seemed unlikely that Pb which formed in a compact mineral would be lost. A number of minerals were dated, yielding ages ranging between 410 and 2200 Ma, but the results were based on the total Pb content of the samples used; as Pb is the stable product of several decay schemes, the ages obtained were unreliable. Despite the limitations of the methods employed by Rutherford and Boltwood, their results clearly indicated that minerals in the Earth's crust have ages that may be reckoned in hundreds or thousands of millions of years. This had already been argued by geologists, contrary to the views of the Church, and even contrary to the limit of tens of millions of years proposed, on the basis of heat loss from the Earth, by Lord Kelvin. It was only a question of time before extension of knowledge of radioactive decay schemes and refinement of analytical procedures allowed radiometric dating to become an invaluable tool for the geologist.

10.2 TIME IN RADIOACTIVE DATING

The essential aspect of any radioactive decay scheme in its application to dating is that the rate of decay is controlled only by the number of radioactive atoms present, and by no other factor; that is, changes in the chemical or physical environment do not produce changes in the rate of decay. This is true for most nuclear reactions because the distance over which nuclear forces act are much smaller than the distances which separate nuclei. Furthermore, the energies involved in nuclear processes are much greater than those normally available in chemical reactions. The two exceptions to this, however, are decay by electron capture and some decays to isomeric states. In particular, capture of an extranuclear electron by a nucleus is influenced by the electron density near the nucleus and this, in turn, may be influenced by the chemical environment. This effect has been observed in the artificial radionuclide ^{7}Be.

If N is the number of atoms remaining at time t,

$$-\frac{dN}{dt} \propto N,$$

where dN/dt is the rate of change in N with time, the negative sign indicating a decrease. Eliminating the proportionality,

$$-\frac{dN}{dt} = \lambda N,$$

where λ is the decay constant. The value of λ represents the probability that an atom will decay in unit time. The mean life τ is the reciprocal of λ and is thus the average life expectancy of an atom:

$$\tau = \frac{1}{N_0}\int_0^{\infty} \lambda N t \, dt,$$

where N_0 is the original number of atoms present in a closed chemical system at the time of its formation ($t_0 = 0$), and N is the number remaining after time t. Furthermore,

$$\int_{N_0}^{N} \frac{dN}{dt} = -\int_0^t \lambda \, dt,$$

$$\log_e\left(\frac{N}{N_0}\right) = -\lambda t,$$

and

$$N = N_0 \, e^{-\lambda t}.$$

Alternatively,

$$N_0 = N \, e^{\lambda t}.$$

As the decay of the parent radionuclide progresses, stable radiogenic daughter atoms are eventually produced. The number D of daughter atoms in a closed chemical system is clearly given by

$$D = N_0 - N.$$

Therefore,

$$D = N_0 - N_0 \, e^{-\lambda t}$$

or

$$D = N \, e^{\lambda t} - N,$$

so that

$$D = N_0(1 - e^{-\lambda t})$$

or

$$D = N(e^{\lambda t} - 1).$$

If D_0 atoms of the radiogenic daughter were already present in the system at time t_0,

$$D_{total} = D_0 + N_0(1 - e^{-\lambda t})$$

or

$$D_{total} = D_0 + N(e^{\lambda t} - 1).$$

In practice, the last equation is used as the basis for radiometric age determination. Rewriting,

$$\frac{D_{\text{total}} - D_0}{N} = e^{\lambda t} - 1;$$

$$t = \frac{1}{\lambda} \log_e \left(\frac{D_{\text{total}} - D_0}{N} + 1 \right).$$

The decay of a radionuclide may also be expressed by reference to its half-life; that is, the time taken for the decay of a given number of atoms to half that number:

$$\frac{N_0}{2} = N_0 \, e^{-\lambda T_{\frac{1}{2}}},$$

where $T_{\frac{1}{2}}$ is the half-life. Thus

$$\log_e \tfrac{1}{2} = -\lambda T_{\frac{1}{2}}$$

and

$$\log_e 2 = \lambda T_{\frac{1}{2}},$$

so that

$$T_{\frac{1}{2}} = \log_e \frac{2}{\lambda},$$

or

$$T_{\frac{1}{2}} = \frac{0.693}{\lambda}.$$

So far, only a simple parent–daughter system has been considered, but several radionuclides produce complex decay chains before a stable daughter is formed.

Consider a parent N_1 which decays with constant λ_1 to daughter N_2, which then decays by constant λ_2 to daughter N_3, which is stable. Let there be $N_{1(\text{original})}$ atoms of the parent present at $t = 0$. The growth of the intermediate daughter is given by

$$\frac{dN_2}{dt} = \lambda_1 N_1 - \lambda_2 N_2,$$

while the decay of the parent is given by

$$N_1 = N_{1(\text{original})} \, e^{-\lambda t}.$$

The number of intermediate daughter atoms is given by

$$N_2 = \frac{\lambda_1}{\lambda_2 - \lambda_1} N_{1(\text{original})} \, (e^{-\lambda_1 t} - e^{-\lambda_2 t}),$$

on the assumption that all atoms of N_2 are radiogenic.

The general solution for a decay series in which all the daughters are radiogenic is given by

$$N_n = \sum_{x=1}^{x=n} C_x \, e^{-\lambda_x t},$$

where

$$C_x = \lambda_1 \lambda_2 \ldots \lambda_{n-1} \frac{N_{1(\text{original})}}{(\lambda_1 - \lambda_x)(\lambda_2 - \lambda_x)\ldots(\lambda_n - \lambda_x)},$$

the denominator excluding the term '$(\lambda_x - \lambda_x)$'. For example (Faure, 1977),

$$N_3 = C_1 \, e^{-\lambda_1 t} + C_2 \, e^{-\lambda_2 t} + C_3 \, e^{-\lambda_3 t},$$

where

$$C_1 = \frac{\lambda_1 \lambda_2 N_{1(\text{original})}}{(\lambda_2 - \lambda_1)(\lambda_3 - \lambda_1)},$$

$$C_2 = \frac{\lambda_1 \lambda_2 N_{1(\text{original})}}{(\lambda_1 - \lambda_2)(\lambda_3 - \lambda_2)}$$

and

$$C_3 = \frac{\lambda_1 \lambda_2 N_{1(\text{original})}}{(\lambda_1 - \lambda_2)(\lambda_2 - \lambda_3)}.$$

The abundance of N_3 as a function of time is given then by

$$N_3 = N_{1(\text{original})} \left(1 + \frac{\lambda_1 \, e^{-\lambda_2 t}}{\lambda_2 - \lambda_1} + \frac{\lambda_2 \, e^{-\lambda_1 t}}{\lambda_2 - \lambda_1}\right).$$

The behaviour of individual decay series will thus be governed by the relative values of λ_1 and λ_2. If λ_1 is greater than λ_2, the abundance of N_1 will decrease exponentially, N_2 will increase to some maximum value less than $N_{1(\text{original})}$ and then decrease, and N_3 will steadily increase until, when N_1 is zero, N_3 will have an abundance that will asymptotically approach the value of $N_{1(\text{original})}$.

In this case, the time t_m at which N_2 reaches a maximum is given by

$$t_m = \frac{1}{(\lambda_1 - \lambda_2) \log_e(\lambda_1/\lambda_2)}.$$

This is known as 'transient equilibrium'. Fig. 10.1 illustrates the variation in activity with time of the decay of a short-lived parent to a longer-lived daughter.

However, when λ_1 is less than λ_2, a rather different picture of changing abundances emerges. From

$$N_2 = \frac{\lambda_1}{\lambda_2 - \lambda_1} N_{1(\text{original})} \, (e^{-\lambda_1 t} - e^{-\lambda_2 t}),$$

it follows that, as t increases, $e^{-\lambda_2 t}$ will decrease more rapidly than $e^{-\lambda_1 t}$. Then, as

$$N_1 = N_{1(\text{original})} \, e^{-\lambda_1 t},$$

$$N_2 \approx \frac{\lambda_1}{\lambda_2 - \lambda_1} N_1$$

or

$$\frac{N_1}{N_2} \approx \frac{\lambda_2 - \lambda_1}{\lambda_1},$$

which is, of course, a constant.

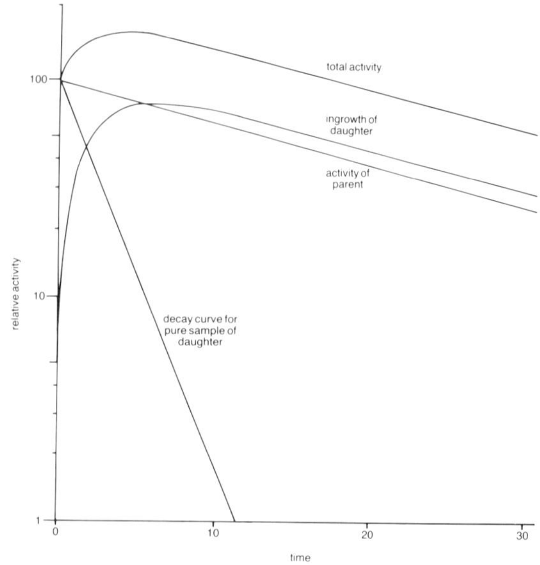

Fig. 10.1 — Transient equilibrium in radioactive decay. (Kathren, 1984, redrawn with permission.)

Thus, after an initial period of time during which growth of N_2 occurs, a condition is established in which the abundances of parent and daughter maintain a constant ratio. Indeed, where the value of λ_1 is very much less than that of λ_2, it follows that

$$\lambda_2 - \lambda_1 \approx \lambda_2,$$

so that

$$\frac{N_1}{N_2} = \frac{\lambda_2}{\lambda_1}$$

and

$$N_1 \lambda_1 = N_2 \lambda_2.$$

This condition is known as 'secular equilibrium' and defines the system in which the rate of decay of the daughter is the same as that of its parent (Fig. 10.2).

The rate at which a daughter approaches equilibrium with its parent is thus dependent upon its own half-life, so that 50% of the equilibrium activity is reached after one half-life. The approach of a daughter to equilibrium with its parent takes the form of a recovery curve as noted above, and full equilibrium is only approached asymptotically; 100% equilibrium is never achieved. The usual reference to equilibrium is, therefore, to either the 95% or the 99% level, which may be considered to be effective equilibrium. The time taken for these equilibrium levels to develop can be determined from

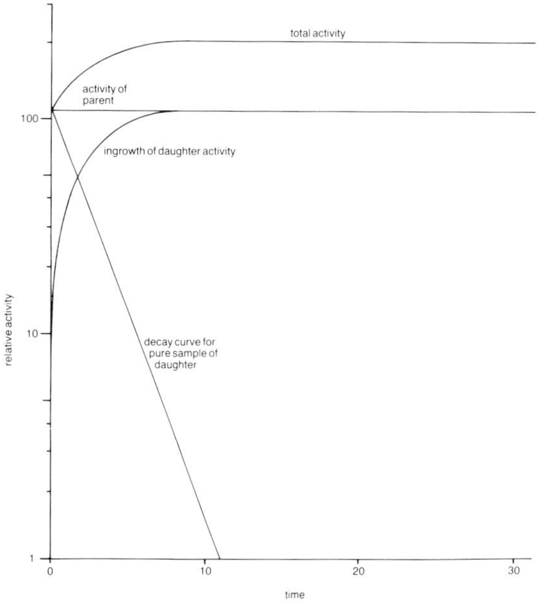

Fig. 10.2 — Secular equilibrium in radioactive decay. (Kathren, 1984, redrawn with permission.)

$$t_{99\%} = \left\{ \frac{\log_e [1/(1-0.99)]}{\log_e 2} \right\} T_{\frac{1}{2}}$$

and

$$t_{95\%} = \left\{ \frac{\log_e [1/(1-0.95)]}{\log_e 2} \right\} T_{\frac{1}{2}},$$

where $T_{\frac{1}{2}}$ is the half-life of the daughter (Table 10.1).

Table 10.1 — Time taken for ingrowth of ^{238}U and ^{232}Th daughters to achieve 99% and 95% secular equilibrium.

Nuclide	99%	95%
^{238}U	–	–
^{234}Th	160.13 days	104.16 days
^{234}Pa	7.84 min	5.10 min
^{234}U	1.661×10^6 years	1.081×10^6 years
^{230}Th	5.315×10^5 years	3.458×10^5 years
^{226}Ra	1.0777×10^4 years	7.011×10^3 years
^{222}Rn	25.41 days	16.53 days
^{218}Po	20.27 min	13.18 min
^{214}Pb	2.97 h	1.93 h
^{214}Bi	2.18 h	1.42 h
^{214}Po	1.09×10^{-3} s	7.09×10^{-4} s
^{210}Pb	146.18 years	95.09 years
^{210}Bi	33.22 days	21.61 days
^{210}Po	2.51 years	1.63 years
^{206}Pb	Stable (24.1% of natural Pb)	
^{232}Th	–	–
^{228}Ra	44.52 years	28.96 years
^{228}Ac	40.73 h	26.49 h
^{228}Th	12.62 years	8.21 years
^{224}Ra	24.19 days	15.73 days
^{220}Rn	6.02 min	3.93 min
^{216}Po	1.05 s	0.68 s
^{212}Pb	2.93 days	1.91 days
^{212}Bi	6.7 h	4.36 h
^{212}Po	1.99×10^{-6} s	1.3×10^{-6} s
^{208}Tl	20.6 min	13.4 min
^{208}Pb	Stable (52.3% of natural Pb)	

In a decay series in which λ_n is always much less than λ_{n+1}, it follows that

$$\lambda_1 N_1 \approx \lambda_2 N_2 \approx \lambda_3 N_3 \approx \lambda_n N_n.$$

After effective equilibrium has been established, the number of atoms of any unstable daughter present is given by

$$N_n = \frac{\lambda_1 N_1}{\lambda_n},$$

while the number of atoms of the stable end product daughter is given by

$$D = N_{1(\text{original})} - N_1 - N_2 \ldots - \mathrm{N_n}.$$

As

$N_2 = (\lambda_1/\lambda_2)N_1, N_3 = (\lambda_1/\lambda_3)N_1, \ldots, N_n = (\lambda_1/\lambda_n)N_1,$

it follows that

$$D = N_{1(\text{original})} - N_1\left(1 - \frac{\lambda_1}{\lambda_2} - \frac{\lambda_1}{\lambda_3} \cdots - \frac{\lambda_1}{\lambda_n}\right),$$

and, if λ_1 is much less than λ_2,

$$D = N_{1(\text{original})} - N_1$$

or

$$D = N_1(\mathrm{e}^{-\lambda_1 t} - 1),$$

so that the same condition is obtained as if the decay had been directly from the radioactive parent to the stable daughter.

10.3 DATING SCHEMES

10.3.1 General principles

From the more or less constant half-life of radioactive decay schemes, it might appear that all the naturally occurring radionuclides could be used for dating purposes. To some extent this is true but, in practice, there are a number of factors which limit those that can be used. Firstly, the radionuclide should have a half-life about the same as that of the materials to be dated. For most geological purposes, therefore, a half-life of the order of 10^8–10^9 years is required. Only comparatively young geological materials can be dated using decay schemes which have half-lives of the order of 10^3–10^4 years, while radionuclides with half-lives of only a few seconds, minutes, hours or days are of no use in geological dating. Conversely, if the half-life of a decay is so long that no appreciable quantity of daughter atoms have been produced since the origin of the Earth, the decay scheme is, again, of little use. Secondly, to be applicable to the dating of various geological environments, the radionuclide needs to be very widely distributed. The major elements of the Earth's crust, considered in Chapter 2, generally have the widest distribution, but of these only K has a naturally occurring radioactive isotope. Some minor elements, however, also possess a wide distribution owing to their chemical similarity with a major element. Rb, for example, is very similar in its chemical properties to K and tends to occur in the same chemical environments as K. If minor elements are to be used, it is necessary that concentrations of both parent and daughter isotopes are sufficient to be detected and to provide results within acceptable errors. In general, measurement errors are dependent upon the current level of available technology, and so decay schemes which cannot be utilised at one time may become available with advances in the sophistication of equipment. Thirdly, to obtain the age of formation of a rock or mineral, the material must have remained a closed chemical system since its formation, with neither gain nor loss of radioactive parent or daughter atoms or, in the case of complex decay chains, of intermediate members. Unfortunately, geological materials and environments do not often meet this requirement. In decay schemes in which gaseous daughters are produced, such as that used in ^{40}K–^{40}Ar age determinations and in the ^{238}U–^{206}Pb dating method (in which ^{222}Rn is a gaseous intermediate daughter), for example, the loss of the daughter by diffusion, even through compact materials, can be considerable. The rate of loss is controlled to some extent by temperature; at elevated temperatures, expansion of crystal lattices allows more rapid diffusion. Episodic heating of minerals, therefore, gives rise to a complex pattern of daughter product retention. Even where the decay products are not gaseous, chemical mobilisation can result from the movement of groundwaters, the extent of the process being influenced by the chemical properties of the decay products, the stability of their host material and the chemical nature of the processes taking place. In some circumstances, however, the rate of daughter mobility can be determined and used to unravel the geological history of samples in a more complex fashion than by simply obtaining a date of initial formation. Finally, where the stable daughter product of a decay scheme is added to an original non-radiogenic content of that isotope, it is necessary to be able to distinguish between the radiogenic and non-radiogenic components. For example, ^{40}K also decays to ^{40}Ca, but ^{40}Ca forms 97% of common Ca, and so it is impossible to distinguish between radiogenic and original Ca except in materials which are very rich in K but very poor in their original Ca content.

10.3.2 The ^{40}K–^{40}Ar method

Of the naturally occurring isotopes of K (^{39}K, ^{40}K and ^{41}K), ^{40}K decays both by electron capture to ^{40}Ar and by β-emission to ^{40}Ca. Measurement of the ^{40}Ar content of a sample can yield its age (Dalrymple and Lanphere, 1969). Because ^{40}K has a branched decay, (11.2% decaying by electron capture) there are two decay constants:

$$\frac{d^{40}K}{dt} = \lambda_{40_\beta}{}^{40}K$$

for the β-particle decay to ^{40}Ca;

$$\frac{d^{40}K}{dt} = \lambda_{40_{ec}}{}^{40}K$$

for the electron capture decay to ^{40}Ar.

The total decay is given by

$$\frac{d^{40}K}{dt} = (\lambda_{40_\beta} + \lambda_{40_{ec}})^{40}K = \lambda_{40}{}^{40}K.$$

Unfortunately, two sets of decay constants have been used:

$$\lambda_{40_{ec}} = 0.585 \times 10^{-10}\ \text{year}^{-1},$$

$$\lambda_{40_\beta} = 4.72 \times 10^{-10}\ \text{year}^{-1},$$

so that $\lambda_{40} = 5.305 \times 10^{-10}$ year^{-1} and the half-life is 1.31×10^9 years, or

$$\lambda_{40_{ec}} = 0.581 \times 10^{-10}\ \text{year}^{-1},$$

$$\lambda_{40_\beta} = 4.962 \times 10^{-10}\ \text{year}^{-1}.$$

in which case $\lambda_{40} = 5.543 \times 10^{-10}$ year^{-1} and the half-life is 1.25×10^9 years. Clearly, the results obtained will depend upon which value for the decay constant is employed (Dalrymple, 1979). The latter set is the most recently determined and so is considered to be the most reliable.

The age of a sample may be determined using the relationship

$$t = \frac{1}{\lambda_{40}} \log_e \left[\frac{^{40}Ar}{^{40}K} \left(\frac{\lambda_{40}}{\lambda_{40_{ec}}} \right) + 1 \right],$$

where ^{40}Ar and ^{40}K are the present-day abundances of these isotopes; it is assumed that there was no initial ^{40}Ar in the sample. Thus, to determine t, it is necessary to measure the ^{40}Ar and ^{40}K contents of a sample. ^{40}K is normally obtained by analytical measurement of total K and the assumption that ^{40}K accounts for a constant 0.0119% of total K. This assumption is considered to be valid in most circumstances but, in aqueous environments where certain exchange processes might operate, deviations can occur. Thus, for measurements of zeolites, for example, mass spectrometry should be employed to determine the isotopic composition of K.

The measurement of ^{40}Ar is most widely carried out by the stable isotope dilution method, in which a very precisely known quantity of a 'spike', of different isotopic composition from that to be determined, is added to the sample. As the composition of the spike is known, measurement of the isotopic composition of the mixture (sample plus spike) on a mass spectrometer enables the quantity of the unknown isotope to be obtained with great accuracy.

Mass spectrometers measure isotopic ratios better than absolute quantities and are well suited to this approach. However, although it would be better to use a different element for the spike, this is prevented because material is ionised in a mass spectrometer, and different elements have different ionisation potentials. It is, therefore, necessary to employ a spike of the same element. This is usually ^{38}Ar with minor amounts of ^{36}Ar and ^{40}Ar.

Corrections must also be made for contamination of the sample by atmospheric ^{40}Ar. Atmospheric Ar consists of 99.600% ^{40}Ar, 0.063% ^{38}Ar and 0.337% ^{36}Ar. Contamination can arise in two ways: by direct absorption during sample preparation, and from the experimental apparatus during measurement. Some atmospheric Ar may also be trapped within the sample at the time of its formation, in which case not all the ^{40}Ar present is of radiogenic origin. This trapped Ar is assumed to have the same isotopic composition as modern atmospheric Ar, but this can be determined separately if required by the isochron method discussed later in this chapter. Knowing the isotopic composition of atmospheric Ar, the radiogenic ^{40}Ar in a sample can be calculated using the equation

$$^{40}\mathrm{Ar}_{\mathrm{radiogenic}} = {}^{38}\mathrm{Ar}_{\mathrm{s}} \left\{ \left(\frac{^{40}\mathrm{Ar}}{^{38}\mathrm{Ar}}\right)_{\mathrm{m}} - \left(\frac{^{40}\mathrm{Ar}}{^{38}\mathrm{Ar}}\right)_{\mathrm{s}} - \left[\frac{1 - (^{38}\mathrm{Ar}/^{36}\mathrm{Ar})_{\mathrm{m}}\,(^{36}\mathrm{Ar}/^{38}\mathrm{Ar})_{\mathrm{s}}}{(^{38}\mathrm{Ar}/^{36}\mathrm{Ar})_{\mathrm{m}}\,(^{36}\mathrm{Ar}/^{38}\mathrm{Ar})_{\mathrm{a}} - 1}\right] \times \left[\left(\frac{^{40}\mathrm{Ar}}{^{38}\mathrm{Ar}}\right)_{\mathrm{a}} - \left(\frac{^{40}\mathrm{Ar}}{^{38}\mathrm{Ar}}\right)_{\mathrm{m}}\right]\right\},$$

where the suffixes s, m and a refer to isotopic ratios of spike, spike–sample mixture and atmosphere respectively (Dalrymple and Lanphere, 1969). $^{38}\mathrm{Ar}_{\mathrm{s}}$ is the quantity of ^{38}Ar spike added (in moles). The Ar is most commonly extracted from the sample by fusing in a vacuum.

The advantages of the ^{40}K–^{40}Ar method are considered by Dalrymple and Lanphere to be as follows:

(1) K is an abundant element in the Earth's crust, occurring in most rock-forming minerals. Even when it is not an essential component, it is often present at concentrations greater than 0.1%.
(2) As the half-life of ^{40}K is long, the age range of materials that can be dated is only limited by the need to be greater than 10^4 years old.
(3) As ^{40}Ar is an inert gas, it can be measured easily and accurately, even in small quantities.
(4) The ^{40}Ar in a mineral can only be derived by decay of ^{40}K or by atmospheric contamination; corrections for the latter are easily made.

The limitations, in contrast, are the following:

(1) Not all materials will completely retain ^{40}Ar; moreover, the ability of a material to retain ^{40}Ar varies with ambient temperature and pressure.
(2) Except in a few cases, the method cannot be applied to sedimentary rocks.
(3) Entrapment of ^{40}Ar lost by diffusion from an older source may occur during crystallisation of younger minerals.

Ar loss from geological materials has been studied in some detail, but the quantitative understanding of the processes involved is still far from complete. Clearly, because ^{40}Ar does not form bonds with other elements in crystal structures, its loss by diffusion from crystals containing the parent ^{40}K is to be expected. Because crystal lattice expansion takes place on heating, the rate of loss by diffusion will be greater at high temperatures; conversely, an increase in external pressure decreases loss by diffusion. Laboratory investigations of diffusion loss have shown variations between individual crystals of the same mineral when the effect of changing temperature and pressure is monitored. It appears that the influence exerted by lattice defects on the rate of ^{40}Ar loss is quite profound. In addition to the physical properties of the host mineral and its environment, the time for which particular circumstances persist must also be considered; for example, loss of ^{40}Ar may be greater from a mineral held at 200 °C for 50 Ma than from the same mineral heated to 500 °C for only 1 or 2 Ma.

Despite these difficulties, laboratory studies have shown that, by and large, none of the materials normally used for ^{40}K–^{40}Ar dating will lose a significant amount of ^{40}Ar at temperatures less than 20 °C maintained over the time scale of the age of the Earth. Indeed, minerals which are relatively good at retaining ^{40}Ar, such as sanidine, could be held at temperatures of 150–200 °C for as much as 10 Ma without appreciable ^{40}Ar loss, and some would lose less than 5% Ar at 200 °C even if held at that temperature for 100 Ma.

The concept of a 'blocking temperature', below which a mineral ceases to have appreciable ^{40}Ar loss, is really only applicable to an individual specimen; even then, some time-dependent variation is possible. Mineral species display a wide range of diffusion rates in practice; so to consider a theoretical blocking temperature is unsound. What has been possible, however, is the field investigation of simple geological systems in which the heating of one rock by another has occurred at a specific place and time. In these studies, the mineralogy of the heated rock has been investigated for ^{40}Ar loss effects at increasing distance (and decreasing temperature) from the point of heating. At temperatures below 250 °C, the order of ^{40}Ar retentivity (Damon, 1968) is

hornblende › muscovite › biotite › phlogopite › sanidine › microcline › glauconite,

while from 600 to 700 °C the order is

hornblende › muscovite › sanidine › orthoclase › biotite ›phlogopite › glauconite.

Diffusion thresholds for biotite are about 150–250 °C, for muscovite and sanidine about 250–350 °C and for hornblende about 400–500 °C. At temperatures in excess of 700 °C, ^{40}Ar loss will occur from even the most retentive mineral if it is held at such an elevated temperature for a sufficient length of time.

Heating from an igneous intrusion will, therefore, cause total or partial ^{40}Ar loss in the minerals of the metamorphic aureole. The magnitude of this effect will vary according to distance from the intrusion, the mineral species present and the retentivity of individuals of those species. Similarly, where plutonic igneous rocks cool slowly after emplacement, or where metamorphic rocks cool slowly during uplift after deep burial, it is possible for different mineral species and individuals to yield different ages. In these situations, the difference between the various ^{40}K–^{40}Ar ages can be used to obtain information on the rate of cooling, provided that the difference in the temperatures at which ^{40}Ar loss became negligible is known. Although, as noted above, this involves many uncertainties, ^{40}K–^{40}Ar ages can be used in a semiquantitative way to study the cooling history by drawing thermochrons (lines of equal age for a particular mineral species). Where a pattern of different ages emerges from such slowly cooled rocks, the term 'discordant' is applied, in contrast with the 'concordant' pattern of similar ages produced from a rapidly cooled body.

Weathering will also lead inevitably to loss of ^{40}Ar from minerals, either through their chemical breakdown or by physical disruption of their structure allowing a greater ease of ^{40}Ar

diffusion. Additionally, fracturing of rock by hammering during sample collection, or even by drilling, can cause ^{40}Ar loss, and many procedures for sample preparation which involve grinding can also exert an unwanted influence on the ^{40}Ar content. These problems can usually be avoided by careful choice of both material and sampling procedure. More problematical and widespread is the effect of migrating groundwater on samples. Water–rock interactions can give rise to very profound physical and chemical effects in the rock. Normally, these effects result in ^{40}Ar loss, and so ages obtained by the method are best considered to be minimum values. Groundwater migration can also effect the ^{40}K content of the rock, however, resulting in either gain or loss of the parent isotope, which again leads to erroneous results. Often overlooked in this respect is the effect of radiation damage on minerals (Mitchell, 1972). In Chapter 4, the effect of α-particles on crystal lattice structures was noted as a cause of chemical and structural breakdown; although ^{40}K is not an α-emitter, it is possible that α emitting nuclides will be nearby, and even minor γ-radiation damage will lead to ^{40}Ar loss.

Although ^{40}Ar loss is the main source of difficulty in the interpretation of ^{40}K–^{40}Ar age data, it is also known for samples to contain excess ^{40}Ar. There are two possible ways in which ^{40}Ar can be taken up by a mineral. These are (1) diffusion into the mineral after its formation and (2) inclusion within the mineral at the time of its formation. In either case, an external source of ^{40}Ar is required. Strong theoretical arguments suggest that, as the diffusion gradient is actually away from any crystal, diffusion into crystals is unlikely to occur. It is, nevertheless, possible in the presence of high P_{Ar} values or if the diffusion is from a high K to a low K mineral. Because these circumstances are unlikely to be common, diffusion in this way is probably a rare phenomenon. Evidence in support of the inclusion of ^{40}Ar during mineral formation is provided by observation of greater ^{40}Ar excess in open crystal structures than in tightly structured minerals. This suggests that the amount of excess ^{40}Ar depends upon the availability of space to accommodate the gas. Normal He/Ar ratios are often found in minerals with an ^{40}Ar excess. As these gases have different diffusion rates, this suggests inclusion of both gases at the time of crystallisation. Moreover, excess ^{40}Ar also occurs in fluid inclusions, and it is unlikely that ^{40}Ar diffusion into a mineral would be concentrated in these pre-existing features. For these reasons, inclusion is generally regarded as the most likely cause of excess ^{40}Ar.

Where does this excess ^{40}Ar come from? Ar is most abundant in those circumstances where a gas phase was active in rock formation. It is therefore supposed, not unreasonably, that the ^{40}Ar was available with other gaseous components. The amount of ^{40}Ar in the Earth's atmosphere is only a small proportion of that which would have been formed by the decay of ^{40}K during the history of the Earth. Most radiogenic ^{40}Ar, therefore, does remain trapped within rocks. Some of the ^{40}Ar lost during metamorphism or melting may have escaped directly to the atmosphere, but a major component will have been available for inclusion within minerals which were cooling elsewhere in the surrounding geological environment (Roddick *et al.*, 1980).

Among the K-bearing minerals, the feldspars orthoclase and microcline are not suitable for ^{40}K–^{40}Ar dating as ^{40}Ar is not retained, even at low temperatures. Perthite is similarly unsuitable. It may be concluded, therefore, that all the low temperature feldspars fail to retain ^{40}Ar. In contrast, the high temperature K-feldspars sanidine and anorthoclase have good ^{40}Ar retentivity (Blyuman, 1975). Similarly, the low temperature plagioclase feldspars (which typically contain up to about 1% K_2O) are unsuitable, while high temperature varieties show good ^{40}Ar retentivity. Biotite is one of the most widely used minerals for ^{40}K–^{40}Ar dating, having a high K content and widespread occurrence. ^{40}Ar is lost from the structure on heating, however, but this allows biotite to be studied as an indicator of the thermal history of a sample. Other micas, such as muscovite, lepidolite and phlogopite, are also used, as they have properties very similar to those of biotite, although with different temperature control on the retention of ^{40}Ar. Feldspathoids provide good materials for dating; for example, both nepheline and leucite show good ^{40}Ar retention. Of all the common minerals, hornblende is the most resistant to ^{40}Ar loss. It also typically contains 0.2–1% K_2O, so analysis can be accurately carried out; it also has a wide distribution.

The minerals considered so far are found mainly in igneous rocks. In sedimentary rocks, only glauconite has proved suitable, although it loses ^{40}Ar fairly easily; a temperature of only 150 °C is sufficient to cause ^{40}Ar loss, and even relatively shallow burial of glauconite-bearing rocks often gives radiometric ages that are too low. However, because glauconite is an authigenic mineral, the time of diagenesis is indicated if the age is correct. If authigenic clay minerals can be separated, these too can be useful for dating sedimentary rocks (Clauer, 1981).

Another approach to the problem of dating is to argue that, if a rock contains those minerals which give good ^{40}K–^{40}Ar ages, the rock as a whole is suitable for ^{40}K–^{40}Ar dating, even though individual minerals may contain only small quantities of K. Basalt, containing high temperature feldspar and pyroxene, is a good example; the K is concentrated in the interstitial feldspar and glass. However, glass is not always a good retainer of ^{40}Ar, especially if devitrification has occurred, and so basalts without glass yield the best results.

As discussed above, the basic assumption that a sample contains no initial ^{40}Ar cannot always be justified. Loss of ^{40}Ar since the formation of a mineral can also produce errors; in this case, ages which are too young are found. Difficulties arising from open system geochemistry can be recognised and, to some extent, overcome by the use of isochron plots (Hayatsu and Carmichael, 1977). The principle involved is that the equation

$$\frac{^{40}\mathrm{Ar}}{^{36}\mathrm{Ar}} = \left(\frac{^{40}\mathrm{Ar}}{^{36}\mathrm{Ar}}\right)_{\mathrm{original}} + \frac{\lambda_{40_{\mathrm{ec}}}}{\lambda_{40}} \frac{^{40}\mathrm{K}}{^{36}\mathrm{Ar}} \left(e^{\lambda_{40}t} - 1\right)$$

represents a straight line (the isochron), with axes the isotopic ratios $^{40}Ar/^{36}Ar$ and $^{40}K/^{36}Ar$. The slope of the isochron is given by

$$\frac{\lambda_{40_{\mathrm{ec}}}}{\lambda_{40}} \left(e^{\lambda_{40}t} - 1\right)$$

in which the only variable is t. If, therefore, several cogenetic minerals or rocks of the same age are analysed, their $^{40}Ar/^{36}Ar$ and $^{40}K/^{36}Ar$ isotopic ratios can be plotted to produce an isochron, the slope of which gives the age of the samples.

To fall upon an isochron, the samples must have remained closed systems with respect to ^{40}K and ^{40}Ar. Statistical treatment of the position of the data points with respect to a best-fit straight line, therefore, gives an indication of how much mobility samples have suffered.

The intercept of the isochron on the $^{40}Ar/^{36}Ar$ axis also gives the $(^{40}Ar/^{36}Ar)_{original}$ isotopic ratio. This is commonly known simply as the 'initial ratio'. In the conventional ^{40}K–^{40}Ar dating method, in which excess ^{40}Ar is not recognised, this is taken to be the same as that of the present-day atmosphere (295.5). However, isochron investigations have shown that the initial $^{40}Ar/^{36}Ar$ ratio can, in fact, differ considerably from the present-day atmospheric value; ratios of up to 5000 have been obtained, indicating the possible extent of inclusion of ^{40}Ar. A modification of the simple isochron plot is a three-dimensional system based on axes ^{40}Ar, ^{39}Ar and ^{40}K, which gives more reliable results (Mussett and McCormack, 1978).

Interestingly, studies of initial $^{40}Ar/^{36}Ar$ ratios have thrown light on the origin of the Earth's atmosphere. The ^{40}Ar in the atmosphere has been produced largely by the decay of

^{40}K in rocks and subsequent degassing, while ^{36}Ar is of primordial origin, degassing probably only having contributed to any extent early in the Earth's history. Similarly, within the Earth, the $^{40}Ar/^{36}Ar$ isotopic ratio will not only be controlled by the formation of radiogenic ^{40}Ar and its loss by degassing; any degassing of ^{36}Ar will also have an effect. Measurements made on glassy submarine basalts (in which the risk of atmospheric contamination is negligible) have shown $^{40}Ar/^{36}Ar$ isotopic ratios which are much greater than the atmospheric value, possibly reflecting high ratios in the mantle. Similar ratios are found in Pre-Cambrian ultrabasic rocks. This suggests that ^{36}Ar was largely degassed from the Earth before much ^{40}Ar had accumulated, possibly within a few hundred million years of the Earth's formation (Faure, 1977).

One of the drawbacks of the ^{40}K–^{40}Ar method is that determination of the ^{40}K and ^{40}Ar contents have to be carried out on different parts of the sample to be dated, and this can lead to problems if samples are not homogeneous. Practical difficulties can also arise where only very small samples are available for dating. $^{39}Ar/^{40}Ar$ measurements, however, provide a means of overcoming this difficulty (Dalrymple and Lanphere, 1974). In this approach, a source of fast neutrons in a nuclear reactor is used to convert ^{39}K to ^{39}Ar. The ^{39}Ar has a half-life of 269 years, and so is effectively stable on the time scale of this analysis. The amount of ^{39}Ar formed is given by

$$^{39}\mathrm{Ar} = {}^{39}\mathrm{K}\,\Delta T \int \phi\sigma\,\mathrm{d}E,$$

where ΔT is the duration of irradiation and ϕ and σ are, respectively, the neutron flux density and the neutron capture cross-section of ^{39}K for neutrons of energy E (Faure, 1977). The integration must be carried out over the range of neutron energies present.

If the ^{40}Ar is entirely radiogenic, then

$$\frac{^{40}\mathrm{Ar}}{^{39}\mathrm{Ar}} = \frac{\lambda_{40_{ec}}}{\lambda_{40}} \frac{^{40}\mathrm{K}}{^{39}\mathrm{K}} \frac{1}{\Delta T} \frac{\mathrm{e}^{\lambda_{40}t} - 1}{\int \phi\sigma\,\mathrm{d}E}.$$

Thus

$$\frac{^{40}\mathrm{Ar}}{^{39}\mathrm{Ar}} = \frac{\mathrm{e}^{\lambda_{40}t} - 1}{J},$$

so that

$$t = \frac{1}{\lambda_{40}} \log_e\left(\frac{^{40}\mathrm{Ar}}{^{39}\mathrm{Ar}} J + 1\right).$$

If a flux monitor of known age t_{fm} is placed in the reactor with the sample to be dated, the value of J can be obtained using the equation

$$J = \frac{\mathrm{e}^{\lambda_{40}t_{fm}} - 1}{\left(\frac{^{40}Ar}{^{39}Ar}\right)_{fm}},$$

where $(^{40}Ar/^{39}Ar)_{fm}$ is the isotopic ratio measured in the flux monitor.

Correction for contamination by atmospheric Ar is still necessary, as are additional corrections for the production of Ar isotopes by other neutron-induced reactions involving elements such as Ca which may be present in the sample. The ^{39}Ar–^{40}Ar step heating technique can usefully reveal the presence of both partial Ar loss and excess Ar. In this method, $^{40}Ar/^{36}Ar$ and $^{39}Ar/^{36}Ar$ isotopic ratios are determined as Ar is released by incremental heating. The slope of the straight line produced by these data gives the $^{40}Ar/^{39}Ar$ ratio for use in the age equation given above (Fig. 10.3).

10.3.3 The ^{87}Rb–^{87}Sr method

^{87}Rb decays by β-particle emission to ^{87}Sr. Because the half-life of this decay is of the order 10^{10} years, it is particularly applicable to the dating of very old Rb-bearing minerals. Early work using this decay scheme was only possible on minerals rich in Rb, but now the method can

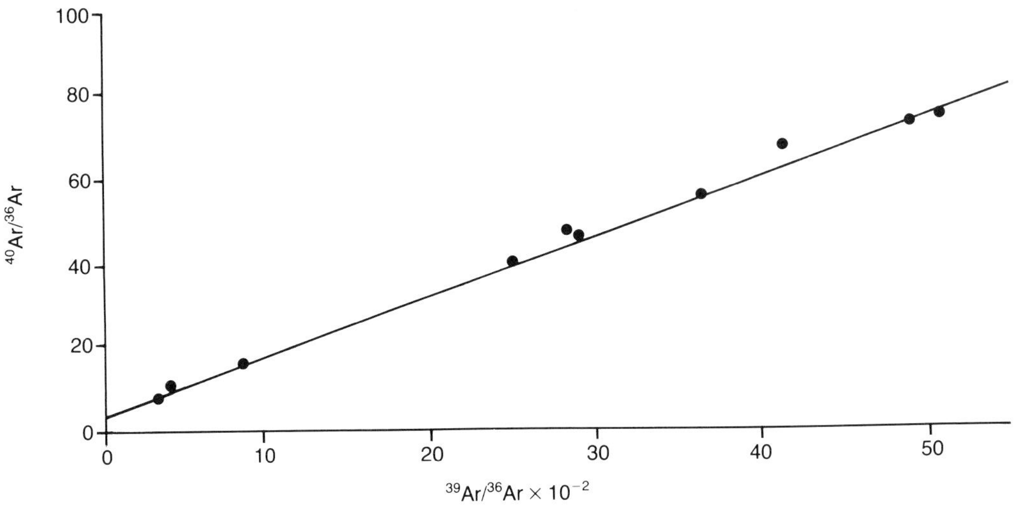

Fig. 10.3 — Ar isotope correlation for a sample of orange glass collected from the Moon on the *Apollo 17* mission. The slope of the line corresponds to an age of 3700 Ma. (Husain and Schaeffer, 1973, redrawn with permission. Copyright: American Association for the Advancement of Science.)

be applied to minerals in which Rb is present only as a trace element.

The decay of ^{87}Rb to ^{87}Sr does not involve branching, and so the equation for the growth of radiogenic ^{87}Sr is simply

$$^{87}\mathrm{Sr} = {}^{87}\mathrm{Rb}\,(e^{\lambda_{87}t} - 1),$$

where ^{87}Rb is the isotopic abundance remaining in a sample t years after its formation and λ_{87} is the decay constant.

As a sample is likely to have contained some initial ^{87}Sr, this must be taken into account:

$$^{87}\mathrm{Sr}_{\mathrm{total}} = {}^{87}\mathrm{Sr}_{\mathrm{original}} + {}^{87}\mathrm{Rb}\,(e^{\lambda_{87}t} - 1).$$

Because a mass spectrometer measures isotopic ratios more accurately than it measures absolute quantities, it is useful to divide this equation through by the non-radiogenic isotope, ^{86}Sr. Thus

$$\left(\frac{^{87}\mathrm{Sr}}{^{86}\mathrm{Sr}}\right)_{\mathrm{total}} = \left(\frac{^{87}\mathrm{Sr}}{^{86}\mathrm{Sr}}\right)_{\mathrm{original}} + \left(\frac{^{87}\mathrm{Rb}}{^{86}\mathrm{Sr}}\right)_{\mathrm{today}} (e^{\lambda_{87}t} - 1).$$

This equation can be rearranged to give t:

$$t = \frac{1}{\lambda_{87}} \log_e \left[\frac{(^{87}\mathrm{Sr}/^{86}\mathrm{Sr})_{\mathrm{total}} - (^{87}\mathrm{Sr}/^{86}\mathrm{Sr})_{\mathrm{original}}}{(^{87}\mathrm{Rb}/^{86}\mathrm{Sr})_{\mathrm{today}}} + 1 \right].$$

Four isotopes of Sr exist naturally: ^{88}Sr, ^{87}Sr, ^{86}Sr and ^{84}Sr. All are stable. Their isotopic abundances are 82.53%, 7.04%, 9.87% and 0.56%, respectively (Faure and Powell, 1972), although the proportion changes as the quantity of ^{87}Sr increases with time.

Unfortunately, because of the low energy of the $\bar{\beta}$-particle emission, the half-life (or decay constant) of the ^{87}Rb–^{87}Sr transition has proved difficult to determine. Most ages obtained by this method have used a half-life of 5.0×10^{10} years, based upon comparisons between Rb–Sr dates for micas with U–Pb dates from coexisting uraninites and monazites (Aldrich *et al.*, 1956), but the most widely accepted direct determination based upon $\bar{\beta}$-particle activity is 4.7×10^{10} years. This shorter half-life has been supported by a number of independent investigations. One study, which involved measurement of the amount of ^{87}Sr produced in a Rb sample over a period of 7 years, yielded a half-life of 4.72×10^{10} years, while another direct determination gave a value of 4.88×10^{10} years (Faure, 1977). Despite the uncertainty, a half-life of 5.0×10^{10} years has been found to give the most consistent results when the ^{87}Rb–^{87}Sr dating method has been used in conjunction with other radiometric dating schemes to investigate geological systems. However, the possibility that ^{87}Rb–^{87}Sr ages are consistently too great and that another factor is involved in achieving the apparent concordance of age dates is, unfortunately, an unresolved problem. The currently used decay constant of 1.39×10^{-11} year^{-1} (a half-life of 5.0×10^{10} years), however, may be replaced by a value of 1.42×10^{-11} year^{-1} (a half-life of 4.88×10^{10} years). Despite this uncertainty in the decay constant, because the half-life is so long, the method is particularly suitable for dating materials older than 10^6 years.

The main advantage that the ^{87}Rb–^{87}Sr method has over ^{40}K–^{40}Ar dating is that, as no gaseous component is involved, loss of daughter product by diffusion is reduced, and atmospheric contamination of samples is not a problem. However, just as there may be an unknown amount of initial ^{40}Ar in a sample, so there will be some initial ^{87}Sr. In order to obtain an age for a sample it is, therefore, necessary to determine the initial $^{87}Sr/^{86}Sr$ isotopic ratio. Indeed, because variation in this value can be great, this is usually carried out for each sample. In cases where the material is greatly enriched in radiogenic ^{87}Sr, however, the age equation is insensitive to small changes in the initial ^{87}Sr value, and a representative value for the particular sample may be chosen.

The $^{87}Rb/^{86}Sr$ isotopic ratio may be obtained from the total Rb/Sr mass ratio of a sample, as it is controlled by the isotopic abundances of the two isotopes and their atomic weights:

$$\frac{^{87}Rb}{^{86}Sr} = \left(\frac{Rb}{Sr}\right)_{total} \left(\frac{^{87}Rb}{^{86}Sr}\right)_{ia} \left(\frac{Sr}{Rb}\right)_{w},$$

where the suffix ia refers to the isotopic abundance and the suffix w to the atomic weight.

The Rb/Sr mass ratio for the sample is obtained by X-ray fluorescence analysis or by isotope dilution techniques on the mass spectrometer. Although, in detail, the abundance of ^{86}Sr and the atomic weight of Sr depend upon the abundances of the other Sr isotopes present, the control exerted by the exact isotopic composition is small because radiogenic ^{87}Sr accounts for less than 10% of total Sr. The total Rb/Sr mass ratio is often simply multiplied by 2.9 to convert to a $^{87}Rb/^{86}Sr$ isotopic ratio.

Where accuracy is required, a mass spectrometer must be used to yield the $^{87}Sr/^{86}Sr$, $^{87}Sr/^{88}Sr$ and $^{84}Sr/^{86}Sr$ isotopic ratios, so that the isotopic abundance of ^{86}Sr and the atomic weight of the actual Sr in the sample can be calculated. However, because ^{88}Sr and ^{84}Sr are stable, and it is unlikely that isotopic fractionation will have occurred, it is often assumed that $^{86}Sr/^{88}Sr = 0.1194$ and $^{84}Sr/^{88}Sr = 0.006\ 75$ (Faure and Powell, 1972).

Various methods have been employed to determine the initial $^{87}Sr/^{86}Sr$ isotopic ratio with variable success. One approach to dating igneous rocks is to determine the $^{87}Sr/^{86}Sr$ isotopic ratio for a mineral in which the initial Rb content can be taken to be zero, so that no radiogenic ^{87}Sr would have formed, and which crystallised from the same source and at the same time as a Rb-bearing mineral that can be dated. Because of its electrochemical properties, Rb normally occurs by substituting for K in K-bearing minerals. Thus a low K mineral, such as Ca-rich plagioclase, may be taken to yield the initial ratio. However, if the ^{87}Sr in a nearby Rb-bearing mineral should be lost from that mineral by, say, fluid migration through the rock, trapping of the liberated ^{87}Sr

is likely in Ca-rich minerals (because the electrochemical properties of Sr and Ca are similar), thereby invalidating the $^{87}Sr/^{86}Sr$ ratio obtained. Such an open chemical system would, of course, also lead to a false $^{87}Sr/^{86}Sr$ ratio in the dated mineral. In passing, it might be noted that, because of the affinity of Sr for Ca, Sr tends to be concentrated in the early formed Ca-rich minerals in igneous rocks (such as Ca-plagioclase) while Rb remains in the liquid phase. Thus, through a crystallisation sequence from a high to low temperature, the Rb/Sr mass ratio will gradually increase and the initial $^{87}Sr/^{86}Sr$ isotopic ratio may also change (McCarthy and Grant, 1980).

Although the problems of loss of daughter product are far less severe in the ^{87}Rb–^{87}Sr method than in ^{40}K–^{40}Ar dating, they do, as shown, still exist. Even in igneous rocks, discordant mineral dates are more often encountered than concordant dates. Gain or loss of ^{87}Sr, caused by reheating and/or the migration of fluids through the rock, is not uncommon. Of course, in metamorphic rocks these effects are even more profound, as a temperature increase of only 200 °C will allow mobility to occur. Moreover, this mobility will not only be of ^{87}Sr; other isotopes of Sr will also be affected, as will Rb. As with the loss of gas in the ^{40}K–^{40}Ar system, the rate of loss of Sr and Rb will vary not only with temperature but also with the mineral species and individuals within mineral species. Again, there is a problem of differential mobility that cannot be expressed in any quantitative way. In general, however, it has been found that the diffusion thresholds for orthoclase, muscovite and biotite lie in the range 200–500 °C with orthoclase›muscovite›biotite.

As K-bearing and Ca-bearing minerals may be present in the same rock, any loss from the former mineral species can lead to gain in the latter. Consequently, it is possible that the rock as a whole remains a closed chemical system.

It is interesting to note here that a wide variety of Rb-bearing minerals can be dated, and even the open lattice structure of orthoclase (which permits loss of ^{40}Ar) is not a significant problem in the case of ^{87}Sr mobility. Indeed, orthoclase is less susceptible to ^{87}Sr loss than either muscovite or biotite.

As with the ^{40}K–^{40}Ar method, the equation for isotope evolution with time takes the form of a straight line when expressed as

$$\left(\frac{^{87}Sr}{^{86}Sr}\right)_{total} = \left(\frac{^{87}Sr}{^{86}Sr}\right)_{original} + \left(\frac{^{87}Rb}{^{86}Sr}\right)_{today}(e^{\lambda_{87}t} - 1),$$

where the axes are the isotopic ratios $(^{87}Sr/^{86}Sr)_{total}$ and $(^{87}Rb/^{86}Sr)_{today}$.

If a suite of cogenetic minerals (and the whole-rock) which have different Rb/Sr mass ratios is analysed, the resulting $^{87}Sr/^{86}Sr$ and $^{87}Rb/^{86}Sr$ values may be plotted to fall on a straight-line isochron, the slope of which yields the age of the suite. Deviation of data points from the straight line provides a means of assessing quantitatively the degree of mobility the elements have undergone. The intercept of the isochron on the $^{87}Sr/^{86}Sr$ axis yields the $(^{87}Sr/^{86}Sr)_{original}$ value. Because whole-rock samples are more likely to have remained closed systems since their formation, an isochron produced by a series of whole-rock samples of the same age is generally considered to be a reliable indication of the age involved.

As noted above, when loss of ^{87}Sr occurs as a result of reheating, trapping of the Sr nearby may result in the rock system as a whole remaining closed, although complete resetting of the mineral system may occur. If the minerals were formed at time t_0, and the reheating occurred at time t_1, the initial $^{87}Sr/^{86}Sr$ ratio at t_0 will have changed as ^{87}Rb decayed to ^{87}Sr until, at time t_1, the mineral can be said to have a new initial $^{87}Sr/^{86}Sr$ ratio. The isochron for a suite of minerals reheated at t_1 will then show the new initial $^{87}Sr/^{86}Sr$ ratio and will have a slope governed by t_1. However, if the rock as a whole remained a closed chemical system, a suite of such rocks would show whole-rock isochrons passing through the original initial $^{87}Sr/^{86}Sr$ ratio and would have a slope governed by t_0 (Fig. 10.4). Note that the whole-rock sample from which the suite of minerals has been collected will lie on both the mineral and the whole-rock isochrons (Gass *et al.*, 1971).

This approach to interpreting isochrons will only be valid if complete homogenisation of the minerals occurred at time t_1, and the whole rock remained a closed chemical system. If only partial homogenisation took place and yet the condition of a whole-rock closed system was satisfied, the data can still be used to yield the time t_0 but will not give t_1. Of course, if the whole rock did not remain a closed system and only partial homogenisation of the minerals took place at the time of reheating, no valid ages can be obtained. These techniques are particularly applicable to the study of metamorphic episodes. Even if a range of ages is produced, none of which has significance individually, useful information may be obtained for the metamorphic petrologist or structural geologist (Cameron *et al.*, 1981).

Age dating of sedimentary rocks is theoretically possible using the ^{87}Rb–^{87}Sr method on authigenic glauconite or on the evaporite minerals sylvite and carnallite. Care must be taken, however, to ensure that samples have not suffered recrystallisation during deep burial. Unfortunately, recrystallisation appears to occur so readily in the evaporite minerals that they do not give reliable isotopic ages. In contrast, glauconite can give results which indicate the time of early diagenesis. In addition, it appears that isotopic homogenisation occurs in fine-grained (clay mineral) rocks during diagenesis (Clauer, 1981). Isochrons can often be obtained from such fine-grained sedimentary rocks and the indicated age is considered to be close to the time of deposition.

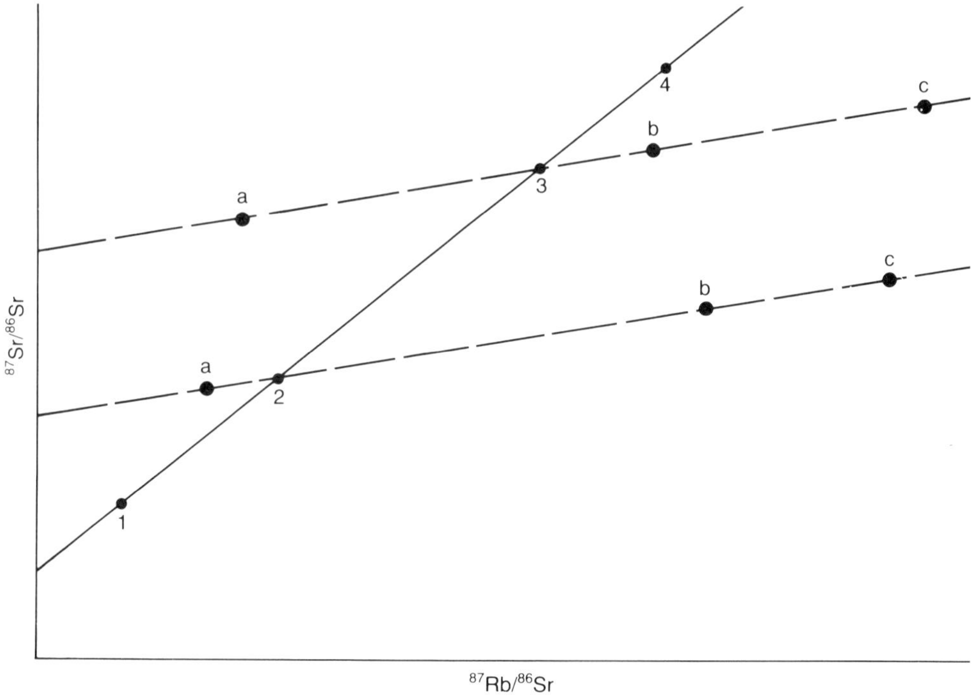

Fig. 10.4 — Individual mineral isochrons from rocks 2 and 3 which are of the same age but have different initial $^{87}Sr/^{86}Sr$ ratios, shown by broken lines (mineral species a, b and c). Whole-rock isochron for older rocks 1, 2, 3 and 4 of the same age, shown by solid lines. The gradients reflect the ages of the mineral open system and the closed whole-rock system.

However, older ages are obtained from detrital minerals that have not undergone complete isotopic homogenisation, and younger ages can be produced by recrystallisation following deep burial.

10.3.4 The U–Pb methods

There are two major naturally occurring radioactive isotopes of U, both of which decay to form stable isotopes of Pb:

$$^{235}U \rightarrow {}^{207}Pb + 7\,^{4}He + 4\beta^-,$$

and

$$^{238}U \rightarrow {}^{206}Pb + 8\,^{4}He + 6\beta^-.$$

Each of these decay schemes can be used for dating. In addition to the radiogenic Pb isotopes, ^{206}Pb and ^{207}Pb, derived from U, another radiogenic isotope, ^{208}Pb, forms from the decay of ^{232}Th. The only naturally occurring non-radiogenic isotope of Pb is ^{204}Pb, which decays to ^{200}Hg by α-particle emission with such a long half-life (1.44×10^{17} years) that it may be considered to be a stable reference isotope.

The U–Pb age equations, based upon numbers of atoms, are similar to those adopted for the ^{40}K–^{40}Ar method:

$$\left(\frac{^{206}Pb}{^{204}Pb}\right)_{total} = \left(\frac{^{206}Pb}{^{204}Pb}\right)_{original} + \left(\frac{^{238}U}{^{204}Pb}\right)_{today} (e^{\lambda_{238}t} - 1),$$

and

$$\left(\frac{^{207}Pb}{^{204}Pb}\right)_{total} = \left(\frac{^{207}Pb}{^{204}Pb}\right)_{original} + \left(\frac{^{235}U}{^{204}Pb}\right)_{today} (e^{\lambda_{235}t} - 1),$$

where λ_{238} and λ_{235} are the decay constants for ^{238}U and ^{235}U, respectively. The half-lives of ^{238}U and ^{235}U are currently accepted as 4.51×10^9 years and 7.129×10^8 years; that is, their decay constants are 1.5369×10^{-10} year^{-1} and 9.7216×10^{-10} year^{-1}, although decay constants of 1.55125×10^{-10} year^{-1} and 9.8485×10^{-10} year^{-1}, have also been obtained (Jaffey *et al.*, 1971).

In these methods, U and Pb concentrations are determined by isotope dilution, and the isotopic ratio of Pb is determined on a mass spectrometer. Rearranging the general equations gives

$$t_{206} = \frac{1}{\lambda_{238}} \log_e \times \left[\frac{(^{206}Pb/^{204}Pb)_{total} - (^{206}Pb/^{204}Pb)_{original}}{(^{238}U/^{204}Pb)_{today}} + 1\right],$$

and

$$t_{207} = \frac{1}{\lambda_{235}} \log_e \times \left[\frac{(^{207}Pb/^{204}Pb)_{total} - (^{207}Pb/^{204}Pb)_{original}}{(^{235}U/^{204}Pb)_{today}} + 1\right],$$

where t_{206} and t_{207} are the ages obtained from the ^{238}U and ^{235}U decay series, respectively. Isochron plots may be used for a suite of samples as in the ^{87}Rb–^{87}Sr method to yield both the age and the initial Pb ratios for both U–Pb systems. As before, if the data points lie close to the best-fit straight line, little Pb or U movement has occurred in the system. The greater the scatter of data points about the best-fit straight line, the more open the system has been. Unfortunately, U–Pb isochrons tend to show a wide scatter and, therefore, have only limited use.

It is possible, however, to combine these two equations for a particular sample:

$$\frac{(^{207}\text{Pb}/^{204}\text{Pb})_{\text{total}} - (^{207}\text{Pb}/^{204}\text{Pb})_{\text{original}}}{(^{206}\text{Pb}/^{204}\text{Pb})_{\text{total}} - (^{206}\text{Pb}/^{204}\text{Pb})_{\text{original}}} = \left(\frac{^{235}\text{U}}{^{238}\text{U}}\right)_{\text{today}} \frac{e^{\lambda_{238}t} - 1}{e^{\lambda_{235}t} - 1}.$$

Isochrons can thus be plotted for ^{207}Pb/^{204}Pb or ^{206}Pb/^{204}Pb isotopic ratios for a suite of samples. The slope of these Pb/Pb isochrons yields an age which is equal to $(e^{\lambda_{235}t} - 1)/(e^{\lambda_{238}t} - 1)$. The value of $(^{235}\text{U}/^{238}\text{U})_{\text{today}}$ is taken to be constant at 0.007 257. Also,

$$\left(\frac{^{207}\text{Pb}}{^{206}\text{Pb}}\right)_{\text{radiogenic}} = \frac{^{235}\text{U}}{^{238}\text{U}} \frac{e^{\lambda_{238}t} - 1}{e^{\lambda_{235}t} - 1},$$

where $(^{207}\text{Pb}/^{206}\text{Pb})_{\text{radiogenic}}$ is the present-day isotopic ratio. Therefore, the age of a sample can be obtained simply from a knowledge of its radiogenic Pb isotopic composition. Correction for reducing the present-day total ^{207}Pb/^{206}Pb isotopic ratio to the radiogenic ^{207}Pb/^{206}Pb isotopic ratio involves subtracting the initial abundances of these isotopes using the ^{207}Pb/^{204}Pb and ^{206}Pb/^{204}Pb isotopic ratios. This equation has to be solved for t by iterative approximations on a computer (Faure, 1977). Moreover, because the present day production rates of ^{207}Pb and ^{206}Pb must be equal to the present-day decay rates of the parents,

$$\left(\frac{^{207}\text{Pb}}{^{206}\text{Pb}}\right)_{\text{radiogenic}} = \frac{^{235}\text{U}}{^{238}\text{U}} \frac{\lambda_{235}}{\lambda_{238}} = 0.046\ 07.$$

The half-lives of both decay series allow their use in the dating of materials older than 10 Ma.

This combination of the ^{235}U and ^{238}U is particularly useful as it is relatively insensitive to loss of Pb. Although values of t_{206} and t_{207} from the same material are often in disagreement, or discordant, because of the differential migration of Pb isotopes during the history of the sample, where Pb loss occurs without isotopic fractionation, the overall ^{207}Pb/^{206}Pb ratio is not affected. When t_{206} and t_{207} do coincide, the dates are said to be concordant, but concordant results are rare. Typically, t_{207} is greater than t_{206}, and both are less than the ^{207}Pb/^{206}Pb age.

It should be noted that the problem of open system chemistry renders many U-bearing minerals unsuitable for age dating. Both of the U decay series include highly mobile daughters and, as U itself is a highly mobile element in many naturally occurring aqueous environments, loss of both parent and daughters frequently occurs. Minerals suitable for U–Pb dating are thus those in which the U and its daughters are trapped within a tight structure. Zircon, monazite, sphene, apatite and uraninite, for example, are likely to prove satisfactory provided that samples are obtained from well below the zone of weathering.

Zircon is often particularly suitable because, although the substitution of Zr^{4+} by U^{4+} is possible (their ionic radii are 0.8×10^{-10} m and 1.1×10^{-10} m, respectively), substitution by Pb is very unlikely to occur as its ionic radius is much greater (1.3×10^{-10} m). Initial Pb concentrations in zircon are, therefore, very low indeed, but the U content can be reasonably high. These features make it very sensitive for dating purposes. Unfortunately, the tightness of the zircon structure that leads to the retention of U and its daughters may allow zircon crystals to survive unchanged through various geological processes. It is even possible that some zircons in igneous rocks may be relics of older materials in which all other minerals melted.

Investigation of Pb loss in the U-Pb decay schemes can be conveniently carried out by means of a concordia–discordia diagram. If undisturbed ^{238}U and ^{235}U systems are considered, the $^{206}\text{Pb}_{\text{radiogenic}}/^{238}\text{U}$ and $^{207}\text{Pb}_{\text{radiogenic}}/^{235}\text{U}$ isotopic ratios measured today will be proportional to the age of the sample. If these ratios are plotted on a graph, a curve (reflecting the different half-lives of ^{238}U and ^{235}U) passing through the origin is obtained. This curve is known as a 'concordia curve'. All closed systems will yield isotope ratios that fall on this curve, those of greater age falling further from the origin (Fig. 10.5).

When, however, Pb loss (or U gain, which has the same effect) has occurred at some specific time during the history of the system, the isotope ratios will fall below the concordia curve, plotting at the origin only if loss of radiogenic Pb had been complete. Now, if a number of individual samples are analysed from a system in which varying degrees of Pb loss are present through the action of modern weathering, all the isotopic ratios will lie scattered about a straight line below the concordia curve, passing through the origin. This line is known as a 'discordia line', the position of which can be used to describe the history of the system. By extrapolating the line back until it cuts the concordia curve, the initial age of formation can be obtained. Moreover, if the Pb loss was not due to modern weathering, but at an earlier date a single event occurred which resulted in the opening of the system, the age of this event can be obtained by extrapolating the discordia line towards the origin until it again intercepts the concordia curve.

It should also be realised that, if the system loses U at some specific time after its initial formation, discordant points will actually plot above the concordia curve. These will still lie on a line extrapolated through the points on the concordia curve determined by the age of initial

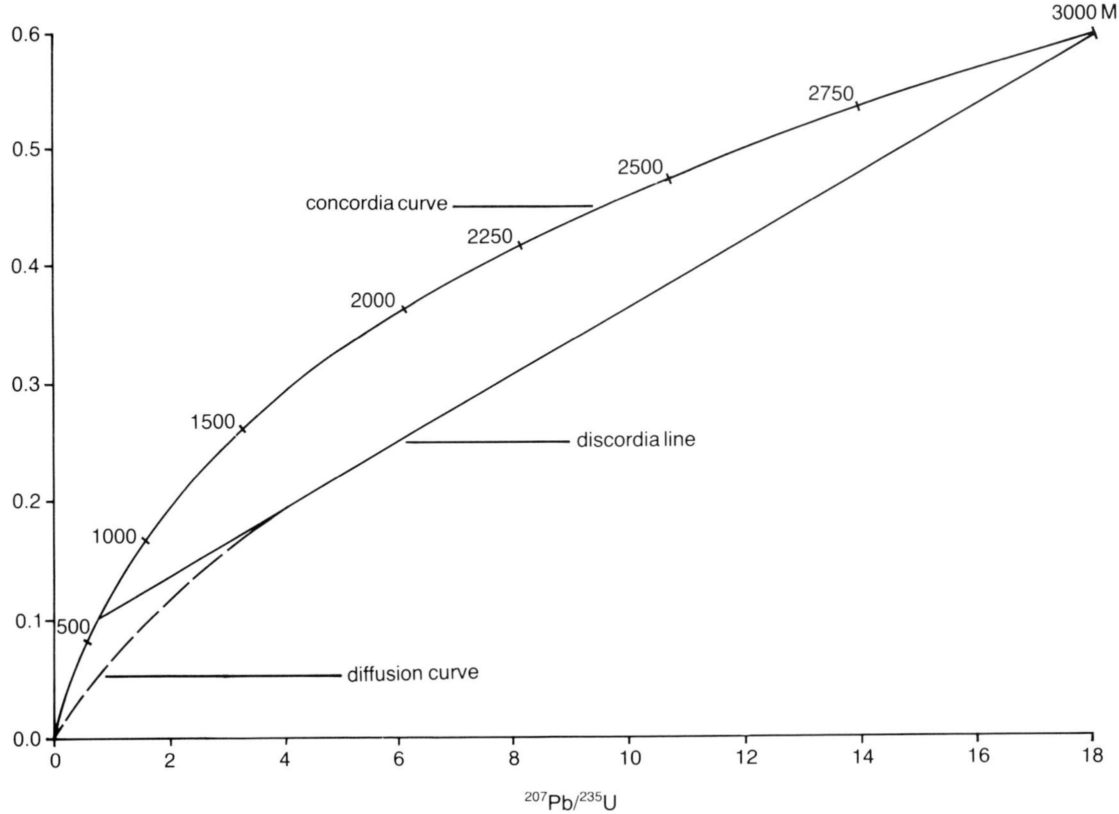

Fig. 10.5 — Concordia–discordia diagram. The concordia curve represents the differential growth of Pb isotopes from U isotopes in a closed system. The intercept of the discordia line with the upper part of the concordia curve indicates the initial age of formation in an open system, while the lower intersection gives the age of opening. The broken section passing from the discordia line to the origin indicates continuous Pb loss by diffusion. (After Faure, 1977.)

formation and the time at which the U loss occurred.

Unfortunately, Pb loss is not always associated with specific events but can also occur simply by continuous diffusion. This results in discordant points which will only lie on a straight line when $d/r^2 < 5 \times 10^{-11}$ year^{-1}, where d is the diffusion coefficient involved and r the effective diffusion radius. When $d/r^2 > 5 \times 10^{-11}$ year^{-1}, the result is a curved discordia line rising from the origin (Tilton, 1960). If, by chance, only those isotopic ratios which produce a straight discordia line are obtained from this type of system, the younger intercept of the discordia line with the concordia curve will therefore yield a meaningless age. The result obtained simply reflects the value of d/r^2. Because of this uncertainty, ages obtained by extrapolating discordia lines towards the origin should always be regarded as fictitious unless there is independent corroborative evidence for an event at that time. A three-dimensional plot of a discordia plane has been used by Wendt (1984) to evaluate samples with non-radiogenic ('common') Pb of unknown isotopic composition.

One interesting aspect of Pb loss in U-bearing minerals concerns the fact that the breakdown of crystalline structures by radiation damage increases in intensity with the age of the minerals and their U content. Moreover, an increase in the water content of the minerals usually accompanies this phenomenon. This water is normally held within the minerals by the confining pressure of deep burial but, on uplift, dilatation occurs, allowing the water to escape and transport Pb with it. The start of Pb loss may, therefore, indicate the uplift of systems that were once deeply buried (Faure, 1977).

10.3.5 U series disequilibrium methods

U–Pb methods which work on the assumption that the entire decay series is in equilibrium are used to date materials older than about 1 Ma. However, this is not the only way in which the decay of U can be employed in geochronology. Following the precipitation of U or the deposition of a U-bearing mineral, it takes a considerable period of time for secular equilibrium to become established between U and the longer-lived members of the decay series. On the assumption that no other processes have subsequently taken place to cause further gain or loss of U, or any of its decay products, the state of disequilibrium can be used to date U deposition that has occurred within the last 1 Ma (Ivanovich, 1982). Simply stated, the greater the degree of disequilibrium, the younger is the material (Fig. 10.6). This is the basis of the ^{230}Th–excess and ^{231}Pa–excess dating methods.

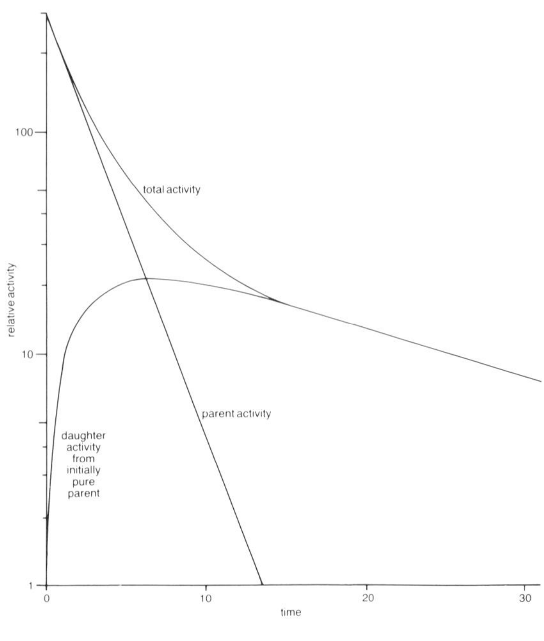

Fig. 10.6 — Disequilibrium in a radioactive decay. (Kathren, 1984, redrawn with permission.)

Processes such as weathering can separate different members of the U decay series by exploiting their very different chemical properties. A member of the decay series may thus become isolated from its parent and reprecipitated at a new site where it will subsequently decay. Again, if this new site remains part of a closed chemical system, the build-up of the daughter products can be used to date the time of precipitation. This is the basis for the ^{230}Th–^{234}U and ^{231}Pa–^{235}U dating methods.

These methods provide a useful tool for dating materials older than the limits of ^{14}C dating and younger than those suitable for methods based on the long half-life radionuclides.

In the following account, it should be noted that the equations are based upon the activity of nuclides rather than on the number of atoms present (which is the case with the radioactive dating methods previously considered in this chapter). However, the relationship between activity and the number of atoms present is straightforward:

$$A = c\,\lambda\,\mathrm{N},$$

where A is the activity, N is the number of atoms present of decay constant λ and c is the counting efficiency of the detector used to determine the activity. Some of the other methods dealt with later in this chapter also employ activity ratios, and in these cases individual attention is drawn to their use.

Methods based upon the accumulation U daughter products

The accumulation of ^{226}Ra

This was the first U series disequilibrium dating method to be used. The accumulation of ^{226}Ra is controlled by the decay of its parent, ^{230}Th, such that the activity ratio of ^{226}Ra with ^{238}U is given by

$$\frac{^{236}\mathrm{Ra}}{^{238}\mathrm{U}} = 1 - \frac{\lambda_{226}}{\lambda_{226} - \lambda_{230}}\,e^{-\lambda_{230}t} + \frac{\lambda_{230}}{\lambda_{226} - \lambda_{230}}\,e^{-\lambda_{238}t},$$

where λ_{226} and λ_{230} are the decay constants for ^{226}Ra and ^{230}Th, respectively. In this relationship, it is assumed that ^{234}U is in secular equilibrium with its parent ^{238}U. However, the ^{234}U/^{238}U activity ratios of solutions which give rise to U deposition are variable, and so the practical use of this equation is limited. A relationship which takes this into account is much more complex (Ivanovich, 1982):

$$\frac{^{226}\mathrm{Ra}}{^{238}\mathrm{U}} = \frac{^{230}\mathrm{Th}}{^{238}\mathrm{U}} - \frac{\lambda_{230}}{\lambda_{226} - \lambda_{230}}\left(e^{-\lambda_{230}t} - e^{-\lambda_{226}t}\right) - \left(\frac{^{234}\mathrm{U}}{^{238}\mathrm{U}} - 1\right) \times \left[\frac{\lambda_{230}\,\lambda_{230}\,e^{-\lambda_{230}t}}{(\lambda_{230} - \lambda_{234})(\lambda_{226} - \lambda_{234})} \times \frac{\lambda_{230}\,\lambda_{234}\,e^{-\lambda_{234}t}}{(\lambda_{230} - \lambda_{234})(\lambda_{226} - \lambda_{234})} - \frac{\lambda_{226}\,\lambda_{230}\,e^{-\lambda_{226}t}}{(\lambda_{226} - \lambda_{234})(\lambda_{226} - \lambda_{230})}\right],$$

but this still involves the assumption that the initial ^{226}Ra content is zero. This condition is unlikely to be satisfied because, with geochemical properties that are similar to those of U, Ra is likely to be present in solutions from which U-bearing minerals precipitate. If zero Ra can correctly be assumed, however, the method can be used to date materials up to 10 ka old.

The ^{230}Th–^{234}U method

At any time after the formation of a sample that contains no initial ^{230}Th, the ^{230}Th/^{234}U activity ratio is given by

$$\frac{^{230}\mathrm{Th}}{^{234}\mathrm{U}} = \frac{1 - e^{-\lambda_{230}t}}{^{234}U/^{238}U} + \left[1 - \frac{1}{(^{234}U/^{238}U)}\right] \frac{\lambda_{230}}{\lambda_{230} - \lambda_{234}} \times [1 - e^{(\lambda_{230} - \lambda_{234})t}].$$

This method can be used to date materials up to about 350 ka old but is most suitable for materials formed during the last 30 ka, as within such materials the ^{230}Th/^{234}U activity ratio varies with age in a manner which is virtually independent of the $^{234}U/^{238}U$ ratio of the sample. In this case

$$\frac{^{230}Th}{^{234}U} = 1 - e^{-\lambda_{238}t},$$

and so t may be easily determined.

The ^{231}Pa–^{235}U method

This method can be used to date materials up to 200 ka old. The half-life of ^{231}Th, which is the immediate daughter of ^{235}U and the parent of ^{231}Pa, is only 25.6 h it can, therefore, be considered to be in secular equilibrium with ^{235}U. This simplifies the age relationship of ^{231}Pa to ^{235}U as

$$\frac{^{231}Pa}{^{235}U} = 1 - e^{-\lambda_{231}t}.$$

Because the $^{235}U/^{238}U$ ratio is taken to be a constant equal to 0.046, ^{235}U is usually measured via ^{238}U. The ^{231}Pa is determined indirectly by measurement of ^{227}Ac or ^{227}Th, which are assumed to be in secular equilibrium with ^{231}Pa.

Combination of ^{230}Th and ^{231}Pa methods

This yields the relationship

$$\frac{^{231}Pa}{^{230}Th} = 1 - e^{-\lambda_{231}t} \Bigg/ \Bigg\{ 21.7\{(1 - e^{-\lambda_{230}t}) + (^{234}U/^{238}U - 1)\left[\frac{\lambda_{230}}{\lambda_{230} - \lambda_{234})}\right](1 - e^{-\lambda_{230} - \lambda_{234})t})\} \Bigg\}$$

which can be applied to minerals up to 300 ka old (Ivanovich, 1982).

The He–U method

Reference has already been made in this chapter to early attempts to use the build-up of ^4He in minerals containing α-particle emitters to date the time of their formation and to the difficulties produced by ^4He leakage.

For materials greater than 1 Ma old, the ^4He production rate can be assumed to be constant and equal to the rate of α-particle emission from the ^{238}U series, if the whole decay series is in secular equilibrium. Clearly, however, for ages less than about 1 Ma, secular equilibrium will not have been established, and so ^4He production from three separate sources must be considered. These are (1) the ^{234}U excess that has decayed from its original activity to its ^{238}U-supported activity; (2) the decay of ^{230}Th and its daughters as they establish equilibrium conditions with ^{234}U and (3) the constant ^{238}U content (Fanale and Schaeffer, 1965).

^4He can also, of course, form by the decay of ^{232}Th and ^{235}U. However, ^4He production from the ^{235}U decay series is considered negligible compared with that from ^{238}U, but the systems chosen for dating must be correctly assumed to have an initial ^{232}Th content of zero. When the different geochemical behaviours of U and Th are borne in mind, this second consideration is not too stringent. Precipitates free of detrital material are, for example, usually suitable.

Nevertheless, as a general rule, it is important to note that, if ^{232}Th is present, then some ^{230}Th is also likely to have been imported into the system at the time of its formation. All ^{238}U series disequilibrium dating methods must, therefore, take into account the possibility of this initial ^{230}Th content. Moreover, for sedimentary precipitates in which the Th is due to the presence of detrital material, there is no fixed initial $^{232}Th/^{230}Th$ ratio, and so this must be determined independently. Similarly, the initial incorporation of ^{234}U not in secular equilibrium with ^{238}U (or, indeed, of any other member of the ^{238}U decay series, such as ^{226}Ra) will give rise to errors in dating (Ivanovich, 1982).

Methods based upon the decay of unsupported (excess) daughter products

The ^{234}U–^{238}U method

This method can yield ages up to about 1 Ma and is based upon the relationship

$$\frac{^{234}\text{U}}{^{238}\text{U}} - 1 = \left[\left(\frac{^{234}\text{U}}{^{238}\text{U}}\right)_{\text{original}} - 1\right] \text{e}^{-\lambda_{234}t},$$

where $^{234}U/^{238}U$ is the U activity at the time of analysis, the suffix 'original' indicates the original $^{234}U/^{238}U$ activity ratio, λ_{234} is the decay constant for ^{234}U, and t is the time that has elapsed since the system became chemically closed.

Because precipitation of U is likely to include both ^{234}U and ^{238}U, no assumptions can be made about the value of the initial isotopic ratio of U and it must be determined by some other technique. However, in applications of this method to the dating of lakes, it has been assumed that the present-day isotopic ratio of U in river water flowing into the lakes has been constant over 200 ka.

The ^{230}Th-excess method

As noted in Chapter 6, at low temperatures the geochemical behaviour of U and Th are often quite different. Under oxidising conditions, for example, U typically forms complex ions which remain in solution, while Th is rapidly precipitated. In sedimentary environments, this may give rise to an excess of ^{230}Th over its parent ^{238}U.

The method was first employed to obtain the age of young sediments using the relationship $t = x/s$, where x is the depth of the sample in the sediment column and s is the sedimentation rate (Piggot and Urry, 1942).

As

$$N = N_0 \text{ e}^{-\lambda t},$$

where N_0 is the original number of atoms of decay constant λ and N is the number remaining after time t, it follows that for ^{230}Th

$$\log_e N = \log_e N_0 - \frac{\lambda_{230}x}{s}.$$

Thus, if results for various values of x are plotted against $\log_e N$, a straight line of slope s and constant N_0 is produced. To obtain N (the ^{230}Th excess in the system), the ^{230}Th activity may be measured directly, but a correction must be made in which the ^{234}U activity is subtracted from the ^{230}Th activity because the U is likely to be associated with detrital clays in which only ^{234}U-supported ^{230}Th is present. Departures from non-linear plots however, have been obtained which may be due to variation in the sedimentation rate. Measurements of the growth of deep-sea manganese nodules have been carried out by this method (Ku *et al.*, 1979).

A similar method, but based upon excess ^{231}Pa, has also been used for dating sediments.

The ^{230}Th–^{232}Th method

Th precipitated from solution will include both ^{230}Th and ^{232}Th. ^{232}Th, however, has a half-life of 1.39×10^{10} years compared with 7.52×10^4 years for ^{230}Th. In young materials, such as Pleistocene sediments, there is a steady reduction in the $^{230}Th/^{232}Th$ activity ratio with increasing age. As before, the relationship is governed by

$$^{230}\text{Th} = {}^{230}\text{Th}_{\text{original}} \text{ e}^{-\lambda_{230}t},$$

which represents the unsupported ^{230}Th. Dividing through by ^{232}Th gives

$$\frac{^{230}\text{Th}}{^{232}\text{Th}} = \left(\frac{^{230}\text{Th}}{^{232}\text{Th}}\right)_{\text{original}} \text{e}^{-\lambda_{230}t}.$$

The assumption that ^{230}Th and ^{232}Th are removed from solution with equal efficiency has been questioned, however, on the grounds that, although they are chemically similar, the geochemical history of these isotopes differs greatly: in marine sedimentary environments, ^{230}Th is produced by the decay of ^{238}U, but ^{232}Th is largely of detrital origin. To overcome this problem, the sediment can be treated with hot HCl solution to liberate sorbed or authigenically incorporated Th but not the Th in detrital minerals (Goldberg and Koide, 1962).

Once again, the total ^{230}Th activity is the sum of supported (indicated by the suffix 'sup') and unsupported (indicated by the suffix 'un') ^{230}Th. The supported ^{230}Th forms by the decay of ^{234}Th so that

$$^{230}\mathrm{Th}_{\mathrm{sup}} = \frac{\lambda_{234}}{\lambda_{230} - \lambda_{234}}\ ^{234}U_{\mathrm{original}} \times (\mathrm{e}^{-\lambda_{234}t} - \mathrm{e}^{-\lambda_{230}t}).$$

If ^{234}U is in secular equilibrium with ^{238}U, its activity will be equal to that of ^{238}U. However, the half-life of ^{238}U is very long, and so $\lambda_{230} - \lambda_{234}$ is approximately equal to λ_{230}, and $\mathrm{e}^{-\lambda_{234}t}$ is approximately equal to 1. Therefore,

$$^{230}\mathrm{Th}_{\mathrm{sup}} = {}^{238}\mathrm{U}_{\mathrm{total}}\ (1 - \mathrm{e}^{-\lambda_{230}t}).$$

Hence

$$^{230}\mathrm{Th}_{\mathrm{total}} = (^{230}\mathrm{Th}_{\mathrm{un}})_{\mathrm{original}}\ \mathrm{e}^{-\lambda_{230}t} + {}^{238}\mathrm{U}_{\mathrm{total}}\ (1 - \mathrm{e}^{-\lambda_{230}t}).$$

Again, dividing by ^{232}Th gives

$$\left(\frac{^{230}\mathrm{Th}}{^{232}\mathrm{Th}}\right)_{\mathrm{total}} = \left(\frac{^{230}\mathrm{Th}}{^{232}\mathrm{Th}}\right)_{\mathrm{original}} \mathrm{e}^{-\lambda_{230}t} + \left(\frac{^{238}U}{^{232}Th}\right)_{total} (1 - \mathrm{e}^{-\lambda_{230}t}).$$

On the assumption that the various phases of a solidifying magma inherit identical ^{230}Th/^{232}Th and ^{234}U/^{238}U isotopic ratios, this equation has been used for dating young igneous rocks. Clearly, to solve this equation, two unknowns have to be determined: the initial Th ratio and the elapsed time (Ivanovich, 1982).

If minerals of the same age have remained chemically closed since their formation, they will plot on a straight line with coordinates ^{230}Th/^{232}Th and ^{238}U/^{232}Th and with a constant slope of $1 - \mathrm{e}^{-\lambda_{230}t}$, giving the initial ^{230}Th/^{232}Th ratio. Deviation of samples from the straight-line plot may indicate different initial ratios, different ages or open chemical systems.

The ^{231}Pa–^{230}Th method

The basis of this method is the formation of ^{231}Pa from ^{235}U and of ^{230}Th from ^{238}U via ^{234}U. It was principally developed for the dating of marine sediments, in which the behaviours of Pa and Th are very similar. As the abundances of ^{238}U, ^{234}U and ^{235}U are relatively constant in sea water, and very little unsupported ^{231}Pa or ^{230}Th is contributed to the sea by rivers, the ^{231}Pa/^{230}Th ratio of sediments at the time of their formation is expected to be close to the calculated value of 0.091. Provided that no post-depositional changes have taken place, measurement of the activity ratio can, therefore, yield the age of sedimentation:

$$\left(\frac{^{231}\mathrm{Pa}_{\mathrm{un}}}{^{230}\mathrm{Th}_{\mathrm{un}}}\right)_{\mathrm{total}} = \left(\frac{^{231}\mathrm{Pa}_{\mathrm{un}}}{^{230}\mathrm{Th}_{\mathrm{un}}}\right)_{\mathrm{original}} \mathrm{e}^{-(\lambda_{230} - \lambda_{231})t},$$

where $(^{231}\mathrm{Pa}_{\mathrm{un}}/^{230}\mathrm{Th}_{\mathrm{un}})_{\mathrm{original}} = 0.091$, $^{231}\mathrm{Pa}_{\mathrm{un}} = {}^{231}\mathrm{Pa} - {}^{235}\mathrm{U}$ and $^{230}\mathrm{Th}_{\mathrm{un}} = {}^{230}\mathrm{Th} - {}^{234}\mathrm{U}$.

Again, difficulties arise because the initial ^{231}Pa/^{230}Th ratio has been found to depart from 0.091. However, if the sediment were deposited at a constant rate s, defined by $s = x/t$, where x is depth and t is time,

$$\log_{\mathrm{e}}\left(\frac{^{230}\mathrm{Th}_{\mathrm{un}}}{^{231}\mathrm{Pa}_{\mathrm{un}}}\right) = \log_{\mathrm{e}}\left(\frac{^{230}\mathrm{Th}_{\mathrm{un}}}{^{231}\mathrm{Pa}_{\mathrm{un}}}\right)_{\mathrm{original}} + \lambda_{231} - \lambda_{230}x,$$

so that a plot of $\log_{\mathrm{e}}(^{230}\mathrm{Th}_{\mathrm{un}}/^{231}\mathrm{Pa}_{\mathrm{un}})$ against x should yield a straight line which gives the age and the original $^{230}\mathrm{Th}_{\mathrm{un}}/^{231}\mathrm{Pa}_{\mathrm{un}}$ ratio (Ivanovich, 1982).

The ^{210}Pb method

The release of ^{222}Rn from U-bearing minerals results in a flux of about 42 atoms min^{-1} of the gas from the surface of the Earth. Most of this

^{222}Rn remains in the atmosphere where it decays through a series of short-lived daughters to ^{210}Pb, which has a half-life of 22.3 years. The residence time of ^{210}Pb in the atmosphere is probably only 1 month at most, however, before it is removed by precipitation and dry fall-out. It then becomes incorporated into snow, ice, lake waters and even organic matter such as trees. The atmospheric ^{210}Pb flux is considered to be constant at any given locality, and the activity of ^{210}Pb in a sample indicates the age of incorporation of the isotope. The method has been used to study rates of deposition of snow and recent sediments. Two different approaches to this method have been adopted. Firstly, if constant initial ^{210}Pb concentrations are assumed, the age difference between the surface and sediment at a depth x is given by

$$t = \log_e \left[\frac{(^{210}\text{Pb}_{\text{un}})_{\text{original}}}{(^{210}\text{Pb}_{\text{un}})_x} \right] \lambda_{210},$$

where

$$^{210}\text{Pb}_{\text{un}} = {}^{210}\text{Pb} - {}^{226}\text{Ra},$$

the value of $(^{210}\text{Pb}_{\text{un}})_{\text{original}}$ coming from the surface sample. Secondly, if a constant rate of sediment supply is assumed throughout time t,

$$\Sigma\ (^{210}\text{Pb}_{\text{un}})_x = F \int_0^t e^{-\lambda_{210} t}\, dt,$$

where $\Sigma\ (^{210}\text{Pb}_{\text{un}})_x$ is the total excess ^{210}Pb activity from the surface to depth x and F is the $^{210}\text{Pb}_{\text{un}}$ flux calculated down to a deposition horizon much older than 100 years (Ivanovich, 1982).

Dating open systems

If reliable radiometric ages are to be obtained using the methods described so far, it is essential that the sample has remained a closed chemical system since its formation, or since some specific point in its history which can be dated. However, in many cases, this condition is not satisfied, chemical exchange processes having commenced at a specific time and having continued ever since. Is it possible to date the time at which these processes started?

Various modifications of the normal decay equations have been proposed to take open system chemistry into account (Hille, 1979). It has been suggested, for example, that fossil shells can be dated if it is assumed that, after death, the shells initially take up U from sea water. This U becomes uniformly distributed throughout the shell, but subsequent alteration by weathering leads to addition of ^{230}Th, ^{231}Pa and ^{234}U, resulting in a change in the activity ratios of these nuclides. On this basis, an equation has been obtained whereby various measured activity ratios can yield an open system age:

$$\frac{^{230}\text{Th}}{^{234}\text{U}} = \frac{^{231}\text{Pa}}{^{234}\text{U}} \frac{(1 - e^{-\lambda_{230} t}) - 1.4352\ [1 - e^{-(\lambda_{230} - \lambda_{234})t}]}{1 - e^{-\lambda_{231} t}} +$$

$$1.4352\ [1 - e^{-(\lambda_{230} - \lambda_{234})t}] \left(\frac{^{231}\text{Pa}/^{238}\text{U}}{1 - e} - 1 \right) \times$$

$$\left(1 - \frac{^{231}\text{Pa}}{^{234}\text{U}} \frac{1 - e^{-\lambda_{234} t}}{1 - e^{-\lambda_{231} t}} \right) +$$

$$1.4352\ [1 - e^{-(\lambda_{230} - \lambda_{234})t}].$$

This method has been applied successfully to the dating of fossil corals (Szabo, 1979), but uncertainty arises because the behaviour of the various daughters involved may not be uniform.

An alternative 'U trend' open system method has been proposed by Rosholt (1980) for dating sediments and minerals. This is an empirical approach which is based upon independent age points within a sequence from about 4 to 720 ka and eliminates the imposition of a rigorous mathematical treatment. Two unknowns need to be determined in this method: the age and the U flux factor (designated λ_0). The value of the flux factor is mainly determined by the quantity of water flowing through the material and the concentration of U in the water.

The technique involves the production of an isochron using the results from three to ten samples from various parts of the unit to be dated. In each sample, the concentrations of ^{238}U, ^{234}U, ^{230}Th and ^{232}Th are determined, and the results plotted on graphs of $(^{234}U–^{238}U)/^{238}U$ activity ratio against $(^{234}U–^{230}Th)/^{238}U$, and of $^{230}Th/^{232}Th$ against $^{238}U/^{232}Th$. The slopes of these lines yield the U trend and the Th index, respectively. The intercept of the U trend line with the $(^{238}U–^{230}Th)/^{238}U$ axis also gives the parameter known as the x intercept. If the value of the x intercept/Th index is known from depositional units of known age, the U flux factor can be assessed. The age of the U trend system can then be calculated using the expression

$$\frac{(^{234}U - {}^{238}U)/^{238}U}{(^{234}U - {}^{230}Th)/^{238}U} = \frac{C_1\ e^{-\lambda_0 t} + C_2\ e^{-\lambda_{234} t}}{C_3\ e^{-\lambda_0 t} + C_4\ e^{-\lambda_{234} t} + C_5\ e^{-\lambda_{230} t}},$$

where

$$C_1 = -\frac{\lambda_0 \lambda_{234}}{\lambda_{234} - \lambda_0}$$

$$C_2 = \frac{\lambda_{234}\lambda_{234}}{\lambda_{234} - \lambda_0} - \lambda_{234}$$

$$C_3 = \frac{3\lambda_0\lambda_{234}\lambda_{230}}{(\lambda_{234} - \lambda_0)(\lambda_{230} - \lambda_0)}$$

$$C_4 = \frac{3\lambda_{234}\lambda_{234}\lambda_{230}}{(\lambda_0 - \lambda_{234})(\lambda_{230} - \lambda_{234})} + 2\lambda_{234}$$

$$C_5 = \frac{3\lambda_{234}\lambda_{230}\lambda_{230}}{(\lambda_0 - \lambda_{230})(\lambda_{234} - \lambda_{230})} - \lambda_{230}.$$

In multiphase sedimentary systems, in which different minerals, cements, etc., are likely to be present, it is necessary to establish λ_0 values for all different transitions. In many cases, however, it is found that one transition is dominant, and so the others can be ignored.

10.3.6 The ^{232}Th–^{208}Pb method

This method is similar in its essential aspects to the U–Pb methods. Here,

$$^{232}Th \rightarrow {}^{208}Pb + 6\,{}^4He + 4\beta^-.$$

The accepted half-life of ^{232}Th is 1.389×10^{10} years (decay constant, 4.987×10^{-11} year^{-1}), but a value of 1.4008×10^{10} years (decay constant 4.9475×10^{-11} year^{-1}) has been obtained recently. Like the U–Pb methods, this method can be used for dating materials older than 10^7 years.

The general equation involving isotopic ratios is similar:

$$\left(\frac{^{208}Pb}{^{204}Pb}\right)_{\text{total}} = \left(\frac{^{208}Pb}{^{204}Pb}\right)_{\text{original}} + \left(\frac{^{232}Th}{^{204}Pb}\right)_{\text{today}} (e^{\lambda_{232} t} - 1),$$

and

$$t_{208} = \frac{1}{\lambda_{232}} \log_e \times \left[\frac{(^{208}Pb/^{204}Pb)_{\text{total}} - (^{208}Pb/^{204}Pb)_{\text{original}}}{(^{232}Th/^{204}Pb)_{\text{today}}} + 1\right].$$

As before, Th is measured by isotope dilution, and the isotopic composition of Pb is determined on a mass spectrometer. ^{232}Th–^{208}Pb dates may be compared with U–Pb values, and discordant results are common, with t_{208} less than t_{206}. Comparison with discordant U–Pb ages leads to the interesting conclusion that, during Pb loss, ^{208}Pb is removed preferentially, ^{206}Pb is less mobile and ^{207}Pb is the least easily removed. Minerals suitable for the U–Pb methods are also suitable for ^{232}Th–^{208}Pb dating. In zircon, for example, substitution of Zr^{4+} by Th^{4+} is only slightly less than by U^{4+}; Th^{4+} has an ionic radius of 1.10×10^{-10} m, and so is almost as readily admitted as U^{4+} (ionic radius, 1.05×10^{-10} m).

Although it is possible for an isochron plot to be successfully used to interpret ^{232}Th–^{208}Pb data, this method is more commonly used in combination with the results of U–Pb studies, by constructing U–Th–Pb concordia–discordia diagrams. Because of the similar half-lives of ^{232}Th and ^{238}U, the concordia plot for these schemes is close to a straight line. Identification of the intersections of the straight discordia line with this nearly straight concordia curve is, therefore, subject to serious error. However, with ^{232}Th and ^{235}U, a curve is produced which is very similar to the U–Pb concordia curve, the positions of points upon which increase in distance from the origin with increasing age. Provided that no fractionation of parent or daughters occurs, these systems plot in the same way as U–Pb concordia–discordia. Thus, if gain of Th or U occurs such that the Th/U ratio remains constant, data points plot on a discordia line beneath the concrodia curve. Extrapolation of this discordia line to intersect the concordia curve will indicate the initial time of formation and the time at which the system became open (if that was due to a specific event). Loss of Pb will have an identical effect. However, the same problem of distinguishing between episodic and continuous diffusion loss applies as in the U–Pb methods. For U or Th loss, data points will plot on the extrapolated discordia line above the concordia curve and, for sudden opening of the system, will lie on a line passing through the origin. In this argument, the absence of fractionation of isotopes is critical. Unfortunately, this is rarely the case, as open systems lead to either gain or loss of parents and daughters, and to changes in the Th/U and Pb isotope ratios. Lower intersections of the discordia line with the concordia curve will thus be produced which have no geological significance (Faure, 1977).

10.3.7 The common Pb method

The isotopic composition of Pb in minerals such as galena, which do not contain significant amounts of U or Th, will have remained unchanged since the mineral formed, provided that no leaching involving isotopic fractionation has occurred. The isotopic ratio of Pb in the mineral will be determined by its age relative to a time when uniform Pb isotopic composition and abundance and uniform Th and U concentrations can be assumed to have been present. This original composition is taken as the intial state of a reservoir from which the Pb can be extracted at any time, the Pb isotopic composition in the reservoir changing as U and Th decay.

If it is assumed that the initial reservoir formed at the time of the origin of the Earth, with isotopic ratios $(^{206}\mathrm{Pb}/^{204}\mathrm{Pb})_0$ and $(^{207}\mathrm{Pb}/^{204}\mathrm{Pb})_0$, and the isotopic ratios $(^{206}\mathrm{Pb}/^{204}\mathrm{Pb})_t$ and $(^{207}\mathrm{Pb}/^{204}\mathrm{Pb})_t$ were abstracted t years ago (obtained when the Pb ratios are measured today), the age of the Pb mineral is given by

$$\frac{(^{207}\mathrm{Pb}/^{204}\mathrm{Pb})_t - (^{207}\mathrm{Pb}/^{204}\mathrm{Pb})_0}{(^{206}\mathrm{Pb}/^{204}\mathrm{Pb})_t - (^{206}\mathrm{Pb}/^{204}\mathrm{Pb})_0} = 0.007\,257\,\frac{e^{\lambda_{235}T} - e^{\lambda_{235}T}}{e^{\lambda_{238}T} - e^{\lambda_{238}t}},$$

where T is the age of the Earth.

This argument has been used to obtain the age of meteorites, on the assumption that they originated in the Primitive Solar Niebula soon after its formation. In this case, $t = 0$ in the above equation, and the measurement of Pb isotopic ratios in meteorites can be plotted as a $^{206}\mathrm{Pb}/^{204}\mathrm{Pb}$–$^{207}\mathrm{Pb}/^{204}\mathrm{Pb}$ isochron, the slope of which yields the age. This is found to be about 4550 Ma. The data points are usually close to the best-fit straight line, indicating the closed nature of these systems since their formation.

Because the average isotopic composition of Pb in the Earth falls on this same straight line, it follows that the Earth is of the same age as the meteorites. Studies of the evolution of Pb during the history of the Earth are considered later in this chapter.

10.3.8 Spontaneous fission of ^{238}U

As a result of spontaneous fission, ^{238}U disintegrates to produce Xe. The isotopic composition of the Xe formed in this way however, is not simple, consisting of ^{129}Xe, ^{131}Xe, ^{132}Xe, ^{134}Xe and ^{136}Xe, although ^{129}Xe is not produced in significant amounts. Radiogenic Xe will, therefore, accumulate in U-bearing minerals at a rate governed by the fission half-life λ_f of ^{238}U, which is 8.19×10^{15} years. However, as Xe is gaseous at ordinary temperatures and pressures, it is very easily lost from most U-bearing minerals, and so the method usually gives unreliable results. However, a modification of the system, in which a tightly structured mineral (such as zircon) is irradiated with thermal neutrons to induce fission of ^{235}U and the consequent production of Xe, has been developed (Teitsma *et al.*, 1975). The Xe produced is released by incremental heating, and its isotopic composition is determined by mass spectrometry. Correction for atmospheric contamination can then be made by reference to the ^{129}Xe and ^{130}Xe contents, and the ^{136}Xe/^{134}Xe ratio of the remaining gas is used to determine the age; this ratio is controlled by the number of both spontaneous and neutron-induced fission events. The neutron-induced fission can, of course, be determined from the known character of the neutron irradiation.

The number S of spontaneous fission events is given by

$$S = \frac{\lambda_f}{\lambda_{238}} \, {}^{238}\mathrm{U}(e^{\lambda_{238}t} - 1),$$

where t is the age of the sample, while the number N of neutron-induced fission events is given by

$$N = {}^{235}\mathrm{U}\sigma d,$$

where σ is the neutron fission cross-section for ^{235}U and d is the neutron dose. Thus

$$\frac{S}{N} = \frac{\lambda_f}{\lambda_{238}} \frac{{}^{238}\mathrm{U}}{{}^{235}\mathrm{U}} \frac{e^{\lambda_{238}t} - 1}{\sigma d}.$$

The ratio S/N will also be given by the ratio R of any two isotopes of Xe (^{I}Xe and ^{II}Xe) produced by fission, in terms of the absolute fission yield Y_n of those isotopes by neutron bombardment and the absolute fission yield Y_s of those isotopes by spontaneous ^{238}U fission such that (Faure, 1977)

$$\frac{S}{N} = \frac{Y_n^{\mathrm{I}} - RY_n^{\mathrm{II}}}{RY_s^{\mathrm{II}} - Y_s^{\mathrm{I}}}.$$

Therefore,

$$t = \frac{1}{\lambda_{238}} \log_e \left(\frac{\lambda_{238}}{\lambda_f} \frac{{}^{235}\mathrm{U}}{{}^{238}\mathrm{U}} \frac{S}{N} \sigma d + 1 \right).$$

As the age determintion is carried out on the gases released at each heating increment, a pattern of values emerges which shows that the highest temperatures release that part of the gas produced by spontaneous fission which has been most effectively trapped within the mineral, and it is this that indicates the time of mineral formation.

Also dependent upon the spontaneous fission of ^{238}U, and utilising thermal neutron bombardment for calibration, is the fission track dating method (Wagner *et al.*, 1975). In this, the fission products of ^{238}U, consisting of high energy charged particles, cause radiation damage in the minerals which contain and surround the U. The lines of radiation damage produced by the passage of these particles can be made visible by etching (Gleadow *et al.*, 1976), and the tracks so revealed will vary in density according to the amount of U present and the age of the material.

The standard procedure is to count a statistically large number of spontaneous fission tracks in a given area, to anneal the sample so that the tracks are destroyed and then to place the sample in a known flux of thermal neutrons to induce fission of ^{238}U. This, in turn, produces new fission tracks that are then counted. The number of tracks is controlled by the amount of U present in the sample and by the neutron dose. As the neutron dose can be calculated, the U content of the sample can be determined. The equation involved is

$$t = 6.446 \times 10^9 \log_e [1 + 7.744 \times 10^{-18} \gamma d]$$

where γ is the ratio of spontaneous fission tracks per unit area to neutron-induced fission tracks per unit area and d is the thermal neutron dose per unit area (Faure, 1977).

Fission tracks produced by the spontaneous fission of ^{235}U and ^{232}Th are negligible and, as the production rate of thermal neutrons in the Earth is usually small, the tracks caused by induced fission of ^{235}U is also considered to be small enough to ignore.

Materials suitable for fission track dating must contain enough U to have produced a track density greater than 10 cm^{-2} since their formation. The U should be uniformly distributed within a material which is free of inclusions, defects, etc., which could make etching and counting of tracks difficult. Finally, the tracks should not self-anneal at low temperatures, or over a time scale equal to their age. Apatite, sphene, zircon, monazite, epidote and natural glass have been found suitable in these respects. All materials, however, will show annealing of the tracks at elevated temperatures and, in some cases, under stress (Macdougall, 1976). Track-loss characteristics vary considerably from one mineral to another, even under the same environmental conditions. Patterns of fission track ages from different minerals in an individual sample may, therefore, provide an accurate picture of its cooling history. Of the materials listed above, epidote has the highest annealing temperature (the temperature at which complete loss of tracks occurs in 1 h at atmospheric pressure) of 715 °C, while sphene, monazite and zircon all have annealing temperatures in the range 600–700 °C. In contrast, apatite and glass have annealing temperatures of only 340 °C and 300 °C, respectively. Apatite is, therefore, a particularly good mineral for indicating in detail the cooling history of a sample at temperatures below the blocking temperatures of the ^{40}K–^{40}Ar and ^{87}Rb–^{87}Sr systems (Naeser and Brookins, 1975).

Because the retention of fission tracks depends upon the rate of cooling, complete track retention will occur at higher temperatures in systems that cool rapidly than in those which cool slowly. Moreover, as the potential for track retention increases from zero to total during cooling, the measured track density actually represents the age of the sample at which its temperature allowed 50% of the tracks to be retained. Clearly, different minerals have different 50% track retention temperatures (for apatite it is about 125 °C, for example), but this temperature increases for materials with higher annealing temperatures. The rate of cooling will also affect the track retention temperature, but the value for apatite appears to be applicable to a rate of temperature decrease ranging between 1 and 100 °C over a period of 1 Ma (Wagner and Reimer, 1972). As well as indicating the presence of mild episodes of heating, the fission track system can also yield useful information on U migration.

10.3.9 The ^{14}C method

As discussed in Chapter 3, ^{14}C is produced in the Earth's atmosphere by the interaction of cosmic ray particles with N, O and C. Of these, interaction with N is the most important in terms of the quantity of ^{14}C produced:

$$^{14}N + {}^{1}n \rightarrow {}^{14}C + {}^{1}H.$$

The ^{14}C produced is quickly oxidised to CO_2,

which disperses throughout the atmosphere to attain a constant level of concentration which represents a balance between its production and its loss by radioactive decay:

$$^{14}C \rightarrow {}^{14}N + \beta^-.$$

^{14}C has a half-life of 5730 years; decay is directly to the ground state, and so is not accompanied by emission of γ-rays.

Because CO_2 containing ^{14}C becomes incorporated into plant tissues as a result of photosynthesis and absorption through the roots, another balance is established between the intake of ^{14}C and its loss by radioactive decay. Similarly, when animals eat these plants or absorb ^{14}C from the atmosphere, they too acquire a ^{14}C content which is a balance between intake and decay.

Atmospheric CO_2 containing ^{14}C also becomes mixed into the hydrosphere and, again, a constant concentration is maintained between supply and loss by decay. Aquatic organisms thus take up ^{14}C in a similar way to terrestrial organisms.

All organisms maintain a steady-state concentration of ^{14}C throughout their lives. When they die, however, the ^{14}C lost by decay is no longer replenished, and so the amount remaining in an organism after its death is proportional to the time that has elapsed since death occurred. The relatively short half-life of ^{14}C, however, limits the use of this system to the dating of materials less than 50 ka old. Nevertheless, it should be noted, that the method is not only applicable to organic matter, as any material which has formed in contact with the atmosphere and has been kept isolated from further reaction with CO_2 can be dated (Libby, 1952).

The ^{14}C activity today, denoted $^{14}C_{today}$, of a sample of either organic or inorganic matter into which ^{14}C was incorporated t years ago is given by

$$^{14}C_{today} = {}^{14}C_0 \; e^{\lambda_{14} t},$$

where $^{14}C_0$ is the original activity; therefore,

$$t = 8266.6 \; log_e \left(\frac{^{14}C_0}{^{14}C_{today}} \right).$$

Here it is assumed that the value of $^{14}C_0$ has not changed with time and that it does not vary with geographical position or climatic conditions. It is also assumed that no isotopic fractionation takes place during the uptake of C by organisms or, if fractionation does occur, it does not vary between one species and another (or, indeed, between individuals).

The assumption that the rate of ^{14}C production is constant, however, cannot be true. Since cosmic rays originate largely in the Sun, and the Sun's activity is known to fluctuate, the production of ^{14}C must also vary. Furthermore, the form of the cosmic ray field in the vicinity of the Earth is controlled by the Earth's magnetic field; as this is also known to change in intensity with time, the rate of ^{14}C production must show a sympathetic variation. Studies of materials that can be dated by other methods, particularly dendrochronology (the use of annual growth rings in trees to obtain dates of up to about 7 ka) and varve chronology (the use of annual deposition of varves in lakes to obtain dates of up to 30 ka), have shown that, prior to 1945, the atmosphere had a depressed ^{14}C content, thought to be due to the burning of fossil fuels in the nineteenth and twentieth centuries. Since 1945, however, testing of nuclear weapons has added considerable amounts of ^{14}C to the atmosphere. In the sixteenth and seventeenth centuries, with fluctuations, the ^{14}C content of the atmosphere was much higher than the mid-twentieth-century value, but similar levels to those of the midtwentieth-century were present in the twelfth to fourteenth centuries (Fig. 10.7). Corrections (of the order of a few per cent) have to be applied, therefore, to ^{14}C ages obtained for this period, but the correction factor is not linear with respect to time (Fig. 10.8). Before historical times, corrections for

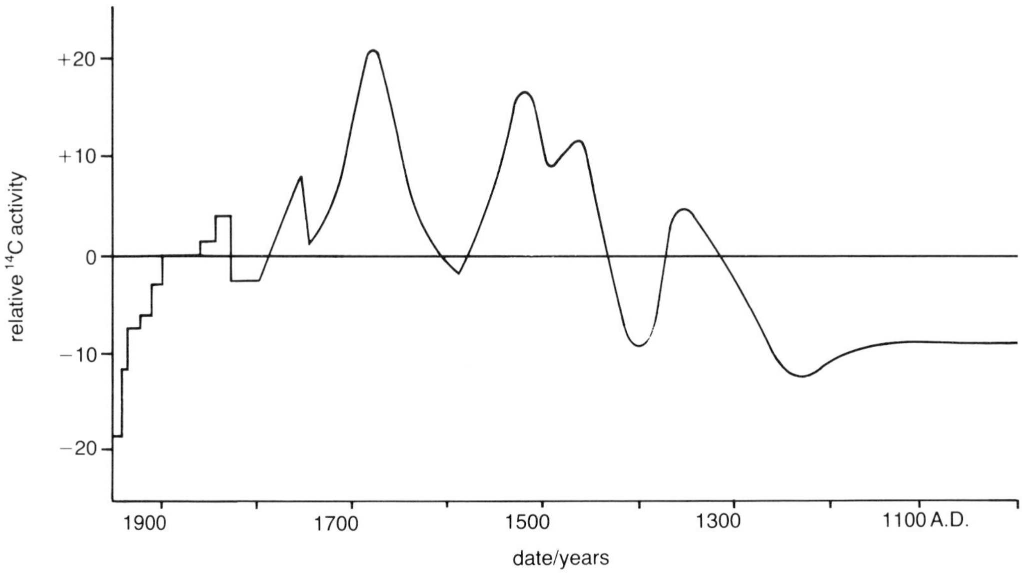

Fig. 10.7 — Variation in ^{14}C activity during the last 10^3 years. (Faure, 1977, redrawn with permission.)

ages up to 4760 BC are reasonably well established, but the comparison between true age and ^{14}C age for older materials is more uncertain. It is clear, however, that ^{14}C ages are about 800 years too low for the interval 3000–10 000 years BP ('before present'; note that ^{14}C ages are mostly referred to the present time, taken conventionally as 1950). Fluctuations in the level of atmospheric ^{14}C within about $\pm 10\%$ of the nineteenth-century value appear to be normal during the period for which the ^{14}C method can be used (Faure, 1977). Wood from the year 1850 is taken as representative of modern atmosphere.

In contrast with problems caused by variation in ^{14}C production with time, the variation that undoubtedly occurs with latitude is negligible. The cosmic ray neutron flux at high geomagnetic latitudes is about four times that at the geomagnetic equator, and ^{14}C production rates are correspondingly greater at high geomagnetic latitudes. However, because maximum ^{14}C production occurs at a high altitude and the mixing of the atmosphere is very effective, the ^{14}C concentration at sea level is effectively independent of geomagnetic latitude.

Organisms are noted for their ability to select preferentially specific isotopes of a given element. Of the two stable isotopes of C, plants take up ^{12}C in preference to ^{13}C; of the radioactive isotopes, ^{12}C is preferred to ^{14}C. Organic materials are, therefore, depleted in both ^{13}C and ^{14}C compared with normal isotopic ratios. Inorganic materials, of course, do not show this fractionation. In order to correct for isotopic fractionation, the isotope composition of the sample must be known so that a fractionation index δ^{13}C can be determined:

$$\delta^{13}C = \frac{(^{13}C/^{12}C)_{sample} - (^{13}C/^{12}C)_{standard}}{(^{13}C/^{12}C)_{standard}} \times 10^3 \ ‰.$$

The original standard ratio used (the 'PDB standard') was based upon CO_2 prepared from specimens of the Cretaceous belemnite *Belemnitella americana* from the Peedee Formation of South Carolina. This is no longer

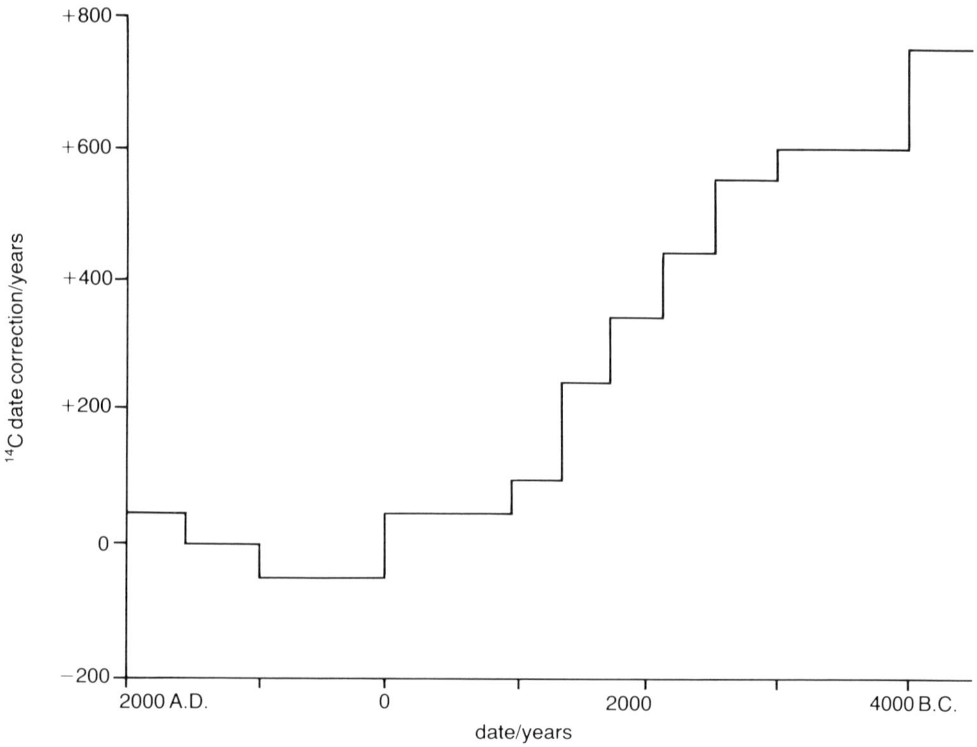

Fig. 10.8 — Corrections to ^{14}C dates obtained by dendrochronology. (Michael and Ralph, 1970, redrawn with permission).

available and a carbonate sample NBS-19 is now employed. Alternatively, a standard based upon CO_2 prepared from oxalic acid is also used. The relationship between these standards is not simple but is agreed to allow interlaboratory correlations to be made. Thus, CO_2 prepared from the oxalic acid has an agreed δ^{13}C value of $-19^{o}/_{oo}$ relative to the PDB standard and NBS-19 has a $\delta\ ^{13}$C value of $+1.95^{o}/_{oo}$. The corrected activity is also based on the assumption that ^{14}C is depleted or enriched twice as much as ^{13}C relative to ^{12}C. Wood has an average δ^{13}C of $-25^{o}/_{oo}$ If a sample to be dated has a different δ^{13}C value, the correction to be applied to the measured ^{14}C activity for isotopic fractionation is, therefore, given by Faure (1977) as

$$^{14}C_{corrected} = {}^{14}C_{measured}\left[1 - \frac{2(25 + \delta^{13}C)}{1000}\right].$$

A wide variety of materials has been found suitable for dating by the ^{14}C method. Wood, seeds, grain, paper, cloth, charcoal, peat and bones have all been found to yield reliable dates, while materials such as pottery and Fe implements may also be satisfactory, having incorporated atmospheric C during manufacture. Shells contain C extracted from sea water, but dates obtained for this material are unreliable because exchange of C with groundwaters may have occurred, and old C may have been incorporated at the time of shell formation. Thus, for cabonate sediments, whether chemically or biologically precipitated, the initial ^{14}C content is largely controlled by the nature of the environment of formation. Where older carbonate material is not present, the precipitated carbonate may truly reflect the character of atmospheric CO_2, and the values obtained may be close to those of plant material. However, where older carbonate material is present, dissolution of this 'dead' C, and its subsequent incorporation into newly precipitated carbonate, thoroughly changes the isotopic composition from that of the atmosphere at the time. Fresh-water conditions, in particular, may have such high levels of dissolved HCO_3^- that the isotopic composition of gastropod shells, for example, can be changed so much that fixation ages much greater than the real ages result. Because the HCO_3^- content of the water in which these animals live will vary spatially in a drainage basin and may even change during the life time of the organism, interpretation of the results obtained can be hazardous. Even in marine waters, the presence of 'dead' C can lead to fictitious ages. Variation in the amount of 'dead' C, due to mixing of surface and deep waters in ocean basins and the mixing typical of coastal waters, is also significant. The conchiolin content of shells, produced directly by the metabolism of an organism, is a more reliable sample material, but large quantities of shell materials (several kilograms) are needed to provide a suitable sample.

In the case of sedimentary materials, $CaCO_3$ can be used for dating, but the problems caused by coprecipitation of 'dead' C and post-depositional changes due to permeating groundwater, discussed in relation to shell material, are also encountered here. Fictitious ages due to these factors are common and may be difficult to evaluate. In addition, the problems due to open system chemistry encountered in other radiometric dating schemes are also encountered. There is also the possibility of contamination by modern ^{14}C and that ^{14}C is produced within the sample by neutron interaction.

Because of the low activity of ^{14}C, sophisticated sample preparation and counting systems have to be employed. The first step in the procedure is pre-treatment of the sample to remove impurities. The C is liberated as CO_2 by burning organic samples with O_2 or treating carbonate samples with acid. Impurities are then removed, and the CO_2 may be converted to other C-bearing gases such as CH_4. A large volume sample is then placed in a carefully shielded counting tube at a pressure of about 3 b. The shielding does not eliminate the background due to cosmic ray mesons, however, and so the apparatus is surrounded by a ring of Geiger–Müller detectors, the response of which allows the background to be calculated and removed. Locally produced neutrons are prevented from entering the sample by means of a 'neutron trap' composed of a mixture of paraffin and boric acid (Aitken, 1974).

10.3.10 The ^{187}Re–^{187}Os method

There are two naturally occurring isotopes of Re: ^{185}Re (37.4%) and ^{187}Re (62.6%). ^{187}Re is radioactive, decaying to ^{187}Os by β^--particle emission with a half-life of 4.3×10^{10} years:

$$^{187}Re \rightarrow {}^{187}Os + \beta^- .$$

^{187}Re occurs principally by trace element substitution in Mo-bearing minerals, in the rare earth mineral gadolinite, in columbite–tantalite and in Cu sulphides.

Although of limited occurrence, ^{187}Re can be used to date Mo mineralisation (Luck and Allegre, 1982) and meteorites. The theoretical basis is the same as that of the ^{87}Rb–^{87}Sr method, in that isotope concentrations are ratioed to the stable isotope ^{187}Os (1.6% of natural Os). Thus

$$\left(\frac{^{187}Os}{^{186}Os}\right)_{total} = \left(\frac{^{187}Os}{^{186}Os}\right)_{original} + \left(\frac{^{187}Re}{^{186}Os}\right)_{today} (e^{\lambda_{187}t} - 1).$$

Again, a series of cogenetic samples will yield an isochron and the original $^{187}Os/^{186}Os$ isotopic ratio if $^{187}Os/^{186}Os$ is plotted against $^{187}Re/^{186}Os$ (Allegre and Luck, 1980). The variation in $^{187}Os/^{186}Os$ isotopic ratios with time for various mantle-derived rocks is shown in Fig. 10.9.

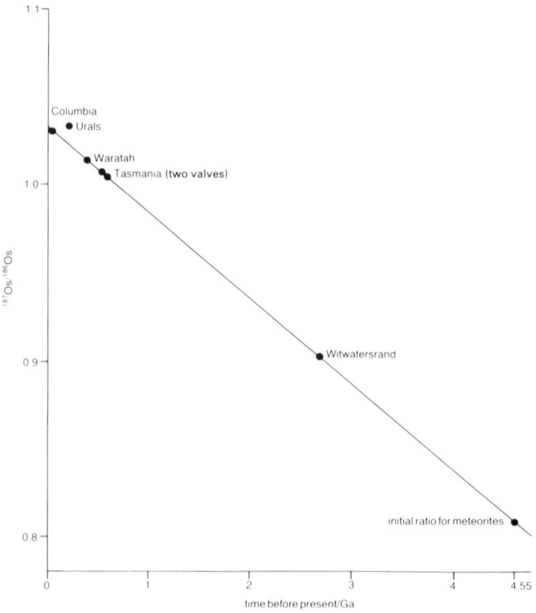

Fig. 10.9 — Variation in $^{187}Os/^{186}Os$ with time in various mantle-derived materials. (Allegre and Luck, 1980, redrawn with permission.)

10.3.11 The ^{176}Lu–^{176}Hf method

Lu is a widely distributed rare earth element with two naturally occurring isotopes: ^{176}Lu (97.4%) and ^{176}Lu (2.6%). ^{176}Lu is radioactive, with a half-life of 3.6×10^{10} years. It has a branched decay by β^--particle emission to ^{176}Hf according to

$$^{176}\mathrm{Lu} \rightarrow {}^{176}\mathrm{Hf} + \beta^-,$$

and by electron capture to ^{176}Yb. Because the electron capture branch accounts for only about 3% of the total decay, it is usually ignored in the theoretical basis of this method.

Many minerals contain Lu, including apatite, zircon, garnet and biotite (by trace element substitution), and the rare earth minerals, in which concentations are usually high. Provided that the mineral does not have a large Hf content, it can be used for age determination; unfortunatcly, zircon is excluded on this basis. Hf has six naturally occurring isotopes, of which the radiogenic ^{176}Hf accounts for 5.2% of the total. The stable isotope ^{177}Hf (18.5%) is used to ratio against ^{176}Hf and ^{176}Lu. The general form of the age equation is, therefore, very similar to that used for the ^{87}Rb–^{87}Sr method:

$$\left(\frac{^{176}\mathrm{Hf}}{^{177}\mathrm{Hf}}\right)_{\mathrm{total}} = \left(\frac{^{176}\mathrm{Hf}}{^{177}\mathrm{Hf}}\right)_{\mathrm{original}} + \left(\frac{^{176}\mathrm{Lu}}{^{177}\mathrm{Hf}}\right)_{\mathrm{today}} (e^{\lambda_{176} t} - 1).$$

Thus a series of cogenetic samples will yield an isochron and the initial $^{176}Hf/^{177}Hf$ ratio when the Hf isotopic ratio $^{176}Hf/^{177}Hf$ is plotted against $^{176}Lu/^{177}Hf$. The main problem with this method is that determination of the isotopic proportions of Hf is difficult. Although Hf concentrates are easily obtained because it separates with Zr, further separation of Hf and Zr is needed if Hf isotopic ratios are wanted from the mass spectrometer. Only where the initial Hf content of a mineral is low can the procedure be easily applied (Faure, 1977).

10.3.12 The ^{147}Sm–^{147}Nd method

This method has been developed during the last 10 years. It is particularly appropriate for old rocks because ^{147}Sm has a long half-life (1.06×10^{11} years). It decays by α-particle emission to ^{143}Nd. The limited enrichment of Sm relative to Nd that occurs during petrogenetic processes also makes the method particularly suitable for old rocks.

However, the ^{147}Sm–^{143}Nd system is different from that of the other long-lived radionuclides in that both the parent and the daughter are refractory in their chemical behaviour, and so their separation is difficult. Both are members of the rare earth element group and are usually found in significant trace quantities in minerals such as zircon, allanite, sphene, biotite, apatite, plagioclase and quartz. Whole-rock systems can also be analysed for their

^{147}Sm–^{143}Nd content. Because of their chemical similarity, separation of Sm and Nd from mineral or rock samples has to be carried out using cation exchange resins. As with other methods of isotopic age determination, the parent and daughter nuclides are ratioed against a non-radiogenic isotope, in this case ^{144}Nd. Isotope dilution methods are used to obtain the isotopic ratios. A suite of cogenetic samples will thus yield an isochron from the equation

$$\left(\frac{^{143}\mathrm{Nd}}{^{144}\mathrm{Nd}}\right)_{\mathrm{total}} = \left(\frac{^{143}\mathrm{Nd}}{^{144}\mathrm{Nd}}\right)_{\mathrm{original}} + \left(\frac{^{147}\mathrm{Sm}}{^{144}\mathrm{Nd}}\right)_{\mathrm{today}} (e^{\lambda_{147}t} - 1) \ ,$$

and ^{143}Nd/^{144}Nd is plotted against ^{147}Sm/^{144}Nd. ^{143}Nd/^{144}Nd ratios are normalised to the ^{146}Nd/^{144}Nd ratio of 0.7219. Model ages for whole-rock samples may be obtained using ^{143}Nd/^{144}Nd and ^{147}Sm/^{144}Nd values of 0.512 636 and 0.1936, respectively, derived from meteorites.

Individual minerals have been shown to become chemically open in whole-rock systems that have been subject to reheating, and so it is difficult to construct simple isochrons for individual minerals. However, pairs of minerals in intimate contact within the rock may form a parallel set of isochrons, indicating that ^{147}Sm and ^{143}Nd only migrate between immediately adjacent minerals (Futa, 1981). In one particular case, sphene and biotite were found to form a linked pair in bands separated by plagioclase and quartz. However, allanite and zircon in the same rock showed isotopic ratios indicative of open system chemistry (possibly related to the extensive radiation damage exhibited by these minerals). It appears that the zircon was contaminated by rare earths derived from the allanite.

Early studies of Nd isotopic evolution appeared to show that Archaean and Lower Proterozoic rocks derived from the mantle possessed approximately chondritic Sm–Nd compositions. This was considered surprising because, although Sm–Nd fractionation was probably very limited during condensation of the Primitive Solar Nebula, some should occur during magmatic processes. Thus, variations in the Sm/Nd ratio in the Earth today should arise through the internal differentiation by which the initial ^{143}Nd/^{144}Nd ratios evolved (Fig. 10.10). Variation in the initial ^{143}Nd/^{144}Nd ratio

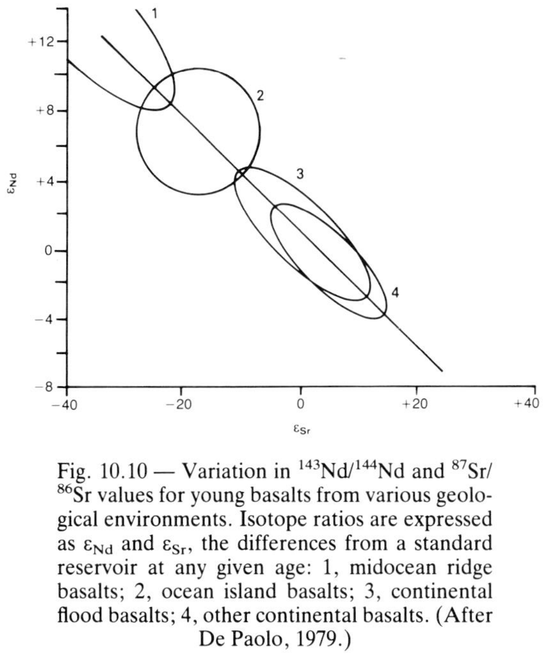

Fig. 10.10 — Variation in ^{143}Nd/^{144}Nd and ^{87}Sr/^{86}Sr values for young basalts from various geological environments. Isotope ratios are expressed as ε_{Nd} and ε_{Sr}, the differences from a standard reservoir at any given age: 1, midocean ridge basalts; 2, ocean island basalts; 3, continental flood basalts; 4, other continental basalts. (After De Paolo, 1979.)

may, therefore, be expected in Pre-Cambrian rocks, and the application of this method should take into account deviations from the initial chondritic ratio. This, however, is not particularly easy because, although the slope of an isochron (which yields the age) is usually subject to reasonable error, the geochemical character of Sm and Nd usually causes the ^{143}Nd/^{144}Nd–^{147}Nd/^{147}Sm data points to plot well away from the origin. In fact, the ^{147}Sm/^{144}Nd ratio is usually in the range 0.09–0.23. When the isochron is extrapolated back to $^{147}Sm/^{144}Nd = 0$, a large error may, therefore, be introduced (Fletcher and Rosmann, 1982). This is especially true for whole-rock samples for which, in order to get a reasonable spread of data points, it is necessary to include specimens of both acid and basic–ultrabasic rocks; in such cases the cogenetic nature of the suite may be in considerable doubt. To minimise errors, a common practice is to translate the origin of the $^{147}Sm/^{144}Nd$ axis from zero to $^{147}Sm/^{144}Nd = 0.1967$, which is the present-day value of the chondritic uniform reservoir, before isochron fitting is carried out.

10.3.13 Other methods of dating geologically young materials

The ^{10}Be method

^{10}Be is produced in the upper atmosphere by cosmic-ray-induced spallation of N and, to a lesser extent, O. Once formed, ^{10}Be decays by β^--particle emission with a half-life of 1.6×10^6 years. The subsequent incorporation of ^{10}Be into sediments may be used as the basis for dating accumulations less than about 10 Ma old (Amin *et al.*, 1975).

For sedimentary deposits, the sedimentation rate s can be calculated by measuring the ^{10}Be concentration at depth x in a layer deposited t years ago, using the expression

$$^{10}\mathrm{Be} = \frac{p}{s}\, e^{\lambda_{10} x/s},$$

where λ_{10} is the decay constant for ^{10}Be (currently taken as 4.33×10^{-7} year^{-1}) and p is the rate of precipitation of ^{10}Be atoms. It is assumed that the sedimentation and precipitation rates have not changed with time.

If x/s is very much less than the decay constant, this equation simplifies to

$$^{10}\mathrm{Be} = p/s.$$

The production rate of ^{10}Be is taken as 1.8×10^{-2} atoms cm^{-2} s^{-1}. If the activity of ^{10}Be is plotted against sedimentation rate, it is found that no unique solution of the sedimentation rate is possible except when $x = 0$. Samples from two depths must, therefore, be analysed to get an average sedimentation rate. Further research, similar to that carried out for ^{14}C methods, is needed to investigate variation in the production rate of ^{10}Be with time and geographical position.

The ^{41}Ca method

^{41} Ca is produced within the Earth by the interaction of thermal neutrons with ^{40}Ca. ^{41}Ca is also another of the spallation reaction products of the cosmic ray flux in the upper atmosphere. However, the parents are Kr, which has a very low abundance in the atmosphere, and Ar, which only acts as a minor source. The contribution of spallogenic ^{41}Ca to the Earth can, therefore, be ignored. The presence of ^{41}Ca in geological materials can thus be taken to indicate the time that the sample has been exposed to a thermal neutron flux, such that

$$\frac{^{41}\mathrm{Ca}}{^{40}\mathrm{Ca}} = \frac{f\sigma T_{\frac{1}{2}}}{0.693}\left(1 - e^{-0.693t/T_{\frac{1}{2}}}\right),$$

where $T_{\frac{1}{2}}$ is the half-life of ^{41}Ca (1.3×10^5 years), f the neutron flux (taken as 3×10^{-3} neutrons cm^{-2} s^{-1}), σ the neutron cross-section (0.44 barns) and t the time that has elapsed since the formation of the material. As the thermal neutron flux is usually only significant near the Earth's surface, the method is restricted to the dating of near-surface materials. Unfortunately, the decay of ^{41}Ca is by electron capture directly to ground state ^{41}K. The only radiation emitted is a 3.3 keV X-ray, and so ^{41}Ca is difficult to detect. However, if a particle accelerator is used, it is theoretically possible to detect radionuclides with a high sensitivity, irrespective of their mode of decay (Raisbeck and Yiou, 1979.

If the method is successful, direct dating of samples that were produced at the surface of the Earth, and then immediately buried, is possible. The shielding necessary to reduce production of ^{41}Ca to negligible levels is about 500 g cm^2 (equivalent to about 1 m of cover). The applications and limitations of the method demand much more detailed investigations.

The ^{129}I method

^{129}I is one of the longest-lived spallogenic radionuclides. It has a half-life of 15.7×10^6 years and decays by β^--particle emission to ^{129}Xe. It is formed in the atmosphere by the interaction of cosmic ray particles with Xe. However, ^{129}Xe is also produced in the Earth by the spontaneous fission of ^{238}U and is a product of nuclear reactors. A tandem accelerator mass spectrometer can be used to give reliable analytical data, but application of the method is limited (Ivanovich, 1982).

The ^{26}Al method

This method is very similar to the ^{10}Be method. Like ^{10}Be, ^{26}Al is produced by spallation in the atmosphere. Because Be and Al possess similar geochemical properties but have different half-lives (^{26}Al has a half-life of 7.3×10^5 years), it has been suggested that the $^{26}Al/^{10}Be$ isotopic ratio provides a time scale which is less dependent upon variations in the cosmic ray flux than the individual methods . Application of the method to sedimentary material dating would seem appropriate. Difficulty in determining ^{26}Al requires the use of a linear accelerator and a cyclotron (Ivanovich, 1982).

The ^{36}Cl method

^{36}Cl is also continuously produced in the atmosphere by cosmic-ray-induced spallation of Ar and ^{35}Cl. It has a half-life of 3.08×10^5 years. Once formed, ^{36}Cl leaves the atmosphere in rain and enters the hydrosphere directly. Because of its low rate of exchange with solid material in the Earth, soluble chlorides containing ^{36}Cl remain in the hydrosphere. Of the possible alternative techniques for detecting ^{36}Cl, an electrostatic accelerator is the best for giving the required sensitivity needed for measuring $^{36}Cl/Cl$ ratios. Unfortunately, ^{36}Cl forms not only from cosmic ray interactions but also from neutron-induced reactions in the Earth and from nuclear explosions. Furthermore, the Cl content of fresh water may be affected by input of marine-derived water ('spray' in rain, or direct spray) or by dissolution of Cl-bearing minerals. To investigate the ^{36}Cl production rate in the atmosphere, old groundwaters (greater that 2 ka) removed from other sources of Cl have, therefore, been studied. These show significantly lower ^{36}Cl values than recent groundwaters and surface waters, where the ^{36}Cl from nuclear explosions is substantially higher. Again, further development of the method is needed (Ivanovich, 1982).

The ^{32}Si method

Cosmic-ray-induced spallation of Ar also results in the formation of ^{32}Si, which can become incorporated into groundwater, polar ice and ocean water. In meteorites, ^{32}Si is produced by spallation from Fe. Dating of materials by this method has only recently been made possible by determination of the half-life of the decay. Unfortunately, the value is disputed, two separate determinations having given half-lives of 108 ± 18 years and 101 ± 18 years (Ivanovich, 1982). Another problem with the use of the method for dating young materials is that ^{32}Si is also produced in nuclear explosions. The spallation production rate has been estimated (on the assumption of a 300 year half-life) at 1.7×10^{-4} atoms cm^{-2} s^{-1}.

10.4 ERRORS

When the result of an isotopic age determination is presented, it is usually accompanied by an error value, no matter which decay scheme has been used; 1234 ± 18 Ma, for example. In conventional use, this error would be determined by many repeat measurements on the same material, or on statistically identical parts of a larger sample, by the same method and operators, and represents error or statistical deviation in the experimental procedure. In the application of isotopic age dating methods, however, it is the usual practice to estimate the repeatability of results from relatively few determinations. This estimate is normally based upon the results obtained when repeatability is specifically investigated. Thus the quoted error is only illustrative of the precision of the result.

High accuracy (which is the approach to the true value) is not the same as high precision, although of course an accurate result must also be precise. Many factors can influence the accuracy of the result obtained. For example, because age is related to isotopic ratios by a logarithmic function, the proportional error in experimental determinations decreases as the age of the sample increases, and so analytical errors do not cause as large an uncertainty in the age of old materials as they do for young materials. Conversely, uncertainties in the value of the half-life produce a time-dependent error which increases with the age of the material to be dated.

In the construction of isochrons, the fitting of a straight line through the data points is always carried out statistically. In practice, some points will be sufficiently close to the best-fit line for their scatter to be attributable to analytical error. However, where larger deviations are apparent, the particular sample may have a different initial isotopic ratio or a different age from that of the rest of the suite, or gain or loss of parent or daughter isotopes may have taken place since the formation of the sample.

10.5 DATING OF GROUNDWATERS

As noted earlier in this chapter, any isotopic dating system requires radioactive decay schemes with half-lives of the same order as the age of the sample to be dated. Any initial isotopic ratio must also be known, and the system must have remained chemically closed throughout its life. In the discussion of isotopic age dating methods applied to rocks and minerals, it was recognised that the condition of a closed chemical system is the most likely source of difficulty, in that episodic or even continuous gain or loss of isotopes may occur and yet not be recognised. The development of open systems in natural waters might, therefore, be considered to render isotopic dating techniques impractical. However, groundwaters develop a chemical equilibrium with their host aquifers and thus prevent dissolution or precipitation which could lead to the development of open systems. The application of dating techniques would, in such cases, appear valid. The use of U decay series measurements in open system dating has been discussed earlier in this chapter.

Variation in the initial isotopic compositions of elements can also present difficulties, but the application of the same procedures used in rock and mineral dating can be successful, and special techniques, such as plotting disequilibrium against distance for waters with a well-defined flow line, may permit extrapolation back to the initial state by following the behaviour of specific nuclides. ^{3}H, ^{4}He, ^{14}C, ^{36}Cl, ^{39}Ar, ^{222}Rn, ^{226}Ra and U have all been used for the dating of groundwaters with varying degrees of success.

10.5.1 Radiogenic gas methods

Rn

^{222}Rn accumulates in groundwater as a function of age and flow rate but, because of the short half-life of this nuclide, it is only of use in short-term or local studies, as considerable variation is possible. ^{222}Rn activities ranging between 32 and 48 Bq l^{-1}, for example, have been reported in waters from the Triassic Sherwood Sandstone of southern England (Andrews, 1982). Moreover, these measured activities are probably less than the true values, because of degassing during sample collection. To some extent, variation in the ^{222}Rn content in groundwater can directly reflect variation in the concentration of its parent ^{226}Ra in the rock matrix. However, because Rn is chemically inert, its growth curve in solution can be used to estimate age.

The ^{222}Rn content of groundwaters can be calculated using the equation

$$^{222}\text{Rn} = \frac{12.338\ A\rho\text{U}}{\phi}\ \text{Bq}^{-1},$$

where A is the fraction of ^{222}Rn generated in the rock that migrates to the groundwater, U (ppm) is the U concentration in the rock matrix, ϕ is the fractional porosity of the rock and ρ (g cm^{-3}) is the bulk density. Unfortunately, while the parameters ρ, U and ϕ may be readily determined for any given rock, the value of A may be difficult to assess. However, if the activity of ^{222}Rn is measured, the value of A can be determined. This can be a useful guide to the ability of ^{222}Rn to migrate through the rock. A value of A close to unity indicates easy ^{222}Rn migration and suggests that the rock has a very open texture or that the parent nuclides are present mainly as coatings on mineral grains or on fracture surfaces, while a low value of A indicates that the parent nuclides are probably tightly bound in resistate mineral phases (Andrews, 1982).

He

He, like Rn, has the advantage of being chemically inert. One atom of ^{4}He is produced by each α-decay in the U and Th series. Its abundance is, therefore, much greater than that of ^{222}Rn; in the ^{238}U decay series, for example, 8 atoms of ^{4}He are produced for every 1 atom of ^{222}Rn. The build-up of ^{4}He in groundwater, like that of ^{222}Rn, may be seen as a function of the residence time although, unlike ^{222}Rn, ^{4}He is stable. That part of the ^{4}He content of groundwater that is in excess of the concentration in air-equilibrated water must be radiogenic in origin. Therefore, if the U and Th contents of the rock matrix are known, the residence time can be calculated by measuring the ^{4}He concentration of the groundwater (Andrews, 1982). Thus

$$T = \frac{\text{He}}{\rho/\phi(1.19\times10^{-13}\text{U} + 2.88\times10^{-14}\text{Th})},$$

where T (years) is the residence time of groundwater flowing through rock of density ρ (g cm^{-3}) and fractional porosity ϕ, and of known U and Th concentrations (ppm). He is the ^{4}He content cm^3 (cm^3 H_2O)$^{-1}$ at standard temperature and pressure, (STP).

In the case of the Sherwood Sandstone, it has been found that $\rho = 2.15$, $\phi = 0.16$, U = 3.4 and Th = 18.2, and so $T = 8.02\times10^{10}$He (Andrews, 1982). As the ^{4}He content averages 3010×10^{-8} cm^3 (cm^3 H_2O)$^{-1}$ at STP, the age is about 2.4 Ma. However, the ^{4}He content of the interstitial water in the Sherwood Sandstone is reported as being considerably greater than that of the flowing water but, if the interstitial water value is multiplied by the fractional porosity, the concentrations are approximately equal. This indicates that ^{4}He is uniformly distributed between the water and the mineral phases of the rock and that its entry into the water-saturated pores is in proportion to their volume. On this basis, the age equation has been modified by Andrews to

$$T = \frac{\text{He}}{\rho(1.19\times10^{-13}\text{U} + 2.88\times10^{-14}\text{Th})}.$$

With a ^{4}He content of $18\,403\times10^{-8}$ cm^3 (cm^3 H_2O)$^{-1}$ at STP in the interstitial water, the age is 15.1 Ma, while the ^{4}He concentration of the flowing water of 3010×10^{-8} cm^3 (cm^3 H_2O)$^{-1}$ at STP is in agreement with an age of 14.8 Ma. This residence time correlates reasonably well with the age of the last main tectonism in southern England (Miocene), suggesting that recharging of the aquifer may be related to opening of fracture-flow systems.

Interestingly, the groundwater in the Sherwood Sandstone aquifer in the area which gave these results is a thermal brine occurring at a depth of about 2.4 km at a temperature of about 90 °C. Temperature and thermal conductivity measurements made in the succession above the Sherwood Sandstone showed that, in the Cretaceous Upper Greensand–Chalk sequence, the heat flow is significantly lower than in the succession between the top of the

Sherwood Sandstone and the base of the Upper Greensand. There is thus a step in heat flow at the base of the Upper Greensand which is probably caused by convection at a velocity of about 4×10^{-7} m s^{-1} (Bloomer *et al.*, 1982). If the convective movement at the base of the Upper Greensand does not involve exchange outside the system, the assumed equilibrium conditions are correct, and the ^{4}He ages determined for groundwater in the Sherwood Sandstone are true ages. However, if exchange is actively taking place, the true age could be significantly different.

He/Rn ratios

Although both ^{4}He and ^{222}Rn are inert gases, the ^{222}Rn concentration of groundwater is primarily dependent upon the ^{226}Ra (and, therefore, ^{238}U) content of the rock matrix, with limited time dependence, while ^{4}He concentrations are governed by both the U and Th concentrations and the age of the water. Variation in the ^{4}He/^{222}Rn ratio may thus indicate changes in parent nuclide concentrations as well as in age.

If changes due to variation in U and Th concentrations can be eliminated, and any fluctuations in the rate at which ^{222}Rn and ^{4}He escape into the groundwater are also allowed for, increases in the ^{4}He/^{222}Rn ratio might indicate sites at which there is localised diffusion and escape of crustal ^{4}He which has been accumulating over long periods of geological time. In one study, for example, it was found that, owing to such enhancement of ^{4}He concentrations, ^{4}He/^{222}Rn ratios for various groundwaters and natural gas fields usually gave pore water ages that were too young. The one exception concerned natural gas fields which are 10^{6}–10^{8} years old (Torgersen, 1980).

Tritium (^{3}H)

^{3}H (half-life, 12.3 years) can also be used for dating groundwaters. Its presence is generally assumed to indicate a proportion of post-1954 recharge; precipitation after that date contained ^{3}H produced by atmospheric testing of thermonuclear weapons (Table 10.2). At its maximum in 1963, the ^{3}H content of rainfall in Britain was about 2000 tritium units (TU). (1 TU is defined as the presence of 1 atom of ^{3}H in 10^{18} atoms of H) but fell to about 60 TU in 1975. Before 1954, the ^{3}H content of precipitation in western Europe was only about 2.5 TU, reflecting the small amount of ^{3}H produced by cosmic ray interaction processes. A groundwater that had complete recharge in 1953 would now only have a ^{3}H level of 0.5 TU, while one that recharged 60 years ago would have ^{3}H at the limit of detection of 0.1 TU. Thus, groundwaters with ^{3}H levels greater than 0.1 TU are less than 60 years old and those with ^{3}H levels greater than 0.5 TU are less than 20 years old.

Table 10.2 — Abundance of hydrogen isotopes in natural water

^{1}H	99.9844%
^{2}H/^{1}H	1.49×10^{-4}
^{3}H/^{1}H	1.3×10^{-18}

(After Matthess, 1982.)

However, it has been suggested that the ^{3}H content (greater than 0.1 TU) of thermal brines in Cornwall, southwest England, could develop in waters more than 60 years old, because of ^{3}H production in the natural environment (Burgess *et al.*, 1981). The only natural nuclear reaction that could produce significant amounts of ^{3}H is

$$^{6}_{3}\text{Li}(n,\ \alpha) \rightarrow {}^{3}_{1}\text{H}.$$

Therefore, only where Li is present in reasonable quantities might ^{3}H be formed. The activity at time t of ^{3}H produced at a constant rate by neutron irradiation is given by

$$\frac{dN}{dt} = \lambda_3 N = p(1 - e^{-\lambda_3 t}),$$

where p is the ^{3}H production rate, λ_3 the decay constant for ^3H and N the number of ^3H atoms present. Thus

$$N = \frac{p}{\lambda_3}(1 - e^{-\lambda_3 t}),$$

and so the limit on the number of ^3H atoms (as t tends towards infinity) is given by p/λ_3.

To estimate the ^3H production rate, the first relationship used is

$$F = \frac{\sigma_{Li} N_{Li}}{\sum_{i=1}^{i=92} \sigma_i N_i},$$

where F is the probability of neutron absorption by Li relative to that by other neutron-absorbing elements, σ_i and N_i are, respectively, the total neutron absorption cross-sections and abundances for all elements in the rock matrix (for atomic numbers 1–92), σ_{Li} is the cross-section for the nuclear reaction given above and N_{Li} is the abundance of Li in the matrix. Only thermal neutrons need to be considered here because fast neutron cross-sections are generally much smaller than those of thermal neutrons, and fast neutrons in most rocks will degrade to thermal neutrons because of the presence of light elements (such as Si and O). For an average granite with a Li content of 40 ppm, the value of F is 0.034 but for the granites of southwest England, which have a Li content of 494 ppm, it is 0.307 (Burgess *et al.*, 1982).

For the second relationship, $p = Fn$, the value of the annual neutron production rate n (neutrons cm^{-3}) is needed. Neutrons are generated by spontaneous fission of ^{238}U and by (α, n) reactions involving light elements, the most important of which are

$$^{28}_{14}Si(\alpha,\ n) \rightarrow {}^{31}_{16}S,$$

$$^{16}_{8}O(\alpha,\ n) \rightarrow {}^{19}_{10}Ne,$$

$$^{24}_{12}Mg(\alpha,\ n) \rightarrow {}^{27}_{14}Si,$$

$$^{27}_{13}Al(\alpha,\ n) \rightarrow {}^{30}_{15}P.$$

The concentrations of these elements in a rock, therefore, control the neutron production rate. In granitic rocks, spontaneous fission of ^{238}U and ^{235}U produces annually 0.4762 neutrons and 0.0002 neutrons, respectively, per microgram of natural U. Neutron production by α-particle irradiation of light elements is a more important process. For example, in a rock with 70% SiO_2, 1.44 neutrons are produced per year per gram of rock per ppm U, and 0.65 neutrons are produced per ppm Th (Burgess *et al.*, 1982).

Thus, for an average granite, about 50 neutrons are produced per cm^3 per year but, in the granites of southwest England, the value has been calculated at just over 100, because of the high U and Th contents (Burgess *et al.*, 1982).

Once formed, ^3H nuclei capture electrons to produce ^3H atoms, then diffuse through the rock matrix and eventually combine to form $^3H_2$ (gas) or, more probably, exchange with H atoms in water to form $^3H^1HO$. If it is assumed that all the ^3H atoms are trapped in the fracture fluids, the ^3H content of the fluids (in TU) has been found to be given by

$$\frac{2.67 \times 10^{-4} p}{\phi},$$

where ϕ is the porosity of the rock and λ_3 is taken as 0.0559 $year^{-1}$. Thus an average granite with a porosity of about 0.01 has a ^3H concentration of 0.05 TU but, a southwest England granite with a porosity of 0.01 has a value of 0.88 TU. Indeed, with porosities of 0.001 and 0.0001, these values become 0.8 TU and 8.8 TU, respectively. Thus, in the special circumstances of high concentrations of U and Th, high Li and low porosity, ^3H concentrations far in excess of those expected in old groundwaters can develop.

However, where ^3H concentrations in excess of 10 TU are encountered in groundwater, it might reasonably be concluded that there is a significant component of very young recharge water and that caution should be exercised in the interpretation of the results of other age dating methods. Indeed, for the Cornish thermal brines, it has been calculated that up to 40% of the thermal spring water is derived from groundwater recharge during the last 25 years, while mixing functions interpret ^{14}C data as indicating an age of at least 50 000 years for the other component (Edmunds *et al.*, 1984).

Other gases

Inert gases, such as Ne, Ar and Kr, have also been used to obtain information about recharge of groundwaters. In particular, if abundances are greater than atmospheric equilibration levels, an additional radiogenic or primordial source is indicated. In this context, the $^{40}Ar/^{36}Ar$ ratio can be very informative, high values being obtained where ^{40}Ar is produced by the decay of ^{40}K. The atmospheric value of $^{40}Ar/^{36}Ar$ is 295.5, but in the thermal brines of Cornwall it is about 312, the additional ^{40}Ar coming from the rock. Direct decay of ^{40}K from K-bearing minerals on fracture surfaces in contact with groundwater is unlikely to be significant over intervals of less than 10 Ma, as a considerable amount of ^{40}Ar must be formed radiogenically to influence the $^{40}Ar/^{36}Ar$ ratio, considering the relatively large amount of dissolved atmospheric Ar. Leakage of radiogenic ^{40}Ar stored in minerals over a very long period of geological time is, therefore, indicated (Burgess *et al.*, 1982). Certainly leakage would be aided by the elevated temperatures at depth and by long groundwater residence times. However, it is very difficult to estimate the time required for diffusive loss of Ar from minerals. Diffusion coefficients for ^{40}Ar depend very much upon the mineralogy of the rock and the presence of crystal imperfections and impurities. Temperature-dependent changes are also notable, ^{40}Ar diffusion increasing by a factor of 10^4 for a temperature increase of 100 °C. Nevertheless, high concentrations of ^{40}Ar can be a useful indicator of long residence times and reveal that at least part of the groundwater is ancient.

High concentrations of primordial gases in groundwaters may be indicated by enhanced levels of ^3He, especially where major fractures occur, but it should be noted that ^3He is also the product of β^--particle decay of ^3H; this may cause some error.

10.5.2 Ra methods

Ra isotopes can be used for dating young groundwaters, in which activities due to Ra isotopes reflect U and Th accumulations in the host rock or on fracture surfaces (Osmond and Cowart, 1982). The short-lived isotope ^{224}Ra reaches its steady-state level within a few weeks; so, if it is assumed that all the Ra isotopes escape from rocks into groundwater with approximately equal ease, their dissolved activity ratios are time dependent and are given by

$$\frac{^{228}\mathrm{Ra}}{^{224}\mathrm{Ra}} = 1 - e^{-\lambda_{228}t}$$

and

$$\frac{^{226}\mathrm{Ra}}{^{224}\mathrm{Ra}} = 1 - e^{-\lambda_{226}t},$$

where λ_{228} and λ_{226} are the decay constants for ^{228}Ra and ^{226}Ra, respectively, and t is the elapsed time. Thus determination of groundwater ages in the range 2–20 years for $^{228}Ra/^{224}Ra$ ratios, and 0.5–5000 years for $^{226}Ra/^{224}Ra$ is theoretically possible, although an irregular distribution of U in the host rocks would cause problems. In a decaying groundwater regime, in which water is flowing away from a radioactive rock (as in geothermal systems), the equations (Osmond and Cowart, 1982) take the form

$$\frac{^{224}\mathrm{Ra}}{^{226}\mathrm{Ra}} = K\ e^{-\lambda_{224}t}$$

and

$$\frac{^{228}\mathrm{Ra}}{^{226}\mathrm{Ra}} = K\ e^{-\lambda_{228}t},$$

where K is the steady-state activity ratio, but as this would take several thousand years to attain, the assumption may be invalid.

Several studies have used Ra isotope ratios qualitatively to estimate flow times. For example, the ^{226}Ra activity of the thermal groundwaters of Cornwall has been found to range between 0.02 and 0.6 Bq l^{-1} in shallow groundwaters, and between 8 and 26 Bq l^{-1} in the deepest, highest-temperature waters (Burgess *et al.*, 1982). Moreover, the $^{226}Ra/^{238}U$ activity ratio shows a marked increase during groundwater evolution. In the shallow groundwaters, the value is usually between 1 and 10, but this rises to several thousand in the deep thermal waters. This is because, in the older groundwater, significant dissolution of ^{226}Ra by α-recoil occurs, while deposition of ^{238}U takes place where groundwater enters reducing environments. Increases in the concentration of ^{226}Ra and in the $^{226}Ra/^{238}U$ ratio are, therefore, good indicators of a groundwater ageing trend.

The activity of ^{226}Ra which enters solution by the α-recoil mechanism should increase with time according to the relationship

$$^{226}\text{Ra}_t = {}^{226}\text{Ra}_\infty\ (1 - e^{-\lambda_{226}t}),$$

where $^{226}Ra_t$ and $^{226}Ra_\infty$ are, respectively, the ^{226}Ra activities in solution after time t and at equilibrium. If the value of $^{226}Ra_\infty$ can be estimated, and provided that ^{226}Ra is not lost from solution, a groundwater age can be calculated from the ^{226}Ra content of the rock matrix. This is because, for a ^{238}U decay series in equilibrium, the activity of ^{226}Ra entering solution reaches the same equilibrium value as that attained by ^{222}Rn. The age is then given by

$$t = 2338\ \log_e\left(\frac{^{226}\text{Ra}_t}{^{222}\text{Rn}} + 1\right),$$

in which the groundwater ^{222}Rn replaces $^{226}Ra_\infty$ (Osmond and Cowart, 1982).

If mixing of groundwaters occurs, the ^{226}Ra activity of one end member can be obtained if the ^{226}Ra activity of the other end member and of the mixture are known. Thus

$$^{226}\text{Ra}_1 = \frac{V_2}{V_1}\ (^{226}\text{Ra}_m - {}^{226}\text{Ra}_2) + {}^{226}\text{Ra}_m,$$

where $^{226}Ra_1$ is the activity of the unknown end member, $^{226}Ra_2$ that of the known end member and $^{226}Ra_m$ that of the mixture. V_1 and V_2 are the relative volumes of the end members in the mixture. Once the end member activity is obtained, the relationship given above can be used to determine the age of the water (Osmond and Cowart, 1982).

10.5.3 U isotope methods

Although the U isotopic composition of old groundwaters has been intensively studied with the aim of developing a reliable dating method, there has been little success. A possible reason for this is that, although U is a conservative constituent of surface and shallow groundwaters, it is less conservative in deep aquifers. Even if a deep aquifer has a reasonably constant U content over a large area, it is usually low (0.001–0.01 ppb); so it is likely that individual atoms are dissolved and reprecipitated in a cyclical fashion. It is, therefore, unlikely that isotopic equilibrium between the groundwater and the rock matrix will be established. However, this consideration does not apply to shallow aquifers, for which the application of an augmenting model is justified. This is governed by the activity relationship

$$\frac{^{234}\text{U}}{^{238}\text{U}} = K(1 - e^{-\lambda_{234}t}),$$

where K is the steady-state dissolved activity ratio, which may be deduced using the shorter-lived daughters which will quickly reach steady state activities (Osmond and Cowart, 1982). Where the dissolved ^{234}U and ^{226}Ra content of groundwater is not limited by solubility but only by the recoil decay process, the relationship may be simplified to

$$\frac{^{234}\text{U}}{^{226}\text{Ra}} = 1 - e^{-\lambda_{234}t},$$

although, if ^{226}Ra data are not available, ^{222}Rn values may be substituted with caution (Andrews *et al.*, 1982).

Various workers have used the decay of dissolved ^{234}U and the consequent decrease in ^{234}U/^{238}U activity ratios to determine groundwater ages. However, caution is needed because, although dissolution of U will occur as groundwater passes down through an oxidising zone, and loss of U will normally occur by deposition at greater depths where reducing conditions are encountered, ^{234}U may still enter solution in the reducing zone by α-recoil. This mechanism will not significantly affect the amount of U remaining in solution but will greatly increase the ^{234}U/^{238}U activity ratio, and so the system is not simply one of decay. Indeed, any increase in the activity ratio will depend also upon the U content of the groundwater, the area of rock surface in contact with the groundwater and the length of time over which the recoil process has acted. Starting with the recoil of ^{234}Th by decay of ^{238}U, it has been noted by Burgess *et al.*, (1982) that

$$^{234}\text{Th} = 0.7336\ \text{U}_r 0.235\ \rho R\ \text{min}^{-1}\ \text{cm}^{-2},$$

where U_r is the natural U content of the rock, ρ is the rock density and R is the recoil range of ^{234}Th in the matrix. The factor 0.235 represents the fraction of recoil ^{234}Th atoms within the recoil range of the surface that actually escapes from the surface, and 0.7336 is the ^{238}U disintegration rate ($\text{min}^{-1}\ \text{g}^{-1}$) for natural U, which is the same for ^{234}U at equilibrium. Any groundwater which contains U in solution will show a change in the ^{234}U/^{238}U activity ratio due to a combination of decay of any initial ^{234}U excess and of ingrowth of ^{234}U duc to ^{234}Th recoil. If the surface area of rock in contact with the groundwater is S cm^2 cm^{-3}, the activity ratio at a later time t is given by

$$\frac{1 + (^{234}\text{U}_w - {}^{238}\text{U}_w)\ e^{-\lambda_{234}t}}{0.7336\ \text{U}_w} = \frac{0.235\ \text{U}_r \rho S R\ (1 - e^{-\lambda_{234}t})}{\text{U}_w}$$

where $^{234}\text{U}_w$ and $^{238}\text{U}_w$ are the activities of these isotopes in solution and λ_{234} is the decay constant for ^{234}U. Clearly, this equation contains a large number of variables but, in practice, it has been found to be mainly dependent upon the values of U_r and S. Thus the amount of U in the rock and the size of the fractures (in the case of fracture permeability) dominate the rate of increase or decrease in activity with time. Unfortunately, it has proved difficult to model conditions for a decrease in activity due to decay when dealing with fractured rock systems if the fracture apertures are less than about 1 mm. When fracture dimensions are of that order, an increase in the activity ratio in the recoil zone is only possible if U deposition has taken place on fracture faces (Burgess *et al.*, 1982).

In the case of groundwaters from the Carnmenellis granite in Cornwall, the ^{234}U/^{238}U activity ratio increases from a near-surface value of 1.0 to about 2.2 at a depth of approximately 250 m but then decreases to a value of about 1.73 at a depth of 700 m (Fig. 10.11). It is possible that these deeper groundwaters evolved through a recoil-dominated zone which produced an activity ratio of greater than 2, with present-day activity resulting either from ageing or from mixing with groundwaters with an equilibrium activity ratio. However, the recoil-dominated zone is of limited extent in fracture systems where U deposition has occurred and is dependent elsewhere upon the U content of the fracture surface in contact with the groundwater. For fresh fractures in granite, upon the walls of which there is no U deposition, therefore, recoil solution of ^{234}U is not significant if the fracture apertures are 1 mm or greater. Groundwater which enters such

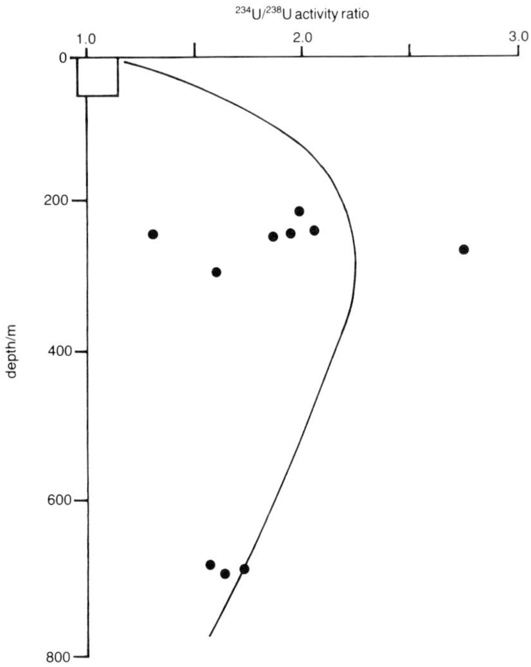

Fig. 10.11 — Variation in $^{234}U/^{238}U$ activity ratio with depth in the groundwaters of the Carnmenellis granite, Cornwall. Shallow groundwater values lie within the open box near the origin. Etch-dominated U solution occurs at shallow depth, passing down into a zone of U deposition and recoil and then to mixing with a deeper component in which the ^{238}U series is in equilibrium. (Edmunds *et al.*, 1984, redrawn with permission.)

fractures thus reaches isotopic equilibrium after about 1 Ma. If mixing then occurs with shallow water from a recoil-dominated zone, the $^{234}U/^{238}U$ activity ratio of the product decreases as the proportion of older water increases. The deep waters from Carnmenellis are either due to a 50%–50% mixture of such components, or if simple decay is assumed, are about 65 ka old (Edmunds *et al.*, 1984).

In conclusion, it may be noted that ages based upon the decay of dissolved ^{234}U and decrease in the $^{234}U/^{238}U$ activity ratio are generally much too old when compared with the results obtained by other methods. For this reason, ages derived by this approach are better interpreted in terms of the migration of U atoms in the flow direction beyond the zone of maximum accumulation. Thus, since U spends much of its time out of solution (adsorbed on the rock), the age obtained is greater than that of the flow-through water (Osmond and Cowart, 1982). Clearly, great uncertainty remains in these aspects of groundwater dating.

10.5.4 ^{14}C dating of groundwater

There are a number of possible ways in which dissolved inorganic C can enter groundwater, the different sources being related to depth. In many shallow groundwaters, CO_2 from the decomposition of organic material in soils or from root respiration in the unsaturated zone can dissolve soil carbonates to produce soluble HCO_3^- ions. Because recent vegetation contains high concentrations of ^{14}C from nuclear weapons testing, high concentrations may also be produced in very young groundwaters by this mechanism. This may be recognised by an associated high ^3H content. Dissolution of soil carbonates by atmospheric CO_2 dissolved in rainwater can also liberate soluble HCO_3^- ions when rain penetrates the ground, but this is a minor source of CO_2 in soil (soil gas usually contains several per cent of CO_2) and is probably negligible except in areas where there is little or no vegetation cover and the groundwater is consequently very soft. Where the soil is free of carbonates, CO_2 in soil gas can alter silicate minerals such as Ca-plagioclase to bring about the formation of HCO_3^- ions but, as this process is very slow, it is usually ignored. Finally, the action of H^+ ions, originating in humic acids in soil, can act on soil carbonate to produce HCO_3^- ions. This mechanism is probably also of minor significance, except in environments, such as peat bogs or in peat layers within the soil profile, in which humic acids are present in large amounts.

During the later stages of movement of groundwater in an aquifer, C can be added by the oxidation of organic material, even though the water itself may be of great age. In contrast, in the absence of O_2, sulphate reduction occurs by reaction with CH_4 to give H_2S and HCO_3^-

ions. In this case, the age of the CH_4 will have an important effect on the C isotope composition of the groundwater. Moreover, if the Ca^{2+} concentration in an aquifer changes due to Ca–Na exchange, some carbonate will be dissolved from the rock to maintain calcite saturation. CO_2 may also be added through volcanic activity. Whether volcanic C is derived from primary material or from nearby limestones, it is unlikely to contain any ^{14}C, and so considerable alteration of the C isotopic composition of the groundwater will occur (Mook, 1980).

The great number of possible sources of C (Fig. 10.12) suggest that ^{14}C dating of groundwaters is unlikely to yield reliable results, especially when it is also considered that there are other processes which, while not changing the total C content of the water, will affect isotopic ratios. These processes include isotopic exchange between dissolved C and C in calcite crystals, and dissolution–precipitation of carbonate minerals. In the isotopic exchange process, ^{14}C in the groundwater exchanges for stable C in a calcite crystal. It might be thought that this would only continue until equilibrium is attained but, in fact, the damage caused in the crystal by the entry of ^{14}C from the surface causes continued diffusion of ^{14}C within the crystal. At elevated temperatures, this is probably a particularly important mechanism. High ^{14}C ages from thermal waters are therefore questionable. Dissolution and precipitation of carbonates is a frequently occurring process in soils in areas where there are long periods of aridity and high air temperatures, but these processes also occur on a seasonal basis in temperate climates. Again, the result is to change C isotopic ratios in groundwater. Despite these difficulties, it is considered that, provided that the dominant mechanism is the production of HCO_3^- ions through CO_2 released during decomposition of soil organic matter or the action of root respiration on soil carbonates, the process is simply

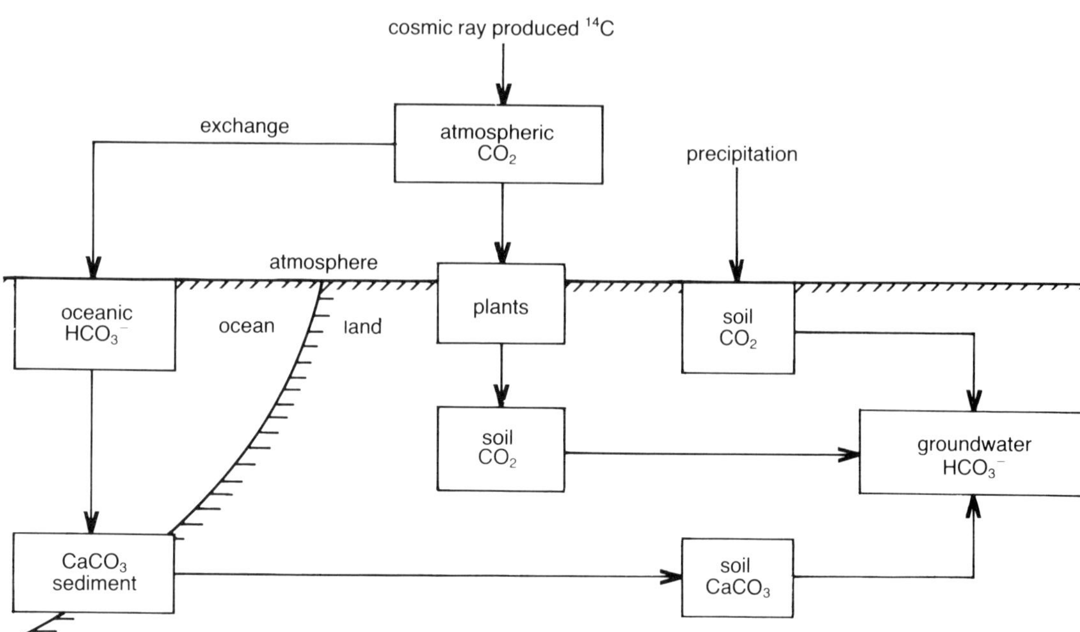

Fig. 10.12 — Simplified geochemical cycle for C illustrating the possible routes of ^{14}C to groundwater. (Mook, 1980, redrawn with permission.)

$$CaCO_3 + CO_2 + H_2O \rightarrow Ca^{2+} + 2HCO_3^-.$$

This seems to be the process which largely controls the C content of groundwater, but it should always be appreciated that the other factors may disturb the system. To check on the possibility of such disturbance, a chemical analysis of the sample is essential before any ^{14}C dating of groundwater is attempted.

Finally, some workers have assumed that any dissolved HCO_3^- is likely to have a ^{14}C content of 50 per cent modern C (pmc) (measured in relation to an activity standard of oxalic acid mulitplied by 0.95; that is, 0.95×13.56 disintegrations min^{-1} g^{-1} in the year 1950 multiplied by 100), so that additional CO_2 increases the ^{14}C content. The age of the water can then be calculated using the relationship

$$t = -\left(\frac{t_{\frac{1}{2}}}{\log_e 2}\right) \log_e\left(\frac{^{14}C_w}{^{14}C_0}\right)$$

$$\approx -8270 \log_e\left(\frac{^{14}C_w}{^{14}C_0}\right),$$

where $^{14}C_w$ is the measured activity of ^{14}C (pmc) in the groundwater and $^{14}C_0$ is the original (or even recent) activity. Flow direction may be determined by noting changes in the apparent ages obtained using this equation; that is, the results are used merely in a relative way. Caution must be exercised if absolute dating is required, and information about whether closed system or open system conditions have prevailed is needed. In closed systems, the carbonate is dissolved by water that has been in contact with the CO_2 reservoir and then infiltrated the system, while in open systems the water is in continuous contact with the CO_2 reservoir at a fixed P_{CO_2}. In practice, it is likely that both systems may have operated during the evolution of a natural groundwater.

If closed systems are assumed, and provided that dissolution of soil carbonates containing no ^{14}C is the only process taking place, the reaction

$$(a + 0.5b)\ CO_2 + 0.5b\ CaCO_3 + H_2O \rightarrow 0.5b\ Ca^{2+} + b\ HCO_3^- + a\ CO_2$$

applies, where a and b are the molar concentrations of dissolved CO_2 and HCO_3^-, respectively. The ^{14}C age is then given by

$$t = -8270\left[\log_e\left(\frac{^{14}C_w}{^{14}C_0}\right) - \log_e\left(\frac{a + 0.5b}{a + b}\right)\right],$$

where $^{14}C_0$ is the original activity. This relationship can be rewritten as

$$t = -8270 \log_e\left(\frac{^{14}C_w}{^{14}C_s}\right),$$

where the ratio $^{14}C_s/^{14}C_0$ $(= (a + 0.5b)/(a + b))$ is known as the 'dilution factor'. Dilution factor values can thus be obtained from measurements of the concentrations of dissolved CO_2 and HCO_3^- (Mook, 1980). An alternative approach works on the basis that the chemical origin of the total dissolved inorganic C determines both the stable C isotope composition and the ^{14}C concentration and that therefore

$$\frac{^{14}C_s}{^{14}C_0} = (^{13}\delta_s - {}^{13}\delta_1)(^{13}\delta_0 - {}^{13}\delta_1),$$

where $^{13}\delta_0$ is the ^{13}C content (in ‰) of soil CO_2, $^{13}\delta_1$ is the ^{13}C content of $CaCO_3$ (also in ‰) and $^{13}\delta_s$ is the ^{13}C content of total dissolved C in the sample.

The value of $^{13}\delta_1$ is not critical, normally ranging only between 0 and 2‰. Occasionally, because of dissolution–precipitation processes, negative values (to -2‰) may be appropriate. In temperate regions, a value of about

$-25‰$ is used safely for $^{13}\delta_0$, although in semi-arid regions there is a great deal of uncertainty and this approach to the determination of dilution factors is not recommended by Mook.

Open systems are more complex in their behaviour and can develop both above and below the water table. Because chemical and isotopic changes in water in contact with atmospheric CO_2 will occur rapidly, it is necessary to ensure that water samples for dating are collected from at least several metres below the water table in open wells and that the samples are sealed as soon as possible to prevent contact with the air.

Once water has entered the saturated zone, changes in its chemical and isotopic composition occur along its flow path as additional C sources and sinks are encountered. These changes will, of course, affect stable isotopes of C as well as ^{14}C. In general, such systems have been modelled as a series of incremental changes in the C content of the water, with associated changes in the isotopic composition based on a mass balance for ^{12}C and ^{13}C, such that

$$d\,\Sigma = \sum_{i=1}^{n} dF_i - \sum_{i=1}^{n} dQ_i$$

and

$$d(R\Sigma) = \sum_{i=1}^{n} R_i\, dF_i - \sum_{i=1}^{n} \alpha_i R\, dQ_i,$$

where, in the first expression, Σ is the total C content of the water and dF_i and dQ_i are, respectively, the carbon input (source) and output (sink) in each incremental step i. The second expression contains the isotopic ratio R of these quantities. α_i is a fractionation factor which assumes isotopic equilibrium between the water and the output C. The model allows correction of measured ^{13}C and ^{14}C values in the groundwater to values that originated by processes taking place in the unsaturated zone, provided that those processes are known from a general chemical analysis of the samples (Mook, 1980).

The problems associated with the determination of the absolute age of groundwaters are considerably reduced in the determination of relative ages, which can be extremely useful in determining groundwater flow direction. For this purpose, it has to be assumed that the recharge area is uniform in terms of vegetation and climate and that these factors have been constant throughout the history of the samples. If this is true, $^{14}C_s$ activity values may be taken as constant for all samples in the groundwater system. The age Δt of sample $k+1$, relative to that of sample k, is then in terms of the activity values $^{14}C_k$ and $^{14}C_{k+1}$ given by Mook as

$$\Delta t = -8270 \log_e\left(\frac{^{14}C_{k+1}}{^{14}C_k}\right).$$

If, however, the C content has increased in the flow direction, then

$$^{14}C_{k+1}\ e^{\lambda_{14}\Delta t} = \frac{(C_{k+1} - C_k)\,^{14}C_{k+1} + C_k\ ^{14}C_k}{C_{k+1}},$$

where λ_{14} is the decay constant for ^{14}C. However, if the additional C is ancient (no ^{14}C),

$$\Delta t = -8270 \log_e\left(\frac{^{14}C_{k+1}}{^{14}C_k}\right)\frac{C_{k+1}}{C_k},$$

where C_k and C_{k+1} are the total C contents of the samples.

10.6 EVOLUTION OF RADIOGENIC NUCLIDES IN THE EARTH

The production of particular isotopes of elements at a known rate by radioactive decay, together with the separation of elements and fractionation of isotopes that can occur by organic and inorganic processes, provides a mechanism for investigating the evolution of the Earth as a chemical system.

10.6.1 Extinct radionuclides

During the supernova that produced the Primitive Solar Nebula from which the Earth formed, short-lived radionuclides would have been produced, just as long-lived radionuclides and stable nuclides were formed. Although these short-lived radionuclides are now extinct (many half-lives having elapsed since their creation), measurement of their stable daughter products can be used to estimate the time that elapsed between the supernova event and the accretion of solid planetesimals.

Certain chondrites contain an excess of ^{129}Xe. As discussed earlier, this isotope is formed by the β^--decay of ^{129}I with a half-life of 15.7×10^6 years, and so its presence in meteorites shows that they must have been formed and sealed (so that Xe was retained) within a very short time of nucleosynthesis.

If trapping of Xe began at time t after synthesis was complete, then

$$\frac{^{129}\mathrm{Xe}}{^{127}\mathrm{I}} = \left(\frac{^{129}\mathrm{I}}{^{127}\mathrm{I}}\right)_{\mathrm{original}} \mathrm{e}^{-\lambda_{129}t},$$

where isotopes are ratioed to the stable nuclide ^{127}I, and λ_{129} is the decay constant for ^{129}I. Moreover, if the length of the nucleosynthesis was T years, and the decay of ^{129}I is expressed in terms of its mean life $(\tau_{129} = \lambda_{129})^{-1})$,

$$\frac{^{129}\mathrm{Xe}}{^{127}\mathrm{I}} = \left(\frac{^{129}\mathrm{I}}{^{127}\mathrm{I}}\right)_{\mathrm{original}} \mathrm{e}^{-t/\tau_{129}}$$

$$\left(\frac{\tau_{129}}{T}\right) \mathrm{e}^{-t/\tau_{129}}$$

and the abundance of radiogenic ^{129}Xe gives the relative magnitudes of T and t. As a result of such studies, it has been suggested by Reynolds (1963) that T is about 2×10^{10} years and t only about 5×10^7 years, but these are maximum values.

An isotope with a shorter half-life than ^{129}I would enable these values to be further refined. ^{26}Al has a half-life of only 7.3×10^5 years and decays to form ^{26}Mg, and so the discovery of excess ^{26}Mg in part of the Allende meteorite demonstrates that the condensation occurred within less than 10^6 years of the completion of nucleosynthesis. Indeed, this short delay between supernova and condensation has led to the suggestion that the supernova produced a compressional shock wave through the Primitive Solar Nebula which set up the mechanical instabilities that, in turn, led to the formation of the Solar System. Astrophysical models also suggest that the life of the Primitive Solar Nebula was less than 10^6 years. Variations in the initial $^{129}I/^{127}I$ isotopic ratio are interpreted as indicating inhomogeneities in the Primitive Solar Nebula before condensation.

Meteorites also show variations in the proportions of ^{131}Xe, ^{132}Xe, ^{134}Xe and ^{136}Xe, all of which are known to be formed by the fission of ^{244}Pu. Thus during nuclueosynthesis the production of superheavy elements occurred; only their decay products are now recognised because, although the spontaneous fission half-life of ^{244}Pu is 2.5×10^{10} years, its decay half-life is only 7.5×10^7 years.

10.6.2 Evolution of Sr

Because ^{87}Sr is continuously produced by the decay of ^{87}Rb, the isotopic composition of Sr has been continuously changing since nucleosynthesis. The ratio $^{87}Sr/^{86}Sr$ (in which ^{86}Sr is a representative of stable non-radiogenic Sr) will, therefore, increase with the passage of time. A plot of $^{87}Sr/^{86}Sr$ against time gives the isotope evolution line. Now, since

$$\left(\frac{^{87}\mathrm{Sr}}{^{86}\mathrm{Sr}}\right)_{\mathrm{total}} = \left(\frac{^{87}\mathrm{Sr}}{^{86}\mathrm{Sr}}\right)_{\mathrm{original}} + \left(\frac{^{87}\mathrm{Rb}}{^{86}\mathrm{Sr}}\right)_{\mathrm{today}} (\mathrm{e}^{\lambda_{87}t} - 1)$$

and as λ_{87} is very small (1.39×10^{-11} year^{-1}), it follows that

$$\left(\frac{^{87}\mathrm{Sr}}{^{86}\mathrm{Sr}}\right)_{\mathrm{total}} \approx \left(\frac{^{87}\mathrm{Sr}}{^{86}\mathrm{Sr}}\right)_{\mathrm{original}} + \left(\frac{^{87}\mathrm{Rb}}{^{86}\mathrm{Sr}}\right)_{\mathrm{today}} \lambda_{97} t.$$

Thus an isotope evolution diagram, in which $(^{87}\mathrm{Sr}/^{86}\mathrm{Sr})_{\mathrm{total}}$ is plotted against t, will follow a straight line with a slope of $\lambda_{87}(^{87}\mathrm{Rb}/^{86}\mathrm{Sr})_{\mathrm{today}}$ and intercept giving $(^{87}\mathrm{Sr}/^{86}\mathrm{Sr})_{\mathrm{original}}$. Moreover, because the ^{87}Rb/^{86}Sr ratio is approximately proportional to the Rb/Sr ratio ($^{87}\mathrm{Rb}/^{86}\mathrm{Sr} \approx 2.9\mathrm{Rb/Sr}$), the slope is also approximately proportional to Rb/Sr.

To obtain the ^{87}Sr/^{86}Sr ratio in the primitive Earth, the ^{87}Rb/^{86}Sr and ^{87}Sr/^{86}Sr ratios in basaltic chondrites have been investigated. These meteorites have very low Rb/Sr ratios and display evidence of crystallisation from a silicate melt; their origin is thus similar to that of igneous rocks on Earth. A $(^{87}\mathrm{Sr}/^{86}\mathrm{Sr})_{\mathrm{original}}$ value of 0.698 99 has been determined, which may indicate the ratio in the solar nebula at a very early stage in the formation of the Solar System, although it is certainly not the ratio of the earliest material. The Allende meteorite, for example, has yielded chondrules which have a ratio of 0.698 87.

If an initial ^{87}Sr/^{86}Sr ratio of 0.699 is taken for 4550 Ma ago (the age of the Earth), it is possible to produce an isotope evolution line for the Earth if we know its Rb/Sr ratio. This is conventionally taken as 0.03, representing approximately the proportion of these elements in the Primitive Solar Nebula. Extrapolation of this line to the present day yields a ^{87}Sr/^{86}Sr value of about 0.705. Now, because Rb and Sr possess different geochemical properties, they become separated during various geological processes. In particular, Rb, with affinities with K, appears as an incompatible element during the partial melting of basic rocks. The Earth's continental crust has, therefore, been progressively enriched in Rb, while the mantle has been depleted. As a consequence, the ^{87}Sr/^{86}Sr composition of the mantle will be lower than if partition of Rb and Sr had not taken place. Thus, modern mantle-derived melts, such as ocean ridge basalts, have ^{87}Sr/^{86}Sr initial ratios of only about 0.703, while young granites derived by remelting of old crustal material can have initial ^{87}Sr/^{86}Sr values as high as 0.740.

If a present-day ^{87}Sr/^{86}Sr ratio of 0.703 and Rb/Sr ratio of about 0.01 are taken for ocean ridge basalts, an isotope evolution line for the mantle can be constructed which intercepts the bulk Earth evolution line at about 2100 Ma ago (Fig. 10.13). This has led to the idea of catastrophic two-stage evolution of the Earth's mantle. In this concept, the isotopic evolution of the mantle followed that of the bulk Earth and remained a closed chemical system with respect to Rb–Sr partition until 2100 Ma ago, when a catastrophic event occurred which depleted the mantle in Rb relative to Sr. Thereafter, the mantle continued to evolve along the new isotope line. This picture, however, appears somewhat unrealistic in that it implies that the major part of the non-recycled continental crust formed 2100 Ma ago. Evidence of young sialic rocks and the youth of ocean basins clearly refute this. Moreover, mantle-derived rocks older than 2100 Ma show initial ^{87}Sr/^{86}Sr values lower than those for the bulk Earth at that time. More probably, a process of continuous abstraction of Rb from the mantle has been going on throughout most of the Earth's history.

The overall composition of the mantle, however, is not simple. Young ocean ridge basalts show a range of ^{87}Sr/^{86}Sr values between 0.702 and 0.705, while young ocean island basalts (intraplate volcanic rocks) show an even wider range, between 0.702 and 0.708. These values indicate a large degree of heterogeneity in the composition of the mantle within the source regions of these basalts. Such heterogeneity is also shown by substantial differences between the major and trace element chemistry of ocean

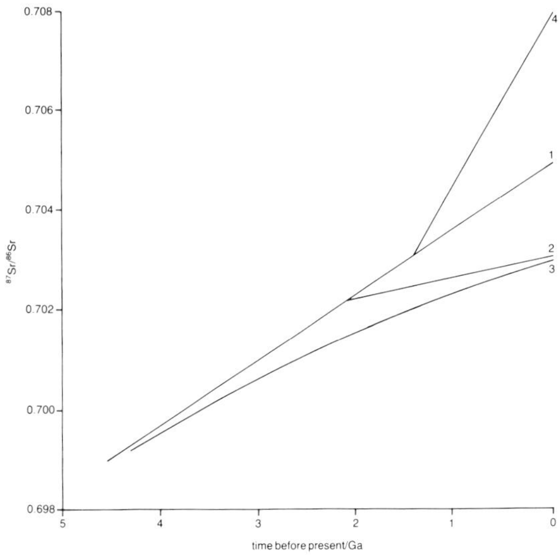

Fig. 10.13 — Sr isotope evolution in the Earth: 1, assuming an average Rb/Sr ratio of 0.03, an initial $^{87}Sr/^{86}Sr$ ratio of 0.699 and an age of 4550 Ma; 2, assuming uniform conditions until 2100 Ma and then catastrophic depletion of Rb in the mantle to give an independent mantle evolution line; 3, continuous depletion of Rb in the mantle resulting in an alternative mantle evolution line; 4, evolution of continental crust (enriched in Rb) about 1500 Ma old.

ridge sources and those of ocean island sources. Ocean island volcanic rocks are distinctly more enriched in incompatible trace elements, for example, than ocean ridge basalts. Recent work has suggested that some of these differences relate to the recycling of subducted older oceanic crust. Another view is that, as the mantle appears to contain substantial quantities of volatile compounds, some of which (such as ^{3}He and ^{129}Xe) are primordial while others (such as H_2O and CO_2) are added by the recycling of crustal materials, volatile transport in the mantle may be sufficient to produce mantle metasomatism as a precursor to partial melting and the formation of basaltic magmas. Other aspects of the chemistry and mineralogy of ocean ridge basalts and ocean island basalts suggest that they are formed by partial melting of the mantle at different depths.

In models of mantle evolution, material of the type represented isotopically by ocean ridge basalts can only account for a small fraction of the whole mantle (De Paolo, 1980a). It seems, therefore, that the mantle could be chemically layered, with a lower part of primitive composition underlying an upper zone in which there is continual recycling of mantle and oceanic crust, and in which the formation of continental crust results in depleted mantle. This model suggests extreme Rb depletion in the upper layer, 80–90% being lost to the continental crust. The boundary between these layers may, in fact, be marked by the seismic boundary which separates the upper mantle from the lower mantle at a depth of about 650 km. Interestingly, because of the similar chemical behaviours of K, U and Th in this process, it is likely that the upper layer of the mantle is strongly depleted in heat-producing elements.

Rb and Sr also display radically different chemical behaviour in near-surface processes. For example, the oceans form a large reservoir of Sr which has been mainly derived from the continental crust. Hydrothermal alteration of basalts at ocean ridges will, therefore, result in a net transfer of Sr to the oceanic crust, from where it may be later returned to the mantle by subduction. It has been suggested that this process involves ^{87}Sr but not ^{86}Sr (De Paolo, 1980b).

The transfer of continental crust back into the mantle by differential solution is not the only mechanism by which reworking will occur. Despite the inability of the subduction process to recycle low density continental crust in a wholesale fashion, erosion of continental crust, followed by deposition of sediments and subsequent subduction will cause some material to be recycled. It is possible that the midocean ridge basalts are derived from a source within the recycled layer in the upper mantle, while the ocean island basalts come from the primitive lower mantle.

The evolution of Sr isotopes also provides useful information on the origin of the continental crust. Study of the initial $^{87}Sr/^{86}Sr$

isotopic ratios from granitic rocks of various ages shows that many areas of continental crust were derived from a source with initial ratios very similar to those of the mantle at the time that those rocks were formed. Continued growth of continental crust from a mantle source can, therefore, be shown to have taken place throughout geological time. Only relatively small volumes of younger continental crust have been shown by their initial $^{87}Sr/^{86}Sr$ isotopic ratios to be formed by reworking of older continental crust. This view of the extent to which continental crust is reworked is not universally accepted, however, and various researchers have argued that recycling of continental crust is very important (Armstrong, 1981). In this alternative model, the formation of continental crust from the mantle was largely accomplished early in the Earth's history, and since then there has merely been recycling of the continental crust with no continental growth. The observed pattern of Sr isotope data for the continental crust can then be explained in terms of a system in which the volume of continental crust recycled remains approximately constant through time, but the rate of recycling decreases steeply from very high values greater than 2500 Ma ago to only a small value today. A recycling rate curve can be constructed on this basis that will exactly fit both the Sr data and any model of the growth of continental crust. Moreover, a decreasing recycling rate produces a distribution of continental crustal ages which closely matches the observed distribution. Thus, while some recycling of continental crust is generally considered to have taken place, much further study is needed to resolve the problem of the degree to which it takes place today and took place in the past.

Sr isotope ratios in the oceans have been subject to much variation (Fig. 10.14), although the modern $^{87}Sr/^{86}Sr$ ratio is generally 0.7090. The Sr in the oceans is mainly derived by weathering of continental crust and transport to the sea in solution; additionally, Sr may be liberated during the diagenesis of marine carbonates and may also be derived directly by input of volcanogenic material. Thus the $^{87}Sr/^{86}Sr$ ratio is an indicator of the composition of the rocks exposed on the surface of continents and in the ocean basins. Studies of the $^{87}Sr/^{86}Sr$ ratios of fossil shells in which no replacement of material has occurred have shown that during the Palaeozoic and early Mesozoic the $^{87}Sr/^{86}Sr$ ratio of sea water fell from about 0.7078 to 0.70675, reaching a minimum during the late Jurassic. Thereafter,

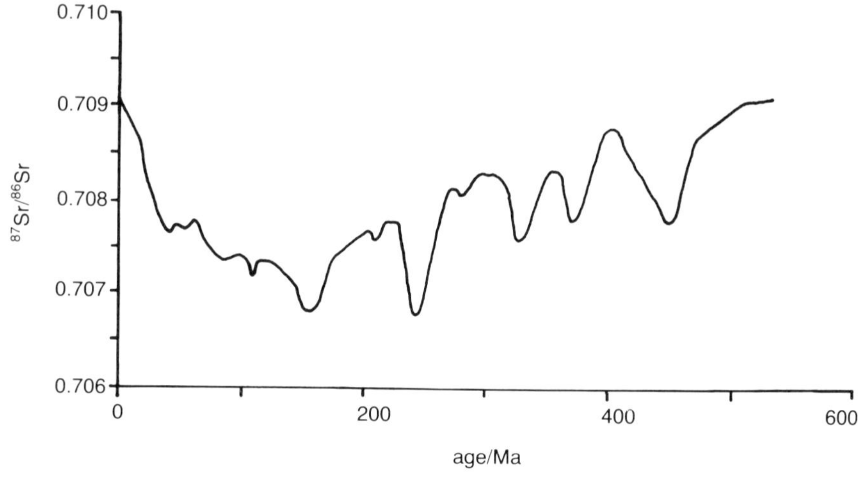

Fig. 10.14 — The variation in $^{87}Sr/^{86}Sr$ in sea water through the Phanerozoic. (After Burke *et al.*, 1982.)

the ratio has risen to its present value. Because, overall, the $^{87}Sr/^{86}Sr$ ratio has been steadily increasing with time as a result of the decay of ^{87}Rb, the fall in this ratio during the Palaeozoic and early Mesozoic is seen as representing a steadily increasing contribution of material with a lower $^{87}Sr/^{86}Sr$ content. The source must have been contemporaneous volcanic activity. Conversely, where the $^{87}Sr/^{86}Sr$ ratio increases beyond the level expected by decay of ^{87}Rb, increased input of Sr eroded from older continental crust may be indicated. It has been suggested that a correlation exists between increases in the intensity of glaciation and increases in the $^{87}Sr/^{86}Sr$ ratio of the oceans (Armstrong, 1971).

10.6.3 Evolution of Pb

Pb has four naturally occurring isotopes: ^{204}Pb, ^{206}Pb, ^{207}Pb and ^{208}Pb. Since ^{204}Pb is not radiogenic, changes in the isotopic composition of Pb with time are reflected in changes in the $^{206}Pb/^{204}Pb$, $^{207}Pb/^{204}Pb$ and $^{208}Pb/^{204}Pb$ isotopic ratios. These ratios generally plot as a curved evolution line for the period since the formation of the Earth. The position of each curve and its form are determined by the abundance of the parent of each radiogenic isotope, and the half-life of the decay involved; the greater the half-life, the nearer to a straight line the evolution curve approximates.

Because Pb minerals such as galena contain very little U or Th, their Pb content will have remained the same since the time of their formation, provided that no open system exchange of Pb has taken place. This argument forms the basis of the U–Pb dating methods described earlier in this chapter. In the case of minerals which initially contain Pb in addition to U and Th (which produce radiogenic Pb), a reasonably uniform picture of Pb evolution is provided, whatever the source of the mineral, and a number of models have been proposed to account for this.

Perhaps the simplest model is based upon the notion that primaeval Pb was originally incorporated with U and Th into a closed chemical system. The way in which the Pb isotope composition then changed with time depended upon the U/Pb and Th/Pb ratios originally present. However, the values of these ratios for the Earth are unknown. Because of differences in the properties of U, Th and Pb, partition of these elements is likely to have occurred within the Primitive Solar Nebula before condensation of the Earth. The problem of determining the isotopic evolution of Pb is thus tackled by working back from present-day Pb isotopic ratios. The $^{206}Pb/^{204}Pb$ and $^{207}Pb/^{204}Pb$ systems are usually employed, as they both involve U, and changes in these ratios can be calculated for different values of the $^{238}U/^{204}Pb$ ratio. The present-day Pb isotope composition can, therefore, be used to derive the U/Pb ratio that gave rise to that composition. The Pb isotope growth curves are taken to originate at the Pb isotope values of iron meteorites (the primaeval Pb composition): 10.29 for $^{207}Pb/^{204}Pb$ and 9.31 for $^{206}Pb/^{204}Pb$. Straight-line isochrons can be constructed from the same origin, the slope of which is governed by the age of the sample (Fig. 10.15). The isochron for $t = 0$ (the present) is also known as the 'geochron' (Faure, 1977).

The rate of increase in the $^{206}Pb/^{204}Pb$ and $^{207}Pb/^{204}Pb$ isotopic ratios along a growth curve thus depends upon the initial U/Pb ratio of the system and its age. Any geochemical process which extracts Pb from a closed U–Pb system and concentrates it into a Pb mineral will, therefore, provide a record of the isotopic composition of the Pb immediately prior to extraction.

Interestingly, the age of Pb extraction from such a closed U–Pb system may not be the date at which the Pb mineral formed. Remobilisation of Pb in successive phases of mineralisation may lead to the reworking of old Pb on a number of occasions. Where 'common lead' ages agree with other evidence for the mineralogical age, the Pb isotopes are said to be 'ordinary' but, where the common lead method gives meaningless results (in terms of the age of

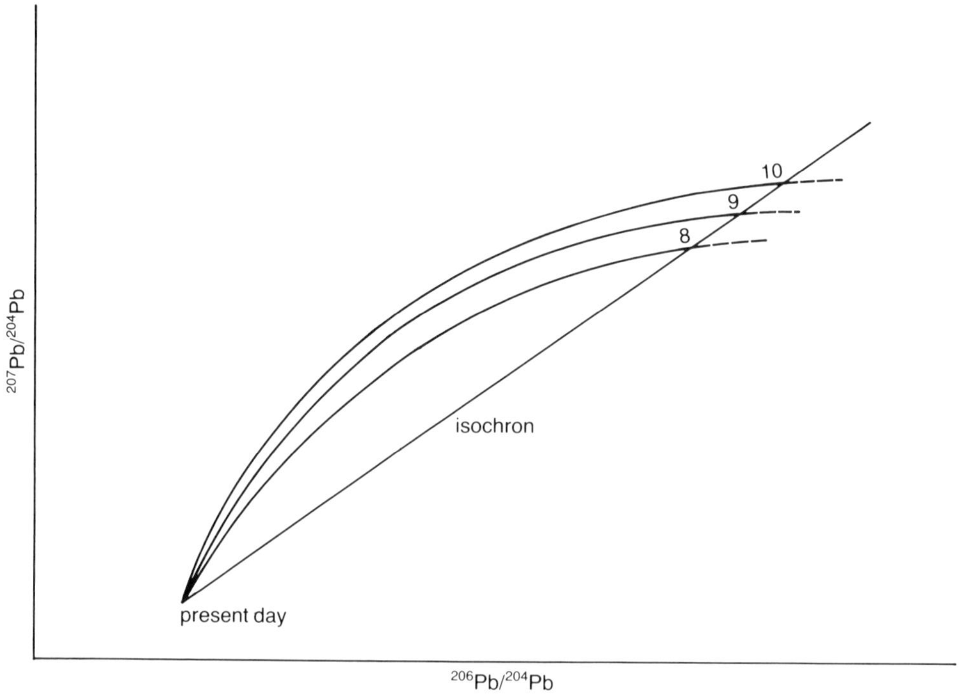

Fig. 10.15 — Growth curves for Pb isotopes in systems where $^{238}U/^{204}Pb$ is 8, 9 and 10. The present day isochron is also shown. (After Faure; 1977.)

mineralisation), the Pb isotopes are known as 'anomalous'. Defining the difference between ordinary and anomalous Pb is at least the first step in the recognition of single-stage Pb. Because of their simple history, these must be created directly from a uniform reservoir. Continental crust, composed of various rocks and minerals of different types and ages, would produce evidence of multistage evolution. The Pb isotope compositions of the upper mantle, and of syngenetic sulphide ore deposits found in sedimentary basins of various ages, with associated mantle-derived volcanic material, have been investigated and found to be on a single growth curve (Stanton and Russell, 1959). On this basis, it has been argued that, if any other Pb isotope data fit this line ($^{238}U/^{204}Pb = 8.99$), the Pb isotopes can be taken as ordinary; anomalous Pb isotopes have compositions which do not fall on this curve.

Anomalous Pb isotopic characteristics are not valueless: they can still yield useful information about the evolution of the system in which they occur. In ore assemblages, for example, the isotopic composition of anomalous Pb is often found to lie close to a straight line on a $^{207}Pb/^{204}Pb$–$^{206}Pb/^{204}Pb$ isotope ratio diagram. The data points may lie to the left of the geochron (in which case apparent ages are in the past) or to the right (the future), but most are found to contain excess radiogenic Pb. The slope of the anomalous Pb line can be used to determine the age of initial withdrawal of Pb from the source, if projected back to the normal growth curve, and the $^{207}Pb/^{206}Pb$ isotopic ratio of the oldest component is indicated by the slope of the line itself.

The excess radiogenic Pb may be considered to have been added at the time t_i, which is indicated by the $^{207}Pb/^{206}Pb$ ratio:

$$t_i = \frac{1}{\lambda_{235} - \lambda_{238}} \log_e\left(\frac{137.8m\lambda_{238}}{\lambda_{235}}\right) ,$$

where m is the slope of the anomalous Pb line (Faure, 1977). However, it is most unlikely that the excess Pb only became available at one specific time, and mixing of Pb of various isotopic compositions according to the time of extraction is more likely.

If t_s is the age of the source rocks from which the Pb is derived, and t is the age of mineralisation, then

$$m = \frac{1}{137.8} \frac{e^{\lambda_{235}t_s} - e^{\lambda_{235}t}}{e^{\lambda_{238}t_s} - e^{\lambda_{235}t}} .$$

If the limiting value of $t = 0$ is taken, this simplifies to

$$m = \frac{1}{137.8} \frac{e^{\lambda_{235}t_s} - 1}{e^{\lambda_{238}t_s} - 1} .$$

but, as t increases, t_s decreases, until they converge at the value t_i. Thus, if the value of t is known from another method, the age of the source rocks from which the mineralisation was derived can be determined.

While the multistage evolution of Pb isotopes may follow a complex route, in the simplest case, Pb merely passes from one system (with a specific $^{238}U/^{204}Pb$ ratio) to another (with a different $^{238}U/^{204}Pb$ ratio). More complex histories can develop by extending this two-stage system to three or more stages. Other possible routes to complexity include changes in the $^{238}U/^{204}Pb$ ratio within a single system by U loss and the addition of Pb of different isotopic composition.

If the first system begins its evolution at time t_0, and, at time t_1, Pb is transferred to the next system, in which it continues to evolve until time t_2, when it is removed from any further association with U or Th, and if $(^{206}Pb/^{204}Pb)_{t_1}$ is the isotopic ratio at time t_1, the Pb isotopic ratio at time t_2 will be given by

$$\left(\frac{^{206}Pb}{^{204}Pb}\right)_{t_2} = \left(\frac{^{206}Pb}{^{204}Pb}\right)_{t_1} + \left(\frac{^{238}U}{^{204}Pb}\right)_2 (e^{\lambda_{238}t_1} - e^{\lambda_{238}t_2}) ,$$

where $(^{238}U/^{204}Pb)_2$ relates to the second system. Also, if $(^{207}Pb/^{204}Pb)_{t_1}$ is the isotopic ratio at time t_1, at time t_2 the Pb ratio will be

$$\left(\frac{^{207}Pb}{^{204}Pb}\right)_{t_2} = \left(\frac{^{207}Pb}{^{204}Pb}\right)_{t_1} + \left(\frac{^{235}U}{^{204}Pb}\right)_2 (e^{\lambda_{235}t_1} - e^{\lambda_{235}t_2}),$$

where $(^{235}U/^{204}Pb)_2$ relates to the second system. However, because $^{235}U/^{204}Pb = {^{238}U}/137.8{^{204}Pb}$, then

$$\frac{(^{207}Pb/^{204}Pb)_{t_2} - (^{207}Pb/^{204}Pb)_{t_1}}{(^{206}Pb/^{204}Pb)_{t_2} - (^{206}Pb/^{204}Pb)_{t_1}} = \frac{1}{137.8} \frac{e^{\lambda_{235}t_1} - e^{\lambda_{235}t_2}}{e^{\lambda_{238}t_1} - e^{\lambda_{235}t_2}} .$$

This equation represents a straight line passing through the point $(^{206}Pb/^{204}Pb)_{t_1}$, $(^{207}Pb/^{204}Pb)_{2_1}$, and has the slope

$$m = \frac{1}{137.8} \frac{e^{\lambda_{235}t_1} - e^{\lambda_{235}t_2}}{e^{\lambda_{238}t_1} - e^{\lambda_{238}t_2}} .$$

Thus, when the growth curve for the first stage of Pb evolution reaches the time t_1, the second stage will continue along a secondary isochron, the slope of which depends upon the time interval $t_1 - t_2$. If the value of t_2 is known independently, a value for t_1 can, of course, be obtained (Faure, 1977).

This argument can be extended to deal with systems of three or more stages. Each stage will produce an isochron of the type described, producing tertiary isochrons, quaternary isochrons and so forth.

Just as the producton of ^{206}Pb and ^{207}Pb from U provides the basis for the $^{206}Pb/^{204}Pb$–$^{207}Pb/^{204}Pb$ method, the growth of ^{208}Pb from Th yields information which can also be interpreted in terms of common Pb, and a diagram based upon $^{208}Pb/^{204}Pb$–$^{207}Pb/^{204}Pb$ isotopic ratios can be used as before. Again, growth curves can be obtained for various values of $^{232}Th/^{238}U$ (instead of $^{238}U/^{204}Pb$). Anomalous Pb values are often obtained which lie on a straight line of slope

$$m = \frac{^{232}\mathrm{Th}}{^{238}\mathrm{U}} \frac{e^{\lambda_{232}t_s} - e^{\lambda_{232}t}}{e^{\lambda_{238}t_s} - e^{\lambda_{238}t}},$$

where t_s may be determined as before. As this system is analogous to the normal common Pb method, the $^{232}Th/^{238}U$ ratio of the source rocks may be obtained.

Multistage Pb isotope patterns appear to be common, even in geological systems which would otherwise seem fairly straightforward. Even young ocean basalts, which might be thought to have originated directly from a uniform mantle source, show evidence of remobilisation of Pb systems and add to the picture of upper mantle recycling already discussed. It follows, therefore, that, with a degree of heterogeneity in the upper mantle, the Pb isotope characteristics from one oceanic basalt area will not be the same as those from another area which may be only a few tens of kilometres away.

An interesting view of the development of upper mantle heterogeneity is provided by two-stage Pb systems. If the concordia–discordia plot is used to interpret the history of young basalts (the U/Pb ratio of which is due to the U/Pb ratio of their source), it is found that Pb isotope ratios fall on a single discordia line. This line intersects the concordia curve at about 4530 Ma and thus agrees with a system commencing with the origin of the Earth. The younger intersection of the discordia line with the concordia curve, however, ranges from over 1000 to 0 Ma. This range of dates points to wide variation in the development of a heterogeneous upper mantle but also suggests that it can be a comparatively recent phenomenon. This view is supported by the $^{238}U/^{204}Pb$ ratios at time t_1 in a two-stage evolutionary system, which are found to be of the order of 8.70. As this is only slightly lower than the value of 8.94 discussed previously, the heterogeneity of $^{238}U/^{204}Pb$ ratios found in the upper mantle seems also to be a relatively recent development.

BIBLIOGRAPHY

Aitken, M. J. (1974) *Physics and archaeology*. Oxford, Oxford University Press.

Albarede, F., and Juteau, M. (1984) Unscrambling the lead model ages. *Geochim. Cosmochim. Acta* **48,** 207.

Aldrich, L. T., Wetherill, G. W., Tilton, G. R., and Davis, G. L. (1956) The half-life of ^{87}Rb. *Phys. Rev.* **104,** 1045.

Allegre, C. J., and Luck, J. M. (1980) Isotopic and chemical effects produced in a continuously differentiating convecting Earth mantle. In Bailey, D. K., Tarney, J. and Dunham, K. (eds), The evidence for chemical heterogeneity in the Earth's mantle. *Phil. Trans. R. Soc. Lond. A* **297.**

Allegre, C. J., and Luck, J. M. (1980) Osmium isotopes as petrogenetic and geological tracers. *Earth Planet. Sci. Lett.* **48,** 148.

Amin, B. S., Lal, D., and Somayajolu, B. L. K. (1975) Chronology of marine sediments using the ^{10}Be method: intercomparisons with other methods. *Geochim. Cosmochim. Acta* **39,** 1187.

Andrews, J. N. (1982) Radiochemical analyses of DST water and core interstial fluids from the Winterborne Kingston borehole. *Rep. Inst. Geol. Sci. Lond.* **81–3,** 161.

Andrews, J. N., Burgess, W. G., Edmunds, W. M., Kay, R. L. F., and Lee, D. J. (1982) The thermal springs of Bath. *Nature* **298,** 339.

Andrews, J. N., Giles, I. S., Kay, R. L. F., Lee, D. J., Osmond, J. K., Cowart, J. B., Fritz, P., Baker, J. F., and Gale, J. (1982) Radioelements, radiogenic helium and age relationships for groundwaters from the granites at Stripa, Sweden. *Geochim. Cosmochim. Acta* **46,** 1533.

Andrews, J. N., and Kay, R. L. F. (1982) Natural production of tritium in permeable rocks. *Nature* **298,** 361.

Armstrong, R. L. (1971) Glacial erosion and the variable isotopic composition of strontium in sea water. *Nature Phys. Sci.* **230,** 132.

Armstrong, R. L. (1978) Removal of atmospheric argon contamination and the use and misuse of the K–Ar isochron method: discussion. *Can. J. Earth Sci.* **15,** 325.

Armstrong, R. L. (1981) Comment on crystal growth and mantle evolution: inferences from models of element transport and Nd and Sr isotopes. *Geochim. Cosmochim. Acta*. **45,** 1251.

Baksi, A. K. (1982) A note on the calculation of errors in conventional K–Ar dating. *Chem. Geol.* **35,** 167.

Bal, K. D., Lal, N., and Nagpaul, K. K. (1983) Zircon and sphene as fission track geochronometer and geothermometer: a reappraisal. *Contrib. Mineral. Petrol.* **83,** 199.

Barbetti, M., and Flude, K. (1979) Geomagnetic variation during the late Pleistocene period and changes in the radiocarbon time scale. *Nature* **279,** 202.

Bennett, C. L. (1979) Radiocarbon dating with accelerators. *Am. Sci.* **67,** 450.

Bentley, H. W., Phillips, F. M., Davis, S. N., Gifford, S., Elmore, D., Tubbs, L. E., and Gove, H. E. (1982) Thermonuclear ^{36}Cl pulse in natural water. *Nature* **300,** 737.

Birck, J. L., and Allegre, C. J., (1979) ^{87}Rb–^{87}Sr chronology of the Binda Howardite. *Nature* **282,** 288.

Bloomer, J. R., Kuckes, A. F., Oxburgh, E. R., and Richardson, S. W., (1982) Heat flow studies in the Winterborne Kingston borehole. *Rep. Inst. Geol. Sci. Lond.* **81–3,** 176.

Blyuman, B. A. (1975) Retention of radiogenic argon in potassic feldspars with various degrees of ordering. *Dokl. Acad. Sci. USSR Earth Sci.* **223,** 146.

Bonhomme, M. G. (1982) The use of Rb–Sr and K–Ar dating methods as a stratigraphic tool applied to sedimentary rocks and minerals. *Precambrian Res.* **18,** 5.

Burgess, W. G., Edmunds, W. M., Andrews, J. N., Kay, R. L. F., and Lee, D. J. (1982) The origin of groundwater in the Carnmenellis granite: the hydrogeochemical evidence. *Rep. Inst. Geol. Sci. Lond.*

Burke, W. H., Denison, R. E., Hetherington, E. A., Koepnick, R. B., Nelson, H. F., and Otto, J. B. (1982) Variation of seawater ^{87}Sr/^{86}Sr throughout Phanerozoic time. *Geology* **10,** 516.

Burleigh, R. (1979) Extending the range of radiocarbon dating. *Nature* **277,** 327.

Butler, J. C. (1982) Artificial isochrons. *Lithos* **15,** 207.

Cameron, M., Collerson, K. D., Compston, W., and Morton, R. (1981) The statistical analysis and interpretation of imperfectly fitted Rb–Sr isochrons from polymetamorphic terrains. *Geochim. Cosmochim. Acta* **45,** 1087.

Capaldi, G., and Pece, R. (1981) On the reliability of the ^{230}Th–^{238}U dating method applied to young volcanic rocks. *J. Volcanol. Geotherm.* **11,** 367.

Clauer, N. (1981) Rb–Sr and K–Ar dating of Precambrian clays and glauconites. *Precambrian Res.* **15,** 331.

Cliff, R. A. (1985) Isotopic dating in metamorphic belts. *J. Geol. Soc.* **142,** 97.

Cohen, R. S., and O'Nions, R. K. (1982) The lead, neodymium and strontium isotope structure of ocean ridge basalts. *J. Petrol.* **23,** 299.

Condomines, M., and Allegre, C. J. (1980) Age and magmatic evolution of Stromboli volcano from ^{230}Th–^{238}U disequilibrium data. *Nature* **288,** 354.

Dalrymple, G. B. (1979) Critical tables for conversion of K–Ar ages from old to new constants. *Geology* **7,** 558.

Dalrymple, G. B., and Lanphere, M. A., (1969) *Potassium–argon dating*. San Francisco, California, Freeman.

Dalrymple, G. B., and Lanphere, M. A. (1974) ^{40}Ar/^{39}Ar age spectra of some undisturbed terrestial samples. *Geochim. Cosmochim. Acta* **38,** 715.

Damon, P. E. (1968) Potassium–argon dating of igneous and metamorphic rocks with applications to the Basin ranges of Arizona and Sonora. In Hamilton, E. I., and Farquhar, R. M. (eds), *Radiometric dating for geologists*. London, Wiley.

Davis, D. W. (1982) Optimum linear regression and error estimation applied to U–Pb data. *Can. J. Earth. Sci.* **19** 2141.

Davis, P. K. (1977) Effect of shock pressure on ^{40}Ar–^{39}Ar radiometric age determination. *Geochim. Cosmochim. Acta* **41,** 195.

De Paolo, D. J. (1979) Implications of correlated Nd and Sr isotopic variations for the chemical evolution of the crust and mantle. *Earth Planet. Sci. Lett.* **43,** 201.

De Paolo, D. J. (1980a) Sources of continental crust: neodymium isotope evidence for the Sierra Nevada and Peninsular Ranges. *Science* **209,** 684.

De Paolo, D. J. (1980b) Crustal and mantle evolution: inferences from models of element transport and Nd and Sr isotopes. *Geochim. Cosmochim. Acta* **44,** 1185.

De Paolo, D. J. (1981) Crustal growth and mantle evolution: inferences from models of element transport and Nd and Sr isotopes (reply to discussion). *Geochim. Cosmochim. Acta* **45,** 1253.

De Paolo, D. J., and Wasserberg, G. J. (1979) Petrogenetic mixing models and Nd–Sr isotopic patterns. *Geochim. Cosmochim. Acta* **43,** 615.

De Witt, E., Armstrong, R. L., Sutter, J. F., and Zartman, R. E. (1984) U–Th–Pb, Rb–Sr and Ar–Ar mineral and whole rock isotopic systematics in a metamorphosed granitic terrane, southeastern California. *Bull. Geol. Soc. Am.* **95,** 723.

Dodson, M. H. (1982) On 'spurious' correlations in Rb–Sr isochron diagrams. *Lithos* **15,** 215.

Duyverman, H. J., Harris, N. B. W., and Hawkesworth, C. J. (1982) Crustal accretion in the Pan African: Nd and Sr isotopic evidence from the Arabian Shield. *Earth Planet. Sci. Lett.* **59,** 315.

Edmunds, W. M., Andrews, J. N., Burgess, W. G., Kay, R. L. F., and Lee, D. J. (1984) The evolution of saline and thermal groundwaters in the Carnmenellis granite. *Mineral Mag.* **48,** 407.

Evans, J. C., Reeves, J. H., and Rancitelli, L. (1980) Terrestial age dating of Antarctic meteorites. *US Antarctic J.* **15,** 53.

Fanale, F. P. (1971) A case for catastrophic early degassing of the *Chem. Geol.* **8,** 79.

Fanale, F. P., and Schaeffer, O. A. (1965) Helium and uranium ratios for Pleistocene and Tertiary fossil aragonites. *Science* **149,** 312.

Faure, G. (1977) *Principles of isotope geology*. New York, Wiley.

Faure, G., and Powell, J. L. (1972) *Strontium isotope geology*. Berlin, Springer.

Field, D., and Raheim, A. (1979) A geologically meaningless Rb–Sr total rock isochron. *Nature* **282,** 497.

Field, D., and Raheim, A. (1980) Secondary geologically

meaningless Rb–Sr isochrons, low $^{87}Sr/^{86}Sr$ initial ratios and crustal residence times of high grade gneisses. *Lithos* **13,** 295.

Fitch, F. J., and Miller, J. A. (1973) Dating granites by the potassium–argon method. *Spec. Publ. Geol. Soc. S.Afr.* **3,** 219.

Fitch, F. J., Miller, J. A., and Hooker, P. J. (1976) Single whole rock K–Ar isochrons. *Geol. Mag.* **113,** 1.

Fletcher, I. R., and Rosman, K. J. R. (1982) Precise determination of initial Nd from Sm–Nd isochron data. *Geochim. Cosmochim. Acta* **46,** 1983.

Foland, K. A. (1979) Limited mobility of argon in metamorphic terrain. *Geochim. Cosmochim. Acta* **43,** 793.

Fontes, J. C. (1980) Environmental isotopes in groundwater hydrology. In Fritz, P., and Fontes, J. C. (eds), *Handbook of environmental isotope geochemistry*. Amsterdam, Elsevier.

Fraundorf, P., Flynn, G. J., Shirck, J. R., and Walker, R. M. (1977) Search for fission tracks from superheavy elements in Allende. *Earth Planet. Sci. Lett.* **37,** 285.

Fullagar, P. D., and Ragland, P. C. (1975) Chemical weathering and Rb–Sr whole rock ages. *Geochim. Cosmochim. Acta* **39,** 1245.

Futa, K. (1981) Sm–Nd systematics of a tonalitic gneiss and its constituent minerals from northern Michigan. *Geochim. Cosmochim. Acta* **45,** 1245.

Galer, S. J. G., and O'Nions, R. K. (1985) Residence time of thorium, uranium and lead in the mantle with implications for mantle convection. *Nature* **316,** 778.

Gass, I. G., Smith, P. J., and Wilson, C. R. L. (1971) *Understanding the Earth*. Milton Keynes, Open University.

Gleadow, A. J. W., Hurford, A. J. and Quaife, R. D., 1976. Fission track dating of zircon: improved etching techniques. *Earth Planet Sci. Lett.* **33,** 273.

Goldberg, E. D., and Koide, M. (1962) Geochronological studies of deep sea sediments by the ionium/thorium method. *Geochim. Cosmochim. Acta.* **26,** 417.

Gopalan, K., and Rao, M. N. (1975) Uranium–xenon dating by thermal neutron irradiation. *Can. J. Earth Sci.* **12,** 887.

Hamilton, P. J., O'Nions, R. K., and Evensen, N. M. (1977) Sm–Nd dating of Archaean basic and ultrabasic volcanics. *Earth Planet. Sci. Lett.* **36,** 263.

Hangini, A., and Sonntag, C. (1977) ^{231}Pa dating of deep sea cores via ^{227}Th counting. *Earth Planet. Sci. Lett.* **37,** 251.

Harmon, R. S. (1979) U-series dating of speleothems and a glacial chronology of western North America. *Bull. Natl. Speleol. Soc.* **41,** 102.

Harmon, R. S., Glazek, J., and Nowbak, K. (1980) $^{230}Th/^{238}U$ dating of travertine from the Bilzingsleben archaeological site. *Nature* **284,** 132.

Harmon, R. S., Ku, T. L., Matthews, R. K., and Smart, P. L. (1979) Limits of U series analysis: phase 1 results of the Uranium Series Intercomparison Project. *Geology* **7,** 405.

Harmon, R. S., Thompson, P., Schwarcz, H. P., and Ford, D. C. (1975) Uranium series dating of speleothems. *Bull. Nat. Speleol. Soc.* **37,** 21.

Harrison, T. M. (1983) Some observations on the interpretation of $^{40}Ar/^{39}Ar$ age spectra. *Isotope Geosci.* **1,** 319.

Harrison, T. M., and McDougall, I. (1981) Excess ^{40}Ar in metamorphic rocks from Broken Hill, New South Wales: implications for $^{40}Ar/^{39}Ar$ age spectra and the thermal history of the region. *Earth Planet. Sci. Lett.* **55,** 123.

Harrison, T. M., and McDougall, I. (1982) The thermal significance of potassium feldspar K–Ar ages inferred from $^{40}Ar/^{39}Ar$ age spectrum results. *Geochim. Cosmochim. Acta* **46,** 1811.

Hayatsu, A., and Carmichael, C. M. (1977) Removal of atmospheric argon contamination and the use and misuse of the K–Ar isochron method. *Can. J. Earth Sci.* **14,** 337.

Hille, P. (1979) An open system model for uranium dating. *Earth Planet. Sci. Lett.* **42,** 138.

Hohenberg, C. M., and Kennedy, B. M. (1981) I–Xe dating: intercomparisons of neutron irradiations and reproducibility of the Bjurbole standard. *Geochim. Cosmochim. Acta* **45,** 251.

Holmes, A. (1937) *The age of the Earth*. Edinburgh, Nelson.

Humphries, F. J., and Cliff, R. A. (1982) Sm–Nd dating and cooling history of Scourian granites, Sutherland. *Nature* **295,** 515.

Hurford, A. J., and Green, P. F. (1982) A user's guide to fission track dating calibration. *Earth Planet. Sci. Lett.* **59,** 343.

Hurford, A. J., and Green, P. F. (1983) The zeta age calibration of fission track dating. *Isotope Geosci.* **1,** 285.

Husain, L., and Schaeffer, O. A. (1973) Lunar volcanism: age of glass in the Apollo 17 orange soil. *Science* **180,** 1338.

Ineson, P. R. (1980) Dating mineralization. *J. Geol. Soc.* **137,** 101.

Ivanovich, M. (1982) Uranium series disequilibria applications in geochronology. In Ivanovich, M. and Harmon, R. S. (eds), *Uranium series disequilibrium: applications to environmental problems*. Oxford, Oxford University Press.

Ivanovich, M., and Harmon, R. S. (1982) *Uranium series disequilibrium: applications to environmental problems*. Oxford, Oxford University Press.

Ivanovich, M., Vita-Finzi, C., and Hennig, G. J. (1983) Uranium series dating of molluscs from uplifted Holocene beaches in the Persian Gulf. *Nature* **302,** 408.

Jacobsen, S. B., and Wasserburg, G. J. (1979) Nd and Sr isotopic study of the Bay of Islands ophiolite complex and the evolution of the source of midocean ridge basalts. *J. Geophys. Res. B* **84,** 7429.

Jaffey, A. H., Flynn, K. F., Glendenin, L. E., Bentley, W. C., and Essling, A. M. (1971) Precision measurement of the half lives and specific activities of U^{235} and U^{238}. *Phys. Rev. C* **4,** 1889.

Johnson, N. M., McGee, V. E., and Naeser, C. W. (1979) A practical method of estimating standard error of age in the fission track dating method. *Nucl. Tracks* **3,** 93.

Kaneoka, I. (1983) Noble gas constraints on the layered structure of the mantle. *Nature* **302,** 698.

Kathren, R. L. (1984) *Radioactivity in the environment*. Chur, Harwood.

Krishanaswami, S., Mangini, A., Thomas, J. H., Sharma,

P., Cochran, J. K., Turekian, K. K., and Parker, P. D. (1982) ^{10}Be and Th isotopes in manganese nodules and adjacent sediments: nodule growth histories and nuclide behaviour. *Earth Planet. Sci. Lett.* **59,** 217.

Ku, T. L., Omura, A., and Chen, P. S. (1979) Be^{10} and U-series isotopes in manganese from the central North Pacific. In Bischoff, J. L. and Puper, D. Z. (eds), *Marine geology and oceanography of the Pacific manganese nodule province*. New York, Plenum.

Lal, N., Parshad, R., and Nagpaul, M. (1977) Fission track etching and annealing of tourmaline. *Nucl. track Detection* **1,** 145.

Lally, A. E. (1982) Chemical procedures. In Ivanovich, M., and Harmon, R. S. (eds), *Uranium series disequilibrium: applications to environmental problems*. Oxford, Oxford University Press.

Libby, W. F. (1952) *Radiocarbon dating*. Chicago, Illinois, University of Chicago Press.

Lofvendahl, R., and Holm, E. (1981) Radioactive disequilibria and apparent ages of secondary uranium minerals from Sweden. *Lithos* **14,** 189.

Loosli, H. H. (1983) A dating method with ^{39}Ar. *Earth Planet. Sci. Lett.* **63,** 51.

Luck, J. M., and Allegre, C. J. (1982) The study of molybdenites through the ^{187}Re–^{187}Os chronometer. *Earth Planet. Sci. Lett.* **61,** 291.

Luck, J. M., and Allegre, C. J. (1983) ^{187}Re–^{187}Os systematics in meteorites and cosmochemical consequences. *Nature* **302,** 130.

Ludwig, K. R. (1977) Effect of initial radioactive daughter disequilibrium on U–Pb isotope apparent ages in young minerals. *J. Res. US Geol. Surv.* **5,** 663.

Ludwig, K. R. (1980) Calculations of uncertainties of U–Pb isotope data. *Earth Planet. Sci. Lett.* **46,** 212.

Macdougall, J. D. (1976) Fission track annealing and correction procedures for oceanic basalt glasses. *Earth Planet. Sci. Lett.* **30,** 19.

Macedo, C. R., Costa, C. V., Ferretra, J. T., Ferretra, M. P., and Reynolds, J. H. (1977) Rare gas dating, III, Evaluation of a double spiking procedure for potassium–argon dating. *Earth Planet. Sci. Lett.* **34,** 411.

Matthess, G. (1982) *The properties of groundwater*. New York, Wiley.

McCarthy, T. S., and Grant, C. R. (1980) Changes in initial $^{87}Sr/^{86}Sr$ ratio during protracted fractionation in igneous complexes. *J. Petrol.* **21,** 245.

McCulloch, M. T., and Wasserburg, G. J. (1980) Early Archaean Sm–Nd model ages from a tonalitic gneiss, northern Michigan. In Morey, G. D. and Hanson, G. N. (eds), Selected studies of Archaean gneisses and lower Proterozoic rocks, southern Canadian Shield. *Spec. Paper. Geol. Soc. Am.* **182,** 135.

McDowell, F. W. (1983) K–Ar dating: incomplete extraction of radiogenic argon from alkali feldspar. *Isotope Geosci.* **1,** 119.

McLennan, S. M. (1982) On the geochemical evolution of sedimentary rocks. *Chem. Geol.* **37,** 335.

Megumi, K. (1978) A problem in ^{210}Pb geochronologies of sediments. *Nature* **274,** 885.

Michael, H. N., and Ralph, E. K. (1970) Corretion factors applied to Egyptian radiocarbon dates from the era before Christ. In Olsson, I. V. (ed.), *Radiocarbon variations and absolute chronology*. New York, Wiley.

Mitchell, J. G. (1972) Potassium-argon dating of gamma-irradiated minerals. *Earth Planet. Sci. Lett.* **14,** 91.

Mook, W. G. (1980) Carbon-14 in hydrogeological studies. In Fritz, P., and Fontes, J. C. (eds). *Handbook of environmental isotope geochemistry*. Amsterdam, Elsevier.

Muller, P. J., and Mangini, A. (1980) Organic carbon decomposition in sediments of the Pacific manganese nodule belt dated by ^{230}Th and ^{231}Pa. *Earth Planet. Sci. Lett.* **51,** 94.

Mussett, A. E., and McCormack, A. G. (1978) On the use of three-dimensional plots in K–Ar dating. *Geochim. Cosmochim. Acta* **42,** 1877.

Naeser, C. W., and Brookins, D. G. (1975) Comparison of fission track, K–Ar and Rb–Sr radiometric age determinations for some granite plutons, Maine. *J. Res. US Geol. Surv.* **3,** 229.

Nishiizumi, K., Elmore, D., Honda, M., Arnold, J. R., and Gove, H. E. (1983) Measurements of ^{129}I in meteorites and lunar rock by tandem accelerator mass spectrometry. *Nature* **305,** 611.

Norman, E. B., and Schramm, D. N. (1983) ^{182}Hf chronometer for the early Solar System. *Nature* **304,** 515.

Nunes, P. D. (1981) The age of the Stillwater Complex — a comparison of U–Pb zircon and Sm–Nd isochron systematics. *Geochim. Cosmochim. Acta* **45,** 1961.

Oldfield, F., and Appleby, P. G. (1984) A combined radiometric and mineral magnetic approach to recent geochronology of lakes affected by catchment disturbance and sediment redistribution. *Chem. Geol.* **44,** 67.

Oldfield, F., Appleby, P. G., and Battarbee, R. W. (1978) Alternative ^{210}Pb dating: results for the New Guinea Highlands and Lough Erne. *Nature* **271,** 339.

O'Nions, R. K., Evensen, N. M., and Hamilton, P. J. (1980) Differentiation and evolution of the mantle. In Bailey, D. K., Tarney, J., and Dunham, K. (eds), *The evidence for chemical heterogeneity in the Earth's mantle., Phil. Trans. R. Soc. Lond. A* **297.**

Osmond, J. K. (1980) Uranium disequilibrium in hydrologic studies. In Fritz, P., and Fontes, J. C. (eds), *Handbook of environmental isotope geochemistry*. Amsterdam, Elsevier.

Osmond, J. K., and Cowart, J. B. (1982) Groundwater. In Ivanovich, M., and Harmon, R. S. (eds), *Uranium series disequilibrium: applications to environmental problems*. Oxford, Oxford University Press.

Ozima, M., Kaneoka, I., and Yanagisawa, M. (1979) Temperature and pressure effects on ^{40}Ar–^{39}Ar systematics. *Earth Planet. Sci. Lett.* **42,** 463.

Patchett, P. J. (1983) The importance of the Lu–Hf isotopic system in studies of planetary chronology and chemical evolution. *Geochim. Cosmochim. Acta* **47,** 81.

Pavich, M. J., Brown, L., Klein, J., and Middleton, R. (1984) ^{10}Be accumulation in a soil chronosequence. *Earth Planet. Sci. Lett.* **68,** 198.

Pettinghill, H. S., and Patchett, P. J. (1981) Lu–Hf total rock age for the Amitsoq gneisses, west Greenland. *Earth Planet. Sci. Lett.* **55,** 150.

Piggott, G. S., and Urry, W. D. (1942) Time relations in oceanic sediments. *Bull. Geol. Soc. Am.* **53,** 1187.

Raisbeck, G. M., and Yiou, F. (1979) Possible use of ^{41}Ca for radiocarbon dating. *Nature* **277,** 42.

Reynolds, J. H. (1963) Xenology. *J. Geophys. Res.* **68,** 2939.

Roddick, J. C. (1978) The application of isochron diagrams in ^{40}Ar–^{39}Ar dating: a discussion. *Earth Planet. Sci. Lett.* **41,** 233.

Roddick, J. C., Cliff, R. A., and Rex, D. C. (1980) The evolution of excess argon in Alpine bodies — a ^{40}Ar–^{39}Ar analysis. *Earth Planet Sci. Lett.* **48,** 185.

Rosenberg, P. E., and Hooper, R. L. (1982) Fission track dating of sandstone-type uranium deposits. *Geology* **10,** 481.

Rosholt, J. N. (1980) Uranium-trend dating of Quaternary sediments. *Rep. US Geol. Surv.* **80–1087.**

Seward, D. (1979) Comparison of zircon and glass fission track ages from tephra horizons. *Geology* **7,** 479.

Shafiqullah, M., and Damon, P. E. (1974) Evaluation of K–Ar isochron methods. *Geochim. Cosmochim. Acta* **38,** 1341.

Sharma, Y. P., Lal, N., Bal, K. D., Parshad, R., and Nagpaul, K. K. (1980) Closing temperatures of different fission track clocks. *Contrib. Mineral. Petrol.* **72,** 335.

Sharma, Y. P., and Somayajulu, B. L. K. (1982) ^{10}Be dating of large manganese nodules from world oceans. *Earth Planet. Sci. Lett.* **59,** 235.

Shepherd, T. J., and Darbyshire, D. P. F. (1981) Fluid inclusion Rb–Sr isochrons for dating mineral deposits. *Nature* **290,** 578.

Stanton, R. L., and Russell, R. D. (1959) Anomalous leads and the emplacement of lead sulphide ores. *Econ. Geol.* **54,** 588.

Stille, P. (1981) The measurement of rubidium and strontium blanks in the geochronological laboratory. *Chem. Geol.* **32,** 303.

Stuiver, M. (1978) Carbon–14 dating: a comparison of beta and ion counting. *Science* **202,** 881.

Stuiver, M. (1978) Radiocarbon timescale tested against magnetic and other dating methods. *Nature* **273,** 271.

Szabo, B. J. (1979) Uranium series age of coral growth on Rottnest Island, Western Australia. *Marine Geol.* **29,** 11.

Tanaka, S., Inoue, T., and Imamura, M. (1977) The ^{10}Be method of dating marine sediments — comparison with the paleomagnetic method. *Earth Planet. Sci. Lett.* **37,** 55.

Tanaka, T., and Masuda, A. (1983) The La–Ce geochronometer: a new dating method. *Nature* **300,** 515.

Taylor, K. S., and Faure, G. (1981) Rb–Sr dating of detrital feldspar: a new method to study till. *J. Geol.* **89,** 97.

Teitsma, A., Clarke, W. B., and Allegre, C. J. (1975) Spontaneous fission — neutron fission xenon: a new technique for dating geological events. *Science* **189,** 878.

Tera, F. (1981) Aspects of isochronism in Pb isotope systematics — application to planetary evolution. *Geochim. Cosmochim. Acta* **45,** 1439.

Thiel, K., Vorwerk, R., Saager, R. and Stupp, H. D. (1983) ^{235}U fission tracks and ^{238}U-series disequilibrium as a means to study recent mobilization of uranium in Archaean pyritic conglomerates. *Earth Planet. Sci. Lett.* **65,** 249.

Tilton, G. R. (1960) Volume diffusion as a mechanism for discordant lead ages. *J. Geophys. Res.* **65,** 2933.

Tilton, G. R. (1983) Evolution of depleted mantle: the lead perspective. *Geochim. Cosmochim. Acta* **47,** 1191.

Torgersen, T. (1980), Controls on pore-fluid concentrations of ^{4}He and ^{222}Rn and the calculation of ^{4}He/^{222}Rn ages. *J. Geochem. Expl.* **13,** 57.

Turner, G., and Cadogan, P. H. (1974) Possible effects of ^{39}Ar recoil in ^{40}Ar–^{39}Ar dating. *Geochim. Cosmochim. Acta Supple.* **5,** 1601.

Van Breeman, O., and Hawkesworth, C. J. (1980) Sm–Nd isotopic study of garnets and their metamorphic hosts. *Trans. R. Soc. Edinburgh.* **71,** 97.

Wagner, G. A., and Reimer, G. M. (1972) Fission track tectonics: the tectonic interpretation of fission track apatite ages. *Earth Planet. Sci. Lett.* **14,** 263.

Wagner, G. A., Reimer, G. M., Carpenter, B. S., Faul, H., Van der Linden, R., and Gijbels, R. (1975) The spontaneous fission rate of U–238 and fission track dating. *Geochim. Cosmochim. Acta* **39,** 1279.

Wakita, H., and Sano, Y. (1983) ^{3}H/^{4}He ratios in CH_4-rich natural gases suggest magmatic origin. *Nature* **305,** 792.

Wang, S., McDougall, I., Tetley, N., and Harrison, T. M. (1980) ^{40}Ar/^{39}Ar age and thermal history of the Kirin chondrite. *Earth Planet. Sci. Lett.* **49,** 117.

Wigley, T. M. L. (1976) A three-dimensional U–Pb discordia plane to evaluate samples with common lead of unknown isotopic composition. *Isotope Geosci.* **2,** 1.

Wendt, I. (1984) A three-dimensional U–Pb discordia plane to evaluate samples with common lead of unknown isotopic composition. *Isotope Geosci.* 2, 1.

Wetherill, G. W. (1966) Radioactive decay constants and energies. *Mem. Geol. Soc. Am.* **97,** 513.

Wilson, G. C., Litherland, A. E., and Rucklidge, J. C. (1984) Dating of sediments using accelerator mass spectrometry. *Chem. Geol.* **44,** 1.

11

Radioactivity and petroleum

The association of U deposits with organic material in sedimentary rocks has been described in Chapter 6. It is not surprising, therefore, that high concentrations of U also occur in asphalt. The chemical association of radionuclides with hydrocarbons is not the only link, however, between radioactivity and petroleum, as there is also a physical link, in the presence of He and Rn in anomalous concentrations in many oil and gas fields. Moreover, there are physiochemical arguments that suggest that radioactivity has a strong (even direct) influence on the origin of petroleum. Various exploration techniques based upon the measurement of radioactivity, or employing artificial radioactive sources, are also of great importance in the petroleum industry.

11.1 CHEMICAL ASSOCIATIONS

The association of radionuclides with hydrocarbons is exemplified in the occurrence of the U hydrocarbon thucolite, the general correlation between U concentration and the organic content of sedimentary rocks, and the concentration of the radionuclide ^{50}V in organic materials. However, radionuclides are only part of a suite of non-essential elements, including stable isotopes of Ni, Co, Cu, As, Pb, Sn, Ag, Au, Mo and W, which occur in this way, and any explanation of the radionuclide–petroleum association must also take into account this assemblage of metals. Moreover, concentrations in crude oil are usually low.

The non-essential metals probably occur as metallo-organic complexes, possibly as porphyrin compounds. Early studies of the association appeared to show that the exceptionally high metal abundances found in asphalt, and in crude oil extracted from petroliferous rock, could not have arisen by leaching of either the source or the host rocks (Erickson *et al.* 1954). Consequently, it was believed that the metals were concentrated by biological activity in the organisms that made up the primary hydrocarbon source. However, this ignores the fact that hydrocarbons tend to separate during oil maturation according to their molecular composition, metallo-organic complexes becoming concentrated in the asphaltic fraction. U, for example, can reach abundances of about 10% in ash from asphalt, and as high as 1% in bulk asphalt. In addition to porphyrins, asphaltenes and naphthenic acids may be important carriers of non-essential metals.

In the light of more recent investigations, it is now considered that the U–petroleum association can be produced in at least three ways, each of which involves groundwater (Bloch and Key, 1981):

(1) H_2S derived from oil and gas may cause reduction of U and other metals in aqueous solution, leading to the precipitation of U and other metal compounds.
(2) Microbiological oxidation of petroleum during upward migration may result in the formation of 'solid bitumens', which then become enriched in U and other metals from circulating groundwater.
(3) Invasion of the reducing environment of a petroleum reservoir by groundwater carrying U and other metals in solution may lead to precipitation of the metal assemblage.

The fact that there is a link between the U–petroleum association and groundwater is suggested by the fact that many oil field brines contain anomalously high concentrations of ^{226}Ra, ^{228}Ra and ^{224}Ra. The first of these isotopes forms from the decay of ^{238}U, but the second and third are produced in the decay of ^{232}Th. It is, therefore, possible that the Ra isotope characteristics can be used to reveal the presence of U mineralisation close to oil pools, as well as indicating the geochemical environment of petroleum reservoirs. Ba–^{228}Ra–^{226}Ra discrimination diagrams (Fig. 11.1), for example, have been used to distinguish between oil field brines associated with U deposits (which plot close to the ^{226}Ra apex) and normal Ra-rich brines (which fall close to the Ba–^{228}Ra side of the diagram).

Each of these views of the origin of the U–hydrocarbon association has been accepted as accounting for the origin of thucolite. However, the present-day occurrence of thucolite seems to be associated with the movement of metallo-organic complexes away from the source environment. Transport *en masse* as colloids or emulsoids in groundwater, or by diffusion, are generally considered to be viable mechanisms (Boyle, 1982). Mobilisation of metallic complexes may be greatly enhanced by the presence of U within asphaltic material, as the effect of radioactivity on organic structures can be profound. It has been

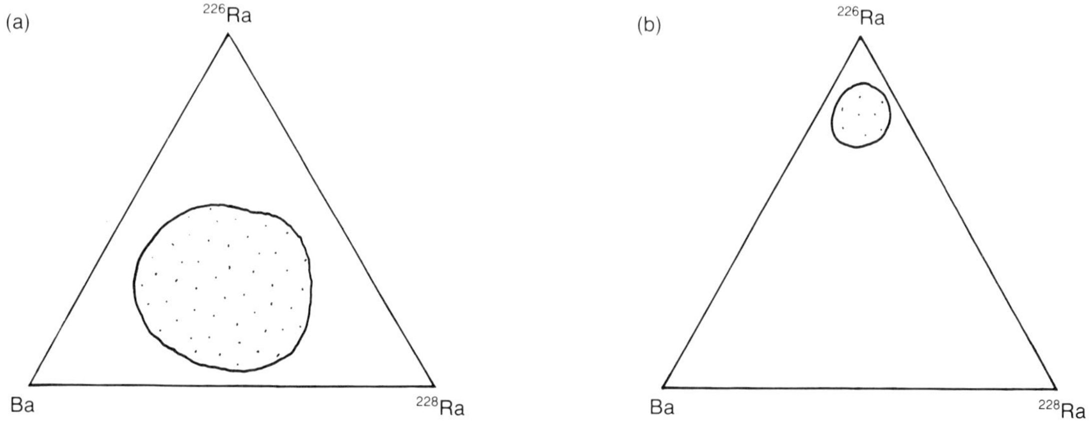

Fig. 11.1 — Ra isotope and Ba ratios for (a) ordinary oil field brines and (b) oil field brines associated with U occurrences. (Bloch and Key, 1981, redrawn with permission.)

suggested, for example, that radioactivity due to the concentration of Rn in crude oil can be so intense that appreciable cracking of heavy hydrocarbon molecules may occur through α-particle bombardment over long periods of time (Bell *et al.*, 1940). The ability of α-particles to break down fatty acids is well known.

The interaction between radioactivity and organic matter during both petroleum formation and reservoir degradation formed the basis of much speculation during the 1940s and early 1950s. It was suggested, for example, that dissociation of water under γ-radiation would release H^+ ions that could then react with free C^{4-} ions to form hydrocarbon compounds. The free C^{4-} was thought to have been provided by organic matter in shaly rocks. Because of the greater abundance of U in shales than in most other sedimentary rocks, a link between high radioactivity, high organic content and the generation and maturation of petroleum was clearly demonstrated. While this association is certainly real, it is not established that petroleum is genetically linked with radioactivity in petroleum source rocks. It can be shown that γ-radiation will split-off H^+ ions from hydrocarbon compounds and, with time, produce progressively heavier hydrocarbon molecules. However, there is no evidence of any trend towards the production of petroleum composed of these heavier compounds. Moreover, in this process, some of the H^+ ions will gain electrons to form H_2 gas. Again, there is no evidence of the generation of large quantities of free H_2 in petroleum source rocks, even though the reducing conditions present would favour its preservation. The argument that radioactivity plays a significant role in petroleum formation is also countered by the observation that some highly radioactive, highly organic black shales do not have associated petroleum deposits (Hedberg, 1964).

Black shales with a high organic content, which are currently considered to be the normal source rocks for petroleum, may nevertheless be modified by high levels of natural radioactivity, and this may have some bearing on the primary migration of petroleum from their source rocks.

11.2 OIL GENERATION AND MIGRATION

11.2.1 Primary migration

Investigation of primary migration has been prompted by the apparently fundamental difficulty of removing large hydrocarbon molecules from rocks of low permeability, even in the presence of moving groundwater. Although shales have reasonable porosity, the packing of clay minerals tends to isolate pores or to reduce their connections to narrow throats. Certainly, continuous flow of liquid hydrocarbons is difficult to envisage where the throats may be significantly smaller than the hydrocarbon molecules. To overcome this problem, many different mechanisms have been proposed, involving either an aqueous medium, in which the petroleum or its precursors are removed from the source rock, or conditions which will allow the migration of petroleum separately from any other phase. In the case of aqueous transport, the petroleum may be as oil or gas in solution, as an emulsion or as a micellar solution. Other transported and diffusive migration systems have also been proposed.

Primary migration mechanisms which involve water as a transport medium face many difficulties. It has been suggested, for example, that an insufficient volume of rising water would be available to account for the accumulations of oil found in large reservoirs because, in subsiding sedimentary basins, the overall water movement will tend to be into the basin and downwards (Jones, 1981). Therefore, although the migration of water may be stratigraphically upwards, it will be spatially downwards. An exception, however, is where heating occurs, in which case an upward flow may be induced.

Perhaps more significant as an argument against aqueous transport is the fact that crude oils do not have high solubility in water. Clearly, if crude oil is transported in aqueous solution, an association of hydrocarbons with similar solubilities should be found in reservoirs. However, hydrocarbon solubility varies not only with composition but also with temperature (Fig. 11.2), so that once again heating may be an important factor (Price, 1976).

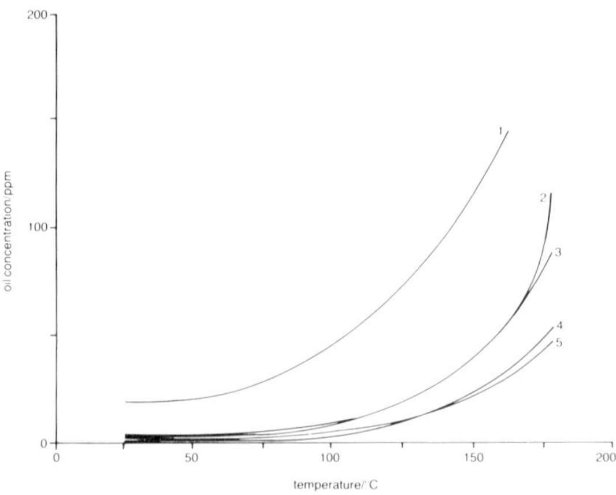

Fig. 11.2 — Variation in solubility of crude oil with temperature (topping temperature, 200 °C): 1, Wyoming Farmers whole oil; 2, Alaska topped oil; 3, Reedy Creek topped oil; 4, Union topped oil; 5, Louisiana State whole oil; 6, Amoseas topped oil. (Price. 1976, redrawn with permission.)

Although most source rocks that generated oil have not been heated to temperatures in excess of 120 °C, some increase in solubility, and change in relative solubility, would have accompanied such a rise in temperature. As a result, it can be argued that, if primary migration occurred at elevated temperatures, the transport of liquid hydrocarbons in aqueous solution is a viable mechanism. However, the availability of the large volume of thermal water necessary for producing large oil accumulations can still be questioned, and specific doubts have been raised about reservoirs which occur above very thick shale successions. In these circumstances, falling hydrocarbon solubility in water rising through the shale would have caused precipitation within the shale long before the reservoir horizons were reached; this is the case in the Tertiary succession of the Gulf Coast of America (Jones, 1981).

Finally, it must be emphasised that liquid hydrocarbon solubilities are particularly low in saline groundwaters. Transport in solution becomes increasingly difficult during rising migration, therefore, as an increasing ratio of saline connate water to fresh meteoric water (or fresh water liberated by the dehydration of clay minerals) is encountered within the shallow sections of marine sedimentary successions.

For these reasons, most arguments which invoke hydrocarbon transport in aqueous solution rely upon the availability of copious quantities of fresh water produced by re-crystallisation of clay minerals (Price, 1976). Unfortunately, some source rocks have never been raised to the temperatures and pressures required for this process. Alternative proposals which seek to increase the ability of cool water to transport hydrocarbons, by producing micelles, for example, also encounter difficulties. In the case of micelles there is no clear source of the large amount of surfactant needed for their production, and there is still a problem concerning their size relative to that of pores and throats.

Aqueous transport has generally been invoked to overcome difficulties in the continuous migration of crude oil as a separate phase, not only because of the large size of oil molecules but also because it is likely that pores contain both oil and water. In water-wet pores, the surface tension between oil and water causes separation of the phases to the extent that capillary effects cannot overcome resistance to movement of discrete droplets of oil. However, if the surfaces of the pores are not water-wet but are oil-wet, this difficulty does not arise. Consequently, much discussion has arisen about whether or not capillary forces alone are sufficient to remove the oil. High differential pressures, for example, which can arise during compaction of a sedimentary pile or as a secondary effect of petroleum generation itself, may be important.

This still leaves the apparently insoluble problem posed by the physical size of the throats between pores. However, it has been shown that an irregular distribution of particle sizes, together with small variations in lithology and the development of microfractures, can lead to the formation of relatively large pores and micropathways, even in shales, which may allow migration of petroleum by channel flow from the source rock (Momper, 1978). Primary migration by continuous oil flow, therefore, rather than in an aqueous solution is the majority view at present.

The role of radioactivity in this process has been largely ignored in the discussion so far, although the effects of radiation in producing alteration in hydrocarbon molecules was mentioned earlier in this chapter. The changes caused by the α-particle bombardment of hydrocarbon molecules, together with the recoil effects on the parent nuclei, are not the same as in crystalline materials. It is interesting, however, to speculate about the degree to which complex liquid hydrocarbons can be considered analogous to solid crystalline structures. It is known, for example, that irradiated hydrocarbons show fragmentation and ionisation of molecular groups in a manner somewhat similar to the metamictisation process (Heinrich, 1958).

Radiation-induced changes in hydrocarbons favour the formation of metallo-organic complexes which, in the reducing environment of a petroleum source rock, would probably be rapidly precipitated. In addition, α-particle bombardment can cause unsaturated hydrocarbons to polymerise and condense, while saturated hydrocarbons become dehydrogenated to form unsaturated hydrocarbons. Subsequently, these newly produced unsaturated hydrocarbons also undergo polymerisation and condense. These changes are in accord with the general view that radioactivity is likely to lead to the formation of the heavier hydrocarbons. Thus, irradiation of hydrocarbons within a source rock would appear to produce heavy, relatively immobile hydrocarbons and to inhibit processes of primary migration. The question arises, therefore, whether or not the high radioactivity found in asphaltic material and petroliferous sedimentary rocks is merely coincidental. Moreover, is the contrast with the generally low radioactivity of light crude oil also coincidental? Clearly, it is possible that the high radioactivity of some source rocks could have effectively immobilised the hydrocarbons to such an extent that primary migration was prevented. If this is, indeed, the case, a petroleum source rock must possess low radioactivity if it is to lead to the formation of oil reservoirs.

11.2.2 Oil generation and maturation

The common association of U with organic material in sedimentary rocks suggests that any model involving only the concentration of radionuclides is not sufficient; it is the total effect of radioactivity on the hydrocarbon molecules that is important, and that will be determined not only by the level of activity but also by its duration and the nature of the material affected. This can be considered as the non-metabolic dose equivalent. A potential source rock containing sufficient organic matter will, under the right conditions, see the generation and maturation of crude oil. If this takes place quickly, it is unlikely that even high levels of radioactivity will have much effect, and the petroleum will be free to migrate to a reservoir. However, if oil generation and maturation take place slowly, the presence of high levels of radioactivity directly associated with the organic matter may result in a preponderance of heavy hydrocarbons which would not be susceptible to migration. It is, therefore, clear that the effects of radioactivity depend upon the rate of petroleum generation and maturation. Interestingly, this is largely controlled by the temperature to which the organic-rich sediment is raised on burial, but that temperature may well be related to the radionuclide content of the crust.

Temperature is not the only factor controlling oil generation, as it is now generallly realised that time also plays a very significant role, and combined time–temperature control modelling has been attempted for the assessment of source rock potential. The basis of all models is very similar; time and temperature are interchangeable, in that either a high temperature for a short time or a low temperature for a long time will lead to oil generation in a suitable source rock.

The relationship between time and oil generation was originally believed to be simply linear, so that not only does oil generation commence after a certain time at a certain temperature, but also doubling the time doubles the maturity of the petroleum (Lopatin, 1971). However, chemical reactions are related to temperature exponentially. If a temperature range of 100–110 °C is given a value of zero, a temperature factor γ can be assigned to each 10 °C interval above and below this range. Thus a range of 90–100 °C becomes expressed as $\gamma = r^{-1}$, 80–90 °C as $\gamma = r^{-2}$, 120–130 °C as $\gamma = r^2$, etc., where r reflects the nature of the reaction. The total maturity is then given by

$$\sum_{n_{\min}}^{n_{\max}} \Delta T_n \, r^n,$$

where ΔT refers to the length of time at a particular temperature interval for which n is the appropriate exponent as defined above (that is $n = -1, -2$ etc.). In the summation, $n_{\min}$ and $n_{\max}$ are the values of n for the lowest and highest temperature ranges experienced (Lopatin, 1971).

Several attempts have been made to solve for the value of r by studying the thermal maturity data of crude oil from many sites. In general, it is found that, if $r = 2$, maturity correlates well with temperature (Waples, 1980). Interestingly, this is the same value as given by the theoretical relationship in which rates of chemical reaction approximately double for each 10 °C rise in temperature. From comparisons of total maturity values with the petroleum maturity indicated by organic and physical factors (such as vitrinite reflectance and thermal alteration), it has been concluded that the onset of oil generation occurs at $\gamma = 15$, peak oil generation occurs at $\gamma = 75$ and oil generation ends at $\gamma = 160$. Values higher than 160 indicate limits for the occurrence of oil (up to 1000), wet gas (1500) and dry gas (6500).

An alternative approach has been based upon the period for which the sediment was held at its maximum temperature (Hood *et al.*, 1975). For this purpose, a 'level of organic metamorphism', on a scale of 0–20, determined by vitrinite reflectance, is employed (Fig. 11.3). In this model, oil begins to form at a value of about 7.8, reaching maturity between 9 and 11.5 and then passing to gas condensates and eventually to high temperature CH_4 generation. The relationship between the effective heating period and temperature is log-normal, with linear trends for different levels of organic metamorphism (Fig. 11.4). For example, a value of 9 (initial oil maturity) is produced at a temperature of 150 °C and a heating time of about 600 ka, or at a temperature of 50 °C with a heating time of 500 Ma. Again, this is consistent with the concept of a doubling of reaction rates for each 10 °C rise in temperature.

In either scheme, it is clear that, theoretically, there can be very wide variation in the time taken for petroleum generation and maturation, the major control being the temperature to which the source rock is raised. This temperature (to the extent that it has a strong radiogenic component) may be related to the radiation dose received by the petroleum which, in turn, will affect the degree to which the formation of heavy hydrocarbon molecules will restrict primary migration.

When considering the thermal history of sedimentary basins, internally generated heat from the decay of radionuclides can be very significant, especially if a subsiding basin develops in an area where the ambient

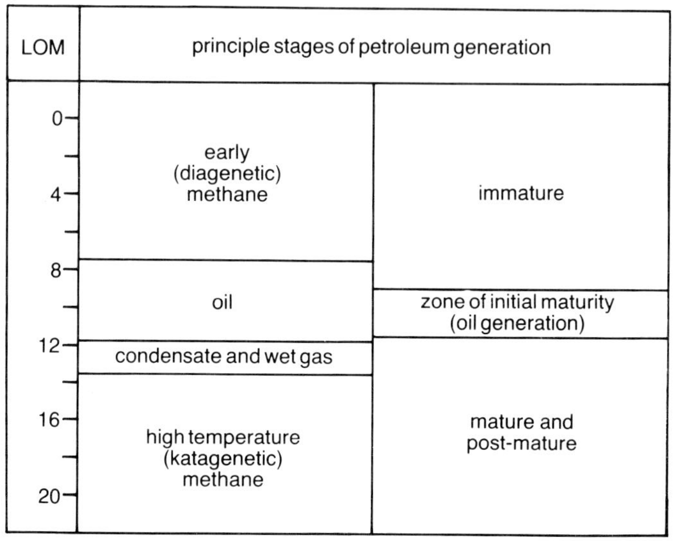

Fig. 11.3 — Level of organic metamorphism (LOM) and petroleum generation and maturation. (After Hood *et al.*, 1975.)

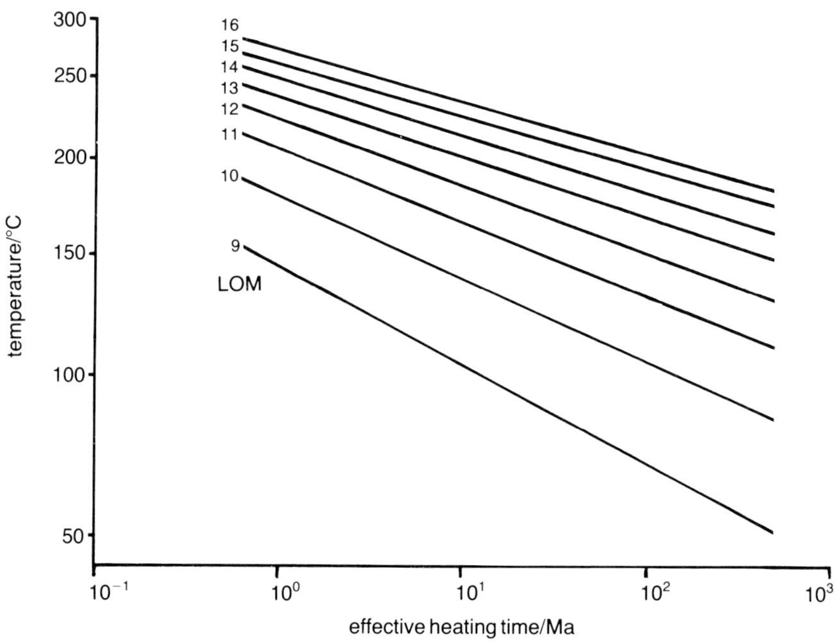

Fig. 11.4 — The relationship between level of organic metamorphism (LOM) and effective heating time. (After Hood *et al.*, 1975.)

geothermal gradient is relatively low. A detailed study of radiogenic heat production in a succession of Jurassic to Tertiary sedimentary rocks in the continental margin of the north-eastern USA, in terms of the implications for petroleum generation, revealed values of heat production (based upon the concentrations of U, Th and K) ranging between 0.3 and 1.8 $\mu W\,m^{-3}$ (Keen and Lewis, 1982). The highest values, between 1.4 and 1.8 $\mu W\,m^{-3}$ were found in shales, while the lowest values of 0.3–0.6 $\mu W\,m^{-3}$ were found to be associated with limestones. The distribution of radionuclides within the succession is sufficient to produce a significant contribution to the surface heat flow, probably somewhere in the region of 8.4 $mW\,m^{-2}$, or about 15% of the total heat flow of about 60 $mW\,m^{-2}$. The effect is thus to enhance the normal geothermal gradient within the sedimentary pile and to increase the temperature by between 4 and 17 °C, depending upon the position in the sequence, the smaller modification occurring near the top of the succession and rapidly increasing with depth to a steady maximum value.

The significance of this increase in temperature can be appreciated when it is recalled that a temperature rise of 10 °C doubles the reaction rate for petroleum generation and maturation. Thus, for this succession, the maturation time is reduced by about one-third. However, the situation is rather more complex than this because the integrated thermal history of the rocks is a function of the lithologies deposited and the structural evolution of the basin. Nevertheless, for any situation, it is likely that the contribution from radiogenic heat will have increased from zero at the time of deposition to a maximum value today.

For the sedimentary succession of the north-eastern seaboard of the USA, temperature–time relationships derived by Keen and Lewis (1982) show that the initial rise in temperature was quite rapid, in that about 80% of the final temperature was reached within 25 Ma for rocks between 100 and 150 Ma old. The final temperatures of strata 100 Ma, 120 Ma and 150 Ma old are 70 °C, 100 °C and 140 °C, respectively. Whether or not these temperatures represent a stable equilibrium between heat production and heat loss is uncertain, as the rate of heat loss can change significantly with time if even low-vigour hydrothermal circulation systems develop. However, it is clear that effective maturation temperatures of about 70–140 °C existed for a period of about 100 Ma, radiogenic heat having raised these temperatures by some 10–20 °C from a range of 60–120 °C. As a result, petroleum generation occurred some tens of millions of years earlier in the Jurassic part of the succession than if the additional radiogenic heat had been absent (the effect on the Cretaceous strata being only marginal). In this case, the 150 Ma old (Jurassic) strata were buried to a depth of about 5 km within 50 Ma. In other circumstances the contribution of radiogenic heat would be greater. The area studied is characterised by rapid subsidence and rifting, however, so the overall temperatures to which the succession has been raised are considerably higher than for successions of similar age deposited in more stable environments.

Two important conclusions emerge from the studies discussed so far. Firstly, radiogenic heat from within a sedimentary sequence can reduce considerably the time required for petroleum generation and maturation. Secondly, radioactivity can lead to the formation of heavy hydrocarbons; in this process, the greater the dose received by petroleum in source rocks, the more difficult primary migration becomes. These effects are, to a certain extent, probably self-regulating, as they act in opposition to each other. Moreover, a source rock with a high concentration of radionuclides will have a high heat production which, in turn, will accelerate petroleum generation. At the same time, the direct effect of radiation on the molecular structure of hydrocarbons decreases with time (as radionuclides decay). It is, therefore, likely that any inhibiting effects that radioactivity may exert upon petroleum migration will only be apparent in environments where oil generation is at the threshold level; specifically, where a source rock has high organic and radionuclide concentrations but is not deeply buried or associated with rifting or other sites of high non-radiogenic heat flow. In such cases, the temperature–time relationship is restricted, and the activity may be such that radiogenic heat production will not overcome the build-up of heavy hydrocarbon molecules.

11.2.3 Secondary migration and trapping

Petroleum transport in solution is generally assumed to be governed by normal hydraulic pressure gradients. Such a passive transport mechanism, involving a finite volume of available water, will certainly encounter the limitations to flow referred to earlier in this chapter. If, however, hydrothermal circulation is taking place, the continuous flushing-through of source rocks by fresh water at a high temperature could produce a very efficient forced transport mechanism. The fact that the passage of thermal waters along microfractures through shaly rocks is a viable mechanism for the transfer of elements on a large scale is shown by metalliferous hydrothermal vein deposits. Certainly, the concept of hydrothermal leaching along microfractures, leading to the concentration of leached material elsewhere in the circulation system, also appears to be applicable to oil migration and reservoir formation. Indeed, the similarity between the hydrothermal charging of an aquifer, in which concentrations of metals in solution moving up fractures encounter a highly permeable porous horizon, and the ingress of oil into a reservoir is marked; if such forced systems are capable of leaching and transporting metals, they are probably also able to effect the migration of oil from source to reservoir.

In this mechanism, the rapid rise of thermal waters along major fractures could transfer oil-bearing solutions to any aquifer encountered *en route*. Discharge at the ground surface and loss of volatile hydrocarbon compounds to the atmosphere is also possible. Charging of aquifers by hydrocarbons in solution requires dynamic flow-through, but a decrease in temperature, or the presence of saline groundwater, will cause the hydrocarbons to be thrown out of solution. Trapping of oil within the aquifer will thus occur, and concentrations will increase as long as the flow regime is maintained. In this way, very large volumes of hydrocarbon can be transferred from source to reservoir. The composition of the hydrocarbons transferred will be determined by the character of the source rock and the temperature and chemical properties of the circulating groundwater. Passive migration of oil due to buoyancy is likely only to cause secondary transfer once precipitation within the aquifer has taken place. The role of radioactivity in this process is evident; it may make a significant contribution to the thermal energy available and will thus determine both the vigour of the circulation system as a whole and the composition of the leached product (Dahlberg, 1982).

11.3 BOREHOLE LOGGING

11.3.1 The γ-ray log

Various geophysical logging devices used by the petroleum industry either measure the natural radioactivity of formations or incorporate a radiation source in order to measure the response of formations.

The most straightforward of these devices is the γ-ray log, in which a scintillation detector measures the total natural γ-ray emissions from the borehole walls. Unlike stationary measurements carried out on the surface, borehole logging systems need to operate while the logging tool is lowered down the hole. As discussed in Chapters 8 and 9, statistical variation in count rates is determined by the magnitude of the count obtained, which, in turn, is partly controlled by the counting period. If a sufficiently long counting period is employed, successive results obtained under identical circumstances will be almost constant. Reducing the counting period introduces a statistical error. For most surface measurements, counting periods usually need to be of the order of a few seconds at least, to get good quality results but, clearly, this is not possible when making continuous measurements in boreholes. The same problem is encountered in surveys using fixed-wing aircraft. While air-borne systems can partly overcome this difficulty by employing detectors of very large volume, the limited space within a normal borehole prevents this solution, and it is necessary to incorporate smoothing circuits into the measuring system to ensure that statistical variations are averaged. The time constants used in these circuits have to be adjusted according to the level of activity recorded (because activity also controls the magnitude of the recorded count). Unfortunately, smoothing circuits cause a lag in recording; so, if the device is run at high logging speed, severe distortion of the response occurs. It has been determined empirically that, to avoid this distortion, the log has to be run at a speed at which the detector does not descend more than about 0.3 m per unit time constant; a typical logging speed is about $0.15\,\mathrm{m\,s^{-1}}$ for a time constant of 2 s (Schlumberger, 1972).

γ-rays emitted from radionuclides in the walls of a borehole will be attenuated as they pass through the rock to the detector by an amount controlled by the composition of the formation and the energy of the radiation. Compositional variations are normally corrected in borehole logging systems by taking the bulk density into account. It is usually assumed that the intensity of γ-radiation per unit mass of a given radioactive mineral is constant. On these bases, the activity is taken to be proportional to the weight concentration of the radioactive mineral present (if only one species is present) such that the measured activity is given by

$$\frac{MA_{\mathrm{m}}}{\rho}$$

where M is the unit volume mass of radioactive mineral of activity A_{m} present and ρ is the bulk density of the rock. M/ρ is, therefore, the concentration of the radioactive mineral. If more than one radioactive mineral species is present, the measured activity will be the sum effect of the individual sources with fractional masses M_1, M, etc., and is thus given by

$$\frac{M_1A_1}{\rho}+\frac{M_2A_2}{\rho}\ldots+\frac{MnAn}{\rho}$$

Measured activity is quoted in 'American Petroleum Institute (API) units', which is a scale based upon a defined difference of 200 API units between the responses of standard high and low activity cements. Day-to-day calibration is carried out by means of small radioactive sources of standard activity.

γ-ray logs are used for identifying the presence of shaly strata or for determining the proportion of the shale within otherwise sandy formations. As such, γ-ray logs have a major use in the determination of variation in grain size within sandstone beds. This parameter is of great importance in the environmental interpretation of sandstone reservoirs and in assessing reservoir production characteristics (Selley, 1985a). Moreover, marine petroleum source rocks which have a high U content can be detected and oil shale and tar sand reserves assayed using the results from a γ-ray log. Unfortunately, however, this is not possible for lacustrine source rocks which typically have low γ-ray activities (Meyer and Nederlof, 1984). High activity can also be used to locate K-rich beds (evaporite sequences), and low activity can help in the identification of humic coal seams (although algal, boghead and cannel coals have high γ-ray activities).

In groundwater flow studies in which radioactive tracers are employed, a log must be run before the tracer is injected to determine the background variation, and then at different times after injection to find the time-response and position of tracer breakthrough. Quantification of the system, involving the addition of a tracer of precisely known activity, also allows the percentage recovery of injected tracer to be determined and permits volumetric flow path characteristics to be analysed.

Conventional γ-ray logging normally involves the use of NaI(Tl) scintillation detectors, but the resolution of these is not really adequate for γ-ray spectrometry, bearing in mind the limitations imposed by short counting periods. Although some single-channel γ-ray spectrometers employing NaI(Tl) scintillation crystals have been developed for use in boreholes, low logging speeds are required in order to obtain statistically acceptable data for the very low count rates normally encountered, at least in the Th window. Semiconductor detectors, which must be cooled to liquid N temperature, are difficult to adapt for extended logging in borehole environments, where the ambient temperature may easily be 100–200 °C. A fairly recent innovation in this field has been the introduction of very high efficiency scintillation detectors which incorporate 'BGO' crystals. These are crystals of $Bi_4Ge_3O_{12}$, which give such a good response that they have increased the statistical reliability of the counts obtained to the extent that determination of the ^{232}Th decay series, as well as the ^{238}U decay series and ^{40}K, is possible. Clearly, the ability to identify the radiation originating from these different nuclides is useful; for example, in distinguishing between the activity emitted by an evaporite sequence (in which the source is ^{40}K) and that arising from shales (the ^{238}U decay series). Reliable data are also important in calculations of radiogenic heat production. Calibration in terms of U (ppm), Th (ppm) and K_2O (%) is possible, if it is assumed that equilibrium is present in the ^{232}Th and ^{238}U decay series (as discussed in Chapter 9).

Another point of interest concerning γ-ray logging in petroleum geology is the fact that phosphate-rich beds, with their high U content, can also be detected. Phosphate-rich beds are often found where there has been sediment reworking. As such, they can be of great value in the reconstruction of sedimentary histories. γ-ray logs may thus allow correlations to be made between separate lithological logs and may even reveal specific cycles of sedimentation not easily recognised in the lithological logs.

Finally, we may note the use of γ-ray activity surveys of core that has been removed from a borehole so that core positions can be tied to specific depths indicated by anomalies on the γ-ray log. This is especially useful where core recovery has been uncertain or where drilling and logging depths do not agree.

11.3.2 The density log

Density logs (also known as γ–γ-logs) are used to measure formation densities *in situ* and, thus, to obtain information on rock porosities (Tittman and Wahl, 1965). The technique is based upon the irradiation of the walls of a borehole with γ-rays. Typical γ-ray sources include ^{60}Co and ^{137}Cs, which emit medium–energy radiation at 1.17 Mev and 0.66 MeV, respectively. On entering the formation, the γ-rays interact with the orbital electrons of atoms in the rock; the extent of the Compton scattering that results depending upon the electron density (the number of electrons per unit volume) of the minerals and fluids present. The electron density is, in turn, dependent upon the bulk density of the formation. In the logging tool, a scintillation detector is placed at a fixed distance from the source and is carefully shielded from direct radiation. This detector measures the intensity of the backscattered radiation; the greater the intensity, the greater is the electron density of the formation.

The relationship between the electron density ρ_e of an element and the bulk density ρ_b of a solid composed of that element as is given by

$$\rho_e = \frac{\rho_b(2Z/W)}{N},$$

where Z is the atomic number of the element and W is its atomic weight. $N(= 6.02 \times 10^{23})$ is Avogadro's number. Interestingly, for a wide range of elements, $2Z/W \approx 1$, for example, in Si (0.9968), O (1.0000) and Ca (0.9980). In such cases, when the proportionality constant N is disregarded, the bulk density is approximately equal to electron density. Rocks are composed of minerals which are the macroscopic expression of clusters of molecules of compounds, and so a similar expression may be given for a mineral:

$$\rho_e = \frac{\rho_b(2\Sigma Z/M)}{N},$$

where ΣZ is the sum of the atomic numbers of the elements in the mineral and M is its molecular weight. For rocks, most of which consist of several minerals, the relative proportions of each mineral phases present must be taken into account:

$$\rho_e = \frac{\rho_b}{N}\left(\frac{2\Sigma Z_1 x_1}{M_1} + \frac{2\Sigma Z_2 x_2}{M_2} \cdots + \frac{2\Sigma Z_i x_i}{M_i}\right),$$

where the suffixes 1, 2, ..., i represent the different mineral phases and x is the proportion of each phase in the rock. Clearly, it does not matter whether these phases are solid minerals or pore fluids (Schlumberger, 1972).

For liquid-filled pores in limestones and sandstones, the electron density and the bulk density are almost identical, whether the liquid is oil or water. Thus, for a non-shaly formation with a porosity ϕ made up of minerals with an average density ρ_{mi} and pores filled with liquid of density ρ_l,

$$\phi = \frac{\rho_{mi} - \rho_b}{\rho_{mi} - \rho_l},$$

where ρ_b can be obtained as ρ_e, and ρ_{mi} and ρ_l can be assigned from a knowledge of the formation as a whole (from the results of other logging methods or core, for example).

There may be some difficulty in assigning a true value of ρ_l, owing to the limited penetration of the γ-rays into the borehole wall and the limited thickness of rock for which ρ_l is, therefore, determined. The logging tool usually employs two detectors placed at different distances from the source (the so-called 'short' and 'long' detectors). Two detectors are needed because, although both source and detector are mounted on a skid which is held against the borehole wall by a caliper arm, there are still borehole effects which must be taken into account in the interpretation of the results; for example, although the skid is designed to cut through the mud cake which coats the borehole wall, some of the mud cake is likely to remain and be recorded as part of the formation. Using two detectors, however, the effect of the mud cake can be allowed for by recording the setting of the caliper arm as the log is run to determine the position of the skid with respect to the borehole wall. However, if the density of the mud cake is known (and it will vary with the different compositions of mud used during drilling), an additional check is possible. Even with no mud cake and the skid right up against the rock, the penetration of the γ-rays is limited; moreover, the return of the scattered radiation is also only possible through a limited thickness of rock. As the typical range for γ-rays in rock is only a few centimetres, it is not surprising that the overall value of ρ_e is only determined for a few centimetres of rock behind the borehole wall. Unfortunately, it is precisely this zone which is most affected by the drilling mud as the borehole is sunk. During drilling, the mud is usually kept at a greater pressure than that of the formation fluids so that drilling is not affected by the ingress of formation fluids into the borehole. There is a tendency, therefore, for the mud to pass from the borehole into the formation. In so doing, the particles of clay, etc., making up the mud get trapped (to form the mud cake), while the mud filtrate passes through to invade the rock. This filtrate is composed of comparatively fresh water and may displace saline formation water or oil. Thus the overall composition of the pore liquids in the rock is changed. Moreover, because the displaced fluids may not be totally removed, the components can be present in various proportions which will change with distance behind the borehole wall. The density of the liquid filling the pores in the invaded zone is thus intermediate between the density of the mud filtrate and the density of the formation fluid, but its actual value cannot be determined directly by a density log. Fortunately, however, very clear information about the character of the invaded zone can be obtained from electrical resistivity logs, and that data may allow a reliable value of ρ_l to be determined (Serra, 1984).

Primary calibration of the device is by reference to fresh-water-saturated limestones of high purity and accurately known density. Secondary calibration standards consist of large Al and S blocks, while day-to-day calibration is performed by checking the response of the detectors to a standard radioactive source. Because there is, again, a need to reduce counting errors to statistically acceptable levels, the input signal is smoothed over a time constant in the same way that the natural γ-ray activity is smoothed. In the case of density logs, the smoothing interval may be 2, 3 or 4 s, depending upon the formation densities encountered. For low density (that is high porosity) rocks, a 2 s interval is satisfactory, but high density (low porosity) rocks may require a 4 s interval. The logging speed must reflect the time interval used and, as before, a speed of about 0.3 m per unit time interval is taken as the maximum.

The output from the log is given directly as a density value and is usually accompanied by a record of the density compensation factor that has been applied in correcting for the mud cake thickness and borehole characteristics, and a caliper log. Interpretation can be made more complex by the presence of hydrocarbons as a pore fluid and of clay or shaley formations. The effect of oil is not usually noticeable. However, the presence of gas (which is considerably less dense than formation water), or of oil saturated with gas, will produce an apparently large porosity which can be used as an indicator of the presence of gas, and from which the gas density can be calculated. Similarly, because clay particles usually have a lower density than other formation minerals, the presence of dispersed clay within an otherwise arenaceous formation will give rise to apparently high porosity values.

11.3.3 The litho-density log

Similar to the conventional density log is the litho-density log, in which a ^{137}Cs source provides 0.66 MeV γ-rays which undergo Compton scattering in the formation and lose energy. The resulting low energy photons, in turn, interact with the formation by photoelectric absorption. The probability that a γ-ray photon will be involved in one of these interactions is governed by the photoelectric cross-section of the material. For a single element, the cross-section σ_e is given by

$$\sigma_e = \frac{KZ^n}{(E_\gamma)^m} \text{ barns atom}^{-1},$$

where K is a constant, the exponents n and m are functions of the energy E_γ of the γ-ray and Z is the atomic number of the target. Empirically,

$$\sigma_e = \frac{12.1Z^{4.6}}{E_\gamma^{3.15}},$$

for ordinary sedimentary rocks. The photoelectric absorption index is given by

$$P_e = \frac{\sigma_e}{KZ},$$

which, under most circumstances, can be taken as

$$P_e = \left(\frac{Z}{10}\right)^{3.6}.$$

However, rocks are not composed of single elements, and so the bulk cross-section P_b is required:

$$P_b = G/\rho_{be},$$

where ρ_{be} is the bulk electron density, made up of individual electron densities ρ_e and

$$G = \Sigma V_i P_{ei} \rho_{ei}$$

for each individual component i, where V is the fractional volume (Serra, 1984).

Two γ-ray detectors are used at different distances from the source, as before. In this case, however, the recorded intensity varies with the γ-ray energy because, although the high energy count is determined only by the electron density of the formation, the low energy count varies with both the electron density and the photoelectric cross-section. The long-spaced detector records both the high and the low energy parts of the signal, while the short-spaced detector records only the high energy signal. The ratio of the two outputs is a function of G/ρ_e. Problems arise from formations rich in U or Th which produce high counts and from the presence of saline formation water. Nevertheless, the method is extremely useful as an indicator of the lithology of the various formations encountered by a borehole.

11.3.4 The neutron log

Neutron logging tools also incorporate a radioactive source (Tittle and Allen, 1966). Typically this is a ^{226}Ra–Be, ^{239}Pu–Be or ^{241}Am–Be system, in which α-particles emitted by the heavy nuclide bombard targets of ^{9}Be, producing ^{12}C and neutrons with a mean energy of 4.5 MeV:

$$^{9}_{4}\mathrm{Be} + ^{4}_{2}\mathrm{He} \rightarrow ^{12}_{6}\mathrm{Be} + ^{1}_{0}\mathrm{n}.$$

Ra–Be sources typically produce 4.5×10^6 neutrons s^{-1}, while Pu–Be sources-produce 8.5×10^6 neutrons s^{-1} and Am–Be sources 10^7 neutrons s^{-1} (Serra, 1984). When these fast neutrons collide with the nuclei of atoms in the formation, they lose some of their energy, the amount depending upon the relative mass of the nuclei involved. Most energy is lost by a neutron when it collides with another object of the same mass. However, nuclei with the same mass as that of a neutron are not common; in fact, only ^{1}H will meet this condition and act as a braking mechanism. Simply stated, it is the amount of H in the formation that controls the neutron energy. In the case of a formation containing a reasonable proportion of H, fast neutrons emitted by the source are soon degraded into thermal neutrons with energies of only a few hundredths of an electronvolt. Subsequently, the thermal neutrons do not lose further energy but move randomly through the formation until they are captured. Any nucleus which captures a thermal neutron becomes excited, causing the emission of a high energy γ-ray. By measuring the γ-ray activity generated in this way (or, alternatively, by measuring the thermal neutrons) it is possible to obtain a means of measuring the amount of H in a formation.

Where the formation contains a large amount of H, the transition from fast to thermal neutrons, and the subsequent capture of the thermal neutrons, takes place close to the neutron source. Thus, if a detector is placed some distance from the source, the count rate recorded will be low. However, if the H concentration in the formation is low, greater penetration of the fast neutrons occurs before the chain of events leading to the count recorded by the detector is set in motion. Consequently, the count will be high.

Different designs of neutron logging system are in operation, some non-directional and some with a directional source mounted on a side-wall pad. Where non-directional systems are used, they are usually run offset from the centre of the borehole so that the effect of the drilling mud is minimised. As noted before, different systems may employ γ-ray or thermal neutron detectors or both. Several detectors at different distances from the source are also employed in the logging tool so that optimum conditions can be chosen for any particular type of formation. The length of borehole logged depends upon the source–detector spacing (typically less than 0.6 m) and the logging speed; at zero logging speed, the length is approximately this distance, but at normal logging speeds the length increases somewhat and resolution is impaired. The thickness of rock investigated depends upon the formation porosity and the H content but is usually less than 0.3 m. Depending upon the type of logging tool, surveys can be run in either cased or uncased holes but, where a side-wall pad is used, the tool can only be run in an uncased hole (Schlumberger, 1972).

Once again, the problem of maximising logging speed while minimising statistical error arises, so that the data collected over a time constant are averaged, and logging speeds are kept below 0.3 m per unit time constant. A time constant of 2 s generally provides satisfactory results in most circumstances but, for large-diameter holes or where the formation response is low, longer time constants are needed.

Primary calibration is based upon the tool's response to a water-saturated limestone with 19% porosity at the American Petroleum Institute neutron pit at Houston. This is defined as giving 1000 API units relative to a zero background. Secondary calibration at well sites employs an environmental calibrator designed to give specific responses to the porosity of pure limestone. The output of the log may, therefore, be either in API units or as a porosity.

In practice, it is the H present in the formation fluid that mainly controls the response of the system, whether the H is within water or oil. Indeed, the hydrogen indices of oil and water are very similar. The hydrogen index HI is defined as 1 for fresh water, and variation in its value can only be used to determine porosity if corrections for salinity and gaseous hydrocarbon content are applied. In uncased holes, invasion of the formation by the mud filtrate means that fresh water does, in fact, saturate the pores, but in cased holes, in which it is possible that saline formation water has returned, the hydrogen index is given by

$$\mathrm{HI} = 1 - 0.4c$$

where c is the weight concentration of NaCl at 24 °C. For gaseous hydrocarbons, the hydrogen index varies with composition. For a molecule CH_n,

$$\mathrm{HI} = \frac{9n}{12+n}\rho_{\mathrm{hy}}$$

where ρ_{hy} is the density of the hydrocarbon concerned. Oil, for example, for which n is, say, 2, has a HI of 1.29 ρ_{oil}, so if $\rho_{\mathrm{oil}} = 0.85$ (a typical figure), HI = 1.09 (Schlumberger, 1972). Where gas is present, the response of the neutron log is such that an apparently very low porosity is indicated. This can be a useful guide to the location of gas zones in a formation known to be uniform, from the results of other logging methods, for example. Another source of complexity is the presence of H in water bound up in clay minerals. Because the response of the system depends upon the total amount of H present, and not just that in the pore fluid, any interpretation in terms of porosity will be in error if the matrix minerals also contain appreciable quantities of H. Clearly, therefore, shales will appear to have a porosity in excess of the true value. The problem is not confined to shales, however, as any mineral which contains H will give rise to error; even gypsum horizons ($CaSO_4{\cdot}2H_2O$) may appear to have significant porosity. Nevertheless, the neutron log is extremely valuable for recording porosity, especially if run with a directional tool in an uncased hole. Used in conjunction with a γ-ray log, γ-ray spectrometry and a density log, it can yield valuable information about lithology and locate the presence of gas. For example, when density and neutron logs are run together and the results displayed on a common calibrated scale for sandstones and limestones, any separation of the results indicates either shale (when the density value is less than the neutron response) or gas (when the density is greater than the neutron response).

11.3.5 The thermal decay time log

Thermal decay time logs also work on the basis of the conversion of fast neutrons to thermal neutrons in the formation, and the subsequent capture of the thermal neutrons (Schlumberger, 1976). In this case, however, it is the nature of the capture that is of interest. Clearly, the rate at which thermal neutrons are captured depends upon the composition of the capturing medium. Where typical rocks and formation fluids are concerned, this is largely determined by the abundance of Cl. Of all the elements likely to be encountered, Cl is considerably the strongest absorber of thermal neutrons. In this logging system, the neutron source is a pulsed generator which emits 14 MeV neutrons. The generator consists typically of an accelerator containing ions of ^{2}H and ^{3}H which collide and are then absorbed onto a C target:

$$^{2}_{1}\mathrm{H} + ^{3}_{1}\mathrm{H} \rightarrow ^{4}_{2}\mathrm{H}_{e} + ^{1}_{0}n.$$

Each burst lasts about 20 μs and is repeated every 1 ms (Serra, 1984). The proportion of thermal neutrons that remain are sampled on a number of occasions after the output burst so that their decay rate can be determined. Alternatively, the γ-rays emitted after each capture may be monitored to provide the same information.

After the neutron burst, a delay time is introduced by the recording circuit before the first count from the detector is obtained. This is to give the neutrons time to reach thermal velocities and to allow the effects of the borehole itself to die away. If γ-ray detectors are used, a shorter delay is needed than when direct detection of thermal neutrons is employed. For this reason, γ-ray detectors are preferred. To avoid spurious results duc to the natural radioactivity of the formation, only γ-rays with energies above 2.2 MeV are counted. Alternative systems which record lower energies need to incorporate a background correction.

Within the formation, the proportion of thermal neutrons that are captured per unit time is given by $v\sigma$, where v is the neutron velocity and σ is the macroscopic neutron absorption cross-section (the linear absorption coefficient) of the formation. σ is, in effect, the sum of the cross-sections of the individual nuclei per unit volume and is expressed in units of cm^2 cm^{-3} or cm^{-1}. Because the value of σ is inversely proportional to v, the quantity $v\sigma$ is constant and defines the formation. Alternatively, its reciprocal, the intrinsic thermal decay time τ_{int}, may be employed. This can be related to the number N of neutrons remaining after time t:

$$N = N_0 \mathrm{e}^{-t/\tau_{\mathrm{int}}},$$

where N_0 is the initial number of neutrons. Taking a number of successive readings yields

$$N_2 = N_1 \mathrm{e}^{(t_2 - t_1)/\tau_{\mathrm{int}}}$$

or

$$\frac{1}{\tau_{\mathrm{int}}} = \frac{1}{t_2 - t_1} \log \mathrm{e} \left(\frac{N_1}{N_2} \right).$$

For a thermal neutron velocity of 2200 m s^{-1},

$$\tau_{\mathrm{int}} = \frac{4.55}{\sigma} \mu\mathrm{s}.$$

If, after the initial borehole disturbance has passed, the relative count rate (on a logarithmic scale) is plotted against time (on an arithmetic scale), a linear relationship is obtained which eventually decays to background. This linear section, which lasts from about 200 to 1000 μs after the neutron burst, yields the decay time. However, the effect of neutron diffusion must be taken into account in understanding the duration of this linear response. When τ_{int} for the formation is greater than that for the borehole mud (that is, the neutron energies decay more rapidly in the mud), the neutron density in the formation builds up and a neutron current flows from the formation into the mud. This neutron current is determined mainly by the value of τ_{int} for the formation, but the nature of the borehole mud, the presence and thickness of casing and so forth all make a contribution to the current. When the values of τ_{int} are reversed, the neutron current flows from the borehole into the formation. A correction for this diffusion current must be applied when determining τ_{int} from the measured decay time (Schlumberger, 1972).

The exponential decay of the count is typically obtained by means of three measurement gates (Fig. 11.5). The first gate opens $2\tau_{int}$ after the neutron burst and remains open for τ_{int}. At the same time as the first gate closes, a second gate opens for a period of $2\tau_{int}$. These two gates measure the decay but, after an interval of τ_{int}, a third gate opens for a period of $3\tau_{int}$ to count the background. Recording thus depends upon the value of τ_{int}, and an optimisation procedure is used which allows recording details to be continuously varied. Typically, three traces are given as the output: the thermal decay time τ, the corresponding value of the absorbtion cross-section σ (determined as $\sigma_{20°C} = 4.55/\tau_{measured}$), and the casing collar log (an electromagnetic system for locating borehole casing collars at which the casing is extra-thick). In addition, the natural γ-ray count may be given as a separate log. Calibration of the system can be carried out using a large tank filled with pure water, which should yield a $\tau_{measured}$ value of 195 μs. For day-to-day checks, a small radionuclide source is used to determine the response of the detectors.

Most sandy lithologies have been found to

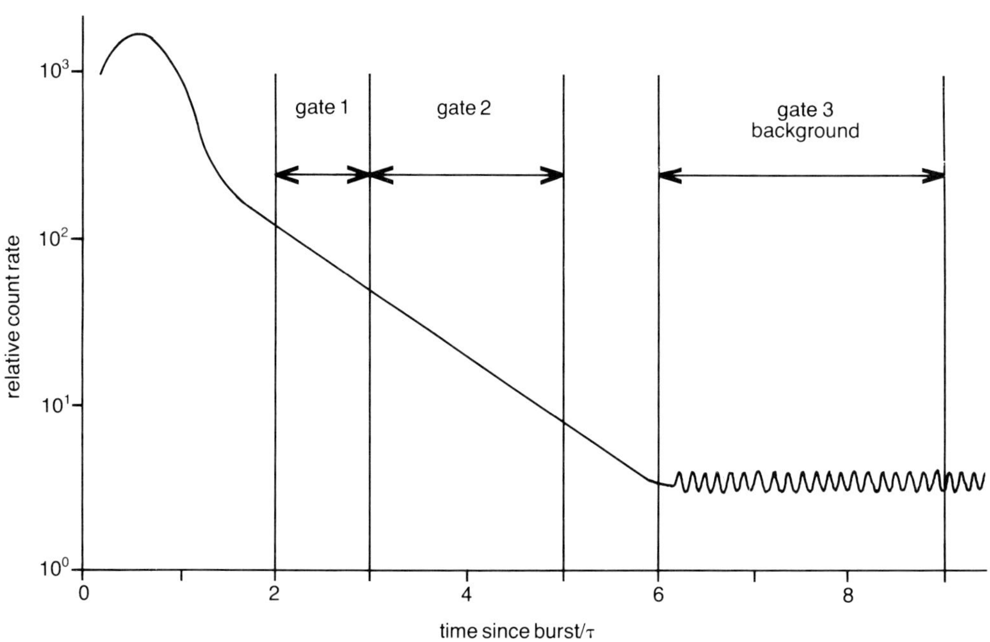

Fig. 11.5 — Characteristic response of the thermal decay time log, NB and the sliding measurement windows. (Serra, 1984, redrawn with permission.)

have a matrix absorption decay time τ_{ma} of about 500 μs, while the most common value for limestones is about 550 μs. The presence of saline formation water, however, dramatically affects the value of $\tau_{measured}$ owing to the presence of Cl, although neither changes in temperature nor changes in pressure produce appreciable differences. Interestingly, the decay time for oil is about the same as that for fresh water (210 μs); in detail, this varies with composition, pressure, temperature and amount of dissolved gas present. Gas on its own has a decay constant which is extremely sensitive to pressure and temperature (Serra, 1984).

The most important use of the thermal decay time log is to distinguish between oil-bearing and saline-water-bearing formations. If the porosity of a formation is known, gas–oil contacts can be located. Water saturation can be determined if the porosity, hydrocarbon composition, pore water salinity and lithology are known. The ability to use the method in cased holes is extremely important, as it permits the behaviour of production wells to be monitored. Problems can arise, however, from the presence of trace quantities of B, Li and rare earth elements, which have large capture cross-sections; 50 ppm B, for example, has the same effect as 6100 ppm NaCl. For this reason, the formation water salinities yielded by this method are always on the high side. Clay minerals also generally contain some B and, in addition, much water. Consequently, they have a high capture cross-section, and so the shaliness of a formation needs to be taken into account before the results can be correctly interpreted.

11.3.6 The induced γ-ray spectrometry log

Although, as noted above, Cl is the main absorber of thermal neutrons in sedimentary formations, many other nuclides also respond in this way, each emitting γ-rays of characteristic energy when they capture a neutron. The characteristic γ-ray spectra of excited nuclei can thus be used to identify and determine the abundance of the nuclides present. H, for example, produces a single peak at 2.2 MeV, while Fe produces a peak at 7.7 MeV. The logging tool is, therefore, very similar to that used for thermal decay time logging, employing a neutron generator which emits 14 MeV neutrons, and a γ-ray detector. Here, however, the energies of the recorded γ-rays are analysed.

Nuclear reactions are produced not only by capture of thermal neutrons but also by inelastic scattering, by activation and by reaction with fast neutrons. Inelastic scattering involves the transfer of energy from an incident neutron to the target nucleus. This excites the target but leaves the neutron to continue. The excited nucleus undergoes rapid return to the ground state with the emission of γ-radiation of specific energy or energy spectrum. For example, ^{12}C emits a 4.44 MeV photon. Activation occurs when the target nucleus changes to that of an unstable nuclide with a relatively long half-life. Again, the product emits a characteristic γ-ray spectrum. The reaction mechanism is similar but, in this case, annihilation of the neutron takes place, and α-particle emission occurs to produce an excited daughter. γ-rays emitted as a result of these fast neutron interactions can be used to determine C, O, Si, Ca, Fe and S, while thermal neutron capture is used for spectral analysis for Cl, H, Si, Ca, Fe and S. Because shaliness also affects the recorded response, automatic correction based upon the ratio of the H concentration relative to Cl is used in the shale-compensated log (Serra, 1984). This is illustrated in Fig. 11.6.

As it is difficult to record full energy spectra using NaI(Tl) detectors, specific energy windows are employed as described in Chapter 9 for conventional γ-ray spectrometry, but, instead of using only a few such windows, an energy analyser with 256 channels is typically employed. Because the timing of the response of the formation following fast neutron interactions and thermal neutron captures are different, the logs for each system must be run separately, with different time constants for the two modes. In the fast neutron mode, measurements are made at the same time as the neutron burst. Even during this short period, however, some thermal neutrons will be formed and captured, and so an early 'capture background' gate is opened immediately. A late capture gate measures the response of the formation and the borehole to capture events. The capture mode operates in a very similar manner to the thermal decay time log. Because of the statistical variation in the response, the logging tools have to be run very slowly, perhaps $0.05\ m\ s^{-1}$ for the capture mode and only $0.0025\ m\ s^{-1}$ for the fast neutron mode. Indeed, the fast neutron mode is often used in a stationary manner, data being collected at point stations over intervals of 5 min. High resolution spectroscopy logs using Ge detectors have to be run at about the same speed as those using NaI(Tl) detectors because of their poorer efficiency but yield a much greater number of elements with greater accuracy. They must, of course, be operated at liquid N temperature.

Variations in the abundances of elements can be used to detect lithological boundaries, variations in porosity and changes in the character of pore fluids and can even yield a mineralogical composition which, in turn, can be used for lithological analysis.

11.3.7 Combination of logs and formation evaluation

We have already noted that the combination of logs of different character can yield information which is not directly obtainable from the results of single logs. In this context, logging systems which either measure the natural radioactivity of the formations or employ a radioactive source in order to measure some other parameter, dealt with in this chapter, can be used in combination with one-another or with systems which are based on different physical or chemical properties.

Where there is no interference between the different logging systems, they may be run at the same time, but, clearly, where one measurement technique affects the results obtained by another, they must be run separately.

The use of data obtained from different logs to derive some additional parameter is known as 'cross-plot' analysis. Apart from the application of cross-plot analysis of neutron and

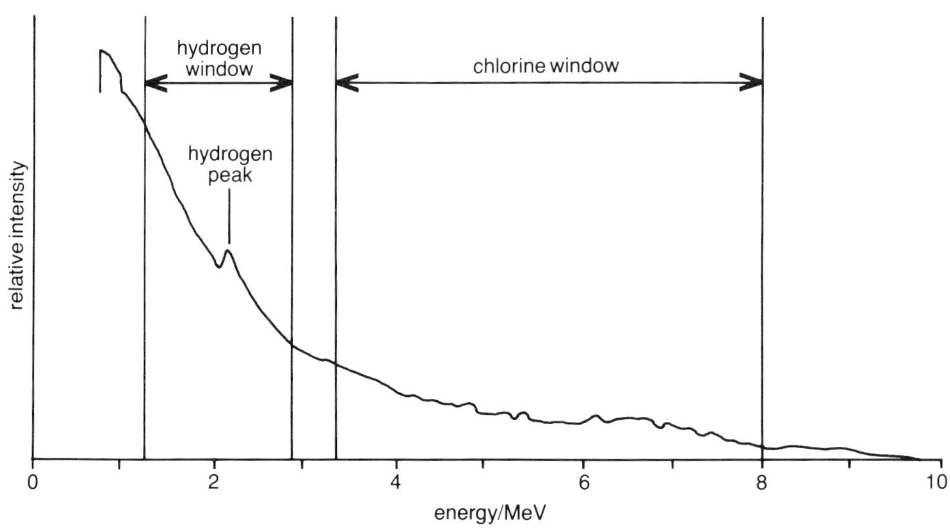

Fig. 11.6 — A typical γ-ray spectrum recorded by a shale-compensated chlorine log, NB and the positions of the H and Cl windows. (After Fletcher and Walter, 1978.)

density logs to yield shaliness or gas content, as already considered, a combination of neutron porosity and bulk density will also permit determination of porosity in relation to lithology. Another especially useful cross-plot, which allows the identification of specific minerals, also involves two of the logging methods considered in this chapter. This cross-plot is based upon the results from a litho-density log giving the photoelectric cross-section and the γ-ray spectrometer log giving the Th/K ratio. A reasonably unique combination of these values (quoted as P_e, $Th/(K \times 10^{-4})$) distinguishes, for example, between glauconite (6.5, 0.9), biotite (6.3, 2.5), chlorite (6.4, 20), feldspar (2.9, 0.65), illite (3.5, 2.5), muscovite (2.5, 2.5) and kaolinite (1.9, 20). In most cases, each of these values is the median of a range of only about 10%. However, montmorillonite (3, 5.5), because of the swelling nature of its lattice in the presence of water, has a more variable photolectric cross-section, with a range of approximately 30% about the median.

Because of the ability to run different logs at the same time to yield additional information, a new nomenclature has arisen. Thus the combination of neutron and litho-density is known by Schlumberger (1983) as the compensated neutron–litho-density log, while the litho-analysis log incorporates γ-ray spectrometry, compensated neutron and litho-density logs. This combination is particularly powerful as these logs allow determination of mineralogy and porosity, and thus interpretation of lithological components within a formation. Similar Schlumberger logs incorporate results from electrical resistivity, sonic transit time and electromagnetic propagation logging systems with γ-ray spectrometry and litho-density to derive dependable porosity values, hydrocarbon type and amount, permeability analysis and volumetric proportions of matrix and pore fluids.

Clearly, in these applications, computerised handling of the data output from the individual logging systems is needed to provide an integrated interpretation, but the complexity of the processing varies very significantly with the overall nature of the tasks involved. For example, in the neutron–litho-density combination the computation progresses step-by-step in the same way as if a cross-plot interpretation were carried out manually. However, where there are many different parameters involved, a more powerful approach which permits simultaneous solution is needed. A good example of this is given by Schlumberger's GLOBAL processing programme (Mayer and Sibbit, 1980), which incorporates, for instance, multidimensional cross-plot analysis. As a result, lithological models are determined which involve any combination of the common petroleum reservoir rocks and clay. In addition, as many as four other non-clay minerals can be included to allow variation in such components as the heavy minerals or micas to be modelled. On the basis of the model that has been chosen, response equations for the different logging systems are defined for the petrophysical properties of interest. Obviously, more response equations than unknowns are needed. After solution of these equations a minimisation routine is used to obtain the results which most closely match the actual measurements made at a particular level in the borehole. These are then output as the most reliable interpretation, although also output is a quality control function which indicates the degree of matching obtained. Where agreement is not good, more data (from additional logging methods) may be needed to permit reasonable interpretations to be made. Sedimentary environment analysis (Selley, 1985b) is thus possible using the results from these geophysical borehole logging methods (Serra, 1985).

BIBLIOGRAPHY

Adams, J. A. S., and Weaver, C. E (1958) Thorium to uranium ratios as indicators of sedimentary processes, example of concept of geochemical facies. *Bull. Am. Assoc. Petrol. Geol.* **42**, 387.

Asquith, G. B., and Gibson, C. R. (1982) *Basic well log analysis for geologists. Tulsa American Association of Petroleum Geologists*

Barker, C. (1977) Aqueous solubility of petroleum as applied to its origin and primary migration. *Bull. Am. Assoc. Petrol. Geol.* **61**, 2146.

Barker, R. D., Lloyd, J. W., and Peach, D. W. (1984) The use of resistivity and gamma logging in lithostratigraphical studies of the Chalk in Lincolnshire and south Humberside. *Q. J. Eng. Geol.* **17**, 71.

Beers, R. F., and Goodman, C. (1944) Distribution of radioactivity in ancient sediments. *Bull. Geol. Soc. Am.* **55**, 1229.

Bell, K. G., Goodman, C., and Whitehead, W. L. (1940) Radioactivity of sedimentary rocks and associated petroleum. *Bull. Am. Assoc. Petrol. Geol.* **24**, 1529.

Bloch, S., and Key, R. M. (1981) Modes of formation of anomalously high radioactivity in oil-field brines. *Bull. Am. Assoc. Petrol. Geol.* **65**, 154.

Boyle, R. W. (1982) *Geochemical prospecting for thorium and uranium deposits*. Amsterdam, Elsevier.

Brandon, H. R., and Osoba, J. S. (1956) Spectral gamma ray logging. *Trans. Am. Inst. Min. Eng.* **207**.

Caldwell, R. L., and Sippel, R. F. (1958) New developments in radioactive well-logging research. *Bull. Am. Assoc. Petrol. Geol.* **42**, 159.

Connan, J. (1974) Time–temperature relation in oil genesis. *Bull. Am. Assoc. Petrol. Geol.* **58**, 2516.

Dahlberg,. E. G. (1982) *Applied hydrodynamics in petroleum exploration*. Berlin, Springer.

Davidson, C. F., and Bowie, S. H. U. (1951) On thucolite and related hydrocarbon–uraninite complexes with a note on the origin of the Witwatersrand gold ores. *Bull. Geol. Surv. Gt. Brit.* **10**, 74.

Edmundson, H., and Raymer, L. L. (1979) Radioactive logging parameters for common minerals. *Log Anal.* **20**, 38.

Erickson, R. L., Myers, A. T., and Horr, C. A. (1954) Association of uranium and other metals with crude oil, asphalt and petroliferous rock. *Bull. Am. Assoc. Petrol. Geol.* **38**, 2200.

Fertl, W. H., Welker, D. W., and Hopkinson, E. C., (1978) The Dresser Atlas Spectralog — a look at basic principles, field applications and interpretative concepts of gamma ray spectral logging. *Dresser Atlas Publ.* **3332.**

Fletcher, J. W., and Walter, J. (1978) A practical shale compensated chlorine log. *Proceedings of the Nineteenth Annual Symposium of Society of Petroleum Well Log Analysts,* Paper GG.

Galloway, W. E., and Hobday, D. K. (1983) *Terrigenous clastic depositional systems: application to petroleum exploration*. New York, Springer.

Giardini, A. A., Melton, C. E., and Mitchell, R. S. (1982) The nature of the upper 400km of the Earth and its potential as the source for non-biogenic petroleum. *J. Petrol. Geol.* **5**, 173.

Givens, W. W., Mills, W. R., Dennis, C. L., and Caldwell, R. L. (1976) Uranium assay logging using a pulsed 14-MeV neutron source and detection of delayed fission neutrons. *Geophysics* **41**, 468.

Glenn, W. E., and Hohmann, G. W. (1981) Well logging and borehole geophysics in mineral exploration. In Skinner, B. J. (ed.), *Econ. Geol., Seventy-Fifth Anniversary Volume (1905–1980)*.

Guyod, H. (1969) *Geophysical well logging*. Houston, Texas, Guyod.

Hedberg, H. D. (1964) Geologic aspects of origin of petroleum. *Bull. Am. Assoc. Petrol. Geol.* **48**, 1755.

Heinrich, E. W. (1958) *Mineralogy and geology of radioactive raw materials*. New York, McGraw-Hill.

Hood, A., Gutjahar, C. C. M., and Heacock, R. L. (1979) Organic metamorphism and the generation of petroleum. *Bull. Am. Assoc. Petrol. Geol.* **59**, 986.

Jones, R. W. (1981) Some mass balance and geological constraints in migration mechanisms. *Bull. Am. Assoc. Petrol. Geol.* **65**, 103.

Keen, C. E., and Lewis, T. (1982) Measured radiogenic heat production in sediments from continental margin of eastern North America: implications for petroleum generation. *Bull. Am. Assoc. Petrol. Geol.* **66**, 1402.

Kraemer, T. F. (1981) ^{234}U and ^{238}U concentration in brine from geopressurised aquifers of the north Gulf of Mexico Basin. *Earth Planet. Sci. Lett.* **56**, 210.

Kraemer, T. F., and Reid, D. F. (1984) The occurrence and behaviour of radium in saline formation water of the US Gulf Coast region. *Isotope Geosci.* **2**, 153.

Landa, E. R., and Reid, D. F. (1983) Sorption of radium 226 from oil-production brine by sediments and soils. *Environ. Geol.* **5**, 1.

Lane Wells (1966) *The gamma ray neutron log*. Lane Wells Technical Bulletin.

Lawrence, T. D. (1981) Continuous carbon/oxygen log interpretation techniques. *J. Petrol. Technol.* **33**, 1394.

Levorsen, A. J. (1954) *Geology of petroleum*. San Francisco, California, Freeman.

Lopatin, W. V. (1971) Temperature and geologic time as factors in coalification. *Dokl. Acad. Sci. USSR Earth Sci.* **3**, 95.

Magara, K. (1978) *Compaction and fluid migration: practical petroleum geology*. Amsterdam, Elsevier.

Mayer, C., and Sibbit, A. (1980) GLOBAL, a new approach to computer-processed log interpretation. *Paper Soc. Petrol. Eng.* **9341**.

Meyer, B. L., and Nederlof, M. H. (1984) Identification of source rocks on wireline logs by density/resistivity and sonic transit time/resistivity crossplots. *Bull. Am. Assoc. Petrol. Geol.* **68**, 121.

McKinlay, P. F., and Tanner, H. L. (1975) The shale compensated chlorine log. *J. Petrol. Technol.* **27**, 164.

Memper, J. A. (1978) Oil migration limitations suggested by geologic and geochemical considerations. *Am. Assoc. Petrol. Geol. Short Course Note Ser.* **8**.

Pacey, N. R. (1984) Some aspects of the natural radioactivity of the English Chalk. *Mod. Geol.* **8**, 199.

Philippi, G. T. (1966) On the depth, time and mechanism of petroleum generation. *Geochim. Cosmochim. Acta* **29**, 1021.

Pirson, S. J. (1963) *Handbook of well log analysis for oil and gas formation evaluation*. New York, Prentice-Hall.

Pirson, S. J. (1983) *Geological well log analysis*. Houston, Texas, Gulf.

Price, L. C. (1976) Aqueous solubility of petroleum as

applied to its origin and primary migration. *Bull. Am. Assoc. Petrol. Geol.* **60**, 213.

Price, L. C. (1977) Aqueous solubility of petroleum as applied to its origin and primary migration (reply). *Bull. Am. Assoc. Petrol. Geol.* **61**, 2149.

Russell, W. L. (1945) Relation of radioactivity, organic content and sedimentation. *Bull. Am. Assoc. Petrol. Geol.* **29**, 1470.

Schlumberger Ltd. (1972) *Log interpretation: principles.* New York, Schlumberger Ltd.

Schlumberger Ltd. (1974) *Log interpretation: applications.* New York, Schlumberger Ltd.

Schlumberger Ltd. (1976) *The essentials of thermal decay time logging.* New York, Schlumberger Ltd.

Schlumberger Ltd. (1983) *Schlumberger open hole services catalogue.* New York, Schlumberger Ltd.

Selley, R. C. (1985a) *Elements of petroleum geology.* New York, Freeman.

Selley, R. C. (1985b) *Ancient sedimentary environments.* London, Chapman and Hall.

Serra, O. (1984) *Fundamentals of well-log interpretation.* Amsterdam, Elsevier.

Serra, O. (1985) *Sedimentary environments from wireline logs.* New York, Schlumberger Ltd.

Smith, N. G., Kyle, J. R., and Magara, K. (1983) Geophysical log documentation of fluid migration from compacting shales: a mineralization model from the Devonian strata of the Pine Point area, Canada. *Econ. Geol.* **78**, 1364.

Stroud, S. G., and Schaller, H. E. (1960) New radiation log for the determination of reservoir salinity. *Petrol. Technol.* **2**, 37.

Sverjensky, D. A. (1984) Oil field brines as ore-forming solutions. *Econ. Geol.* **79**, 23.

Swanson, V. E. (1960) Oil yield and uranium content of black shales. *Prof. Paper US Geol. Surv.* **356**.

Tissot, B. P., and Welte, D. H. (1978) *Petroleum formation and occurrence.* Berlin, Springer.

Tittle, C. W., and Allen, L. S. (1966) The theory of neutron logging. *Geophysics* **31**, 214.

Tittman, J., and Wahl, J. S. (1965) The physical foundation of formation density logging (gamma–gamma). *Geophysics* **30**, 284.

Waples, D. W. (1980) Time and temperature in petroleum formation: application of Lopatin's method to petroleum exploration. *Bull. Am. Assoc. Petrol. Geol.* **64**, 916.

Williamson, W. H., and Wolley, D. H. (1980) *Hydraulic fracturing to improve the yield of bores in fractured rock.* Canberra, Australian Government Publication Service.

Wyllie, A. W. (1985) *Nuclear assaying of mining boreholes.* Amsterdam, Elsevier.

Youngblood, W. (1980) The application of pulsed-neutron decay time logs to monitoring water floods with changing salinity. *J. Petrol. Technol.* **32**, 957.

12

Environmental radioactivity

There is a general awareness of the dangers of radioactivity, and it is often thought that, if radioactivity can be detected at all in an environment, it must be harmful. This public opinion is widespread and takes little regard of the intensity of the activity, the type of emissions involved or the duration of exposure, all of which are important in determining the risk of the harmful biological effects that can result from exposure to radiation arising from many sources, both natural and artificial. Indeed, throughout the history of life on Earth, many organisms have been exposed to natural sources of radioactivity and are, therefore, generally able to tolerate a certain level of activity without noticeably harmful consequences. Occasionally, however, an individual organism may be more susceptible to radiation than the population as a whole and may suffer the damage that leads to illness, death or the creation of mutated offspring. Because of this, in radiological protection it is conventional to postulate that there is no threshold which may be considered the upper limit of a safe level of exposure.

This chapter will review the main sources of natural and artificial radioactivity, the impact of radioactivity on man and possible future developments.

12.1 NATURAL SOURCES OF RADIOACTIVITY

12.1.1 Cosmic rays

This radiation, which originates outside the Earth, is also known as 'galactic radiation'. Although the origin of cosmic rays is not known with certainty, it appears that the main sources are the Sun and the Galaxy. The component which comes from outside the Solar System consists chiefly of high energy protons, while the solar component consists of a continuum flux of low energy particles which, during times of solar flares, becomes supplemented by high energy particles. Apart from protons, which make up almost 85% of cosmic ray particles, α-particles are also important (making up about 14%), and there is a smaller component of heavier atomic nuclei (about 1%). Low energy particles which can be stopped by Pb shielding are referred to as the 'soft' component, while higher energy particles make up the 'hard' component.

Cosmic radiation which has not interacted

with the Earth may be termed 'primary'. 'Secondary' cosmic rays, in contrast, consist principally of subatomic particles such as pions, muons and electrons, produced as a result of interaction with the Earth's atmosphere. The cosmic ray flux at sea level is mainly of secondary origin and comprises about 70% muons and about 30% electrons; less than 1% is of primary protons. As discussed in Chapter 3, the intensity of cosmic radiation increases with altitude and varies with latitude, increasing from the equator to latitudes of about 50–60° and then remaining fairly constant to the poles. This variation with latitude is caused by changes in the strength of the Earth's magnetic field and the energy distribution of the primary cosmic ray particles. Temporal fluctuations also occur in the cosmic ray flux with the solar cycle and because of variations in the magnetic field.

For a variety of reasons, therefore, the cosmic ray dose shows considerable variation over the Earth's surface. In Britain, the average sea level dose equivalent for the directly ionising component of the secondary radiation is about 310 μSv year^{-1} but ranges between 290 and 320 μSv year^{-1} with latitude (National Radiological Protection Board, 1981). The increase with altitude is about 5 μSv year^{-1} for every 100 m. Because only about 0.1% of the population of the UK live above 500 m (and only about 6% live between 200 and 500 m), altitude effects are usually ignored in calculating total dosages. The neutron component of secondary cosmic radiation is more difficult to determine but is found to have a dose equivalent rate of about 20 μSv year^{-1} (Hughes and Roberts, 1984). During solar flares, the cosmic ray intensities may be increased tenfold over the normal rate.

12.1.2 Cosmogenic radionuclides

The interaction of cosmic ray particles with the atmosphere produces a wide range of radionuclides. About 70% of these arise in the stratosphere and about 30% in the troposphere. The quantities of these radionuclides arc generally so small that they can be ignored in calculations of dosage, but it must be recalled that some may be incorporated directly into organisms. Particularly important in this respect are ^{14}C and ^{7}Be (Table 12.1). Radionuclides in extraterrestial dust also make a contribution to dosage from nautral sources, but this is so small that it can also be ignored.

12.1.3 Primordial and radiogenic radionuclides

Of the many radionuclides that must have been present when the Earth accreted about 4500 Ma ago, only the long-lived ones are still present today. The most important of these, as far as the natural radiation dose is concerned, are ^{40}K and members of the decay series headed by ^{238}U and ^{232}Th (Table 12.1). Although many of the U and Th decay series radionuclides have short half-lives, they are still important because they are continuously generated by the decay of their long-lived parents. The concentrations of U and Th series nuclides in crustal rocks are comparatively small, and their presence in organic matter is usually even smaller. ^{40}K, however, is fairly abundant, forming about 0.012% of total K, and is widely distributed throughout many different crustal rocks. Moreover, K is an essential element for most life forms so that reasonable levels of ^{40}K are found in organic matter.

External exposure due to U, Th and K in rocks and soils

Because the concentration of ^{238}U, ^{232}Th and ^{40}K in rocks and soils is variable, it is difficult to calculate an average dose from this source. In Britain, the average dose equivalent due to terrestrial sources has been estimated to lie between 270 and 490 μSv year^{-1}, a range based upon both the measured and the estimated dose equivalent rates in the areas of known geological conditions and population distribution (Taylor and Webb, 1979). Clearly, the actual dose will be determined by the nature of the underlying rock, and any statistics which cover a wide area will average the effects of many different lithologies (Hughes and

Table 12.1 — Estimated average annual dose per person arising from internal sources.

	Average annual dose per person ($\times 10^{-6}$ Sv)		
	Whole body	Endosteal (bone)	Bone marrow
^{3}H	0.04	0.04	0.04
^{7}Be	3	—	—
^{14}C	12	16	16
^{40}K	170	80	150
^{87}Rb	6	4	6
^{210}Po	30	210	30
^{222}Rn	30	30	30
^{226}Ra	—	61	3
^{228}Ra	—	70	3

External dose due to local γ-rays estimated at 500, that due to cosmic ray mesons 280 and that due to cosmic ray neutrons 20. (After Spiers, 1968; Brookins, 1984; Hughes and Roberts, 1984.)

Roberts, 1984). Thus the mean value of the average dose equivalents (Table 12.2) for the whole of England (395 μSv year^{-1}) is only slightly greater than that for Northern Ireland (351 μSv year^{-1}) and only a little less than the values for Wales and Scotland (438 μSv year^{-1} for both). Within each of these areas, however, rocks containing above-average levels of radionuclides, such as the granites of south-west England and the Cairngorm Mountains of Scotland, will contribute average annual dose equivalents considerably in excess of these mean values. Although it is little consolation to the inhabitants of such areas, it is important to appreciate that, because these areas are also characterised by low populations, the effect on the population as a whole is very small.

Table 12.2 — Average annual radiation dose in the UK.

		Average annual radiation dose ($\times 10^{-6}$ Sv)
Natural sources		
Cosmic radiation		300
Terrestial γ-rays		400
^{222}Rn daughters		700
^{220}Rn daughters		100
Other internal sources		370
	Total	1870
Artificial sources		
Medical sources		250
Weapons fall-out		10
Artificial discharges		2
Occupational exposure		9
Miscellaneous sources		11
	Total	282

(After Hughes and Roberts, 1984.)

Statistics are available for different parts of Scotland which illustrate the effect of geological variation. In the Edinburgh area, where the geology consists mainly of sedimentary rocks with relatively low radionuclide contents, the outdoor absorbed dose rate is about 485 μGy year^{-1} while in Aberdeen, where the granite has a high radionuclide content, the rate is 1040 μGy year^{-1}. Because the indoor exposure is made up of γ-rays from the building materials and from the ground outside, it is interesting to note that the mean indoor exposure in Edinburgh is 600 μGy year^{-1} (significantly greater than the outdoor level), but in Aberdeen it is 853 μGy year^{-1} (significantly less than the outdoor level). On the assumption that an individual spends 2.5 h per day outdoors and that the ratio of effective dose to dose in air is 0.7, the average dose equivalent has been calculated as 410 μSv year^{-1} in Edinburgh and 610 μSv year^{-1} in Aberdeen. Perhaps of more interest, however, is the implication that, in Edinburgh, the houses are constructed of materials which enhance the level of radiation exposure while, in Aberdeen, the houses give some protection from the external environment even though many are made of granite. Whether this is true or not is, however, difficult to say. In both cities there will be some attenuation of the γ-rays from the underlying ground by the building materials and a contribution to the γ-radiation by the buildings themselves. Which of these is greater depends on such factors as the thickness of floor and walls, the energy spectrum of the γ-rays and the proportions of U, Th and ^{40}K present in the building materials. Clearly, these values also mask a great range experienced by individuals, between 200 and 1000 μSv year^{-1}, depending upon lifestyle (Taylor and Webb, 1979). In the USA, a considerable range in the average dose equivalent is found between different areas (Table 12.3). These values may be compared with the annual radiation dose limits given by the UK National Radiological Protection Board and the International Commission on Radiological Protection (Table 12.4).

Table 12.3 — Annual radiation dose in the USA.

	Annual radiation dose ($\times 10^{-6}$ Sv)
Estimated individual whole-body external γ-ray dose from natural radioactivity	
Lowest state (Texas)	300
Highest state (South Dakota)	1150
Average of all states	600
Estimated individual whole-body cosmic-ray dose	
Lowest state (Hawaii)	300
Highest state (Wyoming)	1300
Average of all states	450
Estimated individual whole-body dose from gaseous reactor effluents	
Average of all states	0.26

(After Klement 1982a.)

Table 12.4 — Annual radiation dose limits.

	Annual radiation dose limit ($\times 10^{-6}$ Sv)
Effective dose equivalent or uniform dose equivalent to the whole body	
Radiation worker	50 000
General public	1 000
Dose equivalent to single organ or tissue	
Radiation worker	500 000
General public	50 000
Dose equivalent to the eye	
Radiation worker	150 000
General public	15 000

(After National Radiological Protection Board, 1981; International Commission on Radiological Protection, 1985.)

Ingestion of radionuclides

Although external exposure to radiation is obviously one way in which terrestrial sources affect man, another is by ingestion or inhalation of naturally occurring radionuclides. This exposure can also vary considerably according to diet, habitat and lifestyle of the individual. Compared with their parent bedrocks, soils are, with the exception of ^{226}Ra, generally depleted in radionuclides. Recycling of ^{226}Ra by organisms in the soil may explain its persistence. Radionuclides in small soil particles may be transported by the wind as dust or aerosols through the atmosphere to plants and animals. Many plants also trap soil within their leaves or roots during growth. Once on plants, they may be ingested directly by man, or by herbivores which may be later eaten by man. Indeed, there may be a complex sequence, depending upon the number of stages in the food chain. While indirect ingestion often leads to some dilution of the effect, this is not always the case. The best-known example of concentration occurs in people who eat reindeer and caribou meat which contain large amounts of ^{210}Po.

Although herbivores may ingest a wide range of radionuclides, it is mainly ^{40}K, ^{87}Rb, ^{210}Pb, ^{210}Po, ^{224}Ra, ^{226}Ra, ^{228}Ra, ^{230}Th, ^{232}Th, ^{234}U and ^{238}U that are assimilated; most other natural radionuclides are poorly absorbed. For example, K accumulates mainly in the soft tissues, while Ra accumulates in Ca-rich structures such as bones. Consumption of only the soft tissues of animals, therefore, effectively removes the Ra source.

Direct ingestion of potentially high levels of radionuclides is also possible in the case of individuals, particularly young children, who actually eat soil. Finally, the presence of dissolved and, to a lesser extent, suspended material in drinking water must also be considered.

When the population is taken as a whole, there is no clear evidence of harmful effects produced by naturally occurring radionuclides in food and water but, because the levels of activity involved are very low, there have been few measurements made and little attempt to study the statistical relationship between the incidences of morbidity and mortality and different levels of natural radioactivity.

One exception concerns the natural radioactivity of drinking waters. This is extremely variable, depending upon both the source of the water and any treatment which it may receive before consumption. In general, groundwater sources have a higher level of radioactivity than surface waters. The range within groundwaters depends upon the natural radioactivity of the underlying geology and the groundwater residence time.

The results of a study of the distribution of cancer in west Devon (Allen-Price, 1960) suggested that the radioactivity of the drinking water in the area might have been responsible for anomalously high incidences of cancer found in certain localities. The activity of drinking waters in the area was found to be up to 500 Bq l^{-1}, mainly attributable to ^{222}Rn (Abbot *et al.*, 1960). As a result, the most active supplies were abandoned.

An investigation of the natural α-activity of drinking waters from various parts of Great Britain revealed particularly high ^{222}Rn concentrations in Cornwall, with activities of up to 400 Bq l^{-1} at St Ives (Turner *et al.*, 1961). In addition to ^{222}Rn, concentrations of ^{226}Ra and other ^{238}U daughters may contribute to the high radioactivity of groundwaters (Kenny *et al.*, 1966); typically these can be up to 800 Bq l^{-1} in the areas underlain by granite in southwest England, although a total activity of over 1000 Bq l^{-1} has been found in one spring (fortunately this is not now used as a source of drinking water). Generally, however, ^{226}Ra concentrations are low in comparison with ^{222}Rn (a maximum activity of 0.17 Bq l^{-1}, which is still above the World Health Organisation limit for gross α-activity after removal of ^{220}Rn and ^{222}Rn for drinking water), and unrelated to ^{222}Rn concentrations which are a cause of concern. When a large public water supply was developed utilising groundwater within the alluvium of Taw Marsh on the Dartmoor granite in Devon, investigations of ^{222}Rn revealed activities of about 550 Bq l^{-1}. Because of the high levels of ^{222}Rn in this water, a degassing plant was constructed which also removes CO_2 before it enters the public distribution system (removing CO_2 from solution eliminates a major cause of corrosion in the distributing network). As the site is in a National Park, planning conditions required that the plant be constructed underground. Care had to be taken, therefore, that the high levels of ^{222}Rn degassed from the water were not allowed to accumulate in the enclosed environment of the plant, where workers would be in danger of excessive exposure to the gas. The dangers of ^{222}Rn inhalation will be discussed later in this chapter.

Despite such considerations, the effect of Rn in drinking water on human health is uncertain and is made up of both ingested and inhaled components. The half-life of Rn in the body following consumption of well water with an activity of 2500 Bq l^{-1} due to ^{222}Rn has been found to be only 30 min (Anderson and Nilson, 1964); yet, if it is assumed that the whole body is a critical organ, and that there is an average daily intake of 2.2 l of drinking water with an activity of 3300 Bq l^{-1}, a dose equivalent of 5000 μSv $year^{-1}$ results! Nevertheless, it has been argued that, because the mean life of ^{222}Rn in the body is very short, the resulting radiation dose is negligible. Moreover, most drinking water in granite areas comes from surface reservoirs in which the activity is typically less than 10 Bq l^{-1}, and so any source of hazard would be very localised.

The maximum occupational activity level of ^{222}Rn in drinking water has been calculated to be 7400 Bq l^{-1} (Hursh *et al.*, 1965). In this case, the critical organ is the stomach (receiving almost all the radiation dose), to which a maximum exposure limit of 150 000 μSv $year^{-1}$ was set, assuming a daily intake of 1.2 l of tap water. This corresponds to 740 Bq l^{-1} for a small population permitted 15 000 μSv $year^{-1}$ to a single organ. Alternatively, if we work on the assumption of a daily intake of only 300 ml of raw tap water, in two draughts, a maximum permissible activity of 2200 Bq l^{-1} for ^{222}Rn in drinking water has been calculated, with a dose equivalent to the stomach of 15 000 μSv $year^{-1}$ (Hems, 1966). The remainder of the daily intake of water is assumed to be in the form of beverages made with boiled water from which most of the Rn has been degassed. Rn loss during collection of water from the tap and during swallowing are factors also taken into account in these calculations. Another calculation, based upon a daily intake of 1 l of tap water and taking the stomach as the critical organ, suggests a maximum permitted activity of 3700 Bq l^{-1} (Lindell, 1968). On the basis of an annual limit of oral intake for the general public of 2×10^8 Bq, and a daily intake of 2 l of tap water, the UK's National Radiological Protection Board suggests a maximum permitted activity of about 30 000 Bq l^{-1}, giving an average dose equivalent of about 50 000 μSv $year^{-1}$, assuming the whole body to be the critical organ. This is not an official limit, but it may be compared with the average annual effective dose equivalent for radiation of natural origin of 1870 μSv $year^{-1}$ for the inhabitants of the UK.

As far as the natural radioactivity of foodstuffs is concerned, there is, again, relatively little information on the transfer of radionuclides through the various steps of the food chain. The total α-activity of foods varies from less than 0.04 up to 625 Bq kg^{-1} (Table 12.5). In general, the values are low for milk products, fruits and vegetables, but high for cereals and nuts. ^{226}Ra is one of the radionuclides of concern. This mainly enters the body in food and, as noted above, tends to follow Ca in metabolic processes to become concentrated in bones. Typically the α-activity of human bones is between 0.7 and 1.3 Bq for the whole of an average human skeleton; the overall range is assumed to be between about 0.3 and 26 Bq. The concentration of ^{226}Ra in human bones has been found to be independent of age and sex but might correlate with the quantity of ingested ^{226}Ra. In the USA, it is

Table 12.5 — α-radioactivity of various foodstuffs.

	α-radioactivity (Bq kg^{-1})
Brazil nuts	625
Confectionery	<0.04–107
Breakfast cereals	0.3–21
Beverages	<0.04–19
Cockles	11
Bread and flour	0.3–5
Peanuts	4
Mussels	4
Sausages	0.6
Salmon	0.6
Cheese	0.3–0.6
Beef	0.3
Pears	0.3
Lamb	0.04

(After Klement, 1982a).

estimated that about 75% of this ^{226}Ra is derived from cereals, fruits and vegetables, and most of the remainder from animal products (Klement, 1982a).

The decay of ^{226}Ra ingested in foodstuffs produces ^{222}Rn although only about one-third of this is retained within the body. In bone material, ^{222}Rn will provide a dosage to the cells lining the bone and to the bone marrow. ^{214}Bi, the main γ-ray emitter in the ^{238}U decay chain, also lies below ^{226}Ra, and so γ-radiation will penetrate all parts of the body from sources in the bones. However, as γ-radiation is less damaging to tissue than α-particle radiation, this is not considered to be a problem. The whole body dose from ^{222}Rn derived by the decay of ^{226}Ra is estimated to be about 3 μSv year^{-1}. Similarly, low dosages to the whole body of about 3 μSv year^{-1}, and to bone of about 20 μSv year^{-1}, are estimated to arise from ^{210}Pb and ^{210}Po, although it is considered that the main source of these radionuclides is not the decay of ^{226}Ra but direct ingestion of plant and animal foodstuffs.

There is no evidence of the presence of ^{232}Th in foodstuffs, although its daughter, ^{228}Th, has been found in human tissues. As the metabolic uptake of Th is known to be poor, it is assumed that it owes its origin to uptake of ^{228}Ra from ingested foodstuffs, the ^{228}Ra subsequently decaying to ^{228}Th. The estimated whole body and bone dose from ^{228}Ra and its decay products are 8 μSv year^{-1} and 10 μSv year^{-1}, respectively. ^{230}Th, which has formed from the decay of ingested ^{238}U, also contributes a small dose.

Because ^{40}K is present as 0.0118% of all natural K, its presence in food can be readily determined from the total K content. The activity due to ^{40}K intake for an adult male in the USA is estimated at about 85 Bq day^{-1} and leads to an estimated whole body dose for an adult male of about 200 μSv year^{-1}. This radionuclide is thus by far the largest source of the natural radioactivity resulting from ingestion of foodstuffs.

Some specialised diets can lead to particularly high doses from ingested radionuclides. ^{210}Po and ^{210}Pb, for example, may be concentrated by special food chains. Although small quantities of these nuclides are ubiquitous in human bones, gonads and other tissues, we have already noted that consumption of reindeer and caribou meat, in particular, leads to their concentration; apparently the lichen upon which these animals feed is the main concentrating stage. ^{210}Pb is uniformly distributed in soft tissues but may reach high levels in human blood, liver and kidneys. Estimates of the α-particle dose from these sources to Lapps are about 700 μSv year^{-1} to the gonads and 650 μSv year^{-1} to the Haversian canals of the bones (Klement, 1982a).

Inhalation of Rn and its decay products

Although ingestion of radionuclides provides a significant proportion of the natural radiation dose, a more serious source of internal radioactivity is inhalation of Rn and its decay products, as breathing Rn-rich air leads to the build-up of α-particle-emitting daughters in the lungs. There are three major isotopes of Rn: ^{222}Rn, ^{220}Rn and ^{219}Rn. ^{220}Rn is commonly known as 'thoron' (as it is a member of the ^{232}Th decay chain), while ^{219}Rn is known as 'actinon' (as it occurs in the ^{235}U or actinium decay chain). The main source of concern, however, is ^{222}Rn, as it is the most abundant of the isotopes; it has a half-life of 3.825 days. ^{220}Rn, with a half-life of 55.6 s, must also be considered significant. Because the abundance of ^{219}Rn is so low, and as it has a half-life of only 3.92 s, it can be ignored.

Apart from its greater abundance, ^{222}Rn also gives rise to radioactive daughters which can be a serious hazard if lodged in any quantity in the lungs. For some years, the maximum permissible activity of air in industrial environments has been 3.7 Bq l^{-1}, also referred to as one working level (WL). An occupational exposure level of 4 working level months (WLM) for U workers has been introduced in the USA by the Department of the Interior, assuming a working month of 170 h with a maximum air-borne Rn concentration of 1.0 WL at any time (1 WLM is thus 1 WL for 170 h, 2 WL for 85 h, etc.) Recently, however, this exposure level has been questioned and a new limit of 0.7 WL proposed (Dickson, 1980).

As long ago as 1556, Agricola noted that, in some mining areas in Europe, the miners were dying because of the "pestilential air they breath" so that "sometimes their lungs rot away", and between 1877 and 1899 some 75% of deaths among miners at Schneeberg in Germany were due to lung cancer. Similarly, at Jachymov in Czechoslovakia (then known as Joachimsthal), 50% of miners' deaths between 1929 and 1938 were caused by lung cancer. In the 1930s, it was suggested that Rn in the mines might be the reason for this preponderance of lung cancer among causes of death. It was later found that Rn was generally present at activities less than 133 Bq l^{-1}, but occasionally the values were as high as 200 Bq l^{-1}. Experiments were also carried out in which laboratory animals inhaled large Rn doses, but there was little evidence to indicate that Rn caused lung cancer. As a consequence of these studies, it was concluded that, in addition to any effect caused by Rn, factors such as pneumoconiosis, the presence of As, and heredity also played a part. More recent work, however, has shown that, in the Jachymov mines, the distribution of Rn has a range of activities in air of 12–330 Bq l^{-1} and of 2–160 Bq l^{-1} in water. Interestingly, activities in gas within the rock range up to 2700 Bq l^{-1}. With the introduction of mine ventilation in 1930, the air-borne concentration fell below 13 Bq l^{-1}. The average induction time for lung cancer among Jachymov miners was 17 years, with a minimum of 13 years (Cliff, 1980).

When a radioactive aerosol is inhaled, radioactive material is deposited in the respiratory tract, where it has the potential to cause lung cancer. Most radiation-induced tumours occur in the bronchi, and so the most critical cells are assumed to be the stem cells of the bronchial lining; the cells in the pulmonary (or gas exchange) region of the lung are somewhat less sensitive, while the nasal passages are less sensitive still (they also receive less radiation). However, it should be noted that the development of a 'primary' cancer often leads to the development of 'secondary' cancers elsewhere in the body.

The characteristics of the aerosol itself, particularly the fraction of the potential α-particle energy carried by unattached decay products and its particle size distribution, will determine the potential effect of inhalation. In turn, the unattached fraction is largely determined by the ventilation of the area and the nature and concentration of the aerosol present in the air. For example, ventilation rates tend to give relatively high unattached fractions of small particles. However, the breathing rate of the individual, determined by his or her actions, age and health, is also an important factor in deciding the effect of inhalation. Even so, it is important to note that initially the ^{222}Rn decay products are only of atomic size, but they react chemically to form particles with an average diameter of about 0.001 μm. These become attached to solid and liquid particles in the air to form units with a diameter of about 0.1 μm; so, at any one time, most of the decay products are attached. Finally, an effect referred to as 'plate-out', which is the attachment of the decay products to indoor surfaces, may be significant but is as yet difficult to quantify (O'Riordan *et al.*, 1983).

The depth of the stem cells, most of which are located in the basal layer of the bronchial lining, is very important in determining the effect of decay products on the lung; α-particle radiation may be prevented from reaching these cells if there is a thick layer of overlying tissues and mucus. The thickness of this protective zone, the bronchial epithelium, is determined by the age of the person, any condition of respiratory disease, smoking, etc.

In general, for normal young adults in moderately ventilated conditions, a conversion coefficient of 5.5 mSv WLM^{-1} is used for domestic situations (that is, 1 WLM provides an effective dose equivalent of 5 mSv) and a coefficient of 10 mSv WLM^{-1} for miners.

The effects of radiation exposure are considered in terms of risk factors; that is, the probability that a received dose will have a certain effect. These risk factors have been formulated by the International Commission on Radiological Protection on the basis of epidemiological studies of various groups of people exposed to radiation. These studies involved high radiation exposures, but it is assumed that linear extrapolation of the results is appropriate for low level exposure; this implies that any dose, however small, creates a finite risk of cancer. This is a correctly cautious approach, in that it is unlikely to lead to an underestimate of the risk factors for normal circumstances. Risk factors have been calculated for the organs in the body, the risk to the whole body being the sum of the risks to individual organs. A risk factor for serious hereditary effects has also been obtained. A typical example of the application of the risk factor can be illustrated by the case of a dose of 500 Sv received in a year from ^{222}Rn decay products (O'Riordan *et al.*, 1983). In this example, a weighting factor of 0.12 is obtained (derived from a risk factor for fatal lung cancer of 2×10^{-3} Sv^{-1} and the overall risk factor of 1.65×10^{-2} Sv^{-1}), yielding a lifetime risk of fatal lung cancer of $5 \times 10^{-1} \times 2 \times 10^{-3}$, or 10^{-3}; that is, a chance of 1 in 1000.

The definition of the working level is independent of the state of equilibrium between ^{222}Rn and its daughters; 1 WL is effectively defined as any combination of the short-lived daughters which will produce 1.3×10^{-5} MeV of α-particle energy in 1 l of air. In the decay chain, each ^{222}Rn decay produces 5.49 MeV, each ^{218}Po decay produces 6 MeV, and a further 7.68 MeV is produced as ^{214}Pb decays. Because ^{210}Pb has a half-life of greater than 20 years, it is regarded as having been cleared from the lungs before it decays to ^{210}Po, and so both ^{210}Pb and ^{210}Po are ignored in calculating the WL value. When the working level was introduced in 1967, it was chosen as an acceptable convention for mine air, and a maximum permissible annual exposure was set at 12 WLM. However, this was simply based upon the incidence of lung cancer among U miners in the Colorado Plateau area, rather than on strict exposure–dose considerations. In 1969, the maximum permissible exposure was lowered to 4 WLM year^{-1} (Cliff, 1980).

Table 12.6 — Average ^{222}Rn activities in US uranium mines.

	Activity (Bq m^{-3})
Colorado	94 000
Navajo Reservation	13 000
Utah	190 000

(After Cliff, 1980.)

Interestingly, the Colorado Plateau study showed that undifferentiated small cell tumours were prevalent (Saccomanno *et al.*, 1964). Such tumours metastasise very readily, and the first symptoms are typically from metastasis rather than the primary tumour. This also gives a very short survival time; the median survival time from the appearance of the first symptom was only 8 months and from time of diagnosis only 2 months! The results of more general studies of the incidence of lung cancer among U miners in the USA showed that there was a significantly greater risk to miners from ^{222}Rn daughter product exposure if they smoked than if they were non-smokers (Archer *et al.*, 1976). Details of ^{222}Rn activities in US mines are given in Table 12.6.

The danger of contracting lung cancer, however, appears to vary with the locality of the mines. In the Colorado area, for example, the lifetime risk of lung cancer per WLM as an excess per 10^6 persons in the population is 120 ± 50, but in Czechoslovakia it is more than twice this, at 300 ± 100 (Cliff, 1980).

In Czechoslovakian U mines, the population of all miners with a cumulative exposure in excess of 99 WLM had excess lung cancers of most types, and there was a clear linear relationship between cumulative exposure and the frequency of lung cancer. However, the population studied was divided into short (with a mean of 5.6 years), intermediate (9.5 years) and long (14 years) exposure groups (Kunz *et al.*, 1979). For those in the short and intermediate groups with exposure rates in excess of 50 WLM $year^{-1}$, the excess incidence of cancer decreased with increasing exposure. A possible explanation for this is the 'sterilising effect', in which high exposure rates increase cell mortality rather than cause sublethal damage which permits the cell to carry forwards carcinogenic characteristics to its later generations.

In radiological risk assessment, until recently attention focused on ^{222}Rn and its daughters, while ^{220}Rn was largely ignored. This is because, although exposure to ^{220}Rn certainly occurs, there was no data base from which to estimate the danger to health. Because of the short half-lives involved, the member of the ^{220}Rn decay series contributing most to the WL value is ^{212}Pb which, after decaying through ^{212}Bi to ^{212}Po, yields an 8.785 MeV α-particle from the decay of ^{212}Po. With a half-life of 10.64 h, however, it is considered that most ^{212}Pb deposited in the lung will be biologically removed before significant radioactive decay has occurred (Cliff, 1980). Workers engaged in the extraction of Th from its ores, and in the manufacture of certain Th-containing products (such as gas mantles), are most at risk to this nuclide. With its short half-life (55.6 s) there is also some danger of a significant dose from ^{220}Rn itself, as well as from its daughters. However, because ^{220}Rn daughters are, in practice, never in equilibrium, ^{212}Pb, having the largest half-life, contributes 90% to the WL value. 1 WL of ^{220}Rn daughters is produced by an activity of 0.28 Bq l^{-1} for each of the nuclides ^{216}Po, ^{212}Pb and ^{212}Bi. Exposure to ^{220}Rn is now estimated to contribute 100 μSv to the average annual dose of the UK population.

While U miners may be exposed to particularly high ^{222}Rn doses, they are not alone in being at risk. Because U occurs in most rocks, any underground mining operation may be accompanied by high ^{222}Rn levels. The excavation of cavities in mines allows ^{222}Rn to excape into the mine atmosphere, while groundwaters carrying high concentrations of ^{222}Rn may be intersected in mine workings. The Sn mines in the granite of southwest England, for example, have high ^{222}Rn levels because of intersection with hydrothermal groundwater courses.

Table 12.7 — Excess lung cancer incidence in miners per 10^6 persons (lifetime risk per WLM).

Uranium miners, Colorado	120
Uranium miners, Czechoslovakia	300
Fluorspar miners, Newfoundland	100
Swedish miners (excluding uranium)	150
Iron miners, UK	200

(After Cliff, 1980.)

A survey of fluorspar miners in Newfoundland revealed that the excess incidence of lung cancer per 10^6 persons, expressed as lifetime risk per WLM, was approximately 100 (de Villiers and Windish, 1964). This is only slightly lower than that for Colorado U miners (Table 12.7). Indeed, non-U miners in Sweden were found in one survey to have an excess incidence of about 150, while another study showed workers in British Fe mines to have an excess incidence of about 200, both values exceeding the level for Colorado U miners significantly (Cliff, 1980).

Although there is well-established evidence of excess lung cancer among U and other miners exposed to relatively high concentrations of ^{222}Rn decay products, there is some difference of opinion about the risk factors involved. However, they are usually somewhere in the region of 3×10^{-4} WLM^{-1}. Additional information, which suggests that miners are particularly at risk because of the nature of the aerosols inhaled, their smoking habits, breathing pattern, etc., suggests that the risk factor for the population as a whole for ^{222}Rn decay products alone, is about 10^{-4} WLM^{-1}. Interestingly, a lifetime risk of 10^{-4} would result from an effective dose equivalent of 6 mSv; so a conversion factor of 6 mSv WLM^{-1} is obtained, which is in close agreement with the 5–10 mSv WLM^{-1} value given by dosimetric studies. Because safety requires that estimates err on the side of caution, the National Radiological Protection Board uses a reference conversion coefficient of 5 mSv WLM^{-1} for the Rn exposure of the UK population as a whole. Details of limits and mean annual exposure or dose for various categories of workers in Great Britain are given in Tables 12.8 and 12.9, while Table 12.10 shows the risk of death from different types of cancer in the USA and Table 12.11 compares the occupational risk of death in the UK for radiation workers with other industries.

^{222}Rn dosages can, of course, be received by any person through natural sources. Even if U is present in rocks at only 2 ppm, in a volume of rock measuring 1 km × 1 km × 10 m (that is, 10^7 m^3), there will be about 50 t of U, and an activity of about 74 MBq m^{-3} due to ^{226}Ra (Evans, 1969). The latter will decay to ^{222}Rn, a significant proportion of which may escape into the air or into buildings above or below the ground surface. ^{222}Rn which escapes directly to the outdoor air is quickly diluted and dispersed by atmospheric turbulence, but that which enters buildings may accumulate to hazardous levels. Additionally, building materials normally contain some ^{226}Ra and thus act as an extra source of ^{222}Rn (Cliff, 1980).

While ^{222}Rn in the open air may be at low concentrations compared with indoors, interesting variations occur which tie in with the discussion of Rn in soil gas in Chapter 9. The ^{222}Rn concentration in the air is controlled by the exhalation rate from the soil, the rate of vertical mixing, the rate of horizontal movement and the effect of rainfall. It has been found that concentrations are generally greatest during the early morning and fall to a minimum about noon or in the afternoon. High wind speeds cause micro-oscillations in barometric pressure at the surface of the soil which give rise to increased exhalation of ^{222}Rn, while wet ground reduces exhalation by clogging the soil pores with water into which the ^{222}Rn dissolves. The mean value for open air ^{222}Rn activity in the Britain is 2.6 Bq m^{-3} (Cliff, 1980).

One other interesting aspect of ^{222}Rn in the open air concerns its use in exploration work. Because the γ-ray emitters ^{214}Bi and ^{214}Pb form after ^{222}Rn in the ^{238}U decay series, the γ-ray activity of the air is controlled to some extent by these daughters. Thus, part of the radiation detected during air-borne γ-ray surveys is determined by ^{222}Rn. Variation in the concentration of ^{222}Rn or its daughters can, therefore, seriously affect the results of such

Table 12.8 — Radiation exposure of the British workforce.

	Annual exposure limit (WLM)	Annual dose limit (mSv)	Mean annual exposure (WLM)	Mean annual dose (mSv)
Coal miners	4		0.12	
Other miners	4		2.6	
Medical radiographers		50		2
Site radiographers		50		27
Nuclear power workers		50		5.5

(After O'Riordan *et al.*, 1982.)

Table 12.9 — Annual risk of death from mining.

	Coal	Other
Accidents	2.2×10^{-4}	1.5×10^{-3}
Pneumoconiosis	10^{-3}	10^{-3}
Radon daughters	3.6×10^{-5}	7.8×10^{-4}

(After O'Riordan *et al.*, 1982.)

surveys. In particular, when melting of the frozen ground of high latitudes occurs in the spring, the rapid release of ^{222}Rn gives rise to the so-called 'radon-days'. This can be tracked with air-borne γ-ray spectrometer surveys using the ^{214}Bi daughter, and sites of high ^{222}Rn productivity identified. Similarly, holes may be drilled through the frozen soil layer to obtain soil gas samples with enhanced ^{222}Rn concentrations, again facilitating the location of anomalies.

The first study of the concentration of ^{222}Rn in houses was carried out in Sweden in 1956, although similar work in other countries soon followed. It was previously believed that any Rn found inside houses would be due to direct emanation from building materials, but these surveys showed that excess values were so common that the main source was identified as the ground below the houses (Cliff, 1980). Rn can enter houses either directly through the foundations or by migration along the routes used to carry services such as water, electricity, gas and sewage. The potential Rn level that might build up depends upon the Rn emanation characteristics of the ground itself, particularly its composition and permeability, while the actual level reached will also be controlled by the ventilation of each individual room. Ventilation is conveniently expressed in terms of room air changes per hour, or simply air changes per hour (ach), and this varies considerably with the type of construction of the house and the lifestyle of its inhabitants. Measurements in typical modern houses show that, with windows and other controllable openings closed, the average rate is 0.7 ach. An average national reference value of 1 ach is, therefore, believed to be reasonable by the National Radiological Protection Board (Cliff, 1980).

Table 12.10 — Risk of death from cancer of various types for a young adult.

	Dose ($\times10^{-4}$ Sv)
Breast	45
Lung	39
Intestine	30
Leukemia	25
Bone	6
Thyroid	6

(After Cohen, 1982.)

Finally, it should be noted that the dose which an individual receives will also depend upon the occupancy of the building. Again, a common reference value (the occupancy factor) is used. As noted before, it is customary to assume that individuals spend some proportion of the day outdoors. In the UK, this is considered to be 2.5 h per day, giving an occupancy factor of 0.9. This, however, includes a variety of buildings: factories, shops, offices, etc., as well as dwellings. For homes in Britain, a factor of 0.75 is considered more appropriate by the National Radiological Protection Board. Nevertheless, this value should be seen as masking considerable variation between individuals. A person confined to bed will clearly spend virtually no time outdoors, while housewives may have occupancy factors of about 0.9. In contrast, people who work outdoors and pursue outdoor leisure activities may have occupancy factors as low as 0.3.

Measurement methods

There are many different methods for measuring ^{222}Rn, ^{220}Rn and their decay products in air, different techniques being employed for different members of the decay series. The following summary of the main methods is based upon a detailed review given by Cliff (1980).

Table 12.11 — Average annual occupational risk of death in the UK.

Occupation	Risk
Clothing and footware	1 in 250 000
Coal mining	1 in 6 000
Construction	1 in 10 000
Fishing	1 in 800
Food, drink and tobacco	1 in 59 000
Metal manufacturing	1 in 11 000
Timber and woodworking	1 in 34 500
Textiles	1 in 50 000
Radiation workers	1 in 57 000
Average of all occupations	1 in 43 500

(After National Radiological Protection Board, 1986.)

One of the commonest methods for the measurement of Rn gas is the use of the scintillation cell, which consists of an Al alloy shell with a quartz window. The inside of the shell is coated with ZnS scintillator, and the quartz window has the addition of a transparent coat of SnO to reduce variability in the counting efficiency. Attached to the cell is a valve system which allows it to be evacuated and then filled with the atmosphere of interest. After about 3.5 h, the ^{222}Rn in the sample will be in equilibrium with its short-lived daughters. The cell is then placed in a photomultiplier system and the scintillation activity counted. Conversion to absolute activity is made using the response characteristics of the cell determined by calibration against known standards. After use, the cell must be thoroughly flushed with non-radioactive gas and then left until the background activity has diminished to acceptable levels before re-use.

In a modification of this system, two valves are introduced, one an inlet port, the other an outlet port, and a small pump is used to flush the cell with the atmosphere of interest until the original air in the cell has been displaced. Again, after an interval of time, the activity is recorded in a photomultiplier system. With either device, it is possible to avoid waiting for the establishment of equilibrium by noting the time that elapses between collection of the sample and the start of counting, and using this to calculate the degree of daughter equilibrium.

An alternative approach to direct collection of air is to use a filter method. In a typical example, the ^{222}Rn-bearing atmosphere is drawn through a metal tube with filters fitted at each end. The entry filter removes all particulate matter but allows the gas through. As the sample passes down the tube, some of the ^{222}Rn decays to solid daughters, and these are collected by the exit filter. After sampling, the α-activity of the exit filter is counted. This must be carried out very soon after its removal from the apparatus; otherwise too much of the deposited ^{218}Po will have decayed. For this reason, modifications which allow the α-activity of the filter to be measured while sampling is in progress are preferred. Moreover, introducing an aerosol-laden air stream along the periphery of the tube to scavenge daughter products such as ^{218}Po (which would otherwise become attached to the walls of the tube) is an important improvement to the basic design. The aerosol used is NH_4Cl, which is claimed to remove 95% of the Rn daughters.

Another alternative to scintillation cells are thermoluminescent dosimeters, which have the advantage of allowing integration of α-activity over a long period of time. In one such system, ^{222}Rn gas diffuses through SiO_2 crystals and a filter. Above this arrangement is mounted a metal funnel. A large electrical potential (900 V) is applied between the funnel and an insulated electrode inserted in the top of the funnel. This potential causes ^{218}Po ions to migrate to a LiF thermoluminescent chip. A second chip placed outside the funnel is used for correcting for the external γ-ray flux.

More usually used for long-term measurements, however, are devices based upon the damage produced in nuclear track plastics by α-particles. These are similar to the track-etching system described in Chapter 9. The plastic is left exposed to the atmosphere being tested for a long period of time, perhaps several months. It is then recovered, and the tracks etched by a suitable solution or electrochemical process. Counting of the now-visible tracks allows the α-activity of the atmosphere to be determined. The usual material currently used for this purpose is allyl diglycol carbonate (CR39), which has lower background characteristics than other plastics, and responds to α-particles of all energies. It also has a high sensitivity and has been used successfully for integrated measurement of ^{222}Rn concentrations in the environment.

The measurement of ^{222}Rn daughters is generally based upon the activity of the deposits collected by filter systems, either during or after sampling. The most widely used approach for measuring the WL in mines is the Kusnetz method. In theory, this assumes an instantaneous air sample, with equilibrium conditions established; so the measured activity of the filters must be corrected, both for the time taken to collect the sample and for the time delay between collection and measurement. In fact, as long as at least 40 min elapses between collection and analysis, which delay correction factor is selected is virtually immaterial. Much more important is the correction needed for sampling over a finite time interval, and correction factors for sampling intervals of 1, 5 and 20 min have been determined. Moreover, it has been found better to determine counts over a given time interval than to rely on measured count rates. Another important consideration, however, is the assumption of equilibrium conditions. In mines, this condition may not be satisfied, and so the concept of the 'age of air' must be taken into account. Clearly, time is required for ^{222}Rn daughters to approach equilibrium. The initial point can be taken to be the influx of ^{222}Rn (without daughters). As time passes, the daughter concentrations build up, and this is reflected in the ratios of ^{214}Po/^{218}Po and ^{214}Bi/^{218}Po, which develop after specific time intervals. An alternative starting point is when the inflow of ^{222}Rn becomes constant.

The age of the air can very significantly affect the results obtained by the Kusnetz method. With air 3 min old, for example, it has been calculated that a 10 min sample collection period and a 40 min decay time will produce a 25% underestimate in the results. Of course, serious errors may also arise if ^{220}Rn daughters are present, and if the delay between sampling and counting exceeds 90 min. Under these conditions, ^{220}Rn daughters may collect on the filter paper. At typical ^{222}Rn/^{220}Rn ratios, this could cause an overestimate of the true WL value by as much as 100% if the delay is greater than 4 h.

More accurate results are, of course, obtained if the activities of the individual daughters on the filter paper are measured. Typically, this is carried out by α-particle spectrometry. The methods used normally involve passing the air sample through a filter paper which is then used as the source for the spectrometer. The α-particle energy spectrum is then determined at successive intervals. In one approach, a 10 min sample is taken and the spectrum recorded from 120 to 320 s after the end of sampling to give C_1 counts in the ^{218}Po channels and C_2 counts in the ^{214}Bi channels. A further count is taken between 1200 and 1600 s to give C_3 counts in the ^{214}Bi channels (all the ^{218}Po having decayed by this time). For a system with a sampling flow of V l s^{-1} and counting efficiency E, the concentrations of ^{218}Po, ^{214}Pb and ^{214}Bi are given by Cliff (1980) as

$$^{218}\mathrm{Po} = \frac{4.73 \times 10^{-2}}{V\,E} C_1 \ \mathrm{Bq\ m^{-3}},$$

$$^{214}\mathrm{Pb} = \frac{10^{-5}}{V\,E}(-5.783C_1 - 13.70C_2 + 13.67C_3) \ \mathrm{Bq\ m^{-3}},$$

$$^{214}\mathrm{Bi} = \frac{10^{-5}}{V\,E}(0.5525C_1 + 15.45C_2 - 4.192C_3) \ \mathrm{Bq\ m^{-3}}.$$

As described in Chapter 9, α-particle

spectrometry is usually carried out with a vacuum in the α-chamber. Unfortunately, this causes loss of ^{214}Pb atoms from the filter paper as they are generated by decay of ^{218}Po. This is because the recoil energy of ^{214}Pb (110 keV maximum) allows the atoms to move an appreciable distance in a vacuum. The rate of ^{214}Pb loss depends upon the pore size of the filter. In fact, with a pore size of less than 1 μm, the recoil loss rapidly approaches a maximum value of 50%. However, for pore sizes greater than 1 μm, the loss is essentially constant at about 25%. If air is present in the chamber, degradation of energies occurs which results in an overlap of the ^{214}Po and ^{218}Po peaks, requiring corrections to be applied. It should also be noted that some of the ^{222}Rn daughters deposited onto the filter will have penetrated into the filter material, leading to degradation of α-particle energies. This degradation will occur even with the sample in a vacuum, and so corrections for this effect must still be applied. To overcome the problem of the decay of short-lived daughters before analysis, systems have been developed which apply α-particle spectrometry to the filter as the air is drawn through it. Flow rates of about 80 l min^{-1} are used, and count rates are recorded under steady conditions and at the end of the sampling period.

Because α-particle spectrometry has very limited field applications, a number of simple gross α-particle-counting techniques have been developed. These are the most widely used methods now employed for determining the concentration of ^{222}Rn daughter products in the air. In the method of Tsivoglov, the sample is collected on a filter over 5 min, and the α-activity is then recorded 5, 15 and 30 min after the end of sampling. From the decay rates observed, concentrations of ^{218}Po, ^{214}Pb and ^{214}Bi can be determined. A modification of this system involves the recording of counts over intervals of 2–5, 6–20 and 21–30 min after the end of sampling. If these integrated counts are I_1, I_2 and I_3, respectively, V (l min^{-1}) is the sampling flow rate and E the detector efficiency, then, according to Cliff (1980),

$$^{218}\text{Po} = \frac{1}{V\,E}(6.251I_1 - 3.03I_2 + 2.869I_3)\ \text{Bq m}^{-3},$$

$$^{214}\text{Pb} = \frac{1}{V\,E}(0.0451I_1 - 0.7611I_2 + 1.816I_3)\ \text{Bq m}^{-3},$$

$$^{214}\text{Bi} = \frac{1}{V\,E}(-0.8332I_1 + 1.228I_2 - 1.395I_3)\ \text{Bg m}^{-3}.$$

Alternatively, if time intervals of 0–7, 7.5–11, 12–31 and 32–47 min are used to give counts C_1, C_2 and C_3, then Cliff (1980) modifies the equations to

$$^{218}\text{Po} = \frac{1}{V\,E}(2.974C_1 - 1.275C_2 + 1.073C_3)\ \text{Bq m}^{-3},$$

$$^{214}\text{Pb} = \frac{1}{V\,E}(-0.1382C_1 - 0.2212C_2 + 0.624C_3)\ \text{Bq m}^{-3},$$

$$^{214}\text{Bi} = \frac{1}{V\,E}(-0.2938C_1 + 0.5384C_2 - 0.5416C_3)\ \text{Bq m}^{-3},$$

The Rn daughter monitor allows major sources of ^{222}Rn to be located, as well as giving the WL value. It works on the basis that, because of its short half-life, ^{218}Po is usually in equilibrium with ^{222}Rn and allows counting throughout the 15 min sampling interval. Subsequent counts are recorded for time intervals of 16–36 min and 37–57 min. The respective counts, C_1, C_2 and C_3, are then used to determine the concentration of Rn daughters:

$$^{218}\text{Po} = \frac{1}{V\,E}(0.4743C_1 - 0.4188C_2 + 0.2924C_3)\ \text{Bq m}^{-3},$$

$$^{214}\text{Pb} = \frac{1}{V\,E}(-0.01531C_1 - 0.09439C_2 + 0.2165C_3)\ \text{Bq m}^{-3},$$

$$^{214}\text{Bi} = \frac{1}{V\,E}(-0.06257C_1 + 0.228C_2 - 0.1843C_3)\ \text{Bq m}^{-3},$$

The WL is then given by Cliff (1980) as

$$\text{mWL} = \frac{1}{V\,E}(0.00474C_1 - 0.00149C_2 + 0.01913C_3)$$

The use of integrated gross α-particle-counting methods can also be applied to the measurement of ^{220}Rn daughters. In one approach, five counts are made after sampling, extending the total count period to 24 h. Typically, count C_1 is obtained during sampling over 15 min. Further counts are then taken at 16–36, 37–57, 240–360 and 480–1440 min, to give C_2–C_5. From these data (Cliff, 1980) gives

$$^{218}\text{Po} = \frac{1}{V\,E}(0.4756C_1 - 0.4232C_2 + 0.3003C_3 - 0.1124C_4 + 0.0239C_5)\ \text{Bq m}^{-3},$$

$$^{214}\text{Pb} = \frac{1}{V\,E}(-0.0159C_1 - 0.1065C_2 + 0.2378C_3 - 0.3035C_4 + 0.06702C_5)\ \text{Bq m}^{-3},$$

$$^{214}\text{Bc} = \frac{1}{V\,E}(-0.06015C_1 + 0.2201C_2 - 0.1705C_3 - 0.1956C_4 + 0.04668C_5)\ \text{Bq m}^{-3},$$

$$^{212}\text{Pb} = \frac{1}{V\,E}(0.002883C_5)\ \text{Bq m}^{-3},$$

$$^{212}\text{Bi} = \frac{1}{V\,E}(-0.003338C_1 + 0.01086C_2 - 0.01903C_3 + 0.2711C_4 - 0.06273C_5)\ \text{Bq m}^{-3}.$$

The filters used for collecting ^{222}Rn daughters are either of a glass fibre or membrane type. Where the use of electrical pumps is not practical (in coal mines, for example), gravimetric samples are used.

Passive detection of ^{222}Rn daughters using cellulose nitrate film is possible, and active systems which induce a flow of air over the film have also been devised. An alternative detection system operates on the basis of the thermoluminescence of LiF, air being drawn through a membrane filter placed 2 mm behind the LiF detector plate. A second LiF detector plate measures background γ-ray activity so that the activity detected by the first plate can be calibrated in terms of the α-particle flux.

The dose to the body that results from inhalation of ^{222}Rn daughters depends mainly upon the proportion of ^{218}Po that exists as free atoms. This is because the diffusion coefficient of the free atom is much greater than that of attached particles. Free ^{218}Po diffuses rapidly, therefore, to the walls of the upper bronchial airways and, in nasal breathing, most free ^{218}Po will be removed in the nasal–pharyngeal region.

In the systems considered so far, unattached ^{218}Po may well pass through normal filters and not be detected. To trap free ^{218}Po, it is necessary to use a fine wire mesh screen. Such a screen should not trap the attached ^{218}Po, thus allowing normal detection procedures to be used thereafter. However, the trapping efficiency is controlled by the mesh size in that, the smaller the mesh, the more likely is attached ^{218}Po to be trapped in addition to free ^{218}Po. A 60 mesh, for example, collects no attached ^{218}Po and about 77% of unattached ^{218}Po with a flow rate of 2 l min^{-1}, while a 200 mesh with a flow rate of 10 l min^{-1} collects more than 98% of unattached atoms but may trap a proportion of the attached atoms as well.

Another method of measuring the proportion of unattached atoms is to use two monitoring systems, one of which incorporates a diffusion battery to remove the unattached atoms. The difference in output from the two systems gives a measure of the unattached fraction.

In addition to ^{218}Po, unattached atoms of ^{214}Pb are also probably present, and some unattached ^{214}Bi may form as a result of its decay.

Rn in dwellings

In the UK, the average risk of death from all malignancies is about 2×10^{-1}; for lung cancer the risk is about 5×10^{-2}, most of which is attributable to smoking. In comparison with these figures, the risk of lung cancer from exposure to Rn decay products (6×10^{-4}) would not appear to be very high; this represents about 0.3% of the risk from all causes, 1% of the risk of lung cancer. Even if the average overall dose from all sources is considered, the figure is only 0.5% for all causes. There is thus an average risk that 1 fatal malignancy in 200 results from natural radiation (O'Riordan *et al.*, 1983).

Perhaps more easily appreciated than the statistical chance of a fatal malignancy is the average reduction in life expectancy. It has been calculated that the reduced life expectancy from the average dose of Rn decay products in the UK is 5 days. This may be compared with a reduction of about 1000 days for all fatal cancers.

However, the dosage received by the population in the UK varies significantly with the nature of the underlying geology, the location of houses with respect to Rn-conducting fractures in the ground, and the lifestyle of the occupants (Fig. 12.1). One of the areas of highest risk is that underlain by the granites of Devon and Cornwall, where exposure may be as high as 68 mSv year^{-1}. Although this is found in only less than 1% of dwellings, at such dose levels life expectancy is reduced by about 1 year, and the risk of death from a cancer induced by natural radiation increases to about 25%. Indeed, some families may have members which receive as much as 100 mSv year $^{-1}$, leading to a 35% chance of contracting a fatal cancer (7% of death due to all causes), and a

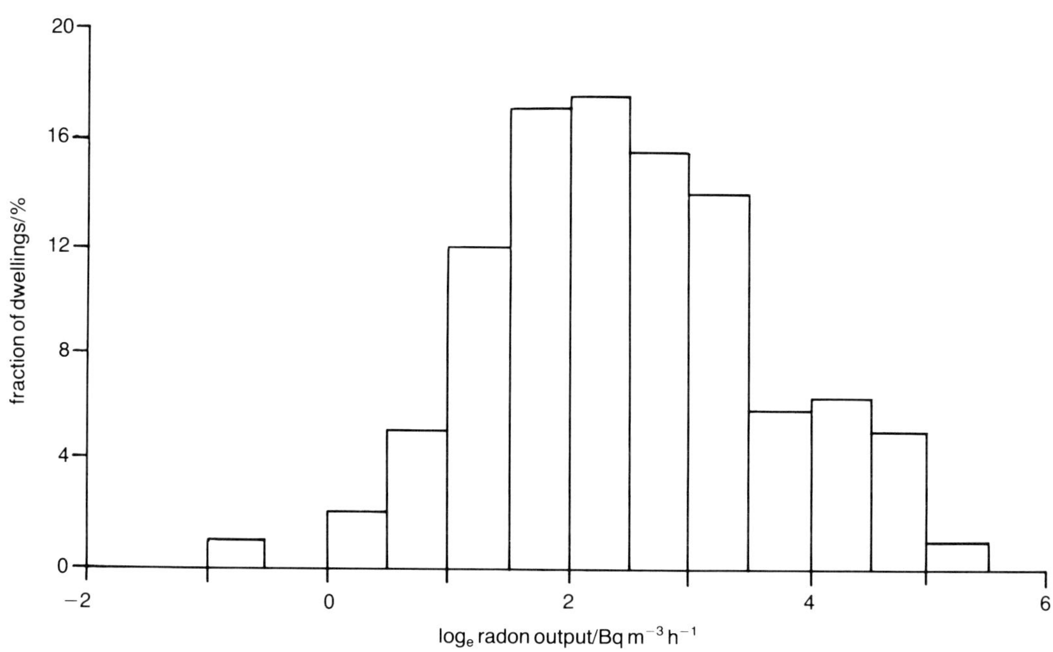

Fig. 12.1 — ^{222}Rn production rates in the living rooms of UK Dwellings. (Cliff, 1980, redrawn with permission.)

decrease in life expectancy of 640 days (O'Riordan *et al.*, 1983). Even when compared with other causes of death, such figures are of concern. For example, the reduction in life expectancy for lung cancer from 10 cigarettes a day is 400 days, road accidents 100 days, accidents in the home 60 days and accidents to men at work 20 days.

The number of dwellings in the UK in which the dose exceeds 25 mSv year^{-1} (a reduction in life expectancy of 160 days) is estimated to be of the order of 1000, and that for doses above 10 mSv year^{-1} (a reduction in life expectancy of 60 days) 10 000. The social impact of the effects of natural radiation are, therefore, significant, with 10 000 families some three times at greater risk of death than through accidents at work. Yet the investment in containing the effects of industrial injury is enormous, while expenditure on controlling exposure to the natural radiation dose in the UK is minimal (O'Riordan *et al.*, 1983).

The problems lie, firstly, in identifying which dwellings pose a health hazard (Tables 12.12 and 12.13) and, secondly, in carrying out remedial work. Associated with these is the problem of recognising potentially hazardous sites, so that future development may either be avoided or constructions made with the appropriate safeguards incorporated. The investigation of a large number of dwelling houses is no simple task. Surveys to date may be considered to represent only pilot studies. Even so, they have identified the fact that alterations to a dwelling can produce significant changes in the dose received by its inhabitants: the blocking-up of fireplaces and removal of chimneys following installation of central heating, the installation of double glazing, the draught-proofing of doors, etc., may result in a dramatic increase in the dose. Change in occupancy from a family with a fondness for open windows to one to which the slightest draught is an abomination can similarly affect the dose received. The design of surveys to permit the investigation of large numbers of

Table 12.12 — Rn decay products and Rn activities in UK dwellings.

	Rn decay products (mWL)	Rn activity (Bq m^{-3})
Granite houses	16	
Brick houses	6	
Houses with solid floors	20	
Houses with suspended wood floors	13	
Average English living rooms		25
Average Scottish living rooms		18
Average Welsh living rooms		21
Average Northern Irish living rooms		13
Average English bedrooms		18
Average Scottish bedrooms		14
Average Welsh bedrooms		16
Average Northern Irish bedrooms		12
Average Cornish living rooms		140

(After Wrixon *et al.*, 1984.)

Table 12.13 — γ-ray dose rates in UK dwellings.

	γ-ray dose rate ($\times 10^{-6}$ Gy h^{-1})
Granite houses	0.12
Brick houses	0.08
Houses with solid floors	0.11
Houses with suspended wood floors	0.11
Average English living rooms	0.07
Average Scottish living rooms	0.07
Average Welsh living rooms	0.07
Average Northern Irish living rooms	0.06
Average English bedrooms	0.06
Average Scottish bedrooms	0.07
Average Welsh bedrooms	0.07
Average Northern Irish bedrooms	0.06
Average Cornish living rooms	0.102

(After Wrixon *et al.*, 1984.)

dwellings in the UK by passive detectors is currently in hand by the National Radiological Protection Board. The initial choice of investigation area, based upon high risk, should, of course, be made by reference to the nature of the geology of the area.

Remedial measures which may be possible in the already existing housing stock are also currently being investigated by the National Radiological Protection Board; it is important to determine the most cost-effective approach. Because Rn may enter the house through service conduits, even in solid floors, one approach is to ensure that such systems are properly sealed. Cracks in foundations can also be sealed at moderate cost. Sheets of thermoplastic such as polyvinyl chloride have been found to act as effective barriers to Rn migration, and polyamide sheet is also very useful for this purpose. Where Rn is exhaled from surfaces, painting or covering the surface with resin will reduce the flux. More extensive measures, such as provision of efficient underfloor ventilation, re-laying of floors, rerouteing of services, etc., may be needed in more difficult cases. Such modification of houses would, of course, give a permanent solution to any Rn problem, but temporary expedients such as the use of electric fans may be necessary. Unfortunately, increased ventilation can add considerably to the cost of home heating. Perhaps the use of combined ventilation–heat exchange systems will overcome the difficulty. Electrostatic precipitators can be used to reduce the Rn daughter concentrations at low running cost. These devices work by charging and collecting the aerosol and are typically about 90% efficient in most normal circumstances. However, they have two disadvantages: firstly, they represent a high capital outlay and, secondly, they can generate O_3 at unacceptable levels unless care is taken in the operation of the equipment (O'Riordan *et al.*, 1983).

In preference to remedial measures, the provision of Rn barriers in new housing should be actively considered for developments in high risk areas identified by Rn measurements carried out as part of normal site investigation procedures. The main feature of any such modified design must be the introduction of a good seal which separates the dwelling area from the ground. The use of impermeable membranes is particularly important in this respect. Also of great importance is the need to eliminate the possibility that Rn enters the dwelling area by service access routes or by cracks and structural discontinuities in the walls and foundations.

Radiation hazards of geological specimens

The collection and storage of rocks and minerals with high concentrations of radionuclides could constitute a radiation hazard. This problem has been investigated by the National Radiological Protection Board by measurement of the activity of a range of samples from Devon and Cornwall, and high grade ore mineral species typically available for purchase in Britain, although they may have an overseas source (Dixon, 1983). Because radiation exposure can arise by direct handling of specimens, by inhalation of air-borne activity and by ingestion of activity transferred from specimens to hands, this study measured the radiation dose rate, the

Rn exhalation rate and the amount of loose activity. From these results an overall assessment of the hazard was made.

The average activity of radioactive geological specimens is 270 Bq g^{-1}, with high activity samples averaging 1853 Bq g^{-1}. Among high activity speimens, the activity of autunite is only 23 Bq g^{-1}, while the activity of uraninite samples can be over 5000 Bq g^{-1}. The average β plus γ absorbed dose rate in air at the surface of the samples is 95 μGy h^{-1}. For specific minerals studied, the rate ranged between 25 μGy h^{-1} for autunite and 3410 μGy h^{-1} for a particularly active uraninite. Surface dose rates, however, can be somewhat misleading, as the rate decreases rapidly with distance from the sample; at a distance of 50 cm, the dose rate is only 0.2% of that at the surface. The Rn exhalation rate was found to be, on average, 397 Bq h^{-1} kg^{-1}. In the high activity group, the average was 1749 Bq h^{-1} kg^{-1} with a single sample maximum of 5760 Bq h^{-1} kg^{-1}. Rn exhalation from autunite was consistently low, the highest values being associated with carnotite and, to a slightly lesser degree, uraninite. The amount of loose activity was assessed in terms of variation in the degree of abrasion which the specimens received. During gentle abrasion, the average activity removed was 0.3 Bq cm^{-2}, but moderate abrasion released 3.4 Bq cm^{-2}. The largest amount removed from ordinary specimens was 3.5 Bq cm^{-2} for gentle abrasion and 38 Bq cm^{-2} for moderate abrasion but, in the high activity samples, as much as 61 Bq cm^{-2} was removed during moderate abrasion (Dixon, 1983).

From these results, it was concluded that the typical activity of U-bearing minerals collected in Britain is likely to be about 300 Bq g^{-1}, while overseas samples of ore minerals are more likely to have activities of around 2000 Bq g^{-1}. The dose rate varies accordingly, with a value of about 100 μGy h^{-1} from the British specimens, and about 500 μGy h^{-1} for the high grade minerals collected elsewhere. Rn exhalation rates also differ between these two categories of sample, with values of about 400 Bq h^{-1} kg^{-1} and 1500 Bq h^{-1} kg^{-1}, respectively, and the loose activity ranges between 0.3 and 3.5 Bq cm^{-2} for gentle to moderate abrasion of British samples and between 3.0 and 30 Bq cm^{-2} for altered high grade minerals. High grade minerals showing no signs of alteration possessed the same loose activity as the British samples.

In assessing the hazards of radioactive minerals in a collection, the presence of two samples of radioactive mineral collected in Britain and three high grade overseas specimens was considered to be typical in a collection of 500 specimens. If 100 h $year^{-1}$ are spent handling such a collection, the average annual handling time of radioactive specimens will be 1 h. Under these conditions, the total annual dose equivalent to the hands is 0.3 mSv, of which nearly 80% is contributed by the high grade minerals. The typical whole body effective dose equivalent is 0.08 mSv. Also on the basis of 100 h handling per year, and storage of the collection in an otherwise unoccupied room with typical ventilation rates, the annual exposure to Rn daughters by inhalation is 0.6 mWLM. At lower ventilation rates, this rises to 8.7 mWLM and, if the room is continuously occupied, the exposure rises to nearly 100 mWLM (Dixon, 1983).

The intake of active material will largely depend upon the type of handling which the material receives. Most handling is considered to be comparable with gentle abrasion but, on about 10% of occasions, the more severe conditions equivalent to moderate abrasion may occur. On the assumption that, on ten occasions during the year, ingestion occurs at the level assumed for the occupational control of contaminated surfaces (ingestion of all activity on a 10 cm^2 surface on each working day) and that only 1% of the transferred activity is ingested on each occasion (equivalent to a 1 cm^2 area), the annual intake was calculated as 2.6 Bq (Dixon, 1983).

The main conclusion drawn from these findings is that radiation dose rates, Rn exhalation rates and the amount of loose contamination from U-bearing minerals can be sufficiently high to warrant the exercise of some caution in their collection and storage. The risks from samples collected in Britain, however, are somewhat less than those from imported specimens. The annual dose arising from the contact times assumed are only 3–5% of the level appropriate for limiting radiation exposure to members of the public, and most of this comes from the imported specimens.

Excessive handling of high grade U-bearing minerals should, therefore, be avoided. Measures to reduce the external exposure can be taken, such as handling specimens only while wearing disposable plastic gloves. If specimens are in contact with the hands, the hands should be thoroughly washed immediately afterwards, and certainly before consuming any hand-held food. Exposure to Rn decay products can be minimised by storing the samples in Rn-tight boxes and in rooms not normally occupied. Care should be taken when handling specimens to avoid abrasion and the detachment of fine loose particles. It is unwise to store any radioactive sample in close proximity to an individual for long periods of time.

12.2 GEOLOGICAL ASPECTS OF THE DISPOSAL OF RADIOACTIVE WASTE

The safe disposal of the potentially harmful radioactive waste products of the nuclear fuel cycle and processing systems is a major concern in countries with a nuclear industry. The relatively long half-lives of some of the radionuclides present in this material, combined with high activity and chemical toxicity, make this a question of long-term importance (Klement, 1982b). Disposal sites must be chosen at which safeguards can be maintained over periods of about 10 ka. Recovery of waste during the first 50 years is also desirable should difficulties arise.

12.1.1 Categories of waste

There are several types of radioactive waste, each having essentially different characteristics (Table 12.14). Unreprocessed spent fuel contains fissile actinides and fission products. This material is extremely hazardous and must be kept under strictly controlled conditions. In the USA, Federal Regulations require that this waste be converted into a dry and stable solid within 5 years and shipped to a repository within

Table 12.14 — Classification of radioactive wastes.

Category	Gaseous ($Bq\ l^{-1}$)	Liquid ($Bq\ l^{-1}$)	Solid ($Bq\ m^{-3}$)
Low level			
α	$<3.7\times10^{-6}$	$<3.7\times10^{7}$	$<10^{10}$
β, photon	$<3.7\times10^{-3}$	$<3.7\times10^{7}$	$<10^{10}$
Intermediate level			
α	$<3.7\times10^{-6}-3.7$	$3.7\times10^{7}-3.7\times10^{10}$	$10^{10}-10^{12}$
β, photon	$3.7\times10^{-3}-3.7\times10^{3}$	$3.7\times10^{7}-3.7\times10^{10}$	$10^{10}-10^{12}$
High level			
α	>3.7	$>3.7\times10^{10}$	$>10^{12}$
β, photon	$>3.7\times10^{3}$	$>3.7\times10^{10}$	$>10^{12}$

(After Kathren, 1984.)

10 years (Cohen, 1982). At this stage, the material is referred to as 'high level waste'. It still has a high heat production and is highly radioactive. As an alternative to disposal, this material may be stored under supervised controlled conditions for a period of several decades until the activity and heat production have sufficiently decayed.

At the end of this period, the term 'intermediate level waste' is applied. Disposal of intermediate level waste is under active consideration, even where the decision to store high level waste has been taken. Many of the problems of the disposal of high level waste also relate to intermediate level waste, but the magnitude of the problem is substantially reduced.

The term 'low level waste' is used to encompass a range of materials which are contaminated with radionuclides from various sources. In the nuclear energy industry, low level wastes comprise filters, ion exchange resins, laboratory wastes, etc.; other sources include hospitals and industry. Curiously, some material may be classed as low level waste even though it does not contain radionuclides, simply to avoid the need for low level checks to be run. The disposal of low level waste is generally by shallow burial and presents the same difficulties as other toxic wastes. High and intermediate level wastes, however, require special consideration, especially for disposal. Many of the problems connected with radioactive waste disposal discussed in this chapter are related to high and intermediate level waste.

12.2.2 Disposal options

In 1984, there were about 1200 m^3 of high level liquid waste in store in the UK, with a total activity of about 2.6×10^{19} Bq. By the year 2030 , this will have risen to about 3400 m^3 and 7.3×10^{19} Bq. After solidification, 3400 m^3 reduces to 1360 m^3 of glass but, even 50 years after vitrification, 1 m^3 of waste would still give as much heat as a 3 kW electric fire! By the year 2030, there will be about 190 000 m^3 of intermediate level waste in the UK. Figures 12.2 and 12.3 illustrate the decay of the γ-ray and thermal power produced by the waste that results from the generation of 400 GW year and 100 MW year of electricity, respectively.

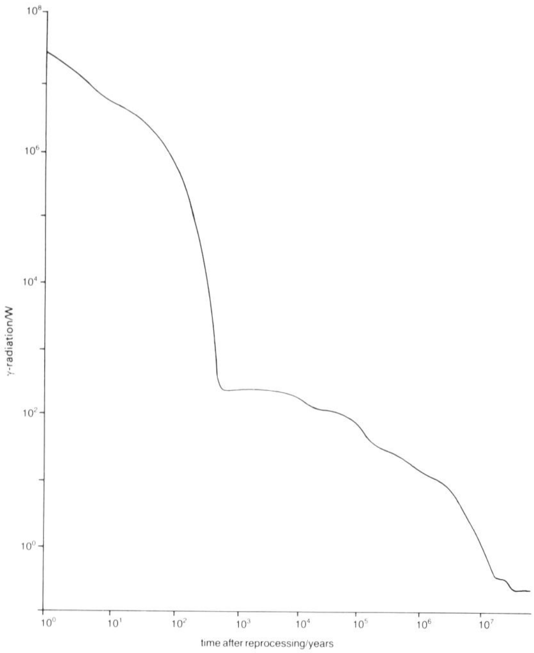

Fig. 12.2 — γ-radiation released by the waste produced from the nuclear power generation of 400 GW years of electricity. (After Cohen, 1982.)

Disposal of waste by burning, by burial on land or by dumping at sea has been practised since man first decided to deal with the garbage that his activities create. Burning of radioactive waste, however, is neither practical nor sensible; the radionuclides would not be destroyed, and radioactive dust would contaminate a huge area. Disposal at sea has been practised but has met much environmentalist opposition. The dangers of marine disposal that are highlighted concern the breakdown of the containment system in which the waste is stabilised, and subsequent leaching by sea water. Indeed, in any disposal system, it must be assumed that the artificial barriers will

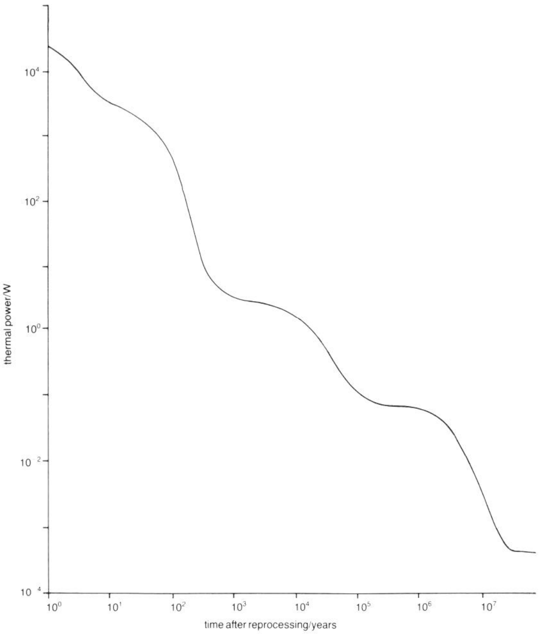

Fig. 12.3 — Thermal power released from the waste produced by the nuclear power generation of 100 MW years of electricity. (After Cohen, 1982.)

break down in a relatively short period of time. The real containment must be provided by natural barriers. Once the radionuclides have dispersed in the sea, it may be thought that the danger is eliminated, but chemical processes and, particularly, biological activity can act to reconcentrate these radionuclides. If this reconcentration occurs in edible marine organisms, direct return to man is possible. Obviously, indirect pathways also occur by which dispersed radionuclides can return through complex food chains. Any material disposed of in the biosphere is, in fact, likely to find its way back to man, and so disposal outside the biosphere is necessary if it is assumed that artificial barriers will fail.

Knowledge of the depth to which the biosphere extends below the surface of the Earth is, therefore, critical. Unfortunately, this is not precisely known. Moreover, it is likely to vary from one area to another, depending upon the geological formations and structures present. One estimate is that the biosphere extends to a depth of 2000 m and that, within the zone between this depth and the Earth's surface, any site can be colonised by organisms, particularly bacteria (Alexandra, 1961). Whether or not such colonisation occurs, however, depends upon the nature of specific environments. If ingress of bacteria is possible along free-flowing groundwater passages, and if there is a ready supply of nutrients, bacteria are likely to thrive and multiply. The degradation of petroleum reservoirs by bacteria clearly demonstrates that the process is not confined merely to the realm of theoretical possibility (Bailey *et al.*, 1974). The fact that bacteria may gain access to radioactive waste is not, in itself, a problem, provided that migration by either direct or indirect pathways does not result in interaction with surface and near-surface organic processes. The recognition of bacteriological acitivity at considerable depth in the Earth is indicative, however, of the more insidious problem of migrating groundwater; the most likely way in which radionuclides from a leaking repository could find their way back to the Earth's surface is as part of a groundwater circulation system.

It is not only the hydrological characteristics of a potential disposal site, and the way in which these could change in the future, that must be known. As the lithologies present at the site may sorb radionuclides transported in solution through them, the mineralogical and chemical behaviour of the rock must also be understood; the geological environment itself can act as the main chemical barrier. Indeed, as most rocks do allow the passage of water over a long period of time, the chemical barrier which they present is possibly more important than any short-term physical barrier which they provide. Nevertheless, as the nature of the contact between the rock and flowing groundwater is determined by the groundwater flow characteristics, the nature of the hydrological system must be understood before the chemical retardation processes can be quantitatively assessed. The results of hydrological investigations will, therefore, be briefly reviewed before dealing with the chemical evidence. At this stage, the discussion will be concerned with general principles rather than with the disposal of specific waste types.

12.2.3 The hydrology of potential radioactive waste repository host rocks (far-field containment)

To the hydrogeologist, rocks are either aquifers or aquicludes. Aquifers have a high porosity and permeability and are potentially suitable for water supply purposes, while aquicludes have a very low permeability and may confine aquifers. Because the physical barrier surrounding any repository needs to have a high integrity, aquifers must be avoided; repositories will need to be sited in aquicludes. However, no rock mass as a whole is totally impermeable to water movement. Even in the case of closely packed and well-cemented sedimentary or crystalline rocks, some water movement along grain boundaries will occur, although the permeabilities involved are very small. More important is the presence of fractures, particularly joints, which are almost ubiquitous in rock masses. The ability of fractures to conduct large volumes of water at high flow rates is well known. Indeed, some aquifers (like the Chalk) only attain that status because of their high fracture porosity and permeability. The problem of a leaking repository is one of fracture flow in the host rock.

In the UK, hydrological investigations of fracture flow have been carried out in the Carnmenellis granite in Cornwall (Heath and Durrance, 1985). Granites are potential repository hosts because they occur in large masses which extend to a very great depth, the intact rock has a low permeability and it has a reasonable chemical stability. Fracture separations in granites are also relatively large, typically of the order of a few metres. The early part of the programme was devoted to detailed mapping of the frequency and orientation of fractures from a number of closely spaced boreholes using diamond-drilled core and geophysical logs. Once this information was available, pump tests were carried out to determine the frequency of water-conducting fractures and their permeabilities. Investigation was then possible of the nature of the general flow regime by interborehole tracing using radionuclides. Finally, the character of flow along single fractures was studied using stable tracers. The detailed results obtained are, of course, only applicable to the specific site investigated, but a number of important general conclusions can be drawn from this work. Firstly, although subhorizontal joints become less frequent with depth, they still occur occasionally at depths of several hundred metres, where they may still form major water-conducting features. Secondly, near-vertical joints occur with well-defined preferred orientations which reflect the stress history of the rock. The frequency of near-vertical joints decreases with depth, areas of relatively intact rock being separated by more highly fractured zones. There are still very important features at depths of several hundred metres. Thirdly, not all fractures intercepted by boreholes are water conducting. The conductivity of fractures appears somewhat bimodal in distribution, with numerous fractures of moderate permeability and occasional important features with a very high permeability. The high permeability fractures have become known as ‘main drains’; these features tend to control the overall hydrology of the site. Fourthly, flow path connections between boreholes are extremely complex, cross-linkages occurring in no obviously systematic way. Moreover, individual flow paths are often not reversible when subject to hydrological testing, and their characteristics change with time. Finally, single fracture experiments indicate that flow within fractures is not planar but is concentrated into linear channels within the fracture. These channels can be highly irregular, and there may be a degree of interconnection of adjacent channel systems. Individual channels can have a very high hydraulic conductivity. It is likely that the variation in fracture permeability observed at the site is simply the product of boreholes which intersect either high conductivity channels in fractures or fractures between channels. In the latter case the fractures appear tight. The flow characteristics of rock masses are thus determined by the frequency, size and orientation of channels within fractures. Interestingly, the orientation of flow paths in the Carnmenellis granite may well be determined by the residual stress anisotropy in the rock. At depth in the granite, both the maximum and the minimum regional compressive stresses are directed horizontally. The maximum horizontal compression is orientated approximately northnorthwest–southsoutheast and appears to allow flow along this direction preferentially; that is, at right angles to the minimum compressive stress.

The geological control of the flow path characteristics seen in granite is perhaps the simplest case. In other rock types, fracture characteristics and the nature of channel flow are likely to be more complex. Clay, shale and slate have all been suggested as suitable repository hosts, but they can possess more discontinuities than granite. Clay formations may be of great thickness as a whole and yet contain numerous silty or sandy strata within them. Such coarser horizons can provide extremely efficient pathways for water flow through what would otherwise be considered an impermeable rock. In mudstones or consolidated clays, joints have small separations owing to the relatively low tensile strength of these rocks. Only if the clay is uniform over a reasonable thickness and has responded to stress in a more or less ductile manner will the conditions suitable for a repository be found. However, in these circumstances, the plastic nature of the clay may, in fact, make repository construction a complex procedure, as creep is likely to occur during excavation. Another difficulty with clays, which are usually deposited in O_2-deficient environments, is that they frequently contain abundant pyrite which will be oxidised on exposure to air. This results in expansion, the physical degradation of the pyrite and the surrounding clay, and the liberation of sulphuric acid. The production of highly aggressive groundwaters by this process not only is a potential hazard to the repository itself but also can cause increased permeability in the host by attacking barriers in groundwater flow channels. The suggestion that pyrite in a clay could be beneficial, in that it would cause reduction of U^{6+} in solution to insoluble U^{4+} and lead to trapping of other radionuclides, is uncertain. The actual response of the chemical system represented by the clay will depend upon the Eh–pH conditions of the leached solutions and the nature of the environment as a whole.

Shales generally fracture parallel to their bedding planes, but some shales are so fine grained that they also fracture across the bedding. The presence of these additional microfractures can increase significantly the permeability of the rock mass as a whole. The presence of silty and sandy layers is also likely to increase overall permeability, as it does in clays.

The cleavage typical of slates can impart a high permeability to the rock mass. Indeed, there may be so many planes (with separations of only centimetres or millimetres) along which water may be able to flow that the rock mass can appear to present the characteristics of continuum flow rather than discrete fracture flow. However, within this three-dimensional network, individual groundwater flow paths will still probably be controlled by channels.

A third rock type considered to provide suitable repository host characteristics is salt (halite). Both bedded salt deposits and salt domes have very low water contents and low porosity and permeability. Moreover, salt anneals easily when fractured. In bedded salt deposits, however, the sedimentary succession normally includes other lithologies, including detrital sedimentary rocks. Also, anhydrite and gypsum are commonly found interbedded with halite. Carbonate layers, too, may be present, together with highly soluble K salts, such as sylvite, carnallite, kainite and polyhalite. Once again, it is the whole environment of the repository that is important and, unless extremely thick halite sequences are utilised, lithological variations around the site may give rise to problems. Anhydrite, for example, is more prone to fracture than halite, and its self-sealing properties are poorer. Interbedded sandstones may be highly permeable, while carbonate rocks may also possess high permeability, either through fracturing or through the development of cavernous structures resulting from dolomitisation–dedolomitisation processes.

The ability of halite to flow under pressure (an advantage in terms of annealing of fractures) may also be disadvantageous during repository construction, when creeep can be a problem in the maintenance of open workings. Moreover, halite is not completely dry, and pools of formation water can be found trapped in the so-called ‘brine pockets’. The significance of brine pockets to a repository is uncertain at present but is not likely to prove a major difficulty (Cohen, 1982).

Salt domes provide an attractive alternative to bedded salt deposits because they can occur as columns 2 km or more in diameter and extend to depths of over 2 km. Halite flows more readily than other evaporites to form diapiric structures, although domes of anhydrite, gypsum, etc., may also develop. The ability of a material to flow is expressed as its rheidity R, which is defined as

$$R = \frac{1000\eta}{\psi}$$

where η is the viscosity and ψ the rigidity.

Interestingly, the rheidity of halite is about 3–30 years, while that of dry anhydrite is about 1000 years (these values can be compared with the rheidity of ice which is about 12 days). Although halite flows more readily than anhydrite, there is no great difference in practical terms, because of the long time scale involved in the development of geological structures.

Volcanic rock sequences, including basalt and pyroclastic rocks such as tuffs, have also been considered as possible repository hosts. Basalts are rather similar to granites in that fracture flow dominates their hydrology. However, volcanic sequences rarely consist simply of basalt flows with no intervening material; pyroclastic layers frequently separate individual flows, while the upper surfaces of flows were often weathered and degraded before being buried by younger lava. For these reasons, volcanic sequences may provide such good water storage characteristics that they are considered as aquifers. Certainly, the flow of water along weathered horizons or tuff bands can be considerable. Unless an individual lava is of such thickness that these problems can be ignored (which is very unlikely), basalt-hosted repositories seem to present fundamental difficulties as far as providing a physical barrier is concerned.

Gneisses and schists possess characteristics which range from those similar to granite to those similar to slate. Gneisses, like granite, generally provide a good physical barrier to water movement except along joints. However, shear zones of high conductivity may occur, and the gneissose foliation itself may allow water movement. As schists possess even more marked foliation, the flow of water is likely to be very high; there will also be flow along joints and shear planes.

In summary, no rock type is likely to provide a perfect physical barrier which isolates the repository from its surrounding environment. However, some rock types will be better than others. The tectonic history of the area of the site is particularly important, as it will have determined the fracture characteristics of the rocks. Some granites, for example, which have suffered little extensional tectonism or uplift since intrusion, may provide good physical barriers, possessing relatively few fractures and low hydraulic conductivities. In contrast, granites in areas with a long history of extensional tectonism and uplift post-dating intrusion are likely to be so intensely fractured that their ability to present a physical barrier must be seriously questioned. High heat production granites, especially, must be viewed with caution, as expansion due to internally generated heat can cause sufficient isostatic uplift to generate fractures even in areas that are otherwise tectonically quiet (Durrance, 1985). Among the metamorphic rocks, only gneisses form good barriers to water movement. Archaean gneisses are particularly suitable, as they tend to occur in relatively stable areas. Of the possible sedimentary rock hosts, thick clays which do not have interbedded silty or sandy beds can provide a suitable host, as can thick salt beds. Salt domes, however, are likely to provide a more coherent barrier than any bedded salt formation. Problems relating to creep during repository construction in clay and salt will need to be overcome.

12.2.4 Geochemical considerations for far-field containment

Geological isolation of radioactive waste will need to be secured in rocks which provide both a physical and a chemical barrier to radionuclide migration (Table 12.15). The chemical barrier will be provided by rocks capable of sorbing radionuclides. Experimental studies of the sorption of various species by rocks of different mineralogy have been carried out (the term 'sorption' is used here to include both absorption and adsorption). The results show that the sorption potential of a rock is determined by the presence of clay minerals and oxyhydroxides (particularly Fe and Mn oxyhydroxides). However, even the most strongly sorptive rock selectively removes some species and leaves others in solution, depending to a large extent upon the chemistry of the solutions involved.

Table 12.15 — Radionuclides in transuranic waste.

	Approximate half-life	Activity ($Bq\ l^{-1}$)
^{90}Sr	28 years	1.3×10^{10}
^{60}Co	5 years	8.1×10^{7}
^{106}Ru	370 days	1.1×10^{8}
^{132}Cs	6.5 days	6.5×10^{7}
^{152}Eu	12 years	1.6×10^{8}
^{154}Eu	16 years	6.5×10^{7}
^{232}Th	1.4×10^{10} years	3.8×10^{4}
^{234}U	2.5×10^{5} years	3.0×10^{2}
^{235}U	7×10^{8} years	1.3×10^{4}
^{238}U	4.5×10^{9} years	2.8×10^{5}
^{238}Pu	86 years	3.4×10^{6}
^{239}Pu	2.4×10^{4} years	3.9×10^{7}
^{240}Pu	6.5×10^{3} years	9.4×10^{6}
^{241}Pu	13 years	2.4×10^{8}
^{241}Am	450 years	6.5×10^{5}
^{244}Cm	18 years	1.6×10^{8}

(After Brookins. 1984.)

12.2.5 Geological analogues

A major limitation of laboratory studies of the sorption of radionuclides is the time scale involved, and there is much uncertainty about the extent to which short-term measurements can be extrapolated to periods of 10^4–10^5 years or so. Fortunately, there exist a number of natural systems which provide analogues of the situation around a leaking repository (Table 12.16).

Perhaps the most important of the natural analogues studied to date is the Oklo reactor (Brookins, 1984). As discussed in Chapter 5, the natural fission reactions at Oklo resulted from the presence of a ^{235}U content of about 3.2% (similar to the level to which modern U ore is enriched for artificial reactors), a favourable U/H_2O ratio, and the absence of neutron poisons. The critical state was attained about 1800 Ma ago, and the reactor continued for about 0.5 Ma during which time about 15 000 MW years of energy was released. About 850 t of U were involved in the reactions, with over 6 t of ^{235}U undergoing fission. Smaller quantities of ^{238}U and ^{239}Pu also fissioned. These reactions resulted in the formation of more than 10 t of fission products identical with those which characterise artificial reactors

Table 12.16 — Radionuclides likely to be associated with radioactive waste disposed of in the earth.

Nuclide	Approximate half-life (years)
^{3}H	12
^{14}C	6×10^{3}
^{36}Cl	3×10^{5}
^{63}Ni	10^{2}
^{79}Se	6.5×10^{4}
^{90}Sr	28
^{93}Mo	3×10^{3}
^{93}Zr	1.5×10^{6}
^{94}Nb	2×10^{4}
^{99}Tc	2×10^{5}
^{106}Ru	1
^{107}Pd	6.5×10^{6}
^{126}Sn	10^{5}
^{129}I	16×10^{6}
^{135}Cs	3×10^{6}
^{137}Cs	30
^{147}Sm	10^{11}
^{151}Sm	90
^{154}Eu	9
^{210}Pb	22
^{226}Ra	1.6×10^{3}
^{227}Ac	22
^{230}Th	8×10^{4}
^{231}Pa	3.3×10^{4}
^{232}Th	14×10^{9}
^{234}U	2.5×10^{5}
^{235}U	7×10^{8}
^{237}Np	2×10^{6}
^{238}U	4.5×10^{9}
^{238}Pu	88
^{239}Pu	2.4×10^{4}
^{241}Am	4.3×10^{2}
^{243}Am	7.4×10^{3}
^{243}Cm	29

However, even though reactor core temperatures of more than 400°C were probably reached, high temperatures do not appear to have been maintained throughout the lifetime of the reactors. On the basis of the normal geothermal gradient in the area and the likely depth at which the reactors developed, it is estimated that the surrounding rocks were probably at a temperature of about 150 °C, rising to 250 °C in the vicinity of the reactors.

Interestingly, the dates obtained by ^{40}K–^{40}Ar methods on 1 Md type illites correspond very well with the results from other isotopic systems. Because the loss of ^{40}Ar from 1 Md illites occurs at temperatures above 125 °C, and transformation to higher illite structures occurs at temperatures of 150–250 °C if sufficient time is available, it is clear that burial of the 1 Md illites cannot have been greater than a few hundred metres and temperatures cannot have exceeded 200 °C. This environment, therefore, provides useful information on the conditions surrounding a repository containing high level waste over a time scale of about 240 ka (ten times the half-life of ^{239}Pu).

Elements in the range $Z=34$ (Se) to $Z=71$ (Lu), the actinides U, Th, Np, Pu and Am, and Pb and Bi (formed by actinide decay) have all been studied in the Oklo investigations. On this basis, it has been concluded that four conditions must be satisfied before elements can migrate (Brookins, 1984).

(1) Transfer of elements from the inside of grains (of pitchblende, in the Oklo system) to the grain surfaces must take place, either through the grains or along cracks.
(2) Desorption of elements concentrated on the grain surfaces must occur, or some other mechanism must bring about their removal.
(3) There must be transport of elements away from the grain surfaces, either as unassociated ions or as complexes.
(4) As the length of transport path depends upon the rock porosity, permeability, tortuosity, degree of saturation, kinetics of nucleation, secondary dissolution–desorption, etc., these factors must be favourable if significant transport is to occur.

Movement of elements to the surface of the pitchblende appears to be largely controlled by diffusion, because elements with similar ionic sizes, ionic charges and electronegativities (such as rare earth elements, Bi, Te and Nb) are retained. Even if fracturing is accompanied by dissolution, these elements would be reprecipitated close to the sites of their release from the pitchblende. Elements such as Mo, Rb, Sr, Cd, Ag, Pd, Sb, Sn, Cs, I, Ba, etc., may be considered to be metastable and may be either removed from the pitchblende or retained depending upon the chemistry of the solution; Ba^{2+}, for example, will not be transported in the presence of SO_4^{2-}, which causes the precipitation of barite (Brookins, 1984).

In the Oklo system, a three-fold division of fission product elements has been recognised: (a) those that have migrated, (b) those that have only migrated locally (that is, over very short distances) and (c) those retained in the system. Clearly, it is those elements which fall into the first category that are important as potential hazards in a leaking repository site. Migration of Rb, Sr, Tc, Cd, I, Xe and Cs has been recognised at Oklo, although each element has some special considerations. Rb, for example, is widely distributed in the reactor system and appears to have undergone long-distance migration. However, when the total amount of Rb present is considered in proportion to the amount of fissiogenic Rb lost from the pitchblende, it appears that up to 60% of the fissiogenic Rb may have been fixed within a few metres of its origin (Brookins, 1981). Similar arguments suggest that Sr may also have been fixed close to the reactor ores. More direct evidence shows that Tc has been confined to a zone extending only a few metres from the reactor ore (Gancarz *et al.*, 1980). The extensive loss of Xe is not surprising, as most gases can be expected to travel freely in rocks of such high porosity and permeability. Migration of Cd, I and Cs also occurred at Oklo but, in the case of Cs, it did not occur until some 25 Ma after shut-down (Brookins, 1984).

There is, therefore, a gradation from local migration to long-distance migration. Indeed, it is likely that a particular element is very mobile under some conditions, while under other conditions it is sorbed or precipitated. Opinions about how 'local migration' should be defined differ. Any movement from the original mineral constitutes migration, and so definition of 'local' is generally just a question of the distance within which fixing occurs. In view of the distances likely to be involved in the transport of radionuclides from a repository back to the surface, a distance of about 10 cm can, perhaps, be considered to be local. Of the fission product elements studied at Oklo, Zr, Mo, Ru, Ag and Sn fall into this category, but they show variation. Zr, for example, appears to have undergone some limited migration, being fixed very near the point of release; most fissiogenic Zr, however, was retained. Mo, in contrast, has been removed from the reactor ore but only locally. Ru also migrated but was fixed within clays that cut the reactor and in sulphides that lie within 10 m of the reactor ore. Like Zr, some of the fissiogenic Ag appears to have migrated locally, while some was retained in the reactor ore. Finally, as shown in Chapter 5, Sn migrated locally.

Fission product elements retained in the reactor ore are Y, Nb, Rh, Pd, In, Te and the rare earth elements. These are elements which are chemically very stable or have properties which allow them to substitute into the pitchblende structure. However, the degree to which the pitchblende structure was maintained in the reactor during the fission reaction is questionable, as expansion and hydration probably occurred under these conditions (Brookins, 1984). Understanding of the chemical substitutions is, therefore, uncertain. The preservation of microcrystalline domains may have exerted important controls, however; without these domains, elements which would normally substitute into the pitchblende structure may have been available for migration.

Unfortunately, there exists another group of fission product elements which cannot be placed into any of these three divisions, simply because there are insufficient data available from Oklo about their migration behaviour. Among these, Br is important because it is unlikely to be retained in pitchblende, irrespective of the degree of coherence of the structure. Sb is also important because of its significance as a fission product in artificial reactions; theoretically, migration of Sb can be expected. Ba is also of interest, and yet little is known in detail about its behaviour in the Oklo system because high background levels of Ba make detection of fissiogenic Ba difficult. More work is needed on these elements before a complete picture of the migration of fission products is available. Furthermore, the actual migration pathways and retention mechanisms, even of those elements which have been studied so far, are still largely unclear and also require further investigation. Particularly important in this respect is the need to acquire a better understanding of the behaviour of I (Brookins, 1984).

Interesting information on element migration can also be obtained from studies of radiogenic and parent nuclides at Oklo. Local migration of radiogenic Pb, for example, has been reported by Gancarz *et al.* (1980), but veins of galena enriched in radiogenic Pb cut the pitchblende. Fixation of Pb near the ore possibly indicates only limited migration, but more study is needed of transport mechanisms and the duration of migration; movement possibly continued well after the reactors ceased to function. Migration of U from both reactor and low grade ores is known to have occurred, but the extent to which U was taken from the reactor ore is unknown. U was probably transported as a carbonate complex, and it is possible that other elements also migrated in this way. However, Np and Pu appear to have been retained within the reactor, possibly because they do not form carbonate complexes. Bi and Th were also retained (Brookins, 1984).

Because the Oklo reactors operated in more or less water-saturated conditions, the system acting in the host rocks has been likened to moderate temperature hydrothermal circulation. Moreover, the porosity and permeability of some of the lithological units present at Oklo are quite high, and the host series has been strongly deformed. Relatively free movement of water can, therefore, be expected to have occurred. Despite these favourable circumstances for migration, fission products have, for the most part, remained close to the reactors and, where migration did occur, it was only over very limited distances. In some cases, elements were retained in the reactors. Some migration only occurred after the reactors had stopped operating. This generally very limited migration suggests that phosphate and carbonate complexes had a very minor role in the movement of any element but U. Although transport of unassociated ions or as oxyions is possible under oxidising conditions, fixing occurs at low temperatures, in reducing conditions, at sorption sites or by precipitation. Transport under reducing conditions could have occurred, but precipitation by dilution or sorption onto, or ion exchange with, clay minerals, may have been effective in removing elements from solution (Brookins, 1984).

On the basis of the very limited and local nature of fission product migration at Oklo, it may be concluded that, for a repository in a rock mass with a lower permeability in a chemically reducing environment, the retention of radioactive waste would probably be better than at Oklo. It is necessary, however, to consider the wider geological characteristics of moderate temperature hydrothermal systems in order to test this conclusion. Unfortunately, a rather different picture then emerges, as such hydrothermal circulation typically gives rise to mineralised vein systems, implying considerable mobility of elements taken into solution. Moreover, circulation systems that give rise to hydrothermal veins typically develop in host rocks very similar to those found at Oklo. Indeed, vein mineralisation does occur at Oklo. The characteristics of hydrothermal mineralisation is thus likely to yield valuable insight into the way in which groundwaters leach, transport and finally deposit radionuclides.

A typical, and well-studied, hydrothermal vein system occurs in southwest England. In the Bodmin Moor–Dartmoor area of east Cornwall and west Devon, the main metalliferous veins typically trend east–west, with a subsidiary set trending north–south (the 'cross-courses'). The east–west veins generally carry ores of Sn, W, As, Cu, Pb and Zn with quartz as the main gangue mineral. Spatially, these ores appear to be arranged in a somewhat concentric fashion, Sn mineralisation being taken to mark the 'centres of emanation' (Dines, 1956). In this concept, hydrothermal fluids largely originated within the granite masses during their emplacement and cooling, the temperature of crystallisation of the minerals controlling their location with distance from the point of origin of the fluid. The highest temperature assemblage (hypothermal, 300–500 °C) contains Sn, W, As and Cu, the former occurring at the highest temperatures. Moderate temperature veins (mesothermal, 200–300 °C) are characterised by U, Ni and Co ores at higher temperatures and Ag, Pb and Zn ores at lower temperatures. Low temperatures (epithermal, 50–200 °C) gave rise to ores of Sb and Fe. In the mesothermal and epithermal veins, gangue minerals such as barite, dolomite and calcite may be present as well as quartz. In some hypothermal veins, feldspar may occur, while fluorite is found in all vein types. The 'centres of emanation' indicated by Sn mineralisation are found either close to the contact of the granite with the Devonian–Carboniferous host rocks or in the interior of granite outcrops. None occurs at any distance from the granite. In contrast with the varied mineralogy of the east–west veins, the north–south trending cross-courses are much less complex. They are mainly characterised by ores of Pb and Zn, typically of mesothermal and epithermal type, and are much more widely distributed away from the granite outcrops than the east–west veins (Beer and Scrivener, 1982).

Although most veins, whether east–west trending or cross-courses, occur over the granite batholith, a number of veins occur well to the north, particularly in Exmoor, where the trend is again either approximately east–west or north–south. Their mineralogy is essentially confined to low temperature hypothermal and mesothermal–epithermal assemblages. As these veins are of the order of 50 km from the granite, it is not thought that they owe their origin to fluids arising from the granite.

In fact, most of the hydrothermal fluids of the veins in the Bodmin Moor–Dartmoor area have now also been shown not to have originated within the granite. Isotope signatures of fluid inclusions and their chemistry characteristically display affinities with formation water or with meteroic water that has undergone interaction with the host rocks, rather than with magmatic water (Rankin and Alderton, 1985). The hydrothermal system that gave rise to the veins appears to have originated in the movement of water in the rocks, principally by fracture flow. On descent to several kilometres, these waters were heated (aided by enhanced geothermal gradients arising from the presence of magmatic heat from the granite intrusion, from the radiogenic heat of the radionuclide-rich granites and from occasional episodes of mantle activity in the area) and then rose as thermal waters. Within this circulation system, leaching of the rocks occurred, followed by transport of the solutions and, finally, deposition of the ore and gangue minerals.

The occurrence of both east–west and north–south trending veins implies that ground-water movement took the form of continuum flow through an interconnected fracture network when viewed on a scale of a few kilometres. However, the cross-course veins are younger that the east–west veins, and the flow that produced the veins seems to have been restricted to simple two-dimensional fracture planes. Moreover, as veins characteristically pinch out and swell within fractures, it is clear that flow was directed along specific channels.

Channel flow thus appears to have provided

a suitable means for transporting large volumes of mineralising fluid carrying large quantities of ore in solution. The system presents profound difficulties, however, when it is remembered that these ore-forming fluids originated by leaching along the earlier part of the flow path. The distribution of the ore minerals suggests that it is the composition of the leached rock that determines vein mineralogy; unless Sn and W are available, for example, ores of these elements cannot be formed, irrespective of the temperature of the fluids. There has been considerable debate about these elements, but the present consensus is that their ores (with those of As) only originate by leaching of granite, and so circulation systems which give rise to these ores must impinge upon the granite to some extent. In contrast, the other ore minerals appear to have formed by leaching of the Devonian and Carboniferous succession, and even As and Sn minerals could be explained in this way. Different members of the succession appear to have contributed different elements; for example, Cu may be related to the presence of basic igneous rocks.

To allow sufficient contact between hydrothermal fluid and rock and to provide sufficient time for the necessary material to be leached, it is clear that flow must have occurred along numerous small fractures over a long period of time. The system favoured is thus one of hydrothermal circulation in which draw-down occurs through a large volume of rock via many small-scale fractures. During draw-down and heating, leaching of the host rocks occurred, probably aided by a progressive decrease in the pH of the water as oxidation of reduced species (particularly pyrite) in the host took place. The fact that draw-down of oxygenated meteoric water occurs to depths of at least 2 km has been shown in deep boreholes in the Carnmenellis granite of Cornwall (Durrance *et al.*, 1982). Expansion of the heated water then caused upward movement to occur. In southwest England, this occurred rapidly along high permeability channels within fractures with a strong preferred orientation. The preferred orientation of the fractures probably reflects the *in situ* stress configuration at the time of vein formation, that is, north–south tension during formation of the east–west veins, and east–west tension when the cross-courses formed. Indeed, because normal hydrothermal circulation is controlled to a greater extent by permeability than by heat flow, the development of vigorous circulation may be related to tectonic activity; episodic tectonic reactivation leading to the episodic mineralisation of the region.

The lesson from hydrothermal vein systems which operated at temperatures of 50–500 °C over long periods of time is that slow movement of hydrothermal fluid through rocks can result in very significant leaching of elements from a wide variety of lithologies. Moreover, the resulting solutions can travel large distances through rocks, such as shales and slates, which should provide good sorption sites. Deposition of material from solution appears to take place only when the ambient temperature has fallen sufficiently to allow direct precipitation.

A possible explanation of transport over long distances is that sorption sites adjacent to flow paths soon become occupied, and diffusion through the rock mass from the passing fluids is too slow to be significant. This appears to contradict the general conclusions drawn from the studies of element leaching and migration from the Oklo reactors. It may reasonably be asked what the true picture would be in the vicinity of a leaking repository, and to what extent the evidence from Oklo and from hydrothermal vein systems can be applied.

Unfortunately, the detailed answer to this question must await the results of further investigations, but there seem to be a number of important controlling factors. Clearly, the chemistry of the invading groundwater is of critical importance in determining the composition of the leached material, its concentration and its form. Also of importance in this respect is the temperature–pressure environment. In addition, the specific nature of the channels in fractures will control the speed of water movement and the path length. Moreover, the nature of channelling will determine the contact area between the groundwater and the rock, and this potentially limits the processes of sorption, irrespective of the nature of the rock.

Each of these factors is site specific, and so very thorough investigations of possible repository sites will be needed before their suitability can be assessed. The evidence from Oklo and hydrothermal veins suggests that rocks which are highly fractured should be avoided. Although they may appear to be tightly closed systems when viewed on a small scale, the presence of sporadic high permeability channels can serve as conduits to transport waste rapidly over large distances. Even the absence of a conduit at the time of site investigation cannot be taken to indicate safety, as stress relief during excavation will change the hydrology of the site. When the probable size of a repository and the need to confine the waste on a scale of 10^2–10^3 m is borne in mind, salt domes and clays are the most promising geological media for the long-term isolation of waste, although even these rocks may prove unsuitable because of the presence of local lithological variation or structural discontinuities. However, there could be political difficulties in the use of such rocks because population density is generally higher in such 'soft rock' areas than in areas underlain by hard fractured rocks which tend to form topographically higher and less productive ground. Clearly, political judgement will play a large part in determining future policy.

12.2.6 Near-field confinement of high level waste

There has been much debate about the best medium for the containment of high level waste, particularly on the relative merits of glass, cement and synthetic rock, for each of which a geological analogue is available.

Low and high melting temperature glasses are available for the containment in vitrified form of high level waste. Low melting glasses, typically borosilicate glasses, are processed at less than 1200 °C. Glass is naturally metastable, but the occurrence of undevitrified natural glass of the order of 10^7 years old suggests that, under suitable conditions, devitrification may not be a problem. However, low melting glasses do contain small-scale fractures and, in the presence of water, devitrification may be more rapid. Interestingly, in the short term, devitrification itself does not lead to loss of stability or to an increase in the potential leach rate, as the crystalline phases that are produced appear to be more resistant to leaching than is the original glass matrix. Nevertheless, it will be recalled from Chapter 4 that the breakdown of crystalline structures by α-particle emitters, that leads to metamictisation, is often accompanied by considerable chemical exchange. Breakdown of the glasses by irradiation does not appear to be a problem, however, as it has been shown that they can withstand 500 ka of α-radiation. Reaction with water is the main concern, with reaction rates of 10^{-7}–10^{-5} g of glass per square centimetre per day at 25 °C, with greater leach rates at higher temperatures. High temperature glasses are richer in SiO_2 and Al_2O_3 and tend to have better resistance to leaching (Brookins, 1984).

'Synroc' is an acronym for synthetic rock in which high level radioactive waste is held in solid solution within natural minerals, particularly perovskite, hollandite and zincolite. The exact nature of the assemblage is determined by that of the waste itself. Thus, for an Al-rich waste, hercynite would also be present, while an Fe-rich waste would contain ulvospinel. These minerals provide suitable host sites for the radionuclides present in the waste. Actinides, rare earth elements, Y, Ru, Tc, Rh, Pd, Sb, Te, In and Sn will, to a greater or lesser extent, be incorporated into zincolite, while nuclides which escape capture and have ionic radii more similar to Ca^{2+} than to Zr^{4+} will be taken up by the perovskite. Larger-radii cations become incorporated into hollandite. Only I is not easily taken into any of these minerals and could be difficult to contain. Clearly, the danger that metamictisation may lead to unacceptable leach rates needs to be considered. However, the fact that Mössbauer spectra have been obtained which suggest that order remains in small structural domains in apparently thoroughly metamict allanite implies that fixing of hazardous radionuclides within synroc minerals could be more secure than in borosilicate glass (Ringwood, 1985).

Cement may be used for containing low level waste, but is not suitable for intermediate level waste. Portland cement, which contains some clay, and bentonite-added cements, in particular, have reasonable retention properties (Rudolph *et al.*., 1981). The main advantages of cement as a means of containing low level waste are its relative stability in near-surface environments and, bearing in mind that large volumes of low level waste are generated each year, its availability and low cost.

12.2.7 Alternatives to excavated repositories

A number of interesting alternatives to conventional excavation of repositories have been put forward, some more practical than others. Very deep burial in large-diameter boreholes drilled in stable crust is perhaps one of the more sensible. In this proposal, boreholes about 2 m in diameter would be drilled to depths of, say, 4 km in areas of high grade metamorphic rocks or igneous rocks which have a low porosity and permeability (Ringwood, 1985). The contained waste would then be deposited in the lower part of the borehole and sealed. This would be practical for relatively small volumes of high level waste, and the temperature increase resulting from the waste itself would be less marked in an environment where temperatures were already probably of the order of 100–150 °C. Although depths of 10 km have been considered by Brookins (1984), the problems of drilling even normal-diameter boreholes to these levels suggest such great depths to be impractical, at least in terms of present-day drilling technology. Ultradeep holes drilled to date also suggest that there will be profound difficulties caused by the presence of high pressure (and, of course, high temperature) fluids and high residual stresses in the rocks at these depths. It might appear that the volume of such a 'repository' is rather restricted. However, this is not a severe limitation. Even for a hole 2 m in diameter drilled to a depth of 4 km, of which only the lower 1 km is used for waste disposal, the available volume is over 3000 m^3, equivalent to an excavation 20 m × 15 m × 10 m. Depending on their diameter and how much of each hole is used, only a few such holes would be needed each year to cope with the waste generated worldwide. A minor disadvantage of this proposal, however, is that, once placed in the hole, the waste would not be easily retrievable. If problems arise after disposal has taken place, solution of those problems is, therefore, likely to be more complex than if access were possible. Whether it is politically advantageous to have a small number of major disposal sites or a large number of smaller repositories is debatable.

Another practical suggestion is disposal within the clays of the abyssal plains of the deep oceans. Various means of emplacing the contained waste within these clays have been proposed, including free-fall penetrators, the drilling of vertical holes, and horizontal trenching using a plough system towed by a surface vessel. The clay layer varies in thickness between 50 and 200 m, but no more than 30 m of this would be available except in drilled holes. Disposal would, therefore, be within the biosphere. Nevertheless, it is possible that the remoteness of the abyssal plain from direct access by man, together with the unfractured nature of the clays and their ability to sorb dangerous nuclides strongly should the containment fail, make this a potentially suitable technique. Moreover, should any waste reach the ocean water, it will be immediately diluted (Brookins, 1984).

The major potential hazard is reconcentration of toxic material by biological or chemical agents with subsequent return to man by indirect routes. However, it is clear that this potential hazard is considerably minimised by burial beneath the sea floor. Recovery of waste disposed of in this way would be difficult, but not impossible, should the need arise.

In addition to possible problems arising from actual disposal, hazards connected with the transport of waste to a disposal site must be considered. Concern has been expressed in connection with oceanic disposal about the safety of the ships that would be used . For this reason, the suggestion that ocean islands be used as repository sites is not as clearly advantageous as a consideration of the disposal environment suggests. The rock of the island itself, although likely to be basaltic and, therefore, to contain permeable zones, is a reasonable containment medium, as groundwater–ocean exchange is very restricted in small islands. Just as with disposal beneath the ocean floor, it can be argued that, should any waste escape from its confinement to reach the ocean, dilution will eliminate the hazard.

Two further proposals for disposal make use of the high heat production of high level waste. The rock melt concept involves the placement of waste as a liquid or slurry into subterranean cavities where the water would soon be lost by evaporation, leaving the radioactive waste to raise the temperature sufficiently to melt the surrounding rock. Mixing of the rock melt with the waste would then occur, with homogenisation and solidification taking place in about 10^3 years. A shaft about 2 km deep to a 6000 m^3 cavity would be suitable. The waste would be emplaced over a period of about 25 years. When filled with waste, the water and steam present would be pumped out and the waste sealed into place. This idea is less practical than other suggestions because of the need for chemical treatment prior to disposal, the problems caused by the pumping of water and steam from the cavity, the irretrievability of the waste and the unknown character of the reactions that might occur between the waste and the host rock (Brookins, 1984).

Allied to the rock melt concept, ice sheet disposal would exploit the heat produced to cause containers of waste buried at shallow levels in an ice sheet to melt their way down to the base of the ice. Clearly, a large ice sheet of great thickness would be needed to ensure optimum conditions for isolation from the surface or margins. Three ways have been described by Brookins (1984) in which the system could operate in practice. In the first, the waste would simply be housed in a suitable cannister and left to melt its way down to the bedrock; the second is similar to the first except that the waste containers would be tethered to the surface and only allowed to sink to a specified depth. The third approach envisages the surface storage of the waste under controlled conditions until a specified mass has been reached, at which time the whole store would be allowed to melt itself down into the ice. This concept, in any of its forms, has several disadvantages over other proposals, particularly because of uncertainties in the behaviour of the radionuclides should the containers fail. For example, water movement within an ice sheet can be concentrated along its base, precisely where, in the untethered systems, the waste would come to rest. Furthermore, ice sheets are not static, movement occurring throughout the sheet can lead to such distortion that overfolding develops. Retrievability of untethered wastes would also be extremely difficult should problems arise. Allied to these difficulties are the shipping problems considered when dealing with ocean and island disposal sites. Furthermore, because sea lanes would be seasonal, additional transport factors argue against implementation of this idea.

For both the rock melt and the ice melt disposal concepts to work, it is important to have a high rate of heat production. High level wastes should, therefore, be processed immediately into a form suitable for disposal, without storage during which the activity decreases. This is counter to the conventional policy of controlled storage of high level waste generally adopted as a sensible precaution during the period of highest activity.

Finally, waste may be disposed of in liquid form by injection into wells at depths of up to 5 km. The waste would be in an acidic form, and sites would be chosen in which the rocks above the injection level were impermeable so that, although the liquid waste would penetrate the rocks, it would be prevented from reaching the surface. Such an approach, however, would be extremely unwise. No ordinary rock is impermeable over the necessary time scale, and fracture flow leading to mixing of the waste with groundwater would certainly occur. Moreover, because of the possible presence of hydrothermal circulation systems of low vigour, even in areas of average and below average geothermal gradient (Durrance and Heath, 1985), there is a great danger that radionuclides would find their way back to the biosphere. Indeed, because the waste would need to be injected under pressure, hydraulic fracturing and enhancement of natural permeability would occur. Injection of the waste at greater depth would only serve to emplace it in a zone of higher temperature. If accompanied by hydraulic fracturing, this could cause hydrothermal circulation to be even more of a problem.

A rather different approach to the use of well injection is the so-called grout injection method, in which a shaly host rock at a depth of 300–500 m would initially be hydraulically fractured. After fracturing, a slurry would be made up of waste, clay and cement and would be injected into the fracture system where it would solidify. In this method, it is important for the fractures to be horizontally rather than vertically orientated. This is necessary to prevent vertical migration of the waste which might bring it back to the surface (Brookins, 1984). As mentioned before, the risk that waste inadvertently reaches highly permeable horizons is one of the general drawbacks to clay or shale as a disposal medium. In the case of liquefied waste in rock that has been deliberately fractured, this problem is considerably enhanced. Only if horizontal fracturing can be produced will the danger be minimised, and that requires a residual stress configuration with a vertical minimum principal stress. Whether or not this requirement is satisfied at any particular depth at any given locality can only be determined by *in situ* stress measurements. The orientation of the residual maximum, intermediate and minimum principal stresses is governed by the tectonic history of the rocks in question. Unfortunately, because recent unloading interacts with the older aspects of the stress history, the relative positions of these stresses are found to change with depth. Clearly, the residual stress must be measured at each horizon being considered for injection. However, even if hydraulic fracturing could be confined to horizontal planes, problems could still arise if lateral lithological variations occurred. Thus the shale must be uniform over as wide an areal extent as possible. Vertical uniformity would also reduce the hazard from any minor vertical fractures that may develop in association with the horizontal fracturing.

Where these conditions are met, the method has a number of advantages over other proposals, especially for the disposal of waste with relatively low levels of activity. This is because shale can be considered to provide good sorption characteristics, should migration of the waste in groundwater take place. The radionuclides would also be reasonably securely held within the cement, contributing to the retention behaviour of the system as a whole.

One word of caution about any method that involves the injection of liquid waste must be given, however, in relation to movement on faults and the generation of earthquakes. The fact that liquid waste injected at depth can actually generate earthquakes is well known in the Denver area of Colorado (Evans, 1966). In this case, the liquid waste involved was contaminated water produced during the manufacture of chemical warfare weapons. In 1961, the Rocky Mountain Arsenal well was drilled to a depth of 3671 m about 19 km northeast of Denver. Up to April 1962, the area had no noticeable seismicity but, in March 1962, high pressure injection of waste at an average rate of 11 000 m^3 $month^{-1}$ commenced and, between April 1962 and September 1965, some

710 earthquakes of magnitude 0.7–4.3 with epicentres in the vicinity of the well were detected. The link between waste injection and earthquake activity became very clear when operation of the well was stopped between September 1963 and September 1964, and the frequency of earthquakes dramatically decreased. On recommencement of waste injection, a low pressure (gravity discharge) system was used which did not noticeably increase the general level of seismic activity but, when high pressure injection started up again in April 1965, once again a noticeable increase in the frequency of earthquakes occurred (Smith, 1973).

The explanation for this is the presence of faults in a stressed state, but not to such a degree that the stress can overcome the frictional resistance along the fault planes to produce movement and to generate earthquakes. However, when the liquid waste is injected, the stress field is changed and resistance to shearing along fault planes is reduced. Thus the injection of liquid waste results in the triggering of fault movements. Now, in terms of waste disposal sites, it should be noted that, except in most unusual circumstances, faults are likely to be present, to a greater or lesser extent, anywhere. In areas of good exposure, even in tectonically quiet ground such as southern England, faults are often seen. Their presence in or near repository sites must, therefore, be expected. The reaction of these faults to liquid waste injected under pressure can only be determined in the light of mesured residual stresses, but it is possible for reactivation of faults to occur during horizontal hydraulic fracturing even if the fault planes are dipping at steep angles. Thus, not only should the possibility of fluid migration be taken into account, but also the possible secondary risk of minor earthquake hazards should be noted. Despite these words of caution, this is a practical proposal and has been used in the United States of America where, at least over a period of several decades, no serious problems have arisen (Brookins, 1984).

12.2.8 Comparisons

Problems that may arise from the interaction of buried radioactive waste with groundwater will be eliminated if burial takes place above the water table. The arid regions of the world would be particularly suitable and tend to have low population densities. However, as far as European radioactive waste is concerned, no simple solution by this means is possible. International agreements may be reached to dispose of radioactive waste on purely commercial terms, but the question is a political mine-field. It is possible, however, that U-bearing countries with suitably arid terrain and low water tables may, as part of a total package, negotiate commercially for the disposal of waste. Again, the political sensitivity of the issue may outweigh any other considerations. Anyway, the difficulties of shipping waste must still be considered and the risks may be unacceptable.

On a time scale of 100 ka to 1 Ma, environmental conditions which are suitable today may not continue unchanged. Although the hazard diminishes with the passage of time, profound environmental changes can occur over relatively short periods. The major changes in glacial activity during the Pleistocene, for example, occurred over time scales of only 1–10 ka, while the climatic fluctuations that gave rise to these changes probably only extended over 100–1000 years. The onset of glaciation in temperate climates containing radioactive waste repositories would probably not present a hazard to mankind. If the glaciation were short lived, it would probably not dramatically alter the geological characteristics of a site; if it was so extensive that it produced major changes, the areas would be largely unpopulated. However, more significantly, general changes in climate may cause profound modification of the groundwater hydrology around a repository, and this should be borne in mind, even where disposal environments today possess suitable groundwater charateristics.

It is claimed that watching our wastes will impose a great burden on our descendants, but the results of detailed analysis of the probabilities of the hazards likely to result from the disposal of high level waste suggest, in fact, that this burden would be trivial (Cohen, 1982). Once a repository has been sealed, surveillance need only consist of periodic inspections to make sure that warning signs were not destroyed or defaced beyond recognition, to prevent drilling or mining in the area and to take water samples from streams and wells for testing. On this basis, it has been concluded that a single part-time employee could supervise the waste. Furthermore, it has been calculated that radioactive wastes are only likely to lead to 0.4 deaths per million years, compared with an estimated 3000 fatalities that are caused each year by natural radiation (Cohen, 1982).

Also of interest is the hazard from U mill tailings. This is the material left after U ore has been processed into 'yellowcake' (which is 70–90% U_3O_8 equivalent), consisting of U daughters together with any ^{232}Th and its daughters, rock fragments, etc. On the basis that, for every tonne of fuel, more than 10 t of U pass through the ore processing mill, not surprisingly it has been calculated that both the ingestion hazard and the inhalation hazard presented by particulate matter from high level waste are higher than for mill tailings. However, because of the different half-lives of these materials, the situation becomes reversed after about 250 years in the case of the ingestion hazard, and after about 10 ka for the inhalation hazard (Fig. 12.4). Moreover, at the greatest difference (zero time), the hazard from freely dispersed high level waste is only about two orders of magnitude greater than for the mill tailings (Cohen, 1982).

At its upper limit, it is estimated that the ingestion hazard would lead to between 10^3 and 10^6 deaths per year if all high level waste were dumped in rivers while, at its upper limit, the mill tailings hazard would cause 10^3 deaths per year. Release as fine particles would also

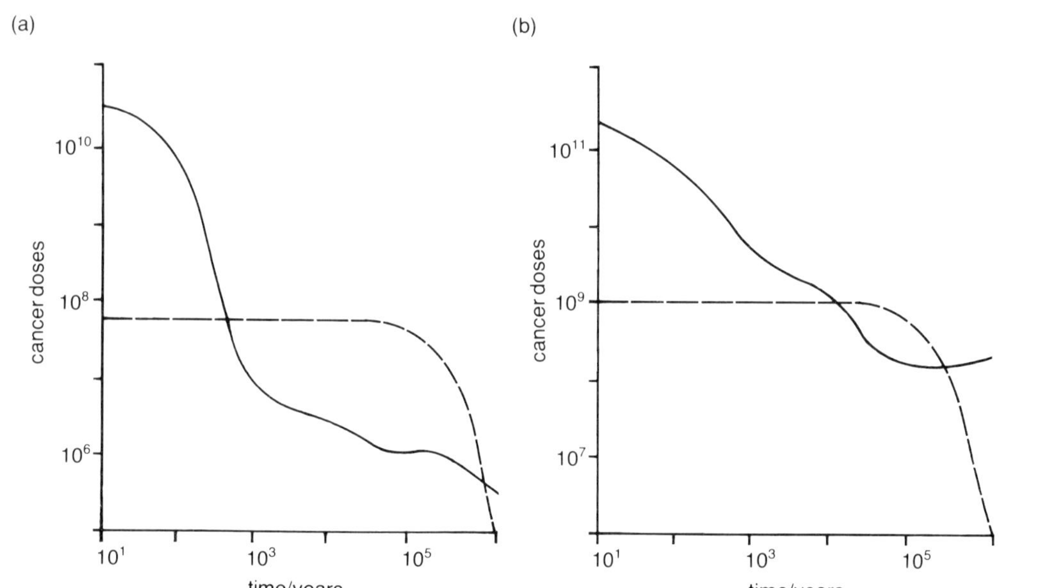

Fig. 12.4 — Comparison of the hazards from U mill tailings (broken lines) and high level radioactive waste (solid lines): (a) as a result of ingestion; (b) as a result of inhalation. (Redrawn with permission from Cohen (1982). Copyright CRC Press, Inc., Boca Raton, Florida, USA.)

produce about 10^6 deaths per year from high level waste and about 10^3 deaths per year from mill tailings. As these calculations ignore the danger of inhalation of ^{222}Rn from mill tailings, it is possible that U mill tailings are a greater hazard than high level waste, and adequate safety precautions should be taken to ensure their safe disposal (Cohen, 1982). Interestingly, the processing of high level waste in a fast reactor can dramatically reduce the short-term radiation hazard (Fig. 12.5). The comparison with mill tailings then becomes even closer.

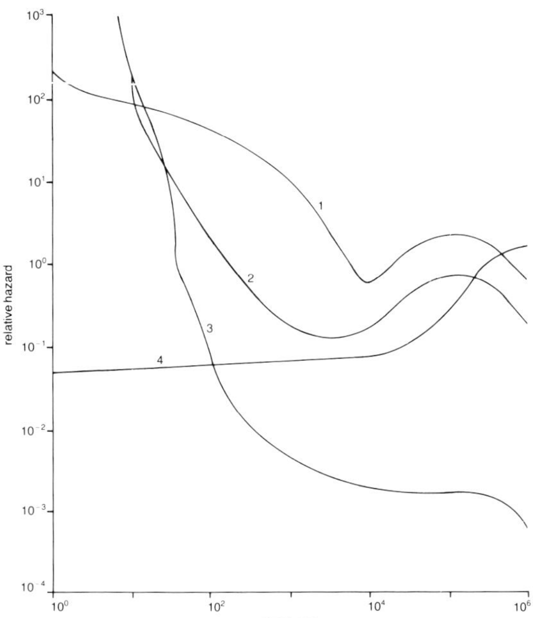

Fig. 12.5 — Long-lived actinides in radioactive waste could be converted to short-lived fission products in a fast reactor The magnitude of the effect depends upon the duration of re-irradiation. Comparison of the radiation hazards due to 12 and 36 year re-irradiated wastes is made with normal waste and that due to unirradiated ^{238}U: 1, waste for reprocessing; 2, waste after 12 years re-irradiation; 3, waste after 36 years re-irradiation; 4, unirradiated ^{238}U. (Clarke *et al.*, 1978, redrawn with permission.)

Although earlier reference was made to the impracticality of disposing of radioactive waste by burning, it should be noted that burning of fossil fuels liberates the radionuclides they contain and disperses them into the atmosphere. In this connection, the radiation exposure resulting from material released by the stack of a 2000 MW coal-fired power station has been investigated (Camplin and Hallam, 1978). These studies revealed that particulate ash contains ^{232}Th, ^{235}U, ^{238}U and ^{40}K, as well as U and Th decays products. Rn, of course, principally occurs as a gaseous phase, but the dose received from Rn is considered negligible at only 0.2% of the total. The main sources of exposure were found to be external radiation from surface deposits originating in the plume, the inhalation of deposited material which becomes resuspended in the air, and the ingestion of food contaminated by surface deposition. For typical conditions, a maximum total individual annual dose of 230 μSv was calculated to arise from these sources. The main contribution (over 95%) to the individual dose was identified by Camplin and Hallam as arising from ingestion of contaminated foodstuffs, with ^{210}Pb (43%) and ^{231}Pa (38%) as the main sources, and ^{210}Po (14%) a secondary source. More recently, however, Corbett (1983) has noted that these values need to be revised to take into account more realistic assumptions about the levels of ^{210}Pb and ^{231}Pa in animal livers and kidneys. An individual maximum annual dose of 50 μSv is now considered more appropriate for ingested material. Corbett has also calculated that the use of waste products from coal-fired generators in building materials contributes a maximum individual dose of about 120 μSv year^{-1}, although some uncertainty exists in that these building materials may actually reduce the dose received from external γ-radiation. Clearly, however, no individual dose is as high as the natural background.

These values may be compared with the individual doses likely to be received, under normal circumstances, in the vicinity of nuclear establishments. For example, individuals living near nuclear fuel preparation plants may expect only about 5 μSv year^{-1} from air-borne discharges and less than 50 μSv year^{-1} from discharges to water. Near reactors, individuals may receive 200 μSv year^{-1} from gaseous discharges and no more than 100 μSv year^{-1} due to water discharges. Near a nuclear fuel reprocessing plant, about 200 μSv year^{-1} may be received through discharges to air and less than 1300 μSv year^{-1} due to the discharge of liquid wastes (a high value owing to a high level for the most exposed individual who consumes much fish containing Cs). Overall, therefore, the most exposed individual living near a nuclear establishment in the UK is likely to receive about 1000 μSv year^{-1}, or only about six times the maximum level that arises from living within the discharge plume of a coal-fired power station in a house containing a high proportion of building materials using ash products.

12.3 UNDERGROUND NUCLEAR EXPLOSIONS

Underground testing of nuclear weapons provides an energy source which can be used for investigating the internal structure of the Earth by seismic methods. Much research has also been undertaken into the recognition of seismic events caused by such tests, and how they may be distinguished from natural earthquakes. The seismic signal from an underground explosion is different from that caused by a natural earthquake. Among the seismic data collected at seismological stations distributed over the Earth's surface is the polarity of the first motion of P and S waves. Whether P waves have a compressional or extensional first motion at a particular seismological station depends upon the nature of the source of the waves and the geometrical relationship of the station to the source. Similarly, the polarity of the first S wave motion is governed by this geometrical factor. The initial direction from which the waves originate is known from the path that they have taken to reach the station, and this can be plotted with the polarity on a suitable projection. P waves generated by a fault movement produce a seismograph which shows a pattern of compression and extension which defines the orientation of the fault (and a complementary direction) and the sense of movement on the fault plane. Fault movements also give rise to the generation of S waves of large amplitude. The pattern of compressional and extensional first movements obtained in this way is known as a 'fault plane solution' and, if a number of such solutions are constructed for a seismically active region, patterns emerge that demonstrate the type of fault displacements taking place; for example, they can distinguish normal from thrust faults, and horizontal displacements from vertical displacements (Bott, 1971).

If fault plane solutions are constructed for seismic waves originating in an underground nuclear explosion, it is found that the pattern of compression and extension characteristic of fault-generated earthquakes does not occur. Instead, there is simply a compressional first-motion field. This arises because an underground explosion is a point source of expansion, the vibrations from which move out as a spherical compressional wave front. For this reason too, S wave motions are much less in evidence than in fault-generated earthquakes. Detailed recording of the wave characteristics thus enables not only the location, time and magnitude of an earthquake source to be determined but also whether the source is natural or man-made (Bolt, 1976).

The peaceful use of underground nuclear explosions has also been proposed for a variety of purposes, ranging from the creation of cavities, which could be exploited as geothermal reservoirs, to improving the recovery of oil and natural gas. The movement of large volumes of barren rock to gain access to valuable deposits, the excavation of depressions for constructional purposes, and even the development of methods to 'blow out' gas and oil well fires have been suggested as possible applications. However, the obvious environmental and social implications of such uses make the majority of ideas impractical.

The use of a deeply buried explosion to

produce a large underground cavity, or to stimulate gas or oil production, is perhaps worthy of further consideration, but the danger of escape of radioactive debris following such an explosion is great, particularly where gaseous radionuclides such as ^{85}Kr, ^{133}Xe and ^{131}I are concerned. Other radionuclides can, of course, be transported to the surface in solution by flowing groundwater. Clearly, the factors to be consisdered in the disposal of high level radioactive waste should also be considered here although, in this case, the source is limited in both size and activity. In fact, a number of experimental determinations of the effect of nuclear explosions have been carried out by the USA and the USSR to test their application in these fields (Kruger, 1975).

The use of underground nuclear explosions in stimulating gas production simply exploits the profound effect that contained detonation has on the surrounding rocks. Essentially, the heat from the explosion is sufficient to melt some of the host rock, and this collects as a pool beneath the point of detonation. In the immediate vicinity of this point, and in a column extending above and around the explosion, fracturing of the host rock is so intense that a rubble chimney is formed (Fig. 12.6). At the top of the chimney, a void is produced which marks the limit of intense fracturing. Around the rubble chimney and beneath the solidified pool, the host rocks display a decreasing intensity of fracturing with distance. Fractures around the point of detonation may be roughly concentric and radial, but, with increasing height above the detonation point, the development of subhorizontal fractures becomes paramount. Close to the rubble chimney, subvertical fractures are also well developed. When a particular gas-bearing horizon cannot be exploited because of low permeability, and conventional methods of hydraulic fracturing cannot successfully improve the rate of gas flow, such an explosion can be detonated close to the base of the horizon (assuming that this level is sufficiently deep to allow containment of the explosion). This is likely to cause sufficient

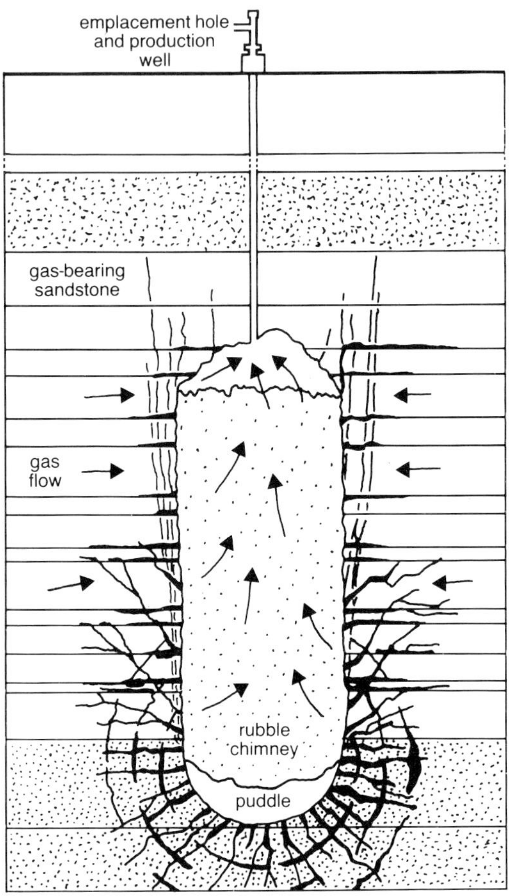

Fig. 12.6 — Fracturing and gas stimulation by an underground nuclear explosion. (Redrawn with permission fron Whicker and Schultz (1982). Copyright CRC Press, Inc., Boca Raton, Florida, USA.)

fracturing of the rock over a suitably large volume to permit seepage of gas into the zone of the rubble chimney, subsequently to collect in the void at the top. A production well sunk to tap the void (now a gas reservoir) can then extract reasonable quantities of gas at acceptable flow rates maintained by recharge from the large volume of fractured rock

(Whicker and Schultz, 1982). Environmental problems associated with this, however, restrict its application. Even if the release of radionuclides to the surface is small, ground movements produced by the detonation may prove unacceptably damaging.

The formation of surface excavations using underground nuclear explosions is technically feasible and has been investigated experimentally. In this case, the detonation point is so shallow that the heave of the ground above the explosion continues to grow until the mound begins to fall over and to collapse on itself. Because venting occurs during the growth of the mound, an extensive cavity is produced around and above the point of detonation. On collapse of the mound, fall-back occurs, leaving a surface crater. The environmental contamination that results from such exercises is unacceptable, however, and seriously limits this application of nuclear explosions (Whicker and Schultz, 1982).

BIBLIOGRAPHY

Abbot, J. D., Lakey, J. R. A., and Mathias, D. J. (1960) Natural radioactivity in west Devon water supplies. *Lancet* **2**, 1272.

Aberg, B., and Hungate, F. P., (1967) *Radioecological concentration processes.* New York, Pergamon.

Adams, J. A. S., and Lowder, W. M. (1964) *The natural radiation environment.* Chicago, Illinois, University of Chicago Press.

Alexandra, A. (1961) *Introduction to soil microbiology.* New York, Wiley.

Allen-Price, E. D. (1960) Uneven distribution of cancer in west Devon with particular reference to the diverse water supplies. *Lancet* **1**, 1235.

Ames, L. L., McGarrah, J. E., Walker, B. A., and Slater, P. F. (1982). Sorption of uranium and cesium by Hanford basalts and associated secondary smectite. *Chem. Geol.* **35**, 205.

Anderson, I. D., and Nilson, I. (1964) Exposure following ingestion of water containing radon-222. *Proceedings of the IAEA Symposium on Assessment of Radioactive Body Burdens in Man II.* Vienna, International Atomic Energy Agency.

Archer, V. E., Gillam, J. D., and Wagoner, J. K. (1976) Respiratory disease mortality among uranium miners. *Anal. New York Acad. Sci.* **271**, 280.

Augier, J. A., Christian, T. J., Jones, T. D., Kerr, G. D., Perdue, P. D., Shinpaugh, W. H., and Thorngate, J. H. (1973) Contribution of natural terrestial sources to the total radiation dose to man. *Rep.* **ORNL TM-4323.** Oak Ridge, Tennessee, Oak Ridge National Laboratory.

Bailey, N. J. L., Evans, C. R., and Milner, C. W. D. (1974) Applying petroleum geochemistry to search for oil: examples from Western Canada Basin. *Bull. Am. Assoc. Petrol. Geol.* **58**, 2284.

Barretto, P. M. C. (1973) Radon-222 emanation characteristics of rocks and minerals. *Proceedings of Panel on Radon in Uranium Mining, IAEA Publ.* **PL-565/1.** Washington, DC, International Atomic Energy Agency.

Beer, K. E., and Scrivener, R. C. (1982) Metalliferous mineralisation. In Durrance, E. M., and Laming, D. J. C. (eds), *The geology of Devon.* Exeter, University of Exeter Press.

Behounek, F. (1970) History of exposure of miners to radon. *Health Phys.* **19**, 56.

Bird, G. W., and Fyfe, W. S. (1982) The nuclear waste disposal problem — an overview from a geological and geochemical perspective. *Chem. Geol.* **36**, 1.

Bolt, B. A. (1976) *Nuclear explosions and earthquakes.* San Francisco, California, Freeman.

Boothe, G. F. (1977) The need for radiation controls in the phosphate and related industries. *Health Phys.* **32**, 285.

Bott, M. H. P. (1971) *The interior of the Earth.* London, Arnold.

Bourke, P. J., Durrance, E. M., Heath, M. J., and Hodgkinson, D. P. (1985) Fracture hydrology relevant to radionuclide transport. *CEC Rep.* **EUR 9854 EN.** Luxembourg, Commission of the European Communities.

Brookins, D. G. (1976) Shale as a repository for radioactive waste: the evidence from Oklo. *Environ. Geol.* **1**, 255.

Brookins, D. G. (1978) Retention of transuranic and actinide elements and bismuth at the Oklo natural reactor, Gabon: application of Eh–pH diagrams. *Chem. Geol.* **23**, 309.

Brookins, D. G. (1981) Alkali and alkaline earth studies at Oklo. In Moore, J. G. (ed.) *The scientific basis for nuclear waste management,* Vol. III. New York, Plenum.

Brookins, D. G. (1982) *The scientific basis for nuclear waste management,* Vol. VI. New York, Plenum.

Brookins, D. G. (1984) *Geochemical aspects of radioactive waste disposal.* New York, Springer.

Burns, W. G., Hughes, A. E., Marples, J. A. C., Nelson, R. S., and Stoneham, M. (1982) Radioactive effects and the leach rates of vitrified radioactive waste. *Nature* **295**, 130.

Busigin, A., Van Der Vooren, A. W., Lin Pai, H., and Phillips, C. R. (1979) Development of instrumentation for the measurement of radon, thoron and their daughters in uranium mines and in the environment. *Min. Sci. Eng.* **11**, 193.

Camplin, W. C., and Hallam, J. (1978) Assessment of the radiation exposure from the radioactivity released from the stack of a 2000 MW coal-fired power station. *Proceedings of the Fifth International IRPA Congress,* Abstract 1114. International Radiation Protection Agency.

Clarke, R. H., Macdonald, H. F., and Wheatley, B. M.

(1978) Nuclear power and the environment. *CEGB Res.* **6**.
Cliff, K. D. (1978) Assessment of airborne radon daughter concentrations in dwellings in Great Britain *Phys. Med. Biol.* **23**, 696.
Cliff, K. D. (1978) Population exposure to short-lived daughters of radon-222 in Great Britain. *Bull. Natl. Radiol. Prot. Board* **22**, 18.
Cliff, K. D. (1980) Radon. Notes for Postgraduate Radiological Protection Course. National Radiological Protection Board.
Cohen, B. L. (1982) High-level radioactive waste from light water reactors. In Klement, A. W. (ed.), *CRC handbook of environmental radiation.* Boca Raton, Florida, CRC.
Cohen, J. J., Lewis, A. E., and Braun, R. L. (1972) *In situ* incorporation of nuclear waste in deep molten silicate rock. *Nucl. Technol.* **14**, 76.
Corbett, J. O. (1983) The radiation dose from coal burning: a review of pathways and data. *Radiat. Prot. Dosim.* **4**, 5.
DeLaeter, J. R., Rosman, J. K. R., and Smith, C. L. (1980) The Oklo natural reactor: cumulative fission yields and retentivity of the symmetric mass region fission products. *Earth Planet. Sci. Lett.* **50**, 239.
Denham, D. H. (1982) Sampling instruments and methods. In Klement, A. W. (ed.), *CRC handbook of environmental radiation.* Boca Raton, Florida, CRC.
Dickin, A. P. (1981) Hydrothermal leaching of rhyolite glass in the environment has implications for nuclear waste disposal. *Nature* **294**, 342.
Dickson, D. (1980) Uranium exposure limits in dispute. *Nature* **286**, 833.
Dines, H. G. (1956) The metalliferous mining region of south west England. *Mem. Geol. Surv. Gr. Brit.*
Dixon, D. W. (1983) Radiation hazards to collectors of geological specimens containing natural radioactivity. *Rep. Natl. Radiol. Prot. Board* **R131**.
Donath, F. A. (1980) Postclosure environmental conditions and waste form stability in a geologic perspective for high level nuclear waste. *Nucl. Chem. Waste Management* **1**, 103.
Down, C. G., and Stocks, J. (1976) *The environmental problems of tailings disposal at metal mines.* London, HMSO.
Dreesen, D. R., Williams, J. M., Marple, M. L., Gladney, E. S., and Perrin, D. R. (1982) Mobility and bioavailability of uranium mill tailings contaminants. *Environ. Sci. Technol.* **16**, 702.
Drozd, R. J.., Hohenberg, C. M., and Morgan, C. J. (1974) Heavy rare gases from Rabbit Lake (Canada) and Oklo mine (Gabon): natural spontaneous chain reactions in old uranium deposits *Earth Planet. Sci. Lett.* **23**, 28.
Duggan, M. J. (1973) Some aspects of the hazards from airborne thoron and its daughter products. *Health Phys.* **24**, 301.
Duguid, J. O. (1978) Radionuclide absorption by manganese oxides and its implications for radioactive waste disposal. *Nature* **274**, 44.
Durrance, E. M. (1985) Hydrothermal circulation and isostasy with particular reference to the granites of southwest England. In *High heat production granites, hydrothermal circulation and ore genesis.* London, Institution of Mining and Metallurgy.
Durrance, E. M., Bromley, A. V., Bristow, C. M., Heath, M. J., and Penman, J. (1982) Hydrothermal circulation and post-magmatic changes in granites of south west England. *Proc. Ussher Soc.* **5**, 304.
Durrance, E. M., and Heath, M. J. (1985) Thermal groundwater movement and radionuclide transport in south west England. *Mineral. Mag.* **49**, 289.
Dyer, A., and Mikhal, K. Y. (1985) The use of zeolites for treatment of radioactive waste. *Mineral. Mag.* **49**, 203.
Edmonds, E. A., McKeown, M. C., and Williams, M. (1975) *British regional geology: south-west England.* London, HMSO.
Eisenbud, M. (1973) *Environmental radioactivity.* New York, Academic Press.
Eisenbud, M. (1981) The status of radioactive waste management: needs for reassessment. *Health Phys.* **40**, 429.
Evans, C. R., Rogers, M. A., and Bailey, N. J. L. (1971) Evolution and alteration of petroleum in western Canada. *Chem. Geol.* **8**, 147.
Evans, D. M. (1966) The Denver area earthquakes and the Rocky Mountain Arsenal disposal well. *Mountain Geol.* **3**, 23.
Evans, R. D. (1969) Engineer's guide to the elementary behaviour of radon daughters. *Health Phys.* **17**, 229.
Fleischer, R. L., Burford-Price, P., and Walker, R. M. (1975) *Nuclear tracks in solids, principles and applications.* Berkeley, California, University of California Press.
Fleischer, R. L., and Mogro-Campero, A. (1978) Mapping of integrated radon emanation for the detection of long-distance migration within the Earth. Techniques and principles. *J. Geophys. Res.* **83**, 3539.
Galkin, N. P., Maiorov, A. A., and Verynatin, U. D. (1963) *The technology of the treatment of uranium concentrates.* Oxford, Pergamon.
Gancarz, A., Cowan, G., Curtis, D., and Maeck, W. (1980) Tc-99, Pb and Ru migration around the Oklo natural fission reactors. In Northrup, C. J. M. (ed.), *The scientific basis for nuclear waste management,* Vol. II. New York, Plenum.
Gera, F. (1972) Review of salt tectonics in relation to the disposal of radioactive wastes in salt formations. *Bull. Geol. Soc. Am.* **83**, 3551.
Gera, F. (1975) Geochemical behaviour of long-lived radioactive wastes. *Rep.* **ORNL TM-4481.** Oak Ridge, Tennessee, Oak Ridge National Laboratory.
Gomez, M. (1982) *Radiation hazards in mining.* New York, Society of Mining Engineers.
Gonzales, S. (1982) Host rocks for radioactive waste disposal. *Am. Sci.* **70**, 191.
Gray, D. A. (1978) Geological aspects of disposal of solid, high-level radioactive waste. In Duff, P. McL. D. (ed.), *Geological aspects of uranium in the environment, Geol. Soc. Lond. Misc. Paper* **7**.
Gray, D. A., Greenwood, P. B., Bisson, G., Cratchley, C. R., Harrison, R. K., Mather, J. D., Ostle, D., Poole, E. G., Taylor, B. J., and Willmore, P. L. (1976) Disposal of highly active solid radioactive wastes into

geological formations — relevant geological criteria for the United Kingdom. *Rep. Inst. Geol. Sci. London.* **76–12**.

Haggerty, S. E., (1983) Radioactive nuclear waste stabilization: aspects of solid-state molecular engineering and applied geochemistry. *Ann. Rev. Earth Planet. Sci.* **11**, 133.

Hamilton, E. I. (1972) The concentration of uranium in man and his diet. *Health Phys.* **22**, 149.

Hamilton, E. I. (1985) The disposal of radioactive wastes into the marine environment: the presence of hot particles containing Pu and Am in the source term. *Mineral. Mag.* **49**, 177.

Harley, N. H., and Pasternack, B. S. (1979) Potential carcinogenic effects of actinides in the environment. *Health Phys.* **37**, 291.

Heath, M. J. (1984) Solute migration experiments in fractured granite, south west England. *Proceedings of the CEA–NEA Workshop on the Design and Instrumentation of In Situ Experiments in Underground Laboratories for Radioactive Waste Disposal.* Brussels.

Heath, M. J. (1985) Geological control of fracture permeability in the Carnmenellis granite Cornwall: implications for radionuclide migration. *Mineral. Mag.* **49**, 233.

Heath, M. J., and Durrance, E. M. (1985) Radionuclide migration in fractured rock: hydrological investigations at an experimental site in the Carnmenellis granite, Cornwall. *CEC Rep.* **EUR 9668 EN**. Luxembourg, Commission of the European Communities.

Hems, G. (1966) Acceptable concentrations of radon in drinking water. *Air Water Pollut.* **10**, 769.

Holliger, P., Devilliers, C., and Retali, G. (1978) Evaluation des temperatures neutroniques dans les zones de reaction d'Oklo par l'etude des rapports isotopiques. In *Natural fission reactors, IAEA Tech. Commun.* **119**. Vienna, International Atomic Energy Agency.

Hollister, C. D., Anderson, D. R., and Heath, G. R. (1981) Subseabed disposal of nuclear wastes. *Science* **213**, 1321.

Howes, M. J., and Jones, M. J. (1984) Third international mine ventilation congress. London, Institution of Mining and Metallurgy.

Hughes, J. S., and Roberts, G. C. (1984) The radiation exposure of the United Kingdom population. *Rep. Natl. Radiol. Prot. Board* **R173**.

Hultquist, B., (1956). Studies on naturally occurring ionizing radiations with special reference to radiation doses in Swedish houses of various types. *Kungl. Sven. Vetenskapsakad. Handl.* **6**, 3.

Hursh, J.B., Morken, D. A., Davis, T. P., and Lovaas, A. (1965) The fate of radon ingested by man. *Health Phys.* **11**, 465.

International Atomic Energy Agency (1980) *Underground disposal of radioactive wastes,* Vol. II, *Otaniemi, Finland 1979.* Vienna, International Atomic Energy Agency.

International Commission on Radiological Protection (1977) *Recommendations, ICRP Publ.* **26**, International Commission on Radiological Protection.

International Commission on Radiological Protection (1985) *Recommendations, ICRP Bull.,* International Commission on Radiological Protection. **65**, 5.

Kathren, R. L. (1984) *Radioactivity in the environment.* Chur, Harwood.

Keen, N. J. (1978) Radioactive waste management in the United Kingdom. In Duff, P. McL. D. (ed.), *Geological aspects of uranium in the environment, Geol. Soc. Lond. Misc. Paper* **7**.

Kenny, A. W., Crooks, R. N., and Kerr, J. R. W. (1966) Radium, radon and daughter products in certain drinking waters in Great Britain. *J. Inst. Water Eng. Sci.* **20**, 123.

Klement, A. W. (1964) Radioactive fallout from nuclear weapons tests. *AEC Rep.* **CONF-765** Washington, DC, Atomic Energy Commission.

Klement, A. W. (1982a) Natural sources of environmental radiation. In Klement, A. W. (ed.), *CRC handbook of environmental radiation.* Boca Raton, Florida, CRC.

Klement, A. W. (1982b) Man-made sources of environmental radiation. In Klement, A. W. (ed.), *CRC handbook of environmental radiation.* Boca Raton, Florida, CRC.

Klement, A. W. (ed.) (1982) *CRC handbook of environmental radiation.* Boca Raton, Florida, CRC.

Kruger, P. (1975) Peaceful nuclear explosives. *ERDA Rep.* **44**, Washington, DC, US Energy Research and Development Administration.

Kunz, E., Serc, J., Placek, V., and Horacek, J. (1979) Lung cancer in man in relation to different time distributions of radiation exposure. *Health Phys.* **36**, 669.

Lewis, J. B. (1985) Radioactive wastes — an overview. *Mineral Mag.* **49**, 153.

Lindell, B. (1968) Ingested radon as a source of human radiation exposure. *Proceedings of International Congress on Radiation Protection, Rome.*

Logan, S. E. (1974) Deep self-burial of wastes by rock melting capsules. *Nucl. Technol.* **21**, 111.

Lorenz, E. (1944) Radioactivity and lung cancer: A critical review of lung cancer in the miners of Scheeberg and Joachimstahl. *J. Natl. Cancer Inst.* **5**, 1.

Mather, J. D. (1978) Proving a potential site for the geological disposal of high level radioactive waste. In Duff, P. McL. D. (ed.), *Geological aspects of uranium in the environment, Geol. Soc. Lond. Misc. Paper* **7**.

McCarthy, G. (1979) *Scientific basis for nuclear waste management,* Vol. I. New York, Plenum.

McCulloch, C. E., Angus, M. J., Crawford, R. W., Rahman, A. A., and Glasser, F. P. (1985) Cements in radioactive waste disposal: some mineralogical considerations. *Mineral Mag.* **49**, 211.

Moghissi, A. A. (1978) Radioactivity in consumer products. *NRC Rep.* **CP-0001**. Washington, DC, US Nuclear Regulatory Commission.

Moore, J. G. (1981) *The scientific basis for nuclear waste management,* Vol. III. New York, Plenum.

Moore, J. McM. (1982) Mineral zonation near the granite batholiths of south west and northern England, and some geothermal analogues. In Evans, A. M. (ed.), *Metallization associated with acid magmatism.* Chichester, West Sussex, Wiley.

Murray, J. (1981) *Uranium and nuclear energy: 1980.* Guildford, Surrey, Westbury House.

Murray, J., and Casteel, D. (1980) *Uranium and nuclear energy. Mining Journal.*

National Council on Radiation Protection and Measurements (1975) Natural background radiations in the United States. *NCRP Rep.* **45**. Washington, DC, National Council on Radiation Protection and Measurements.

National Radiological Protection Board (1981) *Living with radiation.* London, HMSO.

National Radiological Protection Board (1986) *Living with radiation.* London, HMSO.

Northrup, C. J. M. (1980) *Scientific basis for nuclear waste management,* Vol. II. New York, Plenum.

O'Riordan, M. C. (1976) Radiological protection in the mining industry. *Bull. Natl. Radiol. Prot. Board* **15**, 11.

O'Riordan, M. C. (1978) Radiological protection in uranium mining and milling. In Duff, P. McL. D. (ed.), *Geological aspects of uranium in the environment, Geol. Soc. Lond. Misc. Paper* **7**.

O'Riordan, M. C., James, A. C., Rae, S., and Wrixon, A. D. (1983) Human exposure to radon decay products inside dwellings in the United Kingdom. *NRPB Rep.* **R152**. National Radiological Protection Board.

O'Riordan, M. C., Rae, S., and Thomas, G. H. (1982) Radon in British mines — an abridged review. *Bull. Natl. Radiol. Prot. Board* **45**, 8.

Overcamp, T. J. (1982) Low-level radioactive waste disposal by shallow land burial. In Klement, A. W. (ed.), *CRC handbook of environmental radiation.* Boca Raton, Florida, CRC.

Pearson, J. E. (1967) *Natural environmental radioactivity from radon-222, Publ.* **999**, US Department of Health, Education and Welfare.

Perkins, R. W., and Nielson, J. M. (1965) Cosmic-ray produced radionuclides in the environment. *Health Phys.* **11**, 1297.

Philipp, G., Efister, H., and Pauly, H. (1979) Occupational dose from natural radionuclides in phosphate fertilizers. *Radiat. Environ. Biophys.* **16**, 143.

Polikarpov, G. G. (1966) *Radioecology of aquatic organisms.* New York, Reinhold.

Rankin, A. H., and Alderton, D. H. M. (1985) Fluids in granites from southwest England. In *High heat production granites, hydrothermal circulation and ore genesis* London. Institution of Mining and Metallurgy.

Reichle, D. E., Dunaway, P. B., and Nelson, D. J. (1970) Turnover and concentration of radionuclides in food chains. *Nucl. Safety* **11**, 43.

Rimsaite, J. (1982) Mode of occurrence of secondary radionuclide-bearing minerals in natural argillized rocks: a preliminary report related to a barrier clay in nuclear waste disposal. *Paper Geol. Surv. Can.* **82-1A**, 247.

Ringwood, A. E. (1980) Safe disposal of high level radioactive wastes. *Fortschr. Mineral.* **58**, 149.

Ringwood, A. E. (1982) Immobilization of radioactive wastes in Synroc. *Am. Sci.* **70**, 201.

Ringwood, A. E. (1985) Disposal of high level nuclear wastes: a geological perspective. *Mineral. Mag.* **49**, 159.

Ringwood, A. E., Oversby, V. M., Kesson, S. E., Sinclair, W., Ware, N., Hibberson, W., and Major, A. (1981) Immobilization of high level wastes in synroc: a current appraisal. *Nucl. Chem. Waste Management* **2**, 287.

Robson, A. (1983) Comparison of radiological impact of coal and nuclear fuel cycles. *TPRD/GS/E/1.* London, Central Electricity Generating Board.

Robson, A., Manning, P. T., and Corbett, J. O. (1981) Radioactive emissions from coal-fired power stations. *CEGB Newslett.* **114**. London, Central Electricity Generating Board.

Rock, R. L., Lovett, D. B., and Nelson, S. C. (1969) Radon-daughter exposure measurement with track etch films. *Health Phys.* **16**, 617.

Rudolph, G., Vejmelka, P., and Koster, R. (1981) Leach and corrosion tests under normal and accident conditions on cement products from simulated intermediate level evaporator concentrates. In Moore, J. G. (ed.), *The scientific basis for nuclear waste management,* Vol. III. New York, Plenum.

Rundo, J., Marksun, F., and Plondke, N. J. (1979) Observations on high concentrations of radon in certain houses. *Health Phys.,* **36**, 729.

Saccomanno, G., Archer, V. E., Saunders, R. P., James, L. A., and Becker, P. A. (1964) Lung cancer in uranium miners on the Colorado Plateau, *Health Phys.* **10**, 1195.

Savage, D., and Chapman, N. A. (1982) Hydrothermal behaviour of simulated waste glass and waste–rock interactions under repository conditions. *Chem. Geol.* **36**, 59.

Savage, D., Robbins, J. E., and Merriman, R. J. (1985) Hydrothermal crystallization of radioactive waste storage glass. *Mineral. Mag.* **49**, 195.

Schultz, V., and Klement, A. W. (1963) *Radioecology.* New York, Reinhold.

Sholkovitz, E. R., Carey, A. E., and Cochran, J. K. (1982) Aquatic chemistry of plutonium in seasonally anoxic lake waters. *Nature* **300**, 159.

Simmons, J. H., Macedo, P. B., Barkatt, A., and Litovitz, T. A., (1979) Fixation of radioactive waste in high silica glasses. *Nature* **278**, 729.

Smith, P. J. (1973) *Topics in geophysics.* Bletchley, Open University.

Spiers, F. W. (1968) *Radioisotopes in the human body: physical and biological aspects.* New York, Academic Press.

Stranden, E. (1979) Radioactivity of building materials and the gamma radiation in dwellings. *Phys. Med. Biol.* **24**, 921.

Taylor, F. E., and Webb, G. A. M. (1979) Radiation exposure of the UK population — an extended summary. *Bull. Natl. Radiol. Prot. Board* **26**, 8.

Torstenfelt, B., Andersson, K., and Allard, B. (1982) Sorption of strontium and cesium in rocks and minerals. *Chem. Geol.* **36**, 123.

Turekian, K. K., and Rona, P. A. (1977) Eastern Atlantic fracture zones as potential disposal sites for radioactive waste. *Environ. Geol.* **2**, 59.

Turner, R. C., Radley, J. M., and Mayneord, W. V. (1961) Naturally occurring alpha-activity of drinking waters. *Nature* **189**, 348.

Villiers, A. J. de, and Windish, J. P. (1964) Lung cancer in fluorspar mining communities. I, Radiation, dust and mortality experience. *Brit. J. Ind. Med.* **21**, 94.

Wadden, M. M., and Katsube, T. J. (1982) Radionuclide diffusion in igneous crystalline rocks. *Chem. Geol.* **36**, 191.

Waite, D. A. (1982) Recommended and regulatory limits of environmental radioactivity. In Klement, A. W. (ed.), *CRC handbook of environmental radiation.* Boca Raton, Florida, CRC.

Whicker, F. W., and Schultz, V. (1982) *Radioecology: nuclear energy and the environment,* Vol. 1. Boca Rouge, Florida, CRC.

Wrixon, A. D., Brown, L., Cliff, K. D., Driscoll, C. M. H., Green, B. M. R., and Miles, J. C. H. (1984) Indoor radiation surveys in the U.K. *Radiat. Prot. Dosim.* **7**, 321.

Appendix — Radioactive minerals

The formation of radioactive minerals is mainly related to the incorporation of U or Th in the mineral structure, either as an essential component or as a trace element. K-bearing minerals are, of course, also radioactive to some extent and are significant because of the importance of K in the Earth. Those minerals which contain any of the other radioactive elements in their composition will also be slightly radioactive, but the low concentrations of these elements means that they need not be considered here. A simple classification of radioactive minerals into U-bearing, Th-bearing and K-bearing is thus followed in this account.

Within radioactive minerals the possibility that one element substitutes for another is a common feature. The main factor which governs substitution is the ionic size of the elements involved. Total substitution is only possible if the size difference between the ions is less than 15% of the size of the smaller ion. U^{4+} and Th^{4+} both have ionic radii of 0.11×10^{-9} m, but U^{6+} has a radius of only 0.09×10^{-9} m. Complete substitution can therefore only occur between U^{4+} and Th^{4+}. Where the size difference exceeds 15%, partial substitution is often possible, so some substitution of Th^{4+} for U^{6+} may be thought to take place. However, this is unlikely in view of the charge difference of +2 between these ions. Although substitution between ions of different charge is possible, this is usually limited to +1.

Complete substitution between U^{4+} and Th^{4+} may appear to render the classification of mineral species into U-bearing and Th-bearing somewhat artificial. However, if the division is based upon the relative abundances of U and Th in the mineral structure, a geologically sensible division is produced; U^{4+}-rich minerals are highly soluble in most natural waters, while Th^{4+}-rich minerals typically have a low solubility. Indeed, in minerals showing a high degree of U^{4+}–Th^{4+} substitution, the solubility is governed by the amount of U^{4+} present in the structure.

The extent of ionic substitution is also dependent upon temperature. At high temperatures, thermal energy causes expansion of crystal lattices which can then permit exchange with ions of a slightly different size. Even so, substitution is still controlled to some extent by the actual form of the lattice involved. Defect structures, too, are particularly important in allowing ionic substitutions which otherwise may not occur to any significant degree. This is especially true in radioactive minerals which display radiation damage. For these reasons, substitution may occur to such an extent that it is often not possible to give an exact chemical composition for a mineral. Nevertheless, within the three main divisions listed above, the systematics of the mineralogy of U and Th given in this appendix will consider compositions as follows: oxides and hydroxides; arsenates, carbonates, phosphates and sulphates; molybdenates and vanadates; silicates; other minerals. For K-bearing minerals, chlorides and nitrates are important.

Not all radioactive minerals are included, only those which are relatively common and those which illustrate specific features. Within the classification, the individual minerals are listed in alphabetical order.

It is also necessary to mention here that many U-bearing minerals contain water that is not part of their basic atomic structure. Often, some of this water may be lost without structural damage to the mineral concerned, but certain minerals undergo extensive structural reordering on dehydration. In such cases the minerals may exist in two separate phases. These are the normal phase and the metaphase of the mineral. Spontaneous dehydration and conversion from the normal to the metaphase are possible. Finally, it is well to recognise that all U-bearing and Th-bearing minerals will contain Pb as a stable radiogenic product, and other unstable members of the decay series will also be present.

A.1 URANIUM-BEARING MINERALS

A.1.1 Oxides and hydroxides

(i) Aeschynite–priorite series
Composition: (Ce, Y, Ca, Fe, Th) $(Ti, Nb)_2$ $(O, OH)_6$ with Ce greater than Y in aeschynite and Y greater than Ce in priorite.
Variations: U forms up to 12.5%, and Th up to 4.5%. U content increases with transition to priorite.
Other components: Al, Si, He, N, Ra, F, Na, K, Zn and W may be present in moderate amounts. Hydrated to varying degrees.
Varieties: Blomstrandine and blomstrandinite are synonyms of priorite.
Physical properties:
Crystal system, orthorhombic; Specific gravity, 4.5–5.7; hardness, 5–6; colour, brown–black.

(ii) Becquerelite
Composition: $CaU_6O_{19}.11H_20$.
Variations: U may form up to 75.5%.
Other components: Pb has been reported and Ba is possibly present.
Physical properties: Crystal system, orthorhombic; specific gravity, 5.3; hardness, 2–3; colour, yellow–amber.

(iii) Betafite
Composition:
$(U, Ca, Na)_2(Nb, Ta, Ti)_2O_6(OH)$.
Variations: Degree of hydration variable. U up to 24.5%, Th up to 1%. Normally Nb is more abundant than Ta and Ti.
Other components: Mn, Fe^{2+}, Fe^{3+}, Mg, Pb, Y, Ce, Sn, K, Bi, Al and Si may occur.
Varieties: Ti variety is mendeleyevite. Ta variety is blomstrandite.
Physical properties: Crystal system cubic; specific gravity, 3.7–5.25; hardness, 4–5.5; colour, yellow–brown–black.

(iv) Billietite
Composition: $BaU_6O_{19}.11H_2O$.
Variations: U forms up to 69%. Resembles becquerelite.
Physical properties:
Crystal system, orthorhombic; specific gravity, 5.3; hardness, 2–3; colour, colourless–yellow.

(v) Clarkeite
Composition: $(Na, Cu, Pb, Th)_2U_2(O, OH)$.
Variations: U forms up to 68%, and Th up to 2%.
Other components: Mg, Ba, Si, Fe^{3+}, Al and rare earth elements have all been reported in minor amounts.
Physical properties:
Crystal system, orthorhombic; specific gravity, 6.3; hardness, 4–4.5; colour, pale yellow–orange.

(vi) Curite
Composition: $Pb_2U_5O_{17}.4H_2O$.
Variations: U forms up to 62.5%.
Other components: Traces of Fe^{3+}.
Physical properties:
Crystal system, orthorhombic; specific gravity, 7.2; hardness, 4–5; colour, orange–red.

(vii) Davidite
Composition:
$(Fe^{2+}, La, U, Ca)_6(Ti, Fe^{3+})_{15}(O, OH)_{36}$.
Variations: U forms up to 4.5%, and Th less than 1%.
Other components: Nb, Bi, Pb and Al.
Physical properties: Crystal system, hexagonal; specific gravity, 4.0–5.0; hardness, 5–6; colour, black.
Alteration: Alters to give coatings and fracture infillings of carnotite.

(viii) Fergusonite–formanite series
Composition: (Y, Ce, Te, U) (Nb, Ta, Ti)O_4 (fergusonite) to (Y, U, Th, Ca) (Ta, Nb, Ti)O_4 (formanite).
Variations: In fergusonite, U forms up to 7%, and Th up to 6%, but in formanite both Th and U may only be up to 1%.
Other components: Minor amounts of Er may occur with Y.
Physical properties: Crystal system, tetragonal; specific gravity, 5.2–6.2; hardness, 5.5–6.5; colour, brown–black.

(ix) Fourmarierite
Composition: $PbU_4O_{13}.4H_2O$.
Variations: U may form up to 65.5%.
Physical properties:
Crystal system, orthorhombic; specific gravity, 6.0; hardness, 3–4; colour, red–brown.

(x) Ianthinite
Composition: $U_6O_7(OH)_{20}$.
Variations: U may be up to 75.5%.
Other components: Traces of Fe.
Physical properties:
Crystal system, orthorhombic; specific gravity, 5.2; hardness, 3–4; colour, violet–black.
Alteration: Hydrates further, oxidises to U^{6+} and changes in colour from purple to maroon then grey and finally yellow (when it is known as epiianthinite).

(xi) Ishikawaite
Composition: (U, Fe, Y, Ca) (Nb, Ta)O_4.
Variations: U may form up to 19.5%.
Other components: Minor Ca, Mg, Mn, Sn, Ti, Al and Si, with secondary hydration.
Physical properties:
Crystal system, orthorhombic; specific gravity, 6.2–6.4; hardness, 5–6; colour, black.

(xii) Masuyite
Composition: $UO_3.2H_2O$.
Variations: U may be up to 74%.
Other components: Pb may be present in small quantities.
Physical properties:
Crystal system, orthorhombic; colour, orange–red.

(xiii) Polycrase
Composition: (Y,Ca,Ce,U,Th) (Ti,Nb,Ta)O_6.
Variations: U may form up to 12.5%, and Th up to 4.5%. This mineral forms part of the euxinite–polycrase series. As Nd+Ta increases to euxinite the Th content can become greater than the U content.
Other components: Na, K, Pb, Sn, Be, He, N, S, Hf, Ge, Se and Zr have all been reported, and Al and Si may be present. Hydration has usually occurred to some degree.
Physical properties:
Crystal system, orthorhombic; specific gravity, 4.0–6.0; hardness, 5.5–6.5; colour, black.

(xiv) Pyrochlore–microlite series
Composition: End members are $NaCaNb_2O_6F$ (pyrochlore) and $(NaCa)_2Ta_2O_6(O,OH,F)$ (microlite).
Variations: Substitution of K, Mg, Fe^{2+}, Mn, rare earth elements, Zr, U and Th for Na and Ca. Nb and Ta substitution within series. Ti, Fe^{3+}, Sn and W for Na and Ta. Th may form up to 5.5%. U forms up to 17%.
Other components: Small amounts of Pb, Be, Al, Si, Sr, Bi and Hf occur.
Varieties: U-rich varieties are uranian (batchettalite and ellsworthite); Ce varieties

cerian; Ti varieties titanian; Fe varieties ferroan and ferrian. Various degrees of hydration. Y variety of pyrochlore is called obruchevite.
Physical properties: Crystal system, cubic; specific gravity, 4.0–6.5; hardness, 5–5.5; colour, yellow–red–brown–black.

(xv) Samarskite–yttrotantalite series
Composition:
(Y, Ce, U, Ca, Pb) (Nb, Ti, Ta, Sn)$_2$O$_6$ (Nb-rich end member, samarskite) to
(Y, U, Fe)(Ta, Nb)O$_4$ (Ta-rich end member, yttrotantalite).
Variations: In samarskite, U may be up to 16.5% and Th up to 3.5%. In yttrotantalite, U may be up to 5% and Th up to 2%.
Other components: Minor quantities of Mg, Mn, Pb, Al and Zr may occur in samarskite, with trace amounts of Be, Sc, Ba, Sr, Ra, N and He. Hydrated to various degrees. In yttrotantalite, Be, Pb, Al, Si and As occur, with secondary hydration.
Varieties: Nahlite is a Ca, Fe member similar to samarskite, and yttrocolumbite is a synonym of samarskite.
Physical properties:
Crystal system, orthorhombic; specific gravity, 5.4–6.2; hardness, 5–6; colour, dark brown–black.

(xvi) Schoepite
Composition: $UO_3.2H_2O$.
Variations: U may be up to 72%.
Other components: Some Pb, Te and Fe^{3+}.
Varieties: Paraschoepite may be synonymous or a different hydrate. Urgite is $UO_3.nH_2O$.
Physical properties:
Crystal system, orthorhombic, specific gravity, 4.5–4.8; hardness, 2–3; colour, yellow.

(xvii) Studite
Composition: $UO_3.4H_2O$.
Variations: U forms up to 63.5%.
Other components: Pb present.
Physical properties:
Crystal system, orthorhombic; colour, yellow.

(xviii) Uraninite
Composition: UO_2.
Variations: Substitution of Th^{4+} for U^{4+} so that a series exists with thorianite. Th may be up to 45.5%. U content is up to 88%. Division from thorianite taken at U/Th=1.1. U^{4+} auto-oxidised to U^{6+} so that some UO_3 is present, but some U^{6+} may be primary. Degree of oxidation used to define γ-uraninite (U_4O_7), α-uraninite (UO_2) and β-uraninite (U_3O_7).
Other components: Rare earth elements may form up to 10%. He, Ar, N and decay products ending in Pb are also present.
Varieties: Pitchblende is the name most usually given to the massive, botryoidal or colloform occurrences, but definition of pitchblende has also been made upon oxidation state (higher than normal uraninite) and grain size (finer-grained than normal uraninite). Uranium black is the name applied to the hydrated form; more correctly known as hydronasturan.
Physical properties: Crystal system, cubic; specific gravity, 10.95, decreasing with oxidation and Th content to 8.0; hardness, 5–6; colour, brown–black.
Alteration: Outer sections usually more highly oxidised than inner. Hydration to form gummite (which is in fact a fine-grained mixture of various hydrated uranium minerals such as fourmarierite, vandendreisscheite, uranophane, kasolite, soddyite and sklodowskite. Total form a yellow–red to brown mass which varies greatly in physical properties but with specific gravity about 4.0–6.5 and hardness 2.5–5.0.

(xix) Vandenbrandeite
Composition: $Cu(UO_2)(OH)_2$.
Variations: U forms up to 59.5%.
Physical properties: Crystal system, triclinic; specific gravity, 5.0; hardness, 3–4; colour, dark green–black.

(xx) Vandendreisscheite.
Composition: $PbU_7O_{22}.12H_2O$.
Variations: U may form up to 68.5%.
Other components: Some Ba and Ca may substitute for Pb.

Physical properties:
Crystal system, orthorhombic; specific gravity, 5.5; hardness, 3–4; colour, yellow–orange–red.

A.1.2 Arsenates, carbonates, phosphates and sulphates (anhydrous and hydrous)

(i) Abernathyite
Composition: $K(UO_2)(AsO_4).4H_2O$.
Variations: U may form up to 46%.
Other components: Small amounts of P may substitute for As.
Physical properties: Crystal system, tetragonal; specific gravity, 3.7; hardness, 2.5; colour, yellow.

(ii) Andersonite
Composition: $Na_2Ca(UO_2)(CO_3)_3.6H_2O$.
Variations: U may form up to 39%.
Physical properties: Crystal system, hexagonal; specific gravity, 2.8; colour, yellow–green.

(iii) Autunite
Composition: $Ca(UO_2)_2(PO_4)_2.10\text{–}12H_2O$.
Variations: U may form up to 50%. A series exists with metauranocircite.
Other components: Mg and Cu may substitute for Ca in small amounts. Na may also substitute for Ca in the presence of brines.
Varieties:
Na-autunite $(Na,Ca)(UO_2)_2(PO_4)_2.8H_2O$. Dehydration can occur to form meta-autunite $(2\text{–}6H_2O)$.
Physical properties: Crystal system, tetragonal; specific gravity, 3.1–3.2; hardness, 2–2.5; colour, yellow–green.

(iv) Bassetite
Composition: $Fe(UO_2)_2(PO_4)_2.8H_2O$.
Variations: U may form up to 51%
Physical properties:
Crystal system, monoclinic; specific gravity, 3.1–3.4; hardness, 2–2.5; colour, yellow–bronze.

(v) Bayleyite
Composition: $Mg_2(UO_2)(CO_3)_3.18H_2O$.
Variations: U may form up to 29%.
Other components: Traces of Ca, Fe, Al, Si and Sr. Some SO_4.
Varieties: In dry atmospheres it dehydrates to a powder.
Physical properties:
Crystal system, monoclinic; specific gravity, 2.0; hardness, 2–2.5; colour, yellow–white.

(vi) Dewindite
Composition: $Pb(UO_2)_2(PO_4)_2.3H_2O$.
Variations: U may form up to 49.5%.
Physical properties:
Crystal system, orthorhombic; specific gravity, 5.0; colour, yellow.

(vii) Johannite
Composition: $Cu(UO_2)_2(SO_4)_2(OH)_2.6H_2O$.
Variations: U may form up to 51%.
Physical properties: Crystal system, triclinic; specific gravity, 3.3; hardness, 2–2.5; colour, pale green–deep green.

(viii) Kahlerite
Composition: $Fe(UO_2)_2(AsO_4)_2.nH_2O$.
Variations: U may form up to 47%. The value of n is normally 12 but is 8 in the metaphase metakahlerite.
Physical properties: Crystal system, tetragonal; colour, yellow.

(ix) Lermontovite
Composition: $(U,Ca,Ce)_3(PO_4)_4.6H_2O$.
Variations: U may form up to 44.5%. Many other rare earth elements may substitute for Ce.
Physical properties: Specific gravity, 4.5; colour, green–brown.

(x) Liebigite
Composition: $Ca_2(UO_2)(CO_3)_3.10H_2O$.
Variations: U may form up to 33.5%.
Varieties: Uranothallite is a synonym.
Physical properties:
Crystal system, orthorhombic; specific gravity, 2.4; hardness, 2.5–3; colour, green–greenish yellow.

(xi) Rabbittite
Composition:
$Ca_3Mg_3(UO_2)_2(CO_3)_6(OH)_4.18H_2O$.
Variations: U may form up to 31%.
Other components: Traces of Si, Al and Y.

Physical properties:
Crystal system, monoclinic; specific gravity, 2.6; hardness, 2.5; colour, pale green–greenish yellow.

(xii) Renardite–phosphuranylite series
Composition:
$Pb(UO_2)_4(PO_4)_2(OH)_4.7H_2O$(renardite), $Ca(UO_2)_3(PO_4)_2(OH)_2.6H_2O$ (phosphuranylite).
Variations:
U may form up to 57.5% in renardite, and 63.5% in phosphuranylite.
Physical properties:
Crystal system, orthorhombic; specific gravity, 4.1; (phosphuranylite – 4.4 (renardite); hardness, 3–3.5; colour, golden yellow–brown.

(xiii) Rutherfordine
Composition: $(UO_2)(CO_3)$.
Variations: U may form up to 72%.
Other components: Some Pb, Ca and Si usually present.
Varieties: Diderichite is a synonym.
Physical properties:
Crystal system, orthorhombic; specific gravity, 4.8–5.8; hardness, 1; colour, white–pale yellow.

(xiv) Saleeite–novacekite series
Composition:
$Mg(UO_2)_2(PO_4)_2.10H_2O$(saleeite), $Mg(UO_2)_2(AsO_4)_2 12H_2O$(novacekite).
Variations: U may form up to 53% in saleeite, and slightly less in novacekite. Division in the series occurs at a P/As ratio of 1. Novacekite dehydrates to metanovacekite ($4–8H_2O$).
Other components: Small amounts of Pb and Al substitute for Mg in saleeite and traces of Al, Fe, Cu, Ca, Ti, Sr and Ba have been found in novacekite.
Physical properties:
Crystal system, tetragonal; specific gravity, 3.3 (saleeite) – 3.7 (novacekite); hardness, 2.5; colour, yellow–greenish yellow.

(xv) Schroeckingerite
Composition:
$NaCa_3(UO_2)(CO_3)_3(SO_4)F.10H_2O$.
Variations: U may form up to 27%.
Other components: Mg, Al, Fe, K, Si, Sr, Zn, Ti and Mn in minor amounts.
Varieties: Dakeite is a synonym.
Physical properties:
Crystal system, orthorhombic; specific gravity, 2.5; hardness, 2.5; colour, yellow–greenish yellow.

(xvi) Sharpite
Composition: $(UO_2)_6(CO_3)_5(OH)_2.7H_20$.
Variations: U may form up to 68.5%.
Physical properties:
Crystal system, orthorhombic; specific gravity, 3.3; hardness, 2.5; colour, greenish yellow.

(xvii) Torbernite
Composition: $Cu(UO_2)_2(PO_4)_2.8–12H_2O$.
Variations: U may form up to 47%. As may substitute for P and Pb for Cu, in small amounts.
Other components: Na, Ba, Ca, Mg, Mn, Mo, Te, Fe and Si.
Varieties:
Chalcolite and uran-mica are synonyms. Spontaneous dehydration leads to the metaphase metatorbernite ($8H_2O$).
Physical properties: Crystal system, tetragonal; specific gravity, 3.2; hardness, 2.5; colour, pale green–dark green.

(xviii) Troegerite
Composition: $(H_3O)_2(UO_2)_2(AsO_4)_2.6H_2O$.
Variations: U may form up to 55%.
Physical properties: Crystal system, tetragonal; specific gravity, 3.5; hardness, 2–3; colour, yellow.

(xix) Uranocircite
Composition: $Ba(UO_2)_2(PO_4)_2.8–12H_2O$.
Variations: U may form up to 47%. Ca may substitute for Ba and a series exists with autunite.
Varieties: The mineral is only stable in very hydrous environments and does not normally survive. Spontaneous dehydration to the metaphase metauranocircite ($8H_2O$) takes place rapidly.

Physical properties (metauranocircite): Crystalsystem, tetragonal; specific gravity, 3.6–4.1; hardness, 2–3; colour, yellow–green.

(xx) Uranopilite
Composition: $(UO_2)_6(SO_4)(OH)_{10}.12H_2O$.
Variations:
U may form up to 68%. Dehydration to metauranopilite ($5H_2O$).
Physical properties:
Crystal system, monoclinic; specific gravity, 3.7–4.0; hardness, 2–3; colour, yellow.

(xxi) Uranospathite
Composition: $HAl(UO_2)_4(PO_4)_4.4H_2O$.
Variations: U may be up to 46%.
Other components: As may substitute for P as much as As/P of 2.1. Cu may substitute for HAl.
Physical properties:
Crystal system, orthorhombic; specific gravity, 2.5; colour, yellow–pale green.

(xxii) Uranospinite
Composition: $Ca(UO_2)_2(AsO_4)_2.10H_2O$.
Variations:
U may form up to 46%. Na-uranospinite occurs in which Na substitutes for some Ca and PO_4 substitutes for AsO_4. Dehydration to metauranospinite ($8H_2O$).
Physical properties: Crystal system, tetragonal; specific gravity, 3.5; hardness, 2–3; colour, yellow–pale green.

(xxiii) Voglite
Composition: $Ca_2Cu(UO_2)_2(CO_3)_4.6H_2O$.
Variations: U may form up to 31.5%.
Physical properties: Crystal system, triclinic; colour, green.

(xxiv) Walpurgite
Composition: $(BiO_4)(UO_2)(AsO_4)_2.H_2O$.
Variations: U may form up to 16%. Considerable PO_4 may substitute for AsO_4.
Physical properties: Crystal system, triclinic; specific gravity, 5.95; hardness, 3.5; colour, yellow.

(xxv) Xenotime
Composition: YPO_4.
Variations: U may form up to 3.5%. UO_2 and UO_3 present.
Other components: Nd, Pd, Er, Tb, Ho, Ce, La and Sc may be present substituting for Y. Ca, Be, Al, Zr, Mn, Pb and Fe^{2+} present in small amounts. SiO_4 may be present up to several per cent and SO_4 may substitute for PO_4.
Physical properties: Crystal system, tetragonal; specific gravity, 4.4–5.1; hardness, 4–5; colour, red–brown but may be pale red, grey–white or yellow–green.

(xxvi) Zellerite
Composition: $Ca(UO_2)(CO_3)_2.5H_2O$.
Variations:
U may form up to 45%. Dehydration to metazellerite ($3H_2O$).
Physical properties:
Crystal system, orthorhombic; specific gravity, 3.3; colour, yellow.

(xxvii) Zeunerite
Composition: $CU(UO_2)_2(AsO_4)_2.10\text{–}16H_2O$.
Variations:
U may form up to 49.5%. Dehydration to metazeunerite ($8H_2O$). Metazeunerite is the normal form but natural occurrences of zeunerite have been reported. Transition to heinrichite and metaheinrichite occurs as Ba completely substitutes for Cu.
Other components: Some substitution of P for As occurs.
Physical properties: Crystal system, tetragonal; specific gravity, 3.2–3.6; hardness, 2–2.5; colour, green.

(xxviii) Zippeite
Composition: $K_4(UO_2)_6(SO_4)_3(OH)_{10}.4H_2O$.
Variations: U may form up to 63.5%. It is possible that more than one species occurs, with different degrees of hydration.
Physical properties:
Crystal system, orthorhombic (but some varieties are monoclinic); specific gravity, 3.3–3.7; hardness, 2; colour, golden yellow.

A.1.3 Molybdenates and vanadates

(i) Carnotite
Composition: $K_2(UO_2)_2(VO_4)_2.3H_2O$.
Variations: U may form up to 55%.

Other components: Small amounts of Cu, Na, Ba, Sr, Mg, Fe^{2+}, Al and Si.
Varieties: By hydration and base exchange in $CaHCO_3$-rich solutions, carnotite converts to tyuyamunite.
Physical properties:
Crystal system, monoclinic; specific gravity, 4.6; hardness, 2–3; colour, yellow–greenish yellow–orange.

(ii) Ferghanite
Composition: $(UO_2)_2(VO_4)_2.6H_2O$.
Variations: U may form up to 62%.
Other components: Li may be present.
Physical properties: Specific gravity, 3.3; hardness, 2.5; colour, yellow.

(iii) Francevillite
Composition: $(Ba,Pb)(UO_2)_2(VO_4)_2.5H_2O$.
Variations: U may form up to 46%. Pb may be absent entirely or form Ba/Pb in ratio 2.
Physical properties:
Crystal system, orthorhombic; specific gravity, 4.5 (if Pb-bearing); hardness, 3; colour, yellow.

(iv) Fritzschite
Composition: $Mn(UO_2)_2(VO_4)_2.10H_2O$.
Variations: U may form up to 47.5%.
Physical properties:
Specific gravity, 3.5; hardness, 2–3; colour, reddish brown.

(v) Iriginite
Composition: $U(MoO_4)_2(OH)_2.2H_2O$.
Variations: U may form up to 38.5%.
Physical properties:
Crystal system, monoclinic; specific gravity, 3.8; hardness, 4–5; colour, yellow.

(vi) Malunerite
Composition: $H_4U(UO_2)_3(MoO_4)_7.18H_2O$.
Variations: U may form up to 36%.

(vii) Rauvite
Composition: $Ca(UO_2)_2V_{10}O_{28}.16H_2O$.
Variations: U forms up to 26%. U^{4+} and U^{5+} both present.
Physical properties: Specific gravity, 2.9; colour, brown–red–purplish black.

(viii) Sengierite
Composition: $Cu_2(UO_2)_2(VO_4)_2.8\text{–}10H_2O$.
Variations: U may form up to 48.5%.
Other components: Some Fe^{3+} and Al possible.
Physical properties:
Crystal system, monoclinic, specific gravity, 4.4; hardness, 2.5; colour, greenish yellow.

(ix) Tyuyamunite
Composition: $Ca(UO_2)_2(VO_4)_2.5\text{–}8H_2O$.
Variations:
U forms up to 54%. On spontaneous dehydration, tyuyamunite converts to the metaphase metatyuyamunite ($3\text{–}5H_2O$). Some K replaces Ca.
Other components: Small amounts of Ba, Cu, Pb and Mg, some of which may substitute for Ca.
Varieties: By dehydration and increasing exchange of K for Ca, tyuyamunite converts to carnotite.
Physical properties:
Crystal system, orthorhombic; specific gravity, 3.62 (metatyuyamunite 3.8–3.9); hardness, 2; colour, yellow.

(x) Umohoite
Composition: $(UO_2)MoO_4.4H_2O$.
Variations: U may form up to 47.5%.
Physical properties:
Crystal system, monoclinic; specific gravity, 4.6; hardness, 2; colour, dark green–black.

(xi) Uvanite
Composition: $U_2V_6O_{21}.15H_2O$.
Variations: U may form up to 34.5%.
Physical properties:
Crystal system, orthorhombic; colour, brownish yellow.

A.1.4 Silicates

(i) Boltwoodite
Composition: $(H_3O)K(UO_2)(SiO_4).H_2O$.
Variations: U may form up to 58%.
Physical properties:
Crystal system, monoclinic; specific gravity, 3.6; colour, yellow.

(ii) Coffinite
Composition: $U(SiO_4)_{1-x}(OH)_{4x}$.
Variations: U may form up to 60%. Most of the U is U^{4+}.
Other components: As, V and Al usually present in small quantities.
Physical properties: Crystal system, tetragonal; specific gravity, 3.5–3.7; colour, black.

(iii) Cuprosklodowskite
Composition: $Cu(UO_2)_2Si_2O_6(OH)_2.6H_2O$.
Variations: U may form up to 54%.
Physical properties: Crystal system, triclinic; specific gravity, 3.5–3.9; hardness, 3–4; colour, green, possibly yellow margins.

(iv) Kasolite
Composition: $Pb(UO_2)SiO_4.H_2O$.
Variations: U may form up to 40.5%.
Other components: Traces of Fe, Ca, Mg, Na, Ba and Ca.
Physical properties:
Crystal system, monoclinic; specific gravity, 5.8–5.9; hardness, 4–5; colour, yellow.

(v) Opal
Composition: $SiO_2.nH_2O$.
Variations: U may form up to 0.12%. H_2O is usually between 3% and 10%.
Physical properties: Crystal system, amorphous; specific gravity, 2.0–2.2; hardness, 5.5–6.5; colour, white–grey–brown–red or colourless (hyalite variety).

(vi) Sklodowskite
Composition: $Mg(UO_2)_2Si_2O_6(OH)_2.5H_2O$.
Variations: U may form up to 57%.
Other components: Traces of K, Na, Te, Co and Ni.
Varieties: Chinkolobwite is a synonym.
Physical properties:
Crystal system, monoclinic; specific gravity, 3.5–3.6; hardness, 2-3; colour, yellow.

(vii) Soddyite
Composition: $(UO_2)_5Si_2O_9.6H_2O$.
Variations: U may form up to 72%.
Physical properties:
Crystal system, orthorhombic; specific gravity,4.7; hardness, 3–4; colour, greenish yellow–yellow–amber.

(viii) Uranophane
Composition: $Ca(UO_2)_2Si_2O_7.6H_2O$.
Variations: U may form up to 55.5%. Some Mg may substitute for Ca.
Other components: Fair amounts of Cu, Co, Ni and Fe may occur. Minor traces of Al, Pb, Ba and Sr. Some PO_4 may be present.
Varieties:
Uranotile and lambergite are synonyms. Dimorphous with β-uranophane.
Physical properties:
Crystal system, monoclinic; specific gravity, 3.8; hardness, 2–3; colour, yellow.

A.1.5 Other minerals

(i) Demesmaekerite
Composition:
$Pb_2Cu_5(UO_2)_2(SeO_3)_6(OH)_6.2H_2O$.
Variations: U may form up to 22.5%.
Physical properties: Crystal system, triclinic.

(ii) Guilleminite
Composition: $Ba(UO_2)_3(SeO_3)_2(OH)_4.3H_2O$.
Variations: U may form up to 54%.
Physical properties:
Crystal system, orthorhombic; specific gravity, 4.9; colour, yellow.

(iii) Moctezunite
Composition: $Pb(UO_2)(TeO_3)_2$.
Variations: U may form up to 24%.
Physical properties:
Crystal system, monoclinic; specific gravity, 5.7; colour, orange.

(iv) Thucolite
Composition: The composition is very variable. The minerals grouped under this name are all uraniferous hydrocarbons and contain Th, C and O. Hence the name Th–U–C–O–lite. The hydrocarbons may thus be in fact oxyhydrocarbons.
Variations: U_3O_8 content of the ash varies between 1 and 70%; ThO_2 from less than 1 to 48%; rare earth elements from less than 1 to 35%; C ranges between 23 and 61%;

volatiles range between 26 and 40% and ash from 7 to 38%. A Ti-rich variety is known as titanothucolite (TiO_2, 4.5%).
Other components: Traces of Na, K, Ca, Mg, Mn, Fe^{3+}, Al, Si, Ti, B, Co, Ni, Zn, Pb, Cu, Zr, W, Be, V, As, P, Ar, S, Cl and F.
Physical properties: Crystal system, amorphous; specific gravity, 1.5–2.1; hardness, 2.5–4; colour, black.

A.2 THORIUM-BEARING MINERALS

A.2.1 Oxides

(i) Brannerite–thorutite series
Composition:
(U,Ca,Ce)(Ti,Fe)$_2$O$_6$ (brannerite),
(Th,U,Ca)Ti$_2$(O,OH)$_6$ (thorutite.).
Variations: Th may form up to 47.5%, and U up to 43.5%. The Th/U ratio may thus vary considerably. Most U is present as U^{6+}.
Other components: Traces of Ba, Sr, Ni, Bi, Pb, Sn, Ta, Nb, Sc, Fe^{3+}, P, He, Zr and rare earth elements.
Varieties: The Th-rich brannerite has been called absite. Thorutite is the name given to the Fe-poor Th-rich variety while brannerite has been applied to the U-rich Th-poor variety.
Physical properties:
Crystal system, monoclinic; specific gravity, 4.0–4.5; hardness, 4–6; colour, black–reddish brown when U-rich, yellow–yellowish green when Th-rich.

(ii) Cerianite
Composition: (CeTh)O_2.
Variations: Th may form up to 4.5%. Ce/Th ratio 16.
Other components: Nb, Ta, Zr, and other rare earth elements may be present.
Varieties: A series with thorianite depending on the Ce content may exist.
Physical properties: Crystal system, cubic; specific gravity, 3.2; colour, dark green–amber.

(iii) Euxinite
Composition:
(Y,Ca,Ce,U,Th)(Nb,Ta,Ti)$_2$O$_6$.
Variations: Th may form up to 14%, and U up to 12.5%. Part of the euxenite–polycrase series. For euxenite $TiO_2/(Nb_2O_5+Ta_2O_5)$ is less than 3. Normally Nb exceeds Ta but Ta varieties are known.
Other components: Na, K, Pb, Sn, Be, He, N, S, Hf, Ge, Sc and Zr may also be present.
Varieties: Tanteuxenite and eschwegite are Ta varieties.
Physical properties:
Crystal system, orthorhombic; specific gravity, 4.0–5.9; hardness, 5.5–6.5; colour, black.

(iv) Loparite
Composition: (Ce,Th,Na,Ca)$_2$(Ti,Nb)$_2$O$_6$.
Variations: Th may form up to 11.5%, and U in small amounts up to 0.2% substituting for Th.
Other components: Fe^{2+}, Fe^{3+}, and other rare earth elements in small amounts.
Varieties: Loparite is, in fact, a Th, Ce, Na, Nb perovskite. The Th-rich form of loparite is also known as irinite.
Physical properties:
Crystal system, cubic inverted to monoclinic or orthorhombic; specific gravity, 4.8–4.9; hardness, 5.5–6; colour, black.

(v) Polymignite
Composition:
(Ca,Fe,Y,Th)(Nb,Ti,Ta,Zr)O_4.
Variations: Th may form up to 3.5%.
Other components: Other rare earth elements, traces of Mg, Mn, Sn, Hf, Na, K, Al, Si and Pb.
Physical properties:
Crystal system, orthorhombic; specific gravity, 4.8; hardness, 6.5; colour, yellow–brown–black.

(vi) Thorianite
Composition: (ThU)O_2.
Variations: Th can form up to 88%, and U up to 44.5%. U^{4+} is usually partly oxidised to U^{6+}. Rare earth elements substitute for Th.
Other components: Small amounts of Fe^{3+} and Zr may substitute for Th. Traces of Ca, Mn, Mg and He.
Varieties: The transition to uraninite depends upon the Th/U ratio; those near uraninite are

known as uranothorianite. Uranothorianite oxidises and hydrates to thorogummite ($Th(SiO_4)_{1-x}(OH)_{4x}$) which may contain up to 51% Th and 31.5% U.
Physical properties:
Crystal system, cubic; specific gravity, 9.2–9.7; hardness, 5–7; colour, grey–black.

(vii) Yttrocrasite
Composition: $(Y,Th,U,Ca)(Ti,Fe)_2(O,OH)_6$.
Variations: Th may form up to 7.5%, and U up to 3.5%.
Other components: Mn, Pb, Mg, rare earth elements, W, Si, Fe^{3+}, Nb and Ta may be present.
Physical properties:
Crystal system, orthorhombic; specific gravity, 4.8; hardness, 5.5–6; colour, black.

(viii) Zirkelite
Composition: $(Ca,Th,U,Ce)Zr(Ti,Nb)_2O_7$.
Variations: Th may form up to 6.5%, and U up to 1.5%. Th/U and Ti/Zr ratios vary considerably.
Other components: Small amounts of Mn, Mg, Y and Pb occur.
Physical properties: Crystal system, hexagonal; specific gravity, 4.3–5.1; hardness, 5.5–6; colour, black.

A.2.2 Carbonates and phosphates (anhydrous and hydrous)

(i) Apatite
Composition: $Ca_5(PO_4)_3F_2$.
Variations: Th may form up to 3%, and U up to 0.02%. Much substitution can occur. For Ca: rare earth elements, Sr, Ba, Na, Mn, Mg, Pb, Th, some U and some Ra. For F: OH, Cl, S and O. For PO_4: SO_4, AsO_4, VO_4, CrO_4, SiO_4, AlO_4 and CO_3OH.
Varieties: Th-bearing type is abukumalite; U-bearing type is usually a carbonate fluorapatite. Where the material comprises fossil bone it is known as collophane.
Physical properties:
Crystal system, hexagonal; specific gravity, 2.5–3.0; hardness, 5 (by definition, but usually 3–4); colour, white–grey–yellow–green–brown.

(ii) Bastnaesite
Composition: $(Ce,Th,La,Y,Ca)(CO_3)F$.
Variations: Th may form up to 41%, and U up to 0.1%.
Physical properties: Crystal system, hexagonal; specific gravity, 4.9; hardness, 4–4.5; colour, yellow–red–brown.

(iii) Britholite
Composition: $(Ca,Ce)_5(SiO_4,PO_4)_3(OH,F)$.
Variations: Th may form up to 17%, and U up to 2%. P/Si ratio variable.
Physical properties: Crystal system, hexagonal; specific gravity, 3.9–4.7; colour, red–brown.

(iv) Cheralite
Composition: $(Th,Ca,Ce,La,U,Pb)(P,Si)O_4$.
Variations: Th may form up to 27.5%, and U up to 5.5%. P exceeds Si.
Other components: Other rare earth elements, traces of Fe^{3+} and Al.
Varieties: Part of a series from monazite.
Physical properties:
Crystal system, monoclinic; specific gravity, 5.3; hardness, 5; colour, green.

(v) Grayite
Composition: $(Th,Pb,Ca,Ce)PO_4.H_2O$.
Variations: Th may form up to 37%.
Varieties: Brockite.

(vi) Mckelveyite
Composition: $Ba_3Na(Ca,V)Y(CO_3)_6.3H_2O$.
Variations: Th may form up to 4%, and U up to 0.1%.
Physical properties: Crystal system, hexagonal; specific gravity, 3.4; colour, dark green–black.

(vii) Monazite
Composition: $(Ce,La,Nd,Th)PO_4$.
Variations: Th may form up to 26.5%, and U up to 0.1%.
Other components: Other rare earth elements present up to 4%. Small amounts of Ca, Mg, Mn, Fe^{2+}, Fe^{3+}, Pb, Zr, Be, Sn, Al and Si.
Physical properties: Crystal system, monoclinic; specific gravity, 4.6–5.5 increasing with Th content; hardness, 5–5.5; colour, yellow–brown–green–red.

(viii) Streenstrupine
Composition:
$(Ce,La,Th,Na,Mn)_6(Si,P)_6O_{18}(OH)$.
Variations: Th may form up to 10%, and U up to 0.2%. Si/P ratio is variable.

A.2.3 Silicates

(i) Allanite
Composition:
$(Ce,Ca,Y,Th)_2(Al,Fe,Mg)_3(SiO_4)_3(OH)$.
Variations: Th may form up to 4.5%, and U up to 3%.
Other components: Other rare earth elements, some Fe^{2+}, Ti and Be. O and F substitute for OH, PO_4 for SiO_4, and Na and K for Ca.
Varieties; Orthite is a synonym. It is a member of the epidote group.
Physical properties:
Crystal system, monoclinic; specific gravity, 3.9–4.2; hardness, 5.5–6; colour, dark brown–black.

(ii) Chevkinite
Composition:
$(Ca,Ce,Y,U,Th)_4(Fe,Mg)_2(Ti,Fe)_3Si_4O_{22}$.
Variations: Th may form up to 18.5%, and U up to 2.5%.
Other components: Other rare earth elements, traces of Mn, Mg, Be, Al, Na, Ta, Nb, F, Pb, Sn and Fe^{3+}.
Physical properties:
Crystal system, monoclinic; specific gravity, 4.3–4.7; hardness, 5.5–6; colour, colourless-pink, black when heated.

(iii) Ekanite
Composition: $(Th,U)(Ca,Fe,Pb)_2Si_8O_{22}$.
Variations: Th may form up to 24%, and U up to 2%.
Physical properties: Crystal system, tetragonal; specific gravity, 3; hardness, 6–6.5; Colour, brown.

(iv) Gadolinite
Composition: $Be_2FeY_2Si_2O_{10}$.
Variations: Th may form up to 1.5%.
Other components: Other rare earth elements substitute for Y, typically Sm, Nd, Yb, Er, Dy and Gd. Traces of Ho, Tb, Lu, Tm and Eu may be present. Some Mn, Mg, Fe^{3+}, Na, K, Al and Ca.
Varieties: The Ca variety is called calciogadolinite.
Physical properties:
Crystal system, monoclinic; specific gravity, 4.0–4.6; hardness, 6–7; colour, dark green–brown–black.

(v) Huttonite
Composition: $ThSiO_4$.
Variations: Th may form up to 71.5%.
Other components: Minor Fe^{3+} and Ce with traces of Mn and PO_4. Some substitution of OH for O.
Physical properties:
Crystal system, monoclinic; specific gravity, 7.1; colour, colourless–pale cream.

(vi) Melanocerite
Composition:
$(Ce,Th,Ca)_5(Si,B)_3O_{12}(OH,F).nH_2O$.
Variations: Th may be up to 12%.
Other components: Other rare earth elements, Ta.
Physical properties: Crystal system, hexagonal; specific gravity, 4.1; hardness, 5–6; colour, brown–black.

(vii) Perrierite
Composition:
$(Ca,Ce,Th)_4(Mg,Fe^{2+})_2(Ti,Fe^{3+})_3Si_4O_{22}$.
Variations: Th may form up to 3.5%.
Physical properties:
Crystal system, monoclinic; specific gravity, 4.3; colour, brown–black.

(viii) Sphene
Composition: $CaTiSiO_5$.
Variations: Th may be present in small amounts up to about 0.25%.
Other components: Substituting for Ca are Na and the rare earth elements; substituting for Ti are Fe^{3+}, Al and Nb; substituting for O are OH and F. Traces of Fe^{2+}, Mn, Mg, Cr, Sn, Zr, Pb, Sr, V, Cu, Ga, Ta and Ba.

Physical properties:
Crystal system, monoclinic; specific gravity, 3.5; hardness, 6; colour, brown.

(ix) Thalenite
Composition: $Y_2Si_2O_7$.
Variations: Th may form up to 10.5%, and U up to 4%, substituting for Y.
Other components: Ca, Mg, Fe^{2+}, Mn, Na, Al, F, Pb, Zr, Ti, Sn and He.
Varieties: Th thalenite is called yttrialite.
Physical properties:
Crystal system, monoclinic; specific gravity, 4.4; hardness, 5–6.5; colour, pink–red–green.

(x) Thorite
Composition: $ThSiO_4$.
Variations: Th may form up to 64%, and U up to 10% (U varieties are known as uranothorite or enalite).
Other components: Ca, Mg, Fe^{2+}, Na, K, rare earth elements, P, Ta, Ti, Zr, Sn and Al.
Varieties:
If Fe is abundant, the form is ferrothorite. Presence of OH leads to thorogummite. Hydrous form also known as orangite. P varieties include auerlite. Ce varieties are eucrasite and freyalite.
Physical properties: Crystal system, tetragonal; specific gravity, 4.4–6.7; hardness, 4.5–5; colour, brown–black.

(xi) Thorogummite
Composition: $Th(SiO_4)_{1-x}(OH)_{4x}$.
Variations: Th may form up to 51%, U up to 31.5%.
Other components: PO_4, and SO_4 may substitute for SiO_4. Minor amounts of Ce, Y, Ca, Mg, Mn, Pb, Fe^{3+}, Fe^{2+}, Be, Al, Ti, Zr, Hf, Na, K, As, Sb, Sn, Nb and Ta.
Varieties: Chlorothorite, hyblite, hydrothorite, mackintoshite, maitlandite and nicolayite are synonyms.
Physical properties: Crystal system, tetragonal; specific gravity, 4.5; hardness, 3.5–4.5; colour, white–yellow–red–green.

(xii) Tritomite
Composition:
$(Y,Ca,La,Fe)_5(Si,B,Al)_3(O,OH,F)_{13}$.
Variations: Th may form up to 8.5%.
Other components: Other rare earth elements.
Physical properties: Crystal system, hexagonal; specific gravity, 3.3; hardness, 5.5; colour, black.

(xiii) Zircon
Composition: $ZrSiO_2$.
Variations: Th may form up to 13%, and U up to 2.5%.
Other components: Hf usually present and HfO_2 may form up to 16%. Rare earth elements also present, rare earth element oxides up to 16%. Traces of Fe^{3+}, Fe^{2+}, Ca, Mg, Mn, Na, K, Al, Pb, B, Sr, V, Ga, Ti, Ba, Cr, Ag, Sn and Cu. Some PO_4 substitutes for SiO_4.
Varieties: Radioactive pegmatitic varieties are known as cyrtolite and alvite. Malacon is hydrated zircon $(Zr(SiO_4)_{1-x}(OH)_{4x})$.
Physical properties:
Crystal system, tetragonal; specific gravity, 4.6–4.7; hardness, 7.5; colour, colourless–grey–green–reddish brown–brown–violet.

A.3 POTASSIUM-BEARING MINERALS

A.3.1 Chlorides, sulphates, nitrates and vanadates

(i) Alunite
Composition: $KAl_3(SO_4)_2(OH)_6$.
Variations: Na may substitute for K. If Na exceeds K, the mineral is termed natroalunite.
Physical properties: Crystal system, hexagonal; specific gravity, 2.6–2.9; hardness, 3.5–4; colour, white–greyish–reddish–yellowish.

(ii) Carnallite
Composition: $KCl.MgCl_2.6H_2O$.
Other components: Usually some Fe.
Physical properties:
Crystal system, orthorhombic; specific gravity, 1.6; hardness, 2.5; colour, white–pink–red.

(iii) Carnotite
Composition: $K_2(UO_2)_2(VO_4)_2.3H_2O$. See U-bearing minerals.

(iv) Jarosite
Composition: $KFe_3(SO_4)_2(OH)_6$.
Variations: Continuous series with alunite. Na substitutes for K to form natrojarosite. NH_4, Ag, Pb or H_2O may also substitute for K.
Physical properties: Crystal system, hexagonal; specific gravity, 2.9–3.3; hardness, 2.5–3.5; colour, amber–dark brown.

(v) Kainite
Composition: $KCl.MgSO_4.3H_2O$.
Physical properties:
Crystal system, monoclinic; specific gravity, 2.1; hardness, 3; colour, white.

(vi) Langbeinite
Composition: $K_2SO_4.2MgSO_4$.
Physical properties:
Crystal system, cubic; specific gravity, 2.8; hardness, 3–4; colour, colourless.

(vii) Nitre
Composition: KNO_3.
Varieties: Also known as saltpetre.
Physical properties:
Crystal system, orthorhombic; specific gravity, 2.1; hardness, 2; colour, white.

(viii) Picromerite
Composition: $K_2SO_4.MgSO_4.6H_2O$.
Physical properties:
Crystal system, monoclinic; specific gravity, 2; hardness, 2–3; colour, colourless–white.

(ix) Polyhalite
Composition: $K_2SO_4.MgSO_4.2CaSO_4.2H_2O$.
Physical properties: Crystal system, triclinic; specific gravity, 2.8; hardness, 2.5–3; colour, pink–red.

(x) Sylvite
Composition: KCl.
Physical properties:
Crystal system, cubic; specific gravity, 2; hardness, 2; colour, colourless–white.

A.3.2 Silicates

(i) Apophyllite
Composition: $KCa_4(Si_4O_{10})_2F.8H_2O$.
Variations: Some Na replaces K.
Physical properties: Crystal system, tetragonal; specific gravity, 2.35; hardness, 5; colour, colourless–white–pastel pink–yellow–green.

(ii) Biotite–phlogopite series
Composition:
$K(Mg,Fe)_3(AlSi_3O_{10})(OH)_2$ (biotite), $KMg_3(AlSi_3O_{10})(OH)_2$ (phlogopite).
Variations: Position in series depends on presence of Fe. K may be replaced by Na, Ca, Ba, Rb, Cs; Mg can be replaced by Fe^{2+} and Fe^{3+} or by Ti and Mn. Si/Al ratio variable. OH may be replaced by F.
Physical properties:
Crystal system, monoclinic; specific gravity, 2.8–3.4 increasing with the Fe content; hardness, 2.5 (on cleavage); colour, yellow–brown (phlogopite), brown–black (biotite).

(iii) Glauconite
Composition: $K(Fe,Mg,Al)_2(Si_4O_{10})(OH)_2$.
Variations: Replacement of Al by Si or Fe^{3+}.
Varieties: Celadonite is a synonym for blue–green earthy material found filling vesicles in basalt.
Physical properties:
Crystal system, monoclinic; specific gravity, 2.5–2.8; hardness, 2; colour, green–black.

(iv) Hornblende
Composition:
$NaCa_2(Mg,Fe,Al)_5(SiAl)_8O_{22}(OH)_2$.
Variations: Some Na replaced by K. Much interchange of Al, Si, Mg and Fe. Ti may be present.
Physical properties:
Crystal system, monoclinic; specific gravity, 3.0–3.4, increasing with Fe; hardness, 6; colour, dark green–dark brown–black.

(v) Lepidolite
Composition: $KLi_2Al(Si_4O_{10})(OH)_2$.
Variations: K may be partly replaced by Na, Rb and Cs. Li/Al ratio varies. Si may be partly replaced by Al; OH may be partly replaced by F.
Physical properties:
Crystal system, monoclinic; specific gravity, 2.8–2.9; hardness, 2.5 (on cleavage); colour, colourless–pale yellow–pale lilac.

(vi) Leucite
Composition: $K_2O.Al_2O_3.2SiO_2$.
Variations: A small amount of K may be replaced by Na.
Physical properties:
Crystal system, cubic; specific gravity, 2.47; hardness, 6; colour, colourless–white–grey.

(vii) Microcline
Composition: $K_2O.Al_2O_3.6SiO_2$.
Variations: Low temperature polymorph of orthoclase. Na may substitute for K up to a Na/K ratio of 0.1. Al substitution for Si occurs.
Physical properties: Crystal system, triclinic; specific gravity, 2.56; hardness, 6; colour, white–cream–green.

(viii) Muscovite
Composition: $KAl_2(AlSi_3O_{10})(OH)_2$.
Variations: Some Na substitutes for K. Al can be replaced partly by Mg, Fe^{2+}, Cr^{3+} or V^{3+}. The Si/Al ratio varies from 3 to 7. Some F may replace OH.
Varieties: The Cr variety is known as fuchsite; the V variety is known as roscoelite; fine-grained varieties in low grade metamorphic rocks are sericite; in the sedimentary environment the clay mica is known as illite.
Physical properties:
Crystal system, monoclinic; specific gravity, 2.8–2.9; hardness, 2.5 (parallel to cleavage) — 4 (perpendicular to cleavage); colour, colourless.

(ix) Nepheline
Composition: $Na_2O.Al_2O_3.2SiO_2$.
Variations: K substitutes for Na, often approaching a K/Na ratio of 0.3. Some Ca also occurs.
Physical properties: Crystal system, hexagonal; specific gravity, 2.6–2.7; hardness, 6; colour, white–grey–brown.

(x) Orthoclase
Composition: $K_2O.Al_2O_3.6SiO_2$.
Variations: Na may substitute to K up to a ratio of 0.1. Al substitution for Si occurs.
Physical properties:
Crystal system, monoclinic; specific gravity, 2.56; hardness, 6; colour, white–pink.

(xi) Perthite
An intergrowth association between lamellae of Na and K feldspar in which the K feldspar is the more abundant. If Na feldspar is in greater abundance, the term antiperthite is applied.

(xii) Plagioclase series
Composition:
$Na_2O.Al_2O_3.6SiO_2$ (albite),
$CaO.Al_2O_3.2SiO_2$ (anorthite).
Variations: Continuous series between Na and Ca end members. Some K replaces Na.
Physical properties: Crystal system, triclinic; specific gravity, 2.62–2.76; hardness, 6; colour, white–grey.

(xiii) Sanidine
Composition: $K_2O.Al_2O_3.6SiO_2$.
Variations: Some Na may substitute for K up to a Na/K ratio of 1; if Na exceeds K, the term anorthoclase is used. Al substitution for Si occurs.
Varieties; Sanidine is the high temperature polymorphic form of orthoclase.
Physical properties:
Crystal system, monoclinic; specific gravity, 2.56; hardness, 6; colour, colourless.

Index of Places

Index of Rocks

Index of Minerals

Subject Index